HUMAN BIOLOGY

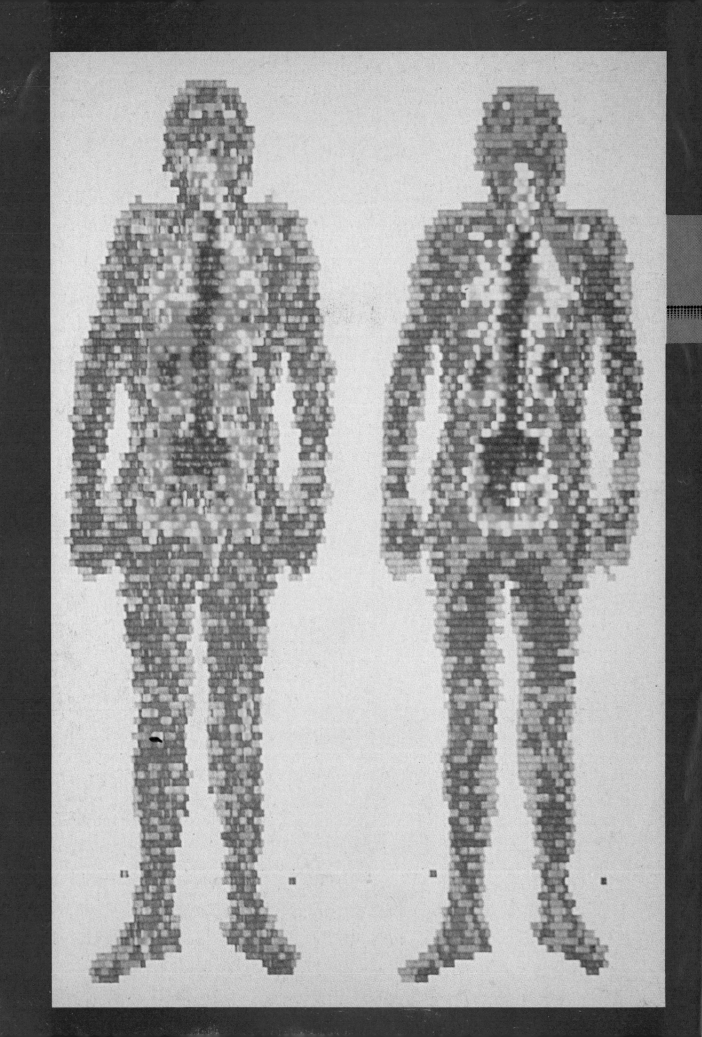

HUMAN BIOLOGY

SECOND EDITION

JOHN D. CUNNINGHAM

Keene State College of the
University System of New Hampshire

1817

HARPER & ROW, PUBLISHERS, New York

Cambridge   Philadelphia   San Francisco   St. Louis
London   São Paulo   Singapore   Sydney   Tokyo

Sponsoring Editor: Sally Cheney
Project Editor: Thomas R. Farrell
Text Design: Laura Ferguson
Cover Design: CIRCA 86, Inc.
Cover Photo and Frontispiece: Diagnostic test. Mildly radioactive body, front
and back. Gamma rays from radioactive isotopes that have been ingested,
collected in different sites in organs. © SPL/Science Source, Photo Researchers.
Text Art: Vantage Art, Inc.
Photo Research: John D. Cunningham/Visuals Unlimited
Production Manager: Kewal K. Sharma
Compositor: Ruttle, Shaw & Wetherill, Inc.
Printer and Binder: Arcata Graphics/Kingsport
Cover Printer: Phoenix Color Corp.

Photo credits for contents: Chapter 1, John D. Cunningham/Visuals Unlimited; Chapter
2, Tom W. French/Visuals Unlimited; Chapter 3, David M. Phillips/Visuals Unlimited;
Chapter 4, George Musil/Visuals Unlimited; Chapter 5, David M. Phillips/Visuals
Unlimited; Chapter 6, John D. Cunningham/Visuals Unlimited; Chapter 7, John Coulter/
Visuals Unlimited; Chapter 8, John D. Cunningham/Visuals Unlimited; Chapter 9,
David M. Phillips/Visuals Unlimited; Chapter 10, K.G. Murti/Visuals Unlimited;
Chapter 11, John D. Cunningham/Visuals Unlimited; Chapter 12, John D. Cunningham/
Visuals Unlimited; Chapter 13, John D. Cunningham/Visuals Unlimited; Chapter 14,
W.H. Hughes/Visuals Unlimited; Chapter 15, Science VU-EPA/Visuals Unlimited;
Chapter 16, David M. Phillips/Visuals Unlimited; Chapter 17, David M. Phillips/Visuals
Unlimited; Chapter 18, David M. Phillips/Visuals Unlimited; Chapter 19, Fred Hossler/
Visuals Unlimited; Chapter 20, William Palmer/Visuals Unlimited; Chapter 21, Eliot C.
Williams/Visuals Unlimited; Chapter 22, E.F. Anderson/Visuals Unlimited; Chapter 23,
Albert J. Copley/Visuals Unlimited; Chapter 24, Frank T. Awbrey/Visuals Unlimited;
Chapter 25, John D. Cunningham/Visuals Unlimited.

A list of credits appears on pages 631–634, which are hereby made part of this
copyright page.

Human Biology, Second Edition
Copyright © 1989 by John D. Cunningham

Library of Congress Cataloging-in-Publication Data
Cunningham, John D.
    Human biology / John D. Cunningham. — 2nd ed.
      p.   cm.
    Includes index.
    ISBN 0-06-041457-X (Student Edition)
    ISBN 0-06-041465-O (Teacher Edition)
    1. Human biology.   I. Title.
QP34.5.C86 1989
612—dc19   88-11159
                                                                           CIP

88 89 90 91 9 8 7 6 5 4 3 2 1

CONTENTS

NOTE TO THE STUDENT

This book is about you and your relationship to other living organisms. It treats both your individuality and the commonality of the structures and functions you share with other humans. Being alive but assembled from nonliving materials is a fascinating circumstance, and we believe you will enjoy the experience of learning more about yourself and about the living world in general.

There are participants and spectators in most aspects of life. Most textbooks are written for spectators in the sense that students are expected to "absorb" the material as if they were sponges. Many texts are also written for passive reading, as if they were novels. You will gain far more from this text by being a participant rather than a spectator, and *Human Biology* has been designed accordingly. For example, there is an old saying that "a picture is worth a thousand words," and illustrations are a very important part of this book. The art, tables, graphs, and photos contain much information and, like the text, should be studied carefully. In some table legends and figure captions you will find questions to help you think about what you have seen and read. There are also end-of-chapter review questions, as well as objective self-evaluation questions and answers for each chapter in Resource 3.

It is assumed that most readers are not preparing for a career in biology. Therefore, the information presented is intended to be relatively nontechnical but useful for any educated person: a reader of current news, a medical patient, a voter, or a shopper.

Like all fields of knowledge, biology uses a specialized vocabulary. Although *detailed* scientific terminology is kept to a minimum in this text, numerous *common* technical terms are used. Many scientific terms have Latin or Greek roots that are used in many combinations. Although the derivations of many terms are given when a concept is introduced, familiarizing yourself with the common prefixes and suffixes in Resource 1 will help you interpret many of the new words you encounter. Many thousands of the long terms in the scientific "word jungle" are combinations of a smaller number of "little" words.

All important concepts and terms (bold-faced items) are defined in the glossary. Other technical terms that may not be used in all courses are included in parentheses for reference or enrichment purposes. Use the glossary frequently, but when looking up terms and concepts, also use the index, which provides page references to each major treatment of a given entry.

The measurement units used throughout most of the world are those of the metric system. However, because the United States is adopting this system slowly, the U.S. customary equivalents of metric units are often given. Where they are not, refer to the information in Resource 2.

If pictures are worth a thousand words, graphs are often worth a thousand pictures. Resource 2 is also designed to help you understand and utilize data presented graphically.

Finally, a word about ecology, the study of the interactions between organisms and their living and nonliving environments. The treatment of ecology is not limited to the final chapter, but is interwoven throughout the text. For example, some population issues are linked with the topic of hunger in the chapter on nutrition, and the health effects of air pollution are discussed in the chapter on our breathing system.

PART ONE

ON SCIENCE AND BIOLOGY

Even to people who rarely travel far, there is no question about how diverse life is. From trees, flowers, and ferns, to mushrooms, mosses, and molds, to birds, fish, and frogs, and to worms, spiders, and humans, life surrounds you. With a microscope, you may see bacteria, algae, and protozoans, and you become aware of even greater biological diversity. During a visit to a zoo or botanical garden you see "stranger" forms of life, often merely because they are from somewhere else. Who can comprehend a million of anything, let alone several million kinds of living organisms as different in appearance and size as amoebas and redwoods? The intricate and diverse structures of these organisms adapt them marvelously to their environments, which for humans is almost everywhere on Earth.

Science is both process and product—the systematic search for an understanding of the natural world and the knowledge gained from the investigations. Specifically, biology attempts to classify, describe, and, most important, to understand the workings of organisms. Obviously, biologists are humans, the only species that explores its environment and attempts to explain it in a systematic way. We still have much to learn about life on Earth, and about ourselves specifically, but you will explore some of the many interesting and provocative areas of human biology in this text.

Although considered simple compared with humans, amoebas, like all organisms, are extremely complex, and we share numerous traits with these and all other forms of life. All species of organisms exhibit variability, but perhaps none so much as humans. In part because of our diversity, we are found almost everywhere on Earth and, on occasion, in nearby space.

CHAPTER 1

Biology: The Study of Life

Discovering, classifying, and understanding the immense variety of life on Earth are the aims of biology. Certain methods and traits aid biologists and other scientists in their work. One basic and necessary mental characteristic is the willingness to alter one's views in the light of new evidence. Thus, the young Charles Darwin held a view of life that was quite different from the one that he held later in life. In more recent times it has become increasingly important for scientists from numerous fields to work together to solve the intricate and intriguing mysteries of life. They are aided in their work by several time-tested concepts, especially those dealing with evolution.

This bust of Charles Darwin resides in the British Museum in London. Darwin was not the first to develop a theory of evolution but his view of life provided us with the basic tenets of the modern theory of evolution that we utilize to understand the workings of all organisms on Earth.

FIGURE 1.1

''You will be a disgrace to yourself and all of your family,'' Darwin's physician father said to young Darwin when Charles seemed more interested in collecting insects than in studying medicine. The prediction hardly materialized: Charles Darwin was destined to shape intellectual history as few have before or since.

DARWIN AND LIFE

Like many other young people, Charles Darwin (1809–1882) was unsure about his future (Figure 1.1). His father and grandfather were well-known and respected British physicians, and the expectation was that Charles would follow suit. At 16, he began medical school, but soon switched to law, which he found equally uninteresting. He finally completed training for the clergy, although he had no intense desire to enter that profession either.

In his spare time, Darwin observed, collected, and studied organisms and rocks. His father once told him, "You care for nothing but shooting, dogs, and ratcatching." While attending divinity school in Cambridge, Darwin found companions in two professors—a botanist named John Henslow and a zoologist named Adam Sedgewick. It has been said that "The sun never set on the British Empire" in the early 1800s, and it was common to assign a naturalist to many government exploratory ships in order to learn more about the natural history of Britain's vast empire. When Professor Henslow could not accept an appointment as a naturalist aboard the British naval survey ship H.M.S. *Beagle,* he recommended Darwin as an alternative. That journey not only provided an opportunity for 22-year-old Darwin to do what he wanted to do, but it also helped to shape an intellectual revolution. The trip was not altogether pleasant, however: Darwin was often very seasick, and he contracted a parasitic disease that greatly weakened him later in life.

Although the 5-year *Beagle* expedition circumnavigated the globe, most of the trip was spent around South America (Figure 1.2). Darwin was able to go ashore on many occasions and often took extended trips inland. In the Argentine Pampas he saw few familiar animals, although the area seemed ideal for many of them. Darwin found the fossils of armadillos and ground sloths that, except for their huge sizes, were not very different from the living species on the same continent. Fossils were not new to Darwin, but these discoveries raised again the question of why some species had become extinct. Were they all simply the remains of those animals unfortunate enough not to have survived the biblical flood?

(A)

FIGURE 1.2

The *Beagle* was a small vessel, and its captain was only one year older than Darwin (A). Captain Fitzroy's official mission was to gather oceanographic information, especially around South America (B). Personally, he also hoped to gather evidence to prove the literal biblical account of creation. Ironically, Charles Darwin was accepted as the ship's naturalist and captain's companion to help him do that. As is true of many volunteer opportunities today, the position of naturalist provided no salary and Darwin had to pay for his own room and board.

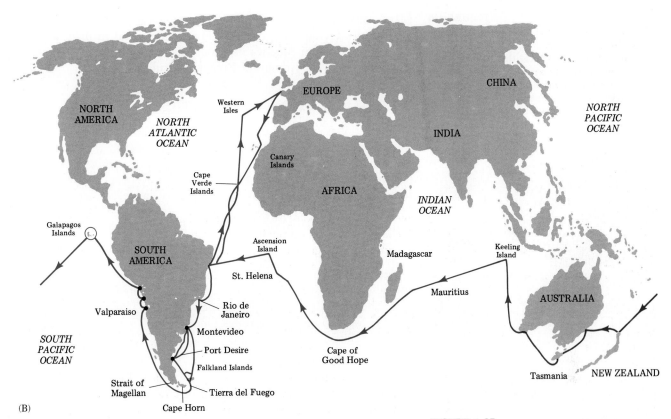

(B)

FIGURE 1.2B

The Galápagos Islands were of special significance to Darwin (Figure 1.3). The shells of Galápagos tortoises that lived in arid habitats were saddleback, and the fronts bent up. However, the shells of the even larger tortoises from relatively moist locations were domed, and the fronts bent down. This suggested to Darwin that the tortoises were related but that their traits had changed after they became isolated on different islands. The several kinds of small finches were similar but somewhat different from those that he had seen on the mainland. Also of interest were cormorants that do not fly, iguanas that dive into the sea to graze underwater on seaweed, and gulls that feed at night. These were clearly not the same kinds of organisms that Darwin had observed on the somewhat similar volcanic Cape Verde Islands off the coast of Africa, nor were they like the species that he would observe on the

FIGURE 1.3
Lying about 900 kilometers (560 miles) off the coast of Ecuador are the relatively young (3.3 million years old) volcanic Galápagos Islands. Habitats on these islands range from dry and bleak to wet and species-rich. New and wonderful species greeted Darwin, as they do modern biologists and tourists who visit the islands: (A) a giant tortoise, (B) marine iguanas, and (C) flightless cormorants.

(A)   (B)   (C)

FIGURE 1.4
"The universal kinship of all generations born from a common mother" was Buffon's way of stating his view that all mammals evolved from a common ancestor. Despite their diverse life-styles, the basic architecture of the mammalian skeleton is similar, as are many other anatomical and physiological traits.

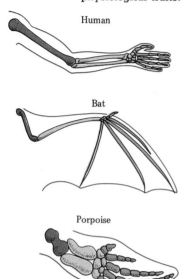

Human

Bat

Porpoise

Horse

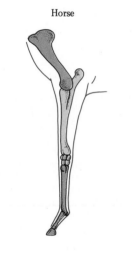

island-continent of Australia. To biologists, isolated islands are natural laboratories for observing changes in organisms and in their environments. Could South America itself once have been isolated, as Darwin theorized, when the isthmus of Panama lay submerged?

SPECIAL CREATION TO EVOLUTION

One traditional view of life, **creationism,** holds that species were created by God individually and in essentially their present form. According to this belief, the ancestry of all present-day organisms can be traced to the individuals created by God, and the essential traits of species are thought to be fixed or immutable. Throughout history, however, many have had doubts. The English philosopher Francis Bacon (1561–1626) wondered about the variations in organisms. The French philosopher René Descartes (1596–1650) believed that all natural events could be explained by physical principles rather than by supernatural forces.

Before the mid-1800s, most biologists as well as others believed in special creation, and their scientific studies were often a way of learning more about the Creator and His works. One such scientist, for example, was the Swedish biologist Carolus Linnaeus (1707–1778), who devised the basic classification system that we still use.

Linnaeus and many other biologists, of course, depended upon organisms for their studies, specimens of which explorers worldwide were bringing back in large numbers. Although many species that were collected or sighted were quite strange, it was clear that organisms fell into seemingly natural groupings. Bats flew, whales swam, moles dug, and buffalo roamed, but all of these animals shared so many features that the respected French biologist Georges Buffon (1707–1788) declared a century before Darwin that mammals evolved from a common ancestor (Figure 1.4). He based his view partially on his study of the numerous fossils that also were accumulating. For example, he noted that the shallow and recently fossilized forms were more modern than the deeper and older fossils.

Within a few decades of Buffon's work, Jean Bapiste Lamarck (1744–1829) proposed the first coherent theory of **evolution,** the view that organisms had changed over the course of time. Lamarck suggested in 1801 that all species, including humans, had descended from other species. He also believed that life evolved progressively from simple to more complex forms. The mechanism for change proposed by Lamarck, now known to be incorrect, was based upon the inheritance of acquired characteristics. The bulging muscles of the weight lifter in Figure 1.5

FIGURE 1.5
The phrase "Use 'em or lose 'em" certainly applies to muscles. Unused muscles atrophy and become weak, whereas those used regularly in exercise or work increase in size and strength. However, the children of this body builder would be no stronger because of the parent's exercising than they would be if the parent had been more sedentary.

obviously resulted from use; unused muscles atrophy or become weak and small. According to Lamarckian theory, the children of weight lifters should be born with larger or stronger muscles than they would have if their parents had not acquired such traits. Although it is now known to be wrong, Lamarck's theory called attention to the study of evolutionary processes and thereby helped to set the stage for the ultimate scientific acceptance of Darwin's view.

Darwin left England on the *Beagle* in 1831 leaning toward creationism, but he returned in 1836 with some doubts. He started his journals on "Transmutation of Species" in 1837 and became an evolutionist over the next several years of studying his specimens and notes, conducting numerous other studies, and discussing ideas with colleagues. Even so, Darwin continued to believe in the existence of a Divine Creator who was personally responsible for all matters. In Darwin's view, however, He had not created Earth's organisms and its natural features to leave them unchanging forever, but rather expressed Himself through the operation of natural laws. The latter could be studied and understood, and that is exactly what Darwin and other scientists tried to do.

Darwin prepared a 35-page sketch of his theory in 1842, but he did not publish it, partly because of concern about the controversy that he knew it would generate. However, because he was often ill, he asked his wife to have it published if he were to die suddenly. In 1859, Darwin published one of the most influential books of all time, *On the Origin of Species by Means of Natural Selection, or the Preservation of Favoured Races in the Struggle for Life.*

Most people of the time interpreted the Bible literally and believed in a fixed and constant world and in supernatural explanations of nature. They found Darwin's work and that of most subsequent biologists troubling, as many still do. (See the essay "Scientific? Creationism.") Darwin's theory represented one of the major intellectual revolutions in human history, and it continues to affect our views of nature. Today, evolution forms the cornerstone of biology.

While sailing on the *Beagle*, Darwin read *Principles of Geology* and other works by the important British geologist Charles Lyell (1797–1895). Especially meaningful to Darwin was Lyell's theory of **uniformitarianism**, through which the past can be understood by studying the present. Uniformitarianism is the theory that the processes now modifying Earth's surface have acted similarly throughout geologic time, although possibly at different rates.

After returning to England, Darwin read and was much influenced by Thomas Malthus's *Essay on the Principles of Population.* Malthus (1766–1834) observed that populations of organisms, including humans, tend to increase geometrically rather than arithmetically. In a **geometric progression**, elements progress by a constant factor, such as 2, 4, 8, 16, and so forth. In an **arithmetic progression**, elements progress by a constant difference, such as 2, 4, 6, 8, and so forth. A major concern of Malthus's was that while the human population increased geometrically, its ability to produce food increased only arithmetically (Figure 1.6).

FIGURE 1.6
Arithmetic and geometric progression compared (A). Although the population size of many species seems to remain relatively stable over time, the human population growth resembles the geometric progression curve (B).

(A)

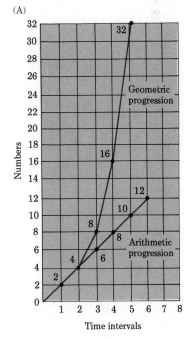

(B)

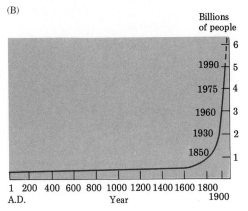

NATURAL SELECTION

The **biotic potential** of most organisms—their ability to reproduce under ideal conditions—is high. If its reproduction and survival are unchecked, every species has the ability to cover Earth or fill its waters in a relatively short period of time. Obviously, this does not happen, although those

Scientific? Creationism

Many creationists are religious fundamentalists who believe that the biblical account of the origin of Earth and its organisms is literally true. According to this view, Earth is relatively young and God created every species of organism separately and in its present form. Fossils are organisms that died in catastrophes, such as Noah's flood. While the ideas of creationism are ancient, and are perfectly acceptable and proper in everyday life, the addition of the term *scientific* to the concept's name was made by some in the 1970s in an attempt to give further credence to a view for which there is no scientific evidence. The lack of scientific evidence, of course, does not make a view wrong, and scientists are not concerned about that question. Scientists work neither to confirm nor to deny the existence of God. Rather, scientists are bothered by the implications of the improper use of the word *scientific*. "Scientific creationism" is a contradiction in terms, much as Groucho Marx joked about the phrase "military intelligence." Science can deal only with the observable, and creationist views are untestable and must be accepted solely on the basis of arbitrary faith, which is an individual matter. As such, however, the views are outside of the realm of science, as are some other aspects of the human experience, because science is a form of human inquiry that depends upon testable hypotheses. Hypotheses are not "believed in" as are some religious creeds; rather they are tentative explanations that are accepted as long as they continue to incorporate known facts satisfactorily. When theories fail to do this, or as we learn more, they are subject to modification.

The scientific evidence against classical creationism that has been gathered during the last few centuries is overwhelming. Earth is clearly very old, about 4.5 billion years old, as opposed to merely a few thousand years, and organisms have occupied it for much of that time. All scientific evidence points to the conclusion that all organisms, including humans, have evolved from earlier, simpler organisms (Chapter 24).

Some creationists who have pressed suits in court attempting to have creationism defined as a science and thereby given "equal time" in school curricula have cited the fact that biologists do not agree on every point. The observation is absolutely correct, just as it is for humans in any other field, including theology. Scientific inquiry must be open and public, and this sometimes draws attention to alternative viewpoints on issues. However, it is erroneous to equate serious controversy about the details of evolutionary mechanisms with any scientific rejection of the concept of evolution itself.

Evolutionary theory provides both a framework within which explanations of biological phenomena fit and a source of testable hypotheses that guides contemporary research. Creationism leads to no testable hypotheses, and it has generated no scientific studies. True science has no need to hypothesize supernatural powers and events. Thus, there need be no basic controversy between religion and science; many scientists are religious, and many who are religious in one form or another are interested in science. The problem is simply that creationist views are not scientific. That does not prohibit many scientists and nonscientists from believing in a God who works through the natural laws of the universe.

FIGURE 1.7

(A) Dandelions are beautiful to some people, partly because they provide such welcome color early in the spring. (B) However, they are despised by others because their biotic potential is very high. To appreciate this, count the seeds on a typical dandelion head before the wind carries them away. Then suppose that every seed survived and grew into a plant that produced as many seeds, and so forth.

FIGURE 1.8
Why is the giraffe's neck so long? This giraffe must stretch its neck to browse upon the uppermost leaves. According to Lamarckian thinking, such stretching would lead to a long neck, and the acquired trait would be passed along to offspring. During a lifetime of neck stretching, the offspring's neck would become even longer and that trait would, in turn, be passed along to the next generation, and so forth. However, acquired traits rarely affect genes and, therefore, are only rarely transmittable to offspring.

who worry about dandelions in their lawns may not be convinced (Figure 1.7). Yet, the population sizes of many nonhuman species tend to remain relatively stable over time. Unfortunately for all life on Earth, the human "population explosion" continues largely unchecked.

Of the large potential number of genetically variable offspring that most organisms can produce, only a few actually survive and reproduce. According to Darwin's key observation, the variant organisms most likely to survive, and thus to have an opportunity to reproduce and pass along a genetic blueprint for their traits to offspring, are those with superior physical and behavioral characteristics. Because the traits of those individuals will increase in the population as more individuals with those traits appear, Darwin called the process **natural selection.** The driving force behind natural selection he called "survival of the fittest."

How can the concept of natural selection account for the giraffe's long neck (Figure 1.8)? Perhaps the ancestors of modern giraffes had relatively short necks, like those of other herbivores with which they competed for limited food. Variation occurs in all species, and any variant giraffe whose neck was longer than usual might have survived better in lean times by being able to reach the uppermost leaves on bushes and, thus, might have produced more offspring than other variant giraffes with short or average necks. Thus, these hypothetical ancestral giraffes already had relatively long necks and did not acquire them through stretching for lofty leaves. While changes such as those described above require long periods of time, many similar changes and examples of natural selection have been observed both in the laboratory and in the field.

The domestication of plants and animals is an old art, and one that Darwin experienced through raising pigeons (Figure 1.9). Through ar-

FIGURE 1.9
Through artificial selection, humans have produced enormous variations in domestic organisms, and continue to do so. If various varieties of dogs (A) were found in the wild, some biologists might classify them, at least initially, as different species. Darwin gained personal experience with artificial selection through his work with domestic pigeons (B).

(A)

(B)

FIGURE 1.10
The finches that live on the Galápagos Islands today belong to several species. A reasonable hypothesis, supported by many studies, suggests that these species evolved from at least one pair of birds (or a single female bearing fertilized eggs) that arrived long ago on the isolated islands from the South American mainland. The so-called woodpecker finch is especially interesting because it uses a cactus spine to prod insects from crevices in bark, much as a true woodpecker does with its long, sharp bill. This presents one of the few instances of nonhuman species using tools.

tificial selection, organisms with traits unfavorable to humans are not allowed to reproduce, while those with favorable traits are. The resulting differences in the traits of domesticated breeds (e.g., dogs, cats, cattle, and horses) often exceed those that separate groups of related species in the wild.

Although some aspects of his theory developed within a few years of his return from the *Beagle* expedition, Darwin worked for nearly two decades perfecting his ideas before publishing *On the Origin of Species*. He was prompted to do so then only when he learned of the work of Alfred Russell Wallace (1823–1913). Wallace, a young British naturalist working in the Malayan archipelago, sent Darwin a concise statement of his own and very similar theory of evolution by natural selection. Wallace, too, had been influenced greatly by Malthus's 1798 essay.

Although his theory was almost completely accepted in scientific circles after the 1860s, Darwin continued to develop his ideas and to write other works, some of them elaborating on the *Origin*. However, he never completed the far more exhaustive treatise on the topic that he had hoped to write. Also, Darwin chose to live quietly on his country estate and to leave the defense of his controversial ideas to others.

Darwin's conception of evolution required an immense period of time, and precise knowledge about Earth's age was unavailable in his lifetime. Another troublesome issue was his inadequate knowledge about heredity, the way by which traits are transmitted to offspring (Chapter 9). Although Darwin and Gregor Mendel (1822–1884), the father of genetics, were contemporaries, they never met and probably were unaware of each other's work. Major development of evolutionary theory was hindered until a satisfactory explanation of the process of heredity was developed about 20 years after Darwin's death. Until the principles of genetics were understood, it was not possible to account for the origin of new variations or for their spread through a population. One could describe aspects of evolution, but one could not explain them without a knowledge of genetics.

Despite these problems, Darwin's work established clearly and firmly that the diversity of life on Earth was due to evolution. In developing his theory, Darwin benefited from a ripening of science and from the work of many other investigators as well as his own observations and ideas. To paraphrase another famous scientist in a similar circumstance, "If I have seen further than others it is because I have stood upon the shoulders of giants." Shaking the foundation of knowledge today is most often a team effort.

SOME METHODS OF BIOLOGY

Defining the scientific method is difficult because there is no single, universal, and prescribed way of conducting scientific research. However, discovering new facts and searching for the mechanisms that explain and integrate them usually begins and ends with observations. Darwin observed that several unique types of finches inhabited the Galápagos Islands, and he also noted that they closely resembled species living on the South American mainland. A common next step is to formulate a hypothesis or tentative explanation of the observed facts. For example, perhaps a few birds arrived on the isolated volcanic islands from the mainland long ago (Figure 1.10). Leading different life-styles in isolation over a long period of time, perhaps the generations of offspring of the ancestral birds gradually became different. Captain Fitzroy of the *Beagle* offered an alternative hypothesis—God created the finches and placed them on the islands because He wanted them there. However, the ideal hypothesis is testable, either directly or indirectly, and, whether it be right or wrong, there is no way to test Fitzroy's hypothesis. Obviously, Darwin could not test his hypothesis directly either. He had no young volcanic islands, no storms to carry birds from a mainland to any isolated and sparsely settled islands, and insufficient time to observe events for hundreds of thousands of years—the *Beagle* had a schedule to maintain, and evolution is a long-term phenomenon. However, Darwin's hypothesis had a characteristic common to good scientific hypotheses—it led to predictions. Both Darwin's and Fitzroy's hypotheses explained events, but only Darwin's could generate testable predictions about evolution and natural selection.

Most hypotheses do not require the immense time factor that Darwin's example did. Suppose a biologist believes that a certain pesticide may have an adverse effect upon the embryonic development of chickens. Many hypotheses are stated in an "If . . . , then . . . ." form. For example, "If I inject the pesticide into an egg, then I will be able to observe whether the bird develops abnormally." This hypothesis suggests an **experiment,** the setting up of the phenomenon of interest under conditions wherein certain observations can be expected, based upon the predictions being tested (Figure 1.11). Ideally, an experiment is set up so there can be only one explanation for the results. We often hear that "the patient died but the operation was a complete success." Similarly, if the embryo in one egg injected with the pesticide failed to develop normally, was this the result of the pesticide, the experimental procedure, or chance? Obviously it is more difficult to draw a meaningful conclusion from an experiment on a single egg than from one that used a larger sample size.

The use of **controls,** or standards of comparison, alleviates most of the problems in the above example. For instance, some eggs may be pierced with a hypodermic needle but injected with nothing. Others may be injected with a solution thought to be unable to affect the outcome, such as sterile, distilled water.

The crucial conditions being observed and measured in experiments are called **variables.** In a controlled experiment, all variables are kept identical except for the one being tested—the experiment uses the same kind of eggs, same incubation time and temperature, and identical treatment in all other ways. Any uncontrolled differences between the control and the experimental groups make a conclusion difficult and tenuous. The conditions that are being manipulated and controlled by the experimenter are called **independent variables,** and those variables that are measured and that are considered to be the experimental results are called **dependent variables.** In this example, the dependent variable is relative growth of the three groups of eggs. Other dependent variables that could be measured include the longevity of the birds; their resistance to disease, cold, or drought; and the number of viable eggs the hens produce. Experimental results are often summarized in graphs, and many graphs are found throughout this text. To learn more about the conventions used in graphing data, see Resource 2, "Measurements and Graphs."

Science may be the search for "truth," but doing an experiment does not necessarily make the results "true." A biologist rarely meets all of his or her colleagues worldwide who are interested in the same topic, although personal contact with some occurs at scientific meetings. However, biologists communicate regularly and formally in scientific journals. The details about experiments and observations are presented in such a

FIGURE 1.11

(A) An uncontrolled experiment to test the effect of a pesticide on the development of a chicken egg. A fertile egg that is injected with a sterile pesticide solution using a sterile needle fails to develop normally. Because one cannot be sure whether the effect was the result of the pesticide, the procedure, or chance, the use of many eggs is advisable, as is the use of controls. (B) In controlled experiments, the groups are treated alike in all ways other than that being tested. For example, the eggs in one group may be pierced and sealed but not injected with anything. The eggs in another group may be injected only with sterile water, or with different concentrations of the pesticides to see whether there is a gradient in the results. The control group consists of a group of untreated eggs that are otherwise handled in the same way as all other eggs.

(A) Uncontrolled experiment

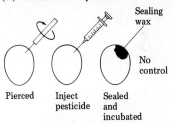

Pierced | Inject pesticide | Sealed and incubated

(B) Controlled experiment

Experimental eggs: Pesticide added to a number of eggs which are then sealed and incubated.

Control eggs:

A number of eggs untreated; a number of eggs pierced but nothing added; a number of eggs injected with sterile water.

FIGURE 1.12
The *Paramecium* (A) conducts all of its activities with its single cell, whereas human activities require trillions of cells. In multicellular organisms, many cells are specialized for different functions. However, although these human skin cells (B) and neurons (C) are somewhat different from each other, all types of cells share many traits.

(A)

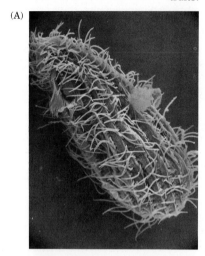

(B)

(C)

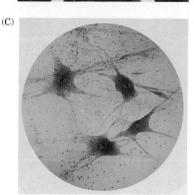

way that they can be repeated by others. Only when experiments and observations have been replicated, reaffirmed or rejected, and refined and altered are the facts considered true. Even then, "tentatively true" is understood in science because all scientific ideas are open to revision when new observations or interpretations warrant changes. Whether research is done by individuals or teams of scientists, the results must be made public in order for them to be validated.

Often a body of facts about a phenomenon is explained by a **theory**. Thus, while evolution is a fact, there are various conjectural and tentative theories about the exact mechanisms that lead to changes in populations during time. All theories have imperfections, and vigorous debate and controversy are the grist for science's mill. As Thomas Jefferson observed, "A little rebellion now and then is a good thing." For example, to question certain *details* about the mechanism of natural selection is not to suggest, as some creationists have, that biologists are rejecting the validity of the theory of evolution itself.

While it is relatively easy to describe the steps used in observing and experimenting, it is more difficult to do this for hypothesizing and theorizing. The mulling over of facts, the surmising, the dreaming, and the speculating occupy much of a scientist's time. The eureka (or "ah-hah") experience of Archimedes is exciting, although it happens all too infrequently to most of us. (See the essay "Serendipity in Science.")

SOME MAJOR CONCEPTS OF BIOLOGY

The bulk of this text concentrates on humans, sometimes telling you more than you ever thought you wanted to know about a topic. This section outlines briefly some of the "big ideas" in biology that will be useful to you in understanding the role of humans in the living world, and our relationship to other organisms. Defining life is not easy, and circular, dictionary-type definitions are of little value—for example, *life* is the condition of not being dead, and *dead* is not being alive.

Evolution: Descent with Modification

You have already been introduced briefly to some notions about evolution, the concept that gives meaning to and integrates all other biological information. In general, by *evolution* we mean changes in the traits of a **population**—an interbreeding group of organisms of the same species—and not changes in the traits of an organism during its lifetime. That is, the traits of the individuals that appear, grow, develop, and die are generally fairly fixed, although some events (mutations) occasionally change them. Populations, and therefore the species of which they are a part, evolve. Thus, over long periods of time, as environmental conditions change, the traits of populations may change because only certain ancestors survived and reproduced.

The Cellular Basis for Life

Most life processes actually occur at the level of **cells** (Figure 1.12), the basic units or building blocks of organisms (Chapter 3). Both health and disease depend upon the structure and function of these tiny units. **Unicellular** organisms such as bacteria and amoebas are composed of single cells, whereas **multicellular** organisms such as sunflowers and humans contain many millions of cells. Understanding cells is essential for an appreciation of biology. The basic matter found in organisms

Serendipity in Science

Unexpected or serendipitous events have played a major role in intellectual history. For example, culture dishes contaminated with molds are a bother to bacteriologists, who generally discard them. However, Alexander Fleming (1881–1955) kept some contaminated dishes in 1928 and, to use Louis Pasteur's phrase, "Chance favors the prepared mind." Fleming's discovery that penicillin from molds killed bacteria started the age of antibiotics (Figure B1.1). For his serendipitous observation and follow-up studies, he was knighted by the British government and was awarded the Nobel Prize in Medicine in 1945.

So admired by Emperor Napoleon III that two laboratories were provided for him, Claude Bernard (1813–1878) was the first scientist to receive a state funeral (Figure B1.2). Like many today who change their fields and areas of interest, Bernard wanted to be one thing—a playwright—but became another instead—a physiological researcher. Bernard's observations on rabbits that he brought into his laboratory from a market serve as another example of serendipity and the prepared mind. Rabbits are generally herbivorous

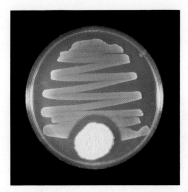

FIGURE B1.1
Bacteria and other microorganisms are often grown on special nutrients in culture dishes like these. Bacterial growths cover the nutrient substance except around the mold. Such an observation led Fleming to conclude that molds produced antibiotic substances that killed the nearby bacteria.

FIGURE B1.2
Claude Bernard.

and usually produce a turbid, alkaline urine, but the urine of these rabbits was clear and acidic, like that of carnivores. Why? One hypothesis was that the starving herbivores had become temporary carnivores, digesting their own tissues. A test of this hypothesis occurred when the rabbits were again fed grass. A few hours after eating their normal diet, the rabbits produced the usual alkaline, turbid urine, but the urine became acid and clear when they fasted again. When fed boiled meat, which the hungry rabbits ate readily, the rabbits continued to produce carnivore-like urine. Thus, Claude Bernard was one of the first to recognize that many processes operate within organisms to maintain internal conditions compatible with life despite external changes.

Serendipitous and other scientific discoveries often lead to the familiar comments about hindsight versus foresight. For example, when Thomas Huxley, one of Darwin's most influential contemporary supporters, was sent a copy of *On the Origin of Species* to review for the London *Times*, he supposedly complained that he was "extremely stupid not to have thought of that."

does not differ from that found in nonliving objects (Chapter 4). In the cell, however, this matter is organized in the unique ways that are characteristic of life.

Organisms Metabolize

Many organisms, called **autotrophs** (or "self-feeders"), take in relatively simple chemicals and use energy such as that in sunlight to produce their own food. However, no amount of sunbathing allows humans to do this. Like many other organisms, we are **heterotrophs** (or "other feeders"), and we require preformed food. Although many of the same chemicals and forms of energy enter different organisms, each organism structures the inputs differently. Thus, Chihuahua and Saint Bernard pups may live in the same home and eat the same dog food, but they obviously do not grow to look alike. Both the dogs and their owners are composed largely of cells different from those they contained last year. Stated very simply for now, the totality of this processing of energy and materials is called **metabolism** (from the Greek for "change"). In other words, the only way to remain in equilibrium with the environment is to change.

The ultimate source of biological energy is the sun. Autotrophs such as green plants absorb the radiant energy of sunlight and convert it into chemical energy (Figure 1.13). Plants use some of the chemical energy for their own growth, maintenance, and reproduction, but some of it is used by plant-eating animals, or **herbivores.** In turn, herbivores such as mice and cows provide the chemical energy for the meat-eating or **carnivorous** animals that feed upon them. Most humans are **omnivores,** eating both plant and animal food. Dead organisms and the wastes of organisms provide energy for **decomposers,** such as bacteria and fungi. Ultimately, energy is degraded to forms not directly usable by organisms. Therefore, energy does not cycle, and life on Earth would cease without a constant influx of solar energy. The materials used to support the metabolism of organisms, however, do cycle. Your body probably contains atoms that were once in stars, earthworms, mushrooms, redwoods, and Julius Caesar. Dust to dust. Some future sponge, amoeba, insect, or human will be the recipient of some of your atoms.

Responsiveness of Organisms

Even rocks respond to environmental changes, as when they expand and contract as the temperature changes. However, the behaviors of organisms are usually far more complicated, and often involve specialized structures to detect environmental changes and respond to them. Nonliving objects simply react passively to environmental stimuli, as when a loosened rock rolls down a hillside because of gravity.

FIGURE 1.13
The supply of materials upon which the metabolism of organisms depends is relatively fixed. Organisms depend upon no regular inflow of materials from space, and few Earth substances are lost to space. Thus, Earth's supply of the calcium for skeletons and the water for numerous metabolic processes, for example, is basically finite. However, the quantity, availability, and purity of materials may change in place and time. In contrast, energy does not cycle constantly between organisms and their environments in the same way. Organisms depend upon a constant influx of solar energy, and life would cease without it.

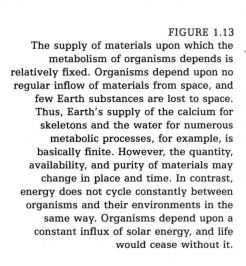
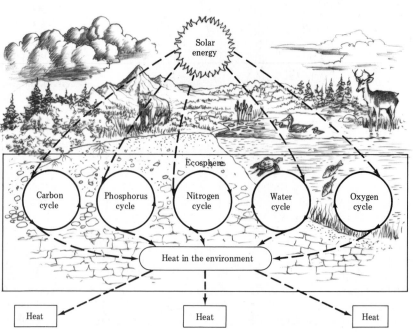

Reproduction and Genetic Programs

The methods of reproduction among organisms are quite varied, with some organisms simply splitting in half occasionally. However, in each case the offspring receive a hereditary blueprint or genetic program from their parents. The **genes**, or basic hereditary units, that constitute the code for the structures of an organism are not always the same as those of the parent or parents. This leads to the variation upon which natural selection acts in the process of evolution.

Adaptations of Organisms

Like evolution, the adaptations of organisms pertain especially to populations rather than to individuals. In general, an organism does not adapt to survive but, rather, it survives if it is adapted. A duck, for example, does not grow bigger and better webs between its toes in order to swim better. However, those ducks who inherited genetic programs that resulted in superior swimming ability have the best chance to survive and reproduce. The adaptations of individuals, such as when mammals grow thick insulative fur in the winter, are changes made possible by their genetic programs.

THE MAJOR GROUPS OF ORGANISMS

A few decades ago, many biologists placed organisms into two large groups—plants and animals—and many nonbiologists still do. Some "troublesome" kinds were simply said to have both plantlike and animallike traits. Now, most biologists classify organisms into at least five large groups or **kingdoms** (Figure 1.14).

FIGURE 1.14
(A) Kingdom Monera: Prokaryotes include the bacteria, such as these *Streptococcus* that cause sore throats and other problems, and the cyanobacteria, such as these *Gleocapsa.* The cyanobacteria were formerly called blue-green algae. **(B) Kingdom Protista:** Protists are unicellular eukaryotes. Some, like *Euglena,* are plantlike in that they may be autotrophic, whereas others, like *Didinium* shown here consuming another protozoan, are animallike. **(C) Kingdom Fungi:** Fungi are generally heterotrophs, like these multicellular mushrooms and unicellular yeast. **(D) Kingdom Plantae:** From microscopic green algae to giant redwoods, plants are multicellular and most are autotrophic. **(E) Kingdom Animalia:** Look in a mirror to see an example of a multicellular, heterotrophic animal. You do not have to look far to find many other examples of animals.

(A)   (B)

(C)   (D)

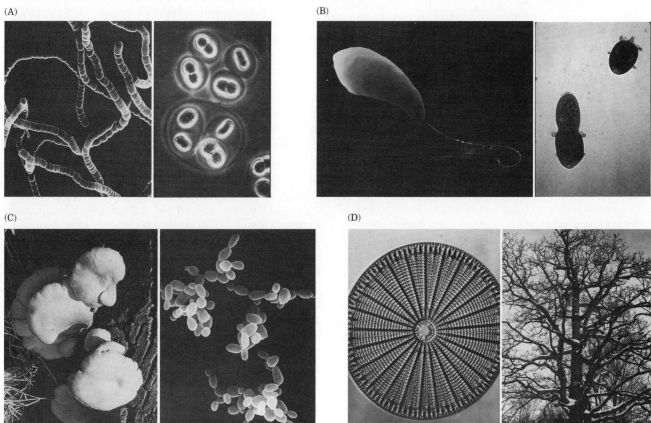

Viruses

Some things are commonly described by telling what they are not. Such is the case with viruses, which neither are cellular nor possess their own metabolic machinery. Even those who consider viruses to be alive do not classify them as true organisms, and viruses are not part of the five-kingdom system. Viruses are simpler than bacteria, which are the simplest organisms known today. Also, viruses are not considered to be an intermediate step in the evolution of life from nonliving substances. On the other hand, viruses do some lifelike things, and many of them have major effects on organisms, to which most people can attest. Today, much information about molecular biology and genetics is derived from the study of viruses.

The existence of viruses was inferred long before they were seen. Because viruses are in the size range of many large molecules, details about their structure can be seen only with an electron microscope (Chapter 3). Prior to the 1930s, when the first viruses were purified and observed, they were known to be agents too small to be filtered by the apparatus used to separate bacteria from solutions.

Viruses cannot release energy or synthesize proteins because they are acellular, and they cannot replicate, metabolize, or grow on their own. Instead, they are parasitic, and their host cells provide the energy, materials, and means (enzymes) to synthesize new proteins. Even crystallized viruses that have been in closed jars for years can infect cells. Although the specifications for the enzymes produced by the host cell are provided by the virus, viruses contain only one to several hundred genes, not the many thousands of genes that most organisms possess. Like a computer program, the virus informs the cell (the "computer") what to do, and utilizes the cell's machinery to perform the tasks. Viruses cannot duplicate themselves outside of a cell any more than a computer program can duplicate itself outside of a computer.

Although they exist in various sizes and shapes, all viruses share many traits. Viruses consist of at least one strand of nucleic acid, which functions as their genetic program. This genetic material is contained in a core that is surrounded by a protein coat and, often, also an outer envelope (Figure B1.3). Most viruses are either threadlike (helical) or roughly spherical with multiple faces (Figure B1.4). Many bacterial viruses, called **bacteriophages** (from the Greek for "bacteria eaters"), or just phages, are even more complex, with a head (within

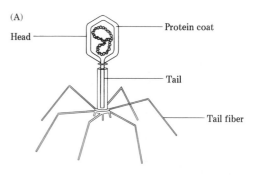

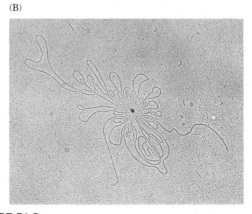

FIGURE B1.3
(A) The structures of a bacteriophage. (B) Viruses consist of genetic material surrounded by a protein coat. The virus seen here has been disrupted so its DNA molecule can be seen. What features of cells do viruses possess? Lack?

which a long DNA molecule is coiled), a neck, a tail, and a base plate. Because phages can be grown relatively easily in bacterial cultures, much information about viruses has come from their study.

The general life cycle of many viruses resembles that of phages (Figure B1.5). The phage attaches itself to the outside of a bacterium by its tail. The phage's genes are then injected into the host cell, although its protein coat remains outside. The phage genes then take over the metabolism of the bacterium, and in 20–30 minutes the cell is filled with phage particles. When the bacterial cell bursts, the phage particles are released and are able to infect other bacteria.

Although viruses are parasitic, not all cause diseases and some may be present in cells without doing apparent harm. However, there are "slow viruses" that take time, sometimes years, to produce symptoms in the

(A)   (B)

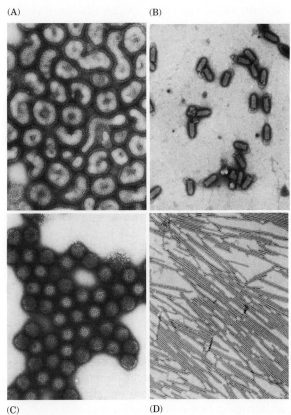

(C)   (D)

FIGURE B1.4
A gallery of viruses. (A) The human influenza virus (282,
100×). (B) A member of the rabies group of viruses, the
vesicular stomatitis virus (100,000×). (C) The human
reovirus (228,000×). (D) A crystalline array of the tobacco
mosaic virus (102,500×).

FIGURE B1.5
The life cycle of a bacteriophage.

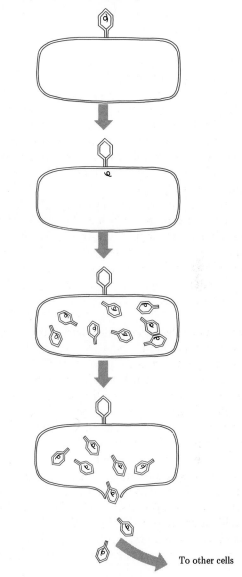

To other cells

victim. Human viral infections include polio, shingles, encephalitis, rabies, influenza, rubella, smallpox, cold sores, AIDS, fever blisters, herpes, mumps, viral pneumonia, the common cold, chicken pox, yellow fever, infectious hepatitis, some warts, measles, infectious mononucleosis, and some cancers. As it does in response to other disease-causing agents, or **pathogens,** the body produces substances called **antibodies** that protect against viruses (Chapter 16). Some antibodies prevent viruses from attaching to host cells, whereas others prevent discharge of the viral genes.

Although the origin of viruses is unknown, the fact that viruses reproduce by providing genes that specify enzymes suggests to some that viruses are degenerate parasitic organisms. Biologically, "degenerate" does not

mean corrupt or perverted, but that an organism has lost structures and functions possessed by an ancestral form. Viruses have also been called "escaped genes" because of their ability to code for protein synthesis coupled with their inability to exist independently of an organism's cellular machinery. Do these traits make viruses simple or complex?

Common criteria used to classify organisms include the structure of their cells, whether they are unicellular or multicellular, and their mode of nutrition (e.g., autotrophic or heterotrophic). Although Linnaeus described the anatomical traits of all the other species that he classified, he simply quoted Thales for humans: "Know thyself."

Two basic types of cells are recognized. **Prokaryotes** are organisms whose cells lack membranes around their nuclei and other cellular structures (Chapter 3). For example, in the bacteria and cyanobacteria (blue-green algae), the nuclear and genetic material is either scattered throughout the cell or concentrated in one area. Prokaryotes (from the Greek for "before kernel") are the most primitive organisms, and the earliest fossils (at least 3.5 billion years old) are of these cells. **Eukaryotes** ("good kernel") are organisms whose cells contain a membrane-bound nucleus and other clearly defined internal structures. Eukaryotes include protists, fungi, plants, and animals.

While most organisms fit nicely into the five-kingdom classification system, some do not. The few "problem" organisms may well be intergrades between our pigeonhole categories. In that sense, our classification system is a progress report on evolution. See the essay "Viruses" for information on those interesting nonorganisms.

WHO ARE BIOLOGISTS?

Biologists are a diverse lot, and it is not hard to find one. One reason is that while Darwin, Mendel, and many thousands of other biologists are dead, the science of biology is a relatively recent human endeavor and most biologists who have ever existed in history are alive and working today! The same is true for most of the other natural sciences.

Some biologists work primarily in the laboratory, whereas others work principally in the field. Why do these men and women of all races study biology? Although we do not know when human thought began, early humans almost certainly wondered, as you undoubtedly do, about themselves and the other life forms with which they shared the Earth. If we are distinct in some ways from other organisms, then perhaps our desire to know, often simply for its own sake, is what sets us apart. Thus, biologists investigate organisms for basically the same reasons that others primarily study music, history, literature, mathematics, or any other subject. Besides, we are alive and most people are inherently interested in and curious about plants, animals, and other critters.

SUMMARY

1. As a young man, <u>Charles Darwin</u> circumnavigated the globe on a 5-year expedition. Many of his observations were important, especially those on the Galápagos Islands, as he developed his theory of **evolution by means of natural selection,** published in 1859. The concept of evolution is the cornerstone of modern biology.

2. Various biologists expressed evolutionary theories before Darwin. **Lamarck's** erroneous view was based upon the **inheritance of acquired characteristics.** Evolutionary changes occur within **populations** of interbreeding individuals because only certain ancestors survived and reproduced.

3. **Creationism** is the traditional view that species or kinds of organisms were created by God individually at about the same time and in essentially their present form. Unlike scientific theories, creationism generates no testable hypotheses.

4. Significant to Darwin's thinking was **Lyell's** concept of **uniformitarianism,** which states that processes now modifying Earth's surface have acted similarly throughout geologic time, except probably at different rates. Also important was Malthus's view that populations tend to increase by **geometric progression.**

5. The ability of organisms to reproduce under ideal conditions—their **biotic potential**—is high, yet the population size of most species, humans excepted, remains relatively constant. Typically, only a few offspring survive and reproduce, usually those variants that have physical and behavioral traits best fitting the immediate environmental conditions. This is the process of natural selection.

6. In **artificial selection,** humans allow only those organisms with selected traits to breed, while they prevent others from reproducing.

7. Although the **scientific method** is difficult to define, there are some common features to scientific research. Most investigations begin and end with **observations.** A common next step is to formulate a **hypothesis,** or tentative explanation of observations. Many hypotheses are tested or are at least testable by **experimentation.** The phenomenon of interest is investigated under conditions that ensure that certain unambiguous observations can be expected based upon the predictions being tested. The conditions being observed and measured in experiments are called **variables.** In **controlled experiments,** all variables are kept identical except for the one being tested. **Independent variables** are the conditions that are being manipulated and controlled by the experimenter. **Dependent variables** are the experimental results. Eventually, a **theory** is developed both to explain a body of facts and to serve as a basis for the further formulation of testable hypotheses.

8. The basic units of organisms are **cells,** and most life processes occur at the cellular level. Some organisms are **unicellular** but most are **multicellular.** It is in the cell that we find matter organized into a state that we call life.

9. **Autotrophs** (such as green plants) are able to produce their own food, whereas **heterotrophs** (such as animals) require preformed food. **Herbivores** eat plants, **carnivores** eat animals, and **omnivores** eat both plants and animals. Ultimately, wastes and dead organisms are acted upon by **decomposers** such as many bacteria and fungi, and the materials of life are recycled. **Metabolism** is the sum of all energy and material processing by organisms.

10. The **behavior** of organisms is far more complex than the relatively simple responses to stimuli made by nonliving objects.

11. **Genes** are the basic hereditary units that constitute the genetic program for the structures and functions of organisms. Offspring often differ genetically from their parent or parents, and natural selection acts upon the resulting variation. Individuals do not change their traits in order to adapt and survive; rather, they survive if their genetic programs provide them with suitable adaptations.

12. Organisms are classified into five **kingdoms** based upon their cell structures, whether they are unicellular or multicellular, and their mode of nutrition. The genetic material and other structures in the cells of **prokaryotes** lack a surrounding membrane, whereas membrane-bound cellular structures are found in **eukaryotes. Viruses** are not alive, although they exhibit some lifelike traits and they affect organisms in major ways.

13. The **Kingdom Monera** includes the bacteria and cyanobacteria, and the **Kingdom Protista** includes some algae and the protozoans. Mushrooms and yeast are common examples of the **Kingdom Fungi,** and familiar plants and animals are members of the **Kingdoms Plantae** and **Animalia,** respectively.

REVIEW QUESTIONS

1. What observations did Charles Darwin make during his expedition on the *Beagle* that influenced the development of his theory of natural selection?

2. Compare and contrast the theories of creationism and evolution.

3. Explain Lamarck's theory of evolution and contrast it with Darwin's theory of natural selection.

4. Define *uniformitarianism* and explain its importance in describing and understanding evolutionary changes.

5. Distinguish between a geometric and an arithmetic progression. Which is characteristic of population increases?

6. If the biotic potential is so high for most species, then why does the population size of most species usually tend to remain relatively constant from year to year?

7. Distinguish between natural selection and artificial selection, using specific examples.

8. Compare and contrast a theory with a hypothesis, using specific examples.

9. Describe examples of controlled and uncontrolled experiments. What is the function of a control in an experiment?

10. State specific examples of dependent and independent variables in a hypothetical experiment.

11. Compare and contrast the general ways that energy and materials are used in metabolic processes.

12. A burning candle flickers, takes in and alters chemicals from its environment, produces and releases wastes, and responds to environmental changes in several ways. How do you distinguish between these activities of burning candles and those of organisms?

13. Distinguish between autotrophs and heterotrophs, using specific examples.

14. State examples of decomposers, herbivores, and carnivores.

15. Compare and contrast the traits of prokaryotes and eukaryotes, and name some examples of each group.

16. Name at least two examples each of the following kingdoms: Plantae, Protista, Animalia, Monera, and Fungi.

17. How do the traits of viruses differ from those of living organisms?

PART TWO

BODY ORGANIZATION AND CONTROL

Conditions on Earth are extremely variable, yet humans and many other animal species have adapted to most of them. We are not surprised to find life almost everywhere we go. Some adaptations are structural, as with the physical organization of the body, and other adaptations are chemical. This part examines both types of adaptations, as well as a general model for understanding how we and other animals respond to changing conditions.

Observing a parrot nesting in Antarctica or seeing a group of kangaroos grazing on a Scottish hillside would be unusual. However, from the constantly warm and humid tropics, to the winter cold and darkness of the Arctic, and to the hot, dry African savannas, humans have adapted and survived quite well, and they exhibit exceptional variation for members of one species. Even regions not inhabited permanently by humans are visited regularly by them.

CHAPTER 2

HOMEOSTASIS AND ADAPTATION: AN INTRODUCTION

Although several million diverse species of organisms live on Earth, all possess one common characteristic—they have mechanisms for survival and reproduction. Organisms survive only if their internal conditions remain within tolerable limits despite changing external conditions. In this chapter, you will study one very important example—temperature regulation—and some of the many body parts involved. Everyone who has shivered or perspired has experienced temperature regulation. Heat is one of the most important physical conditions that affects both organisms and their environments. Temperature regulation (i.e., the way heat is controlled in an organism) presents an excellent example of an important and pervasive regulatory process—homeostasis—that applies to human functions. In other chapters, you will learn about numerous other, but often less obvious, examples of this basic regulatory process.

For this athlete to perform maximally, an extremely wide variety of bodily processes must be working efficiently and in concert. For example, nutrient stores are broken down to release needed energy, the breathing and heart rates increase significantly, the sweat glands pour out fluids that help cool the body, the muscles and tendons send a constant barrage of messages to the brain about their position and condition, and, in contrast, some body systems cease to function. This chapter deals with the general process by which animals maintain their functions at an optimum level and respond to changing internal and external environmental conditions.

FIGURE 2.1
These three animals live in very different environments, all of which are also inhabited by humans. (A) In its nonvarying, almost monotonous tropical rainforest environment, this sloth "hangs around" much of the time, facing few environmental extremes. (B) Lemmings are active throughout the year, surviving the harsh, long arctic winter by using small environments or microclimates that shelter them from environmental extremes. (C) During the desert's searing summer heat, sidewinder rattlesnakes become nocturnal, as do most of their prey. Some of the many homeostatic thermoregulatory mechanisms of humans and other animals are described in this chapter; subsequent chapters describe numerous other homeostatic controls.

(A)

(B)

(C)

HOMEOSTASIS AND ADAPTATION

Organisms that do not survive long enough to reproduce do not pass along a genetic program or "blueprint" to any offspring. **Adaptations** (from the Latin for "to fit") are the heritable traits that help organisms survive and reproduce in their environments. In evolutionary terms, the fittest organisms are generally those that leave behind more fertile offspring with more inherited adaptive traits than their competitors.

Homeostasis (from the Greek for "similar standing") is the process through which organisms maintain their internal conditions within tolerable limits despite changing external conditions. For example, many animals respond to environmental changes so as to keep their body temperatures near an optimum level. Thus, elegant homeostatic controls allow some animals to survive bone-chilling arctic cold and others to tolerate blast-furnace desert heat. We do both, and you would probably not be surprised to find another human anywhere you went on Earth (Figure 2.1). Animals must also have means of securing and distributing food, water, and oxygen to their cells (Chapters 12–15); maintaining stable body conditions when affected by disease-causing organisms or injuries (Chapter 16); and of ridding themselves of wastes (Chapter 17). Naturally, all of these processes must be coordinated (Chapters 18–20), and animals must be able to act upon many of the environmental changes that they sense (Chapter 21). Thus, the self-adjusting balance of homeostasis is a major and intricate refinement of evolution. Walter Cannon (1871–1945), the American physiologist who named the concept during the early twentieth century, called it "the wisdom of the body." In the nineteenth century, the tendency toward maintaining the constancy of the *milieu interieur* or "internal environment" was recognized and studied by the great French physiologist Claude Bernard (1813–1878)(Figure B1.2).

The internal stability of homeostasis does not imply an unchanging condition, and all body functions change to some degree around an optimum level, or set point (Figure 2.2). Thus, the pressure and chemistry of our blood often flucuate above or below the optimum values. The ability to sense and respond to internal and external changes obviously assists an animal in survival.

Homeostatic mechanisms usually have at least three parts: (1) **receptors** that detect changes in the internal and external environments (Chapter 20), (2) **integrators** that coordinate the information received (e.g., the brain and endocrine glands, Chapters 18–19), and (3) **effectors** that produce responses that attempt to return the system to an optimum level (Chapter 21). Receptors in humans include eyes, ears, nose, tongue, and the structures in the skin that sense touch, pain, and heat. Our many internal receptors are not so evident but are equally important. For example, even in the dark you generally know where your body parts are. To demonstrate this, close your eyes and touch with your little finger first the tip of your nose and then your earlobe. Alternatively, close your eyes, extend your arms, and try to touch the tips of your index fingers.

FEEDBACK SYSTEMS

A baby crying in the night could be worrisome to some tired parents. The late, famous cartoonist Rube Goldberg might have solved this problem with one of his intricate inventions that made simple tasks difficult (Figure 2.3). Notice why the phrase "one thing leads to another" describes Goldberg's invention. A breakdown in any one part causes entire

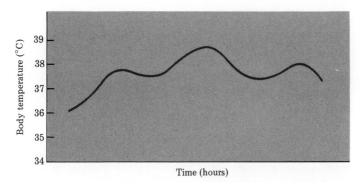

FIGURE 2.2
Over time, human body temperatures
change much as do those in a
thermostatically controlled room. One's
environment, health, body activities, and
even emotions can cause such changes to
occur around the average set point of 37°C.

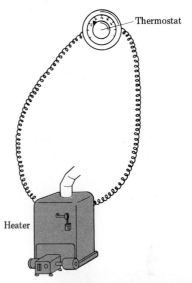

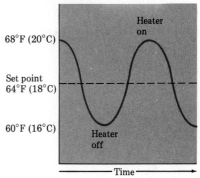

rigid systems like this to fail. For example, the string may break, the
pistol may be empty or not aimed precisely, the electric stove may not
be plugged in or there may be a power failure, there may be insufficient
water in the pot, the candle may be too short, or the crows may not be
hungry. Many familiar devices have similar faults. For example, in most
automobiles, the burning of fuel in the cylinder moves the pistons, the
piston rods move the crankshaft, and so on. As with Goldberg's device,
automobiles are generally built rigidly rather than flexibly, and a failure
of one critical part may cause the entire automobile to stop.

A system is a set of parts that interact with each other for some
purpose. Thermostats are common parts of **feedback systems,** systems
so named because information about a change in the system is fed back
to influence the response. For example, thermostat-heater or thermostat-
heater-cooler systems help keep room temperatures at or near the desired
level even though outside temperatures change.

Elements within thermostats are sensitive to temperature changes,
and a heater is turned on when the room temperature falls below the
set point. When the room temperature rises sufficiently above the set
point, the heater is turned off and/or the cooler is turned on. The room
temperature then falls until it is below the set point, and the process is
repeated. Thus, the room temperature stays close to the desired level.

There are two basic types of feedback systems: negative and positive.
Although we often use the word *negative* in the sense of "bad" or
"wrong," Figure 2.4 shows why the thermostat-heater or thermostat-
heater-cooler systems are called **negative feedback** systems. A decrease
(−) in room temperature leads to an increase (+) in output by the
heater, and an increase (+) in temperature above the set point leads to

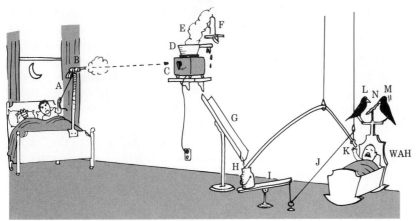

Pull string (A) which discharges pistol (B). Bullet (C) hits switch on electric stove (D), warming
pot of water (E). Water vapor melts candle (F) which drips on handle of pot causing it to upset
and spill water down trough (G) and into can (H). Weight bears down on lever (I), pulling
string (J) which brings nursing nipple (K) within baby's reach. Baby's crying wakens two pet
crows (L & M) and they discover rubber worm (N) which they try to eat. Unable to chew it,
they pull it back and forth causing cradle to rock baby to sleep.

FIGURE 2.3
What to do without having to get up, when
the baby cries in the middle of the night,
with apologies to Rube Goldberg. What
rigid, one-thing-leads-to-another devices do
we use in our lives?

FIGURE 2.4
Negative (A) and positive (B) feedback systems. Note that in negative feedback, the sign of the response is opposite that of the stimulus. In positive feedback, the two signs are identical, which causes a runaway rather than a corrective condition. Some of the main feedback events in the control of room and body temperatures are shown in (C).

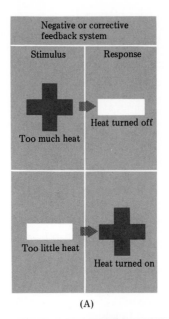

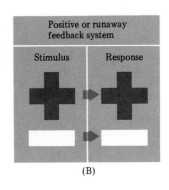

(A)

(B)

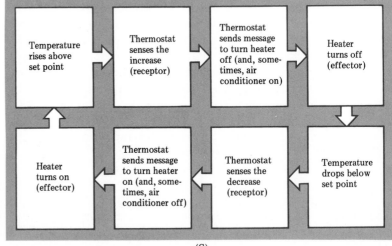

(C)

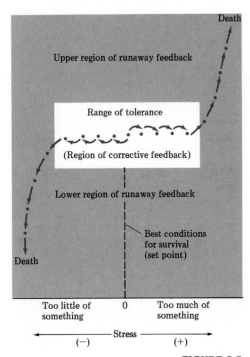

FIGURE 2.5
The general model of homeostasis. External stress may be caused by excessively high or low temperatures as well as many other conditions. Internal stress may occur when chemical substances in the blood, such as glucose, hormones, or dissolved gases, are at too high or too low a level.

a decrease or cessation (−) in heat input. Thus, negative feedback systems operate to keep body processes within tolerable limits because changes lead to events that return the system to its set point or optimal level. For this reason, negative feedback is often called *corrective feedback.* Changes in the size of the pupil in your eye when you are exposed to different levels of light represent another good example of negative feedback—large pupil in dim light and small pupil in bright light.

If an air conditioner further cools an already cold room or a heater stays on when a room is already very warm, you might suspect a faulty thermostat. This is comparable to what happens in **positive feedback,** a chain of events that intensify the original input. Thus, conditions sometimes get worse rather than better, which we commonly call the "snowball effect." Because of this and because deficiencies and excesses are often not corrected, positive feedback is also known as *runaway feedback.* Although it occurs in many disease states and may lead to death, positive feedback is beneficial in some other cases. For example, sexual arousal may lead to increased stimulation, which results in further stimulation until the explosive climax stage is reached (Chapter 5). Another example of positive feedback occurs during childbirth. The pressure of the fetus on the uterus increases the production and secretion of a chemical (the hormone oxytocin), which causes muscles in the uterine wall to contract. This increased pressure on the fetus leads to production of still higher levels of the hormone, and so on, until the fetus is expelled from the mother's body.

Stress occurs when any internal or external condition tends to upset the normal operation of a system. Internal stress may occur when parts age, nutrients are depleted, or wastes accumulate. Although organisms are usually under constant stress, they survive as long as negative feedback systems operate. However, every system has its limits, or its **range of tolerance.** Corrective feedback usually operates well when stress is within the range of tolerance and runaway feedback often occurs when conditions are too stressful. Figure 2.5 shows the general way that body systems react to a changing environment.

In humans and other higher animals, the cells that sense blood temperature and that operate somewhat like a thermostat are found in the brain region called the **hypothalamus** (Figure 2.6). For example, depending upon the stimulus, cells in the hypothalamus may send impulses to sweat glands to increase perspiration, or to appropriate muscles to stimulate shivering. However, extreme body temperatures may cause the thermostat to fail. Humans seldom survive for long with a fever higher than about 43°C (110°F), and even many hours at a lower but above-normal temperature may be fatal. The many thousands of chemical reactions that control body processes are well integrated at the body's normal temperature. While it is true that the higher the temperature the faster many chemical reactions occur, not all chemical reactions are alike. For example, when the temperature increases, one reaction may double its rate but related reactions may increase by factors of 1.6, 3.2, or 0.8. Disability or death may result from the metabolic imbalances. The faster chemical reactions resulting from high body temperatures increase the body temperature still further, which speeds chemical reactions still more, and so on—an example of runaway or positive feedback.

FIGURE 2.6
The tiny hypothalamus is nearly in the middle of your head, and its numerous activities are central to your survival. In its cells are centers that control temperature, thirst, appetite, various emotions; acting through the nearby pituitary gland, it controls many other functions through the hormones and other substances that it produces.

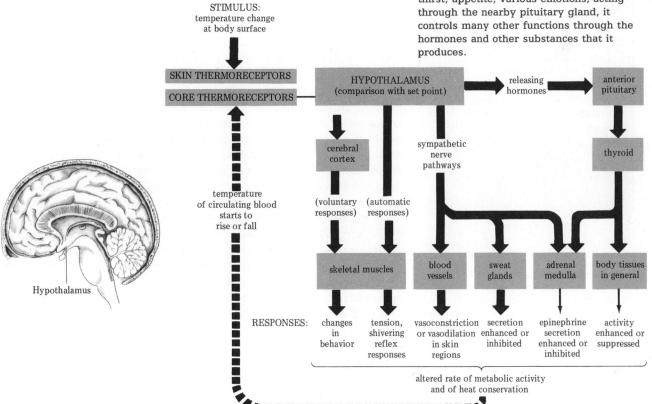

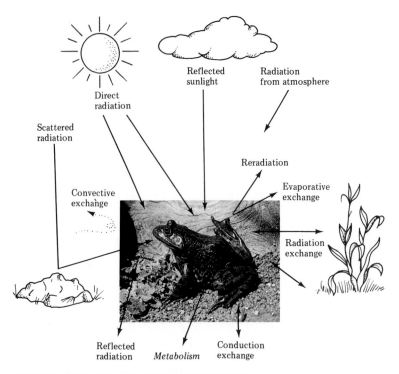

FIGURE 2.7
An animal and its complex thermal environment. How might terrestrial animals gain or lose heat by radiation, evaporation, conduction, or convection? How are thermoregulatory challenges different for aquatic animals?

AN EXAMPLE OF HOMEOSTASIS: TEMPERATURE REGULATION

Almost everything an organism is able to do depends upon its temperature. Temperature extremes on Earth far exceed what organisms can tolerate. Although specialized bacteria thrive in the nearly boiling water of some hot springs, few other organisms survive temperatures over 55°C (131°F). Also, most aquatic organisms cannot tolerate subfreezing water temperatures, as some polar marine fish and crustaceans can. Most of life's organic compounds function only between 0° and 40°C (32°–104°F). Above about 40°C, proteins break down or denature, and this leads to deteriorating chemical functions (Chapter 4). Body temperature control is an excellent example of negative feedback. For example, between blistering summer heat and numbing winter cold, the human body temperature remains about 37°C (98.6°F).

Most of our homeostatic mechanisms operate without our being conscious of them. Thus, the kidneys maintain the quantity and composition of our body fluids, and our nervous system and the diverse hormones from our endocrine system control many metabolic processes. However, we are conscious of many aspects of temperature regulation, such as when we shiver or perspire. Therefore, you will begin the study of homeostasis with this familiar topic and will investigate numerous other examples of homeostasis in other chapters.

To survive, animals must either already exist in suitable thermal environments, move to environments with appropriate external temperatures, or maintain internal temperatures within a range in which their various body systems can operate effectively. Fortunately, environments with tolerable temperatures are commonplace in Earth's waters, and many aquatic animals simply move to an area where suitable temperatures exist. In other cases, the only survivors are those whose early developmental stages settled and grew in a tolerable temperature environment. However, the thermal problems facing terrestrial organisms are more serious because air and surface temperatures often far exceed the tolerances of these organisms.

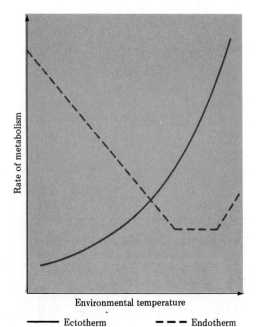

FIGURE 2.8
The general relationship between rate of metabolism and environmental temperature for ectotherms and endotherms. The body temperatures of ectotherms often change directly as their environmental temperatures change. When the body temperatures of endotherms extend beyond the optimal levels, metabolic rates increase as the animals attempt to maintain the desired levels.

Heat Production and Heat Loss

Body temperature results from two processes—heat gain and heat loss. Heat is a by-product of most energy transformations. Living organisms continually produce heat and either gain it from or lose it to their surroundings. The main ways that heat is exchanged between animals and their environments are by conduction, convection, radiation, and evaporation (Figure 2.7). In **conduction,** heat moves because of the contact of objects of different temperatures, as when a spoon warms in a cup of hot soup. **Convection** involves the movement of a mass of air or water, as when air rises after being warmed by a heater. **Radiation** is the transmission of energy through space, as when the sun heats Earth or when food is kept warm under infrared lights in a restaurant. In cooling **evaporation,** heat energy is released to the air from body surfaces, as when liquid water changes to a gas.

Temperature Terms

Humans and most mammals and birds are called **endotherms** because they produce and control most of their heat by means of their metabolisms. These animals are often called *warm-blooded* because they usually maintain fairly high and constant body temperatures despite changing environmental temperatures (Figure 2.8). Most mammals have body temperatures between 36° and 38°C. Average bird temperatures are somewhat higher, between 40° and 42°C.

Most other animals are called **ectotherms** because their body temperatures depend upon an external heat source. Although they are popularly called *cold-blooded,* this term does not describe accurately the body temperature of many ectotherms. For example, active desert iguanas (Figure 2.9) have an average body temperature of about 42°C, a temperature that is lethal to some birds and to most mammals. Also, some ectotherms can temporarily raise their body temperature above environmental levels. This is true of pythons incubating eggs; large, fast-swimming fish such as tuna; and some flying insects. On the other hand, deep-sea fishes have an almost constant body temperature because of their stable thermal environment. See the essay "Temperature Adaptations of Ectotherms."

To maintain their more-or-less constant temperature, endotherms must expend energy constantly. Notice in Figure 2.8 that at some point the amount of energy required for thermoregulation reaches its lowest level. Above or below this optimum level, the metabolic rate increases. Although the only continuously endothermic animals are found among birds and mammals, not all of these animals are endothermic continuously. For example, some endotherms become dormant under harsh conditions and their body temperatures may approximate those of their environments, just as in the case of ectotherms.

The presence of birds and mammals is not considered unusual in deserts but people are often surprised by the numbers of reptiles in such bleak environments. Endotherms use about 80 percent of the energy in their food to regulate their body temperatures. Freed of the necessity to maintain a constantly high body temperature, an ectothermic lizard can survive on much less food and, therefore, can flourish in warm environments where fuel to support their inner furnaces may be inadequate for many endotherms. However, in harsh, cold, high-latitude and high-altitude environments, endothermic vertebrates may abound where few ectothermic vertebrates can.

FIGURE 2.9
(A) The term *cold-blooded* obviously does not apply to active desert iguanas *(Dipsosaurus dorsalis),* which can tolerate a body temperature of 47°C (117°F). (B) Spasmodic contractions of female python muscles produce heat to warm the eggs that she is coiled around. (C) The elevated temperatures of the tuna's deep swimming muscles undoubtedly aid these fast-swimming fish (up to 70 km/hr, or 44 mi/hr) in capturing rapid prey such as squid and herring.

(A)

(B)

(C)

Temperature Adaptations of Ectotherms

Some of the ectotherms shown in Figure B2.1 are sessile, or who live attached to the substratum. Others are **motile** and are able to swim, crawl, or fly through their environments or are transported by them. Sessile ectotherms usually have body temperatures very similar to those of their environments. Living ectotherms always produce some heat metabolically but it is usually lost quickly to their environment, especially if they are aquatic. Animals that are sessile as adults often have motile larval stages. Although many of these larvae do not survive the thermal conditions of their environment, those that do may become attached and develop into adults. Note in Figure B2.1, however, that some motile adults have essentially sessile young.

Many motile ectotherms move back and forth between warm and cool parts of their environments. For example, fiddler crabs return to their cool burrows when surface temperatures are high.

Obviously, subfreezing temperatures may cause death. Ice crystals damage cell structures and may cause dehydration as liquid water turns solid. Metabolism is slowed and disorganized by low temperatures, and some important chemical compounds may become damaged. However, some ectotherms may withstand very low environmental temperatures before death from freezing occurs. Sometimes this occurs because sessile or over-wintering animals produce antifreeze compounds that reduce the freezing point of their body fluids. Motile ectotherms often move to areas where freezing temperatures are not likely. Many freshwater animals such as frogs, turtles, fish, freshwater mussels, and crayfish spend the winter in the bottom mud of lakes and ponds.

Many ectotherms also enter a state of dormancy in response to heat and drought. In desert snails, for example, the inactive period may last for years until a heavy rain occurs. As the unfavorable time approaches, these animals seek refuge in rocks or the soil, and seal the opening of their shell. During the summer months, many kinds of earthworms burrow deep into the soil and line their small chambers with mucus. Similarly, African and South American lungfish may live in a mud cocoon for over a year in times of heat and drought.

FIGURE B2.1
Hydras (A) are common freshwater sessile organisms whose body temperatures are usually identical to those of their environments. In contrast, the motile rattlesnake (B) may move to suitable thermal environments. Although the adult mockingbird (C) is motile and endothermic, its young are essentially sessile and ectothermic for several days after hatching, before they acquire insulating feathers. Parent birds must choose suitable thermal environments for the nest as well as brood the young, much as many mammals do for their hairless and ectothermic young.

(A)

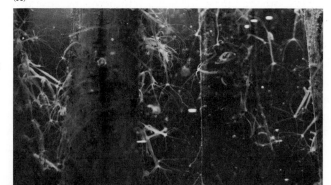

(B)

(C)

Size and Temperature

Body size and shape have much to do with how well an animal adapts to extreme temperatures. Study Figure 2.10 and Table 2.1 to see why. Although most organisms are not cubical, cubes are regularly shaped objects, so calculations about them are simple. Throughout this text, you will study numerous other examples of the importance of **surface area-to-volume relationships.**

Note that small objects have more surface area relative to their volume than do large objects. Figure 2.11 shows that the rate of metabolism in mammals is related closely to body size. The same trend applies when the metabolic rates of young and adults are compared. For example, newborn human infants breathe 50–60 times per minute and have pulse rates of about 120–160 beats per minute. The comparable rates for most resting adults are about 12–16 breaths per minute and 65–85 beats per minute. Despite their differences in metabolism, however, most mammals have life spans that are about equal relative to one another. Thus, the life span of a mouse is not "less" than that of a rhinoceros, it is only faster. Whether large or small, most mammals breathe about the same number of times during their lives.

The amount of heat exchanged between an animal and its environment depends upon the animal's exposed surface area. For example, the body size of many endotherms that live in cold areas tends to be greater than that of their warm-region relatives, the basis for **Bergman's principle** (Figure 2.12). Also, the extremities, such as the ears of cold-region endotherms, tend to be smaller than those of warm-region relatives, a

FIGURE 2.10

The smallest and largest cubes referred to in Table 2.1. Note that while the surface area of cube D increased 100 times more than that of cube A, its volume increased 1000 times. The same patterns apply to objects of other shapes, and humans are actually more like sets of cylinders than cubes.

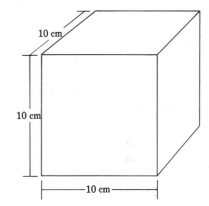

TABLE 2.1

Surface Area-to-Volume Relationships

Cube	Length of Edge of Cube (cm)	Surface Area of Cube (cm²)	Volume of Cube (cm³)	Ratio of Surface Area to Volume
A	1	6	1	6 to 1
B	2	24	8	3 to 1
C	3	54	27	2 to 1
D	10	600	1000	0.6 to 1

Body weight

FIGURE 2.11

The general relationship between body size and metabolic rate. In general, the metabolic rates of large and of mature animals tend to be lower than those of small and of young animals.

(A)

(B)

(C)

FIGURE 2.12
The desert kit fox (A), the temperate forest red fox (B), and the arctic fox (C). Because large animals have a relatively small surface area compared with small animals, they may survive well in cold regions. The large surface area of the extremities of the kit fox provides thermal ''windows'' for the loss of excess body heat to the environment.

Treading Water   Drownproofing*

*Head is raised and lowered alternately for breathing purposes while one floats in a still position.

H.E.L.P.*   Huddling

*Heat
Escape
Lessening
Posture

FIGURE 2.13
Four positions and behaviors for survival in cold water. How does treading water assist survival? Contribute to loss of body heat? In the ''drownproofing'' position, the head is raised and lowered alternately for breathing purposes while one floats in a still position. How could drownproofing assist and hinder survival? The H.E.L.P. and huddling methods are useful while wearing life preservers. How do each of these positions contribute to survival in heat-robbing cold water?

relationship called **Allen's principle.** Large animals with small extremities have less surface area exposed to the environment than do small animals with large extremities. These principles have sometimes been applied to humans in a general way. For example, some human populations native to cold areas are short and plump, whereas some populations native to tropical areas are tall and thin.

In 1985, the remains of the *Titanic* were found. Seventy-three years earlier, on its maiden voyage from Europe to America, this "unsinkable" ocean liner struck an iceberg and sank. Rescuers quickly recovered the dead bodies of over 1500 people. None had drowned, and all wore life preservers. The cause of death was uncontrolled **hypothermia,** a drop in body temperature below the tolerance level. Today, thousands of people die annually from hypothermia, often from falling into cold water, wearing wet clothing during cool weather, or, to reduce heating costs, keeping living areas too cold. As many people who work or play outdoors know, body heat is lost to water about 30 times faster than it is to air.

Figure 2.13 and Table 2.2 show that survival time in cold water depends upon several variables. For example, although the muscular activities of swimming produce about three times as much body heat as floating does, the extra heat is lost rapidly to the water due to increased blood circulation to the arms, legs, and skin. Flotation devices help one remain at the surface without having to tread water or use the drownproofing position, which exposes a large body surface area to the water. Reducing the exposed surface area, as in <u>h</u>eat <u>e</u>scape <u>l</u>essening <u>p</u>osture (H.E.L.P.) and huddling, increases survival time.

Body heat warms the air next to the body, but wind moves the warmed air away from the exposed areas (Figure 2.14). The **wind-chill effect** may cause frostbite and hypothermia in windy weather at seemingly mild temperatures.

ADAPTATIONS OF ENDOTHERMS TO HIGH TEMPERATURES

When too hot, many endotherms either pant or perspire in order to lose heat by evaporative cooling. Hot cats lick their fur. Even before humans begin sweating, their skin may become red due to **vasodilation,** or an increased blood flow in dilated blood vessels to the surface. Vasodilation facilitates the loss of body heat to a cooler environment.

To demonstrate evaporative cooling, an English scientist in 1775 took some friends into a chamber where the air temperature was 126°C (259°F). The humans were unharmed by their 45-minute stay, although a steak they took along was cooked during that time! The chamber also contained two pots of water: The water in the pot that was covered with an oil layer to retard evaporation boiled, but the water in the other pot, which simulated human skin, only became very warm.

Some sweating or insensible perspiration occurs at all times, even when you are quite cold. However, sweating is most noticeable when you work or exercise hard or are exposed to warm, humid air. Although sweating helps cool the body, high rates of sweating cannot continue for long. The body temperature rises about 2°C when water loss reaches about 10 percent of body weight. The dilation of surface blood vessels and the increased viscosity of the blood due to fluid loss strain the heart and many other organs. Naturally, when much blood is near the skin surface, less blood is available for such all-important structures as the brain, heart, lungs, and kidneys.

"Everyday" thirst occurs from drying of the mouth and throat, but the thirst center in the hypothalamus also detects the more serious symptoms of water loss. During dehydration, camels may lose about one-third of their normal weight but they can make up their water deficit in about 10 minutes. Dehydrated donkeys can drink almost 8 L of water per minute. Although the soft drinks and other beverages that many thirsty humans consume do help to make up high water losses, some result in more body heat as their sugars are metabolized.

Bird adaptations to high environmental temperatures are especially interesting. Most birds are small, with high metabolic rates, and relatively few use underground retreats or become active only during the cooler nighttime hours, as many mammals do. However, the high average body temperature of birds gives them an advantage. There are relatively few times when environmental temperatures are higher than bird body temperatures, meaning that birds can lose heat passively. Although birds do not sweat as many mammals do, very hot birds pant or flutter the throat region to lose heat through evaporation from the mouth cavity (Figure 2.15). Overheated storks use an unusual but effective form of evaporative cooling: They urinate periodically on their long legs!

Thick fur keeps some mammals cool just as it keeps others warm. Hair is nonliving and it is undamaged by temperatures that would harm living structures. In some desert mammals, the outer fur gets as hot as 85°C (185°F) although the skin temperature remains normal. Some humans living in deserts wear heavy clothing, partly to take advantage of the same principle. Light-colored fur or clothing reflects sunlight and thus aids in thermoregulation.

TABLE 2.2

Survival Time in Cool Water (10°C or 50°F) of an Average Clothed Human Adult

Situation	Average Survival Time (in Hours)
No flotation	
Drownproofing	1.5
Treading water	2.0
With flotation	
Swimming	2.0
Floating	2.7
H.E.L.P.	4.0
Huddling	4.0

FIGURE 2.14

The wind-chill factor is an index of the cooling power of wind in terms of the equivalent temperature without wind.

Wind Speed (mi/hr)	Actual thermometer reading (°F)											
	50	40	30	20	10	0	−10	−20	−30	−40	−50	−60
	Equivalent temperature (°F)											
0	50	40	30	20	10	0	−10	−20	−30	−40	−50	−60
5	48	37	27	16	6	−5	−15	−26	−36	−47	−57	−68
10	40	28	16	4	−9	−21	−33	−46	−58	−70	−83	−95
15	36	22	9	−5	−18	−36	−45	−58	−72	−85	−99	−112
20	32	18	4	−10	−25	−39	−53	−67	−82	−96	−110	−124
25	30	16	0	−15	−29	−44	−59	−74	−88	−104	−118	−133
30	28	13	−2	−18	−33	−48	−63	−79	−94	−109	−125	−140
35	27	11	−4	−20	−35	−49	−67	−82	−98	−113	−129	−145
40	26	10	−6	−21	−37	−53	−69	−85	−100	−116	−132	−148
Wind speeds greater than 40 mi/hr have little additional effect	Little danger (for properly clothed person)			Increasing danger			Great danger					

FIGURE 2.15 (A) (B)

(A) Sitting under intense sunlight in the open, these boobies pant to reduce excessive body heat through evaporation of moisture from their mouth cavities. Exercising or hard-working humans accomplish the same thing through perspiration. Although copious perspiration for short periods of time can be tolerated, prolonged perspiration may lead to excessive water loss. Humans are generally incapacitated when water loss exceeds 10 percent of their body weight. (B) On the other hand, dehydrated camels may lose over one-third of their body weight before being in danger.

Because endothermy is expensive metabolically, "controlled" hypothermia is often useful. Whereas hibernation (see next section) is common in winter, **estivation** is a response to summer conditions by some endotherms (Figure 2.16). In both cases the body temperature drops, although not as low in estivation as in hibernation. Such body activities as breathing, digestion, kidney functioning, and the heartbeat also decline. These changes reduce significantly the animal's energy needs. The fact that less fluid is lost through breathing and excretion is especially important for small desert mammals. Between the extremes of hibernation and estivation are many cases of daily torpor—having a reduced body temperature and metabolic rate for a few hours daily. Like the seasonal forms of hypothermia (hibernation and estivation), daily torpor leads to reduced energy demands.

FIGURE 2.16

Animals such as the hibernating chipmunk remain endothermic and are still maintaining their body temperatures homeostatically, although at a lower level than when they are active. Because the difference between their body temperatures and those of their immediate environments have been reduced, less body heat is lost to the environment, much like as when your extremities become cool. The curled posture also exposes less body surface area to the environment. Becoming dormant during times of thermal stress reduces metabolic needs. Some animals take advantage of this fact and become dormant (torpid) during a part of each day.

The thermoregulatory challenges presented to many endotherms by the dryness and heat of deserts are often more acute than those presented by cold conditions. This is because the control of heat loss to a cold environment is usually easier than the control of heat gain from a hot environment. It is fortunate, therefore, that birds and mammals have high body temperatures. For example, if body temperatures were only 30°C (86°F), it would be difficult or impossible to produce enough water for evaporative cooling when exposed to very hot air.

Some desert mammals active during the day relax their thermoregulatory controls and allow their body temperatures to rise (Figure 2.17). For example, a camel's body temperature may be 34°C in the morning but about 40°C by evening. When body temperatures exceed environmental levels, body heat is lost passively to the environment without dependence upon evaporative cooling. When the smaller antelope ground squirrel gets too warm, it runs underground for a few minutes. There, it flattens itself against the relatively cool ground of its burrow and loses body heat to it by conduction. Actually, most endotherms are warmer than their environments most of the time. How many minutes during the past year were you exposed to air temperatures in excess of your normal body temperature of 37°C?

Unfortunately, human activities often threaten the homeostatic balances of aquatic animals by changing the thermal environments to which organisms have become adapted. For example, many animals and other organisms are literally in hot water. Factories, power plants, homes, and vehicles all heat the environment. Any major change in the normal environmental temperature is **thermal pollution,** and excess heat presents many challenges to homeostasis.

Power plants, especially nuclear facilities, use enormous amounts of water to keep machinery cool, sometimes several million liters of water a minute. Much of the excess heat ends up in water, causing temperature increases along rivers used by cities and industries. The vertical dotted lines in Figure 2.18 show the distance along a typical river bordered by human cities and industries.

Dissolved gases are needed by aquatic animals and by many decomposers of wastes, such as bacteria. However, as water temperature increases, the amount of dissolved oxygen decreases. Thus, in hot water, animals receive a "double whammy." The warmer the water, the faster the metabolic rate of ectotherms such as fish. However, at the very time when such animals need the most oxygen to support their increased metabolic rate, the available oxygen in their environment is decreasing.

FIGURE 2.17
Comparative daytime body temperatures of camels and antelope ground squirrels. Not only does the relaxation of precise control over temperature help such animals thermoregulate but it also conserves precious body water.

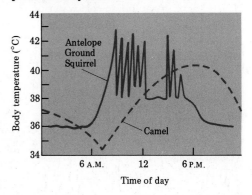

FIGURE 2.18
Water temperatures along a typical industrialized river, with the width of the dotted lines indicating each city. Note the increase in water temperature along each urbanized stretch and some recovery between communities. What does hot water do to the dissolved oxygen content of water? To the metabolic rate of aquatic animals?

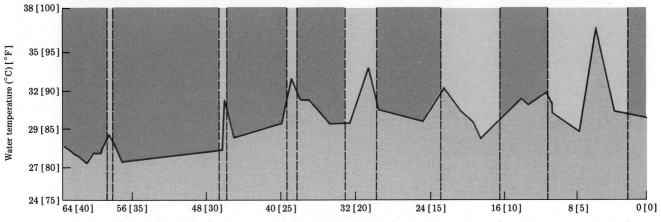

Distance upriver (km) [mi]

FIGURE 2.19
The air trapped beneath the feathers of the lower owl helps to insulate it from the cold. For similar reasons, several layers of lightweight clothing usually keep humans warmer than one layer of heavy clothing. In some mammals, thick fat serves to reduce heat loss to cold water or air.

Warm conditions

Cold conditions

ADAPTATIONS OF ENDOTHERMS TO LOW TEMPERATURES

The range of responses of endotherms to cold is more extensive than the variety of responses to hot conditions. The adaptations fall generally into two categories—increasing heat production and decreasing heat loss. The former often requires an increase in food intake, and this is difficult for many terrestrial birds and mammals during the coldest times. Therefore, conserving body heat is a common adaptation of polar endotherms. For example, air is a good insulator, and air trapped among raised hairs or fluffed feathers provides superb insulation that helps many mammals and birds conserve body heat (Figure 2.19). Because of our minimal body hair, we cannot use this mechanism effectively, except in our choice of clothing. Because insulative air is trapped between layers, several layers of lightweight clothing often keep us warmer than one layer of heavy clothing. One way we raise what little body hair we have is by means of "goose bumps." Shivering is far more effective than goose bumps because more muscles move as we shiver, even though relatively little metabolic energy is used. By shivering, humans may increase heat production by three to four times. Shivering even occurs during sleep, and it need not be visible to be effective. Many other mammals and birds also shiver. Some mammals increase heat production by metabolizing deposits of so-called brown fat (nonshivering thermogenesis) that are found in the shoulder region and around large blood vessels.

Having a cold nose, hands, feet, and ears may be very uncomfortable. However, the **vasoconstriction** of surface blood vessels that accounts for this discomfort helps conserve body heat because heat loss depends upon the difference in temperature between the body and the environment. Thus, as the old saying goes, we "turn blue" in extreme cold. Some body temperatures of cold-region animals that often walk on ice or swim in cold water are shown in Figure 2.20. In herring gulls, part of a nerve that extends from the spinal cord to the feet is exposed simultaneously to temperatures ranging from 6°C to 41°C (43–106°F).

FIGURE 2.20
Some temperatures of body parts of an arctic wolf and a herring gull. Less heat is lost to the environment by cold extremities than by warm ones. This is why it helps us adapt to cold when we "turn blue" due to reduced blood flow to the skin. Physicians sometimes reduce the metabolic rate of patients by placing them in an ice bath before performing certain types of surgery.

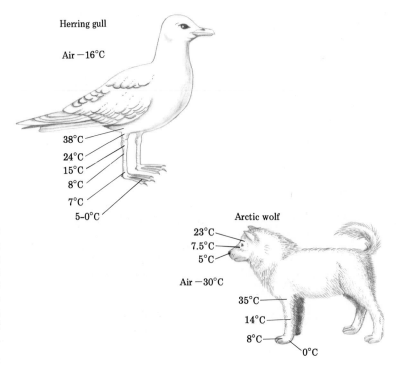

Herring gull

Air −16°C

38°C
24°C
15°C
8°C
7°C
5–0°C

Arctic wolf

23°C
7.5°C
5°C

Air −30°C

35°C
14°C
8°C
0°C

(A)

(B)

The insulative qualities of the polar bear's fur are largely lost when it is wet (Figure 2.21). Thick layers of fat are far more effective than fur in insulating aquatic endotherms against heat loss to cold water. Thus, while whales are largely hairless and their skin temperatures may be about the same as those of the water, there is a sharp upward gradient going in to the mammal's high core body temperature. So effective is fat insulation that seals must have heat-dissipating mechanisms to avoid overheating when they are in warm water or during strenuous activities, as when they crawl about on land to mate. When necessary, blood is shunted into tiny vessels that are not covered by blubber and excess body heat is lost quickly through these "thermal windows" to the environment.

Warm blood in the arteries that carry blood from the heart could

FIGURE 2.21
(A) This polar bear is losing heat quickly to the water because the insulation normally provided by its thick fur is largely lost when the fur is wet. (B) The skin temperature of these walruses closely approximates that of water although the temperature of the inner portion of its insulative blubber is near that of the core body temperature. Special blood vessels (capillary beds) outside of the blubber provide "thermal windows" that allow many marine mammals to dissipate excess body heat when necessary, such as when they must move about on land. We shed heavy coats when we exercise or enter warm areas to accomplish the same goal.

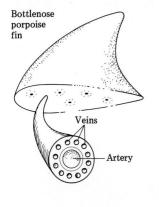

Bottlenose porpoise fin

Veins

Artery

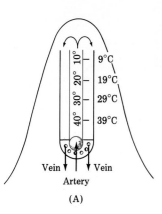

9°C
19°C
29°C
39°C

Vein   Vein

Artery

(A)

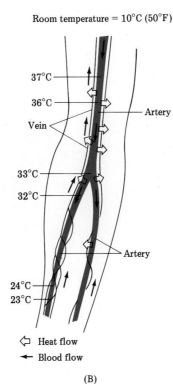

Room temperature = 10°C (50°F)

37°C

36°C

Artery

Vein

33°C

32°C

Artery

24°C

23°C

⇦ Heat flow
← Blood flow

(B)

FIGURE 2.22
(A) Countercurrent circulation in the bottlenose porpoise fin. Some heat from the outflowing arterial blood warms the incoming blood in the nearby veins. This reduces heat loss to the environment and prevents the excessive cooling of the internal body organs. (B) Human countercurrent circulation is not as efficient but exists to some extent in our arms. What thermoregulatory problems could occur without countercurrent circulation?

allow excessive heat loss to the environment. Also, cold blood in the veins that return blood from the skin surface to the heart could chill the body's core. In one type of **countercurrent circulation,** the blood flow in arteries is opposite that in the nearby veins (Figure 2.22). Thus, arterial blood flowing out to the body surface is cooled as some of its heat passes to the venous blood flowing inward from the surface. The cooled arterial blood that reaches the surface loses relatively little heat to the environment, and the venous blood is warmed significantly before it reaches the body's core. While humans have only limited countercurrent circulation, such systems are highly developed in some animals. Other important examples of countercurrent circulation are discussed in other chapters.

It is easier metabolically for a large endotherm to live in a low temperature than it is for a small one to do so. Overall, in fact, life is a low-temperature phenomenon. Low temperatures are generally less damaging to the biochemical integrity of most organisms than high temperatures, unless the water in cells freezes. The evolutionary transition from aquatic to terrestrial life represented a major step in terms of adjustments to temperature.

During **hibernation,** the body temperature often drops to within a few degrees of environmental levels. Breathing and heartbeat rates are also reduced to a fraction of their normal levels. A low body temperature reduces metabolic needs and heat loss to the environment. Many kinds of small ectotherms can go for weeks or months without eating when they are exposed to cold. Many small endotherms in cold could starve

FIGURE 2.23
(A) Comparative metabolic rates during a 24-hour period of time between a shrew and a comparably sized hummingbird. (B) Although they are endothermic when active at night, some white-footed mice *(Peromyscus)* may enter a state of torpor for a few hours during the day, thus reducing energy expenditure for heat production. (C) Some bats not only hibernate during the winter but also exhibit daily torpor during the summer. Science fiction writers often propose controlled hypothermia as a means of reducing the metabolism of astronauts on long space voyages.

(A)

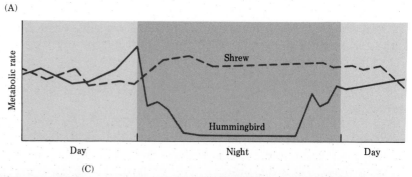

(B)

(C)

in a few hours if they did not eat. These birds and mammals often consume food equal to their own weight daily to maintain endothermy.

Although hibernation is seasonal in some animals, a reduced body temperature called **torpor** may occur daily in such small endotherms as hummingbirds (Figure 2.23). Equally tiny shrews must be active 24 hours a day to feed. However, because hummingbirds do not feed at night to support their very high metabolic rates, they may enter a hibernation-like state daily. These tiny birds are usually just a few hours from death by starvation, and some might not survive the night without a reduction in metabolism. Daily torpor is also common in many small rodents.

BEHAVIORAL THERMOREGULATION

Most examples thus far have emphasized the structural or physiological features of thermoregulation. However, behavioral thermoregulation is also important in both ectotherms and endotherms. For example, many desert reptiles move back and forth between warm and cool areas, and some of these ectotherms can behaviorally maintain their body temperatures during the day almost as precisely as endotherms do physiologically (Figure 2.24). During the long, hot summer, both ectotherms and endotherms may change from daytime (diurnal) to nighttime (nocturnal) activity. Others may migrate, thus avoiding the unfavorable conditions.

The physical conditions immediately surrounding animals, commonly called **microclimates,** are often quite different for small animals compared with those around larger ones (Figure 2.25). Both conditions may differ significantly from those found in a standard weather box, which is a microclimate inhabited by very few organisms, but which is the one from which we secure most of our weather measurements. Because we are large animals, as are many of our domestic animals, we tend to forget that the world of most animals, and of organisms in general, is one of crevices and cracks, holes in logs, burrows in soil, and other small places. It is clear that what appears to us to be a harsh environment may be quite tolerable for the small organisms that live in appropriate microclimates.

An excellent example is found in the fascinating kangaroo rat (*Dipodomys*), a common American desert rodent. Kangaroo rats avoid extreme desert conditions by living in burrows during the day and by becoming nocturnal. If they were exposed to daytime heat, these small endotherms could lose over 10 percent of their body weight hourly to maintain their body temperature by evaporative cooling. This is obviously difficult or impossible, especially since kangaroo rats do not drink! Instead, they get the moisture they need from digesting the seemingly dry seeds they feed upon. Although kangaroo rats have specialized kidneys to reduce urinary water loss (Chapter 17), some water is lost in the feces. Like a number of other mammals, kangaroo rats may reduce this problem somewhat by eating, or recycling, their feces (coprophagy).

Large desert mammals cannot go underground to escape high environmental temperatures, as kangaroo rats can. However, they have increased mobility to find surface water and shelter. Large mammals also have lower surface area-to-volume ratios and lower metabolic rates than smaller mammals. During the afternoon heat, nonworking camels sit on the ground, allowing heat to be conducted to the soil that is kept shaded and thereby cooled by their bodies. They may also huddle, thus reducing the group's total surface area exposed to sunlight and hot air.

FIGURE 2.24
Various positions assumed by the horned lizard *(Phrynosoma)* for thermoregulation during the day. Precise behavioral thermoregulation is often assisted by structural features, such as having lighter-colored body surfaces in the summer than in the winter. What role does behavior play in human thermoregulation?

FIGURE 2.25
Some examples of microclimates. (A) Typical daily pattern of soil and rodent burrow temperatures. The annual range of soil surface temperatures far exceeds the rodent's range of tolerance. In their burrows, however, small mammals live in temperatures far below those on the surface during the summer and far above wintertime surface temperatures. (B) General temperature profile above, in, and under a snow layer during an arctic winter. Some small, nonhibernating mammals live comfortably in the space between the snow and the soil, very close to temperatures far too cold for survival.

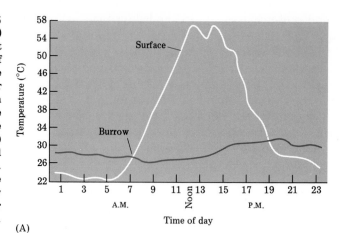

(A)

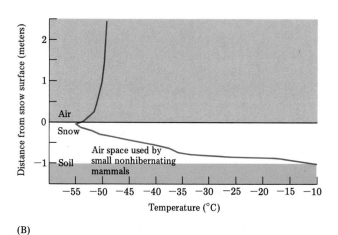

(B)

PERSPECTIVE ON TEMPERATURE REGULATION

A prime advantage of endothermy is the freedom that it allows for birds and mammals to live independently from many external thermal conditions. However, this does not imply that endothermy is superior to ectothermy. The major biological criterion of success is perpetuation of the species, which obviously involves the survival and reproduction of individuals. By this criterion, ectothermy is as successful a mode of life as endothermy. Ectotherms have existed on Earth far longer than endotherms have, and most living species of animals are ectothermic. Endothermy and ectothermy are simply extremes in a continuum of thermal responses that have developed as animal species have adapted to various environments. The earliest type of thermal relationship with the environment was undoubtedly simple ectothermy, but a variety of thermal relationships now exist, from ectotherms controlling their body temperature behaviorally to tightly controlled endothermy.

During evolution, the development of structures such as feathers or hair that served to reduce the conduction of heat to the environment was significant. For example, in a world dominated by reptiles, early mammals probably gained an advantage when they were able to be active at night, without exposure to heat from sunlight. High body temperatures also made possible faster reactions because some substances (neurotransmitters) diffuse faster from neuron to neuron at higher than at lower temperatures (Chapter 19). Endothermy also made it possible for birds and mammals to occupy habitats in cold regions where solar energy was insufficient to warm ectothermic vertebrates. However, modern species often exploit different means of temperature regulation at different times and in different stages of life. Any adaptation that allows sufficient numbers of individuals of a species to survive and reproduce is successful.

In this brief introduction to homeostasis, you have studied several anatomical, physiological, and behavioral aspects of thermoregulation in humans and in animals in general. It is clear that homeostasis involves behavior and the integrated actions of many body structures. Many other examples of homeostasis are discussed throughout the text. The comprehensive feedback control model is one of our most powerful ideas in understanding how we and other organisms adapt to changing conditions.

NORMALITY AND ABNORMALITY: BIOLOGY AND MEDICINE

By *abnormality,* we usually mean a level of functioning somewhat outside the typical range. Because humans are so variable, precise definitions of abnormality are difficult. However, we constantly use and encounter terms for diseases and conditions, including all of those *-itis* terms. Historically, medicine developed as an application of biological, chemical, and physical knowledge to the treatment of malfunctioning. Medicine has often been viewed as an applied field, while the "pure sciences" were seen as constantly pushing outward the boundaries of knowledge for its own sake. This relationship has often been likened to that between technology and science, and in both relationships the boundaries have become blurred. For example, many of the latest discoveries in immunology, genetics, biochemistry, and molecular biology have resulted from research attempting to understand cancer, AIDS, heart disease, and other challenges to homeostasis. However, if we look at the space devoted to science in newspapers (whole sections in some cases), the multitude of science-related magazines, and the popularity of scientific television programs, then most people want more than merely the latest information on the prevention and treatment of human diseases. Rather, they are fascinated with the natural world in general, and especially with our role in natural processes.

SUMMARY

1. Traits that aid organisms in surviving and reproducing in their environments are called **adaptations.**

2. **Homeostasis** is the maintenance of relatively constant internal conditions despite changing external conditions. Temperature is of major importance to both organisms and to events in their environments, and **thermoregulation** is an excellent example of homeostasis.

3. Homeostatic mechanisms usually involve at least three integrated parts: **receptors** that detect changes in the animal's internal and external environments, **integrators** that coordinate the information received, and **effectors** that produce responses to maintain the system at an optimal level.

4. **Feedback systems** are so named because information about a change in the system is fed back to influence the response. The two basic kinds of feedback systems are called **negative or corrective feedback** $(+,-$ or $-,+)$ and **positive or runaway feedback** $(-,-$ or $+,+)$.

5. **Stress** occurs when any internal or external condition tends to upset the normal operation of a system. Every system has an **optimum** level and a **range of tolerance** for each factor.

6. Both the body's thermostat and thirst centers are found in the **hypothalamus** of the brain.

7. Body temperature results from two processes, heat gain and heat loss. Heat transfer between organisms and their environments occurs by means of **conduction** (contact of objects of different temperatures), **convection** (movement of a mass of air or water), **radiation** (transmission of energy through space), and **evaporation** (release of energy as liquid water changes to a gas).

8. **Endotherms,** including most birds and mammals, produce and control their body heat by means of their metabolism. Most other animals, the **ectotherms,** have body temperatures that change as the temperature of their environments change.

9. Small objects have a larger **surface area-to-volume ratio** than do large objects. As objects grow, their volume increases at a faster rate than their surface area. Metabolic rate generally decreases as animals grow and age.

10. Ectotherms living in warm areas tend to be smaller than their relatives living in cold areas—**Bergman's principle.** The size of the extremities of endotherms tends to vary inversely with body size—**Allen's principle.**

11. Adaptations of endotherms to high temperatures include evaporative cooling, **vasodilation** of surface blood vessels, long extremities, fur and feathers, and **estivation. Thermal pollution** is a growing problem for many aquatic organisms.

12. Shivering, **vasoconstriction** of surface blood vessels, **countercurrent circulation,** metabolizing **brown fat,** fluffing of fur or feathers, short extremities, and **hibernation** and **torpor** are common adaptations of endotherms to low temperatures. **Hypothermia** often occurs when the body is exposed to cold water or when moving air removes body heat faster than still air does (**wind-chill effect**).

13. **Behavioral thermoregulation** is important in both ectotherms and endotherms. This often involves the use of **microclimates.**

REVIEW QUESTIONS

1. Name some examples of receptors, integrators, and effectors.

2. Describe several nonthermoregulatory examples of negative and positive feedback.

3. Define *conduction, convection, radiation,* and *evaporation* and describe some everyday examples of each.

4. Name several examples of ectotherms and endotherms. Why are the terms *cold-blooded* and *warm-blooded* inexact in describing the body temperatures of animals?

5. Distinguish among hibernation, torpor, and estivation.

6. Explain the significance of surface area-to-volume relationships to thermoregulation.

7. How does body size and shape affect the metabolic rate?

8. What are some major adaptations of endotherms to warm conditions? To cold conditions?

9. Describe some examples of behavioral thermoregulation.

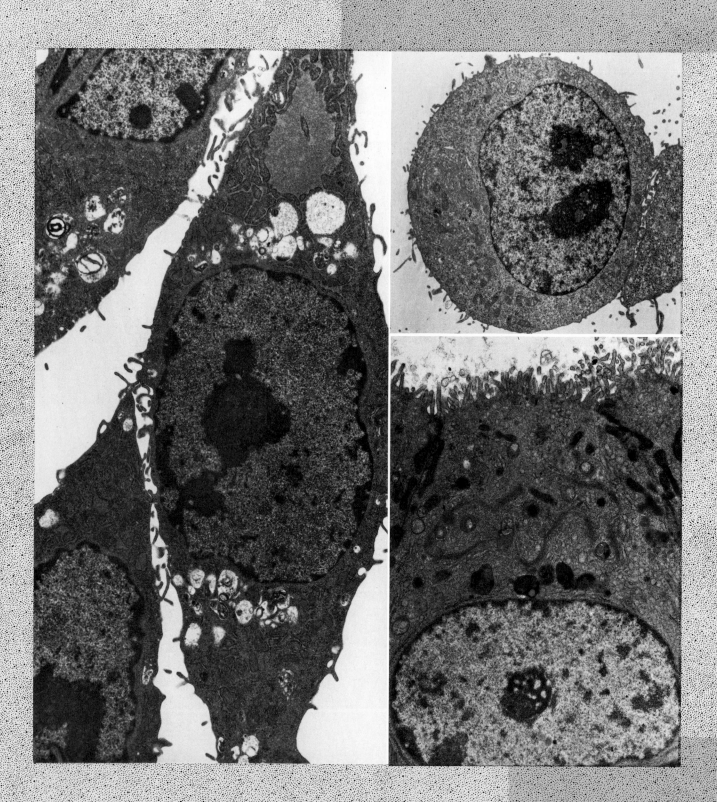

CHAPTER 3

STRUCTURAL ORGANIZATION FOR HOMEOSTASIS

Homeostatic control of body functions depends upon the interaction of many structures. If a stimulus changes the action of any one part, the actions of many other structures also change to keep the entire system integrated. Whether organisms are microscopic bacteria or enormous elephants, homeostatic control and living depend upon the basic structural and functional unit of all organisms—the cell. Understanding cells and how they work and are organized is essential for the study of almost any topic in biology. Both health and disease depend upon the structure and function of cells, and life is a process that occurs only in these units.

A collage of cells, the basic structural units of life. We have come a long way in our understanding of the structures and workings of these tiny units since they were first observed, about 300 years ago.

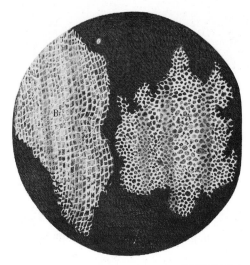

FIGURE 3.1
Cork cells sketched by Robert Hooke in the 1600s. Because cork is dead tissue, Hooke actually saw only the walls of plant cells, not cell contents.

CELLS

In the mid-1600s, Robert Hooke was trying to devise an interesting demonstration for the prestigious Royal Society of London. He happened to cut a thin slice of cork with his sharpened penknife and observed it through a microscope that he had built, one of the world's first. He named the numerous cavities he observed in the dead plant material **cells** (from the Latin for "chamber") because they reminded him of the rows of monk's rooms in a monastery (Figure 3.1). A few years later, the Dutch naturalist Antonie van Leeuwenhoek (1632–1723) observed the first living cells.

The overall significance and importance of cells were not established until two German biologists, Matthias Schleiden and Theodor Schwann, formulated the **cell theory** in the 1830s, almost two centuries after cells were first observed. This principle, which emphasized the basic similarity of all organisms, was broadened in the 1850s by another German biologist, Rudolf Virchow, by the concept that all cells arise from preexisting cells.

Although some cells, such as human eggs, are just visible with the unaided eye, most cells are so small that their details can be seen only with magnification. For example, about 50 of your red blood cells, laid side by side, would equal the width of the period in this sentence. Figure 3.2 shows some of the great variety in our trillions of cells. However, despite the diversity of cellular sizes, shapes, and functions, there are certain main features common to most animal cells, as shown in Figure 3.3.

The cellular material between the cell's bilayer **plasma membrane** and the **nucleus** is called **cytoplasm,** while that within the nucleus is the **nucleoplasm.** The cytoplasm and nucleoplasm are in a semisolid, gelatinlike state, and suspended within them are the specialized **organelles** (from the Latin for "litte tool") that are described below. In addition to the organelles, the cytoplasm also contains a complex mixture of nutrients, wastes, and enzymes (Chapter 4).

FIGURE 3.2
Various human cells. The variety of cells in multicellular organisms allows specialization, each cell having its own abilities and limits. (A) An individual nerve cell or neuron. Such cells are highly irritable to environmental changes and can conduct impulses. (B) A bundle of muscle cells. These muscle fibers are highly contractile and are important in the movement of organisms or their parts. (C) A sperm cell, one of the smallest cells known in the human body. The long tail, or flagellum, aids the sperm in moving. (D) White blood cells have the ability to change their shape, leave blood vessels, and attack foreign materials.

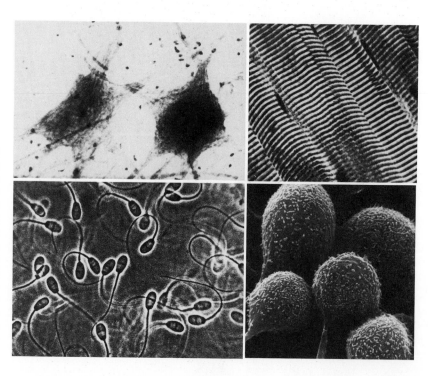

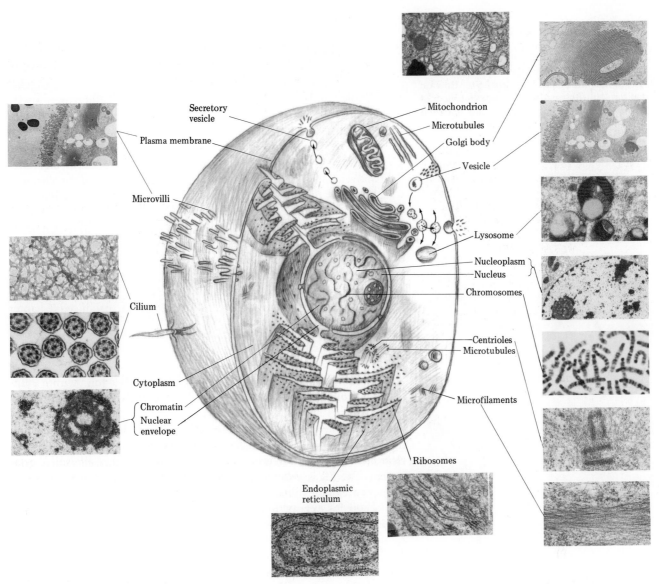

Secretory vesicle

Plasma membrane

Microvilli

Cilium

Cytoplasm

Chromatin
Nuclear
envelope

Endoplasmic
reticulum

Mitochondrion

Microtubules

Golgi body

Vesicle

Lysosome

Nucleoplasm

Nucleus

Chromosomes

Centrioles
Microtubules

Microfilaments

Ribosomes

FIGURE 3.3
A composite animal cell, showing the various organelles and some of the processes involving them. This three-dimensional representation is based upon the study of many photographs taken using electron microscopes, such as those shown for each organelle and structure. Not all animal cells are composed of all of these structures, nor are they in the same proportions.

ORGANELLES—MICROSPECIALISTS

Many cellular substances and functions are compartmentalized in the organelles, which, in a general way, are analogous to the organs of multicellular animals. If a cell is considered a room, then organelles are the closets and cabinets that partition the cytoplasm into functional subcompartments. However, unlike passive cabinets, which merely store items, organelles are active much of the time.

Nucleus

The largest and most prominent part of most cells is the nucleus, the control center of cells (Figure 3.4). In human cells, the nucleus is typically near the cell center and generally spherical in shape. Although one nucleus per cell is typical, some cells have more than one, and mature red blood cells have none. The double membranes of the nucleus are called the **nuclear envelope;** they contain craterlike pores that connect the nucleoplasm and cytoplasm. The pores are not empty, like the holes in donuts, but contain proteins that act somewhat as turnstile gates do,

FIGURE 3.4
(A) A classic experiment to show the importance of the nucleus. In (1), an amoeba is cut in half. The experiment in (2) begins similarly, but the nucleus is then transferred to the half that lacks this organelle. How do you explain the results? (B) Detailed view of the nucleus of a nondividing cell, showing the nuclear envelope with pores, the scattered and indistinct chromosomal material (chromatin, Ch), and a nucleolus, smooth and rough endoplasmic reticulum, lysosomes, and mitochondria.

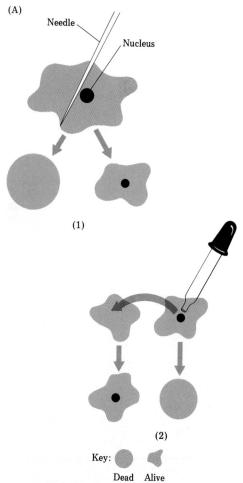

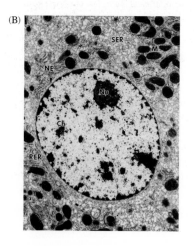

in that they permit the passage of only certain substances, with the right "tickets" or confirmations.

Suspended within the fluid nucleoplasm are the **chromosomes,** which carry the genetic material of the cell. In nondividing cells, the DNA-protein complexes that constitute individual chromosomes are extended and diffuse, and they become readily recognizable only in properly stained dividing cells. Normal human body, or somatic, cells possess 46 chromosomes (23 pairs), but the number of chromosomes differs in many other species (e.g., 64 in some ferns, and 200 in certain crayfish). The condensation of the chromosomes during cell division facilitates their movements (see later section), and the unwinding of the chromosomes at other times in the cell cycle aids in making the genetic blueprint available for directing cell processes. Genes are located within chromosomes and constitute the basic units of hereditary information (Chapters 9 and 10). Genes operate basically by determining the kinds of proteins that cells produce. This, is turn, determines the traits of individual cells and, thus, the characteristics of the organism assembled from them. The one or more small **nucleoli** within the nucleus are the sites where molecules involved in translating the genetic blueprint are produced (Chapter 10).

Endoplasmic Reticulum

The **endoplasmic reticulum** (ER) (from the Greek for "within cytoplasm"; from the Latin for "network") is a very elaborate system of compartments and channels that winds like folded sheets through the cytoplasm. The ER canals are membranous in nature and carry materials through the cytoplasm, connecting the plasma membrane with the nuclear membrane. The channels may be thought of as the cell's circulatory system. The ER is also a manufacturing maze, and on its very extensive surfaces are made numerous key substances, some for use within the cell and others for export. Embedded within its membranes, somewhat like icebergs in the sea, are many **enzymes,** compounds that facilitate or catalyze numerous chemical reactions (Chapter 4). The ER in some cells (e.g., liver) has the capacity to break down potentially toxic substances. The ER membranes are dynamic, not static, and constantly form, dissolve, and reform as they assist with the myriad of chemical reactions occurring in cells. By creating small subcompartments (cisternae) where the membranes fuse at various locations, the ER allows specialized reactions and functions to take place in different locations.

Ribosomes

Ribosomes are small, spherical bodies scattered free in the cytoplasm or attached to ER. Endoplasmic reticulum with attached ribosomes is called rough ER, and that without ribosomes is called smooth ER. Rough ER looks somewhat like the surface of sandpaper, with the grains being ribosomes. Ribosomes are the "molecular machines" that translate copies of genes into proteins. After being synthesized on the ER surfaces, the proteins are transported in ER channels to locations in the cell or to the cell's exterior.

Golgi Bodies

Golgi bodies form from ER membranes that fuse into flattened stacks of membranes. Golgi bodies act as intracellular collecting, modifying, packaging, and distribution centers for substances that cells synthesize at one location and use at another. The packaged substances are transferred in small membrane-bound **vesicles** (from the Latin for "little bladder") that bud off from the edges. Within the flattened, disk-shaped sacs and tubules, various synthetic chemical processes also occur. Golgi bodies are especially prominent in secretory cells, such as those that line the digestive tract. Substances to be released from the cell travel in vesicles to the plasma membrane. After discharging their contents, the membranes of the Golgi vesicles become part of the plasma membrane.

Lysosomes

The discoverer of **lysosomes** called them "suicide bags," although "digestion bags" would be a better description for these mobile vesicles that are derived from the Golgi complex. Lysosomes contain powerful enzymes able to digest cell structures when the organelle's membrane ruptures. One function of lysosomes is to break down and recycle many components of aged, injured, or abnormal cells. Sometimes whole cells are destroyed, as when lysosomes break down the cells that form webbing between the fingers and toes during early embryonic development. Lysosomes also release their products into vesicles for storage, transport, and eventual disposal. Because the enclosed enzymes also decompose the inner part of the lysosome's membrane, energy must be used continually to maintain the membrane.

Some cancers as well as other diseases are associated with abnormal lysosomes. For example, lysosomes missing a crucial enzyme may allow a buildup of harmful chemicals that can result in paralysis, mental retardation, blindness, or death.

Mitochondria

Although scarce in metabolically inactive cells such as those in your skin, fat, and cartilage, the tiny, bacteria-like **mitochondria** are common in active types of cells such as those in muscles and in the liver. There are also more of the long, tubular mitochondria in the working than in the inactive parts of individual cells. Going "where the action is" describes these organelles, because most of a cell's energy results from chemical reactions carried out in these cellular "power plants." The chemical reactions that combine nutrients and oxygen to produce energy occur on the large surface area of the mitochondrion's internal membranes.

Unlike the organelles described thus far, the mitochondria and plastids (see below) are not derivatives of the endoplasmic reticulum. Rather, according to one prevalent theory, these organelles were originally symbionts in ancient cells. **Symbiosis** (from the Greek for "living together") includes relationships whereby two or more different organisms are associated in a way advantageous to all. Some support for the view that the mitochondria were once bacteria taken into a cell comes from the fact that mitochondria, like bacteria, contain a circular DNA molecule. (See the essay "Prokaryotic Cells.")

Plastids

Plastids are found in plant cells but not in animal cells. One type of plastid, the **chloroplast**, contains the green pigment **chlorophyll** that is involved in the process of **photosynthesis** so vital to life on Earth (Figure 3.5). For example, photosynthesis is the ultimate source of our nutrients and oxygen (Chapter 4).

FIGURE 3.5
Not only do plants depend upon the chemical events called *photosynthesis* that occur in chloroplasts like this one, but so do animals and many other organisms, which ultimately secure their food and oxygen from such cellular events. Animals lack chloroplasts and other kinds of plastids.

Prokaryotic Cells

Y ou and most other organisms possess **eukaryotic** cells, and most details in this chapter concern such cells (Chapter 1). However, the cells of some organisms, such as bacteria and cyanobacteria (or blue-green algae) are called **prokaryotic**. Modern prokaryotes trace their evolutionary roots to the most ancient forms of life.

The cells of prokaryotes are generally smaller and simpler than those of eukaryotes (Figure B3.1 and Table B3.1). However, despite their structural simplicity, the prokaryotes are very complex biochemically, and many possess the metabolic machinery to carry out many of the same functions of eukaryotes. Some forms are autotrophic and others are heterotrophic. Many prokaryotes are important pathogens (Chapter 16) and some are very useful in modern genetic engineering (Chapter 10).

FIGURE B3.1
Prokaryotes. (A) A common bacterium, *Streptococcus,* responsible for "strep throats." (B) A well-known cyanobacterium, *Oscillatoria.* (C) Diagrammatic view of the typical features of many kinds of bacteria. Some bacteria also possess a capsule surrounding the cell well.

(A)   (B)   (C)

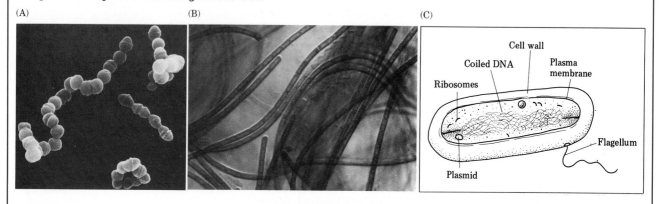

TABLE B3.1
Comparison of Prokaryotic Cells and Those of Typical Eukaryotic Animals

Structure	Prokaryotic Cell	Eukaryotic Animal Cell
Cell membranes	External only	External and internal
Support system	None	Cytoskeleton
Nuclear envelope	None	Present
Chromosomes	Single, circular, composed only of DNA	Multiple, linear, combined with protein
Membrane-bounded organelles	Few	Many
Ribosomes	Smaller, free	Larger, many membrane-bound
Cell wall	Present	Absent
Cilia or flagella (when present)	Solid, rotating	Microtubular (9 + 2 pattern)
Phagocytosis	Absent	Present
Centrioles	Absent	Present

Centrioles

The two tubular **centrioles** found near the nucleus are important in cell division and in the formation of the cilia and flagella described below.

Cilia and Flagella

Many cells have movable hairlike parts on their surfaces. Although otherwise indistinguishable, **flagella** (from the Latin for "whip") are usually few and long, while **cilia** (from the Latin for "eyelash") are numerous and short. These structures either move cells, as in the case of flagella propelling sperm, or move materials past cells, as in the case of cilia beating upward to keep particles from reaching our lungs.

Microfilaments and Microtubules

The delicate threadlike **microfilaments** and the slender and larger tubelike **microtubules** provide a kind of cellular skeleton or **cytoskeleton** that supports the cell and the cilia and flagella. Because these organelles possess the property of contraction, they help determine cellular shape and they assist in moving its structures (e.g., the chromosomes during cell division). Your muscles contain the same contractile microfilaments, and they are allowing your eyes to move from word to word as you read this (Chapter 21).

METHODS OF STUDYING CELLS

The invention of **light microscopes** in the 1500s was very important to biology because many organisms and their parts can be seen clearly only when magnified. Light microscopes can magnify objects a few thousand times, but **electron microscopes** may magnify specimens hundreds of thousands of times (Figure 3.6). However, the resolution, or the ability to distinguish as separate two objects that lie close together, is as important as magnification. For example, notice that when you enlarge a newspaper photograph with a magnifying glass you see more (bigger dots and more space between them) but the "big picture" is no longer recognizable.

Transmission electron microscopes became available in the 1930s, and many of the details of cellular structure were discovered with these instruments. Electron microscopes use beams of electrons instead of beams of light to visualize details. One of their disadvantages is that the specimens being examined must be in a vacuum, and cells cannot live under such conditions. **Scanning electron microscopes** are useful in studying the surfaces of objects.

Cytology, the science of studying cell structures, uses many aspects of microscopy other than magnification. Because most parts of cells are nearly transparent, various stains are useful to make them more visible. Not all organelles have affinity for the same stain, so a variety of stains is often used, frequently in colorful combinations.

If you tip a cup of water slowly, the water spills. However, if you swing the cup fast enough, no water spills even when the cup is upside down. (Try this yourself—carefully!) Figure 3.7 show how **centrifuges** spin materials and separate their parts on the basis of differences in density. If blood is centrifuged rapidly, the heavier cells settle to the bottom of the tube and the less dense plasma, or liquid part of blood, collects near the top. High-speed centrifuges can even separate the cellular molecules that differ only slightly from each other in density.

FIGURE 3.6
(A) Leeuwenhoek's simple microscope led eventually to the modern light microscope (B) familiar in laboratories. Transmission and scanning electron microscopes, shown in cross-section (C) and in use (D), provide both greater magnification and resolving power than light microscopes.

(A)

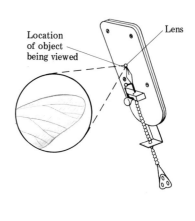

Location of object being viewed

Lens

(B)

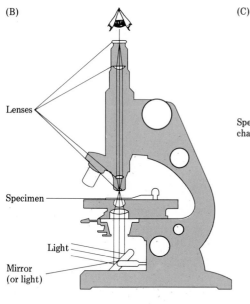

Lenses

Specimen

Light

Mirror (or light)

Compound light microscope

(C)

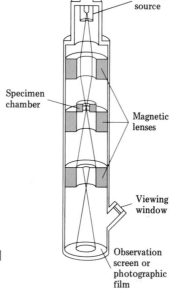

Electron source

Specimen chamber

Magnetic lenses

Viewing window

Observation screen or photographic film

Transmission electron microscope

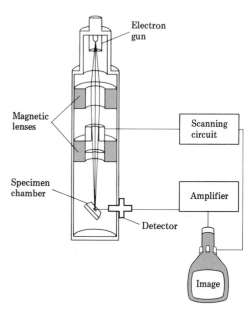

Electron gun

Magnetic lenses

Scanning circuit

Specimen chamber

Amplifier

Detector

Image

Scanning electron microscope

(D)

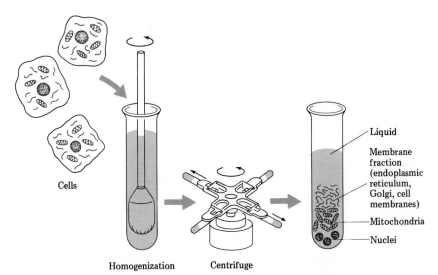

FIGURE 3.7
Centrifuges are used to separate cell parts and other materials with different densities.

Seeing a structure is only a first step, and learning its function is usually more difficult. If something must be magnified many thousands of times just to see it, and in a vacuum at that, imagine the problems of discovering what the structure does. Sometimes certain chemicals (e.g., radioisotopes) can be used as "tags," or tracers that allow one to follow metabolic processes and find out where, when, and how cellular events are occurring.

ON THE SIZES OF CELLS AND ORGANISMS

Have you ever wondered why your body consists of trillions of cells instead of just one large cell? Part of the answer has to do with the **surface area-to-volume ratios** of cells (Chapter 2). The metabolic needs of a cell or an entire multicellular organism depend in part on its volume. The larger the cell or the organism, the more substances it needs and the more wastes it produces. However, the ability of a cell or organism to secure its needs and rid itself of wastes depends in part upon its surface area. All materials that enter or leave the human body must cross one or more membranes.

To understand these relationships, study or review the cubes in Figure 2-10 and the information in Table 2.1. Note that as the objects grow, their volumes are cubed while their surface areas are only squared. Actually, the human body is more like a collection of cylinders than of cubes. Figure 3.8 shows that similar surface area-to-volume relationships exist for cylinders.

A common theme in science fiction has humans getting larger or smaller. Sometimes they are on a different planet or have taken a potent drug. Would it matter if a 68-kg, 175-cm (150-lb, 5-ft 9-in.) person became 5 times larger? The unfortunate person would now be 8.75 meters tall (28 ft 9 in.) and might be considering professional basketball. However, weight is a function of volume, which increases not 5 times but 5^3, or 125, times. Now, perhaps, the 8500-kg (18,750-lb) person is considering becoming a center on the line of a professional football team. If the muscles and bones increase in strength, what is the problem? Bone strength depends upon its cross-sectional area, which has increased by the square, 25 times. But because the load it has to carry has increased by the cube, 125 times, the enlarged human could not stand because the legs would shatter under the enormous weight.

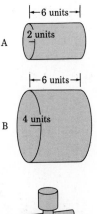

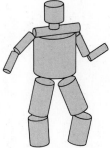

FIGURE 3.8
The surface area-to-volume relationships of cylinders. Surface area = $2\pi rh + 2\pi r^2$; volume = $\pi r^2 h$. Cylinder A: Surface area = 100.5 units²; volume = 75.4 units³. Cylinder B: Surface area = 251.3 units²; volume = 301.6 units³. As with cubes, when cylinders increase in size, their volumes increase at a faster rate than their surface areas.

But suffocation would be as serious a problem as broken bones. The oxygen-absorbing surface area of the lungs would increase 25 times, but the oxygen needs increase as volume increases, by 125 times. Also, our giant human falling off of a building would be smashed, although an ant might survive the fall because its surface area is so large relative to its volume. Would we have problems if our size were reduced by one-fifth?

Small objects have more surface area relative to volume than large objects. For example, imagine a one-celled, full-sized adult human. Because of surface area-to-volume constraints, the surface area would be insufficient for the passage of nutrients and wastes across it. By containing many small cells, multicellular organisms have a very large total surface area of membranes. Because a nucleus cannot control too large a volume of cytoplasm, cells often divide when they grow too large, thus maintaining a favorable surface area-to-volume relationship.

The surface area-to-volume relationships of cells and organisms may also be overcome in other ways, for example, by becoming flattened or elongated as in leaves and flatworms. The flattening of cells or organisms increases their surface area without a corresponding increase in their volume. It also means that all cellular contents are near the cell membrane. Extensions from the surface of cells also increase their surface area with only small increases in volume. Such is the case when absorbing cells contain numerous **microvilli** (Figure 3.9). The immense surface area of the small intestine, where we digest and absorb most of the food we eat, is due mainly to these and larger fingerlike projections (villi, shown in Figure 13.10). Similarly, the large area for gas exchange in the lungs is due mainly to the many small, bubblelike structures (alveoli, shown in Figure 14.4).

MEMBRANES

Most structures in cells are composed of a dynamic system of fluid membranes, and all materials that enter or leave cells pass through membranes. Biological membranes are only a few molecules thick, and it would take about 10,000 membranes to equal the thickness of this page. Light microscopes reveal few of these thin membranes, so most of our knowledge about membrane structure has come from electron microscopy. Not only is the entire cell enclosed by a flexible plasma membrane, but most internal structures are basically membranous or contain internal as well as surrounding membranes. Some of the membrane-bound structures are relatively permanent, such as the nuclear envelope, but many others are transient.

Biological membranes consist of fatty (lipid) bilayers in which numerous large proteins are embedded. Such membranes are freely permeable to water. The solution within the cell consists primarily of water, the **solvent,** and various substances dissolved in it, the **solutes** (Chapter 4).

Cells normally recognize one another by means of receptors on the cell surface (Chapter 16). For example, if cells removed from two different body tissues are mixed, similar cell types often recognize and rejoin one another (Figure 3.10). The unregulated, disorganized cell growth characteristic of cancer results when cell types lose the ability to recognize and respond to each other.

The physical and chemical properties of substances determine their ability to cross membranes. Some materials, such as water, cross membranes easily, some not at all, and some only under certain conditions.

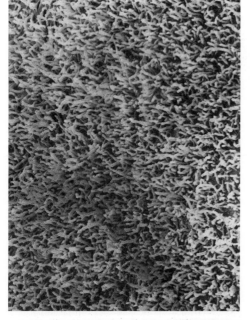

FIGURE 3.9
The numerous fingerlike extensions called *microvilli* are very important in many cells because they greatly increase the surface area without adding significantly to the cell's volume. In many cases they assist the cell in secreting, while in others they facilitate absorption.

The nature of the protein channels or gates in the cell membrane largely determines whether and how materials move into and out of cells. For this reason, cell membranes are called **selectively permeable.** Some people once thought cell membranes to be like screens, allowing small substances to pass through but keeping large particles in or out. However, the fact that different substances of the same size do not always cross membranes in the same way shows that the movement of materials is far from simple.

At first thought, it might seem that eating and drinking have little to do with cells directly. However, cells and cell membranes are involved in every phase of these processes. In digestion, for example, chemicals released by cells in the stomach, small intestine, and other structures break down food. Other cells absorb the nutrients, and still others assist in the transport of the simplified foods throughout the body. Various cells utilize the nutrients to produce energy, respond to stimuli, grow, or reproduce. Incidentally, food and air in the stomach and lungs, respectively, are not *in* the body but are merely *surrounded by* it (Figure 3.11). Components of food and air that we have only surrounded must pass across cell membranes before they are truly inside of cells and therefore inside our living bodies.

To understand how materials cross a membrane, we must know something about its structure. The plasma membrane is composed of two fatty layers with patches of proteins scattered on and/or extending through its surface (Figure 3.12). Studying such a thin structure is difficult, and much remains to be learned about its detailed functions.

Many biological reactions involve chemicals present in small quantities. To interact, these materials must come together, and it is the selectively permeable nature of membranes that makes possible this concentration. Just as many familiar products are manufactured along assembly lines, biochemical substances may be processed similarly along cell membranes.

GETTING MATERIALS ACROSS MEMBRANES

A home would hardly be safe and secure if the owner had no key and anyone could enter and leave freely. Naturally, the wrong key would not benefit even the home owner. So it is with cells, at least in a general

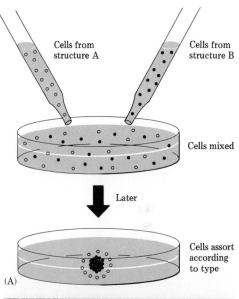

FIGURE 3.10
(A) Isolated cells of like type (either A or B here) can recognize one another as a result of similarities in cell surface receptors. When the two cell types are mixed, they often sort out from each other in accord with their original affinities. (B) Cell cultures like these are often used in studies of cell-recognition phenomena, cancer, and immunology.

Cells from structure A

Cells from structure B

Cells mixed

Later

Cells assort according to type

(A)

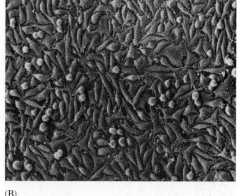

(B)

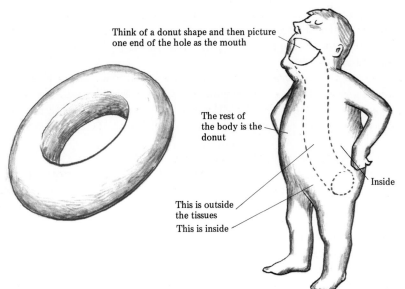

Think of a donut shape and then picture one end of the hole as the mouth

The rest of the body is the donut

Inside

This is outside the tissues

This is inside

FIGURE 3.11
The body of a human or other higher animal can be thought of as a tube within a tube, with the inner tube being the digestive tract. Although the tube is very twisted, its interior is actually outside the body, just as the hole in the donut is outside of but is surrounded by the donut. Much the same analogy applies to substances taken into our lungs—we've only surrounded them until they actually enter cells by crossing the plasma membranes.

FIGURE 3.12
A typical biological membrane is a fatty (lipid) bilayer with scattered embedded proteins that often function as gates or channels. Biological membranes are called *selectively permeable* because some materials, such as water, pass through easily, but others do not, or do so only under certain conditions.

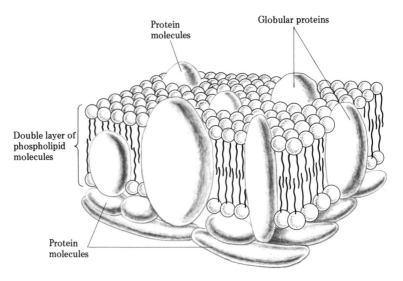

way, because some materials must pass freely in and out and others must not. The plasma membrane, or sections of it, represent the cellular "doors."

Passive Transport

The movement of all materials is affected by **diffusion** (from the Latin for "to pour out"), a process by which substances move from a region of greater concentration to one of lesser concentration (Figure 3.13). Open a jar of perfume or ammonia, and the chemicals will eventually diffuse throughout the room. Add a drop of food coloring to water and, even without your stirring or heating it, the color will spread slowly. Diffusion results from the random movement of individual molecules, and the process tends to distribute molecules uniformly. In the body, diffusion is important in the exchange of oxygen and carbon dioxide in the lungs and between some substances in the blood and neighboring cells. Because materials are moving across the cell membrane without any expenditure of energy by the cell, diffusion is an example of **passive transport**.

FIGURE 3.13
By diffusion, materials move passively from areas of higher concentration to areas of lower concentration. (A) shows the movement of sugar molecules (the solute) as a sugar cube dissolves in water (the solvent), and (B) shows the osmotic movement of water across a plasma membrane until an isotonic condition is reached (see text). Molecular motion does not stop under this condition, but for every movement of a molecule in one direction, there is a movement of another molecule in the opposite direction.

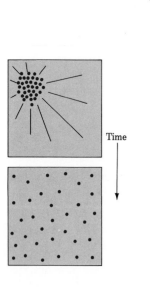

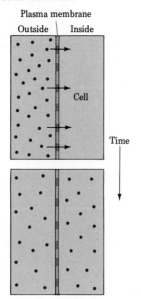

(A)

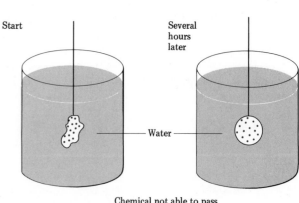

Chemical not able to pass through sac membrane

(B)

A very important type of diffusion is **osmosis,** the diffusion of water across a selectively permeable membrane. The average water content of cells is 70–90 percent, and the human body as a whole is about two-thirds water. Thus, any process that influences water is important to our understanding of life.

Figure 3.14A shows a normal red blood cell in normal blood plasma. On the average, every water molecule entering the cell is balanced by a water molecule leaving the cell, a condition called **isotonic** (from the Greek for "equal tension"). Although water usually moves freely across plasma membranes in both directions, membranes are not as permeable to most other chemicals, such as salt. Notice what happens to the cell placed into pure water (Figure 3.14B), where there is a higher concentration of water molecules outside the cell than inside. Because the cell contains more solutes than its environment, the surrounding solution is called **hypotonic** (from the Greek prefix meaning "less than") with respect to the cell. Water moves into the cell, the opposite of what happens when the cell is placed into salt water (Figure 3.14C). Because the cell then contains a lower concentration of solutes than its environment, the outer solution is called **hypertonic** (from the Greek prefix meaning "more than") relative to the cell. These circumstances are not only important to individual cells but also to entire organisms (Figure 3.15).

FIGURE 3.14

The shape of a red blood cell depends upon its composition relative to that of its environment. While the plasma membrane of the red blood cell is freely permeable to water, it is not permeable to salt. How do you explain the events shown in (B), where the concentration of the cell's cytoplasm is hypertonic to that of its environment (see text), and in (C), where the cell's cytoplasm is hypotonic relative to its environment?

(A)
Red blood cell
in normal blood plasma

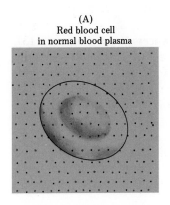

(B)
Red blood cell in pure water

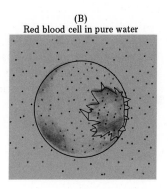

(C)
Red blood cell in saltwater

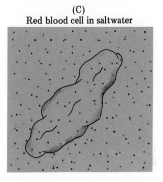

·∴· Water
∷∷ Salt and other
particles

Materials soluble in fats generally pass through cell membranes readily. Figure 3.16A shows one way that some other materials may cross, in either direction, cell membranes that are generally impermeable to them. In **facilitated diffusion,** a "carrier" molecule or transport protein embedded in the membrane may combine temporarily with the substance to be moved. When the substance has passed through the specific membrane channel, the carrier releases it. Like free diffusion, facilitated diffusion does not require metabolic energy. In many cases, needed substances that would otherwise move out of a cell are kept in by becoming attached to molecules that do not pass freely, or not at all, across the cell membrane (e.g., phosphorylation).

Active Transport

Diffusion is often too slow a process for active cells. Also, cells often need more of certain substances from their environment even though the concentration of these chemicals inside the cell is higher than it is in their environment. Thus, substances may be absorbed from your small intestine into your blood even though the concentration of the substances

FIGURE 3.15
Because the gills of fish have a very large surface area and are usually in constant contact with water, we find excellent examples of osmosis, active transport, and homeostasis. (A) Marine bony fish lose much water osmotically across their gills to their hypertonic environment. To compensate for this water loss they must drink salt water, use active transport processes to excrete the excess salt, and produce very little urine. (B) However, because the osmotic concentration of their cells is hypertonic to their environment, freshwater fish usually do not drink. Their kidneys produce large quantities of hypotonic urine from the water that is gained constantly by osmosis across their gills.

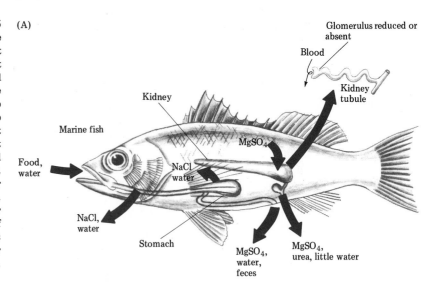

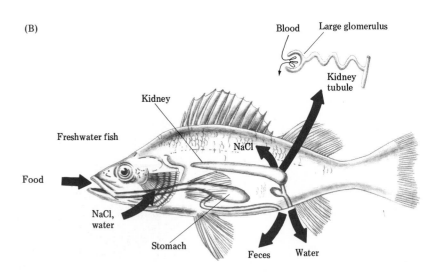

in the blood is already higher. On the other hand, certain chemicals in the environment, such as poisons, may be kept out of a cell even though we might expect them to move in by diffusion. Whereas diffusion is a passive process, these activities require cellular energy. Because materials are being moved "against the tide" and are not moving passively, this is called **active transport** (Figure 3.16B). Defying diffusion requires work, and a good deal of the energy obtained from food is used to move substances across cell membranes against concentration gradients.

Some channels in cell membranes allow numerous small molecules to pass through freely in either direction, at least when the channels are open. Such channels (gap junctions) are important to some cells not adjacent to a blood vessel because they allow substances to move from one cell to another through the channels that cross the intercellular spaces like tunnels. More complex transport channels (so-called pumps) are discussed in chapters that deal with various body systems.

Sometimes large amounts of solid or liquid material enter or leave cells (Figure 3.17). In **phagocytosis** (from the Greek for "cell eating"), external particles become enclosed within vesicles that form from the infolding of the plasma membrane. For example, certain of our white blood cells can engulf bacteria or damaged cells in this way, much as an

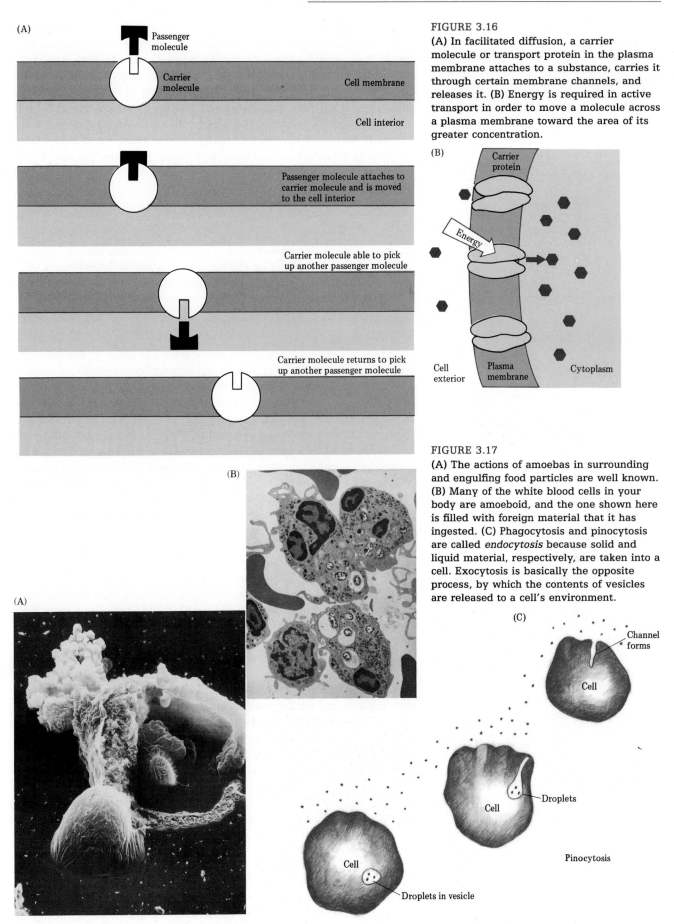

(A)

Passenger molecule

Carrier molecule

Cell membrane

Cell interior

Passenger molecule attaches to carrier molecule and is moved to the cell interior

Carrier molecule able to pick up another passenger molecule

Carrier molecule returns to pick up another passenger molecule

FIGURE 3.16

(A) In facilitated diffusion, a carrier molecule or transport protein in the plasma membrane attaches to a substance, carries it through certain membrane channels, and releases it. (B) Energy is required in active transport in order to move a molecule across a plasma membrane toward the area of its greater concentration.

(B)

Carrier protein

Energy

Cell exterior

Plasma membrane

Cytoplasm

FIGURE 3.17

(A) The actions of amoebas in surrounding and engulfing food particles are well known. (B) Many of the white blood cells in your body are amoeboid, and the one shown here is filled with foreign material that it has ingested. (C) Phagocytosis and pinocytosis are called *endocytosis* because solid and liquid material, respectively, are taken into a cell. Exocytosis is basically the opposite process, by which the contents of vesicles are released to a cell's environment.

(B)

(A)

(C)

Channel forms

Cell

Droplets

Cell

Pinocytosis

Cell

Droplets in vesicle

amoeba feeds. During "cell drinking" or **pinocytosis,** quantities of materials in solution enter cells in vesicles. For example, human eggs secure nutrients secreted by the surrounding "nurse" cells. In both processes, called **endocytosis** generally, the plasma membrane folds inward and the materials become trapped within a vesicle when the edges of the invagination fuse together. **Exocytosis,** the process by which cells rid themselves of materials, occurs in basically the opposite way. Both endocytosis and exocytosis require energy because of the extensive movements of the plasma membrane, and endocytosis does not allow the cell to discriminate between the useful and the potentially harmful materials being taken in.

CELL CYCLES

During a typical second in the life of an average adult human, millions of body cells die, and during the same period about as many new cells form. Because many human cells double in size and divide every few hours, we are not exactly the same as we were an instant ago.

Mitosis

Except for mature muscle and nerve cells, our body or **somatic cells** grow until they reach a certain size and then undergo **mitosis,** or nuclear division (Figure 3.18). The duplicated chromosomes separate and two genetically identical **daughter cells** are formed. These cells grow, their chromosomes duplicate and separate, the cytoplasm divides in turn, and so on, generation after generation—the **cell cycle.** The amount of growth and the time interval between cell divisions vary from one cell type to another. Egg cells grow little between successive divisions, but skin cells grow very rapidly. Intestinal cells last for about a day, white blood cells live for several days to about two weeks, and many red blood cells survive for about four months.

Early cytologists easily saw the cell's nucleus, but only with the development of dyes could other organelles be seen clearly. Some dyes stained the granular material in the nucleus, called **chromatin** (from the Greek for "color"). During cell division this chromatin was seen to condense into the threadlike chromosomes. The suspicion that the chromosomes were involved in the transmission of genetic traits was confirmed early in the twentieth century.

Why cells divide when they do is unknown. However, each cell type has an optimum ratio of cytoplasm to nucleoplasm, and perhaps many cells divide when that ratio is upset. Although the cell cycle is a continuous sequence, with one phase grading smoothly into the next, stages have been named for convenience.

Although **interphase** is sometimes called the "resting stage," the cells are not inactive. During this, the longest part of the cell cycle, the cell grows and all the events that prepare it for division occur. The nucleus is large, and the nucleoplasm consists of a fine network of chromatin. The centriole and the chromosomes duplicate during interphase, although these events do not become evident until later. The replicated and condensing chromosomes are called **chromatids** while they remain attached at a point called the **centromere.**

Mitosis, the shortest part of the cell cycle, begins with **prophase,** when the threadlike chromatids condense further into thicker and denser units. Radiating from the two centrioles are microtubules called **spindle fibers.** The centrioles move in opposite directions until they reach the poles or opposite ends of the cell, and the chromatids move to the center or

(A)

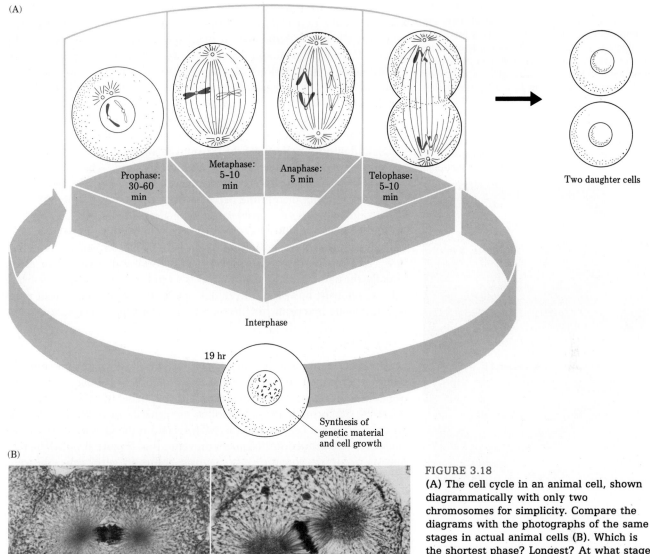

(B)

FIGURE 3.18

(A) The cell cycle in an animal cell, shown diagrammatically with only two chromosomes for simplicity. Compare the diagrams with the photographs of the same stages in actual animal cells (B). Which is the shortest phase? Longest? At what stage are the chromosomes most easily countable? During prophase, what is happening to the nuclear envelope? To the centriole? How does the number of chromosomes in the daughter cells compare with the chromosome number in the parent cell?

equator of the cell. The nuclear envelope breaks down and its components become part of the endoplasmic reticulum.

Metaphase begins when the spindle fibers attach to the centromeres of the chromatids that are at the cell equator. The chromatids and centromeres are usually distinct and may be counted, especially if the cell is treated chemically to stop their movements. Next, each centromere divides, the spindle fibers contract, and the chromatids separate, each with its own centromere.

FIGURE 3.19

Cytokinesis is the cleavage of a cell that has undergone nuclear division. This fertilized egg or zygote from a mammal is dividing for the first time. The two daughter cells produced will divide mitotically, as will their daughter cells, and so on for generation after generation. Although the adult organism that is ultimately produced will have an immense number of cells with the same chromosomal makeup, the cells will have become specialized for a multitude of functions.

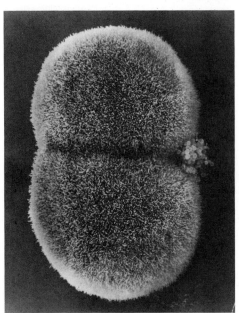

Conditions are now set for the beginning of the short, but beautiful, **anaphase** stage that begins with the separation of the sister chromatids and ends when two dense groups of chromosomes have been moved by microtubules to opposite poles of the cells. When the movement of the chromosomes ends, **telophase** begins. A nuclear envelope re-forms around each set of chromosomes, the spindle fibers dissolve, the microtubules are disassembled, and the chromosomes uncoil to become slender chromatin threads. Basically, the events of telophase are the reverse of those of prophase, and mitosis is complete. However, there is the additional and important event of the cytoplasm constricting into two daughter cells, **cytokinesis.** Cytoplasmic cleavage occurs when the belt of microfilaments around the cell's equator constricts and pinches off the connecting plasma membrane between the daughter cells (Figure 3.19).

Through mitosis, we grow from a single cell, the fertilized egg or **zygote,** to many trillions (60–100) of cells. During our lifetime we also replace by mitosis immense numbers of lost cells. In general, whether new cells become parts of bones, nerves, or muscles, they are alike. Usually the same genetic information is found in big-toe, liver, or brain cells. Obviously, however, a liver cell "pays attention" to different parts of our genetic blueprint for life than a brain or big-toe cell does.

Meiosis

If mitosis were the only type of cell division, the chromosome number would double in each succeeding generation. That is, if each sperm and egg contained 46 chromosomes, then the next generation would contain cells with 92 chromosomes. The following generation would possess 184 chromosomes per cell, and so on. **Meiosis** (from the Greek for "to make smaller") or "reduction division" prevents this (Figure 3.20). Whereas mitosis occurs in somatic cells, meiosis occurs in **germ cells,** the cells that give rise to **gametes** (eggs or sperm). Meiosis makes possible cellular reproduction without increasing the number of chromsomes in the cells of each succeeding generation. Also important is the fact that the genetic messages of the parents may be reassorted and recombined in new ways in the offspring (Chapter 10).

Members of each pair of chromosomes generally are alike in size and shape, and somatic cells that have two of each type of chromosomes are called **diploid** (*2n*). **Haploid** (*n*) cells such as the gametes contain only a single member of each chromosome pair. The letter *n* stands for the number of different chromosomes (Figure 3.21).

PATHOLOGY

Diseased or abnormal cells that change in number, features, functions, or reactions are the units of study in the science of **pathology.** For example, some types of anemia can be diagnosed by counting the number of red blood cells in a given volume of blood. Some allergies or parasites stimulate certain types of white blood cells to increase in number. Rapidly dividing cancer cells may be more susceptible than normal cells to certain chemicals or to radiation (Chapter 16).

Some cells are relatively easy to secure for analysis, such as those in blood, sputum, urine, or feces. Parts of the body with exposed surfaces can be scraped or massaged to secure cells. For example, cells of the vagina and cervix can be scraped (Pap smear) in order to detect cancer in its early stages (Figure 3.22A). Cells from deeper structures can be secured by **biopsy,** the removal of small masses of cells with a scalpel or with special plungerlike devices (Figure 3.22B).

First meiotic division

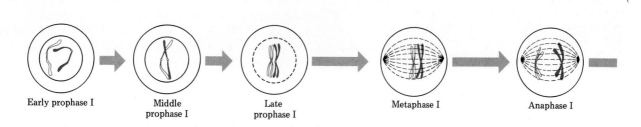

| Early prophase I | Middle prophase I | Late prophase I | Metaphase I | Anaphase I |

Second meiotic division

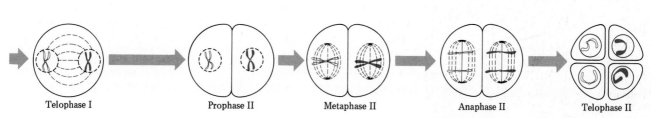

| Telophase I | Prophase II | Metaphase II | Anaphase II | Telophase II |

FIGURE 3.20
Meiosis, shown with only two chromosomes for simplicity. This process occurs only in the germ cells as eggs and sperm are produced. From every immature sperm cell, four haploid sperm are produced. However, only one functional haploid egg results from the two meiotic divisions during egg formation. The small structures called *polar bodies* are not functional, and they disintegrate.

TISSUES

Pioneers were pretty self-sufficient, or they did not survive. Finding and preparing food, caring for medical problems, building shelter, transporting themselves, chopping wood, and fighting enemies were all in a day's work. A contemporary person is usually not such a jack-of-all-trades. One person drives a cement truck, another teaches high school art, and still others stock shelves, grow peanuts, legislate, pump gas, paint houses, set broken bones, or perform endless other specialized tasks. Perhaps we have become too interdependent to survive totally on our own. It is probably as unfair to ask whether the modern human is more or less complex than the pioneer than it is to make that comparison between human and amoeba cells. The amoeba cell does everything for itself since it is the entire organism, whereas the cells of multicellular organisms have become specialized but less flexible in their abilities.

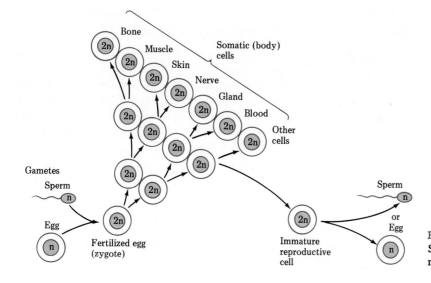

FIGURE 3.21
Summary of the relationship between mitosis and meiosis in an individual's life.

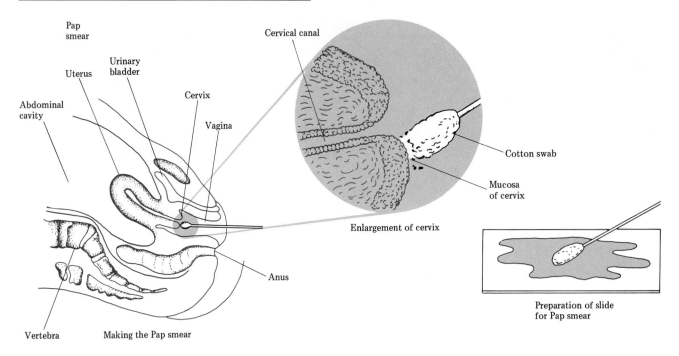

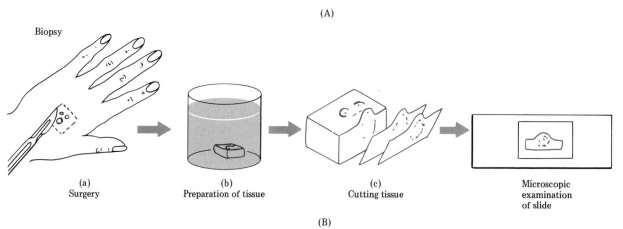

FIGURE 3.22
Two common pathological methods. (A) The Pap smear is named after its developer, Dr. Papanicolaou. An excessive number of white blood cells present on the slide (a common sign of an infection) as well as the presence of tumor cells is cause for concern. (B) Biopsy is a method of removing living cells for examination.

Some cells may be specialized for secretion, others for support, and still others for transmitting nerve impulses. Groups of cells with similar structures and functions make up **tissues.** There are four basic tissue types in humans, and their study is called **histology.**

Epithelial Tissue

Epithelial tissue covers exposed external surfaces and lines internal cavities and tubes, which it protects, repairs, receives stimuli from, and regulates the passage of materials across. Epithelial cells may be flattened (squamous), cube-shaped (cuboidal), or pillarlike (columnar; Figure 3.23). Some epithelial cells absorb (such as those in the stomach, intestines, and kidneys), and others secrete. Some glands are **exocrine,** with ducts to carry their secretions to the exterior. Such is the case with sweat glands, glands of the digestive tract, and mammary glands. The **endocrine** glands are ductless and secrete their products, the **hormones,** into extracellular spaces, from which they diffuse into the circulatory system (Chapter 18). Epithelial cells generally divide regularly, giving the tissue good repair capabilities.

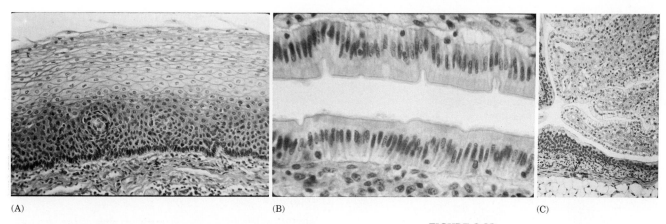

(A)   (B)   (C)

Connective Tissue

Connective tissues fill spaces and bind and support body parts. These tissues contain a variety of cells and intercellular material or **matrix** (Figure 3.24). Fibrous tissues form **tendons,** which attach muscles to bones, and **ligaments,** which hold bones together (Chapter 21). They also wrap muscles, nerves, and other body parts. Supportive connective tissues include **cartilage** (as in your nose and earlobes) and **bone.** Blood may also be supportive when it fills spaces, as during an erection of the penis. Cartilage cells secrete a gristly or rubbery matrix, whereas calcium and other salts make up part of the matrix of bone. **Blood** consists of cells in a liquid matrix (plasma) and is perhaps the most "connective" of our connective tissues because all our cells are near it (Chapter 15). Fat or **adipose** cells and pigment cells are specialized connective cells. Like epithelial cells, connective tissue cells generally divide regularly.

FIGURE 3.23
Some typical shapes and locations of epithelial cells. (A) Cuboidal from kidney. (B) Simple columnar from intestine. (C) Stratified squamous from pharynx. Where else in the body would you expect to find epithelial tissue?

FIGURE 3.24
Examples of human connective tissue. What is the difference between a tendon and a ligament? How does the matrix of cartilage, bone, and blood differ? What is a function of adipose tissue?

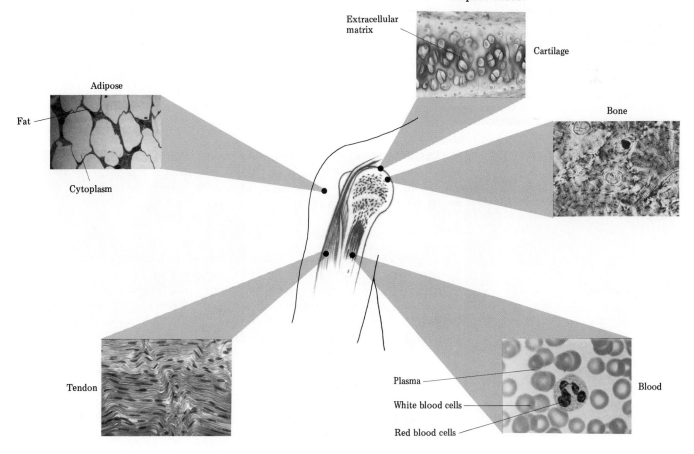

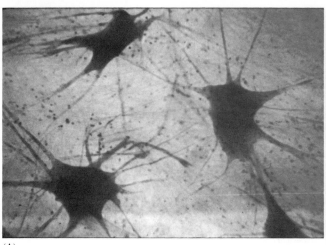

(A)

(B)

FIGURE 3.25
Some examples of neurons, the cells that constitute the nervous system: (A) from the spinal cord and (B) from the brain (cerebellum). How do the structures of neurons relate to their functions?

Nervous Tissue

Nervous tissue is composed of cells called **neurons** that are specialized for initiating and conducting nerve impulses (Figure 3.25; Chapter 19). Most nervous tissue is found in the brain and spinal cord, although many neurons are near most cells. Neurons do not usually divide once they mature, so repair of aged or damaged nervous tissue is minimal.

Muscle Tissue

All movements of the body except those caused by gravity result from the contraction of **muscle fibers,** or individual muscle cells (Chapter 21). The three types of **muscle tissue** are shown in Figure 3.26. **Smooth muscle** occurs in most internal body parts and is usually involuntary, or not controlled consciously. **Skeletal** or **striated muscle** cells have several nuclei and bands or striations. This tissue, found in most voluntary muscles of our body, makes up about 40 percent of our body weight. Striated muscles respond quickly compared with smooth muscles, but they also tire faster. That is, your striated leg muscles become fatigued from a long run, but the smooth muscles in your stomach wall do not tire after

FIGURE 3.26
The three basic types of muscle tissue: (A) striated, (B) smooth and (C) cardiac. Which type acts quickly but also tires easily? Acts and tires relatively slowly? What traits does cardiac muscle share with striated muscle? With smooth muscle? Which muscle types are not under significant conscious control?

(A)   (B)   (C)

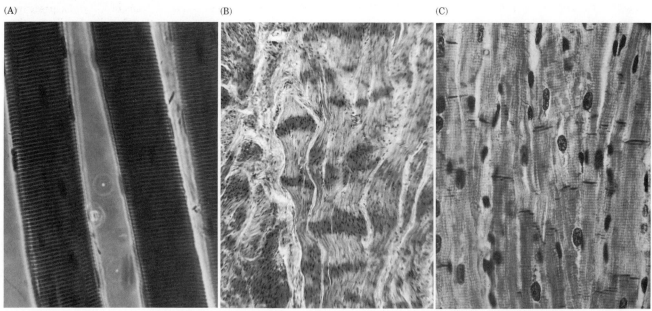

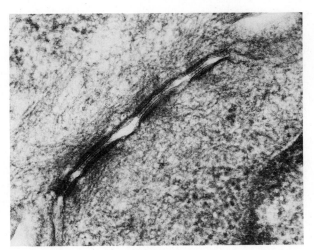

FIGURE 3.27
Some junctions between cells are very tight, such as these. In other cases there may be channels that allow for the movement of certain materials between adjacent cells.

, hEART

you eat a large meal. Although **cardiac muscle** cells are also striated, these cells, found only in the heart, are largely involuntary.

Like neurons, mature muscle fibers do not usually divide. Injured areas of muscles are usually replaced with fibrous connective tissue, or scar tissue, rather than with new muscle cells. Scar tissue in cardiac muscle is especially serious (Chapter 15).

If the cells of our tissues had no adhesiveness they would be loose and might come apart. Various substances act as "glue" or as a "spot weld" to bind certain cells together, especially those subject to much mechanical stress. In some neighboring cells the membranes fuse to form a tight, impermeable junction (Figure 3.27). This explains, for example, why some substances in the blood do not get into the brain. Other junctions not only connect exterior cell membranes but also provide channels between cells to allow for the direct passage of some substances.

ORGANS

Groups of tissues organized into structural and functional units are called **organs.** Such familiar body parts as the stomach, skin, eye, and lung are organs (Figure 3.28). Although all tissue types are usually present in organs, one type may predominate. Thus, the brain is composed primarily of nervous tissue and the heart of muscle tissue.

FIGURE 3.28
Skin is an organ, actually one of the largest organs in the body. Note in the cross-sectional diagram (A) and in the micrograph (B) that each of the four tissue types is represented. Many of the surface cells and those around the base of every hair follicle are epithelial. Connective tissue in the skin is represented by blood and adipose deposits. Fibrous connective tissue (e.g., collagen) is scattered throughout the skin, and as we age our wrinkles will be caused by changes in the cells that produce matrix materials. Many of our sensory neurons are close to our skin surface (e.g., for pressure, temperature, and pain). The small smooth muscle fibers are involved when we shiver and raise "goose bumps."

(A)

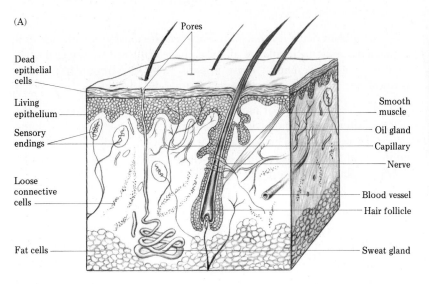

Pores

Dead epithelial cells

Living epithelium

Sensory endings

Loose connective cells

Fat cells

Smooth muscle

Oil gland

Capillary

Nerve

Blood vessel

Hair follicle

Sweat gland

(B)

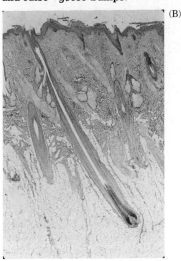

(A)

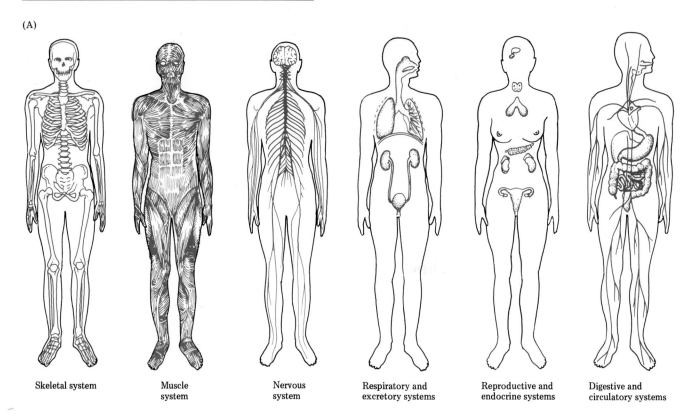

Skeletal system   Muscle system   Nervous system   Respiratory and excretory systems   Reproductive and endocrine systems   Digestive and circulatory systems

(B)

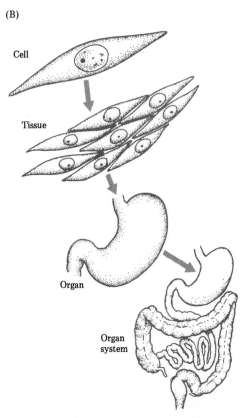

Cell

Tissue

Organ

Organ system

FIGURE 3.29

(A) Some of the major human organ systems. (B) A summary of the organizational hierarchy in the human body.

ORGAN SYSTEMS

Organs that work together to perform an integrated function make up **organ systems.** Figure 3.29 shows the general features of the major human organ systems.

SUMMARY

1. The term **cell** was coined in the mid-1600s by **R. Hooke** when he observed dead cork microscopically. Living cells were observed shortly thereafter by **A. van Leeuwenhoek.** However, it was not until the 1800s that **M. Schleiden, T. Schwann,** and **R. Virchow** developed the **cell theory,** or the generalization that all organisms are composed of cells that arise from preexisting cells.

2. **Cytoplasm** occupies the space between the **plasma membrane** and the **nucleus,** and **nucleoplasm** fills the nucleus. Although animal cells differ in sizes and shapes, most possess the same **organelles,** which partition the cytoplasm into functional subcompartments.

3. The control center of most cells is the large and often centralized nucleus. Pores within the **nuclear envelope** allow for the passage of some substances between the nucleoplasm and the cytoplasm. **Chromosomes** and their constituent **genes** are found within the nucleoplasm, as are one or more **nucleoli.** Human **somatic cells** contain 46 chromosomes (23 pairs).

4. Numerous substances flow through the elaborate channels and compartments of the **endoplasmic reticulum,** which fills the cytoplasm. **Enzymes** embedded within the ER membranes also assist in many synthetic and other chemical reactions.

5. The **ribosomes,** which either float in the cytoplasm or are attached to rough ER, are the organelles that translate copies of genes into the proteins that characterize the cell.

6. **Golgi bodies** are the collection, packaging, and distribution centers for many of the substances synthesized in cells. Transfer from one part of the cell to another occurs by means of small **vesicles.**

7. **Lysosomes** contain enzymes that digest cellular structures in aged, injured, or abnormal cells. The products may be stored and transported in vesicles, and some substances are recycled.

8. **Mitochondria,** the power plants of cells, are especially common in metabolically active cells. The large surface area of mitochondrial membranes facilitates the numerous chemical reactions that occur there. Many biologists theorize that bacteria-like mitochondria originated as **symbionts** in ancient cells.

9. Although animals lack **plastids,** the **chloroplasts** in green plants are the sites of **photosynthesis,** the process upon which animals ultimately depend for food and oxygen.

10. **Centrioles** are important in cell division and in the formation of the hairlike **cilia** (numerous and short) and **flagella** (few and long) that either move materials past cells or propel motile cells.

11. The **cytoskeleton** consists of the contractile **microfilaments** and **microtubules** that give cells their shape and assist in moving their structures.

12. A **light microscope** reveals relatively few cellular features compared with those visible using an **electron microscope.** Scanning electron **microscopes** are used to study surfaces, while **transmission electron microscopes** are used to resolve fine structural details. **Staining** and **centrifugation** are also useful techniques in **cytology,** the study of cellular structure.

13. While a cell's needs are generally proportional to its volume, the cell's ability to satisfy its needs is affected greatly by its surface area. Small objects have a larger **surface area-to-volume ratio** than large objects, and volume increases faster than surface area as objects grow. **Microvilli** increase the surface areas of the plasma membranes of various specialized cells.

14. The materials that enter or leave cells must pass through one or more thin and flexible membranes. **Biological membranes** are freely permeable to water and consist of a fatty bilayer with many embedded proteins, some of which function somewhat like gates. Various **solutes** are dissolved in the cell's watery **solvent.** The chemical properties of these substances determine their ability to cross **selectively permeable** membranes. Some materials cross easily, others only under certain circumstances, and still others not at all. Technically, materials that are breathed or swallowed are not *in* the living body, but are only *surrounded* by it. It is not until they enter cells that they are truly inside of us.

15. **Diffusion** is the **passive movement** of materials from areas of higher concentration to areas of lower concentration. The special case of water diffusing across a selectively permeable membrane is called **osmosis.** The condition that exists when the concentrations of a substance are equal on both sides of a membrane is called **isotonic.** In other

cases, one concentration may by **hypotonic** (less concentrated) or **hypertonic** (more concentrated) relative to another. In **facilitated diffusion,** a transport protein in a membrane combines temporarily with a substance, carries it through a specific channel, and releases it.

16. Passive transport processes are often too slow to satisfy the metabolic needs of a cell. **Active transport** mechanisms that require cellular energy are important in most cells. In such processes, substances may be moved across membranes against concentration gradients.

17. **Endocytosis** includes **phagocytosis** ("cell eating") and **pinocytosis** ("cell drinking"). Cells may rid themselves of materials by **exocytosis,** which is essentially the reverse of the endocytic process.

18. In **mitosis,** somatic cells grow, duplicate their chromosomes, and divide in a manner that normally results in two genetically identical but smaller **daughter cells.** The **cell cycle** is continuous, although it does not occur at the same rate in all cell types. The typical sequence consists of **interphase** (the long "resting" stage), **prophase** (thin **chromatids** connected at a **centromere** undergo condensation, and **spindle fibers** radiate from each centriole), **metaphase** (spindle fibers attach to the centromeres when chromatids are at the equator; chromatids separate after centromeres divide), **anaphase** (sister chromatids are moved to opposite poles), and **telophase** (nuclear envelope reforms around each chromosome set, which again becomes chromatin threads). **Cytokinesis** is the cleavage of the cell into two daughter cells.

19. The **germ cells,** or **gametes,** undergo two nuclear divisions that result in daughter cells with a **haploid** number of chromosomes. This allows the **diploid** number to be reestablished when a **zygote** forms.

20. **Pathology** is the study of abnormal or diseased cells. Surface cells may be secured directly and easily for examination, while **biopsy** is one method of securing cells from deeper structures.

21. **Histology** is the study of **tissues,** or groups of cells with similar structures and functions. There are four tissue types in humans: (a) **epithelial** (covers exposed surfaces and lines internal surfaces; may be protective, absorptive, or secretory, as in **exocrine** and **endocrine** glands), (b) **connective** (fills spaces and supports and binds body parts; consists of cells and an intercellular **matrix;** includes **tendons, ligaments, cartilage, bone, blood,** and **adipose** structures), (c) **nervous** (**neurons** specialized for initiating and conducting impulses), and (d) **muscle** (**muscle fibers** specialized for contraction; includes the involuntary **smooth muscle** cells of most internal organs that act and tire relatively slowly, the voluntary **striated** or **skeletal muscle** cells that act and tire relatively quickly, and the involuntary **cardiac muscle**).

22. **Organs** are groups of tissues organized into structural and functional units (e.g., heart and kidney). **Organ systems** consist of several organs that function in an integrated manner (e.g., digestive, respiratory, and circulatory systems).

REVIEW QUESTIONS

1. Describe the major ideas involved in the cell theory.

2. Describe some of the ways that materials move between the nucleoplasm and the cytoplasm, within the cytoplasm, and from cell to cell.

3. Describe the major functions of the following organelles: nucleus, chromosome, nucleolus, endoplasmic reticulum, ribosome, Golgi body, vesicle, lysosome, mitochondrion, and centriole.

4. How is the process of photosynthesis important to animals, even though they lack the chloroplasts within which the events occur?

5. Distinguish between the structures and functions of the following extensions of or from the plasma membrane: microvilli, cilia, and flagella.

6. What is a cytoskeleton, and what functions do its structures serve?

7. Compare and contrast the traits of prokaryotic and eukaryotic cells, using examples of each.

8. Distinguish between light microscopes and scanning and transmission electron microscopes.

9. What tools and techniques other than microscopy are important to cytologists and histologists?

10. Of what significance are surface area-to-volume ratios to individual cells and to entire organisms? As objects grow, do surface areas or volumes increase at a faster rate, and how does this affect the metabolic rate of living structures?

11. What is the structure of biological membranes, and what is meant by "selectively permeable"?

12. Distinguish between the general processes of passive and active transport and, specifically, among diffusion, osmosis, and facilitated diffusion.

13. Describe circumstances in which substances are isotonic, hypertonic, or hypotonic to one another.

14. What are the functions and key traits of enzymes?

15. Distinguish between endocytosis and exocytosis. What happens to a cell during phagocytosis and pinocytosis?

16. Distinguish between somatic and germinal cells, naming specific examples.

17. Describe the major sequence of events during mitosis and cytokinesis.

18. How does the process of meiosis differ from that of mitosis?

19. Describe the structure, function, and location of epithelial, connective, nervous, and muscle tissues.

20. Distinguish between an exocrine and an endocrine gland.

21. Compare and contrast the structure, function, and location of smooth, striated, and cardiac muscle.

CHAPTER 4

CHEMICAL ORGANIZATION FOR HOMEOSTASIS

The processes of living occur in cells, and it is the chemical reactions in cells that make life possible. Organisms are chemical machines, and the cell is life's chemistry laboratory. Some biologists have little opportunity to study that "lab" directly because they spend most of their time in the field observing animal behavior, or taking body temperatures of porcupines (carefully!), or collecting beautiful blue Brazilian bumblebees, or diving into oceanic trenches to study strange organisms that live in the darkness near volcanic vents. However, to give significance to this fieldwork, biologists must explain observations in cellular and chemical terms. On the other hand, many of the biochemists and molecular biologists who work only in clean laboratories with spotless glassware must be reminded occasionally that some of the chemicals they study come from living organisms.

Molecules were once said jokingly to be too large to interest physicists and too small to concern biologists. Today, many scientists in most fields must enter what was formerly the realm of the chemist. It is possible that one could be a Nobel Prize–winning physicist or chemist without knowing much about biology. However, it is not possible to understand much in biology without some knowledge of chemistry and physics. Although it is possible to describe some aspects of life in nonchemical terms, a knowledge of chemistry is necessary to understand and explain most life processes.

Aghast at studying chemistry when you really wanted to learn more about humans and other organisms? Relax—we need not be mechanics to know, at least in a general way, how automobiles work, nor must we be chemists to gain some insight into some of the ways chemical processes work in organisms.

The chemicals of life, such as this vitamin C molecule, are quite beautiful, at least to chemists and biologists. However, chemicals, by themselves, are not alive. The cell is the structural unit within which the immense variety of chemicals that make up even the simplest organisms are organized.

FIGURE 4.1
Atomic structure. (A) Hydrogen is the smallest atom, and its nucleus contains only a single proton (atomic number = 1). (B) Oxygen contains eight orbiting electrons around its nucleus, which contains eight protons and eight neutrons.

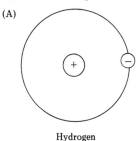

(A)

Hydrogen

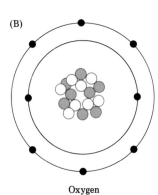

(B)

Oxygen

ATOMS, MOLECULES, AND ELEMENTS

Be it a star, bacterium, rock, or human, all matter is composed of **elements,** substances that cannot be broken down into simpler substances by chemical means. There are 92 naturally occurring elements and over a dozen more human-made elements. The latter elements are unstable, and break down into two or more of the natural elements. Each element has a one- or two-letter symbol that stands for its English or Latin name (Table 4.1). The symbol alone usually means one atom (see below): for example, O designates one atom of oxygen. Three atoms of calcium is symbolized 3Ca.

Atoms are the smallest particles of an element that still possess all of the element's chemical and physical characteristics. Atoms are so small that experiments to determine their structure began only in this century. Atoms consist of a central, dense **nucleus** and a cloud of orbiting, negatively charged (−) **electrons.** Except for hydrogen (the lightest element, which has no neutrons), the nucleus contains positively charged (+) **protons** and, as their name implies, electrically neutral **neutrons** (Figure 4.1). The number of protons in the nucleus, the **atomic number,** is unique to each element. The chemical properties of an atom are largely determined by the number of its electrons, which number in turn depends upon the number of protons.

Isotopes

Isotopes are alternative forms of elements that have the same chemical properties as the elements but whose atoms differ in the number of neutrons they contain (Figure 4.2). Oxygen has 8 isotopes, and some elements have as many as 20 isotopes. Some isotopes are unstable and undergo spontaneous nuclear disintegration or decay, releasing radiation and smaller particles. Although many of these **radioisotopes** may be harmful if not handled properly, some are useful as **tracers** or "tags," allowing us to follow chemical reactions occurring within organisms or in their environments.

Ions

Atoms may also vary in electrical charge. Ordinarily, atoms are electrically balanced, with the same number of positively charged protons as negatively charged electrons. However, atoms may gain or lose electrons, creating **ions.** For example, when a sodium atom (Na) loses an electron, it becomes a sodium ion (Na^+). Many ions are very important in living processes.

FIGURE 4.2
Carbon isotopes. Carbon 14 is often used in tracing chemical reactions in organisms.

Carbon 12

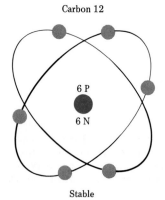

Stable

Carbon 13

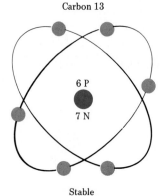

Stable

Carbon 14

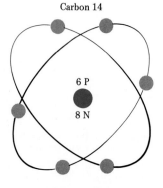

Unstable (radioactive)

TABLE 4.1
Major Chemical Elements in the Human Body

Element	Symbol	Occurrence
Oxygen	O	All tissues
Carbon	C	All tissues
Hydrogen	H	All tissues
Nitrogen	N	Proteins
Calcium	Ca	Bones, teeth; needed for blood clotting
Phosphorus	P	Bones, teeth, some proteins
Potassium	K	Nerves
Sulfur	S	Many proteins
Sodium	Na	Blood, nerves, perspiration
Chlorine	Cl	Perspiration
Magnesium	Mg	Bones, muscles
Iron	Fe	Hemoglobin
Copper	Cu	Some enzymes
Iodine	I	Thyroid hormone
Fluorine	F	Teeth

ELECTRONS AND ATOMIC PROPERTIES

Most of the mass of an atom is in the core nucleus, because the tiny orbiting electrons have almost no mass. The weight of all of your electrons may about equal the weight of your eyelashes. Although it is impossible to know the location of a particular electron, the electrons are often shown in **orbitals** or shells, somewhat like planets in a solar system. Although electrons are said to be maintained in their orbitals by their attraction to the protons in the nucleus, at any given instant a particular electron may be close to or far away from the nucleus. However, when electrons are in outer orbitals they move faster than those in inner orbitals, and therefore have higher energy levels. Although electrons may move from one orbital to another in an atom, an electron that moves from an inner to an outer shell must be "excited" by an external energy source such as light, heat, or electricity (Figure 4.3).

The chemical properties of an atom depend largely upon the number of electrons in the outermost shell. An atom with eight electrons in its outermost orbital is chemically unreactive or inert (e.g., argon and neon). In contrast, an atom with fewer than eight electrons in its outermost shell tends to gain or to lose electrons, and when that orbital is filled or

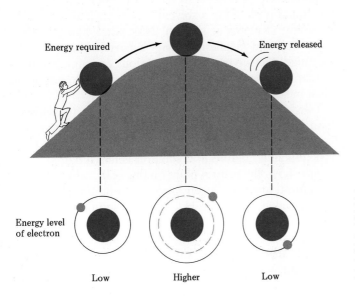

Energy required

Energy released

Energy level
of electron

Low   Higher   Low

FIGURE 4.3
Electrons may exist in different energy levels. Orbitals occur in different shapes, and there may be one to several energy levels in different atoms. Energy is required to move from an inner to an outer energy level.

emptied, the atom becomes stable. However, hydrogen atoms, with only one orbital, need only one additional electron to be stable. All atoms tend to fill their outer energy levels with the maximum number of electrons. Very reactive atoms are those that lack an electron in their outer energy level (e.g., chlorine and fluorine), as are those that possess only one electron in their outer orbital (e.g., potassium and sodium).

Compounds result when the atoms of two or more elements combine chemically in the proper ratios. The chemical properties of a compound differ from those of its constituent atoms. For example, sodium and chlorine are poisonous separately, but together they form table salt. The familiar liquid called *water* is formed from two gaseous elements, hydrogen and oxygen. **Molecules** consist of two or more joined atoms of the same kind and are the smallest units of a compound that retain the compound's physical and chemical properties. The symbol O_2 means that two oxygen atoms are joined to form one molecule of oxygen.

OXIDATION AND REDUCTION

Sometimes an electron gains enough energy to escape from its atom. The loss of an electron by an atom, ion, or molecule is called **oxidation.** Some substances cannot lose electrons unless oxygen is available to accept them. **Reduction,** the gain of an electron, occurs simultaneously with oxidation because an electron lost by one atom is accepted by another. Thus, electron transfer occurs during the oxidation of one substance and the reduction of another. Sometimes an electron is transferred with a proton (as with a hydrogen atom). Thus, when oxygen is reduced by two hydrogen atoms, water is formed. Oxidation-reduction reactions are very important in the transfer of energy within organisms.

CHEMICAL BONDING

The attractive force between the atoms in a molecule is called a **chemical bond,** and each bond contains a certain amount of potential chemical energy. Because atoms of each element have unique bonding traits, only certain kinds of molecules can form.

The process of atoms combining often leads to **ionization** (Figure 4.4). Two ions with opposite electrical charges are attracted to one another, and the chemical bond that holds together the ionic compound is called an **ionic bond.** These bonds tend to break or dissociate when the ionic compound is placed in a solvent (polar) such as water. Such biologically important elements as sodium (Na), potassium (K), chlorine (Cl), calcium (Ca), and magnesium (Mg) behave in this way. Ionic compounds are called **electrolytes** because solutions containing ions conduct electric current. Nonelectrolytes, such as table sugar, do not dissociate into ions when in solution.

Two important classes of ionic compounds are **acids** and **bases.** Acids dissociate in water and donate hydrogen ions (H^+) to solutions, whereas bases dissociate in water and reduce the amount of H^+. The acidic or basic (alkaline) nature of a solution is measured by a 0 to 14 **pH scale** (Figure 4.5). The neutral point on this scale, which measures the relative concentration of hydrogen ions in a solution, is 7. The lower the numerical value, the more H^+ ions the solution contains and the more acidic it is (0–7). Values above 7 indicate an alkaline, or basic, solution. **Buffers** in living systems help to "soak up" excess acid or base ions and keep pH levels constant. This is one of many homeostatic mechanisms that occur within cells.

(A)

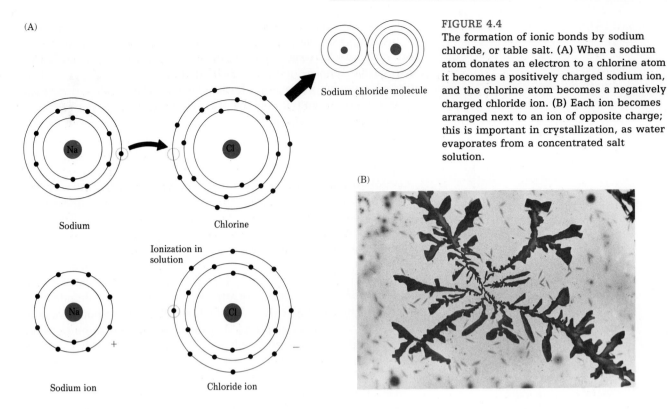

Sodium chloride molecule

Sodium

Chlorine

Ionization in solution

Sodium ion

Chloride ion

FIGURE 4.4
The formation of ionic bonds by sodium chloride, or table salt. (A) When a sodium atom donates an electron to a chlorine atom it becomes a positively charged sodium ion, and the chlorine atom becomes a negatively charged chloride ion. (B) Each ion becomes arranged next to an ion of opposite charge; this is important in crystallization, as water evaporates from a concentrated salt solution.

(B)

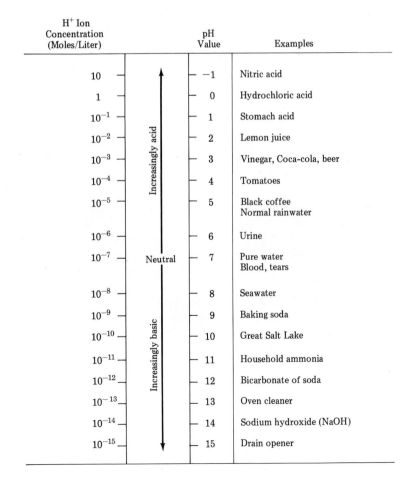

H⁺ Ion Concentration (Moles/Liter)	pH Value	Examples
10	-1	Nitric acid
1	0	Hydrochloric acid
10^{-1}	1	Stomach acid
10^{-2}	2	Lemon juice
10^{-3}	3	Vinegar, Coca-cola, beer
10^{-4}	4	Tomatoes
10^{-5}	5	Black coffee / Normal rainwater
10^{-6}	6	Urine
10^{-7}	7	Pure water / Blood, tears
10^{-8}	8	Seawater
10^{-9}	9	Baking soda
10^{-10}	10	Great Salt Lake
10^{-11}	11	Household ammonia
10^{-12}	12	Bicarbonate of soda
10^{-13}	13	Oven cleaner
10^{-14}	14	Sodium hydroxide (NaOH)
10^{-15}	15	Drain opener

Increasingly acid

Neutral

Increasingly basic

FIGURE 4.5
The pH scale, with examples of common substances. Many acids have a sour taste, while bases tend to taste bitter. When an acid and a base combine, they form a compound plus water, such as HCl (hydrochloric acid) + NaOH (sodium hydroxide) → NaCl (sodium chloride) + H_2O (water).

In contrast to the transfer of electrons from one atom to another in ionic bonds, the stronger **covalent bonds** involve the sharing of one or more pairs of electrons (Figure 4.6). For example, oxygen needs two electrons to complete its outer shell. It may do this by sharing two electrons with hydrogen, thus forming water (H_2O). The strength of covalent bonds depends upon how many electron pairs are shared. Thus, the sharing of two pairs of electrons in oxygen gas (symbolized O=O) represents a stronger bond than the one that involves the sharing of only one pair of electrons, as in hydrogen gas (H—H). More energy is required to break double than single covalent bonds.

Carbon forms covalent bonds with many other elements, and carbon's peculiar bonding abilities enable it to form numerous complex molecules. In this sense, the chemistry of carbon is the chemistry of life. Thus, the study of carbon-containing compounds is called **organic chemistry.** Non-carbon-containing substances are called **inorganic.**

Hydrogen bonds are weaker than ionic or covalent bonds. These bonds, which last for only a tiny fraction of a second, occur when the hydrogen atoms in molecules are attracted electrically to unbonded oxygen and nitrogen (Figure 4.7). The fact that these weak bonds require little energy to form or break them is the basis of their importance to life. For example, hydrogen bonds help give stability to water, an important ingredient of all cells. Thus, as hydrogen bonds break and reform quickly, water "resists" a change in its state.

Because of the bond energies between its atoms, molecules hold energy

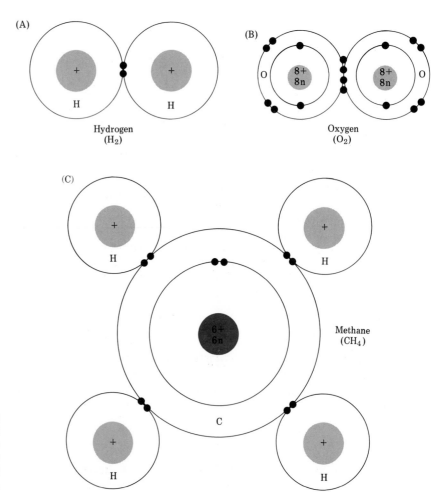

FIGURE 4.6
Covalent bonds. Hydrogen gas (A) and oxygen gas (B) are examples of covalent single and double bonds, respectively. (C) Methane is the simplest combination of hydrogen and oxygen (or hydrocarbons).

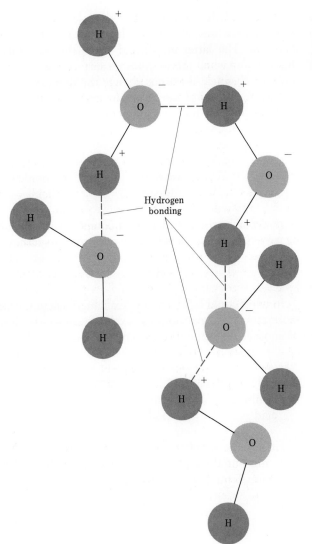

FIGURE 4.7
**Weak hydrogen bonds form when two
electronegative atoms share a single
hydrogen atom, as shown here for water.**

Hydrogen
bonding

"in storage." When the bonds between atoms are broken, the atoms separate and the bond energy is released. The amount of energy released equals the amount supplied to link the atoms. A few atoms, such as helium, cannot form bonds and are therefore called *inert*. Of the main biologically important atoms, hydrogen forms one bond, oxygen two, nitrogen three, and carbon four.

CHEMICAL REACTIONS

Chemical reactions are the making and breaking of chemical bonds between atoms and molecules. Few free atoms exist naturally on Earth. Most molecules in organisms result from contact reactions in which electron sharing or exchange occurs. Many of these chemical reactions are reversible, as shown by the two-headed arrow below:

$$A + B \leftrightarrow C + D$$

Four events often increase the chance of two substances interacting: (1) the concentrations of the reactants (A + B) increases, (2) the concentration of the products (C + D) decreases, (3) the temperature or pressure increases, and (4) the likelihood of collisions by the substances increases. The latter may occur when the reactants are subdivided or dissolved or when there is a large surface area.

Most chemical reactions change the number and/or the types and/or the arrangements of the atoms in molecules. For example, in **synthesis reactions,** two or more simpler substances may combine to form one or more complex substances, as in the following:

$$CO_2 \; + \; H_2O \rightarrow H_2CO_3$$

carbon   water   carbonic
dioxide   acid

In **decomposition reactions,** a complex substance breaks up into two or more simpler substances, the opposite of synthesis. For example,

$$H_2CO_3 \rightarrow CO_2 \; + \; H_2O$$

In what could be called an "exchange reaction" for simplicity, one or more atoms of one molecule exchange places with one or more atoms of another molecule. For example:

$$H - Cl \; + \; Na - O - H \rightarrow H - O - H + Na - Cl$$

hydrochloric   sodium   water   sodium
acid   hydroxide   chloride

Finally, the numbers of types of atoms may remain the same in the molecule but their pattern changes, as follows, and this may lead to different chemical characteristics.

$$
\begin{array}{cc}
\begin{array}{c}
H \\
| \\
H-C-H \\
| \\
Cl-C-O-H \\
| \\
H
\end{array}
&
\begin{array}{c}
H \\
| \\
H-C-Cl \\
| \\
H-O-C-H \\
| \\
H
\end{array}
\end{array}
$$

One way in which two molecules join to form a larger molecule is through a **condensation reaction.** In this reaction, water forms as hydrogen and oxygen atoms are stripped from the reactants. For this reason, condensation reactions are also referred to as *dehydration synthesis.* **Hydrolysis reactions** involve adding hydrogen and oxygen from water to condensation products. Hydrolysis is an important process in digestion because it decomposes a substance into its component subunits (Chapter 12).

The subunits or building blocks of large compounds are called **monomers.** Thus, the sugars that make up carbohydrates and the amino acids (see below) that constitute proteins are monomers. Molecules that consist of many identical or similar molecular subunits or monomers, like the coupled cars that form a long railroad train, are called **polymers.** Thus, starch is a polymer of glucose monomers.

THE CHEMISTRY OF LIFE

Although the labels on many food containers tell us what chemicals enter our bodies, they do not indicate what the chemicals do. Although much still remains to be learned about biological chemistry, we may soon know enough to even explain the chemical basis for thoughts and emotions.

We have long known that the substances of stars have "come alive" in the sense that organisms are composed of many of the same elements as those found in the nonliving world (Table 4.2). Every carbon atom in your body originated in a star. For centuries before that, many assumed that organisms contained some special ingredient or "vital force" that made them alive. By themselves, however, our chemicals are not "alive" until they become organized in cells. The components of life have been synthesized in the laboratory; some new "creatures" have been genetically engineered and even patented (Chapter 10). We cannot yet create life anew, however.

Notice in Table 4.2 that six elements make up about 99 percent of organisms by weight. However, if we measured the *number* of atoms in our bodies instead of their *weight*, only four kinds would exceed 1 percent of the total—nitrogen, oxygen, carbon, and hydrogen (Figure 4.8). The acronym NOCH may help you remember these.

The elements most abundant in Earth's crust are not those most common in organisms. Even as atoms go, however, the NOCH atoms are small and light. In general, smaller atoms react more quickly than larger ones, and life depends upon many integrated and rapid chemical reactions. All of the NOCH atoms form gases and also numerous other compounds by means of covalent bonds, and many of the resulting molecules are water soluble. However, most of the molecules formed are unstable, meaning that the bonds can be broken without the expenditure of excessive energy.

Some elements, such as iron, are present in very small amounts but are essential for life. For example, iron is part of hemoglobin, the chemical in our red blood cells that carries oxygen. Some **trace elements**—those present in extremely small amounts, such as iodine—are also important for normal body activity.

Although organic molecules were first discovered in organisms, many inorganic substances are also important to life. Examples of the latter are water, oxygen, and various minerals. The typical animal cell is composed of almost 90 percent water and about 1 percent minerals, meaning that we are more inorganic than organic.

FIGURE 4.8

The composition of the human body as measured by the number of atoms in elements. The NOCH atoms, and about a dozen other kinds, are of prime importance in the chemistry of life.

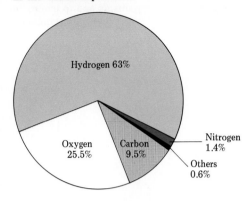

TABLE 4.2

Comparison of Chemicals in Humans and in Earth's Crust

Element	Symbol	Approximate Percent (by Weight) of a Human	Approximate Percent (by Weight) of Earth's Crust
Oxygen	O	65.0	46.6
Carbon	C	18.5	0.03
Hydrogen	H	9.5	0.14
Nitrogen	N	3.3	Trace
Calcium	Ca	1.5	3.6
Phosphorus	P	1.0	0.07
Potassium	K	0.4	2.6
Sulfur	S	0.3	0.03
Sodium	Na	0.2	2.8
Chlorine	Cl	0.2	0.1
Magnesium	Mg	0.1	2.1
Iron	Fe	Trace	5.0
Manganese	Mn	Trace	0.1
Silicon	Si	Trace	27.7

FIGURE 4.9
(A) This view of Earth from space shows clearly why our home is called the "water planet." (B) The fact that water may exist in solid, liquid, and gaseous states within common temperature ranges is of major significance to organisms and to events in their environments.

(A)   (B)

WATER

Cells are largely composed of solutes in water. Water is also an important environment for many organisms and, in its various forms, is a major force in shaping environments (Figure 4.9). Although life may eventually be found elsewhere in space, it is most common in our solar system on the "water planet" Earth. Because cells are mostly water and because life undoubtedly evolved in water, this seemingly simple compound, H_2O, is truly the cradle of life. Where there is insufficient water there is no life, and the chemistry of life is largely the chemistry of water.

Water results from the formation of two single covalent bonds between an oxygen atom and two hydrogen atoms (Figure 4.10). Electrons are attracted more strongly to the water's oxygen atom than they are to its hydrogen atoms, thus giving water electron-rich ($-$) and electron-poor ($+$) ends. Because of this magnetlike trait, such compounds are called **polar,** and water is one of the most polar compounds known. Much of the biological importance of water results from its polarity. For example, numerous polar molecules are soluble in water, as when the negative ends of water molecules face the positive sodium ions and the positive ends of water molecules orient toward the negative chloride ions when table salt dissolves in water.

Oil does not dissolve in water because it is nonpolar. Such nonsoluble compounds are called **hydrophobic** (from the Greek for "water fearing") because they do not form hydrogen bonds with water. The external portions of many biologically important compounds are hydrophobic (e.g., plasma membranes), and this trait also contributes to the unique shapes of many compounds.

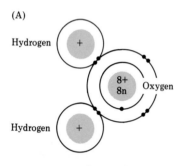

(A)

Hydrogen

8+
8n Oxygen

Hydrogen

(B)

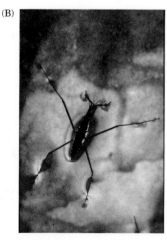

FIGURE 4.10
(A) A water molecule is composed of an oxygen atom that shares one electron with each of two hydrogen atoms. (B) Being polar, water molecules are attracted to other polar molecules. This is called *adhesion* when the other molecules are different and *cohesion* when they are other water molecules. The cohesion of liquid water molecules creates surface tension, which allows some small organisms like this water strider to glide along the water surface without sinking.

ORGANIC COMPOUNDS

In general, organic molecules are larger and more complex than inorganic ones. This is due largely to the properties of carbon atoms that allow them to join readily into multiple covalent bonds with other carbon atoms and with those of many other elements. Large organic molecules, such as most proteins and many carbohydrates, are called **macromolecules.**

Organic compounds have a structural skeleton of carbon atoms bonded together with other atoms. The compounds may be branched, twisted, and looped, giving each molecule a relatively large size and a particular shape. Four main groups of organic compounds are described below.

Carbohydrates

Sugars, starches, glycogen, and cellulose (a major compound in plant fibers) are **carbohydrates,** macromolecules that make up about 1 percent of typical cells. Carbohydrates constitute the largest part of most human diets, and some serve as important cellular fuels, such as the sugar glucose. Carbohydrates store energy in many carbon–hydrogen bonds, which release energy when they are broken. This page of paper is mostly carbohydrate, and although it contains much "food for thought," it is a form (cellulose) that can provide little energy for humans if eaten. Goats and millipedes fare better on such food.

As their name implies, carbohydrates are mainly compounds of carbon, hydrogen, and oxygen—chemically, CH_2O (Figure 4.11). The hydrogen and oxygen are attached to the chain or ring of carbon atoms in a ratio of about $2:1$ (or $1:2:1$ for C, H, and O). However, because more than two carbon atoms are always present, the simple formula is expressed better as $(CH_2O)_n$, where n is 2 or a greater number.

The *saccha-* used in naming sugars is derived from the Latin word for sugar. There are **monosaccharides** or simple sugars (such as glucose), **disaccharides** formed from two simple sugars bonded covalently (such as table sugar or sucrose), and **polysaccharides,** insoluble polymers that consist of long chains of many monosaccharides. Starch, for example, is a polysaccharide formed from glucose. The names of most sugars end in *-ose,* such as glucose, fructose, and galactose. Glucose is the most common simple sugar in our bodies and is the most important energy-storage molecule in our cells. Some carbohydrates also play a structural role in organisms, as in wood (cellulose) and in the outside covering or exoskeleton of insects and crabs (chitin).

Monosaccharides are not common in nature, and sugars in honey, beets, and milk (lactose) are disaccharides. Although sucrose and lactose are soluble in water, their molecular size prevents them from entering cells. To be used for energy, these sugars must first be hydrolized in the small intestine during digestion.

$$C_{12}H_{22}O_{11} + H_2O \rightarrow C_6H_{12}O_6 + C_6H_{12}O_6$$
$$\text{sucrose} \qquad \text{water} \qquad \text{glucose} \qquad \text{fructose}$$

The monosaccharides pass across the intestinal wall, enter the bloodstream, and then enter cells throughout the body. Through the process of cellular respiration discussed in a later section, sugars make energy available to all cells.

Lipids

Fats, oils, steroids, and waxes are examples of **lipids,** which constitute about 2 percent of typical cells. Lipids form part of membranes; some act as chemical messengers (e.g., sex hormones); they serve as cellular fuel; and they form insulation (adipose deposits) in the skin against the loss of body heat. Like carbohydrates, lipids contain carbon, hydrogen, and oxygen. However, lipids have far less oxygen and they are insoluble in water because they are nonpolar. The building blocks of lipids are alcohols (usually glycerol) and fatty acids (Figure 4.12).

Because of their numerous C—H bonds, lipids average about twice as much energy per unit of weight as carbohydrates, making them a concentrated metabolic fuel. Highly saturated (see below) animal fats contain more stored energy than less saturated plant fats. Overweight humans do not necessarily have unusual energy and strength, because

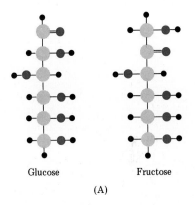

Glucose   Fructose

(A)

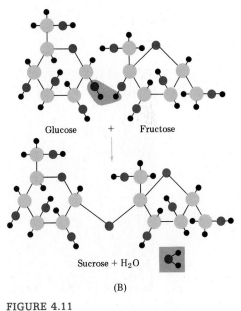

Glucose   +   Fructose

Sucrose + H_2O

(B)

FIGURE 4.11
(A) Glucose and fructose ($C_6H_{12}O_6$), two important simple sugars or monosaccharides, are shown in straight-form diagrams. (B) Disaccharides or double sugars, such as sucrose or table sugar, are formed by the combination of monosaccharides. In this instance, the sugars are shown in ring-form diagrams. This is an example of a condensation reaction, with the H atom coming from fructose and the OH group coming from glucose.

FIGURE 4.12
(A) A fat molecule is composed of the alcohol glycerol joined to fatty acids. The characteristics of lipids depend upon the types of fatty acids, the lengths of the chains, and the ways the carbon atoms are bonded. Note that this is another example of a condensation reaction. (B) Saturated and unsaturated fats. *Unsaturated* means that carbon atoms may form new bonds with other atoms. Are saturated or unsaturated lipids more solid at room temperature?

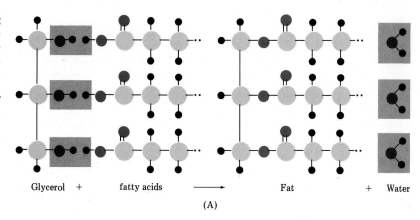

Glycerol +   fatty acids   ⟶   Fat   +   Water

(A)

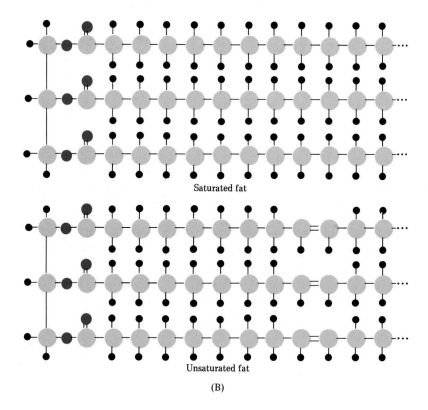

Saturated fat

Unsaturated fat

(B)

lipids are not used as easily by the body as carbohydrates are. Lipids may be consumed as such, or other nutrients such as carbohydrates may be converted to lipids by the body. Most people who "get fat" do so by consuming excess carbohydrates rather than by eating too many lipids. Thus, excess sugar can be converted into fat and stored, or when we diet, body fat may be converted back to sugar.

Advertisements often distinguish between **saturated** and **unsaturated** fats, with the latter usually considered to be better for our health. Chemically, saturated lipids have the carbon atoms that make up the backbone of fatty acid molecules mostly or entirely bonded with hydrogen atoms. These substances (such as butter or lard) are usually solid or semisolid at room temperatures. Unsaturated lipids have many of their carbon atoms double-bonded to one another in the fatty acid molecule, one consequence of which is fewer bonds to hydrogen atoms. These lipids (usually oily) are generally liquid at room temperatures. Many plant fatty acids are unsaturated, whereas animal fats are often saturated.

The accumulation of lipids within the walls of arteries often leads to **atherosclerosis** (one form of arteriosclerosis, or hardening of the arteries). Cholesterol (a steroid common in eggs, butter, and whole milk) has been especially implicated in this problem. However, the problem is complicated by the fact that the body requires cholesterol for the normal functioning of cell membranes.

Lipids that contain phosphorus, **phospholipids,** are important because they are found in all biological membranes (Figure 4.13). Also, part of this molecule is soluble in water, whereas lipids in general are insoluble or only slightly soluble in water. Phospholipids have a "water-loving" **hydrophilic** end or "head" and a hydrophobic "tail." These molecules orient spontaneously in water to form two-layered membranes in which the tails point inward toward one another (Figure 3.12). These traits of phospholipids are very important to cells in terms of the movement of materials across membranes. For example, when phosphate groups are added to sugars, the cell membrane prevents the compounds from diffusing outward in response to a concentration gradient. This occurs because the sugar's negatively charged phosphate group is repelled by the interior of the plasma membrane. Lipid bilayer membranes are impermeable to ions and to most polar molecules.

Proteins

In addition to carbon, hydrogen, and oxygen, **proteins** contain nitrogen (Figure 4.14). Most proteins are immense molecules composed of long chains of **amino acids.** Some polysaccharide molecules may be as large or larger, but they are less complex because they consist of one or a few kinds of sugar units. Proteins, on the other hand, are constructed from 20 different amino acids, which may be thought of as the "alphabet" of proteins. Just as innumerable words are made from the 26 letters of our alphabet, almost an infinite variety of proteins is possible from different combinations of amino acids. Relatively short **peptides** result when amino acids link through covalent bonds (called peptide bonds). Like monosaccharhides, amino acids can form units of different lengths, from very short (as when two amino acids form a dipeptide) to longer chains of amino acids called **polypeptides.** Many proteins, in turn, are

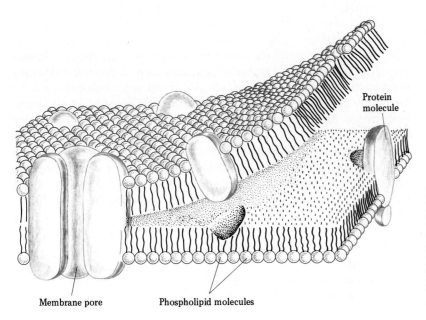

Protein molecule

Membrane pore   Phospholipid molecules

FIGURE 4.13
Phospholipids usually consist of a polar or hydrophilic end or head of alcohol and an end or tail of two fatty acid chains that are nonpolar or hydrophobic. When the tails are long, a bilayer membrane results, with the polar heads in contact with polar water molecules and the tails oriented toward one another.

FIGURE 4.14
(A) The general structure of amino acids is simple to symbolize. Attached to the same carbon atom are an amine group (NH₂) and an acid group (COOH). The carbon chain comes in a variety of forms and is shown as R (for radical). (B) Amino acids may join through condensation reactions to form long polypeptides.

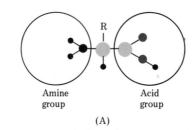

(A)

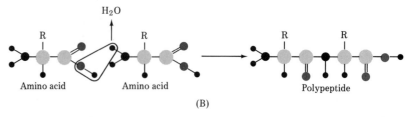

(B)

composed of several polypeptides. Both the sequence of amino acids in the chain and the types of amino acids themselves (there are five classes of them) determine the properties of the protein. Whether cabbage or king, amoeba or human, all organisms are made up of this universal 20-letter alphabet. No two humans, even identical twins, have exactly the same proteins.

The large size of proteins becomes clear when their molecular weights (one way the size of molecules is measured) are compared with those of other nutrients. The proteins insulin (a hormone from the pancreas) and hemoglobin (the red pigment in blood) have molecular weights of about 5,000 and 64,000, respectively. By comparison, the molecular weights of water, glucose, and sucrose are 18, 180, and 342, respectively.

Our body can synthesize only 12 amino acids. The other 8 must come from our diet, and because of this they are called the **essential amino acids.** Most human tissues contain between 10 and 20 percent protein, and proteins make up about 15 percent of typical cells (dry weights). If carbohydrates and lipids are the principal fuels of the body, then proteins are the body's principal structural materials. They are abundant in cells (e.g., in chromosomes and embedded in plasma membranes) and in tissues such as muscles. Fibrous proteins are also found in hair, nails, skin, cartilage, bone, tendons, ligaments, and blood clots.

Proteins also play many regulatory roles in cells, such as forming **enzymes** (see below). These biological catalysts facilitate chemical reactions without being used up or becoming part of the product. A different enzyme catalyzes each reaction in organisms. Some hormones that control various metabolic processes are also composed of proteins.

Usually only during starvation do we use our proteins as fuel. However, excess proteins can provide energy or be converted to fat and stored.

Goof, gold, and *food* have different meanings from the word *good,* even though each differs from it by only one letter. Something very similar can happen with proteins, and error is a real problem in protein structure and function. Even a single change in one amino acid out of a chain of hundreds may produce a serious or fatal problem. The globular protein hemoglobin contains about 300 amino acids, and a change in one particular amino acid may lead to abnormal hemoglobin. When deprived of oxygen, the red blood cells of people with this minor change in an amino acid assume a sickled shape and do not function properly (Chapters 9 and 14).

As some protein chains form, they take on the shape of a circular staircase. This shape is typical of proteins that have structural functions, such as those fibrous forms that hold body parts together or form parts of bones, cartilage, tendons, hair, and nails. Still other proteins, such as enzymes, have more complex shapes, like dense globes. The bonds that hold these proteins in shape are weak and easily broken by agents such as heat. You see this when you cook the high-protein whites of eggs. These **denatured** proteins never regain their former structure and thereby lose their specific biological activity. (Try to unfry an egg to see this for yourself!) Finally, several globular proteins may intertwine to form a very complex structure, which may be required for their function.

Nucleic Acids

Like proteins, **nucleic acids** contain hydrogen, oxygen, carbon, and nitrogen. Amino acids are to proteins what the repeating subunits called **nucleotides** are to nucleic acids (Figure 4.15). The two different types of nucleic acids, ribonucleic acid (RNA) and deoxyribonucleic acid (DNA), consist of chains composed of four different nitrogen-containing bases. The sequence of these base units gives each nucleic acid molecule its unique traits, in a way somewhat analogous to what happens in proteins. Thus, while most computers and the Morse code use two elements to convey messages (0 or 1, and dot or dash, respectively), and our alphabet uses 26 letters, four different molecules suffice to encode hereditary information (Chapter 10).

Chromosomes are combinations of DNA and protein. The precise duplication of DNA when cells divide ensures that the daughter cells produced are genetically identical. DNA provides the master plan of heredity, and RNA is a copy or template of portions of DNA. While DNA remains within the cell's nucleus, certain RNA molecules pass into the cytoplasm, where they specify the amino acid sequence of proteins.

WORKING CELLS

It is obvious that both single cells and large multicellular organisms occupy space and have weight. This is the definition of **matter.** It is also obvious that living requires **energy,** which is the capacity to do work. For example, your heart has been beating from before your birth, and in a 70-year lifetime will have done so about 3 billion times. Unlike matter, energy neither occupies space nor has weight.

Physicists recognize two states of energy, potential and kinetic. Both states are involved if, for example, you wind a wristwatch. **Kinetic energy,** the energy of motion or action, is involved as your muscle cells allow you to turn the knob. The coiled spring in the watch has **potential energy,** stored or inactive energy. In this instance, the work done by the coiled spring is mechanical.

Chemical reactions involve a change from one energy level to another. Sometimes a product of a chemical reaction has more energy than the starting materials, but other times it has less. Both of these types of chemical reactions require **activation energy** in order to start, just as it takes energy to push a boulder uphill. With its high potential energy, the boulder rolling downhill releases kinetic energy, analogous to some chemical reactions (Figure 4.16).

What does energy have to do with life processes? Even as you read this page your cells are working. Cells that cannot obtain and use energy are considered dead. However, just as machines cannot produce energy

FIGURE 4.15
How nucleotides, the subunits of nucleic acids, can attach to form long chains. *S* stands for sugar, *P* for a phosphate group, and *N* for a nitrogenous base.

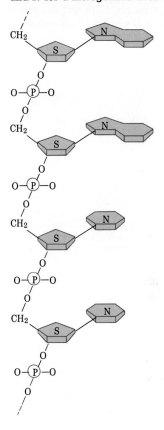

FIGURE 4.16
A molecule must possess sufficient activation energy in order to undergo a specific chemical reaction. The energy of activation is needed to destabilize existing chemical bonds. Note that catalyzed reactions (i.e., those involving enzymes) lower the amount of activation energy needed to initiate a reaction.

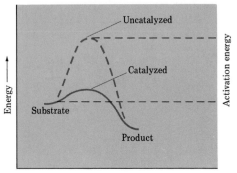

from nothing, cells can only change energy from one form to another. This is the basis of the **first law of thermodynamics,** the concept that energy can be neither created nor destroyed, but only changed in form. The total energy in the universe remains constant, and there is no free lunch. We get no more energy out of the coiled watchspring than we put in. In fact, we usually get less, which is the basis of the **second law of thermodynamics.** This principle states that in any energy transformation, the usable energy is reduced. **Entropy,** a measure of randomness and disorder, increases in a system as its available energy decreases. Thus, because living systems are highly organized, yet our environment is essentially disruptive, energy must constantly be fed into the systems to maintain order and homeostasis. Without the energy input, cells become disorganized, cease to work, and die. (Relax; there is no third law.)

Neither cells nor machines are perfect. That is, the conversion of one form of energy to another is never 100 percent efficient, and some energy is lost in a nonusable form. In biological as well as many mechanical systems much of the unusable energy is in the form of heat. Many motor vehicles have radiators to dissipate the excess heat, while others are air-cooled. As in most automobiles, we lose about two-thirds of the energy derived from our fuel as heat. For most people most of the time, we are warmer than our environment and lose heat *to* it rather than gain heat *from* it.

Somewhat poetically, we could say that life's fires burn with a cool flame. If this page were burned, there would be a sudden increase in temperature. During combustion, the energy stored in many chemical bonds is released suddenly. If you could metabolize the paper as efficiently as a goat does, the bonds would be broken gradually and your body temperature would not change. In other words, we do not get very hot after eating a large meal and cool to the environmental temperature between meals. A chemical mechanism for keeping organisms in a steady-state condition is described below.

ENZYMES

Proteins that are enzymes are so important to life processes that special attention should be given them. When many people think of chemistry they picture someone heating chemicals in a test tube. For chemical reactions to occur, the reactants obviously must come together, and while this can occur slowly at low temperatures, heat increases molecular motion. As a general rule, the rate of many chemical reactions doubles with each 10°C (18°F) increase in temperature. However, the human body temperature is only 37°C (98.6°F), too low for many chemical reactions that our high metabolic rate depends upon. For example, one must heat most food reactants to high temperatures in the laboratory to duplicate digestive reactions that occur in the body at much lower temperatures. The "cold chemistry" of life depends largely upon the work of enzymes that reduce the amount of activation energy required to initiate chemical reactions.

One way that enzymes work is shown in Figure 4.17. In this lock-and-key model, the three-dimensional shape of the large enzyme allows it to bind to the small molecules, called the **substrates,** upon which it works. The substrates or reactants are brought together in clefts or pockets on the surfaces of enzymes, somewhat like fingers fitting into a glove, and it is during this tight fit that the chemical reaction occurs. Once the product has formed, the glove no longer fits properly, the

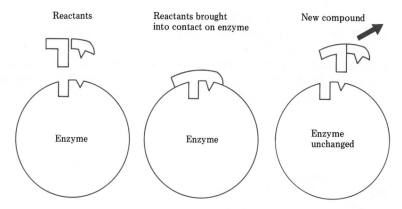

Reactants   Reactants brought into contact on enzyme   New compound

FIGURE 4.17
How enzymes work. Enzymes have characteristic shapes that fit their specific substrates. By combining briefly with the reactant molecules at particular sites, the enzyme brings them together and facilitates their reaction. The unchanged enzyme then dissociates from the new compound, not becoming a part of the product, and it is again available to catalyze similar reactions.

typically weak enzyme-product complex dissociates, and the enzyme is available to catalyze similar reactions again and again. The specific shape of enzymes is important to their work, just as shape is important in our tools. We cannot insert a screw as well with a hammer as we can with a screwdriver, and only certain molecules fit into the enzyme's reactive sites. Although the substrate-binding and the active sites are not the same technically, they are close together on the enzyme's surface. Enzymes do not cause reactions that would not have happened otherwise; they only speed reactions, often a millionfold, that would have occurred slowly. Also, in hastening the inevitable, enzymes do not influence the final proportions of the chemical products.

Most enzymes are very specific in their action and often catalyze only a single reaction or a single step in a complex reaction. This occurs partly because the shapes of enzyme molecules are affected by heat and by other chemicals (such as acids and bases), and there are optimum temperatures and other conditions for enzyme actions. One effect of certain poisons is to prevent enzymes from affecting the substrate (Figure 4.18). Because of the specificity or "pickiness" of enzymes, it is necessary for organisms to possess hundreds to thousands of different enzymes in order to catalyze the numerous and diverse array of metabolic reactions necessary for life.

The names of many enzymes end in -ase. Sometimes they are named for the substrate upon which they work, such as lipase acting upon lipids and sucrase acting upon sucrose. Other enzymes are named for the kind of reaction in which they are involved, such as oxidases facilitating oxidation reactions. However, most metabolic reactions occur in sequences (biochemical pathways) in which the product of one reaction becomes the substrate for another.

Vitamins and minerals are familiar components of food (Chapter 12). Some vitamins are needed in the diet because they function as **cofactors** within the body. Cofactors may act as "adapter molecules" that help bind enzyme and substrate, or they may remove a product of the chemical reaction. Thus, the lack of such vitamins prevents certain chemical reactions and thereby leads to a diseased state. Although only small amounts of vitamins are needed, vitamins must be present in the diet regularly. This is because our cells cannot make most vitamins and vitamins are broken down and excreted rather quickly.

PHOTOSYNTHESIS: GREEN POWER

In the living world the ultimate source of all energy is the sun. Solar energy is captured and changed by green plants and passed along from

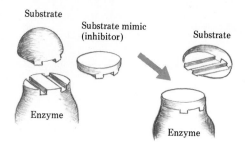

Substrate   Substrate mimic (inhibitor)   Substrate

Enzyme   Enzyme

FIGURE 4.18
Some substances, such as certain poisons, inhibit enzymes. They combine with the enzyme and thus prevent the natural substrate from attaching to the enzyme.

one organism to another, thus reducing entropy locally. After each organism takes its share of energy, such low-energy molecules as carbon dioxide and water are left.

Ever watch a plant eat? Unless you are thinking of an insectivorous plant such as Venus's-flytrap, your answer is probably "no." Although eating, drinking, and other life activities are not as evident in plants as in animals, the fact that plants grow and develop means that they must receive and utilize matter and energy.

One of the first humans to observe plants carefully and record his observations was Aristotle (384–322 B.C.). Although he was uncertain about the nutrients plants used, Aristotle thought that plants got what they needed from the soil through their roots. Because most plants are rooted in soil, this certainly seemed obvious. Stop some "average" people on the street today and ask them where plants get their food and they might express a similar, but erroneous, belief. The general nature of plant nutrition is now clear, and the way we arrived at some of this knowledge is an excellent example of the historical methods of science; see the essay, "How Do Plants Get Food?"

Photosynthesis, the production of food in the presence of light by green plants, is an extremely important biological process, not only for plants themselves but for all organisms (Figure 4.19). The green steroid pigment **chlorophyll** captures the energy of light and makes possible the following general reaction:

$$\begin{array}{ccc} \text{carbon} + \text{water} & \xrightarrow[\text{enzymes}]{\text{chlorophyll}} & \text{glucose} + \text{oxygen} + \text{water} \\ \text{dioxide} \end{array}$$

$$6CO_2 + 12H_2O \longrightarrow C_6H_{12}O_6 + 6O_2 + 6H_2O$$

Much more is known about the detailed chemical steps involved than you probably wish to know.

Ever wonder why most plants are green, instead of blue or red, for example? The answer is that not all wavelengths of visible light contain the same energy and not all objects absorb those wavelengths equally (Figure 4.20). The fact that green wavelengths are reflected indicates that they contain little energy useful in photosynthesis. The colors that we do not see, such as the blue and red wavelengths that are absorbed rather than reflected, are those important in "green power."

Organisms such as green plants that are able to synthesize complex organic molecules from simple inorganic compounds are called **autotrophs** (from the Greek for "self-feeders"). Not all autotrophs are photosynthetic, although the vast majority are. The few **chemoautotrophic** (chemosynthetic) organisms can synthesize food from various ingredients in their environments, often even in the dark. **Heterotrophs** (from the Greek for "other feeders"), which include animals, nongreen plants, fungi, and most bacteria, obtain chemical energy by degrading the organic molecules in the organisms or organic wastes upon which they feed. The energy possessed in organic compounds is often measured by the amount of energy required to break its chemical bonds.

Ultimately, we and other heterotrophs depend upon autotrophs for two basic needs—food and the oxygen that is a by-product of photosynthesis. This is the basis for the pretty ecological posters that are captioned "Have you thanked a green plant today?" We are unlikely to run out of oxygen soon, but sufficient food for the growing human population is a major problem. Exact numbers are unknown but many

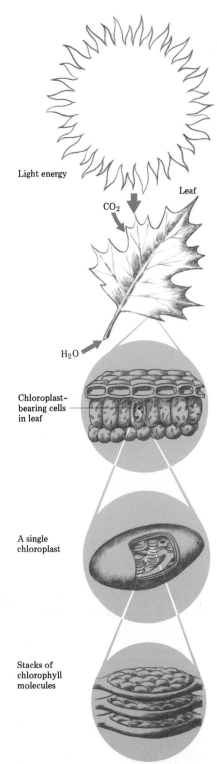

Light energy

CO₂

Leaf

H₂O

Chloroplast-bearing cells in leaf

A single chloroplast

Stacks of chlorophyll molecules

FIGURE 4.19
Photosynthesis occurs when light strikes the chloroplast-bearing cells of green plants. Plant cells usually contain scores of chloroplasts, which hold the chlorophyll molecules stacked in piles. This provides a large surface area for the complex chemical reactions in which the pigment participates.

How Do Plants Get Food?

In Aristotle's time, the curious thought about rather than experimented with natural phenomena. It was not until the early 1600s that the Belgian scientist Jan van Helmont experimented to disprove the "soil eating" theory (Figure B4.1). Van Helmont dried some soil in a furnace, placed it in a large pot, and wet it with rainwater. He planted a small willow tree and covered the pot to keep out dust. In five years the tree grew considerably, yet the weight of the soil decreased only slightly. Because he had added only water to the willow, van Helmont concluded that the plant material arose entirely from the water.

Later in the 1600s an early microscopist (Nehemiah Grew) observed **stomata** (from the Greek for "mouth"), the small openings on leaves, and wondered whether they could allow the plant to exchange materials with the air. However, gases were poorly understood at that time, although by the late 1700s the famous chemist Joseph Priestley had performed the experiment shown in Figure B4.2.

Shortly after Priestley's work, another scientist (Jan Ingenhousz) "saw the light" and discovered that plants cease their "beneficial operation" for animals at night. Plants were observed to "contaminate" the air at night, and even in the daytime in shaded places.

Ingenhousz also learned that not all parts of plants "restored" air (Figure B4.3). Later, others learned that shredded leaves behaved like whole leaves, and this led to the eventual discovery of chloroplasts in living green plant cells. The green pigment chlorophyll was discovered in the chloroplasts, and the nature of the gases exchanged with the environment was identified.

FIGURE B4.1
Van Helmont's experiment. Notice the large increase in the weight of the tree and the small decrease in the weight of the soil. How do you explain the soil loss? How much would the soil have weighed if Aristotle's theory had been correct? What factors did Van Helmont apparently overlook in arriving at his conclusion?

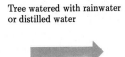

Tree watered with rainwater or distilled water

5 yr

Weight of tree: 5 lb
Weight of soil: 200 lb

Weight of tree: 169 lb 3 oz
Weight of soil: 199 lb 14 oz

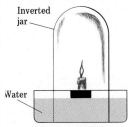

Inverted jar

Water

Floating candle burns

Candle goes out quickly

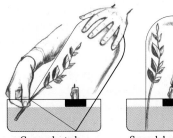

Green plant also placed under jar

Several days later the candle will burn again

Plant and animal live

Animal without plant dies

FIGURE B4.2
Priestley's experiments led to the conclusion that plants reversed the "damaging" effects of fires and animals on air. How do you explain Priestley's results?

Light

Inverted jar

Water

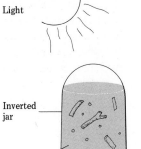

FIGURE B4.3
Ingenhousz discovered that nongreen parts of plants produced less gas (and therefore displaced less water in the inverted jars) than green plant parts. However, he was unsure about why light and the green pigment chlorophyll were important.

Green stems

Green leaves

Nongreen plant parts

FIGURE 4.20
(A) Not all wavelengths of light are absorbed equally by chlorophyll and other common plant pigments. Note that wavelengths near the red and blue ends (the latter being the most important in photosynthesis) are absorbed more than those in the middle range. How do these facts explain the green color of most plants? (B) Like humans, most bacteria depend upon oxygen released by green plants. On a microscope slide, strands of green algae in water also containing bacteria were exposed to different wavelengths of light. Note how the concentration of bacteria along the filament of algae resembles the curve of the graph in (A).

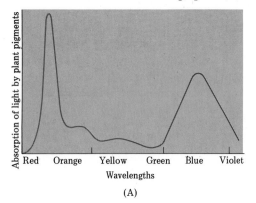

(A)

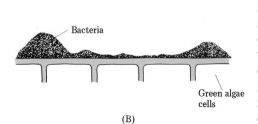

(B)

millions of humans face imminent starvation. Understanding photosynthesis is part of the essential knowledge we need to feed Earth's increasing numbers of hungry humans.

CELLULAR RESPIRATION

Just as we need ways to distribute fuels and sources of energy, such as petroleum and electricity, cells must have an energy-distribution system. As organic compounds are utilized in cells, much of the energy released is used to form **adenosine triphosphate** (ATP). ATP is called the "energy currency" of most cells because it is "spent" for work to be done. Most ATP forms in mitochondria, whose enzymes are specialized to break down glucose and produce ATP. This very reactive molecule is available throughout the cell for nearly all energy-requiring reactions. Release of the energy stored in the chemical bonds of ATP converts the molecule to a less reactive form that can be recharged in the mitochondria and used again.

The cell's currency, ATP, does not just appear but has to be made, and this requires energy. You have seen many ATP makers—they are called green plants. The ultimate sources of energy on Earth are the sugars and polysaccharides made by plants during photosynthesis. This source is "cashed in" for ATP molecules to power the reactions in cells. **Cellular respiration** is the process whereby energy stored in glucose is converted to ATP for use by cells. Plants use some glucose for their own metabolic activities. Energy use continues as **herbivores** eat plants and as **carnivores** consume herbivores and other animals. Eventually, **decomposers** such as many bacteria and fungi secure glucose as they break down dead organisms or their wastes. In this process of harvesting the energy they need for their own metabolism, decomposers also aid in recycling the material substances that organisms require.

You have seen that respiration and combustion are somewhat similar in that both use fuel and release energy (Figure 4.21). However, respiration can be compared to walking down five flights of stairs, whereas combustion is more like jumping out of a fifth-floor window. Either way we end up on the ground, whether by one big step or by many smaller ones. In combustion, many chemical bonds are broken quickly, whereas respiration involves the gradual release of energy by the controlled breaking of chemical bonds. As with photosynthesis, the details of the many chemical steps involved in cellular respiration are complex.

Organic fuel

Combustion: sudden release of heat energy

Respiration: gradual release of heat energy

FIGURE 4.21A

Simplified comparison of combustion and cellular respiration. (A)

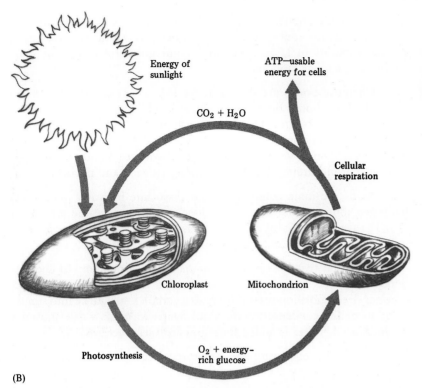

Energy of sunlight

ATP—usable energy for cells

$CO_2 + H_2O$

Cellular respiration

Chloroplast

Mitochondrion

Photosynthesis

O_2 + energy–rich glucose

(B)

FIGURE 4.21B
During photosynthesis, chloroplasts capture solar radiation and produce glucose and oxygen. Glucose is used by the mitochondrion to synthesize ATP. What besides nutrients do autotrophs provide for heterotrophs? What product from cellular respiration is returned to the atmosphere for use by autotrophs in photosynthesis? Thus autotrophs and heterotrophs live happily ever after.

Thus, two of the basic activities that sustain life on Earth depend upon energy from the sun, an object 149 million kilometers (93 million miles) away, and upon two tiny organelles, chloroplasts and mitochondria (Figures 3.3 and 3.5). Again, the processes that define life clearly occur exclusively within cells.

SUMMARY

1. **Elements** are substances that cannot be broken down into simpler substances by chemical means. Over 100 elements are known, of which 92 occur naturally. Each element has a different symbol, such as Fe for iron and N for nitrogen.

2. **Atoms** are the smallest particles of an element that still possess all of the element's chemical and physical properties. Atoms consist of negatively charged **electrons** orbiting a **nucleus** made up of positively charged **protons** and electrically neutral **neutrons.** **Isotopes** are alternative forms of elements that have the same chemical traits but differ in number of neutrons. Atoms that gain or lose electrons are called **ions.** **Oxidation** is the loss of an electron, and **reduction** is the simultaneous gain of the electron by another atom, ion, or molecule.

3. Electrons may exist at various distances from the nucleus and possess different energy levels. Electrons in outer **orbitals** possess a higher energy level than those in inner shells, and it requires external energy for an electron to move from an inner to an outer energy level. The number of electrons in the outermost shell largely determines the chemical behavior of an atom.

4. When atoms of two or more elements combine chemically in the proper ratios, a **compound** forms. The smallest unit of a compound that

retains the compound's physical and chemical traits is a **molecule. Organic** compounds contain carbon; all others are **inorganic. Monomers** are the subunits of large compounds, and **polymers** consist of many identical or similar monomers.

5. **Chemical bonds** are the attractive forces between the atoms in a molecule. **Ionic bonds** bind ionic molecules, which are called **electrolytes** because they dissociate when placed in appropriate **solutions. Covalent bonds** differ from ionic bonds; they involve the sharing of one or more pairs of electrons. Weak and short-term **hydrogen bonds** occur when hydrogen atoms are attracted to unbonded oxygen and nitrogen atoms.

6. **Acids** are ionic compounds that dissociate in water and donate hydrogen ions to solutions; **bases** dissociate to reduce the amount of H^+. The neutral point on the **pH scale** is 7; acidic solutions are assigned lower numbers (to 0) and basic solutions higher numbers (to 14).

7. **Chemical reactions** involve the making and breaking of chemical bonds. **Synthesis reactions** combine two or more simpler substances, whereas **decomposition reactions** break a complex substance into simpler components. In **condensation reactions,** water forms as a new substance is synthesized; water is added during **hydrolysis reactions.**

8. By weight, organisms are composed largely of O, C, H, N, Ca, and P atoms; by their number, N, O, C, and H atoms are the most important. At least a dozen **trace elements** are also important in the chemistry of life.

9. **Cells** are mostly solutes in water, and H_2O is also important in the environments of organisms. Water is a **polar compound** and the H and O atoms are linked by covalent bonds. Because they do not form hydrogen bonds with water, nonsoluble compounds are called **hydrophobic. Hydrophilic** molecules react readily with H_2O.

10. **Macromolecules** are large organic molecules (skeleton of C atoms bonded with other atoms). **Carbohydrate** molecules may be small, such as mono- and disaccharides, or larger polysaccharides. **Lipids** are composed of alcohols and fatty acids, and may be **saturated** (most C atoms bonded with H atoms) or unsaturated (many C atoms unbonded with H atoms). **Phospholipids** are important parts of biological membranes. The monomers of the relatively immense **protein** molecules are the 20 different **amino acids.** Certain covalent bonds link amino acids in **peptides,** which may link to form **polypeptides,** which may link to form proteins. Protein molecules occur in various sizes and shapes. The monomers of **nucleic acids** (DNA and RNA) are **nucleotides.**

11. Some proteins are **enzymes,** which catalyze or speed chemical reactions without being used up or becoming part of the product. The unique shape of each enzyme allows it to combine with only certain substrates. **Cofactors** may assist enzyme reactions, and some substances may interfere with or prevent them (e.g., certain poisons).

12. Living requires energy and material substances. **Kinetic energy** is the energy of motion; **potential energy** is stored or inactive energy. **Activation energy** is required for chemical reactions, and enzymes lower activation energy. The **first law of thermodynamics** states that while energy cannot be created or destroyed, its form can be changed. The **second law of thermodynamics** states that the amount of usable energy declines with each energy transformation.

13. The ultimate source of biological energy is the sun. Solar energy becomes chemical energy through **photosynthesis,** which releases oxygen in the process. **Autotrophs** are organisms capable of synthesizing organic compounds from simple inorganic ingredients. The **heterotrophs** that obtain chemical energy by degrading organic compounds may be herbivores, carnivores, or decomposers.

14. In **cellular respiration,** the energy stored in glucose is transferred to adenosine triphosphate for use by cells.

REVIEW QUESTIONS

1. Define the terms *atom, element, molecule,* and *compound* and state examples of each.

2. Describe the structure of an atom and the traits of electrons, protons, and neutrons.

3. Which atoms are of special importance to living processes?

4. What are orbitals and energy levels, and of what importance are they for life?

5. Distinguish between an ion and an isotope.

6. What determines the chemical and physical properties of each atom?

7. Distinguish between organic and inorganic substances.

8. Define *chemical bond,* and describe the nature of ionic, covalent, and hydrogen bonds, using specific examples.

9. Describe the pH scale and name examples of acidic, neutral, and basic substances.

10. Compare and contrast condensation and hydrolysis reactions.

11. What traits of water contribute to its importance in the chemical reactions within organisms and the events in their environments?

12. Define *monomer* and *polymer,* and describe examples of each for carbohydrates, lipids, proteins, and nucleic acids.

13. Define *enzyme* and describe how enzymes work. What is a cofactor?

14. Describe the biological importance of the first and second laws of thermodynamics.

15. Compare and contrast kinetic and potential energy.

16. Outline the main features of photosynthesis and describe its importance to life.

17. Compare and contrast autotrophic and heterotrophic modes of life.

18. Distinguish between combustion and cellular respiration.

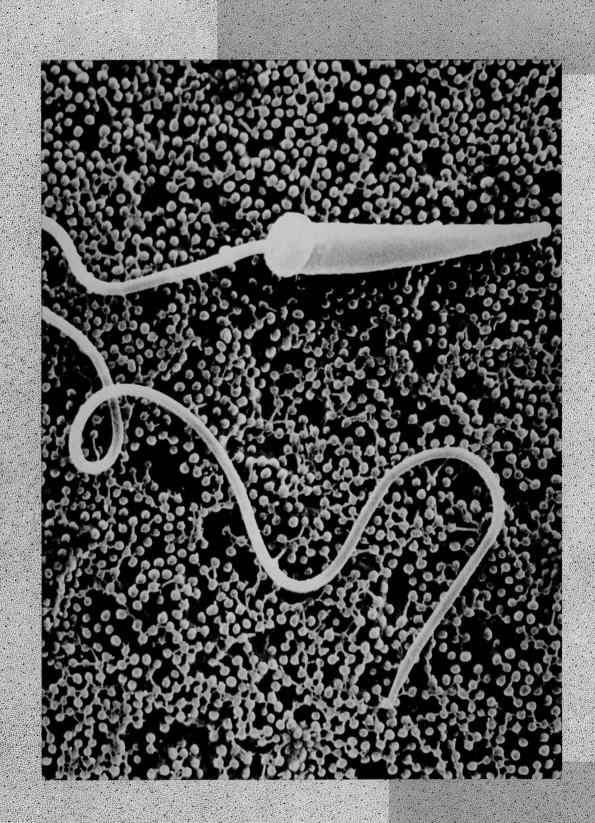

PART THREE

BECOMING

To survive, every individual must secure and process nutrients and gases, circulate needed materials, collect and excrete wastes, sense its internal and external environments, act upon the information, and control its body processes. It is not necessary or even advantageous for each person to reproduce. However, if no members of a species reproduced, the species would become extinct. In this part you will explore the structures and processes involved in reproduction and, thereby, the perpetuation of the species. The English satirist Samuel Butler said that ''a hen is merely an egg's way of making another egg.'' You will also study the mechanisms by which genetic blue-prints are transmitted from one generation to another, and learn how the directions are translated into functional structures. Finally, you will investigate our development, both prenatally and postnatally, and explore how our genetics and our environment interact during our life span.

Eggs are among the largest of cells, and sperm among the smallest. Typically, the number of sperm produced is enormous compared with the number of eggs, but it takes only one sperm to fertilize the egg.

CHAPTER 5

THE REPRODUCTIVE SYSTEM

The thread of life is sustained from one generation to another through reproduction, and this perpetuates the species. Although part of an individual's future is determined genetically when an egg is fertilized, the environment into which people are born also affects them in major ways. Humans have long been aware of the relationship between the birth of a child and sexual intercourse about 40 weeks before, although the significance of copulation remained obscure until after sperm were observed in the 1600s and the cell theory was proposed in the 1800s.

Although most mammals produce gametes only once or a few times each year, humans are atypical and produce them continuously. In this chapter, you will examine the structures and functions of the reproductive system.

Because they are relatively easy to secure, animal sperm have been observed for centuries. Although they exist in a variety of shapes and sizes, as this potpourri illustrates, the function of all these tiny cells is to fertilize the egg, which is usually far larger and less numerous.

MALE REPRODUCTIVE SYSTEM

Gamete-producing organs are called **gonads,** and the male gonad is the egg-shaped **testis** that is usually located within the thin, loose, wrinkled, and pendulous **scrotum** (Figure 5.1). There is usually no medical significance to the fact that one testicle often hangs slightly lower than the other, and it has been suggested that this arrangement reduces the chance for crushing. The scrotum is only sparsely covered with hair, and its numerous nerve endings make it highly sensitive to touch, pressure, and temperature.

The external location of the testis is important in regulating temperature. Sperm do not develop normally at internal body temperatures, and the testes average about 3°C (5°–6°F) cooler. Partial or complete sterility develops if the testes are too warm, which happens sometimes even from wearing tight clothing. However, by being close to the body the testes are also protected against the inhibitory effect of cold upon sperm production. (Mature sperm can survive very low temperatures and may be stored for long periods under such conditions, as they are in sperm banks.) Involuntary muscular actions move the testes closer to or further away from the body depending upon the external temperature. The tightening of the scrotum during physical exercise or sexual arousal may be a protective reflex that lessens potential injury to the testes.

During prenatal development the testes form within the abdominal cavity and move downward slowly, usually descending into the scrotal sacs about the time of birth (Figure 5.2A). **Cryptorchidism,** the condition of undescended testes, must usually be treated hormonally and/or surgically before puberty in order to prevent sterility or low fertility.

The connective tissue that usually covers the opening from the abdomen through which the testes descended may be weak or incomplete, and it sometimes breaks when it is strained, as when a heavy weight is lifted. This relatively common event is called an **inguinal hernia** (Figure 5.2B). If the opening is large it must be repaired surgically to prevent part of the intestine from slipping into the scrotal sac. If that happens, the intestine may be squeezed tightly, circulation to the cells may be reduced, and the cells may die.

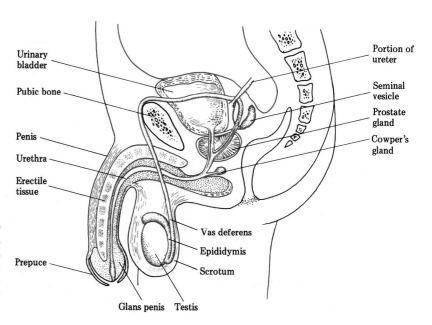

FIGURE 5.1

The male gonad, the testis, is located in the scrotum. Sperm and sex hormones form in the testis, mature in the epididymis, and travel through the vas deferens to the urethra. Various semen-producing glands empty into the ejaculatory duct and the urethra.

Urinary bladder

Pubic bone

Penis

Urethra

Erectile tissue

Prepuce

Portion of ureter

Seminal vesicle

Prostate gland

Cowper's gland

Vas deferens

Epididymis

Scrotum

Glans penis   Testis

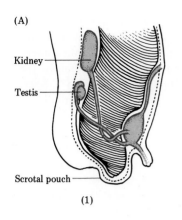

Kidney

Testis

Scrotal pouch

(1)

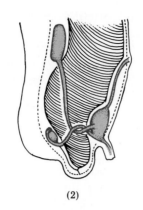

(2)

FIGURE 5.2

(A) The testes form embryologically near the kidneys and descend gradually into the scrotum prenatally or occasionally postnatally. Sperm form normally only in the cooler temperature of the scrotum, and partial or complete sterility results from untreated cryptorchidism. (B) In inguinal hernias, the tissue that covers the opening through which the testes descended ruptures. Sometimes an organ, such as a loop of the small intestine, protrudes into the opening and must be treated medically.

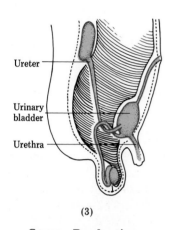

Ureter

Urinary bladder

Urethra

(3)

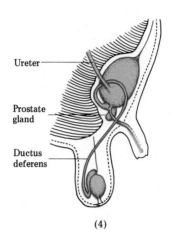

Ureter

Prostate gland

Ductus deferens

(4)

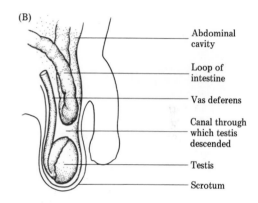

Abdominal cavity

Loop of intestine

Vas deferens

Canal through which testis descended

Testis

Scrotum

Sperm Production

The testis has two major functions (Figure 5.3). They are sperm formation within the **seminiferous tubules,** and production of sex hormones such as **testosterone** by the **interstitial** or **Leydig cells.** Testosterone is important in sexual development and in the sex drive, or **libido.** Sperm production begins at puberty and usually continues throughout a male's life. Healthy males may produce 300–500 million or more sperm daily. Each testis is a maze of hundreds of seminiferous tubules, each about the diameter of a coarse thread, whose combined length is about 500 m (0.3 mi).

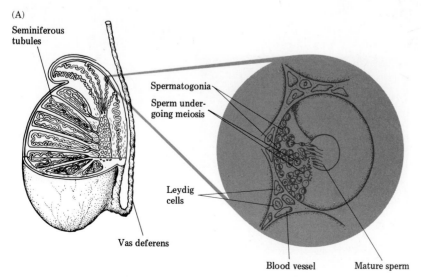

(A)

Seminiferous tubules

Spermatogonia

Sperm undergoing meiosis

Leydig cells

Vas deferens

Blood vessel   Mature sperm

FIGURE 5.3

The dual functions of the testis. Sperm are produced in the seminiferous tubules (A) and sex hormones such as testosterone are released by the Leydig cells in the interstitial tissue (B).

(B)

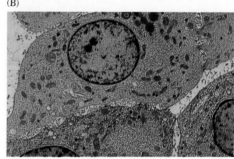

The sequence of events in the formation of sperm is called **spermatogenesis.** Beneath the connective tissue sheaths of seminiferous tubules are diploid (2*n*) cells (spermatogonia and spermatocytes) that undergo two meiotic divisions and a series of structural changes before they become mature, haploid (*n*) **sperm** (from the Greek for "seed"). Sperm are among the smallest of human cells, and about 100,000 tightly packed sperm would be barely visible to the unaided eye. During maturation, most of the sperm's cytoplasm is lost and mature sperm are little more than tailed masses of DNA (Figure 5.4). In addition to containing the haploid nucleus, the head of the sperm also bears a small vesicle called an **acrosome,** which contains enzymes that assist in fertilization by dissolving the numerous cellular and noncellular layers that surround the egg. The sperm's midpiece contains two centrioles, which generate the microtubules of the tail, and one or more mitochondria, which provide energy for the movements of the flagellum.

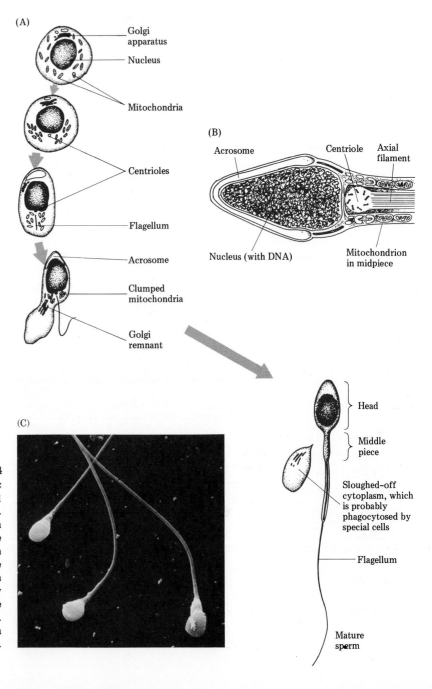

FIGURE 5.4
During spermatogenesis, two meiotic divisions typically result in four haploid sperm from each diploid spermatogonium. (A) Summary of the maturational changes in a sperm, some of which occur in the epididymis. (B) Mature human spermatozoa consist mainly of a haploid nucleus in the head, mitochondria in the midpiece, and a flagellated tail. The tip of the head usually bears an acrosome, whose digestive enzymes assist in the fertilization process. (C) An electron micrograph of three human sperm, magnified 2300 times.

The mature sperm released into the openings of the seminiferous tubules are carried to the **epididymis** located on the back surface of each testis. There, sperm mature, and fluids and cell debris are removed from them. Sperm may spend several weeks in this thin, highly coiled, 6-m (20-ft) network of tubes. Excess sperm may be destroyed here and their molecules recycled.

Semen-Producing Glands

From the epididymis, sperm travel through the 40-cm (16-in.) **vas deferens,** which leaves the scrotum and curves alongside and behind the bladder. **Cowper's gland** and the **seminal vesicles** are found near the junction of the vas deferens and the **urethra,** the tube from the bladder. Together with sperm, fluid secretions from these glands and from the chestnut-sized **prostate gland** make up **semen.** About 70 percent of the volume of semen comes from the seminal vesicles, with most of the remainder being produced by the prostate. The few drops of seminal fluid from Cowper's gland sometimes appear at the tip of the penis before ejaculation, and the fluid may contain some viable sperm. Seminal fluid varies in color from whitish to yellowish or grayish.

The prostate gland, located just below the bladder, often enlarges in older men and is also a common site of cancer. Because the prostate gland surrounds the urethra, much as a bead surrounds a thread, the pressure of an enlarged prostate may make urination difficult, and a part of the gland must sometimes be removed surgically. Retention of urine may result in bladder damage and infections, and this sometimes also damages the kidneys and the tubes (ureters) that carry urine from them to the bladder. Because the prostate is directly in front of the rectum, it can be felt by a physician during a rectal examination.

Semen is a complex, thick, mucuslike fluid with several important functions. Because sperm contain little cytoplasm, they obtain energy by metabolizing the sugar fructose that is found in seminal fluid. Sperm are unusual in this respect because most cells use glucose for their metabolic energy. Seminal fluid also transports sperm and lubricates the urethra, and its slight alkalinity protects sperm from the typically acidic vaginal secretions. The secretion from the prostate stimulates the quiescent sperm as they are released during an ejaculation.

Penis

During sexual excitement, the arteries leading to the limp (average 9.5-cm or 4-in.) **penis** dilate, and blood fills the three cylinders of spongy tissue that are bound in thick connective tissue sheaths. The increased blood pressure closes the veins, and the penis becomes stiff and erect (Figure 5.5). This facilitates coitus, although there is no physiological relationship between normal penis sizes and either virility or sterility. The dimensions of flaccid penises vary considerably, but such sizes are largely of psychological importance. Whatever their flaccid sizes, most erect penises are about 12–18 cm (5–7 in.) in length. Only in fantasy do penises come in three sizes—large, gigantic, and absolutely enormous—and only rarely are penises too small to be functional. Erection has been called "the great equalizer," since small penises enlarge relatively more during erection than large penises do. Erection also occurs in male babies and during certain stages of sleep in most males. The six or so erections during sleep usually last 5–10 minutes each and are not controlled by the specific content of dreams.

The tip or **glans** of the penis consists of an enlarged portion of the spongy tissue mass (corpus spongiosum) that surrounds the urethra on

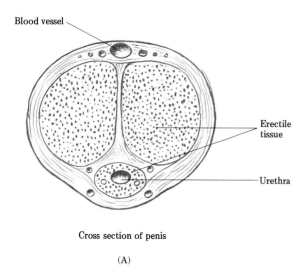

Blood vessel

Erectile tissue

Urethra

Cross section of penis

(A)

(B)

FIGURE 5.5

The cross-sectional structure of the penis. Erect penises are more alike in size than are the variable sizes of flaccid penises.

the underside of the penis shaft. Sensory nerve endings are more concentrated on the glans than they are on the shaft. The foreskin or **prepuce** of the penis sometimes forms such a tight collar over the glans that it cannot be retracted. This condition is one reason for the common practice of **circumcision** in young males, the surgical removal of all or part of the prepuce. Circumcision is also often a religious practice, as in Islam and Judaism. Many physicians believe that circumcision reduces infections and irritations due to the accumulation of **smegma,** a mixture of oily secretions, sweat, bacteria, dirt particles, mucus, dead cells, and semen between the glans and prepuce. Circumcision does not improve or detract from male sexual function, and uncircumcized males who practice good hygiene are not at any major health disadvantage.

Ejaculation

When the penis is stimulated by friction during coitus or masturbation, reflex reactions lead to **ejaculation** and orgasm. Wavelike (peristaltic) smooth muscle contractions involve the epididymis, vas deferens, seminal vesicles, and prostate. A muscle (sphincter) also closes the opening from the bladder to the urethra. Although the urethra is a common canal for semen and urine, the two fluids do not occur together. Before ejaculation, sperm are kept inactive by the thick, mucuslike seminal fluid.

About 3–5 ml of semen are released in a typical ejaculation by a normal male. (The volume of 5 milliliters equals about 1 teaspoonful.) The 60–600 million sperm themselves are so small that they contribute little to the volume of an ejaculate, and a healthy male replaces them in a few days. If there is no ejaculation, the unused sperm are reabsorbed. Infertile males release fewer than about 35–50 million sperm per ejaculation, and/or more than about 25 percent of these sperm are abnormal.

As seminal fluid thins quickly after ejaculation, sperm become very active and swim several millimeters per minute, no small feat for such tiny cells. Movement within the vagina is slower (because of acidic secretions) than in the uterus and fallopian tubes (Chapter 6). Although sperm are motile for only a few hours within the vagina, some may remain viable in the oviducts for several days. Rhythmic secretions of the uterus aid sperm transport, and if fertilization occurs, it usually takes place in the upper third of the fallopian tubes (Figure 5.6). Certain hormones in semen (called *prostaglandins* because they were discovered

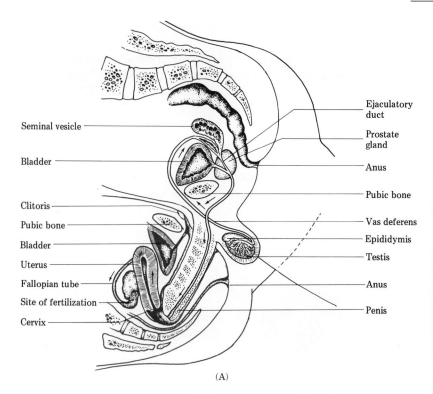

Seminal vesicle

Bladder

Clitoris

Pubic bone

Bladder

Uterus

Fallopian tube

Site of fertilization

Cervix

Ejaculatory duct

Prostate gland

Anus

Pubic bone

Vas deferens

Epididymis

Testis

Anus

Penis

(A)

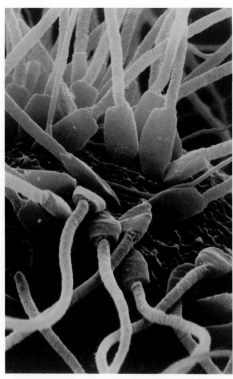

FIGURE 5.6
(A) The pathway of sperm during coitus to the typical site of fertilization. (B) Although only a single sperm fertilizes an egg, it takes the combined actions of many sperm to first loosen the mass of cells that surround the egg's plasma membrane.

(B)

initially in the prostate) cause contractions in the uterus and oviducts and these may help to propel sperm upward. Sperm have their greatest fertilizing ability during the first 24–36 hours after ejaculation.

FEMALE REPRODUCTIVE SYSTEM

The main features of the female reproductive system are shown in Figure 5.7. Note that the reproductive and urinary tracts are separate in females. Like the male genitals, those of females show much variation.

External Genitals

The external genitals, or **vulva,** consist of the mons, labia, clitoris, and the perineum. The **mons veneris** (from the Latin for "mound of Venus": Venus was the Roman goddess of love) is a fatty cushion over the pubic bone that becomes covered with hair at puberty. The region contains many nerve endings sensitive to touch and pressure. Surrounding the vaginal and urethral openings are folds called the outer and inner **labial lips.** Sweat and oil glands are common in the fatty outer labia and pubic hair grows along their sides. The inner lips consist of a vascular core of nonfatty spongy tissue rich in sensory nerve endings. **Bartholin's glands** open through small ducts into the inner lips next to the vaginal opening. The few drops of secretion produced by these glands during sexual arousal moisten the labia.

The inner lips meet above the urethral and vaginal openings to create a fold of skin or hood over the **clitoris.** This small erectile organ forms from embryonic tissue similar, or homologous, to that which gives rise to the glans of the penis in the male. Somewhat like the penis, the clitoris consists of a spongy shaft and a bumplike glans, and it becomes engorged with blood during sexual excitement. It contains many nerve endings sensitive to touch, pressure, and temperature and is a major site of stimulation during coitus or masturbation. Unlike the penis, the clitoris usually does not lengthen when it is stimulated, and it has no reproductive or urinary function. It is the only organ in either sex whose only known function is to concentrate sexual sensations.

The **perineum** is the hairless region between the lower labia and the anus. The sensory nerve endings in the area may be another source of sexual arousal.

Vagina

The tubular **vagina,** with its highly elastic, muscular, and folded walls, is a receptacle for the penis during sexual intercourse and is the birth canal during childbirth. The walls of the relaxed vagina are collapsed, and they vary considerably in length (about 8–15 cm or 3–6 in.). The

FIGURE 5.7
The major structures of the female reproductive system: (A) external genitals, (B) internal organs, and (C) side view.

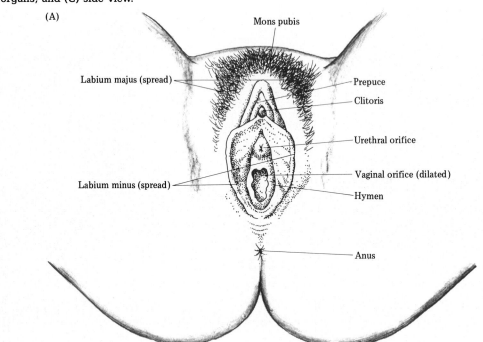

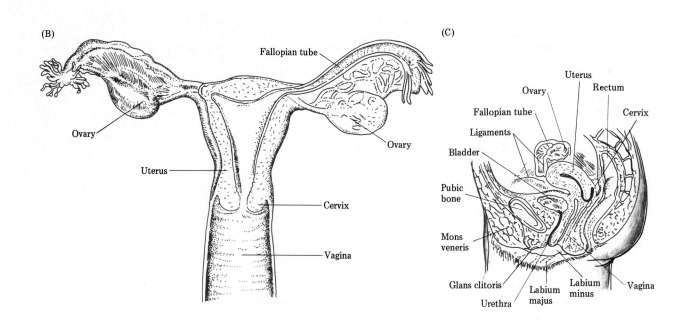

vagina adjusts in size to a small or large penis, and it is unusual for size differences of male and female sex organs to create difficulties. After childbirth, the vagina usually enlarges somewhat and loses some elasticity.

The outer third is highly sensitive but the remainder has relatively few sensory neurons. The vaginal walls are usually kept moist by acidic mucus secretions, and the quantity of these lubricating secretions increases during sexual excitement. Vaginal infections are most common when the acidity decreases.

The vaginal opening of young females may be partially closed by a thin and usually perforated membrane called the **hymen.** Although the hymen is often regarded as a symbol of virginity, it is absent in many females and in others is sometimes torn by exercise, petting, infection, the use of tampons, or surgery. In some females, part of the hymen persists throughout life despite frequent coitus.

Uterus

The fist-sized, pear-shaped, hollow, and very muscular **uterus,** or womb, usually extends upward and forward at nearly a right angle from the vagina. Sometimes it is tipped backward (in about 25 percent of women), and sometimes it tilts even farther forward (in about 10 percent). Several ligaments hold the uterus loosely in place in the pelvic cavity.

The lower end of the uterus is called the **cervix,** and it protrudes into the upper vagina. The opening (cervical os) into the uterus is about the diameter of a thin straw. The numerous glands within the cervical canal produce mucus that varies considerably during the menstrual cycle. Near the time of ovulation, the mucus becomes thin and watery, while at other times it is thicker and forms a plug that blocks the entrance to the cervix. Few sensory neurons are found on the surface of the cervix and it contributes little to sexual feelings, as shown by the fact that its surgical removal leads to no loss of sexual responsivity.

The cervix may tear during childbirth and sometimes becomes infected then as well as at other times. Cervical cancer occurs in about 2 percent of American women, and a regular Pap test is wise. Cervical cancer may occur at any age, but the peak incidence is between the ages of 35 and 45 years. It develops most often in women in poor health, those who have had many children, those with chronic cervical irritation or infections, and those whose mate is not circumcised. Virgins rarely develop cervical cancer. Although the cure rate is high if detection is early, chances of survival decrease dramatically as the cancer passes into the more advanced stages.

The uterus is about 7.5 cm (3 in.) long and 5 cm (2 in.) wide, somewhat flattened from front to back, and its two major parts have separate and distinct functions. The glandular inner lining or **endometrium** changes considerably during menstrual cycles (see below) and is the site where embryos develop. The thick and very muscular outer region is called the **myometrium.** Contractions of the myometrium facilitate labor and delivery and are involved in sexual responsiveness.

Oviducts

The 10-cm (4-in.) **oviducts,** or **fallopian tubes,** extend outward and back from the sides of the uterus. The openings into the fallopian tubes from the uterus are very small, roughly needle-sized. The distant ends of the oviducts are funnel-shaped with fingerlike extensions (fimbria)

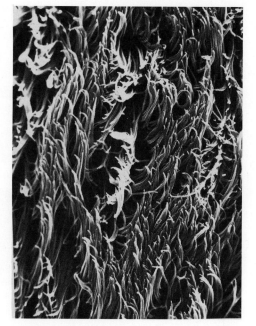

FIGURE 5.8
Scanning electron micrograph of the ciliated lining of the oviduct. Both the beating of cilia and the peristaltic movements of the fallopian tubes propel eggs toward the uterus.

that partly surround the ovaries but are not connected to them. The oviducts are lined with long, thin folds and beating cilia that act with peristaltic movements of the tubes to move the egg toward the uterus (Figure 5.8 on page 109).

Ovaries

The female gonads are the two almond-shaped **ovaries** found on either side and a little behind the uterus. Connective tissue that attaches to the broad ligament of the uterus holds the ovaries in place, and deposits of fat protect them. Like male gonads, ovaries have two functions, to produce gametes and sex hormones.

Oogenesis is the process by which ova form from diploid cells (oogonia) in the ovary (Figure 5.9). After growing, the nucleus of each $2n$

FIGURE 5.9

(A) Cross-sectional view through the ovary showing the maturation of an ovum and the development of the corpus luteum and the corpus albicans. (B) Summary of oogenesis. (C) A tiny nonfunctional polar body shown pinching off of the far larger oocyte during meiosis. (D) Apex of mammalian follicle. (E) Follicle rupturing. (F) Monkey ovary. (G) Ovary primary and secondary oocytes.

(A)

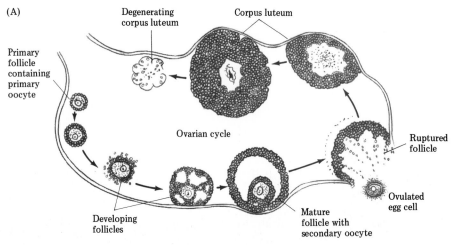

Degenerating corpus luteum

Corpus luteum

Primary follicle containing primary oocyte

Ovarian cycle

Ruptured follicle

Ovulated egg cell

Developing follicles

Mature follicle with secondary oocyte

(B)

Mitosis

Oogonium

Primary oocyte

First polar body

First meiotic division

Secondary oocyte

Second polar body

Second meiotic division

Ootid

Development

Ovum

(C)

(D)

(E)

(F)

(G)

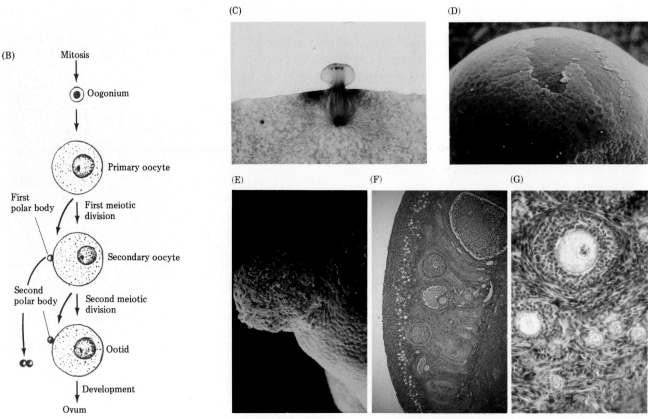

cell (primary oocyte) undergoes a meiotic division. Most of its cytoplasm goes to one cell (secondary oocyte), and very little goes to the tiny adjacent cell called the **polar body.** In most vertebrates the polar body serves no function and disintegrates. The maturing egg cells enlarge before birth, and other cells grow around them to form thin, spherical **follicles** (from the Latin for "small ball").

About halfway through fetal development, a female's ovaries contain 6–7 million small oocytes. Most of these degenerate, and only about 400,000 immature eggs are present at birth. No new eggs form thereafter, in contrast to sperm, which are produced by the billions annually from puberty on. Continued degeneration of eggs occurs during childhood, and by puberty the number is reduced to perhaps 200,000–300,000 or less. The primitive eggs remain inactive between the time of their formation prenatally and puberty, about a dozen years later, at which time some begin further development. Typically, only about 400 oocytes in the ovaries mature during a woman's 30–40 fertile years.

Beginning around puberty, the usual pattern is for one ovum to mature in one ovary during each average 28-day cycle. The egg protrudes inward from the follicle wall into a large fluid-filled space. As follicles mature they grow to be marble-sized **Graafian follicles** and move until they project like blisters from the ovary's surface. Mature eggs are also surrounded by a jellylike material (zona pellucida)(Figure 5.10). At the time of **ovulation,** when the ripened follicle ruptures, the ovum is released along with many surrounding follicle cells. The hundreds of other follicles that underwent various degrees of maturation usually disintegrate. Although the ovum is the largest cell in the human body, it is smaller than the period in this sentence. Bird eggs are far larger because of the yolk they contain.

The epithelium of the funnellike end of the fallopian tube is ciliated, and currents are set up that aid in carrying the egg into the oviduct. If sufficient sperm are present, the egg may be fertilized within the upper third of the fallopian tube. Unfertilized eggs usually disintegrate. Although ovulation usually goes unnoticed, it sometimes causes a minor cramp or abdominal pain. There is no nausea or abdominal tenderness, but there may be some bloody discharge or "spotting" from the vagina.

The cells of the old Graafian follicle multiply and enlarge to form the **corpus luteum** (from the Latin for "yellowish body"). Because of the hormones it produces, the corpus luteum is important in the menstrual cycle (see below) and during pregnancy. The inner part of the ovary contains no follicles but is composed of loose connective tissues and blood vessels that supply the ovaries.

Breasts

Although not reproductive organs, the breasts are usually important female **secondary sex traits.** In many societies, the breasts symbolize femininity, attractiveness, and sexuality, and this is emphasized in advertising, film and video, men's magazines, and clothing styles. However, although breast size may be important psychologically to both males and females, there is no evidence that breast size is related physiologically to sexual interest, libido, or the capacity for sexual responsiveness. In many women, in fact, these modified sweat glands generate few important sexual sensations when they are caressed or fondled. However, some women seek breast augmentation surgery, which formerly involved the injection of liquid silicone but now more commonly utilizes implants of soft plastic pouches of silicone gel. Some other women have surgery to reduce the size of breasts that they consider to be too large.

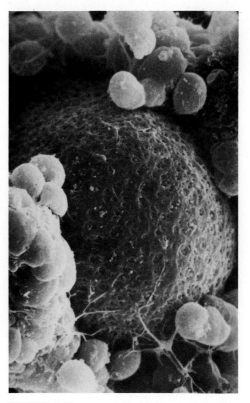

FIGURE 5.10
Mature human eggs are usually surrounded by jellylike material and many nutritional cells.

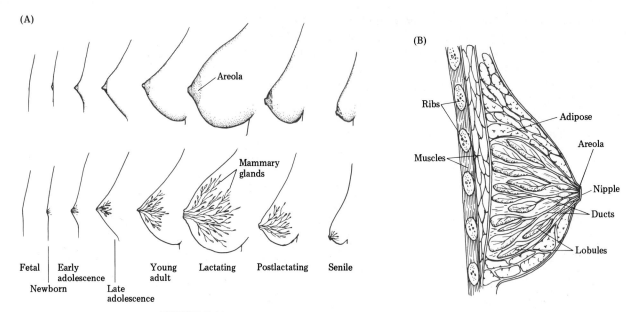

(A)

Areola

Mammary glands

| Fetal | Early adolescence | | Young adult | Lactating | Postlactating | Senile |

Newborn   Late adolescence

(B)

Ribs

Muscles

Adipose

Areola

Nipple

Ducts

Lobules

FIGURE 5.11
(A) Developmental changes in the breast through a female's life. (B) The structure of a mature breast.

The developmental changes in the breasts are dramatic (Figure 5.11). At puberty, when the breasts become conical or hemispherical, the left breast usually grows slightly larger than the right. Most of the breast consists of adipose and fibrous tissues. The 15–20 **milk glands** arranged in a grapelike cluster in each breast drain by ducts to the surface of the **nipple.** Nerve endings in the nipple make it sensitive to touch and temperature. Extending 1–2 cm around each nipple is the darker, wrinkled **areola,** which contains nerve fibers and smooth muscles that cause erection of the nipple. Inverted nipples are common, and these do not interfere with nursing, and extra nipples, while perhaps embarrassing, have no adverse health effects.

THE MENSTRUAL CYCLE

Puberty is the age range when secondary sex traits such as breast enlargement and growth of body and pubic hair develop. Many of the body changes are caused by the steroid hormones called **estrogens** (from the Greek for "frenzy origin"). Estrogens come from the ovary, but their production is controlled by hormones from the pituitary and hypothalamus (Chapter 18).

The first menstrual flow is called **menarche;** it usually begins around the age of 11–12 years (range, 9–18 years). **Menstrual cycles** continue for about the next 30–40 years, usually interrupted only by pregnancies, until the time of **menopause,** at the age of about 47–49 years (typical range, 40–55 years). The length of normal menstrual cycles typically ranges from about 20–36 days, with the average being 28 days. However, even "regular" women may vary from cycle to cycle, and stress and health have major effects upon the cycle. Almost any pattern may be considered normal if it recurs on a regular basis.

In most mammals the blood and tissue associated with the breakdown of the endometrium is reabsorbed. Humans are almost unique in discharging this bloody tissue through the vagina during **menstruation,** or a "period," which usually takes 3–6 days. The menstrual fluid loss sometimes taxes the blood-forming bone marrow and leads to mild anemia and/or iron deficiency. Muscular cramps, nausea, breast tenderness, and pain (dysmenorrhea) may occur during menstruation. This is most common in women under age 25 years and least common in older women who have had children. Sometimes changes in the hormone balance cause emotional depression or other forms of "menstrual blues." At the end of menstruation the endometrium is smooth and thin. Tampons and/or sanitary napkins are most commonly used to absorb the menstrual fluid. The opening of the hymen is normally large enough in all females to accommodate a tampon. Occasional small discharges may occur between periods. The absence of menstrual periods (amenorrhea) may also be a problem. This may involve a delayed start of periods or having periods stop after menarche.

Hormonal Control of the Menstrual Cycle

The precise control of the menstrual cycle by four main hormones, two each from the ovary and the pituitary, is an excellent example of negative feedback (Figure 5.12). The relative levels of these hormones and the associated changes in the ovary and uterus are shown in Figure 5.13.

Although many small eggs may begin to develop during the **follicular phase** under the influence of **follicle-stimulating hormone** (FSH) from the pituitary gland, usually only one egg in one ovary matures fully in each cycle. As the ovum matures, the surrounding follicle cells secrete estrogen, which stimulates the endometrium to thicken due to the growth of glands, connective tissue, and blood vessels. The increasing concentration of estrogen in the blood usually inhibits the release of FSH, and this generally prevents other oocytes from maturing. While estrogen inhibits the release of FSH, it stimulates the pituitary to release increasing amounts of **luteinizing hormone** (LH), and this begins the **luteal phase.**

Midway in the menstrual cycle (days 9–19), ovulation usually follows the peak in LH by 12–24 hours. LH also causes the remains of the Graafian follicle to transform into the corpus luteum. The mature corpus luteum secretes estrogen, although in smaller amounts than that secreted by the follicle cells before ovulation. The corpus luteum also secretes large amounts of the steroid hormone **progesterone**, the function of which is to finalize the preparation of the uterus for the possible implantation of an embryo. The endometrium thickens further, becomes pitted, acquires more blood vessels, and releases various glandular secretions.

What happens next depends upon whether an egg is fertilized. If fertilization does not occur, the corpus luteum continues to secrete progesterone for about a week after ovulation, and this high level of progesterone inhibits the secretion of LH and FSH by the pituitary. Because the levels of estrogen and progesterone that keep the uterus in its "ready state" are declining late in this phase, the endometrium begins to disintegrate. Fragments of the blood-enriched endometrium along with blood and mucus from the uterine glands are shed during menstruation (55–85 g, or 2–3 oz). Because menstrual blood lacks certain factors, it does not clot like other blood.

If an egg is not fertilized, the corpus luteum gradually becomes inactive and changes into a mass of scar tissue called the **corpus albicans,** which decreases gradually during the next several months. However, if fertilization occurs and an embryo becomes implanted in the uterus, the corpus luteum remains functional for a number of weeks and the progesterone it secretes keeps the endometrium intact and inhibits the secretion of FSH. Therefore, no other follicles are usually (but not always) stimulated to develop and menstrual cycles generally cease. The details of pregnancy are discussed in Chapter 6.

FIGURE 5.12

The major negative feedback relationships of the menstrual cycle. Note that high levels of one hormone may inhibit the release of another hormone and/or stimulate the release of still another hormone.

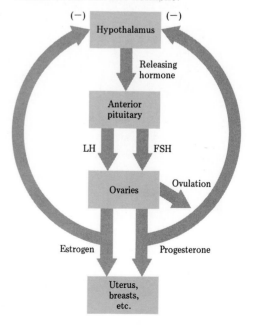

Several important changes also occur in the cervix during the menstrual cycle. Estrogen causes the cervical mucus to become wetter and thinner, thus aiding the entrance of sperm into the uterus. Both estrogen and progesterone affect the chemical content of the cervical mucus around the time of ovulation, and these changes apparently increase sperm survival. They may also be used in some tests to determine whether a woman has ovulated. After ovulation, progesterone causes the cervical mucus to become dryer and thicker.

After the uterus returns to its original state, the cycle begins again.

FIGURE 5.13
Ovarian and uterine changes during the typical 28-day menstrual cycle under the influence of various hormones. In medical practice, day 1 is the beginning of menstruation.

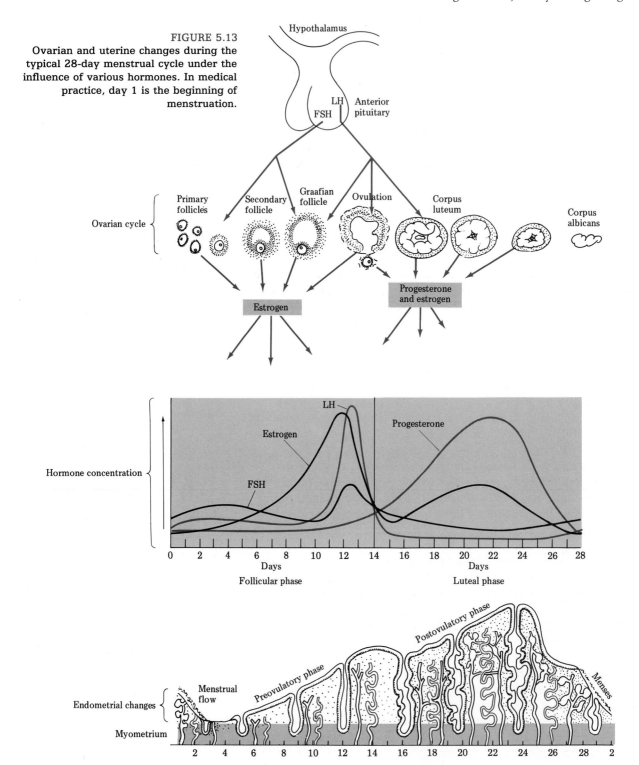

Menopause

The regular release of eggs from the ovaries stops at menopause (from the Greek for "month cease"). At that time the remaining immature eggs in the ovaries disintegrate and a woman's reproductive life ends. The physiological and psychological changes of menopause usually occur over a period of several years, and they range from very mild to occasionally severe. Although ovulation ceases, a postmenopausal woman's sexual life continues and often improves.

THE SEXUAL RESPONSE CYCLE

For most humans the sex drive is one of the strongest drives that influences their behavior. Sexual intercourse is also one of the most pleasurable physical activities, provided that affection and tenderness and not aggression and rage are involved. Although males and females differ in coital reactions, the similarities are greater than the differences. In both sexes, four stages of the **sexual response cycle** are commonly recognized (Figure 5.14). Like many cycles, there are no absolute divisions between the stages, and the responses vary considerably between people and in one person from time to time.

Foreplay, which may involve a wide range of sensory experiences, serves to arouse and prepare a couple for coitus. Its expression varies widely between cultures, individuals, and even from time to time between established sexual partners. The body has many areas besides the genitals that are possible sources of sexual arousal. The largest of these **erogenous zones** is the skin, especially the insides of the thighs, the perineum, and the neck. Touching, caressing, stroking, and massaging are important sources of sensual pleasure and nonverbal communication, including invitations or declines to further sexual activities. The lips and tongue contain concentrated sensory neurons, and kissing and oral exploration, including oral-genital contacts, are common during foreplay. Sexual stimulation may also involve the buttocks, anus, and rectum. Some cultures regard the buttocks as symbols of female sexuality in the way others regard the breasts. In tight jeans and swim suits, the female buttocks are sexually stimulating to many males. Stroking hair is enticing to many, and the presence of body hair and well-developed muscles may make males more sexually exciting to some females.

One result of sensory stimulation during the initial **excitement stage** is the swelling of erectile tissues such as the nipples, breasts, clitoris, and penis. Vasodilation also commonly causes the cervix and vagina walls to secrete moistening or lubricatory fluids, and Bartholin's glands may also secrete some liquid. Another result of sexual stimulation is **myotonia,** an increased muscular irritability and tendency to contract throughout the body, not only in the genital region.

During the excitement stage in the female, the inner two-thirds of the vagina expands, the outer labial lips flatten and move apart, the inner lips dilate, the clitoris enlarges, and the cervix and uterus are pulled upward. In males, besides erection of the penis, the testes are commonly drawn toward the body and may increase in size. Nipple erection may also occur in some males.

In the **plateau stage** the events of the excitement stage are maintained and intensified. In both sexes, the blood pressure and the heart and breathing rates increase. The outer third of the vagina swells, but the size of the vaginal opening decreases by about 30 percent. This ability to "grip" the penis is a major reason that penis size is generally unimportant biologically. Although engorged, the clitoris now retracts be-

(A)

Female:

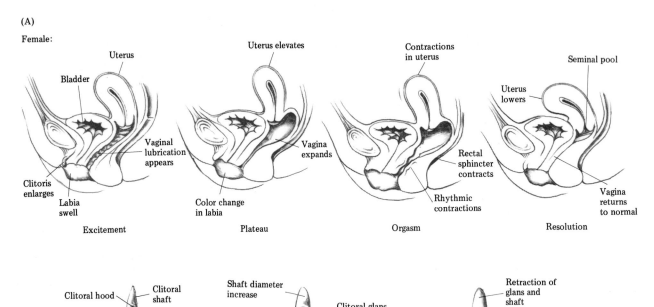

Excitement | Plateau | Orgasm | Resolution

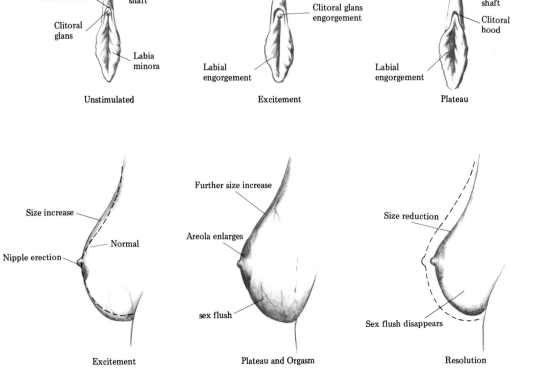

Unstimulated | Excitement | Plateau

Excitement | Plateau and Orgasm | Resolution

Male:

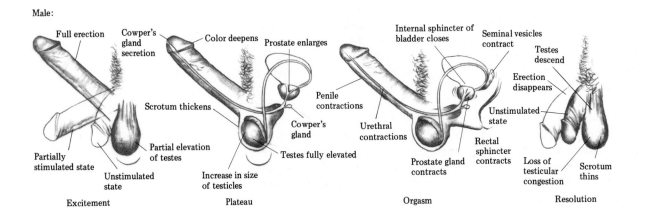

Excitement | Plateau | Orgasm | Resolution

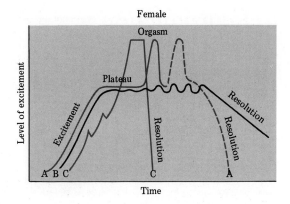

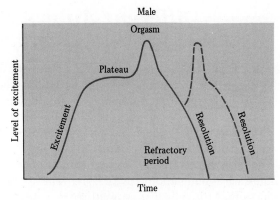

(B)

FIGURE 5.14

(A, opposite page) Common patterns of changes in females and in males during the stages of the sexual response cycle. (B, above) Although both females and males typically progress through the same four stages (solid lines), there are some differences, such as the wider range of patterns in females. While most males may experience another orgasm only following the refractory period (dotted lines), many females may achieve another orgasm quickly, and some may experience a series of orgasms (dashed lines).

neath the clitoral hood and this protects the glans from direct stimulation, although there is no loss of clitoral sensation from its indirect stimulation. Vasodilation causes further enlargement of the inner labial lips, sometimes doubling or even tripling them in thickness. The increased vascular supply causes characteristic color changes in the inner lips (pinkish or reddish to brighter or deeper red). The entire breast and areolar area usually enlarge, and this may obscure nipple erection. Early in this stage or late in the excitement phase, skin color changes occur in 50–75 percent of females and about 25 percent of males due to vasodilation of surface blood vessels. This reddish, spotty "sex flush" usually begins at the base of the breastbone and then spreads rapidly to many parts of the body.

The ridge of the glans of the penis enlarges during the plateau stage and often deepens in color. Vasodilation also swells the testes, sometimes to double their unstimulated size, and they continue to elevate. A little clear liquid from Cowper's glands may be expelled from the urethra.

The **orgasm stage** is the shortest phase of the cycle. At its onset in the female, the inner labial lips, the outer third of the vagina, uterus, and anal sphincter contract forcefully and at intervals less than 1 second apart. As orgasm progresses, the contractions diminish in intensity and length. An intense orgasm may involve 10–15 contractions, while a mild one may produce only 3–5. Biologically, the contractions may aid ascent of sperm toward the oviducts. Although stimulation of the clitoris is a major cause of female orgasm, no particular orgasmic changes occur there. However, muscles elsewhere in the body also contract, brain wave patterns change, and the sex flush may intensify. Although sensory awareness is lessened or lost during orgasm, there is a diffuse feeling of pleasure during those 3–10 seconds.

Conception is possible without orgasm, but it is not known for certain whether orgasm facilitates fertilization. Orgasm must be learned by some females, and some women never experience it despite frequent coitus. Many females are capable of several orgasms, one immediately after the other, and orgasms may last longer than those in most males. Multiple orgasms seem to occur more often during masturbation than during intercourse.

Orgasm and ejaculation usually occur together in males. Early in orgasm, contractions of the vas deferens, prostate, and seminal vesicles force semen into the bulb of the urethra. Next, contractions of the urethra and penis, coupled with continued contractions of the prostate, lead to the spurting of semen from the penis. After the first 3–4 contractions of the penis, the intensity declines and the intervals between contractions lengthen.

Different Approaches to Reproduction

Starfish eat oysters and can become a pest to those who harvest these shellfish. Hoping to eliminate the problem, some people catch numbers of troublesome starfish, chop them into pieces, and throw the chunks back into the sea. These people are often surprised and dismayed when they learn that some starfish pieces can grow back, or **regenerate,** the lost parts and thus multiply the starfish population (Figure B5.1). The asexual reproduction of the cells that grow to repair injuries is familiar, as when a wound heals or a lizard grows a new tail after a narrow escape from a predator.

Most people think even less about flatworms than they do about starfish, unless they have a tapeworm or similar parasite. However, flatworms and some of the other animals shown in Figure B5.2 can do something we cannot—reproduce **asexually** without meiosis, gametes, mating, or fertilization. Some flatworms sometimes simply pinch themselves midway along their lengths, and the missing parts of each half regenerate. Because cells in organisms are usually specialized, it is obvious that for regeneration to occur, cells must despecialize and then respecialize to perform new functions. In asexual reproduction, offspring are usually genetically identical to the parents and to each other.

Budding is another means of asexual reproduction in some organisms. In this process a new organism originates as an outgrowth of the parent and is usually well developed when it breaks away from the parent. Perhaps this is the basis for the creature in the Greek myth of Hydra, which could grow two heads for every one that was cut off. Alternatively, some buds may remain attached to form the basis for a colony of connected individuals.

Parthenogenesis (from the Greek for "virgin birth") is a form of asexual reproduction that occurs when an

FIGURE B5.1
Regeneration. This starfish is regenerating the remainder of its body from one arm. This example of asexual cellular reproduction does not produce a new individual but serves to repair damaged tissues.

animal develops from an unfertilized egg (Figure B5.3). This process occurs widely among arthropods and some other invertebrates, and in some fishes, amphibians, and lizards. For example, aphids multiply rapidly by parthenogenesis in the spring and summer, when environmental conditions are favorable, thus frustrating many gardeners. Their unfertilized eggs develop into both male and female aphids, which mate later in the season. Only females hatch from the fertilized eggs; they then reproduce parthenogenetically.

Asexual reproduction seems so simple and efficient. Why, then, do so many organisms, including humans, rely mostly or solely on sexual reproduction? It seems such a waste to produce the enormous numbers of gametes when only a minuscule few will result in fertilization. Also, some sexually reproducing animals may not encounter an appropriate sexual partner or, if they do, they may be unsuccessful in courtship and mating. Sexual activities require time and effort, and in nature

FIGURE B5.2
Some examples of asexual reproduction. (A) Several buds are forming on this *Hydra*. These may be released to grow, and they may reproduce in a similar manner, as the budding yeast in (B) are. (C) The head of this flatworm was cut and each half regenerated the missing portion.

(A)

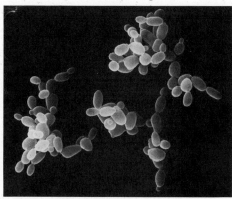

(B)

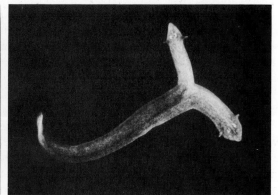

(C)

FIGURE B5.3
Parthenogenesis. Like many animals capable of reproduction by parthenogenesis, aphids also reproduce sexually. During the favorable season, parthenogenetic aphids can produce large numbers of male and female individuals. Sexual reproduction occurs when conditions are less favorable, but all offspring from fertilized eggs are females.

also often entail the increased risks of predation and spread of contagious diseases.

As long as environmental conditions remain within an organism's range of tolerance, asexual reproduction would seem adequate. If the asexually reproducing individuals are well adapted to their environments, then some of their identical offspring may also survive and reproduce. However, if the environmental conditions to which they are adapted change and exceed their ranges of tolerances, as conditions often do over long periods of time, then the population of identical individuals may perish. One analysis compared this with a lottery, wherein the prize is genetic representation in the future and the tickets are offspring. Although asexual females can buy a lot of tickets, almost all the tickets have the same number. The tickets of the sexual female are usually fewer but they are generally different.

Sexually reproducing organisms have often benefited in the long term from the variation in their offspring. Thus, whereas asexual reproduction leads to uniformity, sexual reproduction produces diversity. Meiosis results in gametes that carry different sets of genes, and chance determines which gametes unite during fertilization. The new **genome,** or total genetic makeup of an individual, generally contains the same DNA as the parents had, but variation results from the genetic recombination. Many individual sexually reproducing organisms may not tolerate changing environmental conditions any better than asexually reproducing ones. However, some variants may survive and allow the species to perpetuate itself. Thus, sexual reproduction does not merely preserve old patterns but creates new combinations of traits. Probably for this reason, many species that reproduce asexually can usually reproduce sexually as well.

Although most animals are either male (♂) or female (♀), some (e.g., various sponges, worms, mollusks, and deep-sea fish) are **hermaphroditic** and contain both gonad types. Hermaphroditism occurs most often among sessile and parasitic animals, whose circumstances may sometimes make finding a mate uncertain. However, a disadvantage of self-fertilization is the loss of some genetic diversity, somewhat as with self-pollination in plants. Thus, there are advantages for hermaphrodites to **outbreed,** or mate with other individuals, in order to maintain or increase genetic diversity within a population (Figure B5.4). In fact, even though hermaphrodites have both female and male genitalia, they may be unable to self-fertilize because, for example, eggs and sperm may not mature at the same time.

In nature, copulation has many functions. For example, dominant males in some groups of monkeys may mount low-ranking males, who often present their rears as a sign of submission. Usually there is no penetration, although the dominant male may make several pelvic thrusts. Sexual receptivity among female chimpanzees is "announced" by a vaginal odor and a reddened and enlarged area around the rump. These "pink ladies" may copulate with most or all adult males in the group. Because only one male is necessary for impregnation, this behavior is thought to develop social bonds within the group. Apparently, nymphomania occurs in wild chimpanzees, also indicating the probability that orgasmic reactions to copulation are not strictly human. At the time they are most likely to conceive, female baboons may accept only high-ranking males. At other times, however, they may copulate with males of any rank, perhaps reinforcing group bonds, as with chimpanzees.

FIGURE B5.4
These hermaphroditic earthworms are outbreeding, transferring sperm to one another. Compared with self-fertilization, outbreeding has the advantage of maintaining or increasing genetic diversity within a population.

Although the contraction of the various smooth muscles is controlled by a spinal reflex in the lower back, brain centers may alter the response. Ejaculation may occur without tactile stimulation of the penis, for instance, by erotic thoughts or petting. Ejaculation during sleep is called a nocturnal emission, or "wet dream." In psychological **impotence**, brain centers may also inhibit erection or ejaculation.

In both sexes during the **resolution stage,** the anatomical and physiological changes that occurred during the earlier stages are reversed. Events that elicited excitement earlier may be irritating or distasteful during this phase.

In females, the labial color changes disappear, the clitoris returns to its usual position and size, the vagina shortens, the uterus moves back to its resting position, and the breasts decrease in size.

During this stage, the penis usually becomes flaccid quickly and the testes usually decrease in size and descend into the scrotum. Most males are unable to achieve another erection for a time, or **refractory period,** that may last for a few minutes to many hours. Most males cannot experience multiple orgasms within a short period of time as is possible by some females some of the time.

There are numerous coital positions, and variety is mainly important psychologically. Positions differ in the degree of movement possible, the distribution of body weight, the freedom of the hands, the portion of the partner's body that is seen, and the amount of genital contact.

See the bioperspective essay "Different Approaches to Reproduction" for a summary of some variant ways that animals reproduce.

SEX AND REPRODUCTION

The paired paramecia shown in Figure 5.15 are engaging in **sex,** or the exchange of genetic material, rather than reproducing. The two individuals will soon disattach and go their separate ways without leaving behind any offspring. However, sex will have changed these organisms, as it does most, but for reasons of genetics and, probably, a greater chance for survival.

As is discussed in the bioperspective essay "Different Approaches to Reproduction," reproduction may occur asexually, without any exchange of genetic material. Some species of animals even reproduce asexually as larvae, before they develop into adults. For some species this occurs because food is very scattered and ephemeral, and animals must take advantage of the generally scarce but locally abundant resource quickly. After gorging themselves and reproducing quickly and prolifically, however, the animals may be ready for sex before they disperse in search of new food. Some of the variable individuals that result from sexual reproduction may be better able to cope with the vagaries of their environment and survive than would those individuals who are identical to one another.

Mutations, or random changes in the genetic blueprint, do not occur often, and most are harmful (Chapter 10). However, the beneficial mutations tend to accumulate and recombine at higher rates in sexual than in nonsexual populations. The genetic diversity created by sexual reproduction clearly has a multitude of potential and often real advantages. There is a reduced chance that all individuals will perish when a pond dries, or an early frost occurs, or as a result of any number of other environmental changes. Alert and hungry predators may not detect all variant individuals, as they might if they had "tuned in" to the uniform traits of asexually produced individuals. Because not all sexually pro-

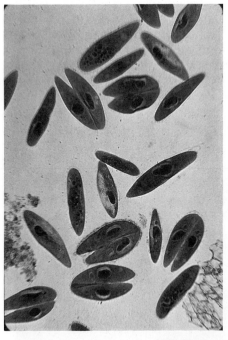

FIGURE 5.15
Several pairs of paramecia are exchanging genetic material (by conjugation), or sex, and they will soon separate without reproducing. In humans and in many other animals, sex and reproduction occur together.

duced offspring are alike, competition among them for resources may be lessened. Finally, some variant individuals may be able to occupy and survive in habitats or in ways quite different from the places and means used by others.

SUMMARY

1. **Gonads** (**testes** and **ovaries**) produce **gametes** (**sperm** and **eggs**) and **sex hormones** (e.g., **testosterone** and **estrogen**). During **spermatogenesis** and **oogenesis**, **diploid** cells become **haploid** gametes. Sperm are among the smallest and eggs among the largest of human cells.

2. Sperm form abnormally at core body temperature, and the temperature of the testis in the **scrotum** is usually about 3°C cooler.

3. **Inguinal hernias** result when the tissue that covers the opening to the abdomen through which the testes descended breaks. A loop of the intestine then may protrude into the scrotum and surgery may be required to correct the condition.

4. Sperm form in the **seminiferous tubules** and testosterone is produced in the interstitial **Leydig cells**. Spermatogenesis, which begins at **puberty** and usually continues throughout life, typically results in the production of millions of sperm daily. After completing maturation in the **epididymis**, mature sperm consist of a **nucleus, acrosome, midpiece,** and **flagellum.**

5. Sperm travel through the **vas deferens** to a region of several **semen-producing glands.** The large **prostate** surrounds the **urethra,** and the **seminal vesicles** and **Cowper's gland** empty into the nearby **ejaculatory duct** and urethra, respectively. Semen transports sperm, lubricates the urethra, and provides **fructose** for metabolic energy; also, its alkalinity buffers the acidic vaginal secretions.

6. **Erection** occurs when the spongy tissues of the **penis** fill with blood. Variations in penis size are of little biological importance. Small penises enlarge more during erection than large penises, and changes in the **vagina** accommodate most penises. The enlarged **glans** contains many sensory neurons. **Circumcision** removes part or all of the **prepuce.**

7. **Ejaculation** typically results in the release of about 3–5 ml of semen, which usually contains over 120 million sperm in fertile males. Sperm move faster and live longer in the **uterus** and **oviducts** than in the vagina. Rhythmic contractions of the uterus, some stimulated by seminal substances, help propel sperm, and most **fertilizations** occur in the upper **fallopian tubes.**

8. The reproductive and urinary tracts are separate in females. The **vulva** includes the **mons, labia, clitoris,** and **perineum.** Ducts from **Bartholin's glands** lead into the inner labial lips. The clitoris develops from embryonic tissue similar to that from which the penis forms in males and, like the penis, consists of a shaft and a sensory glans.

9. The vagina is both the receptacle for the penis during **coitus** and the **birth canal** during delivery. The vaginal opening of young females is often covered partially by a thin **hymen.**

10. Extending slightly into the upper vagina is the **cervix. Cervical mucus** changes from watery near the time of **ovulation** to thicker at

other times. The thick **myometrium** of the uterus is muscular, whereas the thinner **endometrium** is glandular and changes regularly during **menstrual cycles.** The oviducts, which extend from the uterus, have funnellike ends that partially surround the ovaries.

11. Like the testes, the ovaries produce both gametes and sex hormones. Unlike spermatogenesis, oogenesis typically results in only one (not four) haploid gamete from each diploid cell. The small **polar bodies** that form in each **meiotic division** are nonfunctional. Numerous cells form around the **ovum** and create **follicles.** Only a few hundred thousand ova out of several million that form during a woman's embryonic development are present at the time of her birth, and many of those degenerate before puberty. An average woman releases only about 400 eggs in her 30- to 40-year reproductive lifetime. After ovulation, the **corpus luteum** forms from cells in the old **Graafian follicle.**

12. Scattered in grapelike clusters within the adipose tissue that makes up most of each **breast** are the **milk glands,** which drain by ducts to the **nipple.** Breast development is one **secondary sex trait.**

13. Menstrual cycles occur regularly (averaging 28 days) between puberty and **menopause,** usually interrupted only by pregnancies. The cyclical changes are controlled primarily by two hormones from the **pituitary (FSH—follicle-stimulating hormone—**and **LH—luteinizing hormone)** and two from the ovary (estrogen and **progesterone**). In the early **follicular stage,** FSH causes many immature eggs to begin maturation, although only one usually becomes an ovum. Follicle cells secrete estrogen, which stimulates the endometrium to begin redeveloping, inhibits the release of FSH, and stimulates the release of LH. Midway in the cycle, ovulation follows shortly after the peak of LH. LH stimulates the formation of the corpus luteum, which secretes progesterone that finalizes the development of the endometrium. If fertilization occurs, the corpus luteum remains functional for several weeks, but if no fertilization occurs the corpus luteum becomes a **corpus albicans.** Low levels of estrogen and progesterone late in the **luteal phase** lead to **menstruation,** and allow FSH to stimulate the ripening of more eggs in the ovaries.

14. Four stages are commonly recognized in the **sexual response cycle,** with **foreplay** being important in preparing a couple initially. During the **excitement stage,** bodywide **myotonia** occurs, and vasodilation causes erection of the penis, clitoris, and female (and sometimes male) nipples; secretion of vaginal, cervical, and other fluids; and enlargement and movement of various organs in both sexes. These changes are intensified during the **plateau stage,** when a **sex flush** may also develop. During the short **orgasm stage,** rhythmical contractions occur in various organs, and some of these may aid sperm ascent. Many females are multiorgasmic, but males generally experience a **refractory period** of minutes to hours following orgasm. In general, the events of the excitement and plateau stages are reversed during the **resolution stage.**

15. **Sex** is the exchange of genetic material, whereas **reproduction** results in the production of new individuals. In many animals, including humans, sex and reproduction occur together. **Asexual reproduction** (e.g., **budding** and **parthenogenesis**) usually results in individuals identical to the parent(s), whereas **sexual reproduction** usually results in more variation among offspring, and this is often advantageous over long periods of time.

REVIEW QUESTIONS

1. What purpose is served by the location of the testes in the scrotum? What other structures and functions are involved in allowing sperm to form properly?

2. What are the two functions of the gonads, and what structures perform each function?

3. Describe the process of spermatogenesis, including the structural changes that occur in sperm as they mature.

4. Distinguish among sterility, fertility, and impotence.

5. What circulatory processes are involved in erection of the penis and in the return of the erect penis to its flaccid state?

6. What is circumcision, and for what purposes is it performed?

7. Where is semen produced and what are its properties and functions?

8. Describe the pathway of sperm during coitus from the testis to the usual site of fertilization. Which events or conditions slow or prevent sperm from uniting with an egg, and which assist in the process?

9. Name the structures, locations, and functions of the female external genitalia and internal reproductive organs.

10. What are some important differences and similarities between the male and female reproductive tracts?

11. Compare and contrast the processes of oogenesis and spermatogenesis.

12. Describe the hormonal control of the menstrual cycle, specifically explaining the sources and roles of follicle-stimulating hormone, estrogen, luteinizing hormone, and progesterone.

13. What bodily changes occur during menopause, and how do these differ from those during menarche?

14. Compare and contrast the advantages and disadvantages of asexual and sexual reproduction.

15. Distinguish between sex and reproduction.

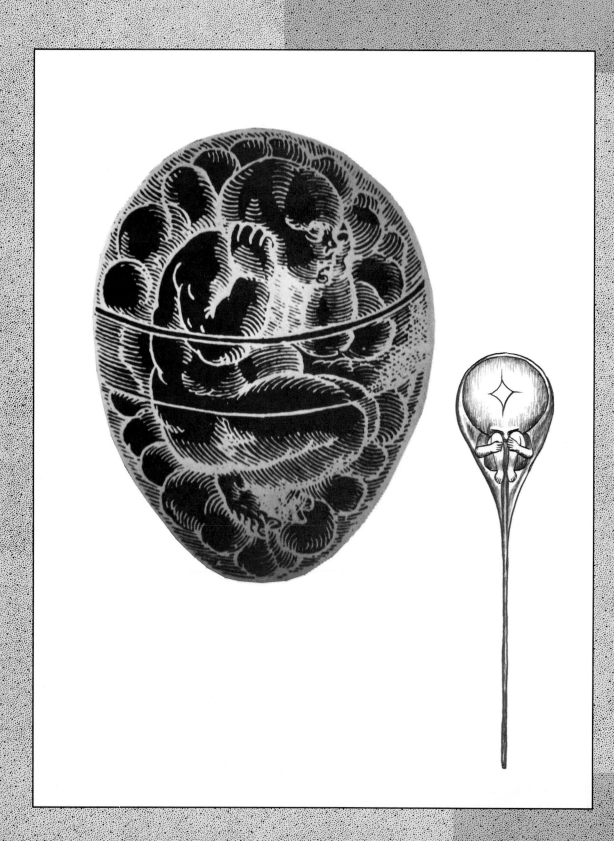

CHAPTER 6

PREGNANCY AND BIRTH

Aristotle thought that semen was the seed that gave rise to a new individual, and that the female's body was the site where the seed was nourished. Perhaps the phrase "Mother Earth" resulted from this type of thinking.

Some early microscopists even believed that they saw a miniature human, called a homunculus, crouched within a sperm. Others felt that preformed humans were found in eggs, and one theory proposed that Eve contained all future generations wrapped one within the other in her eggs, somewhat like the Russian dolls-within-dolls. Supposedly, humans would become extinct when all encased homunculi had been released!

Although sperm were seen by early microscopists, human eggs were not observed for about another 300 years, and it was not until the 1940s that the fertilization of a human egg was seen. Thus, much of our knowledge about human reproduction is recent and many unanswered questions remain. In many respects, we know more about the reproduction of various laboratory organisms than we know about our own reproductive processes.

Belief often affects perception, and some early microscopists were convinced that they saw tiny humans, or homunculi, crouched within the gametes.

FIGURE 6.1
Cross-section through a wandering white blood cell (macrophage), showing sperm that it has engulfed. Like various pathogens, sperm are foreign substances that these specialized and protective leukocytes recognize and phagocytize. Many sperm are also immobilized by inadequate nutrients and by the effects of acidic vaginal secretions.

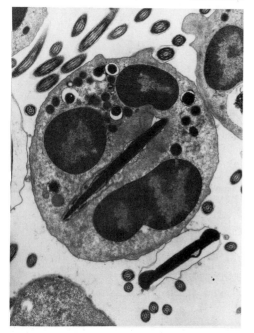

FERTILIZATION AND EARLY DEVELOPMENT

Eggs are most likely to be fertilized within about 12 hours after ovulation, and it is probable that few eggs older than a day are fertilized successfully. Some sperm may survive for three or four days inside the female reproductive tract, but most die within a day or two after ejaculation.

The chance is small that many of the half-billion or so sperm ejaculated into the vagina during coitus will encounter an egg. Perhaps 100 million leak from the vagina or are killed by acidic vaginal fluids. Those that survive must swim upstream toward channels in the cervical mucus, and with little cytoplasmic food reserves, thousands become exhausted quickly. Some are destroyed by the female's wandering white blood cells, much as foreign bacteria are (Figure 6.1). Perhaps only a few thousand enter the uterus and only a few hundred, or even only a dozen, survive to complete the voyage to the egg in the upper oviduct. However, the upward movement of sperm is helped by the rhythmic contractions of the uterus.

Whereas a sperm is small and motile, an egg is enormous and must be moved by other cells. Cilia within the fallopian tubes beat toward the uterus, aiding the peristaltic contractions of the oviducts in moving the egg along. Because the passage of the egg is slow it must usually be fertilized in the upper third of a fallopian tube, before it has become too old for fertilization (Figure 6.2).

Sometimes an egg is fertilized outside of a fallopian tube. Some zygotes settle within the fallopian tube instead of moving along to the uterus, and occasionally embryos attach to the cervix or to structures in the abdominal cavity. These **ectopic pregnancies** (Figure 6.3) are potentially dangerous because structures other than the uterus are unable to support a developing embryo for long. They may rupture and lead to serious internal hemorrhaging, and about one-tenth of maternal fatalities are caused by ectopic pregnancies. Therefore, women of reproductive age who may be pregnant and who feel abdominal pain should be examined quickly for a possible ectopic pregnancy.

The egg is surrounded by many follicle or nurse cells when the bulletlike sperm reach it (Figure 6.4). The egg's volume is huge compared with that of the tiny sperm, about 90,000 times larger. Within the caplike **acrosomes** of sperm are enzymes that digest a path through the egg's surrounding nurse cells. It takes the combined action of several sperm

FIGURE 6.2
The early stages of human embryonic development. Fertilization usually occurs in an upper oviduct, and it takes several days for the young embryo to reach the uterus. The solid morula is about the same size as the zygote, and growth and development into the blastocyst stage typically occur only after the embryo reaches the uterus.

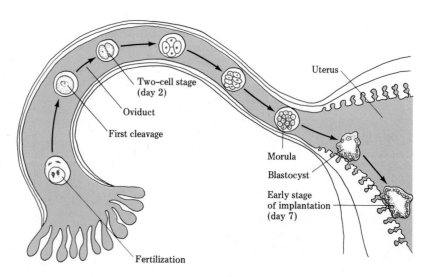

Two–cell stage
(day 2)

Oviduct

First cleavage

Uterus

Morula

Blastocyst

Early stage
of implantation
(day 7)

Fertilization

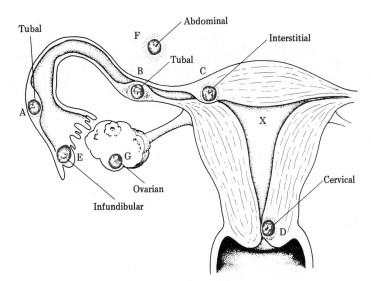

Tubal

Abdominal

F

Interstitial

Tubal

B

C

A

X

E

G

Cervical

Ovarian

Infundibular

D

FIGURE 6.3
Ectopic pregnancies. Improper implantation sites are shown by letters A to G, which indicate the approximate order of frequency of extrauterine implantations. By contrast, typical normal sites for implantation are in the upper half of the uterus (position X).

to accomplish this, which largely explains why so many sperm are needed for a successful fertilization. The surviving sperm then encounter a jellylike layer (zona pellucida) that surrounds the egg, and only one or a few sperm move through it to reach still another layer (vitelline membrane), before one sperm finally reaches the egg's plasma membrane. Electrical changes spread across the egg's membranes and microvilli extend to surround the sperm that enters the egg. Within about a minute, the outer egg membrane lifts slightly from the egg's surface and usually becomes impermeable to other sperm. Multiple fertilizations probably occur occasionally, but the extra set or sets of chromosomes lead to abnormal and short-lived embryonic development.

The cytoplasmic activity within the egg is high when the sperm enters. The egg completes its second meiotic division, and the second tiny, nonfunctional polar body is expelled (Figure 5.9). Within 10–12 hours a new nuclear membrane forms around the two fragile sets of genetic material, and fertilization is complete. Within about 15–20 hours, **cleavage** occurs after the zygote divides mitotically. Succeeding cell divisions follow quickly—two cells become four, four become eight, and so on. About three days after the first cell division, the **embryo** is a solid, raspberrylike mass of 16 cells called a **morula**. Although the number of cells has increased, however, the total mass of the morula is about the same as that of the zygote. Only after the embryo reaches the uterus several days later does any significant cellular growth occur.

FIGURE 6.4
The process of fertilization. Several sperm generally reach an egg (A), and the enzymes released from their acrosomes digest nurse cells and allow one sperm to fertilize the egg (B).

(B)

(A)

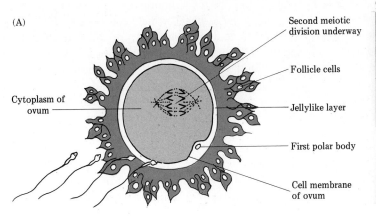

Cytoplasm of ovum

Second meiotic division underway

Follicle cells

Jellylike layer

First polar body

Cell membrane of ovum

Sometimes cells from the early cell divisions separate, and two or more embryos may develop from the original zygote. This is one way identical multiple births occur. However, at a relatively early stage in embryonic development each cell is usually "committed" to form various parts of the body. If these cells separate, they generally do not divide to give rise to complete embryos.

About four days after fertilization and soon after the morula passes from the oviduct into the uterus, fluid enters through the surrounding jellylike layer and the solid morula becomes a hollow, fluid-filled **blastocyst.** The cells that will form the embryo develop from the thick inner cell mass on one side of the blastocyst, whereas the thinner outer cells, or **trophoblast,** form the extraembryonic membranes that protect and nourish the embryo and attach it to the endometrium (Figure 6.5). Both the outer **chorion** and the inner **amnion** surround liquid-filled sacs, and these membranes eventually fuse and fill the womb. The fluid that will cushion the embryo throughout pregnancy begins to flow into the amniotic sac about eight days after fertilization.

IMPLANTATION

The blastocyst usually spends three or four days floating free in the uterus. **Implantation** usually occurs on an upper surface of the swollen and folded endometrial walls of the uterus about six to nine days after fertilization, and a woman is then pregnant. Enzymes released by the trophoblast break down the endometrial cells, and the materials released nourish the blastocyst, which quickly becomes buried in the endometrium.

Fertilization and early embryonic development are chancy events. Of the estimated 80 or so eggs out of every 100 that encounter sperm and are fertilized successfully, only about 60–70 survive to encounter the uterine wall. Perhaps only 40 or so are still alive after seven days, and some of those blastocysts do not become implanted but disintegrate or are expelled through the vagina. These early **miscarriages** or **spontaneous abortions** usually occur without the female's knowledge. Exposure to certain chemicals, pathogens, or radiation causes miscarriages, developmental problems, or genetic changes.

EMBRYONIC DEVELOPMENT

The developing human is called an embryo until about the end of the seventh week. After eight or nine weeks, when bone begins to replace cartilage, the term **fetus** is used. At two months into development, the fetus has a disproportionally large head, which constitutes about 50 percent of the body. Growth and development are rapid during the embryonic period, and all of the basic organs and systems form quickly (Figure 6.6). Blood and blood vessels develop during the third week, when the embryo is still smaller than a sesame seed, and the tiny heart throbs soon thereafter. The nervous system also begins its development during the third week and the bulges and lobes on the upper end of the neural tube will swell and grow into the brain. The eyes are relatively large but visionless at this early stage.

Limb buds and facial features appear during the fifth week, and the embryo can respond to touch. The tail that is present usually disappears but a remnant occasionally persists, to be snipped off shortly after birth.

The liver is large and is the blood-forming organ at this stage, whereas in the adult it is the blood "recycling" center (Chapter 15). It is the yolk sac that produced blood cells earlier in development, and the bone marrow will take over this important function after the liver. By seven weeks the primitive gonads begin to form, and fingers and toes are recognizable on the paddlelike limbs.

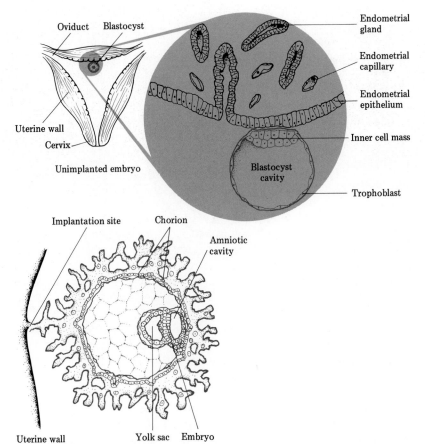

FIGURE 6.5
The process of implantation. At first free in the uterine cavity, the embryo usually contacts an upper wall surface and soon becomes completely embedded in the endometrium.

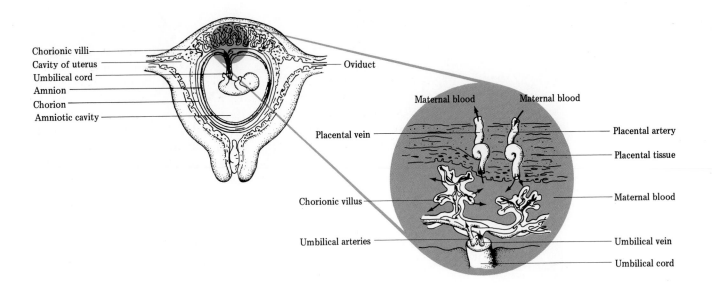

FIGURE 6.6

Some major events in embryonic development. (A) Photos of the (1) 5-week, (2) 8-week, (3) 10-week, and (4) 18-week stages. (B) Diagrammatic views of some embryo stages drawn to the same scale in order to show the relative proportions of the body regions. The 3-week embryo is only about the size of a small seed but its blood and blood vessels are forming rapidly. Note the movements of the eye cells during development, the early presence of gill slits and a tail, and the gradual development of separated fingers and toes from initially fused structures.

(B)

Third week

Fourth week

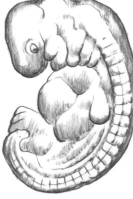

Fifth week

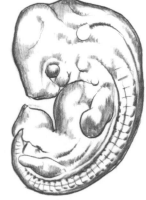

Sixth week

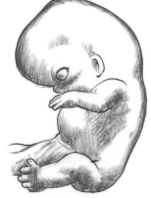

Seventh week

(A)

(1)

(2)

(3)

(4)

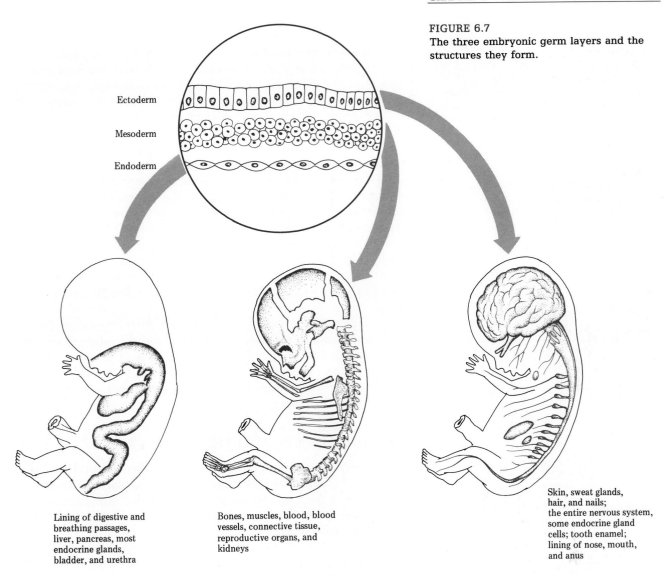

FIGURE 6.7
The three embryonic germ layers and the structures they form.

Ectoderm

Mesoderm

Endoderm

Lining of digestive and
breathing passages,
liver, pancreas, most
endocrine glands,
bladder, and urethra

Bones, muscles, blood, blood
vessels, connective tissue,
reproductive organs, and
kidneys

Skin, sweat glands,
hair, and nails;
the entire nervous system,
some endocrine gland
cells; tooth enamel;
lining of nose, mouth,
and anus

The cells that will form the placenta later (see below) are now secreting **human chorionic gonadotropin** (HCG). This hormone stimulates the corpus luteum to remain functional, which is necessary for pregnancy to continue. Looking for HCG in the urine is the basis for some pregnancy tests.

The three distinctive embryonic **germ layers** from which all body tissues and organs are fated to develop in an orderly and precise fashion are present by the second week (Figure 6.7). The **ectoderm** (from the Greek for "outside skin") gives rise to the outer epithelium (e.g., skin, hair, and nails) and to the nervous system. The cells that form a neural groove along the embryo's back soon grow together to form a tube, first in the midsection and then progressing zipperlike in both directions. The cells of the **mesoderm** or "middle skin" develop into muscle, bone, and other connective tissues, as well as giving rise to the membranous **peritoneum** that lines the body cavity and covers the organs within it, the circulatory system, and most of the reproductive and excretory systems. From the "within skin" or **endoderm** cells develop the epithelial lining of some internal structures such as the digestive tract, most of the respiratory tract, and the bladder, liver, pancreas, and some endocrine glands.

PLACENTATION

The pancake-shaped **placenta** (from the Latin for "flat cake") is the organ that exchanges material between the embryo or the fetus and the pregnant female. Soon after the blastocyst becomes implanted, the placenta begins to form as fingerlike projections or villi of the chorion grow into and break down the surrounding endometrium. As endometrial capillaries degenerate, sinuses that contain maternal blood form around the villi. At the same time blood vessels from the developing embryo grow into the villi. Prior to the development of the placenta, the embryo's energy supplies and the raw materials for its growth came from its own yolk and uterine secretions. Essentially, the placenta functions as the digestive tract, lungs, and kidneys of the fetus. It provides the fetus with oxygen and predigested nutrients, and it eliminates fetal wastes. These materials are generally transferred between the pregnant female and the fetus by diffusion and osmosis. To facilitate these exchanges, the surface area of the placenta is large and grows as the fetus grows, covering about 20 percent of the endometrium by 4 weeks and 50 percent by 20 weeks.

The placenta is also an endocrine structure. After about three months the corpus luteum in the ovary degenerates and the estrogen and progesterone needed to continue the pregnancy then come from the placenta. The placenta also secretes a growth hormone and a thyroid-stimulating hormone. Placental hormones may appear in the urine, and their presence also serve as a pregnancy test.

HARMFUL AGENTS AND THE PLACENTA

Unfortunately, many drugs, medications, and other chemicals, radioactive substances, and some pathogens also cross the placenta. These may kill the embryo or fetus (miscarriage or **stillbirth**), slow development, cause diseases in the developing human, or lead to birth defects and/or mental retardation. Agents that cause birth defects, that alter or mutate genetic material, and that cause cancer are called **teratogens, mutagens,** and **carcinogens,** respectively. Most substances that enter the body of a pregnant female may also enter the embryo or fetus.

Carbon monoxide (CO) results from the incomplete combustion of organic compounds. CO, a common air pollutant and major ingredient of tobacco smoke, greatly reduces the ability of blood to carry oxygen. Pregnant females who smoke heavily and/or who breathe CO-polluted air tend to have **low-birth-weight** (LBW) infants (Figure 6.8). LBW infants, once known as "premature" infants, have lower survival rates and more defects than normal-weight infants.

The rubella, or German measles, virus may cross the placenta and cause eye, ear, heart, or tooth defects and/or mental retardation. However, the nature of the problem also depends upon when the infection occurs (Figure 6.9). For example, heart defects often occur from infections during the fifth week, blindness (e.g, cataracts) from infections during the sixth week, and deafness from ninth-week infections. Thus, the specific effect of an agent depends in part upon the developmental changes that are occurring at the time the agent affects the embryo or fetus.

Obviously, many regulatory mechanisms are involved in development, and the sequencing of development is critical. Before the nerves that influence the heart function, there must be a heart, blood, blood vessels, and somewhere for the blood to flow. Because all cells that developed from the zygote usually have the same genetic blueprint, some cells must

FIGURE 6.8
The effect of maternal diet and smoking behavior upon the weight gain of fetuses. Much of the retardation in growth that results from smoking is due to CO, which greatly reduces the oxygen-carrying ability of blood.

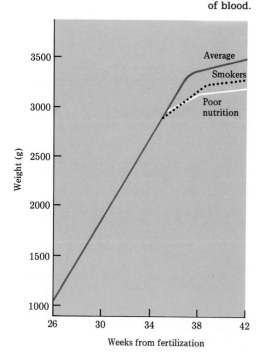

"listen" to certain genes while ignoring others. Also, some cells must "tune out" most directions for a time, until it is their "turn" to become active, multiply, grow, and differentiate. In fact, some cells must die, such as those that form the webbing between the developing fingers and toes.

UMBILICAL CORD

Anchored to the life-supporting placenta, the **umbilical cord** averages 1– 2 cm (0.03–1 in.) in diameter and 50–55 cm (20–22 in.) in length. The internal blood vessels are protected by a jellylike substance. Near the time of birth, about 285 L (75 gal) of blood pass through the placenta daily, and a round trip between the fetus and the placenta takes only a few seconds.

Twisting and bending of the cord are common, but true knots are rare. The cord may become looped around the fetus, and this sometimes complicates the birth process, although it is uncommon for a fetus to be strangled by the cord. Although the umbilical cord looks slack after birth, like a piece of rope, it is less flexible when blood is flowing through it, much as a garden hose with flowing water is more rigid than an empty one.

FIGURE 6.9

Some critical periods in human embryonic development. The colored bars indicate highly sensitive periods, and the open bars denote stages less sensitive to damage by external agents. Note that many agents produce their major effects early in development, when the cells of structures are dividing rapidly.

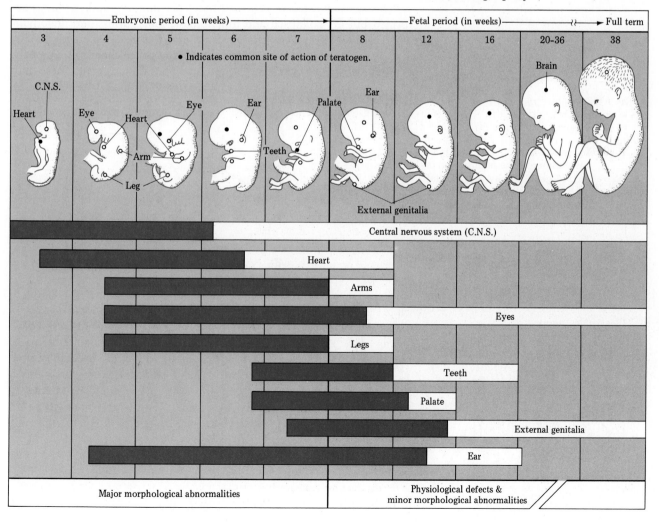

FETAL DEVELOPMENT

Although the early fetus weighs far less than a gram and is only about 3.2 cm (1.3 in.) long (top of head to buttocks), growth occurs rapidly. The face of the two-month fetus is already somewhat humanlike, with slitlike eyes, and nose, ears, and jaws forming. The skeleton, which was formerly cartilage, is beginning to be replaced with bone, and the external sex organs are visible. Most major organs have formed, and subsequent development consists mainly of refining these structures. Figure 6.10 and the descriptions that follow summarize fetal growth.

Third Month

During the third month the fetus grows to a length of about 8 cm (3 in.) and a weight of about 15 g (0.5 oz). Because the head no longer bends forward, the posture becomes more upright. Blood, which formed earlier in the liver and spleen, now forms in the bone marrow. Nails, hair, and tooth buds appear, the kidneys begin to function, and swallowing motions are distinct. Males have an external penis, and females form a vagina and uterus. The long bones form independently, and only later will joints connect them (Figure 6.11).

Fourth Month

Rapid growth during the fourth month increases the fetal length to fist size (about 16 cm, or 6 in.) and the weight to about 113 g (4 oz). The heart beats rapidly (120–160 times per minute), and the lungs are formed but are collapsed and functionless. Fingerprints and fingernails appear. Fetal movements are now commonly felt by the pregnant female.

Fifth Month

The fetus now weighs about 226 g (0.5 lb) and is almost 20 cm (8 in.) long. Movements are more vigorous, several reflexes are present,

FIGURE 6.10
The main external changes during fetal development. The fetal period is characterized generally by growth and elaboration of basic structures and systems formed during the embryo stage.

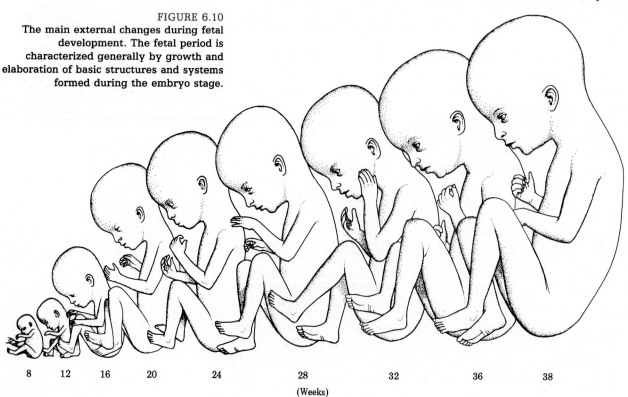

8   12   16   20   24   28   32   36   38

(Weeks)

and the fetus may sleep and hiccup. The eyes are light sensitive, but the ears do not yet hear. Fine hair and a cheesy exfoliation of cells cover the body. During the second **trimester,** or middle third of pregnancy, the body grows rapidly and begins to catch up in size with the head.

Sixth Month

The 600-g (1.5-lb), 28-cm (11-in.) fetus looks like a miniature human. Scalp hair is abundant, eyebrows and lashes form, and the eyelids separate. The fetus may drink amniotic fluid and suck or chew a thumb. Pastelike cellular debris begins to accumulate within the intestine, where it remains until birth. A fetus born at this stage rarely survives because of a lack of temperature regulation and because the lungs and digestive organs are usually nonfunctional.

Seventh Month

The fetus now weighs about 0.09–1.3 kg (2–3 lb) and is about 30.5 cm (12 in.) long. The fine body hair is reduced to the back and shoulders. The brain grows, its surface becomes grooved, and localization of function is evident.

Eighth Month

Growth slows during the eighth month, although fat is deposited under the skin and the fetus appears chubby in contrast to its earlier wrinkled look. Weight is about 2.3–2.7 kg (4–5 lb), and the length is about 33 cm (13 in.). The fetus perceives light, responds to various stimuli, and can taste chemicals, but remains deaf. About three of four fetuses can survive if born at this stage, although the lungs and digestive system are still immature. The fetus usually assumes an upside-down position, partly because of the shape of the uterus and partly because the head is the heaviest body part. During the third trimester the fetus comes to fill the amniotic cavity rather than float freely within it.

Ninth Month

The weight of a late fetus averages 2.7–3.6 kg (6–8 lb), with males generally weighing more than females, while average length is 48–51 cm (19–20 in.). Activity decreases as the fetus fills the amniotic cavity, and growth slows as the placenta becomes fibrous and less functional. Antibodies (protective substances discussed in Chapter 16) are secured from the maternal blood supply, and the immunities thus acquired usually last for the first several months following birth. The chest is prominent, mammary glands protrude in both sexes, the skin is whitish to bluish-pink, and the eyes are blue.

The changing proportions shown in Figure 6.12 illustrate the head-to-tail or **cephalocaudal developmental direction.** Another developmental direction is called **proximodistal,** because the middle part of the body develops earlier and faster than the extremities.

THE PREGNANT FEMALE

The first sign of pregnancy is usually the absence of menstruation. However, some women have been known to continue menstruating at reduced levels for two or three months after conception. The breasts grow, veins become more prominent on the breast surfaces, and the nipples and surrounding areolae may darken and become broader. Some-

FIGURE 6.11
Views of some fetal developments. The lungs (A) and a developing bone (B) are shown in a young fetus; the skin, with sweat glands (C), and the lips, upper and lower incisors, and mouth cavity (D) are shown in an older fetus.

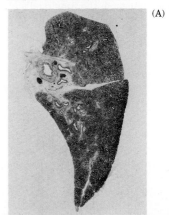
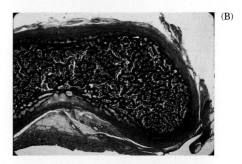
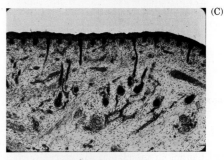
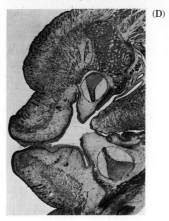

FIGURE 6.12
The changing proportions of the body during the fetal period, with the stages drawn to the same scale for comparison purposes. Note the decrease in the relative size of the head, a prime example of the cephalocaudal developmental direction. What changes illustrate the proximodistal developmental direction?

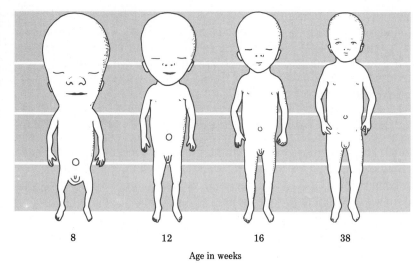

8   12   16   38

Age in weeks

times the breasts tingle, throb, or are painful. Pigment may also darken the eye region and the line from the navel to the pubic area.

Thirst often increases, and the average pregnant woman retains about 3–6 L (3.1–6.3 qt) of additional body fluid (edema), half of this during the last 10 weeks of pregnancy. Ankle swelling is especially common. The frequency of urination may increase, especially as the growing uterus presses against the bladder. Pregnant females are often more susceptible to urinary tract infections, bowel movements may become irregular, and indigestion and/or constipation may occur. Various bodily changes (e.g., stretching of the uterus) may cause occasional nausea or "morning sickness," especially during the first four to six weeks of pregnancy. Vaginal secretions may increase, and a general feeling of tiredness is common. The joints of the pelvic bones usually widen late in the first trimester, the blood volume increases, there is a slight increase in heart size, and the bone marrow becomes more active.

The outward appearance of the pregnant female changes as the uterus grows (Figure 6.13). The uterus is the most extendable organ in the

(B)

FIGURE 6.13
(A) Changes in the size of the uterus during pregnancy. Note that the uterus drops late in pregnancy (dashed lines). (B) By late in pregnancy, the fetus comes to fill most of the amniotic cavity, typically with the head in the cervical region.

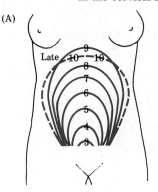

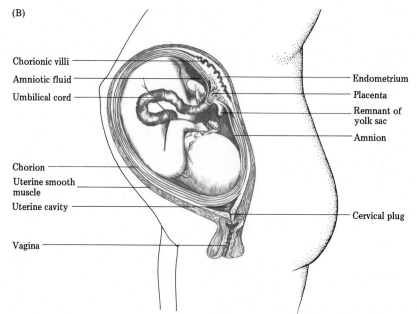

Chorionic villi

Amniotic fluid

Umbilical cord

Endometrium

Placenta

Remnant of yolk sac

Amnion

Chorion

Uterine smooth muscle

Uterine cavity

Cervical plug

Vagina

human body, and its weight normally increases 16–18 times during pregnancy. Early changes include color changes in and a softening of the uterus and the cervix. The uterus, which can be felt through the abdominal wall from about the third month on, crowds many abdominal organs and may affect breathing by making movements of the diaphragm difficult. Pressure upon the abdominal veins may cause enlarged or varicose veins and/or hemorrhoids. Backaches and muscular cramps are also common. The dropping of the fetus during the last few weeks of pregnancy reduces some pressure on the pregnant female's lungs.

Because the pregnant female is "eating for two," dietary needs are increased and a poor diet may adversely affect both the developing human and the pregnant female. For example, during the third trimester, about 85 percent of the calcium and iron the pregnant female consumes goes to form the fetal skeleton and blood cells, respectively. Her protein intake is very important for the proper development of the fetal nervous system. Actually, the pregnant female is doing almost everything for two. As a result, her breathing and heart rates usually increase and her kidneys work harder.

Early and frequent medical attention during pregnancy is wise, and such care may prevent numerous problems. Poor kidney function, high blood pressure (hypertension), and malnutrition adversely affect both the health of the fetus and the pregnant woman.

Weight gain during pregnancy is somewhat less restricted than it was a decade or so ago. Some common minimum weight gains (total = 12.0 kg, or 26.4 lb) during pregnancy are: body fat reserves, 4.0 kg (8.8 lb); the fetus, 3.5 kg (7.7 lb); blood and extracellular fluid, 2.5 kg (5.5 lb); uterus, 1.0 kg (2.2 lb); amniotic fluid, 0.5 kg (1.1 lb); and the placenta, 0.5 kg (1.1 lb).

BIRTH

Trillions of cells have now developed from the single-celled zygote. The average pregnancy lasts 280 days (normal range, 265–295 days) from the first day of the last menstrual period. Many changes occur as **labor** begins. Estrogen levels increase and progesterone levels decrease shortly before birth. The hormone **oxytocin** from the maternal pituitary gland, along with estrogen and several prostaglandins (Chapter 18), stimulates uterine contractions, whereas progesterone inhibits them. The uterus is the largest and most powerful human muscular structure, far stronger than a boxer's biceps. As the uterus and abdominal muscles contract rhythmically, the cervix thins and opens (Figure 6.14). Uterine contractions usually begin at about 30-minute intervals and gradually increase in frequency.

The first stage of labor involves the opening of the cervix to enable the fetus to enter the birth canal. A pink vaginal discharge and the rupture of the amniotic sac indicate cervical dilation. This is the most variable and longest part of labor, 8–24 hours (average 12 hours) for the first pregnancy and 3–12 hours for later pregnancies. Labor pains are caused mainly by the stretching of the vaginal wall and of the area between the vagina and the anus. Because learning to breathe properly may greatly reduce the discomfort, this exercise is common in natural childbirth classes.

Although some fetuses are in breech positions (buttocks or feet first), over 90 percent are head down. If there is difficulty during the birth process, such as a transverse fetal position, the child may be delivered by **caesarean section**. This abdominal surgery involves cutting directly

FIGURE 6.14

(A) Diagrammatic views of the birth process. Most infants are born head first and rotate somewhat during the birth process. The squeezing of the head as it passes through the birth canal seldom results in damage because of the movable skull sutures. (B) The culmination of about 280 days of pregnancy.

(A)

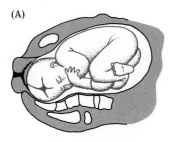

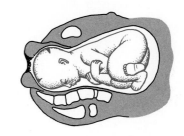

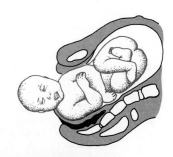

(B)

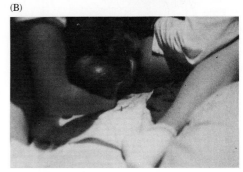

into the womb to remove the fetus. The procedure may also be used when the pelvis is too small, when the placenta is in the lower part of the uterus, or when the pregnant female has died. Survival rates of babies delivered in this way increase from about 10 percent in the seventh month to about 70 percent and 95 percent in the eighth and ninth months, respectively.

When the cervix dilates to a diameter of about 10 cm (4 in.) the second stage of labor, or **delivery,** begins. Learning how and when to push helps to move the child through the birth canal, reduces the discomfort, and often reduces or eliminates the need for anesthesia. This stage may last 20 minutes to two hours, and as the child moves through the birth canal it is usually rotated by the bones of the mother's pelvis. The obstetrician or midwife may assist in guiding the child, and occasionally special forceps are used.

The child's head is squeezed considerably as it passes through the birth canal. However, the brain is rarely damaged because the skull bones are pliable and are separated by loose sutures. The head returns to its normal shape within a few days.

With the expulsion of the child, the uterus shrinks and the major labor contractions cease. The third stage involves delivery of the placenta and associated membranes, the **afterbirth.** This may occur within a few minutes to a half hour following delivery. The placenta averages about 23 cm (9 in.) in diameter and 2.5 cm (1 in.) in width, about the size of a pie pan. The constriction of blood vessels by the shrinking uterus reduces blood loss, and the umbilical cord collapses and is cut and clamped.

FIGURE 6.15
The main features of the circulatory system in (A) the fetus and (B) the neonate. What changes are evident?

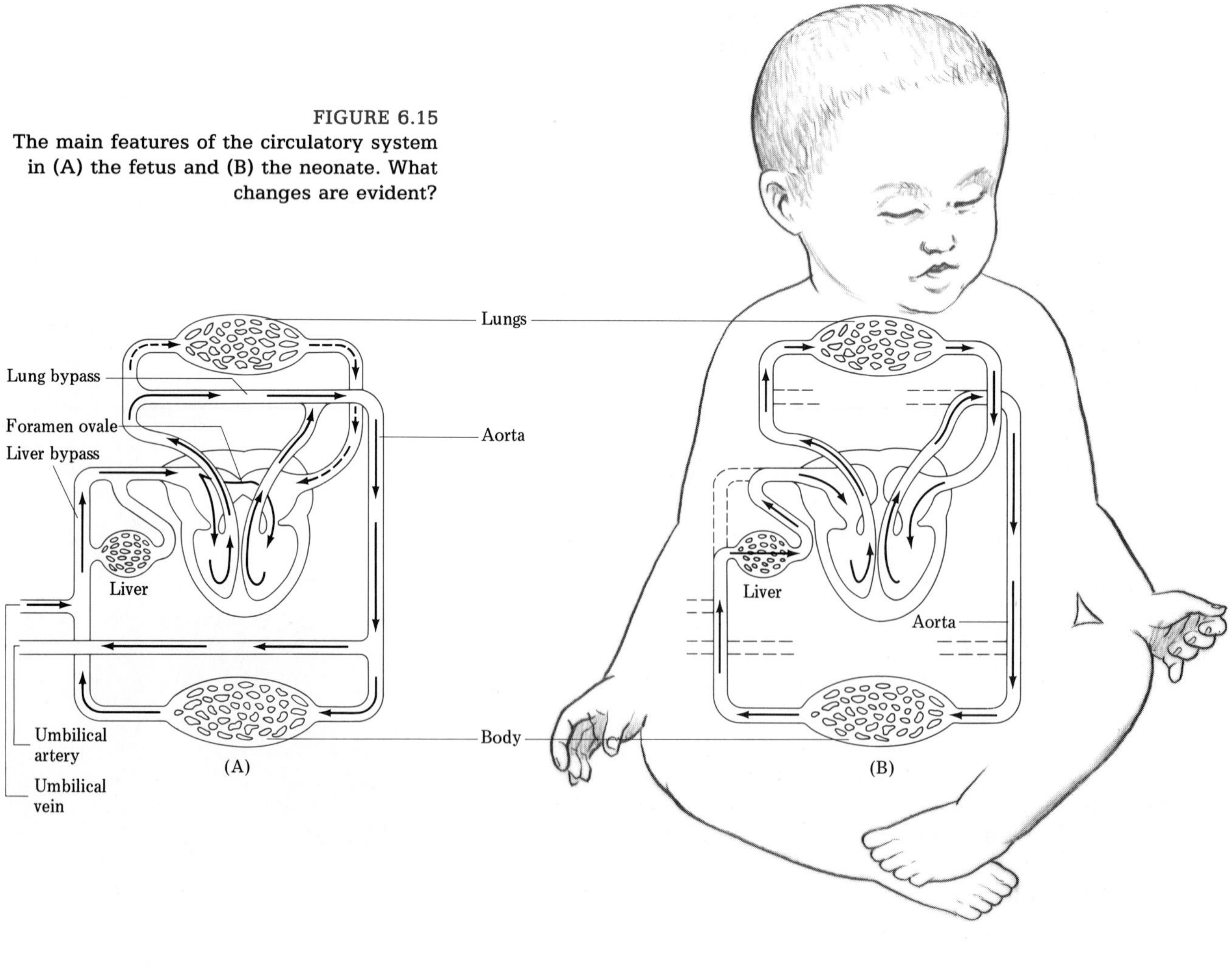

Although newborn infants or **neonates** may cry during and after the birth process, there are usually no tears because the tear glands do not generally function for another few weeks. The skin color of babies of all races is light at first, just as the eye color of most neonates is blue initially. Changes in the skin and eye pigmentation often take several weeks.

Until birth, the pregnant female has done almost everything for the child. Once the neonate leaves the protected, warm, shock-absorbing sea of amniotic fluid, however, it must begin to breathe, or 9 months of development will have been to no avail. Both mechanical and chemical events probably impel the child to take its first breath. As the child passes through the birth canal, pressure forces fluid out through its mouth and nostrils and, after birth, the neonate's lungs fill as air rushes into the partial vacuum. Carbon dioxide (CO_2) is an important chemical factor, because when the umbilical cord is cut, CO_2 accumulates in the infant's blood and quickly stimulates its brain's breathing center. According to one view, the "shock" of moving from a warm, relatively dark, and watery world to a cooler, brighter, and dry one also helps trigger the first breath.

The fetal lungs are shriveled, fluid-filled, and generally idle. However, dramatic changes in the child's circulatory and respiratory systems occur soon after birth (Figure 6.15). Only about 10 percent of the fetus's blood goes to the lungs. Most blood (about 55 percent) flows to the placenta for gas exchange, and the remainder circulates through the other blood vessels. Because the neonate no longer depends upon the placenta for nutrients and gas exchange, its blood must flow through the lungs and liver. Therefore, the heart-lung bypass (foramen ovale) usually closes quickly after birth. If it does not and if other circulatory system changes do not occur, a **blue baby** results and surgery may be necessary.

LOW BIRTH WEIGHT

So-called "prematurity" is usually defined medically by birth weight rather than by length of pregnancy. Low-birth-weight (LBW) infants weigh less than 2.5 kg (5.5 lb), whereas the average birth weight is about 3.2–3.6 kg (7–8 lb). Low birth weight accounts for more than half of the deaths of neonates in the United States, and the lower the birth weight the greater the risk. A common problem in "preemies" is that a film of sticky material covers the air sacs and tubes in the lungs and prevents or hinders normal gas exchange.

Poor prenatal maternal nutrition, alcoholism, hypertension, heavy smoking, and the lack of medical attention contribute to the incidence of LBW infants. Premature labor may also be caused by infections (e.g., syphilis), diabetes, an abnormal placenta or fetus, and thyroid and/or kidney problems. Too rapid a succession of pregnancies also contributes to the problem.

Although most multiple births are LBW infants, they have a better chance for survival than infants of comparable weight that are born singly, **singletons** (Figure 6.16). That is, twins and triplets are relatively more developed than singletons of the same birth weight.

LBW infants are often placed in high-tech, climate-controlled incubators because of their susceptibility to infections. Body temperature regulation may be inadequate, and feeding is a problem because of the underdeveloped digestive tract. Use of the mother's milk and close mother-infant contact contribute to survival.

One of the most extreme cases of a LBW infant surviving involved

FIGURE 6.16
Comparison of weights of average singletons and twins. Most multiple births are LBW infants.

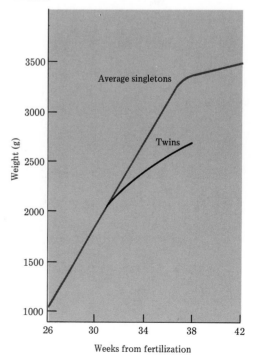

Weeks from fertilization

FIGURE 6.17
The painting shows an infant held so its head is on the left side of the mother's body, a position where the mother's heartbeat can be heard and felt most easily.

an infant born in the sixth month of pregnancy, weighing only 340 g (12 oz). After spending over 4 months in an incubator, the child grew and developed normally.

Suppose you were suddenly handed an infant. Would you hold it so its head was on your right or left side? What stimulus do you think is most prominent to the late fetus? These questions relate to an experiment conducted to improve survival of LBW infants. Because the pregnant female's heartbeat may be the loudest and most regular sound heard by the late fetus, a recording of a heartbeat was played 24 hours a day in one ward of LBW infants but not in another ward of matched infants. Infants in the former ward gained weight faster, cried less, slept better, and breathed more deeply and regularly than those in the latter ward. Assisting the survival of LBW infants with such methods is now common, and some manufacturers make cuddly toys that play a heartbeat recording when turned on.

Imprinting is a form of quick learning, and although often visual, it may involve other senses. Could we become imprinted to our mother's heartbeat during pregnancy? Several studies have found that high percentages of both left-handed and right-handed mothers tend to hold babies close to their left sides. In contrast, inanimate packages are held on the right side about as often as on the left side. Next time you observe museum art, note how the mothers depicted in paintings and sculptures are holding their infants (Figure 6.17).

PHYSICAL CHANGES AFTER BIRTH

Many major physical changes occur after childbirth. The uterus shrinks to about one-twentieth its size at delivery, aided in this process by breast-feeding. Some of the protein in the uterine wall is broken down, absorbed, and eliminated in the urine, and some of the endometrium is discharged through the vagina. The abdominal wall shrinks within a few weeks, although some silky stretch marks may remain. Exercise helps the abdomen return to its nonpregnant state. Postdelivery weight decline is due mainly to fluid loss and to an increased rate of urination.

The vagina may not return completely to its nonpregnant size, and the labia may be looser than before childbirth. The genitals must be kept clean to avoid postbirth infections. Although low estrogen levels sometimes reduce the sex drive, intercourse may usually be resumed safely within three to six weeks. Because of blood loss, anemia is common but usually mild. A proper diet, including adequate iron, is as important postnatally as it is prenatally.

In women who do not nurse, menstruation resumes in about eight weeks; it often does not occur while a woman breast-feeds her infant. There is great variation among women who nurse, however, and menstruation may resume as early as two months later or, most commonly, at about four months. Although some women ovulate while nursing, most do not.

See the bioperspective essay "Altricial and Precocial" for a discussion about the possibility that humans are born too early.

LACTATION

The female breasts or **mammary glands** begin to enlarge at about age 10 years (Figure 5.11). Estrogen increases the connective tissue and ducts, and progesterone causes growth of the glandular structures. After menarche, the breasts enlarge and shrink slightly during each menstrual

Altricial and Precocial

Some mammals give birth to numerous **altricial** young that are small, poorly developed, hairless, and helpless (Figure B6.1). The **precocial** young of other mammals are relatively well developed at birth, and can care for themselves to a significant degree. Primates (e.g., apes, monkeys, chimpanzees, and gorillas) are good examples of mammals with precocial young, with one notable exception—humans. Like our primate relatives (Chapter 24), humans produce few young and these offspring have large brains and relatively long life spans. Could it be, as some have proposed, that human neonates are so altricial because they are born too soon, as essentially young fetuses?

Humans have a lengthy pregnancy, or gestation period—the longest among the primates. However, human brains and bones, for example, develop relatively slowly, and both our childhood and our life spans are comparatively long. The young of many primates are born with bones ossified to a degree not reached by human infants until years after birth. While the brains of many mammal young are fully or largely formed at birth, human brains continue to grow postnatally at rapid, fetal rates. At birth, the brain of a human neonate is less than one-fourth its ultimate size. Even by the age of 3 years, the brain is less than three-fourths of its ultimate size.

While the human brain is disproportionately large, there is a finite size to the female pelvis. It has been suggested that we are born "early" in order to allow our brain the room it needs to develop fully. About one-third of our life span is devoted to growing, and the attainment of sexual maturity occurs relatively late. Our lengthy period of attachment to parents and others increases our learning period and the time available for the transference of culture. Thus, our extended childhood is an important aspect of human social evolution.

FIGURE B6.1
Altricial young of (A) mice and, even more so, of (B) opposums are very helpless and poorly developed at birth. They are small and blind, and, being hairless, have poor thermoregulatory abilities. In contrast are the young of precocial animals, such as these (C) killdeer and (D) white-tailed deer, which are able to walk and function well soon after being hatched or born, respectively.

cycle and may become quite sensitive. Although external breast size differs greatly from one female to another due to differences in the amount of adipose tissue, the amount of glandular tissue is fairly uniform.

During pregnancy the breast size increases markedly and the lobes grow and become more complex. Oil glands in the darkening areolae enlarge, and their secretions keep the nipples supple. The hormones that enlarge the pregnant female's breasts may do the same to the fetal breasts, including those of males. The neonate may even secrete some milk until the hormone effects wear off in a few days and the breasts subside.

When estrogen and progesterone levels drop during birth, milk secretion and release, **lactation,** is no longer inhibited. Lactation is stimulated by so-called **lactotropic hormones** (e.g., **prolactin**) from the anterior pituitary (Chapter 18). The secretion during the first few days after birth is a watery, yellowish-white fluid called **colostrum.** Colostrum is similar to milk but contains numerous white blood cells and little or no fat. Its laxative effect helps clear mucus and debris from the infant's digestive tract.

Although true lactation begins about two to four days after birth, milk remains within the glands and sinuses until the breast is suckled. The **letdown reflex** is triggered by suckling at the nipples, and the resulting nerve impulses cause the hypothalamus to release hormones that cause contraction of the small muscles that surround the milk lobules. The baby's sucking reflex is fully developed at birth, and the more milk is removed, the more hormones are released and the more milk is secreted. Once started, milk secretion may continue for months, and in some human populations children are breast-fed for several years. Milk production averages 1.0–1.8 L (1–2 qt) per day.

Although some antibodies from the mother move across the placenta before birth, colostrum and milk also contain them. These protective substances may confer short-term immunity to the infant against such infectious diseases as measles, scarlet fever, and diphtheria. Breast-fed infants usually suffer fewer gastrointestinal, respiratory, and urinary tract infections than formula-fed infants.

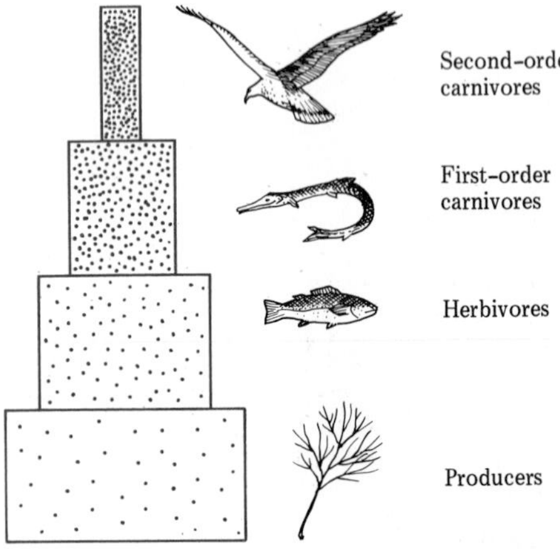

The high level of prolactin during nursing usually, but not always, inhibits the menstrual cycle. This inhibition is short-lived, however, and nursing mothers are not immune from pregnancy. If the mother does not nurse, milk production stops within one to two weeks and any stored milk is reabsorbed.

Multinational corporations often advertise their products worldwide. Thus, baby formulas may be promoted as much in developing as in developed countries. Unfortunately, parents who can ill afford the cost sometimes resort to artificial milk, believing it is better for the baby because of "prestige" (e.g., such behavior mimics that of more affluent parents). However, bottle-feeding not only removes a natural contraceptive effect and a source of useful antibodies but may also increase exposure to disease. Open formulas must be refrigerated to avoid spoilage, and refrigeration is often scarce in developing nations. Also, many parents are not aware of the sanitation needed to keep bottles and nipples safe and clean.

Breast-feeding is sometimes dangerous if milk contains a contaminant such as a pesticide. This is the basis of an ecological poster showing an infant at its mother's breast and the caption "Danger—Keep Out of Reach of Young Children." The problem is a good example of **bioconcentration.** Small amounts of pesticides, radioactive substances, heavy

Second-order carnivores

First-order carnivores

Herbivores

Producers

FIGURE 6.18
Bioconcentration is illustrated in this example of a simple food chain. Pollutants, especially lipid-soluble forms, that enter plants from water, soil, or air may become more highly concentrated in the bodies of the herbivores that eat the plants. In turn, the concentrations may become even higher in the herbivore-eating carnivores. A fetus or neonate is at a higher feeding level than a pregnant or lactating female, and the concentration of some pollutants in its body may reflect this.

metals, or other harmful contaminants in the environment may become part of plant tissues. The concentrations become higher in the bodies of herbivores who consume large amounts of plant material, and still higher in carnivores who eat other animals to survive (Figure 6.18). Similarly, the concentration of a contaminant may become higher in a fetus or in a breast-feeding neonate than in the pregnant or lactating female, respectively.

MULTIPLE BIRTHS

Twins may be **identical** (**monozygotic**) or **fraternal** (**dizygotic**). Monozygotic twins develop from one egg fertilized by one sperm; fraternal individuals develop when two or more eggs are each fertilized by a different sperm (Figure 6.19). In the United States, twins occur in about 1 of 85 and triplets in about 1 of 7200 pregnancies. Fraternal twins are about two to three times more common than identical twins, depending

FIGURE 6.19
Formation of (A) identical and (B) fraternal twins. Monozygotic twins have separate amnions, one chorion, and a common placenta. Dizygotic twins have two amnions and two chorions, and the placentas may be separate or fused. Thus, dizygotic twins not only have different genetics but their prenatal environments may differ somewhat.

(A)

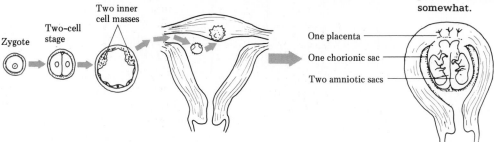

(B)

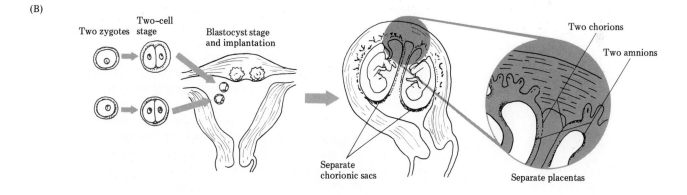

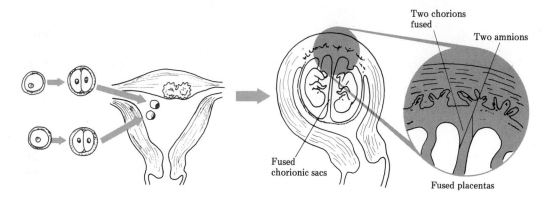

FIGURE 6.20
The relationship between the age and childbearing experience of the mother and the incidence of monozygotic and dizygotic twins. Note that the rates of monozygotic twinning are about equal in all age categories. The age of the father has no influence on rates of twinning.

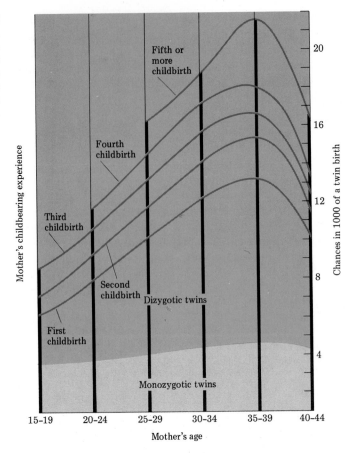

upon the nature of the subculture. Quadruplets, quintuplets, and sextuplets may be either monozygotic, multizygotic, or a mixture of the two.

Three important factors that affect the chance of having multiple births are the age, race, and childbearing experience of the mother (Figure 6.20). Although the incidence of monozygotic twins is similar in all races, Negroid woman have the highest rates of dizygotic twinning, Mongoloid women the lowest, and Caucasoid women intermediate. Once a woman has produced multiple births, the probability of having subsequent multiple births is about 10 times the average chance. Females who have had fraternal twins are more likely than average to have identical twins. One reason for the increase in the incidence of multiple births may be the increased use of FSH-containing fertility drugs.

SUMMARY

1. **Fertilization** generally occurs in an upper **oviduct** within a day following ovulation. Although exhaustion and the effects of acidic vaginal secretions destroy many sperm, peristaltic movements of the uterus help propel other sperm along.

2. **Ectopic pregnancies** occur when **zygotes** implant on structures that cannot support proper development. Serious hemorrhaging and occasional death occur when the inappropriate structure ruptures.

3. Eggs are enormous compared with tiny sperm, and they are surrounded by many **nurse cells.** Enzymes released from the acrosomes of sperm digest the egg's follicle cells, and usually only one sperm fertilizes the egg. The egg's second meiotic division occurs after fertilization, and within a few hours the genetic blueprints from the two gametes unite.

4. **Cleavage** occurs within about 15–30 hours; subsequent cell divisions lead to the solid **morula stage** (16 cells) about three days later. Growth of the **embryo** and its development into the **blastocyst stage** occur only after it reaches the **uterus** about four days after **conception.** **Implantation** occurs about six to nine days following fertilization, usually on an upper surface of the uterine walls. Early **miscarriages** or **spontaneous abortions** often occur without the female's knowledge.

5. The term **embryo** is used for about the first two months of development, and the term **fetus** is used thereafter, following the time when **bone** begins to replace **cartilage.** All basic organs and systems develop during the embryo stage (e.g., the heart beats and the gonads and nervous system begin to form during the first month). At the beginning of the fetal stage, the tailed human is about 50 percent head.

6. The three **embryonic germ layers** develop by the end of the second week. The outer epithelium, the nervous system, and some other internal structures develop from the **ectoderm.** The **mesoderm** gives rise to muscle, bone, the peritoneum, the circulatory system, and most of the reproductive and excretory systems. The linings of many internal structures (e.g., much of the digestive and respiratory tracts) and the liver, pancreas, bladder, and some endocrine glands develop from the **endoderm.**

7. Exchanges of nutrients, gases, and wastes between the developing human and the pregnant female occur through the large surface area of the **placenta** via the **umbilical cord.** The placenta produces **estrogen** and **progesterone,** which are needed to continue **pregnancy** after the ovary's **corpus luteum** degenerates, and it also secretes growth- and thyroid-stimulating hormones.

8. Most substances that enter the body of a pregnant female can also enter the embryo or fetus; some of these may be **teratogenic** (causing birth defects), **mutagenic** (altering genetic material), or **carcinogenic** (producing cancer). Some agents (e.g., CO) slow development and result in **low-birth-weight infants** (<2.5 kg). The specific effect of an agent depends in part upon the developmental stage of the embryo or fetus, and in general, substances are most harmful in the early stages of embryonic development, while structures are still actively forming.

9. Much of fetal development consists of the refinement and growth of the basic organs and systems that formed earlier. Development is often divided into **trimesters;** during the last trimester the fetus grows to occupy most of the **amniotic cavity.** Both the **cephalocaudal** (head-to-tail) and **proximodistal** (middle-to-peripheral) **developmental directions** are very evident.

10. From the usual cessation of menstrual cycles following conception to the extended abdomen later, many physical changes occur in the pregnant female. The stretching of the uterus and other body changes sometimes cause "morning sickness." The breasts grow, thirst increases, and edema is common. Because the pregnant female is performing most physiological functions for two individuals, diet is especially important. Typical minimum weight gains during pregnancy equal about 12 kilograms.

11. Trillions of cells develop from the zygote during pregnancy, which usually lasts 280 days or so. Estrogen, oxytocin, and some prostaglandin levels increase during **labor,** and progesterone levels decrease. The first part of labor includes the dilation of the **cervix,** and the second stage involves **delivery.** The **afterbirth** is usually expelled soon after delivery.

12. **Neonates** must breathe soon after delivery, and breathing occurs because of both mechanical and chemical events and stimuli (e.g., buildup of CO_2 in the blood). The opening between the upper heart chambers that allows most blood to bypass the fetal lungs usually closes quickly after birth; a **blue baby** results if the opening remains.

13. Low birth weight ("prematurity") is a major cause of neonatal death. Among the factors that increase the incidence of LBW infants are heavy smoking, alcoholism, poor prenatal nutrition, lack of medical care, infections, diabetes, hypertension, other metabolic problems, abnormal placentas and fetuses, and too rapid a succession of pregnancies. LBW **singletons** survive less well than LBW **multiple-birth** infants of the same weight. Incubators, the use of the mother's milk, and close mother-infant contact (especially with the infant's head on the left side of the mother's chest) aid survival of LBW infants.

14. Mammalian young that are small, hairless, helpless, and poorly developed at birth are called **altricial,** whereas those that are relatively well developed and can care for themselves to a significant degree are called **precocial.** Unlike most other primates, humans produce essentially altricial young, which according to some are really only young fetuses. About one-third of our comparatively long life span is spent in childhood, which allows for the lengthy learning period necessary for us to utilize the potential of our disproportionately large brain.

15. After birth, the uterus and abdominal wall shrink quickly, various changes occur in the uterine wall and endometrium, and some weight loss occurs from increased urination and loss of tissue fluid. Some structures (e.g., the vagina and labia) usually do not return completely to their prepregnant conditions. Menstruation resumes most quickly in women who do not **breast-feed** their infants.

16. Both estrogen and progesterone stimulate the growth and development of the **mammary glands.** The amount of glandular tissue is fairly uniform from female to female irrespective of external breast size, which is due largely to the quantity of adipose tissue. **Lactotropic hormones** from the hypothalamus affect the letdown reflex. In general, breast-fed infants grow and develop better than formula-fed infants.

17. **Bioconcentration** is the phenomenon by which certain contaminants become more concentrated with each transfer in a food chain. Thus, a pollutant that enters a plant may reach a higher concentration in herbivores and a still higher level in carnivores. The concentration of a contaminant may be higher in a breast-feeding infant than in its lactating mother.

18. Because **monozygotic twins** develop from one egg fertilized by one sperm, they are popularly called **identical twins.** The more common **dizygotic twins** that result from two eggs that are each fertilized by a different sperm are called **fraternal.** Other **multiple births** may be monozygotic, multizygotic, or a mixture of the two. The probability of dizygotic twinning increases with the age of the mother and with childbearing experience. It is also higher in some races (e.g., Negroid) than in others (e.g., Mongoloid).

REVIEW QUESTIONS

1. State when and where eggs are generally fertilized.

2. Explain why so many sperm are necessary for a successful fertilization.

3. Define *ectopic pregnancy* and explain why the condition is potentially life-threatening.

4. Explain the events during the process of fertilization.

5. Define the terms *cleavage, morula,* and *blastocyst,* and state where in the female reproductive tract each stage occurs.

6. Describe the process of implantation.

7. What events lead to spontaneous abortions and stillbirths?

8. Distinguish between an embryo and a fetus. What major developmental events occur during the embryo stage?

9. Describe the structures and functions of the placenta and umbilical cord.

10. Name the three embryonic germ layers and the structures that develop from each.

11. Distinguish among teratogens, mutagens, and carcinogens.

12. How does the stage of embryonic development influence the effect of an environmental agent?

13. Define *low birth weight* and state some of the common causes.

14. Describe the major events that occur during fetal development.

15. Define the *cephalocaudal* and *proximodistal developmental directions,* and state examples of each.

16. Describe the major changes that occur in the organs and systems of the pregnant female.

17. Describe the three stages of labor.

18. Define a *blue baby* and describe the events that generally occur to prevent the development of this condition.

19. Define *imprinting* and describe a possible human example.

20. Distinguish between precocial and altricial young.

21. Describe the major causes and events associated with lactation and the letdown reflex.

22. Explain how lactation influences the resumption of menstrual cycles.

23. Define and explain *bioconcentration,* and state a human example.

24. Distinguish between monozygotic and dizygotic twins, and explain how other types of multiple births may occur.

25. Describe the major variables that affect the incidence of dizygotic twinning.

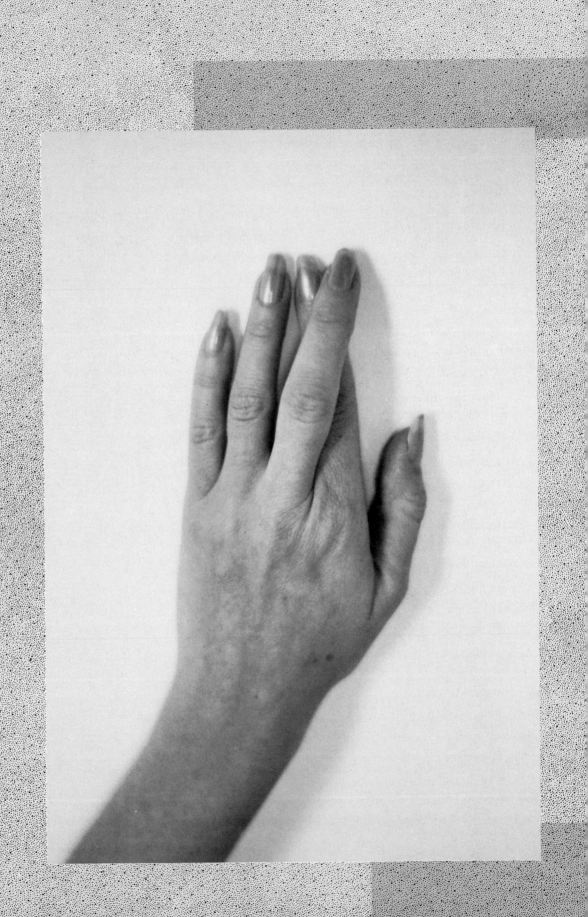

CHAPTER 7

CONTRACEPTION

For many humans, pregnancy and reproduction are largely matters of chance. Others avoid parenthood by practicing contraception—the use of methods and materials that interfere with fertilization or implantation—or by abstinence. Perhaps less than half of Earth's human population has access to modern contraceptive and family-planning information. Folklore methods are relatively ineffective, and for many humans who give thought to birth control the method shown in the chapter opener may be the most common.

Copulation is associated solely with reproduction in many animal species. In humans, reliance on "hope" or "luck" for contraception is a risky form of reproductive roulette.

BIRTH CONTROL

Family planning is an issue of great personal as well as societal concern. The personal concern is important not only to potential parents but also to the child. Although every child should be wanted, perhaps a million unintended and often unwanted children are born annually in the United States. Often, unwanted pregnancies occur in young teenagers or women over 35, ages when the health risks of pregnancy are relatively high.

Why do some couples decide not to have children? Some cite age, future plans, the relational status, the high cost of raising children, the reduced personal freedom that results from parenthood, or the chances that childbearing might adversely affect their physical and/or mental health. Some people label such couples as selfish, if not immoral.

Earth's human population now exceeds 5 billion and it is increasing geometrically (Figure 7.1). The human population reached 1 billion in about 1830 and it took a century to add the second billion. Only 30 years were needed to add the third billion, and less than 15 more years to reach the 4-billion level. At current rates of reproduction, Earth's population could increase by another billion in less than a decade. Human dignity and even human survival are threatened by irresponsible or unplanned reproduction. Chronic hunger and even starvation threaten many millions of humans on Spaceship Earth. Also, the term "popullution" is used by some to describe the relationship between our growing population and the increasing concentrations of environmental contaminants. Numerous other biological and sociological problems are related directly to the explosive growth of the human population. Americans make up about 5 percent of Earth's population, yet we use about half of many of Earth's nonrenewable resources. It is often said that the "burden" of an American child on Earth is many times greater than that of a child born in a developing nation. "The rich get richer and the poor get poorer" is a common phrase, especially when discussing population issues. Over three-fourths of population growth occurs in developing nations.

The method shown in the chapter opener, or no method, works for perhaps 10 percent of American couples, either because one or both are sterile or because they are biochemically incompatible. An example of the latter occurs when the chemistry of the uterus prevents implantation. Otherwise, abstinence from sexual intercourse and sterilization are the only sure ways of avoiding pregnancy.

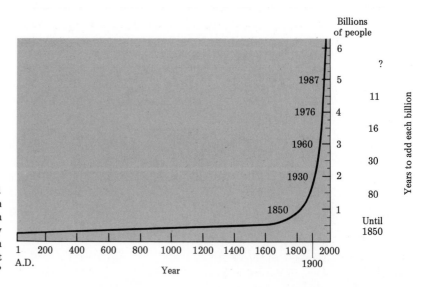

FIGURE 7.1
The growth of the human population, and a graphical representation of the "population explosion." At about what time in history did the J-curve of the population explosion begin? If past trends continue, what might the world population be in the year 2000?

The two most common concerns about contraception relate to effectiveness and safety. No one contraceptive method is always safest or best.

METHODS THAT DESTROY SPERM OR PREVENT THEIR ENTRY

Several varieties of modern contraceptive methods have long been used to prevent the entry of sperm. In numerous cultures, for example, objects such as cloth, cotton, sponges, or plant material have been used to block the cervix or to absorb sperm. The contraceptive failure of mechanical devices may occur from breaks or improper use (e.g., when they are used too early or too late).

Coitus Interruptus

A male using **coitus interruptus** attempts to withdraw his penis from the vagina prior to ejaculation. This ancient and seemingly simple method, referred to in the Old Testament, demands practice, self-control, and high motivation. The high failure rate occurs from the release of semen before ejaculation, from delay in withdrawal, and from semen deposited near the vaginal opening. Even so, the method is superior to using no contraception when nothing else is available.

Condoms

Condoms are cylindrical sheaths of rubber, plastic, or animal membranes that envelop the erect penis (Figure 7.2). The closed end is either plain or tipped with a reservoir end to catch semen. "Rubbers," as they are commonly called, are simple, inexpensive, widely available without prescription, and about 80–90 percent effective in preventing the release of sperm to the vagina when used properly and consistently. Most failures result from tears, from the condom being put on too shortly before ejaculation (e.g., after some preejaculatory fluid containing live sperm has already entered the vagina), from the condom slipping off during coitus, or from leakage when the penis becomes flaccid or when the condom is removed. Condoms of various colors, designs, textures, and thicknesses are available for psychological reasons. Lubricants are sometimes needed when condoms are used, and prelubricated types, some combined with a spermicide, are available.

The condom is also an effective **prophylaxis** or disease-preventing agent against sexually transmitted diseases and infections. In fact, condoms were invented centuries ago for this purpose, not as contraceptives.

Diaphragms and Cervical Caps

Like condoms, **diaphragms** act as barriers to prevent sperm from reaching the egg. Diaphragms are shallow, cup-shaped rubber domes with strong but flexible rims that cover the cervix (Figure 7.3). Because vaginal and cervical dimensions vary, diaphragms must be fitted properly, and as the body changes due to age, pregnancy, and/or weight gain or loss, a refitting may be necessary. Some authorities recommend an annual examination and possible refitting. Diaphragms should be inserted carefully, before any semen is deposited in the vagina but generally not longer than two hours before intercourse, and should usually be left in place for six or more hours after coitus. Coitus during menstruation is facilitated by the use of a diaphragm, which acts as a "reverse barrier."

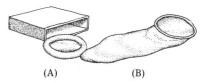

(A)   (B)

FIGURE 7.2
Condoms, (A) rolled and (B) unrolled. Condoms commonly serve both contraceptive and prophylactic functions.

FIGURE 7.3
The diaphragm and its insertion. Cervical
caps are smaller than diaphragms and cover
only the immediate region of the cervical
opening. Contraceptive sponges also fit into
the cervical region.

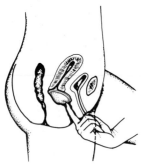

The effectiveness of diaphragms, lower than that of birth control pills or intrauterine devices, is improved greatly when they are used with spermicides (see below).

Plastic or metal **cervical caps** are smaller than diaphragms, fit tightly over the cervical opening, and stay in place by suction. A cap may be used briefly or left in place for weeks, sometimes being removed only for menstruation. The effectiveness is not well known but it appears to be lower than that for the diaphragm.

Spermicides

Many **spermicides** kill or immobilize sperm and help to block the cervical opening. Spermicides may also provide some protection against sexually transmitted infections. Spermicides are widely available in the form of creams, jellies, aerosols, tablets, suppositories, or foams. Although spermicides may be used alone, they are more often used with mechanical devices. Some brands of spermicides irritate the genitals, and other potential disadvantages include the leakage of fluid from the vagina, excessive vaginal lubrication, and the time required for some chemicals to dissolve. Feminine hygiene products have no spermicidal effectiveness.

Contraceptive Sponges

Contraceptive sponges were approved by the U.S. Food and Drug Administration in the early 1980s. These small, soft, disposable, nonprescription devices, which contain a spermicide, kill, slow, and absorb sperm and help block the cervical canal. The sponge is inserted into the vagina prior to intercourse and is removed by pulling a small ribbon attached to one of its sides. Its apparent effectiveness is about the same as the diaphragm but it can be inserted earlier and retained longer.

Postcoital Douche

Douching (from the French for "wash") with water or a spermicidal solution after coitus has a low effectiveness. Many sperm will have entered the uterus by the time the douche is used, and douching may force sperm further into the reproductive tract.

Sterilization

Sterilization operations cut or block the tubes through which gametes must pass and, if successful, are 100 percent effective. While the permanence of sterilization is what appeals to many, this can also be a drawback, as when a couple later separates or a child or mate dies. In general, it is easier to reverse male than female sterilization procedures. One alternative to reversal operations in males is to have semen samples frozen and stored before the surgery, for possible use later.

Males.   The sterility surgery in the male is known as a **vasectomy** (Figure 7.4A). Local anesthesia is adequate, and usually the cutting of each vas deferens is accomplished quickly in a clinic or doctor's office, and postvasectomy problems are usually minimal. Because many sperm are still in his system, a male is not sterile immediately after a vasectomy. Semen samples are usually examined for several weeks to determine when sperm are absent. Sperm are still produced in the testes but they are phagocytized or they disintegrate and are reabsorbed. Seminal fluid is produced and ejaculated as usual, and vasectomy does not affect sex

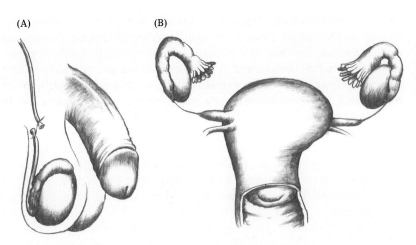

FIGURE 7.4
(A) Sterilization in males—vasectomy—involves cutting the vasa deferentia. Sperm are still produced but they cannot be released during ejaculation, although seminal fluids are. (B) A common sterilization procedure in females, salpingectomy, involves cutting the oviducts. The same result can be achieved by blocking or tying the fallopian tubes. In neither sex do these procedures remove the sources of sex hormones. However, castration, hysterectomy, and ovariectomy not only cause sterility but also eliminate the main sex hormones.

hormone production. About 10 percent of American couples in the United States rely on vasectomy for contraception.

Vasectomy is not the same as **castration**, when the entire testes are severed surgically or accidentally. In castration, both the sources of sperm and the sex hormones are removed; castrated males are called **eunuchs.** If castration occurs before puberty, the limbs grow relatively long, fat deposits form as in females, most secondary sex traits do not develop, the libido is commonly low or absent, and the male is sterile. As late as the eighteenth century, some European choirboys were castrated in order to maintain their high-pitched singing voices into adulthood. Interest in singing to this degree has declined significantly during the past century! If an adult male is castrated, the libido often remains even though no sperm are produced.

Females.   Most of the many female sterilization procedures involve blocking (e.g., by cauterization), cutting, tying, or clipping the fallopian tubes (Figure 7.4B). Formerly, these operations were expensive and often involved major surgery under a general anesthesia and hospitalization. Newer **salpingectomy** methods require only small incisions in the lower abdomen, in the naval, or in the wall of the vagina or uterus, and these relatively inexpensive "Band-Aid" operations usually require no hospitalization. A small tubelike optical instrument called a laparoscope is inserted through the opening to perform the surgery. These procedures cause no change in libido or production of eggs or sex hormones (except when the ovaries are removed).

A **hysterectomy** is the surgical removal of the uterus, and an **ovariectomy** removes the ovaries. Although these operations are rarely performed for contraceptive purposes only, sterility is obviously the result.

METHODS THAT AVOID OR SUPPRESS OVULATION

Rhythm

The **rhythm method** involves periodic abstinence from coitus when an egg is thought to be present and fertilizable. In women who have a regular and "perfect" 28-day cycle, coitus is usually avoided from about days 10 through 17. The other days in the cycle are considered "safe" for coitus. However, ovulation does not always occur on day 14 but may occur, for example, as early as day 12 or as late as day 16. The method must also consider the fact that sperm may survive within the female tract for several days.

Many women have longer, shorter, or variable cycles. For these women as well as for those with regular cycles, the minor temperature decline around the time of ovulation may be useful in predicting ovulation (Figure 7.5). Intercourse is often considered safe two to four days after the postovulation temperature rise. Temperatures are usually taken immediately after waking in the morning, when the body is at a low or basal level of activity.

The rhythm method requires no unusual materials, produces no side effects, and is sanctioned by the Roman Catholic church because it is considered to be "natural." Its disadvantages are the careful calculations, the high motivation required, and the periods of abstinence. Its failure rate is often relatively high, and users of the method have sometimes been defined jokingly as "parents." However, much research is underway to improve its success, including studies of changes in electrical charges in the body and changes in the quantity and quality of cervical mucus, the Billings method. The cervical mucus usually becomes clear and more abundant at or near the time of ovulation. Intercourse is usually considered safe a few days later when the mucus returns to a cloudy color and has a tacky consistency.

"The Pill"

Oral contraceptives were developed in the 1950s after sex hormones were synthesized; such contraceptives became generally available around 1960. Most birth control pills contain various combinations of natural or synthetic estrogen and a synthetic progesterone-like hormone called progestogen.

The so-called combination pills are taken for 20–21 days after menstruation. Manufacturers also market 28-pill packs that contain seven placebos, or inactive pills, so that women get used to taking a pill daily. The period usually starts one to three days after the last "active" pill is taken. Thus, the pill establishes a 28-day menstrual cycle by substituting hormones for those normally produced by the ovary. Both hormones act upon the pituitary (Chapter 5). Estrogen inhibits follicle-stimulating hormone (FSH), and progestogen both inhibits luteinizing hormone (LH) and causes the cervical mucus to thicken, making it hard for sperm to penetrate. The pill often reduces menstrual discomfort in women who experience such problems. Used properly, birth control pills are over 99 percent effective.

In the early 1970s, a pill became available that contained a small dose of progestogen but no estrogen. Unlike the combination pills, these "minipills" do not always prevent ovulation, which leads to a higher failure rate, but their side effects are reduced. Besides thickening the

FIGURE 7.5
(A) Body temperature tends to decline slightly around the time of ovulation and to increase relatively quickly shortly thereafter. Unfortunately, these changes average less than 1 Celsius degree and are often difficult to distinguish from body temperature changes caused by other conditions, such as activity, emotions, and infections. (B) These are strands of cervical mucus as seen through an electron microscope. Changes in the color and texture of cervical mucus can also aid in identifying the time of ovulation.

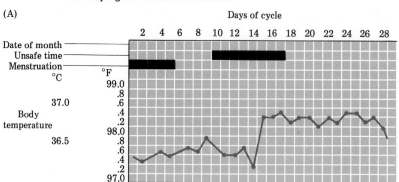

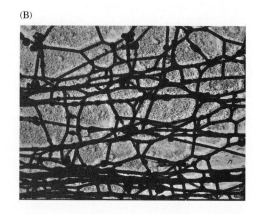

cervical mucus, the minipills make the uterine lining less receptive to implantation.

Some common and bothersome side effects are similar to the bodily changes that occur during pregnancy. For example, some women starting to use the pill experience nausea, vomiting, breast tenderness, irritability, fluid retention, weight gain, minor increases in blood pressure, headache, constipation, and skin rashes. Oral contraceptives also tend to increase the volume of vaginal discharges and increase the susceptibility to vaginal infections (vaginitis), just as pregnancy does. However, menstruation in pill users is often lighter, shorter, and more regular than in non-pill users, and menstrual cramping is often lessened.

Almost 200 million women worldwide have used the pill, and oral contraceptives are among the most studied medications in history. However, the exact evaluation of risks is complex because most problems associated with the pill also occur among non-pill users. Also, the hormone composition of pills has changed over the years since their introduction. For example, the early "sequential pills" were banned because they were associated with uterine cancer. Certain cancers take decades to develop, so it is not known whether any currently used pills will cause cancer, although it appears that they actually decrease the incidence of ovarian and endometrial cancers. Pill users also develop fewer benign or noncancerous breast tumors (one-fourth), ovarian cysts (one-fourteenth), rheumatoid arthritis (one-half), pelvic inflammatory disease (one-half), and iron deficiency anemia (two-thirds) than non-pill users. The latter may be due partly to reduced menstruation.

One potentially fatal side effect in susceptible women is the formation of blood clots. A **thrombus** is an internal blood clot that remains at the site where it forms. A loosened blood clot carried to another site, where it blocks circulation, is called an **embolus.** The combination of these two circulatory disorders is often called thromboembolic disease. Blood clots in the lungs or brain (stroke) and heart attacks may also occur among certain pill users.

Many physicians suggest that women over 35–40 years old and those who smoke heavily choose a form of contraception other than the pill. Similar advice is often given to diabetic and obese women and to those with circulatory, kidney, liver, hypertension, or other serious health problems. Information about family histories of these problems is also useful to physicians in judging the potential risk of using the pill. Pill use during pregnancy may lead to birth defects but prepregnancy pill use apparently does not.

It is likely that some sex hormones in the pill are passed to infants by nursing mothers, but this is not thought to be a serious problem. However, progestogens may cause masculinization of female fetuses, and estrogen may inhibit bone growth in infants of both sexes.

Experimentation with different brands of the pill may be necessary to find the best combination of estrogen and progestogen for each individual, and supervision by a physician is highly desirable. Some side effects are partly psychogenic, as with other medications—that is, symptoms occur mainly in patients who have been told to expect them.

In theory, the pill should be completely effective in preventing pregnancy, and pregnancies among pill users are due more often to misuse than to failure of the contraceptive. No medication is 100 percent safe, and the side effects and/or failure of the pill must be weighed against the chance of dying from complications of pregnancy or childbirth. For the vast majority of women the risk of injury or death from using the pill is far less than that associated with a full-term pregnancy.

(A)

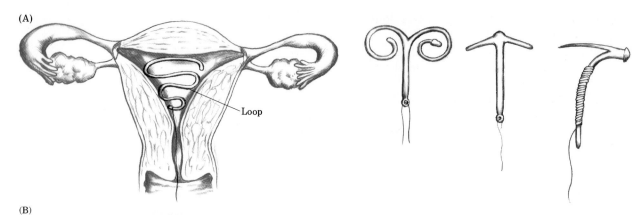

Loop

(B)

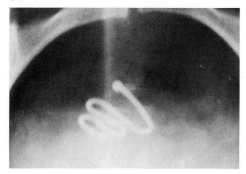

FIGURE 7.6

(A) Intrauterine devices come in various sizes, shapes, and compositions, and they are placed into the uterus for short to long periods of time. Strings that extend into the upper vagina aid the user in detecting loss or other problems. (B) An IUD shown in place in an X-ray photo. How the IUD works is unknown; one theory suggests that it may prevent implantation.

METHODS THAT PREVENT IMPLANTATION

Intrauterine Devices

Intrauterine devices (IUDs) are small devices of various sizes, shapes, and compositions (e.g., some contain copper filaments as well as plastic, and others contain a synthetic form of progesterone) (Figure 7.6). IUDs are inserted into the uterus through the vagina and cervix, and are generally left in place except when pregnancy occurs. Although some human IUDs were developed early in the twentieth century, their use has become common only since the early 1960s. The concept is related to the ancient practice of camel drivers placing large pebbles into the uteruses of their camels to prevent pregnancies during long desert caravan trips. Most modern IUDs are made of plastic or metal, and often include threads that extend into the vagina so the user can check on the device.

IUDs are usually inserted by a professional, usually during a menstrual period (because this generally means that a women is not pregnant) and/ or after a negative pregnancy test. The woman is usually also tested for various infections before insertion in order to reduce the chance of spreading infections into the uterus or other structures.

It is not known for certain how IUDs prevent pregnancy, but some suggested possibilities are that they (1) change the way sperm move through the uterus; (2) prevent implantation of the embryo; (3) increase contractions in the fallopian tubes and uterus, causing embryos to move too quickly and be expelled through the vagina; (4) change the condition of the uterine endometrium; and/or (5) cause the uterine wall to release certain white blood cells (Chapter 16) that devour sperm or the embryo. IUDs sometimes cause discomfort, cramps, and/or bleeding, and expulsion is also a problem for some women (especially young women and women who have had no children). Some expulsions go unnoticed (perhaps 20 percent), and this accounts for about one-third of the pregnancies among IUD users. Experimentation is often necessary to find the best size and shape, and IUDs are most successful after the first pregnancy.

There is no evidence that IUDs cause uterine cancer, and, in fact, the regular IUD checkup examinations may help to detect and treat cancer in an early stage. Rarely, the uterus is perforated by an IUD, usually because of faulty insertion or an improper type of IUD. **Pelvic inflammatory disease** (PID) and ectopic pregnancies are problems for some users. Although IUDs do not cause permanent sterility after removal, women who develop PID while using an IUD may be sterile if their fallopian tubes close as a result of the infection. Infections may enter the uterus during insertion or may "climb" the tail of the IUD. IUDs do not preclude the use of tampons during menstruation but they sometimes cause pain during intercourse.

To the individual, IUDs have several advantages over many other contraceptives. High motivation is not necessary, long-term protection is provided without the regular intake of chemicals, and its use is private and involves no precoital preparation. IUDs also have advantages in family-planning programs because they are simple and inexpensive and can be inserted by trained paraprofessionals. IUDs are not infallible, however, and the failure rate is about 3–6 percent per year.

"Morning After" Pills

Several types of high-estrogen **"morning after" pills** can be used shortly after coitus to prevent conception, as in cases of rape, leakage from a condom or diaphragm, or unprotected intercourse. These preparations are not safe for routine use, although the short-term effects on the user are usually limited to relatively minor nausea, breast tenderness, and changes in the menstrual cycle. Use of the pills later in pregnancy has been associated with abnormalities in the reproductive systems of children born to the user.

METHODS THAT PREVENT BIRTH IF PREGNANCY OCCURS

Abortion

Induced abortions, the intentional termination of pregnancies before the fetuses can live independently outside of the womb, have been a common method of birth control throughout history. **Spontaneous abortions** or **miscarriages** occur through natural causes (Chapter 6). Induced abortions are usually implied when the term *abortion* is used without qualification. Many spontaneous abortions are due to abnormal fetuses and/or placentas, and perhaps 10–20 percent of pregnancies end in miscarriages, usually within the first trimester.

Today, as from earliest times, opinions concerning abortions vary. Plato and Aristotle supported the practice, whereas Hippocrates opposed it. "Thou shalt give life to life" was part of Jewish law, and Roman law saw the fetus as part of a woman's body rather than as a separate human being. However, a woman who aborted against her husband's will was guilty of disobedience! In the United States, there were no laws against abortion in 1800 and by the mid-nineteenth century, abortion services were advertised widely in newspapers, religious magazines, and journals. However, by 1900, abortion was illegal everywhere in the country.

Theologians have long differed in their views of when the "soul" enters the fetus. St. Augustine distinguished between "formed" and "nonformed" fetuses, and did not consider abortion of the latter murder. St. Thomas thought that life began at the time of detectable movement ("quickening"), not at conception. Others argue that "life" begins at some other specific point, such as the time of the first heartbeat or the time when the fetus can survive after birth. The Roman Catholic church takes the position that the soul enters the zygote at conception; the church therefore considers performing abortion to be a sin at any time. Pope Pius IX issued a decree in 1869 that declared abortion sinful and banned the practice. It is very clear that "life" is a matter of definition, not a matter of fact.

Members of the class may respond anonymously to a questionnaire such as this one in order to judge the diversity of opinions present in the group. (Such a questionnaire may also be administered to many others of diverse backgrounds, ages, sexes, and religions in order to identify possible correlations with the opinions.)

Questionnaire on Beliefs About Abortion

Sex: Male   Female   (circle one)
Age:_____   Religion:_____

1. Is an induced abortion wrong under all circumstances?
   Yes   No   (circle one) Why?

2. Should abortion be strictly a medical matter, not a legal one, or should the law be involved in some way?

3. Check those circumstances below for which you think an abortion would be justified. (Naturally, skip this section if you answered "Yes" to Question 1.)

   _____ A. When the pregnancy resulted from rape.
   _____ B. When the pregnancy resulted from incest.
   _____ C. When continuation of the pregnancy would result in a child with major physical deformities.
   _____ D. When continuation of the pregnancy would result in a child with serious mental retardation.
   _____ E. When continuation of the pregnancy would threaten a woman's life or seriously impair her health.
   _____ F. When continuation of the pregnancy would threaten a woman's mental and emotional health.
   _____ G. When the pregnant female is unwed.
   _____ H. When the pregnancy was unplanned and unwanted.
   _____ I. When the family suffers from extreme poverty.
   _____ J. Other circumstances (specify):

The 1973 *Roe* v. *Wade* decision of the Supreme Court was based upon women's rights and concern over mortality and morbidity stemming from improper abortions. The essence of the 7–2 decision is as follows:

1. During the first trimester of the pregnancy, the abortion decision is between a woman and her physician.

2. During the second trimester, a state may not prohibit abortion but may regulate its practice in the interest of protecting the woman's health. (For example, states may regulate who may perform abortions and specify where they may be done.)

3. During the last trimester, the state may choose to protect the potential life of the fetus by prohibiting abortion except when necessary to preserve the life or health of the woman.

This policy may not mean that the rate of abortions is higher now than previously, but simply that illegal abortions are much less common. Rates of abortions in the United States are far lower (currently about 350 per 1000 live births) than in some countries, where they equal or exceed the birthrate. About two-thirds of the almost 1.3 million reported abortions in the United States annually are performed on women under age 25.

Several abortion techniques are available, depending largely upon the stage of pregnancy. **Suction currettage** is the most common method during the first trimester. Suction apparatus is used to remove the embryonic and placental materials, usually within a few minutes. Blood loss is small, and the risk of infection is low.

Later in pregnancy (after 12–13 weeks), the **dilatation and evacuation** (D&E) method is commonly used (Figure 7.7). (This is similar to the dilatation and currettage, or D&C, procedure that is used for various gynecological purposes.) This abortion method involves dilatation of the cervix, the vacuum aspiration of most of the contents of the uterus, and, often, the use of forceps and/or a currette (spoonlike instrument with a long handle) to remove any remaining embryonic materials. Local or general anesthesia is used, and sterile conditions are essential. Unless all embryonic and placental tissue is removed, infection or hemorrhage may result, and perforation of the uterine wall is also a risk because it becomes softer with pregnancy.

The technical difficulties of an abortion increase after about the twelfth week. Second-trimester abortions sometimes use chemicals such as prostaglandins to stimulate contraction of the uterus, which causes the expulsion of the fetus and associated tissues. Another method sometimes used at this time is **hysterotomy,** or removal of the fetus and placenta through an incision in the abdomen and uterus, much as in a caesarean section. (If sterilization is desired, it is usually easily and quickly performed at this time.) **Saline-induced abortions** involve the injection of a hypertonic salt solution into the uterine cavity through the abdominal wall. Like the prostaglandin-induced abortion, this causes labor and subsequent expulsion of the nonliving fetus and placenta through the vagina. One risk is the possibility of injecting salt solution

FIGURE 7.7
Two common methods of abortion: (A) suction currettage, (B) dilatation and evacuation.

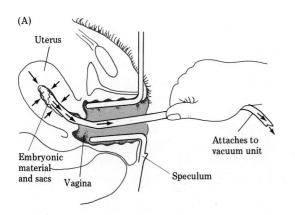

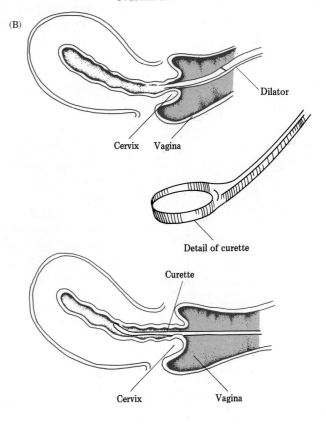

into the bloodstream, which can be fatal. Even when saline injection is successful, the high blood sodium level may cause headaches, nausea, numbed fingers, and sometimes coma. Occasionally, retention of the placenta and difficulties in delivery occur.

Under proper medical conditions, abortion during about the first 12 weeks is usually safe and uncomplicated, with a mortality rate less than that for tonsillectomies. Abortions after the first trimester become progressively more dangerous and complicated. Furthermore, abortion provides no permanent protection against pregnancy, except when sterility occurs as a side effect. Repeated abortions may increase the chance of ectopic pregnancies, predispose some women to miscarriages and/or premature labor, and increase the possibility of low-birth-weight infants.

Self-employed methods are very hazardous, as are those used by the "back-alley butchers." For example, deaths have occurred when untrained people have attempted to use the principles of D&E with inappropriate devices such as vacuum cleaners and coat hangers. It is far safer to have a baby than to abort a fetus without proper medical help.

Menstrual Extraction

Menstrual extraction methods were pioneered by the women's self-help movement. When a period is expected, a thin, flexible plastic tube is inserted into the uterus through the vagina and cervix, and suction (motor-powered or hand-operated) removes the endometrial lining. If the woman is pregnant, the tiny embryo is removed along with the tissue that would normally be lost in several days of menstruation. Current research indicates few major adverse health effects when this method is used professionally.

NEW AND FUTURE CONTRACEPTIVE DEVELOPMENTS

Refinements of current contraceptive methods as well as totally new approaches are possible. For example, major research on chemical contraceptives for males is under way. One problem is that there are fewer "links" or vulnerable points in the reproductive chain of events in males than in females. However, substances or agents (e.g., ultrasound treatment) may be perfected to suppress spermatogenesis without affecting libido or causing serious side effects, interfering with sperm maturation (e.g., acrosome development), or impairing sperm transport. As in females, any compounds may take the form of a pill, an injection, or an implant.

A vaccination that prevents pregnancy is an attractive idea to many people, especially to those who consider pregnancy to be a dreaded disease. One possibility is to develop a vaccine that disrupts the action of a critical hormone needed to maintain pregnancy (e.g., HCG). Another possible vaccine might affect the corpus luteum and block the release of progesterone and thus prevent implantation. Injections of progestogens at intervals are being tested, and implants of these "time capsules" might assure a proper and gradual release of synthetic hormones. Hormones might also be absorbed through the vaginal lining, such as from progestogen-containing plastic rings (similar to the rim of a diaphragm) or treated tampons. Intracervical devices that release substances to cause thickening of the cervical mucus are also being studied, as are ways to block the openings of the oviducts into the uterus non-surgically. The knowledge that sex hormones can be absorbed through the skin has long been used in the manufacture of cosmetics such as estrogenic face creams. Perhaps contraceptive hormone-containing, snugly fitting arm bracelets could be worn.

Chemical antagonists have great potential as contraceptives by providing "fake messages" that could affect the pituitary, the gonads, and/or the gametes. For example, a substance might "trick" an egg membrane into changing so as to prevent sperm penetration, as it does normally when it is fertilized.

SUMMARY

1. Unplanned pregnancies have major effects on many children, parents, and society in general. Earth's **human population** exceeds 5 billion and is growing geometrically. Hunger, starvation, and the concentrations of environmental contaminants could decrease with better family planning, and many of Earth's finite resources would be available longer. Unfortunately, **contraception** is poorly understood by many, and is often not used properly and consistently by those who do use it.

2. Various contraceptive methods are designed to destroy sperm or prevent sperm from reaching an egg. In using **coitus interruptus,** a male attempts to withdraw his penis from the vagina prior to ejaculation. **Condoms, diaphragms, cervical caps,** and **contraceptive sponges** are used to prevent sperm from entering the cervix. **Spermicides** may be used alone but are more effective when used with a mechanical device. **Post-coital douches** have very low effectiveness.

3. Like total **abstinence** from sexual intercourse, **sterilization** is 100 percent effective if performed properly. In males, each vas deferens is severed (**vasectomy**), and in females most procedures involve blocking, cutting, tying, or clipping the oviducts (**salpingectomy**). **Castration, hysterectomy,** and **ovariectomy** also cause sterility, although removal of the gonads also eliminates the major sources of sex hormones.

4. The **rhythm method** involves abstinence from sexual intercourse during times when conception is possible. Failure occurs primarily because of difficulty in predicting the exact time of ovulation. Knowing the pattern of past menstrual cycles, measuring small changes in body temperature, and observing the quantity and quality of the cervical mucus aid in estimating the time of ovulation.

5. Many **oral contraceptives** contain combinations of synthetic estrogen and progesterone, although minipills contain only progestogen. Birth control pills function by inhibiting FSH and thickening the cervical mucus. Most side effects are relatively minor, and resemble the bodily changes associated with pregnancy, but potentially serious **thromboembolic problems,** strokes, or heart attacks may occur in susceptible women.

6. **Intrauterine devices** are small plastic objects (sometimes also containing other substances) that are placed into the uterus. The mechanism(s) by which IUDs work is unknown but some common theories propose that they interfere with implantation, affect the movement of sperm and/or embryos, or stimulate phagocytic cells.

7. **"Morning after" pills** are not safe for routine use because of their high estrogen content and teratogenic potential. They are used primarily in cases of rape and, on occasion, in instances of unprotected intercourse or failed contraception.

8. **Spontaneous abortions** are miscarriages, whereas **induced abortions** are intentional. Abortions are now legal under all circumstances during the first trimester, but states may regulate abortions performed later. The abortion method that is most appropriate depends largely upon the stage of pregnancy. Thus, **vacuum currettage** and **D&E methods** are used in earlier stages than prostaglandin- or **saline-induced abortions** or **hysterotomies.**

9. Somewhat like D&E procedures, **menstrual extraction** uses suction to quickly remove the endometrium and any embryo present at the time of menstruation.

10. Future developments in contraception will include refinements of current methods as well as totally new techniques. Vaccines and the use of implants of synthetic hormones in both sexes are two likely possibilities.

REVIEW QUESTIONS

1. Define the phrase *population explosion* and describe the factors that contribute to it. Historically, when did the population explosion begin?

2. Describe the major factors that lead to failure or reduced effectiveness of coitus interruptus, condoms, diaphragms, cervical caps, contraceptive sponges, spermicides, and postcoital douching.

3. Describe the major methods of sterilization in males and females.

4. Describe the disadvantages of removing the gonads as a way to achieve sterility.

5. Describe ways that sterility procedures might affect people's attitudes about their sexuality.

6. Discuss why someone might desire to reverse a sterility procedure. Describe some alternative methods to reversal operations.

7. Are there reasons against legalizing voluntary sterilization? Against legalizing involuntary sterilization?

8. Describe the composition of oral contraceptives, explain how they work, and describe some common side effects.

9. Describe some theories about how IUDs prevent pregnancy.

10. Discuss the purposes, mechanism, and safety of "morning after" pills.

11. Distinguish between spontaneous and induced abortions.

12. Describe the major methods of induced abortions and relate these to the time of pregnancy when they are most appropriate.

13. Describe some of the reasons that people agree and disagree with all or parts of the Supreme Court decision regarding the legality of abortions.

14. Do you think the term *health* in discussing abortions should include mental as well as physical conditions? Should the father have a role in an abortion decision?

15. Discuss some promising future contraceptive developments.

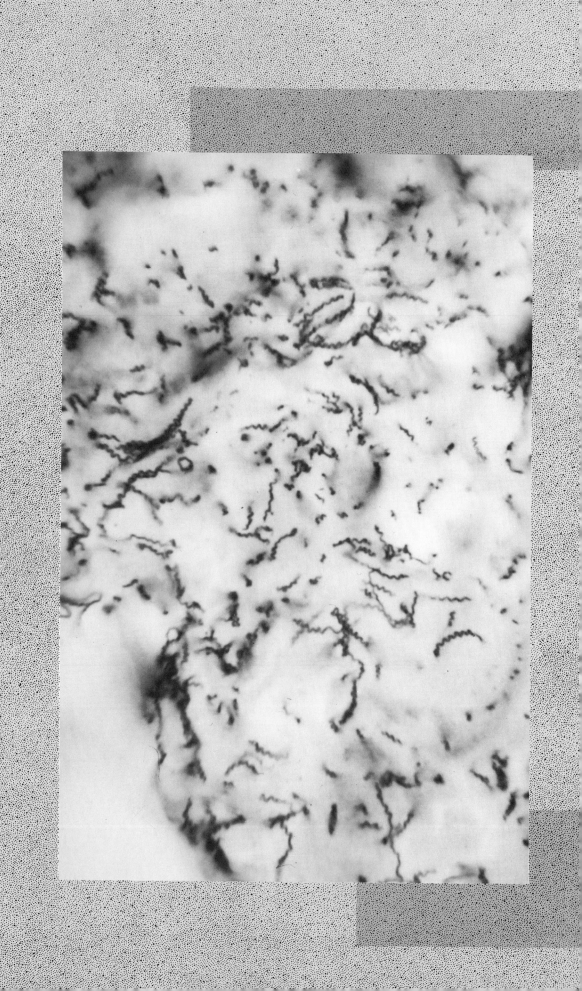

CHAPTER 8

SEXUALLY TRANSMITTED DISEASES

"Thy bones are hollow, impiety has made a feast of thee," wrote Shakespeare, probably referring to the consequences of sexually transmitted diseases. Originally pertaining to Venus, the Roman goddess of love, the venereal, or sexually transmitted, diseases have an ancient history, and some historians trace the origin of syphilis to native Americans. If that is true, Columbus discovered not only the New World but also a new disease, one he eventually died of. Other historians trace syphilis to biblical times. Whatever its origin, syphilis spread throughout the world during the ages of exploration and discovery. Gonorrhea has an even older history and may well have been one of the many plagues described in the Bible.

The tissue shown in the photograph is from the brain of a stillborn infant who contracted syphilis from its untreated mother during pregnancy. The syphilis pathogens are the numerous spiral-shaped structures. Fortunately, most sexually transmitted diseases are easy to treat if they are diagnosed early.

FIGURE 8.1
The syphilis (A) and gonorrhea (B) pathogens are both species of bacteria that usually die quickly outside of the body. The syphilis pathogens shown were secured from the brain of a neonate who contracted the disease from its mother during its embryonic development.

(A)

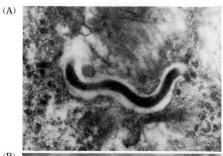

(B)

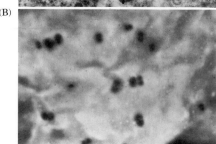

THE SEXUALLY TRANSMITTED DISEASES

Syphilis and **gonorrhea** are not the only **sexually transmitted diseases** (STDs) but they are the best known. Both are caused by species of bacteria (Figure 8.1). The syphilis pathogen is a spiral-shaped **spirochete** (*Treponema pallidum*), whereas the **gonococcus** (*Neisseria gonorrhoeae*) that causes gonorrhea looks like two spheres. These delicate organisms afflict many millions worldwide and have greatly affected human history. They survive best under warm, moist conditions and usually die quickly when exposed to dry air. Thus, it is rare for these diseases to be transmitted by toilet seats, cups, towels, or other objects used by infected persons, although the pathogens sometimes survive for up to two hours in suitable environments. Syphilis and gonorrhea are transmitted, sometimes together, primarily by sexual contact—vaginal and anal intercourse, fellatio, cunnilingus, and sometimes even kissing. Although much less common than gonorrhea, syphilis is far more serious.

Despite knowledge about their causes and treatments, these "social diseases" flourish, and STDs have reached epidemic levels in many areas. A case of an STD occurs every few seconds in the United States. Although most of the reported communicable diseases in American adults are STDs (over 10 million cases annually), perhaps only one-fifth of STD cases are reported. About one-fourth of new infections occur in persons under 22 years of age, and over 85 percent of all reported STDs occur in people between 14 and 30 years of age. There is no exempt age, and STD infections may occur time and time again.

Although some problems of STDs during pregnancy are described below, it is important to note that several forms of STDs produce few if any symptoms in adults but may be serious for neonates who contract them from their mothers. Important pathogens include a type of streptococcus bacterium (type B) and a virus (cytomegalovirus, or CMV). Half of infants infected with the streptococcus die soon after birth. Perhaps 25 percent of all serious infant retardation is caused by the CMV virus, more than that caused by the dreaded rubella virus.

Like most other diseases, many STDs are usually cured easily when diagnosed and treated early. AIDS and genital herpes are exceptions (see descriptions below). Treatment often consists of antibiotics, and the symptoms described below should be familiar to everyone. Following diagnosis and treatment, some infected people may be interviewed by a social worker or epidemiologist in order to obtain confidential information used to discover and contact other infected individuals (Figure 8.2). In public health language, the phrase "sexually active" often applies to someone who has sex with several partners rather than merely frequent sexual intercourse. Generally, the more sexually active one is, the greater the chance of getting STDs, and it is especially important for sexually active individuals to learn to recognize symptoms of STD in their partners as well as in themselves. Regular STD checkups are as important for sexually active individuals as they are for pregnant females, and many local health departments provide confidential and free or inexpensive tests and treatment for STD. If STD is diagnosed, informing one's sexual partners is a responsible action, as is refraining from sexual contacts until the disease is treated and cured. See the essay "How to Study a Disease."

SYPHILIS

Syphilis is usually transmitted during coitus, but the pathogen may be carried by almost any body fluid, including saliva. About twice as many

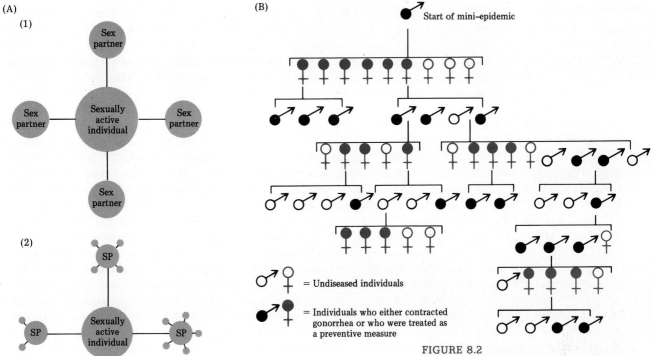

FIGURE 8.2

(A) (1) may represent the "group" about which the sexually active person is aware. However, because sexually active people tend to have partners who are also sexually active, (2) may more accurately describe the circumstance. If the individual indicated by the arrow gets an STD, the condition may spread quickly through the larger group. (B) A mini-epidemic of gonorrhea. Why do so many infected individuals fail to take preventative measures, secure treatment, or notify their sexual partner(s) about an infection?

males as females contract syphilis, and about one-half of men with syphilis are homosexual or bisexual. If the sexual partner is infected, there is no sure way to avoid contracting the disease, although condoms and washing with soap and water give partial protection. Urinating as soon as possible after coitus may also "flush out" some syphilis pathogens. It is common medically to identify four stages of syphilis.

Primary Stage

The first symptoms appear between 10 days and 3 months (typically 2–4 weeks) after infection. These are usually one or more dull red spots that develop pimples and later become ulcerated sores with hard raised edges called **chancres** (pronounced "shankers") at the site(s) of the pathogen's invasion.

In males, infections often occur on the penis or scrotum, and in females they generally develop in the vagina, on the cervix, or on the vulva. Because chancres are usually painless, some infected females with internal chancres may be unaware of their infections, or are **asymptomatic** in medical terms. Chancres in or near the anus may result from anal intercourse. Transmitted through fellatio or cunnilingus, syphilis chancres may form on the lips, tongue, or in the throat. The breasts and even fingers may also be sites of infection.

Untreated chancres disappear within a few weeks, usually leaving no scar. Although many affected people assume erroneously that their problems are over, the pathogen continues to spread within the body and the victim remains contagious.

How to Study a Disease

What causes disease X and how does a human contract it? Where does the causative agent contact or enter the body, and what is the body's response to it? Answering these types of questions has been called the **basic science approach** to the study of disease. The **clinical approach** studies the impact on the victim, exploring how to diagnose and treat the problem. **Epidemiology**, a third approach, is the study of the incidence, distribution, and control of disease within a population. "Who gets disease X and how does it spread through a population?" is a typical question asked.

Actually, all approaches work together. Epidemiologists may discover that those exposed to a certain agent and within a certain age group have relatively high rates of disease X. Basic scientists may expose experimental organisms or cells in tissue culture to the suspected agent to confirm or deny the possible connection. Clinicians may use these findings to learn how to diagnose the disease earlier or treat it more effectively.

The magnitude of a problem is important to an epidemiologist. For example, 400 deaths per year may be normal for a population of 100,000 but low for a population of 1 million and very high for a group of 1000.

Suppose exposure to an industrial chemical seems to be related to a high rate of incidence of a certain type of cancer. Several logical possibilities exist: (1) perhaps the relationship is incorrect due to the small sample size or poor sampling methods, (2) exposure to the chemical may cause the cancer, (3) cancer may cause exposure to the chemical, and (4) another factor may cause both the cancer and the circumstances that bring about exposure to the chemical. Possibility (1) is often solved by better and larger sampling, (2) often takes a long time to prove, and (3) is often, but not always, ruled out. As for (4), perhaps the age, life-style, or health of victims differs from that of those not exposed to the chemical. Thus, **controls** are useful here as they are in most scientific research (Chapter 1). However, humans do not lend themselves to the same type of controlled experimentation used with tissue-cultured cells or laboratory organisms—that is, we do not intentionally expose one group of humans to a chemical but not another group to see which matched group develops the problems or dies first. How would you set up a controlled experiment on humans to determine the possible link between an industrial or environmental chemical and cancer? How does our relatively long life span complicate such research? What ethical constraints should be used with both human subjects and with other organisms?

Some epidemiological research is **retrospective** and looks back, and other work is **prospective** and looks forward. Suppose, again, that a relationship was suspected between a particular chemical and a form of cancer. One might identify as many people as possible with the cancer, as well as a number of matched people without the cancer. The histories of the people could be examined to determine the proportion of those with the cancer that were exposed to the chemical. Also, one could watch groups of exposed and nonexposed people over time to see how many of each group developed the cancer. Which of these studies is retrospective and which is prospective? Which groups are controls?

Secondary Stage

One week to 6 months after the chancres disappear, new symptoms may develop in untreated people (Figure 8.3). The nonitching rash that occurs in many victims may cover much of the body or be limited to small isolated areas. Unlike many other rashes, this type may occur on the palms of the hand and the soles of the feet. It may also cause patchy hair loss on the scalp. The rash may occur as flat pinkish or pale red spots, or as raised bumps. The clear fluid that forms in some patches, usually within and around the genitals, contains many pathogens, making the affected person highly contagious during this stage. Other symptoms may include fever, headache, sore throat, poor appetite, weight loss, and joint pains. Even if the infected person is untreated, these symptoms usually disappear within a few months, although they may recur on and off for several years. Unfortunately, syphilis is often called "the great imitator" because the symptoms it produces resemble those of numerous other common conditions.

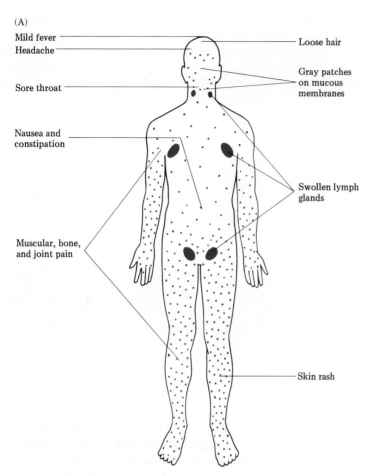

(A)

Mild fever

Headache

Sore throat

Nausea and constipation

Muscular, bone, and joint pain

Loose hair

Gray patches on mucous membranes

Swollen lymph glands

Skin rash

FIGURE 8.3
(A) Possible symptoms of the secondary stage of syphilis, although not all victims experience all symptoms. (B) A skin rash from secondary-stage syphilis.

(B)

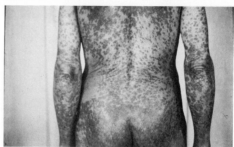

Latent Stage

This "resting" stage may last for years or even decades. Except for an occasional recurrence of a rash, there are usually no major symptoms during this stage. Persons may be infectious early in this stage but usually are not later. Affected people may feel fine, still not realizing that they are infected, and that the pathogens are now deep within their tissues, including the brain and spinal cord, bones, and blood vessels. In about 50–70 percent of cases the disease develops no further.

Tertiary Stage

Symptoms in the advanced stage may vary in time of appearance and in severity, from relatively minor to severe. Large, destructive ulcers may develop on the skin or within internal organs (after 3–7 years). Other tertiary syphilis victims suffer damage, sometimes fatal, to the heart and major blood vessels (after 10–40 years), and the brain and spinal cord may deteriorate (after 10–20 years). Sensory loss, paralysis, and mental illness (e.g., paresis) often precede death. Several thousand victims of advanced syphilis die annually in the United States. The tertiary-stage individual is not infectious, except in the case of pregnant females.

Syphilis and Pregnancy

The syphilis pathogen may reach the fetus of an infected woman through the placenta, although this usually does not occur before the second trimester. Thus, if the infected woman is treated before about the fourth month of pregnancy, the fetus will usually not be infected.

Children born to untreated women almost always have congenital syphilis. If the woman's infection is in the primary or secondary stages, the infant is often stillborn or dies soon after birth. Surviving infants may be blind, deaf, anemic, or mentally retarded, and may have various abnormal organs and/or deformed teeth and bones. Many infants born to mothers with tertiary-stage syphilis develop no obvious symptoms, but others are born with latent syphilis and develop symptoms many years later.

Diagnosis and Treatment

Because most external symptoms of syphilis are similar to those of many other diseases, blood samples and the fluid from chancres must usually be analyzed. Most tests look for antibodies to syphilis rather than for the pathogen itself. No single test is always accurate, so several different tests are often used to avoid "false positive" or "false negative" results.

Antibiotics are often used to treat syphilis, and many primary- and secondary-stage patients are cured quickly. However, the more advanced the disease, the more prolonged the treatment. Although treatment halts progress of the disease, damage done prior to the treatment may be permanent. Some strains of syphilis have become resistant to most antibotics, a common and difficult problem with many pathogens. These resistant strains may require larger doses, longer treatment, and/or different medications.

GONORRHEA

Far more common than syphilis, gonorrhea usually causes less bodily damage. Also, its effects are often restricted mainly to the urinary and reproductive systems, whereas syphilis may affect many organ systems. However, advanced cases of gonorrhea may damage the heart valves and cause arthritis, blindness, or death.

Symptoms

Symptoms of gonorrhea may appear from two days to three weeks after infection. However, because less than 50 percent of infected females experience obvious early symptoms, treatment may be delayed. Many women do not discover their infections until the more obvious symptoms appear in a male partner. Gonorrheal infections in females usually begin (in about 90 percent of cases) in the cervix, but the pus discharged from the cervix within three to eight days after the infection may not reach the vaginal opening. When it does it may go unrecognized or it may be yellow-green and irritating to the vulva. Infections may also occur in the urethra (70 percent), rectum (30–40 percent), throat (10 percent), or in a combination of sites. Anal and rectal gonorrhea is called **proctitis**; and it may cause pain, especially during defecation, and pus in the feces. Some infected women experience a persistent low backache, abdominal pains, painful urination, or abnormal menstruation.

Use of birth control pills increases the chance of getting gonorrhea, apparently because the chemical changes in the genital organs provide a more suitable environment for the pathogen. The slight acidity of the normal vagina in women of reproductive age gives some protection against the gonococcus.

The first symptom in males is often a thick, whitish, yellowish, or clear discharge from the penis. The discharge results from inflammation of the urethra, called **urethritis**, and it contains dead urethral cells, bacteria, and white blood cells. It is irritating and leads to a burning sensation during urination. The urine becomes cloudy with pus and may contain some blood, and the tip of the penis may swell and become inflamed. Even though 10–30 percent of infected males are asymptomatic, they remain contagious.

Because most males experience painful or irritating symptoms, they usually seek treatment and are cured quickly. The longer that treatment is delayed, however, the farther the pathogens spread. If they reach the prostate, an abcess may form, making urination and defecation painful. The pathogens may also infect the epididymis in one or both testicles, and the scar tissue that forms may slow or prevent sperm from passing through the vas deferens. Sterility is possible but is relatively infrequent.

In untreated women, gonorrhea may cause serious complications. For example, the pathogens may spread to the cervix, oviducts, and ovaries and cause **pelvic inflammatory disease** (PID). PID, which is not always caused by gonorrheal or other pathogens, is a common cause of female sterility. Common symptoms include dull pain in the lower abdomen, backache, nausea, vomiting, headache, fever, and pain during intercourse. Serious infections may disrupt the menstrual cycle and make menstruation painful, and the severe symptoms resemble those of appendicitis and ectopic pregnancy.

PID sometimes causes irreversible damage to the fallopian tubes. The blocked tubes enlarge, the scar tissue distorts the reproductive organs, and about 25 percent of victims become sterile (over 100,000 American women annually). Partially blocked oviducts also increase the chance of ectopic pregnancies. Even if a gonorrheal infection is cured, PID may predispose the woman to repeated cases of inflammation caused by a variety of organisms.

A relatively rare but potentially serious condition in both sexes is **septicemia,** when bacteria or their products escape from the infected area and enter the bloodstream. Symptoms include a high fever, chills, and painful joints, and the heart lining (endocardium) and brain coverings (meninges) may also be damaged. Immediate treatment must be received to avoid permanent physical damage or death.

Infant blindness is a possible and tragic side effect of untreated gonorrhea. The gonococcus may be transmitted to the fetus during pregnancy, or it may be acquired during the birth process. Among the most vulnerable organs of the fetus or neonate are its eyes. At least one gonorrhea test is wise during pregnancy because damage can be avoided if the disease is treated early enough. Another preventative measure is to place bacteriocidal drops in the neonate's eyes. Although only a small percentage of pregnant American women have untreated gonorrhea, preventative eyedrops are given routinely to most infants. Gonorrheal eye infections are uncommon in adults but sometimes occur when the eyes are touched with a contaminated finger.

Diagnosis and Treatment

Gonorrhea is diagnosed by means of blood tests and the microscopic examination of samples taken with cotton swabs from the vagina, cervix, anus, urethra, or throat, and grown on a culture medium. In females, the vagina, cervix, anus, and mouth may also be examined directly.

As with syphilis, antibiotics are usually successful in treating gonorrhea. Except for those allergic to it or in cases of resistant strains of the gonococcus, penicillin is a commonly used antibiotic. If treatment occurs within a week following the infection, the disease is often cured within a few days. However, no immunity develops and reinfection may occur repeatedly.

ACQUIRED IMMUNE DEFICIENCY SYNDROME

Because of its devastating and highly lethal effects, **acquired immune deficiency syndrome** (AIDS) has gained widespread attention. AIDS was documented only in 1981, but millions of Americans have already been exposed to the pathogen, as have many millions more in scores of countries, especially in Africa. Tens of thousands of Americans have developed the disease, and the number of confirmed cases has increased dramatically each year.

Most Americans afflicted with AIDS in the early 1980s were males, the majority of whom were homosexual or bisexual. In Africa, however, it seems that most victims are heterosexual, and the disease affects women as often as men. In the United States, blacks and Hispanics contract AIDS at several times the rate for whites, but there is no known genetic predisposition to AIDS. The disparity may result instead from generalized differences in access to health care, educational and other cultural factors, and the prevalence of high-risk behaviors, especially in metropolitan areas.

AIDS is an unusually fatal disease, and some researchers believe that few survive a full-blown case. When their immune system dysfunctions, AIDS patients often succumb to secondary or opportunistic infections, some of which are quite rare in the general population and are seen mainly in certain cancer and transplant patients.

AIDS is spread primarily through sexual contacts and by contaminated needles and other apparatus used by intravenous drug users. Some people (e.g., hemophiliacs) have contracted AIDS upon contact with infected blood, for example, during transfusions or transplants. Now, rapid-screening tests of donated blood, tissues, and sperm (e.g., for artificial insemination purposes) significantly reduce those risks. One cannot contract AIDS from *donating* blood or tissues properly. Although the AIDS pathogen can be carried in tears, saliva, urine, vaginal and cervical secretions, and mother's milk, the prime body fluids that transmit the disease are blood and semen. Despite opinion, or hysteria, to the contrary, AIDS is not known to be transmitted by being in the environment of an infected person or by casual contact with the person. Nor is AIDS known to be spread by insects, although the virus is found in a variety of African arthropods.

Protection for sexual contacts is important in avoiding AIDS, as it is for other STDs. Protective methods include the use of condoms, reduction of the number of different sexual partners, and avoidance of anal intercourse and other sexual activities that result in tissue damage and bleeding. The use of sterile apparatus and products is essential in transfusions and injections. As with many other STDs, AIDS can be transmitted to a fetus by an infected pregnant female, and infected women are usually urged not to breast-feed their infants. Malnutrition, stress, and repeated infections (especially of STDs) lower one's resistance to pathogens and are added risk factors. Also, substance abuse lowers one's inhibitions, and increases the chance of risk-taking behaviors.

The AIDS pathogens are viruses (e.g., various human immunodeficiency viruses, or HIV) (Figure 8.4) that cause the immune system to fail, thus rendering the person defenseless against infections that a healthy body could control easily (Chapter 16). The time from exposure to the development of symptoms (incubation time) may be several years, during which time a person may be unknowingly infectious. The long latent period has allowed the virus to spread widely and rapidly among both males and females. Diagnosis includes blood tests for the antibodies that a person exposed to the virus produces over a period of time.

Although early symptoms are diverse, they often include progressive weight loss; major fatigue; swollen lymph glands; a persistent, thick, whitish coating on the tongue or in the throat that is associated with soreness (a condition known as thrush); shingles (a virus-caused infection); persistent diarrhea; a persistent dry cough not associated with smoking; loss of memory; changes in the sense of equilibrium; easy bruisability and unusual bleeding; and coin-sized red-purple skin spots. Often one infection after another ravages the patient, and death frequently follows diagnosis within 2 or 3 years. However, in the future, death may not be the inevitable result if it turns out that most cases diagnosed during the early years following identification of the disease represented only the most severe forms of the disease. Large numbers of scientists are devoting their talents to research on AIDS, but no cure is known, although various treatments are available that show promise of prolonging the lives of AIDS patients.

OTHER SEXUALLY TRANSMITTED DISEASES

Nonspecific Urethritis and Cervicitis

Nonspecific urethritis (NSU) in males and **cervicitis** in females resemble mild gonorrhea in their symptoms, and their modes of transmission and treatment are similar. In both sexes, a common symptom is frequent and painful urination. Women may also develop PID, although some females are asymptomatic while carrying the pathogens, one of which is the bacterium *Chlamydia trachomatis* (Figure 8.5). However, some urethral and cervical inflammations may be caused by chemicals, allergic reactions, or other pathogens. The causal pathogens can also infect an infant during the birth process and result in eye infection or pneumonia. Millions of Americans are affected by these conditions, which are increasing far faster than gonorrhea. Antibiotic treatment (e.g., tetracycline) is common.

Venereal Warts

Certain viruses cause **venereal warts** to appear within 1–6 months after coitus with an infected person. The glans, foreskin, and shaft of the penis are commonly affected sites in men, although the scrotum and anus may also become infected. In women, warts usually appear near or in the vagina or on the cervix. Like "ordinary" warts, some are dry, small, hard, and yellow-gray, while others are soft and pinkish to reddish in color, with a rough cauliflower-like appearance. These warts, which are usually painless and more of a nuisance than a serious health problem, are usually treated easily with appropriate medications when small but must be removed surgically if they grow large. Treatment is wise, because such warts sometimes become malignant.

Genital Herpes

Genital herpes is caused by one of the herpes family of viruses. In many cases the herpes simplex II form is the pathogen, whereas in others it is the herpes simplex I virus, which commonly causes oral infections such as cold sores or fever blisters. Herpes infections have been known for centuries and were named by ancient Greek physicians from the word "to creep" because of the appearance of the characteristic skin rash.

All forms of sexual contact are the common but not the sole means of transmission. In some cases the virus is transmitted through breaks

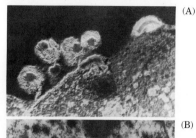
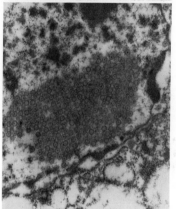

FIGURE 8.4
Virus-caused STDs: (A) A virus associated with AIDS (HIV) and (B) a herpes virus. Most viral diseases are more difficult to treat than bacterial diseases, and antibiotics are ineffective against viruses.

(A)

(B)

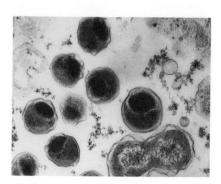

FIGURE 8.5
Chlamydia trachomatis infections are extremely prevalent in the United States.

in the skin. Also, herpes viruses can live for several hours on objects, although this indirect transmission is probably uncommon. Clusters of small, painful blisters are early symptoms in both sexes, appearing between about 2 and 20 days after exposure. In females, the blisters are usually on the vulva, labia, vagina, anus, urethra, and/or cervix. The penis, urethra, or rectum are the most common sites for blisters in males.

The blisters may form reddish open sores when they rupture, except for those on the cervix. Secondary bacterial infections may occur, causing pus to form. Other symptoms may include painful urination, tender and swollen lymph glands in the groin, vaginal or urethral discharges, fever, headache, and general muscle aches. Even without treatment the sores usually begin to heal after 2 or 3 weeks, and little or no scarring remains when healing is complete. However, the pathogens may remain and the blisters may reappear in some people, especially when they are fatigued, ill, or physically or emotionally stressed. Menstruation and lengthy exposure to sunlight may also trigger a recurrence, and infected women are unusually susceptible to cervical cancer and cancer of the vulva. Regular Pap smears and pelvic examinations are useful in detecting these forms of cancer early. Recurrences are usually less severe than the initial infection, and for many people the repeat attacks diminish over a period of years. In some cases the recurrences are preceded by symptoms 1 or 2 days before the active outbreak of blisters.

Only about 1 in 2000 women who have a history of genital herpes are likely to transmit the infection to their babies during a vaginal birth. However, many of the infants born to infected females may die or be blind or brain damaged. This generally results from the infant's having become infected from the cervix or vagina during delivery rather than from the virus having crossed the placenta. Neonates sometimes become infected by the virus transmitted in breast milk or from parents or others who have oral lesions. The discovery of a herpes infection in a pregnant female has often led to the decision to deliver the infant by caesarean section.

Various laboratory tests are used to diagnose herpes—Pap smears in women, blood tests to identify antibodies against the virus, and cultures grown from tissues. As with many other viral diseases, treatment is relatively difficult, although various drugs and radiation are often useful. Because there is no certain cure, victims may carry the virus for life. Most treatment is intended to relieve the pain and itching of active sores and to prevent their becoming infected further. With both oral and genital herpes, people should prevent others from contacting the infected sores. This may involve abstinence from kissing, intercourse, and other sexual contacts during attacks. Unfortunately, some people release the viruses all of the time, not just when symptoms are evident. To prevent severe eye infections, extreme care is needed to avoid transferring the virus to the eyes after touching an open sore or even after sharing a towel or washcloth used by an infected person.

The exact number of infected Americans is unknown; it is estimated to be at least 15–20 million, with perhaps 500,000 new cases occurring annually. These numbers far exceed those for victims of gonorrhea and syphilis combined.

Viral Hepatitis

Viral hepatitis causes a liver infection that may range in severity from symptomless or mild (e.g., indigestion, diarrhea, and poor appetite) to very serious and debilitating. Symptoms of the more serious form include vomiting, fever, jaundice, and sometimes death.

Like herpes, the hepatitis virus exists in several forms (e.g., A and B). Oral-anal contacts, primarily among homosexual males, are a route of transmission of hepatitis A. Perhaps 250,000 cases of hepatitis B occur annually in the United States as a result of sexual contacts. Although the rates of incidence in homosexuals are higher than in heterosexuals, most cases overall occur in heterosexuals because they outnumber homosexuals by about ten to one. Perhaps 200 million humans worldwide are carriers of the hepatitis B virus, including 400,000–800,000 Americans. Carriers may retain the virus for years or for a lifetime, and although they may not be ill they may both remain contagious and have an increased risk of developing liver cancer. Vaccines are being tested, but most treatment is aimed at relieving symptoms and there is no cure yet.

Tropical STDs

Several forms of bacteria-caused STDs are relatively rare in the United States but are common in warm, humid climates. These include chancroid, lymphogranuloma venereum, and granuloma inguinale. Common symptoms include pimplelike sores in the genital region and swollen lymph glands. Travelers to tropical regions who suspect one of these infections should contact a physician for treatment, which usually consists of an antibiotic.

CYSTITIS AND VAGINITIS

Cystitis is a general term for inflammation of the bladder, and most forms of STD may cause cystitis. However, other pathogens may also cause the condition, such as the common bacteria *Escherichia coli,* found in immense numbers in the normal healthy intestine (Chapter 13). In females, *E. coli* may reach the urethra due to wiping from back to front instead of from front to back after defecation.

Symptoms include a frequent desire to urinate and a burning sensation during the process. The urine is often hazy and may contain blood. Antibiotics are usually successful, although a cure may require prolonged treatment.

One of the most common diseases of the female reproductive system is **vaginitis,** and it may occur without sexual contact. This irritating condition may be caused by a variety of organisms, including bacteria, fungi, and protozoans (Figure 8.6). All common types of vaginitis have similar symptoms, although examination of vaginal secretions is necessary to discover the exact pathogen and to plan proper treatment.

The parasitic protozoan *Trichomonas vaginitis* can survive for several hours on moist objects. Thus, "trich" can be transmitted by objects such as shared towels and toilet seats. Although both males and females may be infected, most males have few if any symptoms. The protozoan usually dies quickly in males but may survive for days beneath the foreskin of an uncircumcised penis. Sometimes, however, the trichomonads swim up the urethra to the prostate gland, making a male a carrier of the parasites. During coitus, the protozoans are introduced into the female's vagina as part of the ejaculate. Symptoms in infected women include an itchy inflammation of the vagina and a frothy white or yellow vaginal discharge. Although the symptoms may become less severe even without treatment, they usually do not disappear. In time, the cervix is damaged and the chance of cervical cancer increases. Examination involves sampling vaginal fluid or fluid from the urethra or from beneath the penis's foreskin. Drugs are available to treat this problem, which affects about 3 million Americans annually.

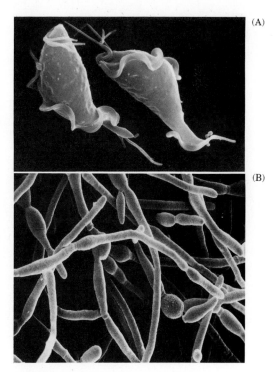

(A)

(B)

FIGURE 8.6
Among the wide variety of organisms that may cause vaginitis (as well as some problems in males) are (A) the protozoan *Trichomonas vaginalis* and (B) the yeast *Candida albicans.*

FIGURE 8.7
Pubic lice often lay their eggs on pubic hairs. The parasitic insects feed upon blood and this may cause itching, an allergic rash, and/or a secondary infection when a person scratches intensely.

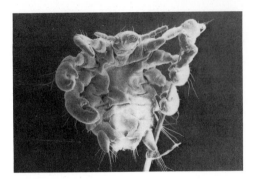

The yeast *Candida albicans* (Figure 8.6B), which causes **monilial vaginitis,** is found on healthy skin, and why it sometimes becomes pathogenic is not known. However, diabetic and pregnant women and those using birth control pills are unusually susceptible to this problem. Symptoms include vaginal itching and a thick, whitish discharge. Coitus is often painful when the vagina wall becomes dry and inflamed. Although the disease is treated successfully with certain antibiotics, accurate diagnosis is important. If the disease is mistaken for a bacterial infection, the wrong antibiotic may be prescribed and the problem may worsen. For example, some antibiotics kill competing microorganisms and allow monilial populations to grow larger.

Several types of bacteria may infect the vagina, and some produce symptoms similar to those of early gonorrhea. Treatment is generally conducted with antibiotics.

PUBIC LICE

Pubic lice (*Phthirus pubis*), or "crabs," are often transmitted during coitus as well as by infected clothing, towels, or bedding (Figure 8.7). These tiny parasitic insects cling to the base of pubic hairs, pierce the skin, and feed upon blood. The lice eggs become cemented to pubic hairs and are very difficult to wash off. Their effects vary from minor to severe itching, and some people develop an allergic rash that may become infected following intense scratching. Various medications usually treat the lice quickly and effectively.

SUMMARY

1. By definition, **sexually transmitted diseases** are spread primarily by sexual contacts rather than by the touching of objects used by infected persons, although some STDs can be transmitted in the latter manner. Millions of Americans have one or more STD infections despite general knowledge about the causes, symptoms, and treatment of most of these communicable diseases.

2. There are several approaches to the study of diseases. The **basic science approach** attempts to identify the pathogen, the way(s) it contacts or enters the body, and the body's reaction to it. The **clinical approach** focuses on the diagnosis and treatment of diseases. **Epidemiologists** study the incidence and control of a disease within a population. Some of the latter research is **retrospective** (e.g., compares people who have a disease with those who do not) and other research is **prospective** (e.g., compares the development of a disease in groups with different histories).

3. **Syphilis** is caused by a spiral-shaped species of bacteria (**spirochete**), whereas the **gonorrhea** pathogen is spherical in shape (**coccus**). Both STD pathogens live best in the body and usually die quickly if they become cool or dry. Syphilis is generally far more serious than gonorrhea but is less common. Antibiotic treatment is typical for both conditions.

4. Medically, syphilis is often divided into four stages. **Chancres** are common symptoms of the **primary stage,** although some infected females are **asymptomatic** if the chancres are internal and painless. **Secondary-stage** symptoms often include common conditions such as a rash, headache, fever, and/or sore throat. No external symptoms appear during the long **latent stage,** when the pathogens spread to many internal organs. Internal damage, mental changes, and death may result from **tertiary-stage** infections. A major concern is transmission to an embryo or fetus carried by an untreated woman.

5. Gonorrhea primarily affects the urogenital system. Females are

commonly asymptomatic, whereas early symptoms in males often include **urethritis.** In untreated people, the pathogens may spread and cause **proctitis, pelvic inflammatory disease,** and/or **septicemia.** Transmission of the pathogens to a fetus during pregnancy or birth is a major concern. Two factors that have contributed to the increased incidence of gonorrhea are the use of oral contraceptives and the development of antibiotic-resistant strains of pathogens.

6. Viral-caused STDs include **acquired immune deficiency syndrome, genital herpes, venereal warts,** and **viral hepatitis.** AIDS is usually fatal and its incubation period of several years allows it to be spread rapidly and widely by sexual contact, contaminated intravenous needles, or contact with infected blood. Herpes blisters form in various locations and may ulcerate and become infected with bacteria when they open, for example, after being scratched. Attacks come and go, often triggered by physical and emotional stresses.

7. The bacterium *Chlamydia trachomatis* infects millions and, among other diseases, causes **nonspecific urethritis** in males and **cervicitis** in females.

8. Bladder infections are called **cystitis** and the common *E. coli* bacterium is one of many causal pathogens. Urination may become painful and the frequency of urination increases. Many pathogens (e.g., protozoans, bacteria, and fungi) cause **vaginitis.** The symptoms of all forms are similar and include itching and unusual vaginal discharges. Treatment generally consists of antibiotics; if the condition is untreated, it may develop into serious internal infections or increase the chance for cancer (e.g., in the cervix).

9. **Pubic lice** are usually only irritating, although some people develop allergic rashes and secondary infections.

REVIEW QUESTIONS

1. Explain why sexually transmitted diseases are still so widespread despite known causes and cures for most forms of STDs.

2. Describe those contraceptive methods that are most and least likely to be effective in reducing the incidence of STDs.

3. Do you think that American society views male and female victims of STD differently? If so, why?

4. Compare and contrast the basic science, clinical, and epidemiological approaches to the study of diseases.

5. Describe how prospective epidemiological studies differ from retrospective investigations.

6. Describe several examples of how STDs affect males and females differently.

7. Compare and contrast the pathogens that cause syphilis with those that cause gonorrhea.

8. Describe how sterility may result from untreated cases of some STDs.

9. Describe how STD pathogens may infect an embryo, fetus, or neonate. Also describe how such cases can be avoided and how they are treated when they occur.

10. Compare and contrast the causes, symptoms, and treatments for nonspecific urethritis and cervicitis.

11. Define *cystitis, proctitis, septicemia,* and *vaginitis,* and describe the respective causes, symptoms, and treatments.

BASIC HUMAN GENETICS

The familiar phrase "a chip off the old block" implies that offspring are like their parents in many ways. Genetics, the study of heritable traits, tries to explain both likeness and variation. That is, offspring are usually like their parents in many major traits but different in numerous minor ways. Understanding genetics is important because it helps us explain many birth defects, the breeding of domesticated organisms, and the evolution of humans and other species.

Genetics began formally with the work of the monk Gregor Mendel in the 1860s in what is now Czechoslovakia. Mendel discovered the basic patterns of inheritance from carefully planned experiments on the garden pea. His success was due partly to his choice of the pea, because those plants differ in some easily observed discontinuous traits such as height (tall or short), pod color (yellow or green), seed color (yellow or green), and seed form (smooth or wrinkled). Most human traits are not of this either-or type and the development of genetics based upon the observation of humans would have been very difficult.

Besides showing clear-cut traits, peas were also self-fertilizing. Thus, true-breeding pea plants produced seeds that grew into plants with certain traits exactly like those of their parents. Mendel crossed (transferred sperm nuclei in the pollen grains) one true-breeding plant with another plant that bred true for the alternate trait. He then collected the seeds produced and observed the results when he planted them the following year. Again, humans would have made poor subjects for the development of major genetic principles.

The height, weight, skin color, and many other traits of the males and females in this group vary, as do their many internal traits. Descriptive genetics attempts to find patterns to such variations and to determine how such traits are passed from parents to their offspring.

FIGURE 9.1

The basic units of heredity, the genes, are found at particular loci on chromosomes. Traits depend upon the nature of the allelic genes, such as whether both are dominant or only one is. The use of dyes and other techniques leads to the banding patterns of chromosomes that assist in specifying the loci of genes.

BASIC GENETIC CONCEPTS AND TERMS

Inherited traits are transmitted by **genes,** which are sections of the DNA molecules that make up part of the chromosomes (Chapter 10) (Figure 9.1). The genes for a particular trait are found at certain locations, or **loci,** on **homologous chromosomes,** one from each parent. The two or more different forms of a gene that may occupy a particular locus are called **alleles,** and traits depend upon the nature of the allelic genes (e.g., a gene for brown eyes vs. a gene for blue eyes). Often one allelic gene is expressed—the **dominant** form—but the other, **recessive** form is not. For example, the gene for normally pigmented skin is dominant whereas the recessive allele for this gene causes **albinism,** or the inability to make the normal skin pigment melanin (Figure 9.2). A capital letter is often used to designate a dominant trait (e.g., *A*) and a lowercase letter to symbolize the recessive trait (*a*).

Individuals in whom the two alleles of a pair are the same are called **homozygotes** (e.g., *AA* or *aa*). **Heterozygotes** are those in whom the two alleles of a pair are different (e.g., *Aa*). **Phenotype** refers to an individual's appearance, whereas **genotype** means the organism's genetic makeup. Thus, both *AA* and *Aa* individuals show the same phenotype for normal skin pigmentation although they differ in genotype.

Albinos are homozygous recessive (*aa*) for skin pigmentation because they received a recessive gene for this trait from each parent. For a recessive gene to be expressed it must occupy the proper locus in both homologous chromosomes. The recessive trait is usually "masked" when it is paired with its dominant allele (*Aa*).

Figure 9.3 shows a **Punnett square,** a simple checkerboard way to illustrate matings and crosses. Notice that capital letters are always written first. The offspring from Mating 1 may also be called **hybrids** as well as heterozygotes. Like the two normally pigmented parents in Mating 2, they are **carriers** of the albino gene. The typical ratio of normally pigmented-to-albino offspring born to carrier (hybrid or heterozygous) parents is 3 : 1. This does not mean that if the parents had one albino child their next three children would have normal pigmentation, because each child results from the chance union of gametes.

Wavy hair is usually determined by a dominant gene (*W*) and straight hair by the recessive allele (*w*). A dominant gene (*B*) causes one type of short fingers, and the recessive allele (*b*) leads to normal finger length. Punnett squares are especially helpful when studying two traits determined by alleles on different chromosomes. For example, the Punnett

FIGURE 9.2
The inheritance of albinism.

Normally pigmented parents who carry the recessive gene for albinism

Male is a normally pigmented carrier. Female is albino.

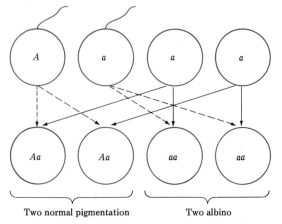

square in Figure 9.4 shows a **dihybrid cross** between a male with normal-length fingers and straight hair (*bw*) and a female with short fingers and wavy hair (*BW*).

Because few families include 16 children, the "ideal" ratio seldom results. The chance that a particular phenotype will result from a single pregnancy is found by dividing each number in the ratio by 16. For example, the chance of a child with short fingers and wavy hair is slightly greater than 50 : 50 (9/16, or 56 percent).

Figure 9.5 and Table 9.1 show some other alternate, discontinuous, or either-or human traits, although not all are inherited in a simple manner.

TABLE 9.1
Some Dominant and Recessive Human Traits

Trait	Comments
Little fingers	Hold your hands with the palms toward you and press the little fingers together. Bent finger tips are dominant over parallel fingers.
Fingers and toes	Having extra fingers and toes is dominant over the normal number of digits. Short fingers are dominant over those of normal length.
Finger hair	Hair on the middle segment of fingers is dominant over hairless fingers.
Nose	A large, convex nose is dominant over a smaller, straight one.
Chin dimples or fissures	Chin dimples or fissures are dominant over no chin dimples or fissures.
Eye region	Eyelashes over 1 cm appear to be dominant over shorter eyelashes. Drooping eyelids and the condition of having the inner corner of the eye bridged by an arching fold of skin extending from the bridge of the nose are dominant traits.
Freckles	Freckles appear to be dominant over no freckles.
Ear traits	Free ear lobes are dominant over attached ones. Shallow ear pits (one or both ears) are dominant over nonpitted ears. A small tubercle on the upper rim of the ear is also a dominant trait.
Skin	Scaly and/or thickened skin may be dominant over normal skin. Albinism is recessive to normal skin pigmentation.
Tongue-rolling	Tongue-rolling is dominant over the lack of this ability.
Common baldness	The **M**-shaped hairline that recedes with age is recessive to normal hair.

FIGURE 9.3
Punnett squares depicting the inheritance of albinism. What are the phenotypes for skin color of each parent in Mating 1? Their offspring? What ratio of genotypes results from Mating 2? Phenotypes? Note the scientific symbols for males and females.

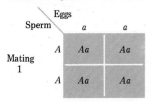

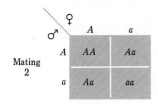

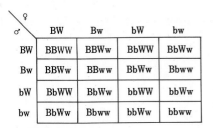

FIGURE 9.4
The inheritance of hair texture and finger length. Describe each different phenotype resulting from such a mating. What is the ratio of the different phenotypes? What is the chance (e.g., the number out of 16) of producing a normal-fingered, straight-haired child? A short-fingered, straight-haired child?

FIGURE 9.5
Some common human discontinuous (either-or) traits (see Table 9.1). Are the most common traits in a population always determined by dominant genes?

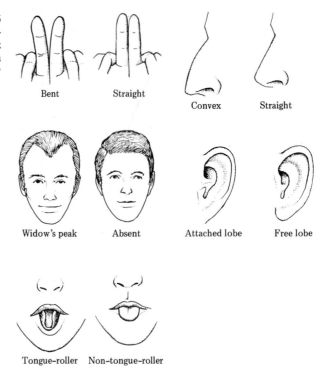

Bent   Straight

Convex   Straight

Widow's peak   Absent   Attached lobe   Free lobe

Tongue-roller   Non-tongue-roller

HUMAN CHROMOSOMES

The fact that Mendel knew nothing of chromosomes makes us appreciate his theory of inheritance and his intellectual accomplishment even more. Although published in 1866, his ideas were largely ignored until about the turn of the century. Perhaps this was because Mendel's "factors" were rather mysterious units for which no physical basis was known. However, by 1900, enough was known of the behavior of chromosomes for Mendel's theory to be acceptable. Mendel's "factors" were genes located on the chromosomes. During meiosis, the allelic genes are segregated or separated so that a gamete gets only one member of each allelic pair. This is Mendel's **principle of segregation** (or "first law"). Mendel also formulated a **principle of independent assortment** ("second law"), which states that when two or more pairs of genes are involved in a cross, the members of one pair usually segregate independently of the members of all other pairs (Figure 9.6). The establishment of this principle was possible because most of the traits he studied were determined by genes found on different pairs of chromosomes. This principle does not always apply when the genes of two or more alleles occur close together on the same chromosome.

The independent assortment of chromosomes during meiosis accounts for much variation in the gametes each individual produces. The number of possible combinations of 23 chromosomes with random assortment is 8,388,608 (2^{23}), and because this variability occurs in each sex, the diversity of zygotes is even greater. The chance of two identical sperm fertilizing two identical eggs is about 1 in 64 trillion.

The exact number of human chromosomes was not determined until 1956. Certain white blood cells are often used in chromosome studies. The blood sample is placed in a culture dish containing chemicals that stimulate the blood cells to divide, and the chromosomes become separate and visible during cell division. After other chemicals are added to make the chromosomes swell and to stop the process of cell division,

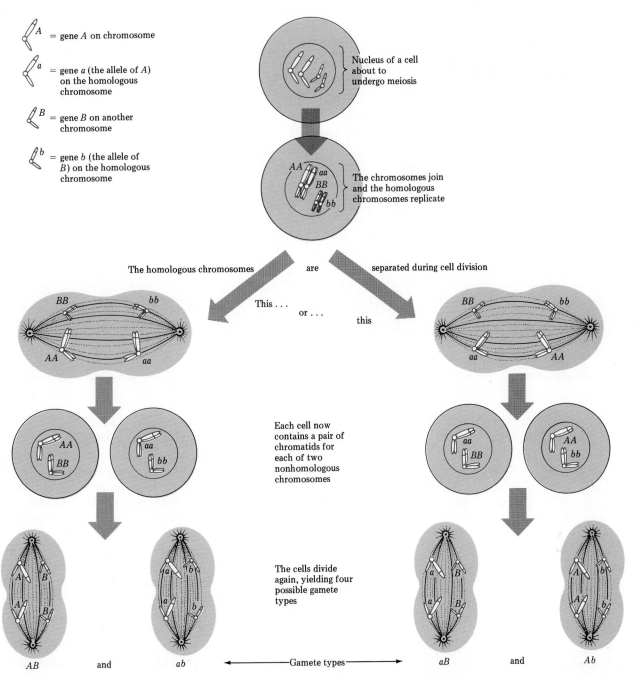

= gene *A* on chromosome

= gene *a* (the allele of *A*) on the homologous chromosome

= gene *B* on another chromosome

= gene *b* (the allele of *B*) on the homologous chromosome

Nucleus of a cell about to undergo meiosis

The chromosomes join and the homologous chromosomes replicate

The homologous chromosomes   are   separated during cell division

This . . .   or . . .   this

Each cell now contains a pair of chromatids for each of two nonhomologous chromosomes

The cells divide again, yielding four possible gamete types

AB   and   *ab*   ←———Gamete types———→   *aB*   and   *Ab*

FIGURE 9.6
Segregation and independent assortment in meiosis. Why do you think the principle of independent assortment does not always apply when the genes of two or more alleles occur close together on the same chromosome?

the cells are broken open so the chromosomes will spread out. They may then be photographed through a microscope (Figure 9.7). Each chromosome can be cut from the photo and lined up with one another in matching pairs to form a **karyotype.** Computers are often used to speed and automate this process. Twenty-two pairs of chromosomes, the **autosomes,** are matched in both sexes. The remaining two chromosomes are called the **sex chromosomes,** and they are matched in females (XX) but unmatched (XY) in males.

DOMINANT AUTOSOMAL TRAITS

Woody Guthrie (1912–1967), the American balladeer ("This Land Is Your Land"), suffered from **Huntington disease** and spent 15 years in a

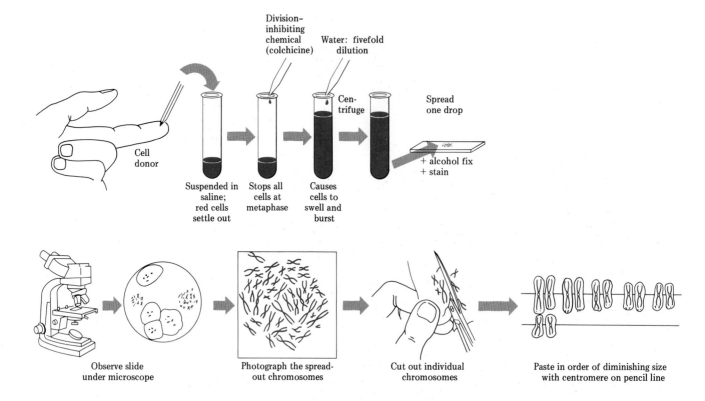

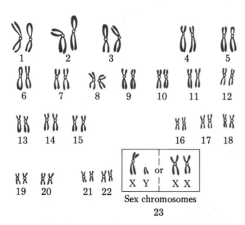

FIGURE 9.7

Method of preparing a human karyotype. How many pairs of autosomes do human body (somatic) cells contain? What traits seem to be used in placing chromosomes into groups? How do the sex chromosomes in males differ from those in females?

mental hospital before his death. Unfortunately, the symptoms of this dominant trait usually do not appear until the affected person approaches middle age, after the typical time when decisions are made about whether to have children. Any child, male or female, of an affected parent has a 50 percent chance of developing this severe neurological disease. Symptoms include mental disturbance and gradual loss of muscular control; the affected people eventually become helpless, bedridden invalids until death.

Many hundreds of other human traits are also determined by dominant genes located on autosomes. Although many dominant genes produce a specific normal substance, some cause undesirable traits. Most humans who show the unfavorable trait are heterozygotes.

The effects of many dominant autosomal genes vary in **expressivity** from person to person. Many variations in gene expression are probably due to the effects of other genes that influence the same trait. For example, the gangly appearance of Abraham Lincoln may have resulted from only a mild case of **Marfan syndrome.** However, the expression of the syndrome in the violinist Nicoli Paganini (1782–1840) was more serious, but perhaps contributed to his extraordinary talent. People with severe symptoms not only have excessively long hands, arms, feet, and legs but also have a defective aorta (large artery that leaves the heart) and an abnormal eye lens.

Human inheritance is often shown by **pedigrees,** such as that in Figure 9.8 for "woolly hair." Affected people have short, brittle, tightly kinked hair. On pedigree charts, females are shown by circles and males by squares, and individuals possessing the trait being studied are shown by darkened symbols. Horizontal lines between individuals indicate matings, and their offspring are shown at the end of short vertical lines. Other common pedigree symbols are shown in Figure 9.9. You may wish to draw family pedigrees for each trait that you can find sufficient information about, such as some of the traits in Table 9.1.

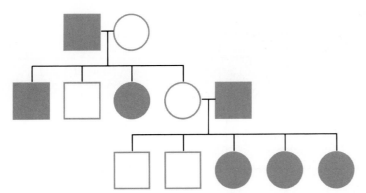

FIGURE 9.8
A pedigree for "wooly hair." What are some possible genotypes for affected humans? Nonaffected people?

RECESSIVE AUTOSOMAL TRAITS

Several hundred other human traits are known to be caused by recessive autosomal genes. Many recessive genes fail to produce a normal product, or the product is inferior to that produced by the dominant allele. Most of these traits are rather rare, and their effects are usually less severe than those with homozygous dominant autosomal traits. Also, the expressivity of these traits, one example of which is the inability to roll one's tongue, is about the same for all affected people.

The more closely related two people are, the greater the chance that they are carriers of the same gene. **Consanguineous matings** are those between close relatives. Although only about 0.1 percent of marriages in the United States are between first cousins, about 8 percent of albino children result from such matings. Figure 9.10 and Table 9.2 show how the incidence of autosomal recessive traits relates to the degree that affected person's parents are related. Although most human societies have strict cultural prohibitions against incestuous matings, incest taboos seem to result as much or more from sociological factors as from applied genetics.

Genetic counseling helps people learn whether they are carriers of autosomal recessive genes, and couples often wish to know the chances of their having a child that is homozygous recessive for a seriously negative trait. However, the advice given by a genetic counselor is usually sought only after the couple or a close relative has had an affected child. For example, people affected by **phenylketonuria** (PKU) cannot break down phenylalanine, a common amino acid in many protein foods. Children born with PKU (about 1 in 15,000 births in the United States) may become severely mentally retarded without proper care, which

FIGURE 9.9
Some common pedigree symbols.

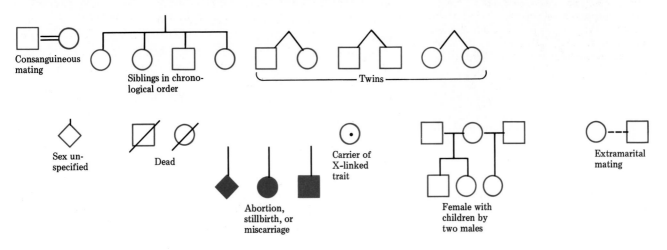

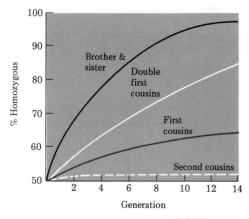

FIGURE 9.10
The relationship between the incidence of autosomal recessive traits and the degree of relationship of the affected person's parents.

TABLE 9.2
Chance of Expression of Recessive Traits in First-Cousin Matings

Trait	Frequency in Nonrelated Matings	Frequency in First-Cousin Matings	Ratio of Increase
Albinism	1 in 20,000	1 in 2,272	8.1:1
Phenylketonuria	1 in 10,000	1 in 1,600	6.2:1
Cystic fibrosis	1 in 1,600	1 in 640	2.2:1

involves a diet low in phenylalanine. Carriers can be identified by giving them unusually large but safe amounts of phenylalanine. Heterozygotes with only one dominant gene cannot break down the phenylalanine as rapidly as those with two such genes. (Incidentally, note the warning labels about the phenylalanine content of some products that are sweetened with certain synthetic compounds.)

INTERMEDIATE INHERITANCE OR INCOMPLETE DOMINANCE

It should be no surprise that there are exceptions to most principles that apply to living organisms. In **intermediate inheritance** or **incomplete dominance**, traits are not expressed in the dominant-recessive form. As before, one allele of the gene is contributed by each parent, but in this

FIGURE 9.11
(A) The sickled red blood cells are quite different from the somewhat more normal, disk-shaped red blood cells in this view. The inheritance pattern of sickle-cell trait and sickle-cell anemia is shown in (B). What is the ratio of normal, sickle-cell trait, and sickle-cell anemia among the offspring of two carrier parents? What is the genotype for sickle-cell anemia? For sickle-cell trait? For normality?

(A)

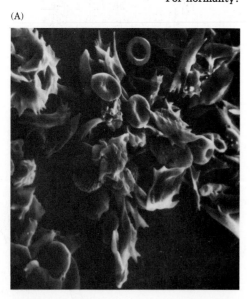

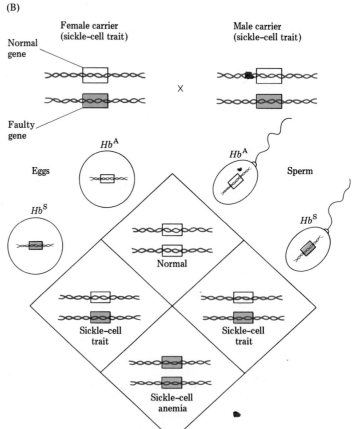

case neither allele "gives way" to the other. For instance, both members of an allelic pair are sometimes expressed to some degree in heterozygous individuals, such as in **sickle-cell anemia** and **sickle-cell trait.** Millions of Americans, mostly blacks, are troubled with these serious problems. Sickle-shaped blood cells with abnormal hemoglobin do not carry normal amounts of oxygen and may clog small blood vessels (Figure 9.11). Although much research is being done on this problem, most of those affected by sickle-cell anemia die by their early teens. People with sickle-cell trait usually appear normal, but if they are exposed to low oxygen levels (such as while flying or traveling in high mountains), their red blood cells may sickle.

Capital and lowercase letters imply dominant and recessive, respectively. The problem this presents in choosing letter symbols for genes with intermediate inheritance is solved by using superscripts. Thus, Hb^A symbolizes the gene for normal hemoglobin and Hb^S the gene for abnormal hemoglobin.

CODOMINANCE

The traits discussed thus far have depended upon a single pair of alleles, such as *A* and *a*. However, some common human traits are caused by **multiple alleles,** such as the set of alleles that determines the **ABO blood group.** (More information on this and other blood groups is found in Chapter 16.) The three different alleles that determine one's ABO blood type are shown in Table 9.3. Both the alleles *A* and *B* are dominant to *O*. Because *A* and *B* both have an effect on the surface of red blood cells when they occur together, they are said to show **codominance.**

Knowledge of the genetics of blood type inheritance has often been used in determining paternity, although new methods of matching tissue types are more accurate. Suppose a woman of blood type B seeks support for her child, blood type AB, from a man of blood type O. Using such genetic information alone, can you prove that this male is *not* the father of the child? That he is the child's father?

EXTRA CHROMOSOMES

Three instead of two chromosomes of pair 21, or **trisomy-21,** cause **Down syndrome** (Figure 9.12). Affected people (about 1 in 800 live-born children) are short, have malformed hearts, and are usually severely mentally retarded. Because of their distinctive face and eyelids, those affected were once said to exhibit "mongolism." However, the eyelids of Mongoloids differ from those with Down syndrome, and people with Down syndrome are not more common among Mongoloids than among other races.

Although the incidence of Down syndrome is not related to race, it is related to the mother's age (but not to the father's age; Figure 9.13). Young parents who have a child with Down syndrome may be advised to have their chromosomes examined to see if they are carriers of abnormal chromosomes.

Trisomy also occurs in other autosomes, usually leading to a miscarriage or to the early death of an infant. Also, the deletion of even a part of an autosome may lead to similar problems. Most of the 10–15 percent of spontaneous abortions that occur before the twentieth week of pregnancy may result from abnormal chromosomes.

TABLE 9.3
Inheritance of ABO Blood Types[a]

Blood Type	Genotype(s)
A	$I^A I^A$, $I^A I^O$
B	$I^B I^B$, $I^B I^O$
AB	$I^A I^B$
O	$I^O I^O$

[a] The three alleles at a single locus on chromosome 9 that cause the ABO blood types are designated I^A, I^B, and I^O.

(A)

(B)

XX   XXX   XX   XX
20   21   22   X Y

FIGURE 9.12
A child with Down syndrome (A). Most cases of Down syndrome result from trisomy-21 (B).

FIGURE 9.13
Comparison of the ages of all mothers with the ages of mothers who bore children with Down syndrome.

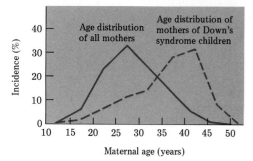

CONTINUOUS GRADATION AND POLYGENIC INHERITANCE

Many human traits show **continuous gradation** rather than expression in an either-or or discontinuous manner. That is, humans do not come in only two sizes, large and small; instead, they differ widely in height, weight, shape, skin color, and mental ability. More than one allele is involved in **polygenic inheritance.** For example, four genes operate when two loci are involved.

Members of all human races contain several different skin pigments, and perhaps five or six alleles are involved in the inheritance of skin color. However, the intensity of pigmentation depends mostly upon the concentration of two dark brown pigments (melanin and melanoid). "Pure Negroid skin color" may be symbolized as *AABBCCDDEE* and "pure Caucasoid skin color" as *aabbccddee*. The nature of the two dark brown pigments appears to be identical in all races, although the amount of pigments produced by "white" genes is less than that produced by "black" genes.

The offspring of "pure Negroids" and "pure Caucasoids" are called **mulattos,** and their skin color is intermediate between that of their parents. Figure 9.14 shows the mating of two mulattos. If independent assortment of skin-color genes occurs, 32 (2^4) different kinds of gametes can be made by the heterozygotes.

Because we do not see just short and tall people, simple dominant-recessive inheritance is not indicated in explaining body height. Also, intermediate inheritance is not determined by one allele, because in that case we would expect short, medium, and tall people with discontinuous variation between the three groups. Figure 9.15 and Table 9.4 show how several pairs of alleles probably control height.

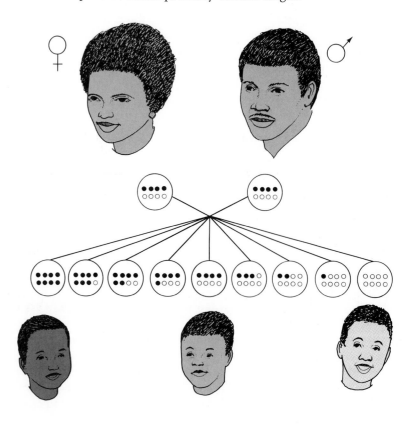

FIGURE 9.14
Because several alleles affect skin pigmentation, a couple may have children of various skin colors. How can some offspring be darker or lighter than their parents?

Very dark — Very light

TABLE 9.4
The Polygenic Inheritance of Height

| | ♀ Gametes | | | |
	AB	Ab	aB	ab
Tall Father ♂ Gamete AB	AABB 188 cm (74 in.)	AABb 180 cm (70.8 in.)	AaBB 180 cm (70.8 in.)	AaBb 172 cm (67.7 in.)
	♀ Gametes			
	AB	Ab	aB	ab
Short Father ♂ Gamete ab	AaBb 172 cm (67.7 in.)	Aabb 164 cm (64.6 in.)	aaBb 164 cm (64.6 in.)	aabb 156 cm (61.4 in.)

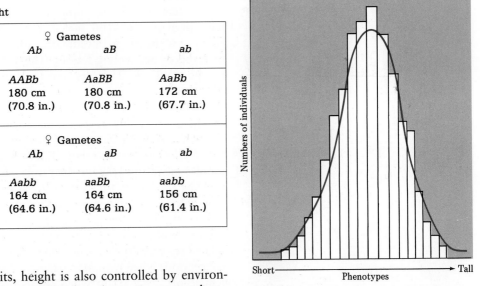

FIGURE 9.15
The range of phenotypes in a population will assume a bell-shaped curve when a trait is controlled by several alleles. How could the environment affect one's height? Other traits?

Like many other human traits, height is also controlled by environmental factors. Genes provide a potential, but the environment determines the degree to which that potential is attained.

The many shades of eye color and the fact that blue-eyed parents frequently have brown-eyed children suggest polygenic inheritance. Except for albinos, blue-eyed people have the least amount of melanin. A little more melanin causes green eyes, still more causes hazel eyes, and even larger melanin deposits cause brown to almost black eyes. In Figure 9.16 all genes for deposition of melanin in excess of that needed for blue eyes are grouped as B.

SEX DETERMINATION

The twenty-third pair of chromosomes, or the sex chromosomes, match in females (XX) but not in males (XY) (Figure 9.17). Although it seems logical that there would be a 50:50 chance for a male or a female child, in most countries more males than females are born alive. For example,

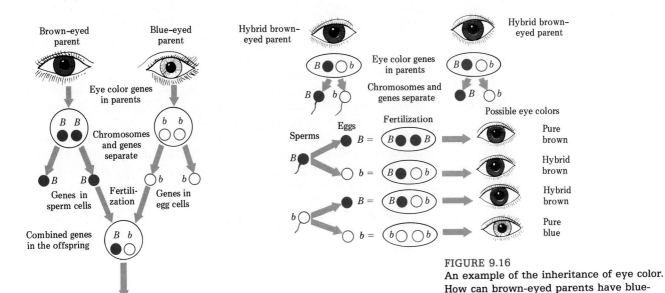

FIGURE 9.16
An example of the inheritance of eye color. How can brown-eyed parents have blue-eyed offspring?

FIGURE 9.17
Sex determination.

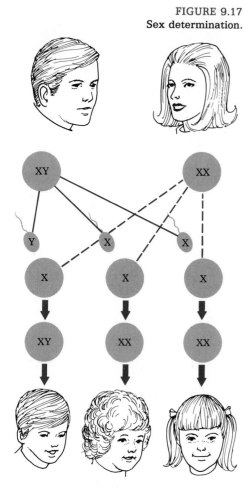

the Caucasoid sex ratio at birth in the United States is about 106 males to 100 females. The sex ratio at the time of fertilization is even higher, with perhaps 130–160 males conceived for every 100 females. One reason for the range in estimates is the difficulty of determining the sex of embryos and young fetuses. The sex ratio is equal between the ages of about 18 and 25 years, but thereafter there are more females than males. About twice as many females than males survive to age 85, and over the age of 100 there are about five times as many females as males.

There are many theories about why more males are conceived. Whereas the **X chromosome** appears to have a normal number of genes, the **Y chromosome** is relatively blank genetically. (Thus, females have more genetic material in their cells than males do.) Because of this, perhaps the Y-carrying sperm are able to reach the egg first. Perhaps conditions within the female reproductive tract favor the survival of one type of sperm, or maybe the egg is more easily fertilized by Y-carrying sperm. Also, the timing of coitus relative to ovulation appears to be related statistically to sex determination. More males seem to result from conceptions when ovulation occurs within a day before coitus, and more female offspring result when coitus occurs three or more days before ovulation. See the essay "Sex Preselection" for a discussion about how this information is sometimes used.

Abnormal Numbers of Sex Chromosomes

Abnormal numbers of sex chromosomes are found in about 1 of every 300 births. For example, some instances are known of females with only one X chromosome (XO or 45,X) or with as many as five or six X chromosomes, and some males have extra X or Y chromosomes. Although one might predict that XXX (47,XXXX) females or XYY (47,XYY) males would be "sexier" than normal, missing or extra sex chromosomes can result in serious problems, although not in these two examples. In many other cases, however, sterility or low fertility may result. Two interesting and relatively common examples of abnormal sex chromosomes are **Klinefelter syndrome** (1 of 500 births) and **Turner syndrome** (1 of 3500 births) (Figure 9.18). Those with Klinefelter syndrome are phenotypic males, and those with Turner syndrome are phenotypic females.

Most Klinefelter males are long-legged, have small sex organs, are sterile, show feminine muscular development, and often have femalelike breasts. Turner females have infantile sex organs, most are sterile, their physical growth is retarded, and they often have a large fold of skin along the sides of the neck. Males comparable to Turner females (i.e., YO) are not known except in aborted fetuses. Apparently at least one X chromosome is necessary for survival.

Some people known as **genetic mosaics** contain some cells with one chromosome number and other cells with a different chromosome number. For example, because the first known Klinefelter male had some XX cells and some XXY cells, the person's sex chromosomes pattern was symbolized XX/XXY. Other examples of phenotypic male mosaics include XY/XXY and XXY/XXXXY, and some phenotypic female mosaics include XO/XX, XX/XXX, and XXX/XXXXX individuals. Mosaics involving the autosomes are also known to exist, although mosaics of the sex chromosomes seem to be more common. Mosaics probably result from **nondisjunction** or from the loss or gain of chromosomes due to improper separation during cell division.

Mosaics whose bodies contain the sex chromosomes of both sexes

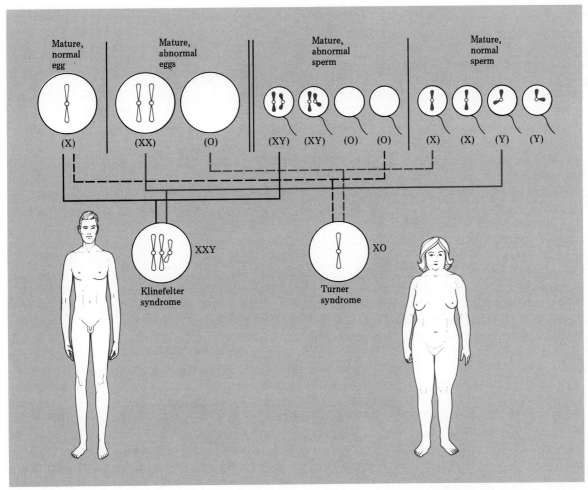

FIGURE 9.18
Klinefelter and Turner syndromes and some
ways they may form. What is the genotype
of Klinefelter males? How many total
chromosomes do Turner-syndrome females
have?

are called **hermaphrodites,** and some primary and secondary sex traits of both sexes may be present. For example, an XX/XXY person may have one breast, growth of hair on part of the face, and some testicular and ovarian tissue. About one-third of human hermaphrodites have a testis on one side and an ovary on the other. About one-fifth have **ovotestes** on both sides, and almost half have an ovary or a testis on one side and an ovotestis on the other. The term *hermaphrodite* is derived from Hermaphroditis, the son of the Greek deities Hermes and Aphrodite; his body coalesces with that of a maiden who is in love with him. (See the essay "Sex Reversal, Genetics, and Sports.")

There was once considerable interest in the possibility that some XYY males are predisposed to criminal or antisocial behavior. XYY individuals occur in about 1 of 1000 live male births. Most such males are tall (over 183 cm, or 6 ft), some have serious acne, and some reach sexual maturity at an earlier age than normal. XYY males are generally fertile and usually produce normal sons. Although most XYY males are virtually normal, a 1965 report that the frequency of XYY males in a Scottish mental-penal institution was far higher than that of the general population created considerable controversy and study. Similar reports appeared from around the world, although most inmates of such facilities are XY males, and the vast majority of XYY males are not institutionalized. Some researchers wondered whether the aggressive, sometimes criminal behavior reported for some XYY males might be due to

Sex Preselection

For centuries, many prospective parents have wanted to choose the sex of their offspring rather than leave it to chance. Most methods of **sex preselection** used in the past are used today only rarely in some developing nations: (1) using special chants during coitus; (2) timing coitus relative to an external condition (e.g., wind direction, rainfall, temperature, tides, or moon's phases); (3) eating sweet foods to produce males; (4) having females wear male clothing and/or males wear boots before coitus to produce sons; (5) having males hang their pants on the right side of the bed to have a son and on the left side to have a daughter; (6) burying the placenta from the last child under nut trees to ensure that the next child will be a male; and (7) engaging in coitus on even-numbered days of the menstrual cycle to produce sons, or on odd-numbered days to produce daughters. Selective infanticide has been used by many cultures to achieve the desired ratio of the sexes within the population. Two additional theories: (1) Early Greeks thought that substances from the right testicle produced males, whereas those from the left testicle produced females. (One testicle could be tied off before coitus and some European noblemen had their left testicles removed in order to ensure them an heir.) (2) "Unripe" eggs produce males, whereas "ripe" eggs produce females. Eating protein before ovulation was believed to help eggs ripen. Why do you think that infanticide—a failure-proof method—was used more often against females? Can you think of any scientific basis for these beliefs?

Some reported, apparently reliable but slight correlations are that (1) the percentage of sons rises during and for a time after wars; (2) in some populations the percentage of males increases with the age of the father, the mother, or both; (3) the ratio of sons to daughters is higher for couples with higher coital rates; (4) a higher percentage of sons is born to couples of higher socioeconomic levels; (5) first births are more likely than subsequent births to be males; (6) the sex ratio varies with the season (e.g., in the United States the ratio of male to female births is highest in summer); and (7) the ratio of male to female births is lower after certain disasters (e.g., floods, serious multiday smog episodes, and hepatitis epidemics). Can you think of any scientific basis for these correlations?

Current research on sex preselection is being conducted in at least four areas.

1. *Timing of coitus relative to ovulation.* One theory is that coitus near the time of ovulation favors the conception of males. Y-carrying sperm are presumed to die sooner than X-carrying sperm. Therefore, coitus several days before ovulation results in more female conceptions.

2. *Sperm separation and artificial insemination.* Until recently, this method was difficult because X- and Y-carrying sperm could not be separated *in vitro* (from the Latin for "in glass"). Now, properly stained Y-carrying sperm are easily identified. Because of their weight difference, X- and Y-carrying sperm can be separated by centrifugation. The success of this method depends in part upon the damage done to the sperm by the process.

Artificial insemination differs from **artificial fertilization.** The former process involves introducing sperm into the female's body so that fertilization occurs internally or *in vivo* (from the Latin for "in life"); fertilization is external in the latter method. If the zygote or young embryo looks normal, it is placed in the female's uterus for implantation. What possible objections might some people have to the use of these methods?

Invariably, there are discarded embryos when artificial fertilization is attempted, and there are currently few legal and moral guidelines that relate to these embryos. Do they belong to the woman, the sperm donor, the obstetrician, the clinic or hospital, or a research-funding group? How should abnormal embryos be discarded? Is the discarding abortion, murder, or embryocide?

Another sex preselection method utilizes the fact that Y-carrying sperm in a mild saline solution with charged wires appear to move toward one pole, while X-carrying sperm move toward the other pole. The desired sperm type can then be collected, and some practitioners claim high success rates, while others have little success.

3. *Immunologic methods.* Because sperm are foreign, protein-containing objects, they should theoretically cause antibody formation (Chapter 16). Normally this does not happen, but perhaps in the future females can be sensitized differentially against X- or Y-carrying sperm.

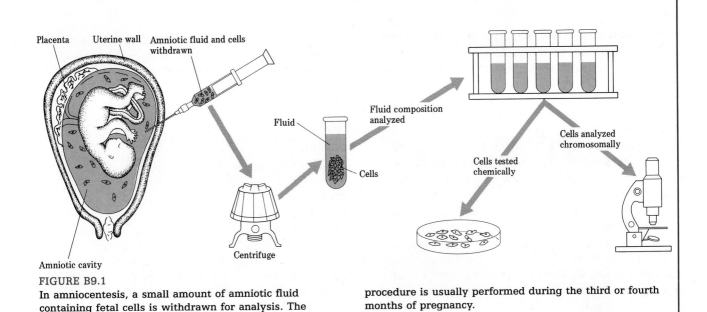

FIGURE B9.1

In amniocentesis, a small amount of amniotic fluid containing fetal cells is withdrawn for analysis. The procedure is usually performed during the third or fourth months of pregnancy.

4. *Induced abortion after fetal sex determination.* **Amniocentesis** (Figure B9.1) makes it possible to identify the sex of the fetus *in utero*, and if the fetus is of the unwanted sex, it can be aborted. However, amniocentesis is safest and most reliable after 16 weeks of gestation, whereas abortions are generally safest and simpler earlier (Chapter 7).

The sex of one's children is important to many people, and the desire for freedom of choice is well established. How would sex preselection affect society if it were practiced widely? Many people responding to questionnaires such as that below indicate either no preference or a desire for equal numbers of males and females. However, if one sex is desired over another in numbers or in birth order it tends to be males. How do you explain these opinions?

On the basis of some questionnaire results, it appears that the sex ratio with sex preselection in practice could be about 120 males to 100 females. Because many forms of social behavior are sex correlated, any significant changes in sex composition could have many sociolog-ical implications. Specific societal effects would depend upon how long the preference for a particular sex lasted.

You may wish to administer a questionnaire like that below to a number of people, instructing them to assume that a safe, reliable, and morally and legally acceptable method of sex preselection is possible.

Sex Preselection Questionnaire

> Sex: Male   Female   (circle one)   Age:_____
>
> Family size: Brothers____   Sisters____
> What do you consider to be the ideal family size?___
> Proportion of sexes wanted in your family: (check one)
>   More males
>   More females
>   Equal numbers of males and females
>   No preference
> Sex of the first child wanted: (check one)
>   Male
>   Female
>   No preference

Sex Reversal, Genetics, and Sports

Although many people had **sex-reversal operations** previously, attention was drawn to this procedure when Richard Raskind, a physician and former captain of Yale University's male tennis team, underwent such an operation (Figure B9.2). This procedure consisted of the surgical removal of the male sex organs, the surgical creation of an artificial vagina, and female hormone treatment. The latter led to redistribution of fat according to a typically female pattern, breast development, and inhibition of male body hair growth. Dr. Richard

FIGURE B9.2
Dr. Richard Raskind and Renee Richards. If you were a ''normal'' female athlete, would you feel that it was unfair to compete against a genetic male who had undergone a sex-reversal operation? Against a genetic female who had received male-hormone treatment? Do you think there should be any restrictions on the right of humans to undergo sex-reversal procedures?

the extra Y chromosome. If so, could an XYY criminal plead innocent to a crime due to genetics, just as some do on the basis of insanity? Or could the behavior of a few XYY males have resulted in part from the social effects of their size and early sexual maturity? This case highlights the problem of what, if anything, the parents of an XYY baby, or one with many other abnormal sets of chromosomes, should be told.

The Weaker Sex?

Although males are often thought to be stronger physically than females, they are clearly weaker in terms of survival. Despite being called "the weaker sex," females typically live longer and most sex-linked genetic conditions (see below) occur in males. Table 9.5 shows the death rates from a number of human problems.

In 1930, there were about 1.5 million more American men than American women. Now there are over 6.5 million more women than men in the United States. Although this difference is partly due to changing immigration patterns, women outlive men on the average and

Raskind became Renee Richards, who looks, thinks, and behaves like a woman. Genetically, however, Dr. Richards is still an XY male.

Female athletes in some countries use male sex hormones to build up their musculature. These females, of course, pass the chromosome analysis test although they are physiologically more male than Renee Richards.

The presence or absence of **Barr bodies** (Figure B9.3) makes it possible to detect a number of genetic abnormalities early. Another useful test involves detection of a small "drumstick" attached to the nucleus of white blood cells in females but usually not in males.

Genetic analyses of occasional "women" athletes who show unusual skill in certain events show that the individuals lack Barr bodies and drumsticks in their cells. Because the "females" in some cases have been found to be genetic males, it is now common to analyze the chromosomes of athletes, especially if records are set. Generally, world and other major records apply only to XX and XY individuals. Do you think there should be separate categories and records for XO, XXY, XYY, and other "abnormal" sex-chromosome patterns?

Actually, there are several arbitrary aspects to athletics because of the way rules are established. For example, few humans average in height and weight are found on most professional basketball and football teams. Is it discrimination that Pygmies hold no world records in the high hurdles? Would "the tables be turned" if the rules were to run under rather than jump over the hurdles? Are there other aspects of athletics that discriminate against those who have inherited certain sets of traits? If so, are there ways (e.g., Special Olympics) to overcome the problems?

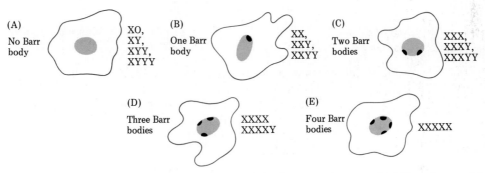

FIGURE B9.3

Barr bodies and their use in diagnosing various abnormal sex chromosome patterns. How many Barr bodies do the cells of women with Turner syndrome have? XXX females? Klinefelter-syndrome males? How do you think the levels of G6PD enzymes compare in Turner females and Klinefelter males?

TABLE 9.5
Some Causes of Death in the United States

Cause of Death (in Rank Order)	Approximate Ratio of Male to Female
Heart diseases	2.0
Cancers	1.5
Cerebrovascular problems	1.2
Accidents	2.9
Chronic obstructive pulmonary diseases	2.2
Pneumonia and influenza	1.8
Diabetes mellitus	1.0
Chronic liver disease and cirrhosis	2.2
Atherosclerosis	1.3
Suicide	3.2
Homicide	3.9
Congenital anomalies	1.1
Kidney problems	1.6
Septicemia	1.4

the trend is increasing. American women outlived their male counter-parts by about 2.5 years in 1900 but now do so by more than 7 years. In fact, American women now have one of the longest life expectancies of any women on Earth, whereas American men rank a relatively poor eighth.

In terms of physical strength, males are generally stronger than fe-males. Although the average protein content of the skeletal muscles of males is higher than in females, the same hormones that give males greater physical strength also contribute to their higher metabolic rate. According to one view, males "burn themselves out" faster than females, and perhaps heterozygote superiority is also important. Many harmful genes are recessive, and males express all such X-linked genes. Also, females produce little testosterone, a hormone that some relate to male vulnerability to cardiovascular diseases.

The ratio of males to females at birth differs from area to area. Often, a lower ratio of male births occurs among low socioeconomic groups than among more affluent groups. This may indicate that male fetuses do not survive the stresses of poor prenatal conditions as well as females. The sex ratio in developed nations is generally higher than in developing nations. The male birth ratio is also lower for multiple births than for single births. In the United States, the birth ratios of males to females for twins, triplets, and quadruplets are about 103:100, 98:100, and 70:100, respectively.

Perhaps cultural factors are as important as biological factors in explaining better female survival. More men are murdered, over three times as many commit suicide, and more die of accidents than females (Table 9.5). Is this due to the traditional male bread-winning role and to a life-style geared to competition, aggressiveness, and both physical and psychological stresses? If women are now freer to engage in roles that were formerly exclusively male, will their life expectancies decline or stabilize?

EPISTASIS

Some albinos have inherited genes for heavy skin melanin deposits as well as a pair of recessive genes for albinism. The latter are **epistatic** to the former because they lead to a failure to produce an enzyme needed to make melanin. Similarly, many genes control our complex hearing apparatus. Any gene that fails to produce a vital structure is epistatic to all other genes, and deafness may result. Some little people ("dwarfs") carry genes for a very tall stature, but if a recessive gene fails to lead to the development of that part of the pituitary gland that produces the growth hormone, the person does not grow normally. If this person mates with a person of average height, their offspring may be above average in height due to the "tall" genes inherited from their short parent.

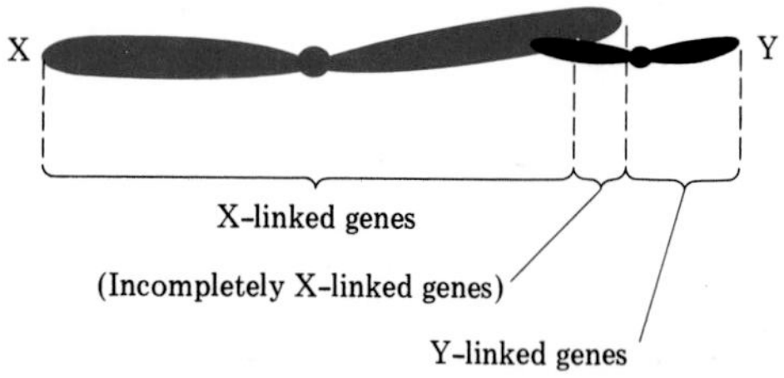

X-linked genes

(Incompletely X-linked genes)

Y-linked genes

FIGURE 9.19
Terms used to designate genes on the X and Y chromosomes. Can males be unaffected carriers of X-linked traits? Does a male receive X-linked genes from his father or from his mother? How does the inheritance of X-linked traits differ from the inheritance of autosomal traits?

SEX-LINKED INHERITANCE

Non-sex-determining genes found in the sex chromosomes are involved in **sex-linked inheritance.** X chromosomes probably have about as many genes as autosomes of similar length, but the Y chromosome is relatively blank for genes not involved in sex determination (Figure 9.19).

X-linked Inheritance

Over 100 abnormal traits are determined by genes found on the X chromosome. The X chromosome is probably the best-known human chromosome, partly because the inheritance patterns of X-linked traits are so distinctive.

X-linked genes may be either dominant or recessive, although most appear to be the latter. Because there are two X chromosomes in normal females but only one in normal males, females may be either heterozygous or homozygous for an abnormal X-linked allele. A woman heterozygous for the trait will express it if it is dominant but will be a normal-appearing carrier if it is recessive. Because males have only one X chromosome, they are affected by the abnormal X-linked allele whether it is dominant or recessive. Note that a male inherits his X chromosome from his mother and transmits it only to his daughters.

"If you can't beat 'em, join 'em" is a common phrase that describes some of the marriage alliances between the royal families of Europe. **Hemophilia,** or "bleeder's disease," is a well-known recessive X-linked trait so named because the blood of affected individuals does not clot properly. "Classical hemophilia" was common among the royal families of Europe (Figure 9.20). Queen Victoria (1819–1901) was a carrier of

(A)

FIGURE 9.20
A royal family portrait (A) and the incidence of hemophilia in the royal families of Europe (B). Why is the present royal family of England free of hemophilia? Why are female hemophiliacs so rare?

(B)

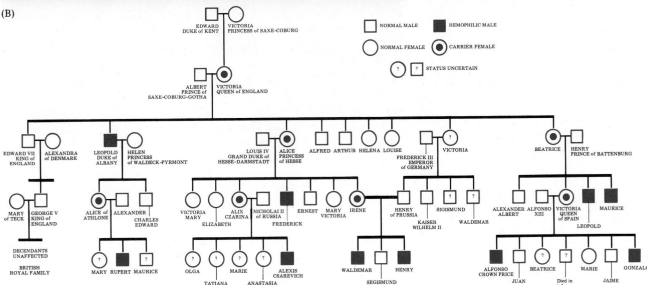

the gene, which may have resulted from a mutation. Queen Victoria's granddaughter Alexandra married Czar Nicholas of Russia. Their son Alexis, who was a hemophiliac, was killed at the age of 14 during the Russian Revolution.

Fortunately, hemophilia is rather rare and not all cases are equally serious. About 1 male in 10,000 has the condition, but only 1 in many millions of females is a hemophiliac. While male hemophiliacs may survive to reproduce, most females do not, many dying following the first menstrual cycle if not before.

A common form of red-green **color blindness** is caused by a recessive X-linked gene (Figure 9.21). About 1 male in 12 has this condition, but only about 1 female in 250 is color-blind.

The distinction between "normal" and "abnormal" X-linked genes is not always clear. For example, some people lack an enzyme involved in carbohydrate metabolism, a condition known as **G6PD deficiency**. Affected people appear normal until they inhale or ingest agents as diverse as some kinds of pollen or certain chemicals. Then their red blood cells break open and severe anemia may result.

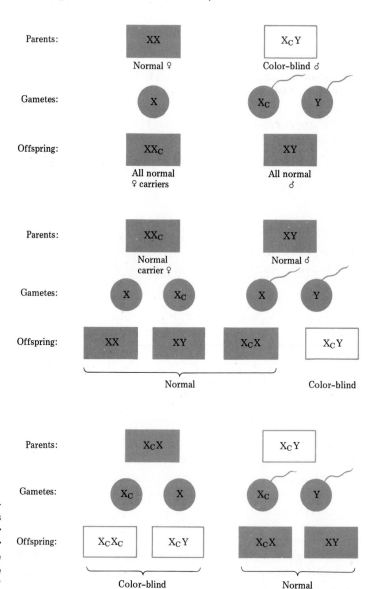

FIGURE 9.21
Some examples of the inheritance of red-green color blindness. Do color-blind males transmit the gene to their sons or to their daughters? Can color-blind males father normal-vision sons? How can a female be color-blind? Could color-blind females be born to non-color-blind parents?

Dosage Compensation

Women, with their two X chromosomes, do not produce more G6PD enzyme than males with their single X chromosome. Could the male's X chromosome "work" twice as hard as the female's X chromosomes? Or could the activity of one of the female's X chromosomes be reduced? The latter explanation appears to be true, and it is known as the **Lyon hypothesis of dosage compensation.**

The nondividing cells of normal human females contain **Barr bodies** within their nuclei, and the Lyon hypothesis holds that the Barr body is a tightly coiled and inactivated X chromosome. Thus, males and females appear to have about the same level of X-linked gene products because one X chromosome is "turned off" in normal women.

Other Sex-linked Inheritance

Most X-linked traits can be expressed in both sexes, although they are more common in males. Some X-linked traits, however, are found only in males because those who express them do not survive long enough to reproduce. An example is one form of **muscular dystrophy** (Duchenne type). Affected males appear normal until their muscles begin to deteriorate by about 6 or 7 years of age, and when they die a few years later they are almost literally skin and bones.

In contrast, some dominant X-linked traits are more common in females than in males. Defective tooth dentin, which leads to rapid tooth wear, is one example.

Although the Y chromosome contains relatively few genes, one cannot be a male without this chromosome because it contains genes that activate the testicular part of the embryonic sex organs. The few genes found on the Y chromosome cause **Y-linked inheritance,** one example of which is to have long hairs on the ears.

SEX-LIMITED AND SEX-INFLUENCED TRAITS

Sex-limited genes are carried by both sexes but are usually expressed in only one sex. Most sex-limited genes are autosomal and should not be confused with sex-linked genes. However, most sex-limited genes affect primary and secondary sex characteristics. For example, most women have little facial hair even though they carry genes needed to produce a beard and the facial hair on their sons is as affected by their genes as by those from the father. Rarely, women are bearded and men develop large breasts. These correctable problems result from genes that produce abnormal secretions of sex hormones and related hormones.

The genes that cause **sex-influenced traits** are recessive in one sex but dominant in the other. For example, an index finger longer than the fourth finger seems to be dominant in females but recessive in males (Figure 9.22). Men have the long index finger only if they are homozygous for the gene. Pattern baldness is another common sex-influenced trait (Figure 9.23).

LETHAL GENES

Genes that produce phenotypes that fail to survive are called **lethal genes.** Not all lethal genes express themselves at the same time during development. Some lethal genes act so early that zygotes fail to divide even once, whereas others may produce abnormal enzymes that prevent im-

FIGURE 9.22
The length of the index finger relative to the length of the fourth finger is a sex-influenced trait. Will a heterozygous woman have a long index finger? If a man has a long index finger and a woman has a short index finger, what will the lengths of the index fingers be in their sons and daughters?

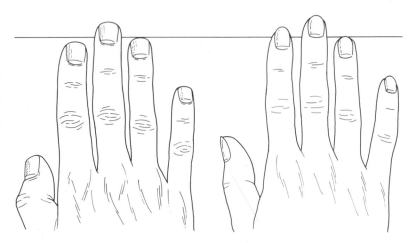

plantation. Still others result in death depending upon when a structure or function becomes vital. Thus, lethal genes that result in abnormal hearts kill embryos when they are about three weeks old. However, genes causing abnormal kidneys may not kill until a child is born, because the mother's kidneys handle the fetal wastes during pregnancy.

Whereas many lethal genes express themselves early, some exert their deadly effect late in life. For example, Huntington disease involves the slow destruction of brain tissue. Early in life when the brain is developing there are no symptoms, but brain deterioration often begins between 30 and 40 years of age, and death usually occurs within a decade.

Some lethal genes cause death when homozygous, but produce non-lethal effects when heterozygous. Because humans have so many genes, most people probably carry some recessive lethal genes, although the chance of mating with someone with identical genes is quite remote.

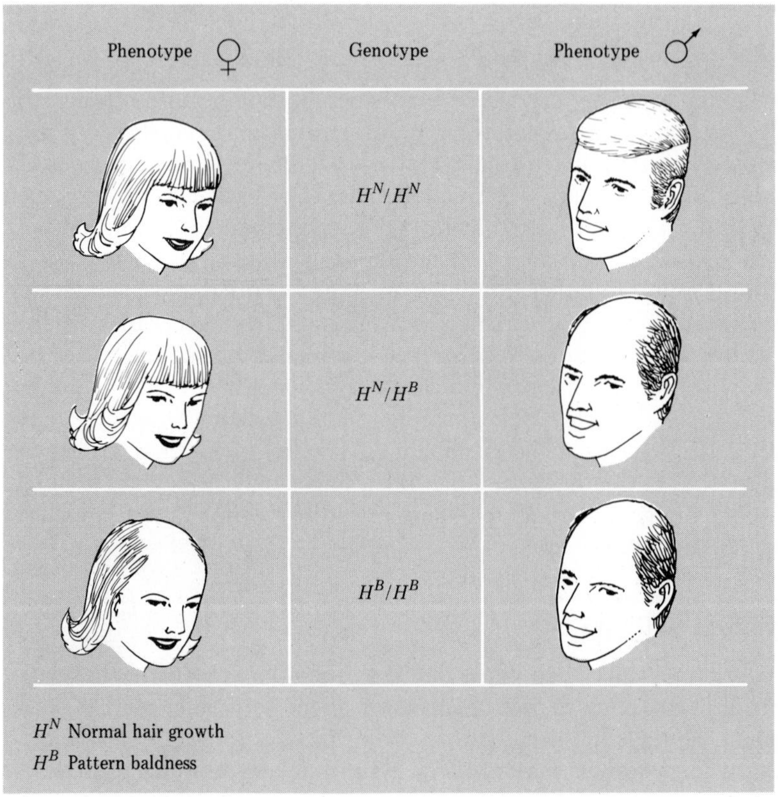

Phenotype ♀	Genotype	Phenotype ♂
	H^N/H^N	
	H^N/H^B	
	H^B/H^B	

H^N Normal hair growth
H^B Pattern baldness

FIGURE 9.23
The inheritance of pattern baldness. How does the inheritance of this trait differ in males and in females?

SUMMARY

1. The two or more forms of a gene are called **alleles. Dominant** allelic genes are usually expressed, whereas **recessive** genes are usually not. Dominant genes are symbolized by a capital letter and recessive genes by a lowercase letter. The environment determines the degree to which the potential provided by genes is attained.

2. **Homozygotes** contain two identical alleles of a pair (e.g., *RR* or *rr*), whereas **heterozygotes** contain two different allelic genes (e.g., *Rr*). The individual's appearance is called its **phenotype,** and its genetic makeup is termed its **genotype.** The phenotype of heterozygous individuals is normally the same as that of organisms that are homozygous dominant for the trait. Heterozygotes are also often called **hybrids** and **carriers.**

3. **Punnett squares,** mathematical calculations, and **pedigrees** may be used to follow the results of **crosses** or matings.

4. **Gregor Mendel** developed his theory of **genetics** in the 1860s based upon experiments with the garden pea. Mendel's **principle of segregation** specifies that pairs of alleles segregate during the formation of gametes and each gamete receives only one member of each allelic pair. His **principle of independent assortment** states that each pair of alleles segregates independently during gamete formation.

5. Humans normally possess 22 pairs of **autosomes** and one pair of **sex chromosomes.** Relatively few human traits are of the **discontinuous gradation** type. Rather, most important human traits result from **polygenic inheritance** and exhibit **continuous gradation.**

6. Hundreds of cases each of dominant autosomal traits (e.g., **Huntington disease** and **Marfan syndrome**) and of recessive autosomal traits (e.g., **PKU** and tongue rolling) are known. Not all traits exhibit the same **expressivity.** Individuals who are homozygous recessive for a trait are especially common among the offspring of **consanguineous matings.**

7. In **incomplete dominance,** heterozygous offspring may exhibit traits intermediate between those of parents homozygous for the two alleles. Thus, $Hb^A Hb^A$ individuals are normal, $Hb^A Hb^S$ people have **sickle-cell trait,** and those characterized by $Hb^S Hb^S$ suffer from **sickle-cell anemia.**

8. In **codominance,** two or more alleles are expressed in heterozygotes, as observed in the inheritance pattern of the **ABO blood group.**

9. **Trisomy** may occur in either autosomes or sex chromosomes, such as trisomy-21, or **Down syndrome.** As with missing chromosomes, extra chromosomes usually lead to problems, including miscarriages.

10. The sex chromosomes (twenty-third pair) differ in males (XY) and females (XX). Typically, more males are conceived and born than females, although women generally survive longer than men. The **X chromosome** is relatively large and contains a normal number of genes, but the **Y chromosome** is small and contains few genes. Perhaps this leads to differences in the speed or survival of the X- and Y-carrying sperm.

11. **Turner** (XO) and **Klinefelter** (XXY) **syndromes** are among the conditions that result from abnormal numbers of sex chromosomes. **Genetic mosaics,** as in **hermaphrodites,** occur when not all cells contain

the same assemblage of chromosomes (e.g., XX/XXY). **Nondisjunction** during cell divisions explains many of these conditions. Early detection (e.g., observing the number of **Barr bodies**) of many chromosomal abnormalities is possible. **Sex reversal** operations change the phenotype but not the person's genotype.

12. Although **sex preselection** has long been of interest, many old methods were based upon faulty assumptions. Modern techniques basically attempt to separate X- and Y-carrying sperm and use these in **artificial insemination** (*in vivo*) or **artificial fertilization** (*in vitro*) methods. **Amniocentesis** is a method of identifying the sex and numerous other traits of fetuses *in utero*.

13. **Epistatic genes** may modify or prevent the expression of nonallelic genes.

14. Genes located on the sex chromosomes are involved in **sex-linked inheritance**. Males may express a recessive gene located on their single X chromosome, but females do so only when the allele is homozygous. **Hemophilia** and a common form of red-green **color blindness** are caused by X-linked recessive genes. Y-linked inheritance is rare in humans.

15. **Sex-limited genes** are usually expressed in only one sex although they are carried, usually on an autosome, by both sexes. Many such genes influence the primary and secondary sex traits. **Sex-influenced traits** are caused by genes that are recessive in one sex but dominant in the other (e.g., pattern baldness).

16. **Lethal genes** influence the survival, either prenatally or postnatally, of an individual.

REVIEW QUESTIONS

1. Define the following and state an example of each: gene, chromosome, allele, hybrid, nondisjunction, epistasis, dosage compensation, consanguineous, and lethal gene.

2. Compare and contrast the following pairs of words, and state an example of each: dominant and recessive, homozygous and heterozygous, phenotype and genotype, sickle-cell trait and sickle-cell anemia, continuous and discontinuous variation, and sex-limited and sex-influenced traits.

3. Distinguish among the patterns of dominant-recessive, incomplete dominance, codominance, and sex-linked inheritance.

4. Suppose two normally pigmented humans who are carriers of the gene for albinism have children. What is the chance of the couple producing albino daughters?

5. Suppose that an apparently healthy, 20-year-old man learns that his father has just been diagnosed as having Huntington disease. His mother is normal. What are the chances that the son will develop the disease? If the son does not develop the disease, what are the chances that his children will have Huntington disease?

6. Total color blindness is a rare condition caused by a recessive autosomal gene. Suppose that a normally seeing man whose mother is totally color-blind and a non-color-blind woman whose father is totally

color-blind have children. What are the chances that their offspring will be normal? Carriers? Totally color-blind?

7. A woman of blood type A and a man of blood type B produce a child whose blood type is O. What are the genotypes of the three individuals? What other blood types might children produced by these parents have?

8. Suppose that you inherit a dog valuable because of its particular coat, which is controlled by a dominant autosomal gene. You decide to provide the dog for breeding purposes, but while checking its pedigree you discover that some of its ancestors possessed a coat defect. How could you determine whether your dog is homozygous dominant or is heterozygous for the coat-controlling gene?

9. A normally seeing man whose mother is red-green color-blind and a non-color-blind woman whose father is red-green color-blind have children. What are the chances that their offspring will be normal? Carriers? Red-green color-blind?

10. A woman with sickle-cell trait and a man whose mother has sickle-cell disease have children. What proportion of their offspring will be normal? Have sickle-cell trait? Have sickle-cell disease?

11. Some males are affected by X-linked genes. Explain why (1) their offspring are unaffected and (2) the sons of their daughters, and those of their sisters, are affected by the traits.

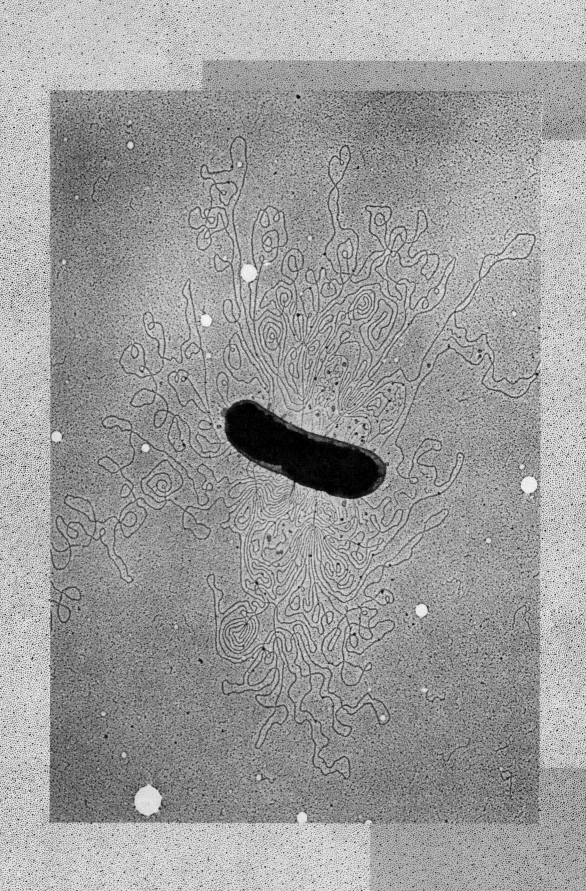

CHAPTER 10

HEREDITARY MECHANISMS

Describing the patterns of inheritance is one thing; explaining them is another. "Like begets like" is an adage that is old and familiar, as are many ideas about how people inherit "their mother's eyes" or "their grandfather's nose." Aristotle believed that particles from all parts of the body came together to become part of the gametes, and Charles Darwin accepted this view. While the theory was wrong, the concept that "particles" are important in inheritance survives. Today, the units of heredity are called genes *and the abbreviation for their chemical composition is known to most educated people—DNA. In the nineteenth century, Mendel predicted heritable factors, and the chromosomal theory of inheritance in the early 1900s added support to the view. The theory of genes was developed in this century, and details are accumulating rapidly as thousands of scientists delve deeper into the mechanisms of heredity.*

Many studies of genetic mechanisms utilize microorganisms, especially the bacterium *Escherichia coli,* which exists in such immense numbers in our intestines. Although this tiny bacterium has only a single chromosome, we can see the large extent of its DNA exposed here. Genes are various sections of the DNA molecule that encode for proteins.

THE SEARCH FOR GENETIC UNITS

Patience is important in studying the inheritance of peas or other large organisms. For example, the pea seeds that form in one summer must be stored until the following spring for planting, and one must wait additional weeks to note the plant's height, the color of its flowers, or the color and texture of its seeds and seed pods. However, many bacteria and viruses reproduce very quickly, sometimes 2 or 3 times per hour, and the answers to many questions about inheritance can be gained rapidly through the study of such subjects.

One significant experiment in 1928 was performed on the bacterium that causes pneumonia in mammals, *Streptococcus pneumoniae* (Figure 10.1). Two forms of colonies, smooth (S) and rough (R), were observed on growth media by Frederick Griffith, a British scientist. The S cells produced a polysaccharide coat that the R cells did not. Mice injected with the living S-cell strain died, while those injected with live R cells

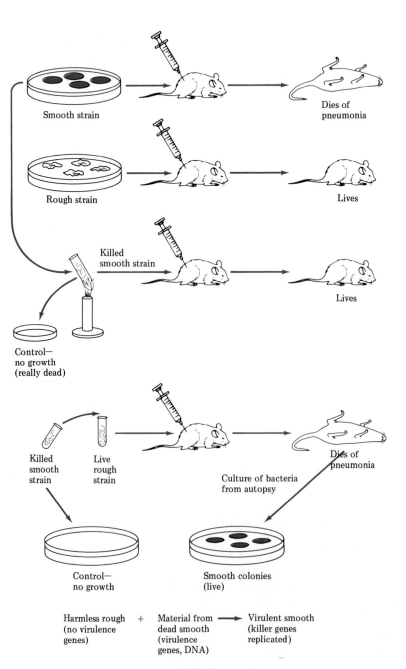

FIGURE 10.1
Griffith's experiment on transformation of bacteria. Mice injected with living rough (R) strains of *Streptococcus pneumoniae* lived, whereas those injected with the encapsulated smooth (S) forms died of pneumonia. To discover whether the coat or the cell was pathogenic, Griffith injected mice with heat-killed S-strain bacteria. Surprisingly, however, mice died when injected with nonpathogenic living R cells and heat-killed S cells. This and much further research by others led to the discovery of transformation, or the fact that genetic information in the dead cells had passed to the live ones.

survived. However, S cells that were killed by heat were harmless to mice, indicating that it was not the cellular coat that was pathogenic. Then why did mice die when injected with harmless heat-killed S cells mixed with harmless live R cells, and why did their blood contain living S cells? These and other observations led to the discovery of **transformation**, or the assimilation of external genetic material by a cell. The R cells had taken in segments of the S-cell DNA and become pathogenic smooth cells.

By the 1940s it was known that chromosomes consisted of protein and a seemingly simple molecule called **deoxyribonucleic acid,** or **DNA.** The existence of DNA had been known since 1869, a time when both Mendel and Darwin were active researchers. Knowing that proteins are extremely varied in forms and functions, many scientists at the time suspected that the protein complement of chromosomes was the material of genes. However, heat denatures or inactivates most proteins, and Griffith's heat-killed S cells were still capable of killing mice when they were mixed with live R cells. This and much other work led ultimately to the conclusion that DNA, not protein, is the genetic material.

For example, most viruses are largely DNA enclosed in a protein coat (Chapter 1). **Bacteriophages** (from the Greek for "bacteria eaters") are extensively studied, and these viruses, often just called *phages,* were used in experimental work published in 1952 by Alfred Hershey and Martha Chase, shown in Figure 10.2. Looking somewhat like space vehicles

FIGURE 10.2
The Hershey-Chase experiment. The fact that protein but not DNA contains sulfur and that most of a phage's phosphorus is in its DNA made it possible to tag the two components with different radioactive isotopes.

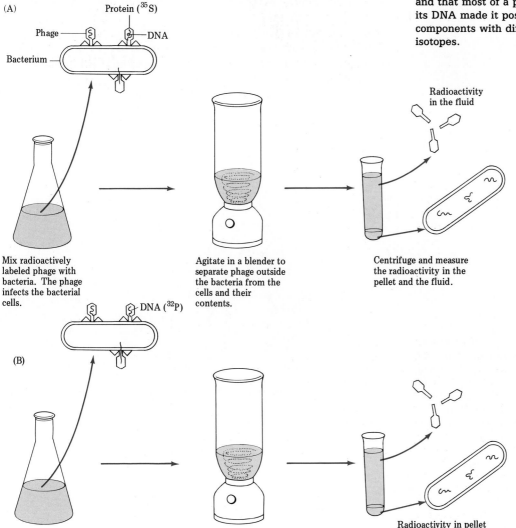

landing on the moon, phages contact cells and inject their DNA into these cells as a hypodermic needle does, with the protein coats remaining outside. The phage's DNA directs the cells to synthesize numerous new phages, which are released when the bacteria rupture.

Different radioactive isotopes were used to tag the two viral components—sulfur in the protein coat, and phosphorus in the DNA. The batches of tagged phages were then allowed to infect the common intestinal bacterium *Escherichia coli*. The mixture was then agitated and centrifuged to separate the heavier bacterial cells from the lighter viral particles and some of their empty protein coats. Most of the radioactive sulfur was found in the solution, and the pellet of bacteria contained most of the radioactive phosphorus.

DNA AND ITS REPLICATION

By the 1950s, the chemical nature of DNA was well established (Figure 10.3). DNA is a polymer synthesized from monomers of four different nucleotides, each consisting of a nitrogenous base, a sugar, and a phosphate group (Chapter 4). Each nucleotide may consist of any one of four different organic bases: adenine (A), cytosine (C), guanine (G), or thymine (T). (The origin of names is often of interest. *Cyto-* means cell, and cytosine is found in all cells. *Adeno-* means gland, and adenine and thymine were first isolated from the thymus gland. Guanine was first isolated from guano, the deposits of fecal material from seabird and bat colonies.)

What was not known until the work of the American James Watson and the Englishman Francis Crick in the early 1950s was the three-dimensional structure of DNA. They proposed that the shape of the DNA molecule was helical, somewhat like a rope ladder with rigid rungs

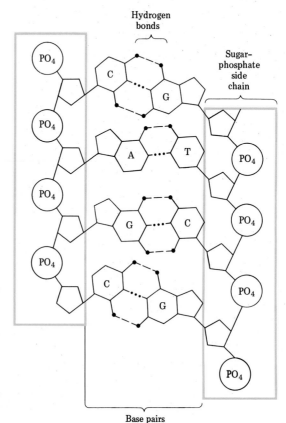

FIGURE 10.3

The chemical structure of DNA. The DNA molecule consists of phosphate and five-carbon sugar (deoxyribose) side chains with one of four ring-shaped organic bases protruding inward from each sugar. The amounts of adenine plus thymine are equal in all DNA molecules, as are the amounts of guanine plus cytosine. However, the observation that the amounts of cytosine plus guanine often differed greatly from the amounts of adenine plus thymine led eventually to the notion that the base-pair pattern was important in inheritance.

that was twisted into a spiral. They believed that DNA consisted of two strands, the now-famous **double helix** (Figure 10.4). The "ropes" were the molecule's sugar-phosphate component and the "rungs" were pairs of nitrogenous bases, with A always coupled with T and C with G, by means of hydrogen bonds. Because DNA molecules can be extremely long, the sequence of the four bases used in their assembly can be varied in an infinite number of combinations. This is a key to the coding of genetic information. **Genes** within DNA typically are hundreds or thousands of nucleotides in length, with each gene having a specific sequence of bases.

Obviously, genetic information must be capable of being duplicated in order for copies of an organism's blueprint to be transmitted to the next generation. During **replication,** the two strands of DNA separate, and each strand becomes a template for the assembly of another, complementary strand (Figure 10.5). Each daughter molecule has one strand derived from the parent molecule and one new strand. The specific chemical events in this process are not only very complex but also very rapid, with about 50 nucleotides per second being added during the replication of your DNA. Most of the time, the events are also very accurate, although occasional mistakes occur (see later section on mutations). Scores of enzymes have been discovered that identify and correct most errors, many of which result from our exposure to such potentially damaging agents as mutagenic chemicals, radioactive substances, X-rays, and ultraviolet light.

Many details about replication are unknown, including what initiates it, why it stops forever in certain cell types that become very specialized (e.g., mature neurons and muscle fibers), and why it becomes uncontrolled in cancer (Chapter 16).

FIGURE 10.4
Using a twisted-ladder analogy, the double helix consists of sugar-phosphate "ropes" and of "rungs" of nitrogenous bases. The base pairs of the rungs are always arranged with A-T and C-G pairs.

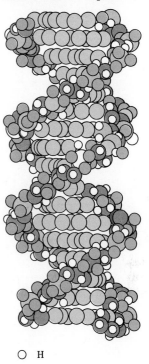

○   H

⬤   O

⬤   C in phosphate–sugar chain

◯   C and N in bases

◯   P

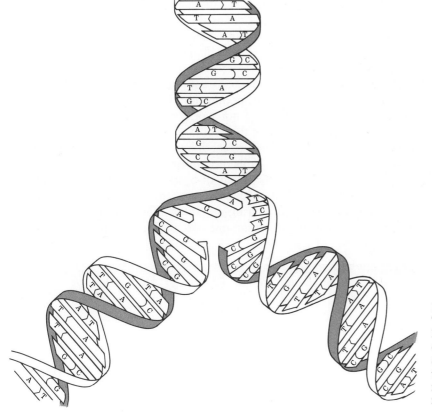

FIGURE 10.5
The general process of DNA replication. Each daughter molecule has one strand derived from the parent molecule (tinted) and one new strand (untinted). Note that the sequence of base pairs in the daughter molecule is identical to that in the parent molecule.

EXPRESSING GENES

Replicating genes and transmitting them during reproduction are obviously important, but genes must also be expressed. The metabolic activities of a cell depend largely upon its enzymes and other proteins, and it is the synthesis of these substances that genes specify. Yet, most of the metabolic "action" in a cell occurs in its cytoplasm, and the genes remain in the nucleus. In this section you will explore the flow of molecular or genetic information from gene to protein. Keep in mind that genes do not construct proteins but, rather, they provide the blueprint that guides the synthesis.

Controlling Metabolism

One of the earliest accurate theories about gene action was proposed in 1908 by the British physician A. E. Garrod. A number of inherited diseases were then called "inborn errors of metabolism." Garrod suggested that individuals who expressed these metabolic defects or phenotypes lacked a particular enzyme. Eventually, this led to the insightful but oversimplified hypothesis of "one gene–one enzyme." For example, while all enzymes are proteins, not all proteins are enzymes: Keratin in hair and the hormone insulin are common nonenzyme proteins that are gene products. However, a one gene–one protein hypothesis is not satisfactory either, because many proteins are composed of two or more different polypeptide chains. Therefore, a one gene–one polypeptide hypothesis is more accurate for the general process of gene action.

Transcription

An early step in the genetic specification of an amino acid sequence for a particular polypeptide is the transfer of information from double-stranded DNA to single-stranded **ribonucleic acid,** or **RNA.** RNA, the other form of nucleic acid, contains the sugar ribose instead of deoxyribose, and uracil (U) rather than thymine pairs with adenine. Also, the RNA molecule is usually far smaller than DNA, usually being a thousand to a million times shorter.

Because there are 20 amino acids and only 4 nucleotides, there must be a pattern or code for such specification. Obviously, a one nucleotide-one amino acid system would not work, nor would a two nucleotide-one amino acid code because this would produce only 16 (4^2) of the 20 amino acids. Therefore, much research has shown that triplets of bases, or **codons,** are the smallest units of uniform length that can code for all of the amino acids. In fact, triplets of bases can specify 64 (4^3) combinations, providing useful redundancy, but no ambiguity. The **genetic code** itself was "cracked" during the 1960s (Figure 10.6). Most amino acids are encoded by more than one codon. For example, GCU, GCC,

FIGURE 10.6

The triplet genetic code. Each amino acid in a polypeptide is encoded by a series of three nucleotides (codon) in one strand of DNA.

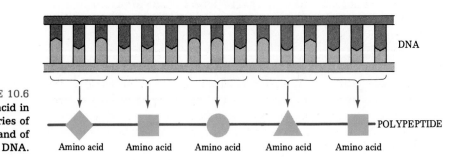

DNA

POLYPEPTIDE

Amino acid   Amino acid   Amino acid   Amino acid   Amino acid

GCA, and GCG all encode for alanine. Three codons act as "stop" signals that, like the period at the end of a sentence, specify the end of a polypeptide.

The so-called **messenger RNA** (mRNA) is **transcribed** from a DNA template by a process similar to that which synthesizes a new DNA strand during DNA replication. Large and special enzymes (RNA polymerase) govern the process of forming a gene copy, and particular sequences of nucleotides provide the "start" (promotor) and "stop" (terminator) signals. Only portions of one strand of DNA are transcribed, although it is usual for many mRNA molecules to be transcribed from different parts of the same gene simultaneously. The final mRNA molecule or transcript then crosses the nuclear membrane and enters the cytoplasm, where it associates with ribosomes, and protein synthesis ensues (Chapter 3).

Translation

The amino acids in the cytoplasm cannot recognize the codons on mRNA directly. Instead, they require an "interpreter" called **transfer RNA** (tRNA). Enzyme-assisted tRNA molecules pick up appropriate amino acids and bring them to mRNA for alignment to form the new polypeptide (Figure 10.7). This process is called **translation** because the nucleotide sequence is like a language using one set of alphabetic symbols that are translated into another language, resulting in a predetermined amino acid sequence.

Ribosomes are important organelles in the process of coupling tRNAs to the codons on mRNAs. In the subunits of ribosomes is another specialized form of RNA, **ribosomal RNA** (rRNA). rRNA is actually the most abundant form of RNA because of the thousands of ribosomes found in most cells. The ribosomes that are bound to the membrane of the endoplasmic reticulum (rough ER) are the sites for production of the proteins that will become part of the cell's membranes or that are packaged for export from the cell (e.g., certain hormones). Free ribosomes are associated with the production of most of the proteins that function within the cytoplasm. One minute or less is enough time for an average-sized protein to be produced. This occurs as the ribosome binds to one end of the mRNA transcript and then moves down the mRNA in increments of three nucleotides and adds appropriate amino acids to the end of a lengthening chain. A "stop" signal indicates the end of the protein, and the ribosome then disengages from the mRNA and releases the new protein.

MUTATIONS

Sometimes a heritable and uncorrected change occurs in the relatively stable DNA molecule, and this is called a **mutation**. For example, substitution in a single base pair in part of the hemoglobin gene (the amino acid valine inserted instead of the normal glutamic acid) causes sickle-cell anemia (Chapter 9). Mutations contribute to genetic diversity, and evolution depends upon genetic variation in organisms. Sometimes these rare changes lead to an improved condition and increase survival and/or reproductive capabilities. In most cases, however, mutations are detrimental. The structure of the DNA molecule and its several associated molecules protects it from many mutations, and numerous enzymes act constantly to find and correct errors.

FIGURE 10.7

Overview of gene actions. Genes are specific sections of the DNA molecules located in the cell's nucleoplasm. When genetic information is expressed, it is via DNA to mRNA (the process of transcription) and then from mRNA to polypeptide (translation). During translation, tRNA brings appropriate amino acids to particular codons on mRNA, and rRNA coordinates the construction of a protein.

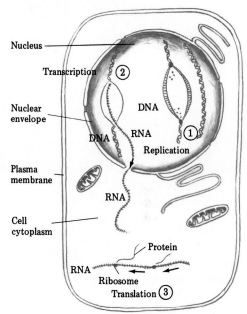

Some mutations affect your somatic cells, such as those found in the skin, lungs, liver, or intestines. These mutations may have major, even fatal, effects, but only in your body. In contrast, the mutations that affect the germ cells may be passed along to offspring. The frequency of such mutations is not known for certain, but a common estimate is that a particular gene is altered in about one out of a million gametes.

The analogy between a healthy organism and a functioning wristwatch has been made to explain several features of mutations. Probing the interior of a wristwatch with an instrument while you are blindfolded is highly unlikely to make the watch keep better time, any more than a random change in the genetic blueprint of an organism will always improve its functioning. The removal of a few teeth or a minor gear may have devastating results on the timepiece, just as a small change in nucleotide sequence may produce a serious metabolic disorder.

Agents that cause mutations are called **mutagens,** and common mutagens include ionizing and ultraviolet radiation, heat, and various chemicals. Mutagens may change the genetic message itself, for instance, by altering the sequence of DNA nucleotides. Ultraviolet light produces a "kink" in the double helix by affecting the way base pairs link, and this prevents the proper replication of the duplex. Still other mutagens alter the way the genetic message is organized.

Actually, changes in organisms can occur in ways other than through mutations, such as in the various forms of **recombination** when existing elements of the genetic message change. Thus, a mutation may change a "letter" or gene in the message, leading to a nonsense "word" or nonfunctional enzyme. However, even if all of the letters and "pages" (or chromosomes) are correct but the order of the pages is wrong, as when chromosomes are rearranged, then the meaning of the message

FIGURE 10.8
The amount of DNA in your cells is enormous, and some of the ways the genetic blueprint is folded and packaged are summarized here.

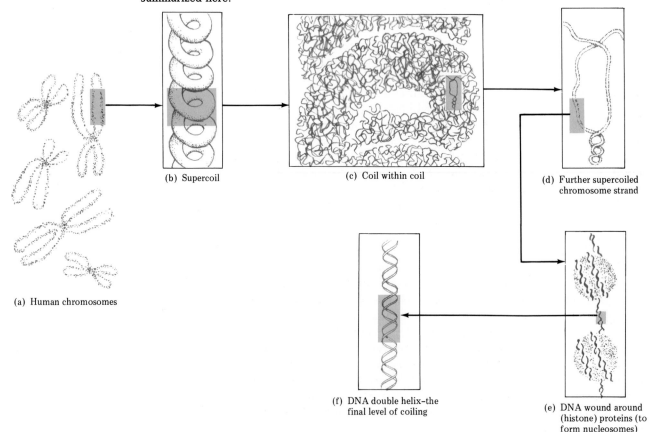

(a) Human chromosomes

(b) Supercoil

(c) Coil within coil

(d) Further supercoiled chromosome strand

(e) DNA wound around (histone) proteins (to form nucleosomes)

(f) DNA double helix–the final level of coiling

may be altered or destroyed. Sometimes two chromosomes trade segments, or one chromosome "donates" a segment (transposon) to another, and, of course, there is chromosomal assortment during meiosis (Chapter 3). Whatever its source, however, genetic change is the raw material for evolution.

CONTROL OF GENE EXPRESSION

"Metabolic anarchy" is a phrase some have used to describe what would happen if a cell expressed all of its genetic information continually. For proper homeostasis, genes must be switched on and off as external and internal conditions change. For example, it would do no good for a cell to produce an enzyme if its substrate were not present. At any given time, a cell expresses only about 1 percent of its genome, which is considerable in the higher organisms. For example, packed into the 46 chromosomes of each human somatic cell is about 2 m (6 ft) of DNA. The average human DNA molecule consists of about 160 million nucleotide pairs, or is about 5 cm (2 in.) long if stretched out. This large amount of DNA fits into the nucleus through an elaborate system of packaging (Figure 10.8). While many details about gene expression are unknown, the way DNA is folded undoubtedly has much to do with the activity of its genes.

So-called **regulatory proteins** act on specific nucleotide sequences to activate or repress them. Because much of our DNA is usually inactive, most of the regulatory proteins may function to turn on transcription of particular genes at appropriate times. For example, the biochemical processes of development must take place at just the right time and in just the right sequence (Chapter 11).

RECOMBINANT DNA

Recombinant DNA technology began in the mid-1970s, creating unprecedented opportunities, challenges, and potential problems. The basic technology is to combine genes from different sources *in vitro* and transfer the recombinant DNA into cells where it is expressed (Figure 10.9). It is now possible to isolate specific genes and to produce large amounts of them and their products. The first introduction of a human gene into another organism occurred in 1980, when the gene that encodes for interferon, a substance that increases human resistance to viral infection, was transferred into a bacterial cell (Chapter 16).

The use of living organisms or their components to do practical tasks is called **biotechnology,** and the growth of biotechnology companies or subsidiaries is phenomenal. It is big business. However, not all such uses of organisms are new: Various microorganisms have been used for many centuries to make products such as cheese and alcoholic beverages. Also, since the advent of the Agricultural Revolution about 11,000 years ago, there has been selective breeding or artificial selection of crop and other plants and of domestic animals. More recently, antibiotics have been produced from microorganisms, again largely through a selection process of finding or creating (e.g., by using a mutagen) the initial organisms. However, biotechnology based upon recombinant DNA technology is a far more precise and powerful tool for the improvement of food production and human health.

FIGURE 10.9

Summary of a common genetic engineering practice. In this example, restriction enzymes are used to cut genes. The "sticky ends" of one segment allow another segment to be inserted, resulting in recombinant DNA. A vector (e.g., a virus or plasmid) is used to carry the altered DNA into a host such as an *E. coli* cell, which then reproduces according to its engineered genetic blueprint.

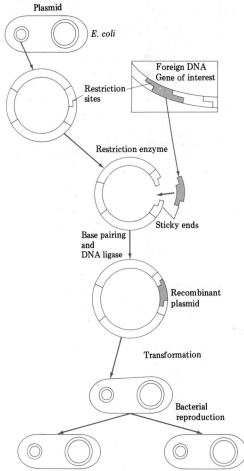

Bacterial clone carrying many copies of the foreign gene

Cloning Is Not Clowning

Geraniums and many other house plants are propagated asexually by means of cuttings, and no gametes need be involved. Asexual reproduction at the cellular level is common in humans as we grow, heal wounds, and replace old cells. Could entire new humans and not just patches of cells be produced asexually, as geraniums are, without any reproductive act on the part of parents? In theory they *can* be produced, but many wonder whether the so-called test-tube babies, somewhat like those of Huxley's *Brave New World, should* be produced.

Cloning is the asexual duplication of cells or organisms and conceptually (not a pun!) is relatively simple, as illustrated by the following scenario, which is possible for some animals. First, a mature, unfertilized egg is taken from the ovary by means of a needlelike instrument (Figure B10.1). After the egg's nucleus is removed (for instance, by microsurgery or radiation), it is replaced with a nucleus from a normal somatic cell. The egg responds by dividing as if it had been fertilized normally. This is possible because all the genetic information needed to produce a human (i.e., the full diploid set of chromosomes) is encoded in the chromosomes of every body cell.

Normal fertilization is called *in vivo* because it occurs in the body. Just as in **artificial fertilization,** cloning occurs *in vitro* or outside of the body. In the latter cases, the embryo may be grown for a time in the laboratory to ensure that early development is normal and then is allowed to implant itself into a properly prepared uterus for later development.

Annually, tens of thousands of American women whose mates are either sterile or carry genetic flaws use **artificial insemination.** In many cases the sperm are provided by anonymous donors whose pedigrees have been screened carefully for genetic defects. Do you see any legal, ethical, or moral problems with artificial insemination? Should potential sperm donors be allowed to advertise their services? Should donors be paid for their services as physicians are for theirs? What are sperm worth?

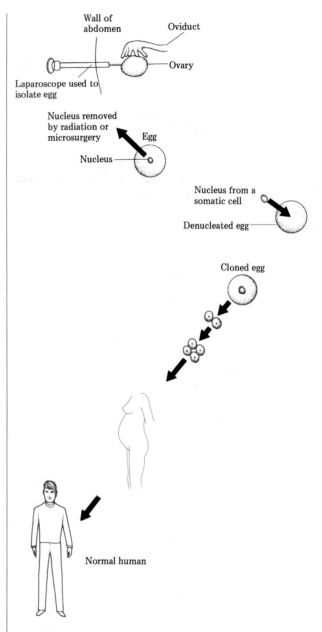

FIGURE B10.1
The general methodology of cloning a human.

Note that artificial insemination is not the same as artificial fertilization. The former occurs *in vivo*, whereas the latter takes place *in vitro*. *In vitro* fertilization necessarily involves human manipulation and the potential risk of damage to the embryo. Who "owns" the embryo produced by artificial insemination? Artificial fertilization? How should damaged or faulty embryos be disposed of, and what legal or moral guidelines should relate to the practice?

In some couples, both of whom are fertile and want a child of their own, as opposed to an adopted child, pregnancy is not appropriate for the woman for one reason or another. Artificial fertilization may be the answer, and the couple may provide the gametes for *in vitro* fertilization and hire a surrogate "mother" for the pregnancy. This practice occurs today, and perhaps when it becomes even more common we will see a new section in the business pages of telephone directories entitled "Wombs for Rent." How much are the services of a surrogate mother worth?

Although some women are sterile, most produce far more eggs than they can ever "use." Just as some males provide sperm for artificial insemination, some females may provide eggs for artificial fertilization. If this practice becomes more common and acceptable, still another new section in the telephone directory might read "Eggs for Sale: Human." How much is an egg worth? A dozen?

Actually, the fertile female who desires a child of her own, totally her own, may eventually be able to produce one through cloning. She has both the eggs and the diploid nuclei to transplant into them. She also has the choice of using her own body for the pregnancy or the body of a surrogate mother. (In terms of reproduction, males could then be irrelevant, although females might wish to keep some males around, even if only for display in "zoos.") Although the person produced would have a genetic blueprint identical to that of its mother, or to any other donor of the somatic nucleus, genes only provide the potential for what we may become; it is the interaction of genes and the environment that determines what we do become. Presumably, however, if you were interested enough to clone a person, then you would also probably attempt to duplicate the "parent's" upbringing so that the clone would be as psychologically like the donor of the somatic nucleus as it would be physically.

Would literature, science, humanitarianism, art, and politics be improved by more Shakespeares, Einsteins, Schweitzers, Picassos, and Lincolns, respectively? Should society select the perfect soldier and clone an army? Or clone small astronauts who take up less room in their space capsules and require fewer nutrients on long voyages? Or, besides the Clone Ranger and Bozo the Clone, who else, living or dead, might improve various fields of human endeavor if cloned in proper numbers? Should permits be required for cloning, and should there be limits on the number of clones allowed?

It has often happened that a person's genius or contribution to a discipline was not recognized until after the individual's death. We cannot exhume the dead and use their genes, but some have proposed that we begin now to establish human "cell banks" wherein some gametes and somatic cells of every person might be preserved indefinitely along with information about their traits. Most such genes existed in the human gene pool before the individual occurred on Earth, and most will persist when the person dies; individuals are merely unique but temporary reservoirs of these genes. Does society, then, have a right to use these genes as and when it sees fit, perhaps long after your death? Do you own your genes or does society? If the latter, then which individuals or agencies are guardians of the gene bank?

Genetic and Biomedical Engineering

Perhaps genetic engineering will someday provide us with the ability to correct or eliminate genetic disorders, not merely to treat their symptoms. Some genetic syndromes involve many genes but others are traceable to a single defective gene. Theoretically, some of the latter might soon be replaced with a normal gene developed through recombinant DNA technology.

Perhaps the easiest cells to begin with are somatic cells with a major genetic defect. If such cells were of an actively reproducing type, such as bone marrow as opposed to nerve or muscle cells, then the normal gene might be replicated quickly within an individual. Obviously, the protein product of the normal genes would have to correct the problem, even though the protein would be absent in most other tissues. Researchers will undoubtedly try to discover how early a normal gene can be introduced, for example, into undifferentiated embryonic cells. Research will also almost certainly reach the stage of attempting to treat germ cells in hope of reducing the defect in future generations.

How much *should* we tinker and tamper with our germ lines, even though we *can*? Could the technology be misused, if, for instance, some people arbitrarily decided which genes were good and which bad? Or could we accidentally create a genetic monster?

Safety is another important issue in biotechnology. While most laboratory procedures are standardized and well tested, accidents happen occasionally in any setting. One concern is that an engineered "organism," some of which have been patented, might escape from the laboratory and be pathogenic, perhaps causing new or more virulent forms of cancer or uncontrollable plagues. A more recent worry concerns scenarios involving engineered organisms designed to be released into the environment, such as those whose laudable functions are to help plants fix atmospheric nitrogen or to degrade environmental pollutants. Naturally, most hope that genetic engineering will not be used in warfare, but it is all too easy today to produce enormous quantities of botulin and other toxins, for example.

Translated literally from its roots, **eugenics** means "good genetics," and who could be against that? As implied above, however, one concern centers on the definitions of "good" and "bad"; we have seen all too many cases in history of one group of humans defining themselves as superior, and annihilating those with "inferior" genes so as not to "dilute" or "weaken" the gene pool. Therefore, some favor **euthenics** as a method of improving the human condition quickly. Euthenics generally implies the intensive use of our extensive knowledge in education, public health, and "traditional" medicine, applied universally on Earth to improve the lives of people.

It is routine today for remote sensing devices to measure the vital signs of hospital patients. One doctor or nurse can monitor many patients, even some outside the hospital, from a central control panel. With closed circuit TV it is even possible to view one of the most sensitive indicators of all—the patient's face. Significantly stimulated by America's space program, **biomedical engineering** affects every area of medicine and is a major thrust of euthenics. Perhaps the most dramatic

---

Restriction enzymes are major tools of recombinant DNA technology. The normal function of these numerous bacterial enzymes seems to be to protect bacteria against the intrusion of DNA from other organisms. The special enzymes do this by cutting, or restricting, specific nucleotide sequences at particular points. In the laboratory, restriction enzymes can be used to cut desired regions of DNA, leaving fragments with "sticky ends," and to insert and join enzymatically DNA pieces originating from different sources. The result, recombinant DNA, is transferred into an infective agent or vector, such as a virus or an extrachromosomal segment of DNA (plasmid), and then into the target cell. As the latter divides, any foreign genes carried by the recombinant DNA are replicated in the **clone** of genetically identical cells that result.

See the essay "Cloning Is Not Clowning" for a discussion of a slightly different form of cloning, as well as some related issues.

Although desired genes may be secured from cells in various ways, some genes whose nucleotide sequence is known can also be chemically synthesized in the laboratory. Techniques for determining sequences of

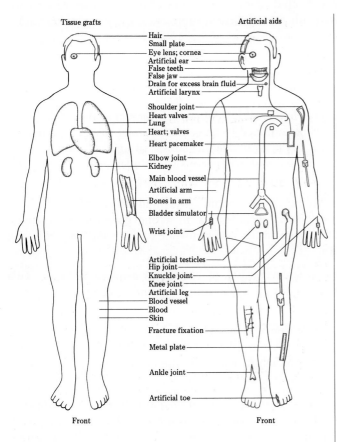

Tissue grafts

Artificial aids

Hair
Small plate
Eye lens; cornea
Artificial ear
False teeth
False jaw
Drain for excess brain fluid
Artificial larynx

Shoulder joint
Heart valves
Lung
Heart; valves
Heart pacemaker

Elbow joint
Kidney

Main blood vessel
Artificial arm
Bones in arm
Bladder simulator

Wrist joint

Artificial testicles
Hip joint
Knuckle joint
Knee joint
Artificial leg
Blood vessel
Blood
Skin

Fracture fixation

Metal plate

Ankle joint

Artificial toe

Front

Front

FIGURE B10.2

Replaceable humans. What other body parts do you think we may soon be able to replace?

euthenic advances to many people have been artificial limbs and internal organs (Figure B10.2). **Bionics** is the relatively new field of designing replacement and sometimes superior body parts, and the term **cyborg** has been coined to name a human-machine combination. Is there a limit to the amount of plumbing and hardware we can fill ourselves with and still be human? Could people become the "living dead," and could robots reach a stage of development when we can call them "alive"? Can we ever replace or transplant morality, will, judgment, compassion, empathy, intelligence, and humanitarianism? Or can we eliminate greed, aggression, and arrogance?

Some propose that we have moved through at least four major periods in recent decades—the Atomic Age, the Computer Age, the Space Age, and, now, the Genetic Engineering (or DNA) Stage. Although all have been telescoped into a few years, each is as significant as the Bronze Age, the Renaissance, and the Industrial Revolution. The Atomic Age has certainly given us veto power over human evolution by means of our ability to bring on a nuclear holocaust. Computers have greatly expanded the faculties of the human mind, and the Space Age has freed us of Earth's gravitational "fence." However, although we exploded "the bomb" decades ago, we have yet to resolve the military, political, economic, and ethical issues of nuclear energy. Similarly, are we closing in on the most intimate secrets of life, and utilizing some of those findings too rapidly? Is technological progress always automatically good?

genes are developing so rapidly, and so many scientists are working in this exciting field, that many parts of the human genome, with its millions of nucleotide pairs, may be sequenced within a few years.

Before 1982, a major source of insulin used to treat diabetics was the pancreatic tissues obtained from cattle and pigs in slaughterhouses (Chapter 18). Now it is possible to insert the human insulin gene into genetically engineered bacteria and produce insulin quickly and in needed quantities. Furthermore, the insulin produced is human insulin, eliminating the adverse side effects experienced by some diabetics from the use of similar but nonhuman insulin from other mammals.

Some dwarfs develop because their pituitary glands produce insufficient amounts of growth hormone (Chapter 18). Unfortunately, growth hormone is generally quite species-specific and it was not possible to treat this condition in humans by use of growth hormone secured from domestic animals. Beginning in 1985, however, genetically engineered human growth hormone became available for clinical uses. See the essay "Genetic and Biomedical Engineering."

SUMMARY

1. Although many geneticists are very interested in large organisms, the small size and rapid reproductive rates of many viruses and microorganisms have made them excellent subjects for the study of hereditary mechanisms.

2. **Transformation,** the assimilation of external genetic material by a cell, provided early insights into the important genetic role of **deoxyribonucleic acid,** or **DNA.**

3. The chemical nature of DNA was well known in the 1950s when Watson and Crick determined its three-dimensional structure, a **double helix.** The "ropes" of the twisted ladder consist of sugar-phosphate units, and the "rungs" are pairs of nitrogenous bases, adenine-forming hydrogen bonds with thymine, and cytosine doing the same with guanine.

4. Structurally, **genes** are specific sequential arrays of nucleotides. Genes are now viewed as operating in a one gene–one polypeptide manner, modified from earlier one gene–one enzyme and one gene–one protein theories.

5. The **genetic code** consists of **codons** or triplets of bases. This provides redundancy for specifying the 20 amino acids and several regulatory signals.

6. During chromosomal **replication,** the two DNA strands separate and each strand becomes a template for the assembly of a complementary strand. However, while DNA remains in the nucleus, most cellular metabolic processes occur in the cytoplasm.

7. **Messenger RNA** molecules are **transcribed** from DNA templates and then move from the nucleus to the ribosomes in the cytoplasm, where polypeptides are synthesized. During the process of **translation, transfer RNA** brings the proper amino acids to the ribosomes, and **ribosomal RNA** coordinates the protein-synthesizing processes.

8. Heritable changes in the genetic blueprint are called **mutations.** Common **mutagens** include heat and various chemicals and forms of radiation. Mutations are rare, random, and usually detrimental, although most are corrected. Those occurring in somatic cells affect only the individual, but those occurring in germinal cells may be passed to future generations.

9. The genetic variation that is the raw material for evolution results from various types of **genetic recombinations** as well as from mutations.

10. **Biotechnology,** the use of organisms or their products for practical ends, is both old and, in the form of **recombinant DNA technologies,** new. Through genetic engineering, genes from various organisms may be combined in desired ways.

11. Part of **euthenics** is **biomedical engineering,** such as **bionics,** the development of replaceable body parts. A **cyborg** is a human-machine combination. **Eugenics** attempts to improve humans through the specific application of genetic knowledge.

12. **Cloning** techniques may eventually allow humans to reproduce asexually, as is now possible with some animals. Eggs may now be **artificially fertilized** *in vitro* and the embryo implanted into a surrogate mother. *In vivo* fertilization is typical in **artificial insemination.**

REVIEW QUESTIONS

1. In what ways were the Griffith and the Hershey–Chase experiments controlled?

2. Define *transformation* and describe an example.

3. If the base sequence of one chain of the DNA molecule is TAACGTA, then what sequence must its partner in the duplex have?

4. Describe the process by which DNA replicates itself.

5. Distinguish between the processes of transcription and translation.

6. Why are the codons of the genetic code composed of three nucleotides rather than of one or two?

7. Compare and contrast the roles of mRNA, tRNA, and rRNA.

8. Explain why mutations and genetic recombinations are called the raw materials for evolution.

9. Why are regulatory proteins so important in homeostasis?

10. Describe some typical methods of recombinant DNA technology.

11. What do you foresee as the promise of genetic engineering? The potential pitfalls?

12. Distinguish between artificial insemination and artificial fertilization, and between *in vivo* and *in vitro* fertilizations.

13. Compare and contrast the aims and methods of eugenics and euthenics.

CHAPTER 11

GROWTH AND DEVELOPMENT

▓▓▓

Why we live so long, with a current life expectancy of between 70 and 80 years, is a legitimate question. Many small mammals live for just a few years, and even elephants usually survive for only 50–60 years. Theoretically, we might expect that 20–30 years or so would be adequate to bear and raise offspring, and thereby perpetuate the species. But we typically live for half a century longer.

In its broadest sense, development includes the results of all our interactions with the environment from conception to death. Chapter 6 deals with major events to the time of birth and this chapter examines some aspects of the postnatal changes during our relatively long life.

There is no mistaking the signs of growth, development, and aging here. In time, we grow, new structures appear, and the abilities of those structures gradually fade. The inevitable end of development, at present at least, is death.

GROWTH AND DEVELOPMENT

Growth involves either an increase in the size of existing cells or the production of new ones. The hundred-trillion cells in the adult body were all derived from the single-celled zygote. About 200 million new cells are produced in your body each minute, mainly to replace worn-out and injured cells. A baby's heart contains about the same number of cells as an adult's heart, although it is only one-sixteenth the size. Thus, the heart grows mainly by an increase in cell size. **Development** means the appearance of new structures and/or functions (Figure 11.1). Thus, a baby's proportions are very different from those of an adult. By age 1 year, babies generally weigh about three times their birth weight and they are usually about 50 percent taller. A rough estimate of final height can be made in young children. For example, by age 18 months in girls and 2 years in boys, most individuals are about half their mature height.

Human growth and development are often uneven, and it is common for persons to be above and below the various norms at different times during their development. Although the so-called typical **growth curve** is roughly S-shaped, not all body parts or systems grow at the same rate (Figure 11.2), and males and females do not grow at equal rates (Figure 11.3). The adolescent growth spurt occurs earlier in females but it lasts longer in males. In general, males increase more in height and females relatively more in weight. Women may develop 30 percent of their weight as fat, about twice that of males. For a period of time, sometimes awkward for some of both sexes, females are taller and heavier than males of the same age. Females typically reach their full height by about age 18, while males often continue to grow to about 20–21.

Several factors control growth and development. Genetic factors are especially important between conception and about 2–3 years. Later the growth hormone from the pituitary and the hormone thyroxin from the thyroid assist in growth and development, and elongation of the long bones is very evident during childhood (Chapters 18 and 21). At the age

(A)

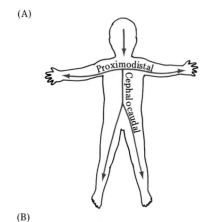

(B)

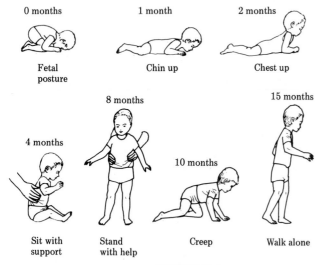

0 months — Fetal posture
1 month — Chin up
2 months — Chest up
4 months — Sit with support
8 months — Stand with help
10 months — Creep
15 months — Walk alone

(C)

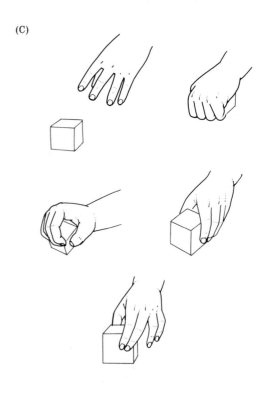

FIGURE 11.1
The two main developmental directions (A) and some changing abilities in locomotion (B) and grasping (C) during early childhood. What proximodistal developmental directions are shown in (B) and (C)? Cephalocaudal?

of about 9–10 years in females but nearer 12 years in males, the gonad-stimulating hormones begin to contribute to the adolescent growth spurt, when individuals may grow 10–15 cm (4–6 in.) and develop both the primary and secondary sex traits. Although these increases are often spectacular, they are much less so than the changes that occurred unobserved prenatally. From conception to birth, our weight increased about 2.4 billion times!

Diet is also obviously related to growth and development. Inadequate protein retards growth, and lack of iron adversely affects the production of hemoglobin. As a result, most body cells are affected because of the reduced ability of red blood cells to carry oxygen. Proper zinc is needed for sexual maturation, calcium and phosphorus for bone development, iodine for thyroxin production in the thyroid, and vitamins for the healthy development of many tissues. Extended illness and lack of exercise during childhood may also slow growth and development, although there is often a significant rebound after recovery. Even the seasons play a role, yet to be explained in detail: Children increase in height about twice as fast in the spring as in the fall, although they gain more weight in the latter season.

It is also evident that many psychological and emotional factors affect growth and development, and so-called psychosocial dwarfism and motor retardation have long been known. Children in certain institutions, such as some orphanages, or in homes where they are not wanted and are unloved may not walk until they are 5 or 6 years old. Removed to more favorable conditions, some of these children have grown 12–25 cm (5–10 in.) within a single year. However, there appear to be various **critical periods** or "points of no return" in development, such as when early deprivation of affection leaves some children permanently changed physically and psychologically. This is somewhat analogous to prenatal development, when each developmental stage must occur in the proper sequence.

FIGURE 11.2

(A) The "typical" growth curve, and some common percentiles. Note that human variation increases with age, as we all begin life within a few centimeters of each other in length and a few kilograms in weight. **(B)** All growth curves are not the same, however, and this graph shows those for the heart and brain compared with that for the body as a whole. What line would describe the growth curve for the reproductive system?

(A)

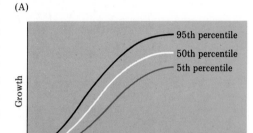

(B)

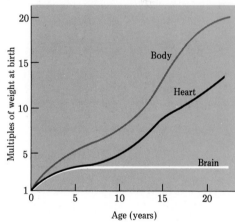

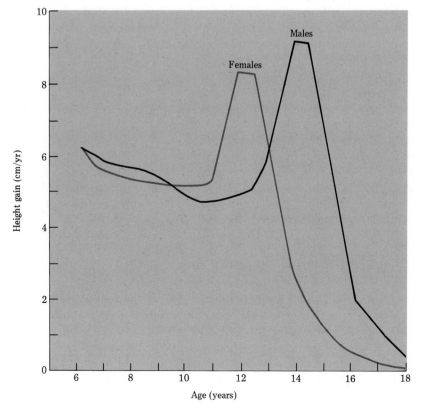

FIGURE 11.3

The relative growth in height of males and females. Do males or females reach their maximum height sooner? In what age range may some females be taller than many males?

Naturally, young children are not separated intentionally from other humans to observe how they develop in isolation. However, natural experiments on early deprivation are provided by several dozen somewhat controversial cases of "wild children" who, when discovered, had apparently been living in isolation or with animals. Although often grouped as "wolf children," some children were also associated with various other animals. Carolus Linnaeus (1707–1778), the Swedish biologist who devised the basis for assigning scientific names to species, knew of some cases and classified them as "feral man." The study of these children helps us understand how some behaviors develop without normal human stimulation.

In captivity, many wild children were mute and quadrupedal. Sensory abilities were often normal although some children were relatively insensitive to pain and temperature changes. Few affective behaviors such as crying and laughter were noted, and sexual interests and interpersonal interactions were minimal.

Some researchers believe that many wild children were mentally retarded. This, they feel, might explain many of their deficiencies as well as provide a possible reason for child abandonment in many cultures. Indeed, an autopsy by neurosurgeons on Ramu, one of the "wolf children" from India, when he died after 14 years "in captivity" showed extensive brain damage. Other researchers wonder how high-I.Q. children, let alone those who are mentally retarded, could survive such living conditions.

ARE PHYSICAL AND PSYCHOLOGICAL TRAITS RELATED?

Of perennial interest to researchers in many fields, as well as to people in general, has been the possibility that physical traits are related to psychological characteristics. Are redheads truly hot-tempered, and do blonds really have more fun? Literature and art have done much to perpetuate many social stereotypes. However, even real as opposed to fancied correlations describe but do not explain relationships. If physique and temperament are related, then the causal relations involved may be of three types:

1. *Psychological traits influence physical traits.* Habitual activity often leaves its mark, from the powerful shoulder muscles of a swimmer to the facial creases of frequent smiling or frowning. Anxiety and stress not only affect most illnesses but also cause many, such as some cases of ulcers, hypertension, heart attacks, asthma, allergies, and skin problems. There is an abundance of evidence that personality affects the susceptibility to, and ability to recover from, many diseases. The hypochondriac's imagined problems represent still another example of these **psychosomatic relationships.**

2. *Physical traits influence psychological traits.* Neural or glandular abnormalities may affect behavior in major ways. The hypoactive thyroid of cretins or the small brain of microcephalics are examples. Being blind, deaf, extremely fat or thin or tall or short, or being afflicted by any chronic ailment commonly affects one's self-image and, often, the way one interacts with others. The term **somatopsychological** describes physical traits influencing psychological traits.

3. *The common influence of a third factor.* Sometimes two factors show a high correlation not because one causes the other but because both are affected by a third factor. For example, one's socioeconomic level has pervasive effects upon both one's physical and psychological

development. The child reared in a "superior" home may have more intellectual stimulation as well as a better diet, hygiene, and medical care than some children raised in a ghetto, barrio, slum, or poor rural area.

Males and females are often said to differ psychologically, although the biological bases for the differences are still largely obscure. However, parents often have different behavioral expectations for boys and girls and therefore raise them differently, often without intending to do so. Sometimes boys are allowed or even encouraged to be aggressive, at least outside the family.

Some behaviors are related to the levels of sex hormones. However, the most dominant male laboratory animals in an experimental group may not have the highest levels of testosterone. Testosterone levels may increase with sexual activity, and the most dominant males often have more sexual opportunities. Therefore, it may be that testosterone, as merely one example of a hormonal effect, reflects behavior in addition to influencing it.

Sex hormones may have similar effects in females, and there are often marked changes in the emotions as the levels of these hormones change. For example, many women experience tension when their estrogen and progesterone levels drop during menstruation. New mothers are psychologically vulnerable for similar reasons, as are some women during the menopausal years.

Dermatoglyphics

Fortune-tellers have long used palm prints to analyze one's personality and/or to predict future events. Although palm reading as a form of predicting the future is as phony as it is ancient, the lines on one's hands and feet are used in a modern field of medical diagnosis called **dermatoglyphics.** Abnormal print patterns were noted in people with Down syndrome over 40 years ago, and today scores of other specific disorders are correlated with print abnormalities. These include PKU and certain heart, brain, kidney, and other defects. Because the analysis of large numbers of individuals helps in identifying correlations, the children born to mothers suffering from epidemic diseases such as rubella during pregnancy have been important subjects. Most print abnormalities are minor, and the fact that no two people have identical prints further complicates the study. However, several signs now looked for routinely by physicians are shown in Figure 11.4 and described below.

1. *Simian line.* Two creases are normal, whereas one crease is abnormal.

2. *Axial triradii.* A triradius is where three lines join. A normal axial triradius is found just above the first crease where the hand joins the wrist, and a higher position on the palm may indicate an abnormality. Other triradii occur near the bases of the index and little fingers. The normal angle between the three triradii (about 48°) is smaller than that found in those with Down syndrome (about 80°), and the angle is even larger in those with some other abnormalities.

3. *Radial loops.* The loop is a common fingerprint pattern, and the open end of a loop usually faces down or points away from the thumb. Various anatomical and physiological abnormalities are most common in those whose loops open toward the thumb.

Dermatoglyphic analysis is not easy because some people with serious problems have normal prints and some apparently normal humans have

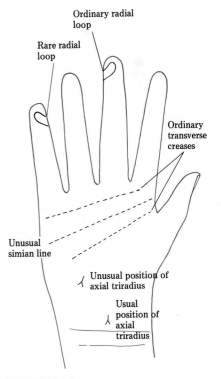

FIGURE 11.4

Some normal and abnormal palm and print patterns looked for in a dermatoglyphic examination.

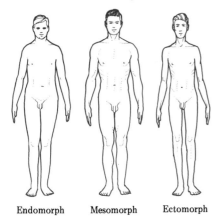

Endomorph   Mesomorph   Ectomorph

abnormal prints. Therefore, abnormal print patterns cannot usually indicate a specific problem with certainty but may prompt a more thorough examination than usual. Also, the earlier a problem is detected, the better the chance that it can be treated successfully. Prints form during the first four months of gestation when the embryo and fetus are particularly vulnerable to damage by external agents. Presumably, the harmful agent both causes the abnormal print and damages internal organs, which then result in abnormal functioning and/or behavior.

Somatotypology

One comprehensive attempt to relate physical and psychological traits was made by W. H. Sheldon, an American physician. Sheldon's theory is neither being promoted nor attacked here but is offered merely as one example of research in this field. Make up your own mind as you relate the following material with observations of humans.

First using college students and later older adults and children, Sheldon proposed the three basic body forms, or **somatotypes,** shown in Figure 11.5 and described below:

1. *Extreme endomorphy.* Soft, roundish body; relatively underdeveloped bone and muscle but highly developed digestive organs; short legs, thick neck, and full face; relatively weak physically.

2. *Extreme mesomorphy.* Hard, rectangular body; relatively well developed bone, muscle, and connective tissue; relatively strong physically.

3. *Extreme ectomorphy.* Delicate, lean, linear body; relatively light musculature, large skin surface, and large brain and central nervous system.

Sheldon developed a seven-point scale for each somatotype, and felt that most individuals had some traits of each group (e.g., 2-5-1). Sheldon's next step was to identify the three basic psychological groups described below and to correlate them with the somatotypes, as shown in Table 11.1:

A. Relatively relaxed and even-tempered; likes food and physical comfort; highly sociable; slow reactions; deep sleeper.
B. Assertive, energetic, and competitive; enjoys risk and power; often extroverted.
C. Restrained posture and movement; fast reactions; inhibited socially; poor sleeper; introverted and quiet; enjoys privacy.

Although Sheldon's theory has been highly criticized and few investigators have found correlations as high, many researchers feel that *some* relationship may exist between physique and temperament. If the relationship is identified, then the next step would be to determine the roles of genetic and cultural factors.

TABLE 11.1
Sheldon's Correlations Between Body and
Temperament Types[a]

Somatotype	Temperament Type[b]		
	A	B	C
Endomorphy	+0.79	−0.29	−0.32
Mesomorphy	−0.23	+0.82	−0.58
Ectomorphy	−0.40	−0.53	+0.83

[a]Mathematically, correlations vary from 0.0 to 1.0. Positive correlations indicate direct relationships and negative correlations indicate inverse relationships. Correlations close to 0.0 indicate little or no relationship between the two variables.

[b]See descriptions in text.

CHANGING TRAITS

There has been a steady rise in the physical dimensions of individuals in most human populations during the past two centuries (Figure 11.6). The height of children between the ages of 5 and 7 has been increasing at the rate of about 1.3 cm (0.5 in.) per decade since 1900. However, in the United States and many other developed nations, the rate of increase is apparently leveling off. During the past half-century, the average height of American 18-year-old men and women increased by about 5 cm and 3 cm, respectively, and the average weight increased by over 8 kg and by almost 5 kg, respectively.

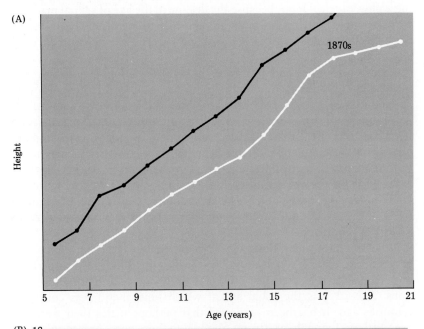

(A)

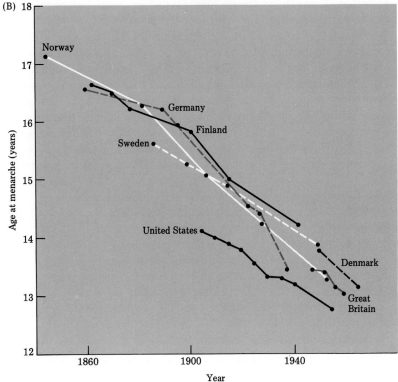

FIGURE 11.6

(A) Evidence from many sources and areas shows that human dimensions are increasing. These data show the changes in height of English males during approximately the century following the 1870s. (B) The steady decline in the age of menarche in Americans and Western Europeans is clear during the past century. How do you explain the changes illustrated in these two graphs?

The age of sexual maturity has also been decreasing. For American males, puberty now averages slightly below age 14, whereas it averaged over 16 years in 1900. The average age of menarche for American females was over 14 years in 1900 but it is now under 13 years. European records show that menarche occurred between the ages of 16 and 17 years in many countries during the mid-1800s. It still is in that age range among some human populations living in harsh environments and among various subsistence-level groups. In subsistence cultures, many women may experience menarche halfway through their lives, pregnancies are few and spaced, and menopause occurs early. Various cultures practice "fattening the bride" customs to improve the chances of conception and successful pregnancies. However, perhaps we are approaching a physiological limit, because the decline in the age of sexual maturity is apparently stabilizing in many developed nations. It has been known since the Renaissance that urban women mature earlier than their rural counterparts.

Almost certainly, better nutrition, the control of disease, and improved medical care have played major roles in these changes. However, a very specific influence is the proportion of body fat referred to above. The general relationship between body fat and reproductive processes in mammals is well established. For most human populations throughout history, the accumulation of sufficient body fat for early menarche, as well as for successful pregnancies, was probably difficult. Perhaps this is the basis for the numerous references in literature (e.g., "the fat of the land" in Genesis) to the link between fat and fertility. It may also explain many of the early statues and fertility figures of obese women with large breasts and buttocks.

It is clear that today's average 18-year-olds are larger, more mature, and better educated than average 21-year-olds of 1900. This was perhaps one reason for reducing the voting age in federal and many other elections. Age 21 was the usual age at which knighthood was conferred centuries ago. It has been suggested that perhaps males then were not usually large enough or strong enough to hop onto horses with full armor to rescue damsels in distress until about age 21!

Menopause is largely a modern phenomenon, because until this century the life expectancy of American women was only about 45 years. Today, however, the average woman in developed nations outlives her ovaries by about 25–40 years, although it is unclear why the cessation of ovulation occurs. Although the ovarian decline produces few symptoms in many women, the symptoms, which are usually treatable, are at least troublesome in other women. Besides the missing menstrual periods, there may also be hot flashes, palpitations, and nighttime sweats. The estrogen deficiency may also soften or decalcify the bones (osteoporosis) and cause a variety of changes in the reproductive and urinary tracts. Common emotional changes associated with these physical changes include depression, fatigue, anxiety, headaches, vertigo, and, sometimes, a severe personality crisis. Major psychological problems, however, are decreasing, perhaps in part because of the realization that a woman's worthiness far exceeds her maternal role.

Estrogen replacement therapy has been used frequently in the past to reduce or eliminate these symptoms. However, there is apparently a chance that estrogen therapy increases the risk of uterine and breast cancer in at least some women. Therefore, estrogen use should be supervised medically, and as with most treatment, second opinions are useful.

Unlike females, males continue to produce gametes and sex hormones as they age, although in slightly decreasing quantities. However, some males in their 40s and 50s may experience a syndrome of psychological traits similar to those of menopause, the so-called male **climacteric.** These changes appear to be culturally based and due, perhaps, to males' experiencing a sense of the loss of their youth.

NATURE AND NURTURE

The question of the relative roles of nature or genetics and nurture or the environment in determining what we become has long been debated. Today, there is agreement that both are important because we inherit genes, not traits, and traits must *develop* under the control of genes and within the limits imposed by the environment. Genes determine what organisms may become, and the environment influences what organisms *do* become.

If you have studied Chapter 9, you have seen several examples of traits that show continuous variation and learned that the several alleles that control these traits interact both with the individual's other genes and with the environment. For example, height is controlled not only by one's "tall" and "short" genes but also by the growth hormone, the hormone thyroxin, the enzymes that digest food, and the genes that control the deposition of calcium in the long bones. Chronically poor nutrition in childhood and exposure to inadequate sunlight for the proper synthesis of vitamin D may also slow growth. However, environmental factors do not affect all traits the same way. For example, an inadequate environment has relatively little effect on eye color as compared with its effect on height and, especially, weight.

Children who inherit genes for phenylketonuria (PKU) do not inherit a low I.Q. but, rather, inherit an inability to break down an amino acid (phenylalanine) (Figure 11.7). The buildup of the amino acid and its byproducts causes brain damage, which results in a low I.Q. A simple test to identify infants with this condition is now available, and eliminating the amino acid from the diet allows normal development to occur.

Phenocopies are environmental duplications of genetic traits, and the concept helps to separate genetic and environmental factors. For example, the light-skinned Hawaiian vacationer can acquire about the same skin color as a native in time. Also, **phocomelia** was known as a recessive abnormality long before the use of certain medications increased its incidence (Figure 11.8). Similarly, the **hydrocephalic** ("water on the brain") individual normally has a low I.Q. because of the effects of one or more recessive genes. However, a viral infection in the pregnant female may also cause hydrocephalus in the fetus (i.e., a phenocopy). Also, hydrocephalus can be prevented by draining the excess fluid. Even though the genes of such children "say" that they should be hydrocephalics, they are phenocopies of normal children.

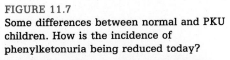

FIGURE 11.7
Some differences between normal and PKU children. How is the incidence of phenylketonuria being reduced today?

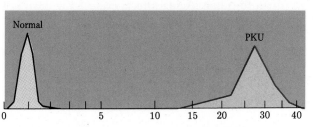

Phenylalanine in blood plasma (mg %)

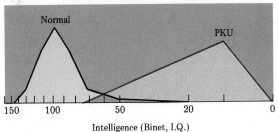

Intelligence (Binet, I.Q.)

FIGURE 11.8
Humans with phocomelia are often armless. This infant acquired its deformity because its mother took a drug (the sedative thalidomide) during her second month of pregnancy. Explain how this instance of phocomelia is a phenocopy of a genetically based case of phocomelia.

TWIN STUDIES

Because monozygotic twins have identical genetic composition, any variation between them is usually due to environmental factors. Although most identical twins are reared together and treated similarly, some sets are raised apart, making interesting comparisons possible. This is called the **co-twin research method** because each twin is the control or comparison of the other. Of course, siblings and fraternal twins (of the same or of different sexes) may also be raised together or apart, and Table 11.2 shows results of various comparative studies.

Heritability in humans is often expressed by the degree of **concordance**. When both twins have a particular trait they are concordant; when only one does they are discordant. Concordance is usually expressed as a percentage of similarity with respect to the presence of a trait. A high concordance in identical twins with a much lower percentage in fraternal twins usually indicates a strong genetic influence. For example, albinism in monozygotic twins shows a concordance of 100 percent, whereas it is only about 25 percent in dizygotic twins.

A similar concordance value in the two kinds of twins may indicate a high environmental influence. For example, deafness resulting from rubella infection of the mother shows a concordance of 86 percent and 88 percent for identical and fraternal twins, respectively. Sometimes there is a low percentage in both types of twins. For example, harelip (incomplete fusion of the upper lip) shows a concordance of 33 percent in identical twins and 5 percent in fraternal twins. This shows a strong environmental influence as well as some genetic effect.

Although most diseases are not inherited, people may inherit a susceptibility to get a certain disease. Table 11.3 shows the concordance between identical and fraternal twins for various diseases and conditions.

In general, we know more about the genetics of physical traits than psychological ones. Behavior is one of the most variable of all human traits, and not everyone agrees on how to describe it, let alone explain it. We know little about either the genetics or the environmental factors of leadership, empathy, curiosity, or most other mental abilities and personality traits.

Some genes that cause physical problems also cause mental retardation. However, the latter can also be affected by environmental factors such as accidents, certain diseases, and poor prenatal and postnatal nutrition. Although the development of intelligence has received much study, about all that can be said for certain is that mental abilities depend upon the effects of many genes located on many chromosomes, as well as upon numerous environmental influences.

AGING

Between adolescence and about age 25–30 years, most body systems operate at peak levels. Prior to about age 25, more cells form than die, but thereafter the number that die outnumbers the new ones, although the rate varies from organ to organ. **Aging** affects every body part, not

TABLE 11.2
Differences Between Twins and Siblings

| Trait | Reared Together | | | Reared Apart |
	Monozygotic Twins	Dizygotic Twins (same sex)	Siblings (same sex)	Monozygotic Twins
Height (cm)	1.7	4.4	4.5	1.8
Weight (kg)	1.9	4.6	4.7	4.5
I.Q. (points)	5.9	9.9	9.8	8.2

TABLE 11.3
Concordance

Disease or Abnormality	Concordance (percentage)	
	Identical Twins	Fraternal Twins
Mental retardation	97	37
Site of cancer (when both have cancer)	95	58
Measles	95	87
Down syndrome	89	6
Rickets	88	22
Schizophrenia	86	15
Diabetes mellitus	84	37
Handedness (right or left)	79	77
Manic-depressive psychosis	77	19
Epilepsy	72	15
Criminal record	68	28
Tuberculosis	65	25
Cancer	61	44
Congenital hip dislocation	42	3
Clubfoot	32	3

always gently, and Figure 11.9 shows some typical changes during aging between about ages 25 and 70 years. In general, changes in females occur somewhat more slowly than in males. As Shakespeare wrote about aging in *As You Like It,* "Sans teeth, sans eyes, sans taste, sans everything."

Some other changes that occur during aging include the following: the skin thins, sags, and becomes less elastic; the bones become brittle; the height declines as the spongy disks between the vertebrae shrink, especially in women; the number of taste buds declines; the number of abnormal chromosomes increases; hearing and vision decline; vocal cords stiffen; cells become more pigmented ("age spots"); muscle tissue is lost and is often replaced by fat; the diameter of hairs declines and color changes such as graying often occur; the abilities to maintain body temperature and blood sugar decline; response time increases, but usually without a decline in actual mental capacity; the blood supply to the brain and to the filtering units in the kidneys decreases; blood pressure increases; the basal metabolic rate declines; breathing becomes shallower; the arteries harden; and the breasts sag in women. The death rate for most major diseases also increases with age, partially because the immune system becomes less efficient.

However, while many structures decline in size, cartilage is one of our few tissues that continues to grow as we age. Because of this, the nose may widen and lengthen and the drooping of the earlobes may enlarge the ears. Although tooth decay often declines with age as tooth enamel hardens, the typical increase in periodontal problems more than offsets that apparent advantage (Chapter 13).

Some scientists prefer to use the term *aging* for the passage of time and the term **senescence** for the body deterioration that accompanies aging. The formal study of aging and senescence is called **gerontology.** The term "prolongevity" has been coined for the increasingly fascinating field of life-extension possibilities.

The 25 million or so Americans over 65 years old use a major portion of our medical skills, and about half of all nonsurgical hospital beds are filled by the elderly. Because so many senior citizens feel rejected and useless, the suicide rate is also very high. Yet one of our greatest reservoirs of talent is our prematurely retired population.

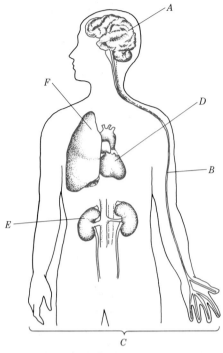

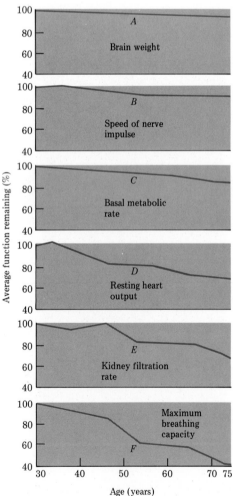

FIGURE 11.9

To illustrate the loss of various functions with age, the average level at age 25–30 years is assigned a value of 100 percent. How do you explain these changes?

THEORIES OF SENESCENCE

Although many people die from a specific disease, condition, or accident, senescence is clearly multifactored, or due to more than one change in structure or function. A few of the dozens of theories of senescence proposed by gerontologists are described briefly below.

Cross-linkage

The structure and function of parts change when certain proteins **cross-link** due to the buildup of metabolites that the body is slow in excreting. Linked proteins become deactivated and relatively useless, and only one cross-link for every 30,000 units of a large molecule will change its traits. **Collagen** fibers make up about one-third of the body protein, and these often cross-link. When the rate of cross-linkage increases, soft connective tissue becomes denser and less elastic. This slows the movement of materials to and from nearby structures, and cross-linked connective tissue probably also damages blood vessels. Muscle activity is also impaired by altered collagen fibers.

Some harmful metabolites are called **free radicals,** and these may link with collagen fibers to produce some of the effects just outlined. These products of numerous metabolic reactions destroy cells, trigger many destructive chemical reactions, and damage DNA to change the genetic code. RNA then "reads" a false message and produces a "nonsense" structural or enzymatic protein.

Somatic Mutation

Every human contains some **mutations,** or changes in genetic material. The body is constantly in contact with **mutagens** such as radiation and certain chemicals and pathogens. A mutation or random change in a well-functioning organism is not likely to benefit it immediately, and most mutations are harmful. Mutations need not be lethal or major to produce harmful effects. For example, a mutation that makes a slight change in the rate of hemoglobin production may adversely affect all body tissues because of the small decline in oxygen-carrying ability.

Immunological Theory

Paradoxically, the body increases its attack against its own cells at the same time that the efficiency of the immune system is declining. For example, damaged or abnormal cells are potential antigens that may stimulate antibody production (Chapter 16). The condition that describes the body's failure to recognize and accept itself is called the **autoimmune response.** Normal as well as abnormal cells are attacked as the body produces antibodies against itself. Some types of arthritis, rheumatism, multiple sclerosis, and anemia are or may be autoimmune problems.

Master Timing Device

"A time to be born, and a time to die" is a well-known biblical passage, and some gerontologists believe that the body has a master timing device, perhaps in the brain, that programs our death. Mature neurons do not reproduce, many neurons die as we age (about 50,000 daily at age 30), and the breakdown of such a brain center could be important in senescence. Perhaps the operation of an automobile is analogous in the sense that even if all its other parts are in good condition, a car will "die" or run poorly if its timing device is off.

There is evidence that whatever the cellular death warrant is, it has a genetic component. For example, the life spans of identical twins coincide more closely than those of fraternal twins of the same sex. Similarly, there is a relatively high correlation between the life spans of children and their parents, and longevity has long been observed to run in certain families. Some researchers propose specific genes that switch off essential cellular functions at a particular time, lead to faulty or nonsense enzymes, or even code for a "death hormone" that interferes with many aspects of metabolism.

Stress

The stress theory proposes that life's physical and emotional stresses lead to wear and tear that is never fully repaired. Much evidence shows that stress disrupts sleep patterns, changes hormone balances, inhibits lymphocyte production, impairs cell division, slows DNA and protein synthesis, and influences examination scores. Stress may also lead some to eat and exercise improperly and to use or abuse alcohol, tobacco, and other substances, all of which may adversely affect one's health.

LIFE EXPECTANCY

The **life span** is the age that a person actually attains, whereas **life expectancy** is the statistical probability of living to a particular age. When the Bible spoke of some people living threescore and ten years 2000 years ago, most humans were fortunate to reach their early 30s. Life expectancies had increased to the late 30s by the Middle Ages. Life expectancy for all Americans born in 1900 averaged about 47 years; it is now about 75 years. On the average, females live longer than males and the difference between the sexes is increasing. Life expectancy for all male American infants today is about 71 years, but it is almost 79 years for females. Lower life expectancies are found among various American subcultures and in most populations in developing nations.

As shown in Figure 11.10, death rates are not constant throughout the life span. Also note the relative unimportance of infectious diseases compared with degenerative disease after the age of 4 years. Although the three most common causes of death in the United States in 1900 were infectious diseases, no infectious diseases are among the top four killers now (Figure 11.11). Part of the increase in life expectancy is also due to a reduction in infant mortality. Unfortunately, and despite our generally sophisticated medicine, the United States has one of the highest infant mortality rates of any developed nation.

INCREASING LIFE EXPECTANCY

Until recently, apathy or resignation to death has been a major obstacle to research on aging, although some have long sought elixirs to produce perpetual youth. For example, in 1492, Pope Innocent VIII supposedly drank the blood of three donors to stay alive, but the attempt failed. Millions of dollars are spent annually on substances designed to slow aging, or at least its signs, from megadoses of vitamins and minerals to hormonal skin creams. Whether the materials being marketed or promoted are miracle cures or "snake oils" remains to be seen. Some products obviously have a major psychosomatic effect, however, even if they are otherwise placebos at best, and because of this effect may be worth the monetary expenditure in terms of mental health.

Improvements in medicine and public health will probably continue to increase life expectancy, especially for those populations that have not benefited greatly from such advances in the past. Yet some of the

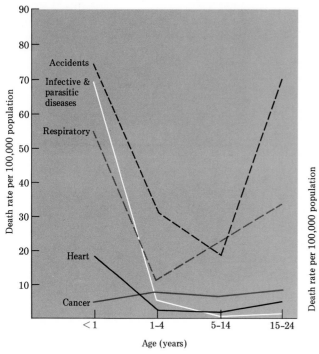

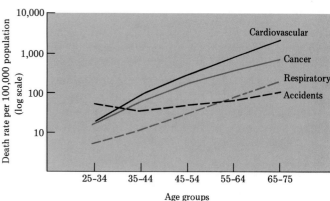

FIGURE 11.10

Death due to various causes over the typical life span. (The right scale is logarithmic because the figures would go far off the page if plotted in conventional form.) How do you explain the trends? Do you think the type of accident that is fatal changes over the life span?

apparent, but controversial, records for longevity come from several developing areas of the world. The several life-extending areas of research discussed below are in most cases either theoretical possibilities or studies mainly on laboratory animals.

Nutrition

Any excess weight taxes most body systems, and death rates are generally lowest for those somewhat below the average weight and highest for those far above the average weight.

Antioxidants remove free radicals and stabilize cell structures and functions. Vitamins C, the B complex, and E, and minerals such as selenium are important antioxidants. Some amino acids and certain food additives also appear to be antioxidants. Although deficiencies of antioxidants hasten aging, little is known yet about optimum amounts or whether excesses lengthen the life span.

Some researchers have found that laboratory rats fed nucleic acids (RNA and DNA) look younger and sometimes live longer than control rats. Clinical tests on humans are under way but the results are inconclusive, in part because of our already long life span.

If enzymes can be considered nutrients, then perhaps some could be administered that would reduce cross-linkage. To facilitate this, the materials needed to maintain and repair the body's tissues must also be provided.

Exercise and Stress

Many symptoms of aging are reduced by regular exercise, although it is unclear whether exercise *per se* increases life expectancy. When people exercise, they feel better, their cardiovascular system is usually healthier, and they are less likely to be overweight than those who are sedentary. Similarly, those who learn to avoid stress usually feel better and may well live longer also. People in good physical condition generally resist the harmful effects of stress and other challenges to survival better than those in poor physical condition. Even having a pet seems to reduce stress in many, and pets certainly add further meaning, purpose, and interest to the lives of many lonely elderly people.

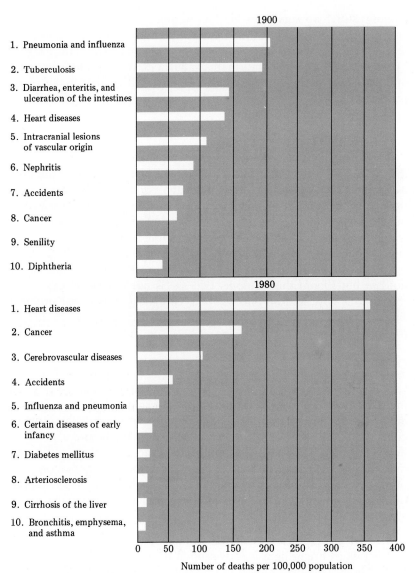

1900

1. Pneumonia and influenza
2. Tuberculosis
3. Diarrhea, enteritis, and ulceration of the intestines
4. Heart diseases
5. Intracranial lesions of vascular origin
6. Nephritis
7. Accidents
8. Cancer
9. Senility
10. Diphtheria

1980

1. Heart diseases
2. Cancer
3. Cerebrovascular diseases
4. Accidents
5. Influenza and pneumonia
6. Certain diseases of early infancy
7. Diabetes mellitus
8. Arteriosclerosis
9. Cirrhosis of the liver
10. Bronchitis, emphysema, and asthma

Number of deaths per 100,000 population

FIGURE 11.11
Changes in the causes of death since 1900.
How do you explain these trends?

Hormone Therapy

Synthetic sex hormones given to menopausal and postmenopausal women and elderly men appear to slow some symptoms of aging. However, hormone therapy is being approached cautiously because of concern about possibly increasing the incidence of certain cancers.

Transplants

Transplants of major organs such as hearts, kidneys, and livers have saved many lives and received much publicity. However, a less spectacular transplant may save even more lives in the future. Some biologists believe that we have a finite antibody-producing ability. If so, transplanting properly matched and compatible juvenile bone marrow and possibly spleen and/or thymus gland cells into the elderly may prolong life. For example, injections of fetal lamb cells were tried by Charlie Chaplin and Winston Churchill.

Reducing Body Temperature

Just as average chemical reactions are slowed by reduced temperatures, the metabolic rate declines somewhat when the body temperature drops. This is why some operations are performed on patients in an ice bath. Volunteers taking experimental temperature-reducing drugs apparently suffer no significant decline in abilities such as reaction time and problem solving when their body temperatures are reduced by a few degrees. Some scientists believe that years could be added to the life span by reducing the metabolic rate in this manner, even if hypothermia occurs only for short periods of time daily. It may well be that many of the benefits of caloric restriction occur because they involve a reduction in body temperature.

Not Smoking

The American Cancer Society posters emphasize that people can harm themselves in many ways, but few common behaviors affect the body as adversely as smoking. Although many associate lung cancer with males, lung cancer is also a major cause of death in women. Fortunately, hordes of people are heeding the advice not to smoke, as well as to refrain from the use and abuse of various other substances.

In summary, it seems likely that a combination of methods could make possible many more centenarians (Figure 11.12). The question is not *can* but *should* this be done. Old age is not a pleasant experience for millions now, and no one wants to complicate this major problem. *In addition to adding years to our life, we must learn how to add life to our years.*

DEATH

Like life, death is a matter of definition rather than a matter of fact, and in these days of transplants and life-support systems, definitions of death are changing. A single cell may be declared dead or alive, but humans contain trillions of cells. How many cells must die before a person is dead? If it is not the absolute number of dead cells, then which cells must be dead? It is clear that dying is now looked upon more as a process than as an event, and definitions of death usually imply a point of irreversibility in the dying process. Some or all of the following criteria are included in many definitions:

1. Lack of response to external stimuli, such as the pupils not contracting when bright light is shown upon them.
2. Absence of any spontaneous muscular movements, notably breathing and the heartbeat.
3. Flat electroencephalogram (EEG) or absence of brain waves ("brain death").

However, the individual's circumstance is often also important. For example, the EEG of a victim of barbiturate poisoning may remain flat for many hours and yet the person may recover. A flat EEG for a few minutes might be an adequate criterion of death for a person with massive brain damage as a result of a serious automobile accident. The death decision is obviously important in cases where an organ is to be

FIGURE 11.12
The proportion of the elderly in the total population is changing. How might increased life spans complicate Earth's population, pollution, and resource problems? What societal changes may be necessary because of these trends?

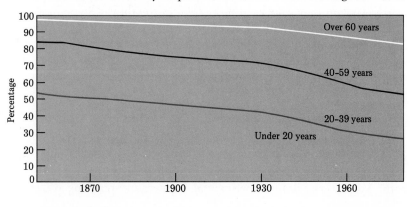

used for a transplant operation. Few want to sacrifice one person in order to obtain an organ for transplant to another.

Although we often hear that "old Joe Smith" or "widow Jones" simply died of old age, old age does not kill people directly. It is simply that specific disorders such as heart attacks, strokes, cancer, or pneumonia or nonspecific assemblages of less recognizable conditions occur more often with advancing age. Many believe that life is shortened when people retire prematurely from living, not merely from a job. Grandma Moses began to paint at the age of 78, and she painted over two dozen picture after she turned 100.

Today, many dead people receive some form of **autopsy** or postmortem examination. At least two main reasons for this are (1) the desire of the family to know the exact cause of death, and (2) the fact that increased medical knowledge results. Because of the important moral and legal restrictions on human experimentation, much of our knowledge of pathology comes from autopsies. This fact prompts many people to donate their bodies to medical schools and/or donate certain organs for possible transplantation.

"Never say die" is the motto of some people who die today of an incurable ailment. These individuals have themselves preserved by chemicals and are stored in liquid nitrogen, so-called **cryonic burial.** These optimists hope that medicine will eventually find a cure for their problems, thaw them, and treat them for a return to normal life, like a modern Rip Van Winkle. However, the success of human cryonic burial and revival is yet untested.

Euthanasia and Mercy Killing

In the past, swift death from **acute** illness was common. Now, medical care prolongs life so that more and more people die from degenerative or **chronic** (long-lasting) diseases, such as cancer. Over 70 percent of Americans die in hospitals or nursing homes. **Euthanasia** (from the Greek for "good death") is not the same as "mercy killing." The latter implies administering or doing something to end a person's life and is a form of homicide. Euthanasia is withholding treatment that is viewed as "artificially" keeping a person alive, such as a drug or mechanical device. Proponents of euthanasia speak of "death with dignity" and object to prolonging the life of those with terminal illnesses, sometimes seemingly just for the sake of doing so. Many who believe in euthanasia sign a "living will," such as the one on the following page, and these documents are recognized legally in many states.

Suicide

Discussions of death often include the issue of **suicide,** now a leading cause of death (Figure 11.13). Suicide or attempted suicide is often a criminal offense, and is commonly viewed as a sign of mental problems, raising the questions about society's obligation or right to prevent individuals from taking their own lives. Whose body is it?

FIGURE 11.13

The death rate for suicide in various age groups. How do you explain the trends? Should society have the right to prevent individuals from taking their own lives?

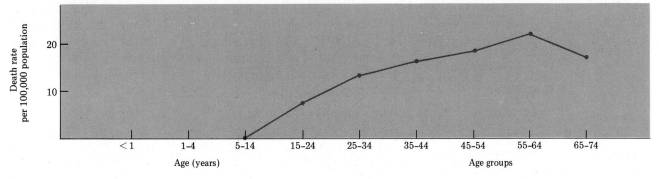

A LIVING WILL

TO MY FAMILY, MY PHYSICIAN, MY LAWYER, MY CLERGYMAN
TO ANY MEDICAL FACILITY IN WHOSE CARE I HAPPEN TO BE
TO ANY INDIVIDUAL WHO MAY BECOME RESPONSIBLE FOR MY
HEALTH, WELFARE, OR AFFAIRS

Death is as much a reality as birth, growth, maturity, and old age—it is the one certainty of life. If the time comes when I, _____, can no longer take part in decisions for my own future, let this statement stand as an expression of my wishes, while I am still of sound mind.

If the situation should arise in which there is no reasonable expectation of my recovery from physical or mental disability, I request that I be allowed to die and not be kept alive by artificial means or "heroic measures." I do not fear death itself as much as the indignities of deterioration, dependence, and hopeless pain. I therefore ask that medication be mercifully administered to me to alleviate suffering even though this may hasten the moment of death.

This request is made after careful consideration. I hope you who care for me will feel morally bound to follow its mandate. I recognize that this appears to place a heavy responsibility upon you, but it is with the intention of relieving you of such responsibility and of placing it upon myself in accordance with my strong convictions that this statement is made.

Date_____   Signed _____

Witness _____

Witness _____

Copies of this request have been given to:

_____

_____

SUMMARY

1. During **growth,** new cells are produced or increase in number and/or size. The appearance of new structures and/or functions occurs during **development.** Although not all structures grow at the same rate, the generalized **growth curve** is S-shaped.

2. Growth and development are especially noticeable around puberty, when secondary sex traits develop. The adolescent growth spurt is earlier in females but greater in males. Genes control one's potential but environmental conditions affect one's realized growth and development. **Critical periods** occur postnatally as they did prenatally.

3. In **psychosomatic** relationships, psychological traits influence physical traits, as when stress affects the development or severity of ulcers, hypertension, heart attacks, asthma, and various other conditions. The term **somatopsychological** means that physical traits influence psychological traits, as when emotional changes occur following a serious handicap or the development of a glandular problem. Both categories of relationships may be affected by a third variable, such as one's socioeconomic situation.

4. **Dermatoglyphics** is the field of medical diagnosis that utilizes patterns of prints on the fingers, palms, and soles to correlate with possible abnormalities as an aid to early treatment.

5. **Somatotypology** is the concept that sets of body traits, or somatotypes, are correlated with sets of psychological traits (e.g., Sheldon's theory of endomorphs, mesomorphs, and ectomorphs).

6. Several human traits have changed during historic time—average height and weight have increased, and the age of puberty has decreased. Because of increased **life expectancy,** women now long outlive their reproductive years, and menopause typically occurs during the late 40s. Life expectancy is longer for females than for males.

7. Determining the relative roles of nature (genetics) and nurture (environment) in what we become is not easy. "Wild children" have intrigued some, but more are interested in the **co-twin control method.** Environmental duplicates of genetic traits are called **phenocopies.**

8. **Gerontologists** study **senescence,** the physical and mental changes that accompany **aging.** There are numerous theories of senescence (e.g., **cross-linkage, somatic mutation, immunological changes, stress,** and the possibility of a **master timing device**) as well as ideas about extending the human life (e.g., diet, exercise, stress, hormonal treatment, transplants, body temperature level, and behavior).

9. Like life, **death** is a matter of definition rather than a matter of fact, and opinions vary considerably. **Mercy killing, euthanasia,** and **suicide** remain controversial.

REVIEW QUESTIONS

1. Describe several important examples of the cephalocaudal and proximodistal developmental directions.

2. Distinguish between growth and development, using specific examples.

3. Describe several ways in which the patterns of growth and development in males differ from those in females.

4. Describe and evaluate the relationships between somatotypes and temperament types proposed by Sheldon.

5. Describe and explain (a) the changing nature of the average height and weight of most populations during the past century, (b) the changing age of menarche, and (c) the differences in development between rural and urban populations.

6. Explain phenocopy and state at least two examples.

7. Describe and explain many of the obvious external and internal changes in structures and functions associated with aging.

8. Would you like to live to be 100–200 years old through the use of one or more of the life-extension possibilities described?

9. What societal changes would make being elderly a more enjoyable and satisfying experience?

10. How do you distinguish between "ordinary" and "extraordinary" means of prolonging life?

11. What are your opinions about mercy killing and euthanasia? Who should decide if and when euthanasia is advisable?

PART FOUR

FUNCTIONING

The pole climbers are obviously functioning. There is an enormous amount of activity in each man's brain from past experiences as well as from the incoming sensory information about the external world and about the state of internal body conditions. The breathing and heart rates are rapid, and the nervous and endocrine systems have primed the person for maximum performance. Good nutrition has helped the contestant become fit for this moment of competition, although the digestive and some other systems are relatively quiet at this moment. Although all organ systems work together in a functioning organism, it is often useful to analyze each system separately, as is done in this part.

Sound digestive, respiratory, circulatory, excretory, and control systems are important for the survival of the individual. Reproduction is critical for the survival of the species. This part deals with the body systems that allow sufficient numbers of individuals to survive and reproduce.

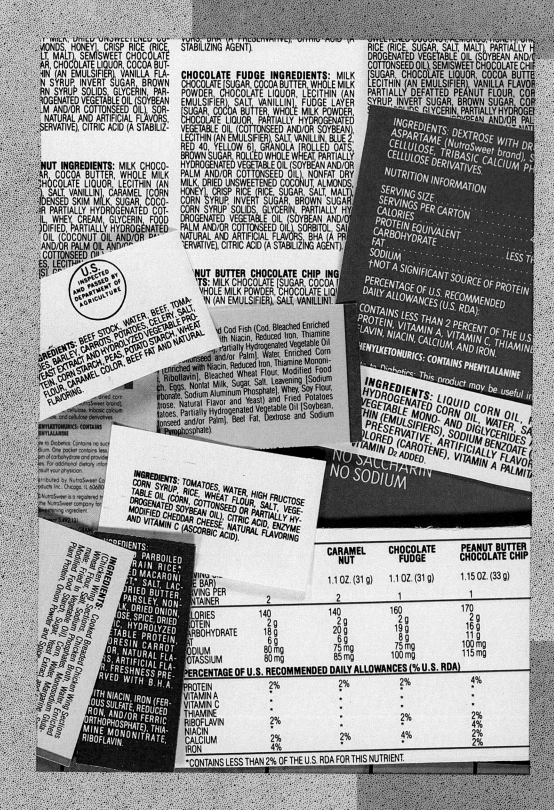

MILK, DRIED UNSWEETENED CO—
MONDS, HONEY], CRISP RICE (RICE,
LT, MALT), SEMISWEET CHOCOLATE
AR, CHOCOLATE LIQUOR, COCOA BUT-
HIN (AN EMULSIFIER), VANILLA FLA-
N SYRUP, INVERT SUGAR, BROWN
RN SYRUP SOLIDS, GLYCERIN, PAR-
ROGENATED VEGETABLE OIL (SOYBEAN
LM AND/OR COTTONSEED OIL), SOR-
NATURAL AND ARTIFICIAL FLAVORS,
ERVATIVE), CITRIC ACID (A STABILIZ-

NUT INGREDIENTS: MILK CHOCO-
AR, COCOA BUTTER, WHOLE MILK
CHOCOLATE LIQUOR, LECITHIN (AN
SALT, VANILLIN], CARAMEL [CORN
DENSED SKIM MILK, SUGAR, COCO-
R PARTIALLY HYDROGENATED COT-
L WHEY, CREAM, GLYCERIN, FOOD
DIFIED, PARTIALLY HYDROGENATED
OIL (COCONUT OIL AND/OR PA—
AND/OR PALM OIL AND/O— PA—
COTTONSEED OIL...
ES, LECITHIN...
SI—

U.S.
INSPECTED
AND PASSED BY
DEPARTMENT OF
AGRICULTURE

REDIENTS: BEEF STOCK, WATER, BEEF, TOMA-
JES, BARLEY, CARROTS, POTATOES, CELERY, SALT,
EAST EXTRACT AND HYDROLYZED VEGETABLE PRO-
TEIN, CORN STARCH, PEAS, POTATO STARCH, WHEAT
FLOUR, CARAMEL COLOR, BEEF FAT AND NATURAL
FLAVORING.

STABILIZING AGENT).

CHOCOLATE FUDGE INGREDIENTS: MILK
CHOCOLATE [SUGAR, COCOA BUTTER, WHOLE MILK
POWDER, CHOCOLATE LIQUOR, LECITHIN (AN
EMULSIFIER), SALT, VANILLIN], FUDGE LAYER
[SUGAR, COCOA BUTTER, WHOLE MILK POWDER,
CHOCOLATE LIQUOR, PARTIALLY HYDROGENATED
VEGETABLE OIL (COTTONSEED AND/OR SOYBEAN),
LECITHIN (AN EMULSIFIER), SALT, VANILLIN, BLUE 2,
RED 40, YELLOW 6], GRANOLA [ROLLED OATS],
BROWN SUGAR, ROLLED WHOLE WHEAT, PARTIALLY
HYDROGENATED VEGETABLE OIL (SOYBEAN AND/OR
PALM AND/OR COTTONSEED OIL), NONFAT DRY
MILK, DRIED UNSWEETENED COCONUT, ALMONDS,
HONEY], CRISP RICE (RICE, SUGAR, SALT, MALT),
CORN SYRUP, INVERT SUGAR, BROWN SUGAR
CORN SYRUP SOLIDS, GLYCERIN, PARTIALLY HY-
DROGENATED VEGETABLE OIL (SOYBEAN AND/O—
PALM AND/OR COTTONSEED OIL), SORBITOL, SA—
NATURAL AND ARTIFICIAL FLAVORS, BHA (A PR—
ERVATIVE), CITRIC ACID (A STABILIZING AGENT).

NUT BUTTER CHOCOLATE CHIP ING
NTS: MILK CHOCOLATE [SUGAR, COCOA
WHOLE MILK POWDER, CHOCOLATE LIQU—
IN (AN EMULSIFIER), SALT, VANILLIN].

d Cod Fish (Cod, Bleached Enriched
with Niacin, Reduced Iron, Thiamine
Partially Hydrogenated Vegetable Oil
ttonseed and/or Palm], Water, Enriched Corn
[enriched with Niacin, Reduced Iron, Thiamine Mononi-
Riboflavin], Bleached Wheat Flour, Modified Food
ch, Eggs, Nonfat Milk, Sugar, Salt, Leavening [Sodium
rbonate, Sodium Aluminum Phosphate], Whey, Soy Flour,
trose, Natural Flavor and Yeast) and Fried Potatoes
tatoes, Partially Hydrogenated Vegetable Oil [Soybean,
onseed and/or Palm], Beef Fat, Dextrose and Sodium
Pyrophosphate).

SWEETENED COCONUT, ALMONDS, HONEY [HP—
RICE (RICE, SUGAR, SALT, MALT), PARTIALLY H—
DROGENATED VEGETABLE OIL (SOYBEAN AND/C
COTTONSEED OIL), SEMISWEET CHOCOLATE CHIP
[SUGAR, CHOCOLATE LIQUOR, COCOA BUTTE
LECITHIN (AN EMULSIFIER), VANILLA FLAVOR
PARTIALLY DEFATTED PEANUT FLOUR, COR
SYRUP, INVERT SUGAR, BROWN SUGAR, COR
SYRUP SOLIDS GLYCERIN, PARTIALLY HYDROGEN
ATED VEGETABLE OIL (SOYBEAN AND/OR PAL—
AND/OR

INGREDIENTS: DEXTROSE WITH DR—
ASPARTAME (NutraSweet brand), S—
CELLULOSE, TRIBASIC CALCIUM PH—
CELLULOSE DERIVATIVES.

NUTRITION INFORMATION

SERVING SIZE ....
SERVINGS PER CARTON ....
CALORIES ....
PROTEIN EQUIVALENT ....
CARBOHYDRATE ....
FAT ....
SODIUM .... LESS TH—
†NOT A SIGNIFICANT SOURCE OF PROTEIN

PERCENTAGE OF U.S. RECOMMENDED
DAILY ALLOWANCES (U.S. RDA):

CONTAINS LESS THAN 2 PERCENT OF THE U.S
PROTEIN, VITAMIN A, VITAMIN C, THIAMINE
FLAVIN, NIACIN, CALCIUM, AND IRON.

PHENYLKETONURICS: CONTAINS PHENYLALANINE

te to Diabetics: This product may be useful i—

INGREDIENTS: LIQUID CORN OIL, P—
HYDROGENATED CORN OIL, WATER, SA—
VEGETABLE MONO- AND DIGLYCERIDES A
THIN (EMULSIFIERS), SODIUM BENZOATE (
PRESERVATIVE, ARTIFICIALLY FLAVOR
OLORED (CAROTENE), ARTIFICIALLY FLAVOR
VITAMIN D₂ ADDED), VITAMIN A PALMIT—

NO SACCHARIN
NO SODIUM

INGREDIENTS: TOMATOES, WATER, HIGH FRUCTOSE
CORN SYRUP, RICE, WHEAT FLOUR, SALT, VEGE-
TABLE OIL (CORN, COTTONSEED OR PARTIALLY HY-
DROGENATED SOYBEAN OIL), CITRIC ACID, ENZYME
MODIFIED CHEDDAR CHEESE, NATURAL FLAVORING
AND VITAMIN C (ASCORBIC ACID).

INGREDIENTS:
(Chicken Wing Sections, Cooked Breaded Chicken Wing Sections
Wheat Flour, Salt, Sodium Phosphates, Chicken Broth, Water,
Modified Food Starch, Sugar, Corn, Water, Monosodium Glu-
Plant Protein, Onion Powder and Spice, Yeast Extract, Margarine (—
mate, Fried in Vegetable Oil, —
PARBOILED
RAIN RICE,
D MACARONI
," SALT, LAC-
RIED BUTTER,
PARSLEY, NON-
LK, DRIED ONION,
OSE, SPICE, DRIED
C, HYDROLYZED
TABLE PROTEIN,
RESIN CARROT
R, NATURAL FLA-
R, ARTIFICIAL FLA-
R, FRESHNESS PRE-
RVED WITH B.H.A.

TH NIACIN, IRON (FER-
OUS SULFATE, REDUCED
RON, AND/OR FERRIC
ORTHOPHOSPHATE], THIA-
MINE MONONITRATE,
RIBOFLAVIN.

	CARAMEL NUT	CHOCOLATE FUDGE	PEANUT BUTTER CHOCOLATE CHIP
	1.1 OZ. (31 g)	1.1 OZ. (31 g)	1.15 OZ. (33 g)
VING PER NTAINER	2	2	1	1
LORIES	140	140	160	170
OTEIN	2 g	2 g	2 g	2 g
ARBOHYDRATE	18 g	20 g	19 g	16 g
T	6 g	6 g	8 g	11 g
ODIUM	80 mg	75 mg	75 mg	100 mg
OTASSIUM	80 mg	85 mg	100 mg	115 mg

PERCENTAGE OF U.S. RECOMMENDED DAILY ALLOWANCES (% U.S. RDA)

PROTEIN	2%	2%	2%	4%
VITAMIN A	*	*	*	*
VITAMIN C	*	*	*	*
THIAMINE	*	*	*	2%
RIBOFLAVIN	2%	*	2%	4%
NIACIN	*	*	*	2%
CALCIUM	2%	2%	4%	2%
IRON	4%	*	*	2%

*CONTAINS LESS THAN 2% OF THE U.S. RDA FOR THIS NUTRIENT.

CHAPTER 12

NUTRITION: FUELING THE BODY

Although humans and other animals do many things that plants do not, plants do two extremely important things that animals cannot— they produce their own food and they release oxygen in the process. Most animals either eat plants (herbivores), eat other animals (carnivores), eat both plants and flesh (omnivores), or feed on the wastes or remains of other organisms (detritivores). For the above reasons, plants, some protistans, and some bacteria are called autotrophs or "self-feeders" because they can synthesize all required compounds from a few basic inorganic substances. Most animals, fungi, many protistans, and most bacteria are called heterotrophs, or "other feeders," because the many compounds that they cannot synthesize must be present in their food.

In this chapter, you will study the nature of nutrients; Chapter 13 deals with the adaptations for securing and digesting food.

Knowing that more and more people are concerned about their diets, advertisements are increasingly providing more information for consumers. Products are often promoted as being high in some substances (such as iron, calcium, or fiber) and low in others (such as cholesterol, sodium, or calories). This chapter will help you understand more about which nutrients are good and which are bad for us; Chapter 13 deals with how we process our nutrients.

OVERVIEW OF NUTRITIONAL METABOLISM

Thousands of different chemical reactions occur simultaneously in any living organism, and **metabolism** (from the Greek for "change") is the sum of these chemical activities. One part of nutritional metabolism involves the breaking down (catabolism) of complex materials into simpler ones, often for the release of energy. Another aspect of metabolism is constructive (anabolism), whereby small molecules are formed into larger ones for producing new tissues or for storing energy.

Figure 12.1 summarizes the flow of biological energy. Photosynthetic organisms capture radiant energy from sunlight and use it to convert water and carbon dioxide into carbohydrates and other nutrients. The oxygen that is released during the photosynthetic process is used by animals during the breakdown and utilization of the nutrients that they consume. During these processes of cellular respiration, CO_2 and H_2O are produced. Because some heat is lost to the environment following each energy transformation, only a constant input of solar energy allows biological processes to continue. The one-way flow of energy is quite different from the cycling of the materials of life (Chapter 1).

Nutrients are the substances used by organisms as chemical building blocks for growth and repair and as energy sources for their activities. Among the substances included in the general term *food* are carbohydrates, proteins, and lipids (Chapter 4). From digesting these substances, we obtain energy trapped in chemical bonds, and the carbon skeletons and other chemical groups that result are used to make an enormous variety of needed compounds (Figure 12.2). Also important are vitamins, minerals, and water. Whereas plants can make all of the amino acids needed for their proteins, animals cannot synthesize all they need. Those that must be present in our diet are called **essential amino acids,** and we need eight such compounds.

Some nutrients are needed in large quantities, the **macronutrients,** but only small amounts of the **micronutrients** or **trace elements** are required. What we make of these substances after we digest and absorb them is determined by our DNA. When we eat soybeans, for example, we must obviously convert the plant's proteins to their constituent amino acids and then reconstitute human proteins from them. Various cats and dogs may eat the same food but their common food source does not cause them to grow alike; the Chihuahua stays small and the Saint Bernard remains huge, as determined by their genetic composition. It is the

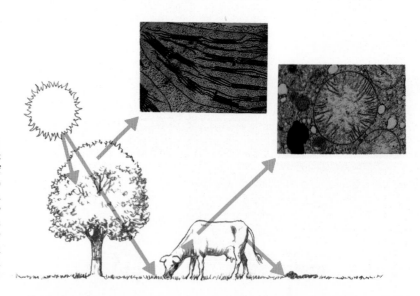

FIGURE 12.1

An overview of energy flow. The chloroplast organelle is central to the process of photosynthesis, as the mitochondrion is to cellular respiration. The mitochondrion carries out the final stages of cellular respiration whereby the energy in nutrients such as carbohydrates is converted into ATP molecules. Have you thanked the sun and green plants today, as various ecological posters ask?

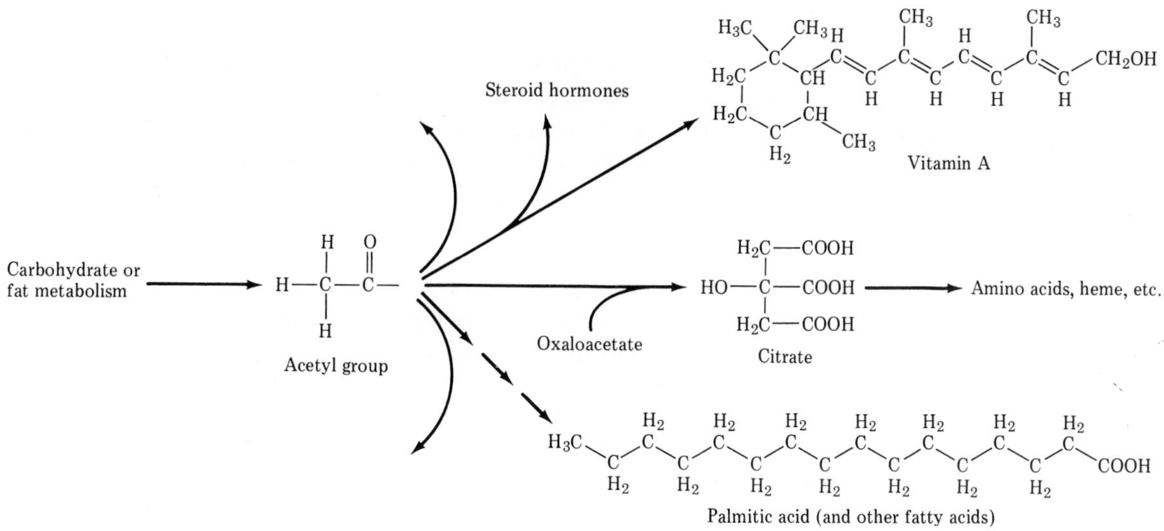

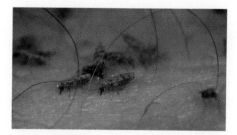

genetic blueprint inherent to each type of organism that determines how the amino acid building blocks are assembled.

CALORIES: FOOD ENERGY

Food provides us with both structural materials and energy, the latter being expressed in calories. A **calorie** is the heat energy needed to raise the temperature of 1 g of water 1°C at about 15°C. This small unit is useful for studies of the energy an aphid gets from sucking plant sap, a mosquito secures from digesting your blood, or a caterpillar secures from chewing leaves (Figure 12.3). However, because the calorie is too small a unit for nutritional studies of humans, the **Calorie** (capital C) is usually used. One Calorie equals 1000 calories. To avoid the confusion of lowercase and uppercase letters, the calorie and Calorie are also called a gram calorie (gcal) and a kilocalorie (kcal), respectively.

The heat energy of food is often measured in a **calorimeter** (Figure 12.4). A dehydrated food sample is ignited by an electric current and the heat produced is measured by the change in water temperature. Indirect calorimetry involves measuring an animal's oxygen consumption and/or CO_2 production. The average energy value for lipids (9 kcal/g) is about twice that of an equal weight of carbohydrates or proteins (4–5 kcal/g).

FIGURE 12.2
The acetyl group is an excellent example of how animals produce a wealth of biological molecules from the carbon skeletons that they secure from the carbohydrates, proteins, and lipids that they consume.

FIGURE 12.3
Most animals are quite small. Some, like these mosquitoes, live on a liquid diet. The gram calorie is adequate for studies of their nutrition, whereas the kilocalorie is more useful for larger animals. In most cases only a small fraction of the energy available at one feeding or trophic level is transferred to a higher feeding level.

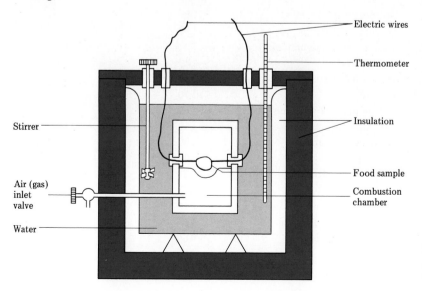

FIGURE 12.4
A calorimeter for measuring the energy value of food.

METABOLIC RATES

The rate of metabolism when you are at complete rest is called the **basal metabolic rate** (BMR). The BMR includes the energy needed for the lungs, heart, and other organs to work even at rest; it is about 10 percent lower when you sleep. To perform any activities, you require additional calories (Table 12.1)

Among the variables that commonly affect the rate of metabolism are age, sex, health, activity level, environmental conditions, and surface area. For example, although the human body is irregular in shape, its average surface area can be estimated from Figure 12.5 (also see Figure 3.8). Use that information and that in Table 12.2 to calculate your possible BMR. The formula is surface area in square meters × heat production per hour per square meter × 24 hours. The average surface areas for adult American males and females are about 1.6 m² and 1.8 m², respectively. Note in Table 12.2 that the metabolic rate (expressed as heat production) is higher for males than it is for females and that it decreases with age.

Certain hormones also affect the BMR (Chapter 18). For example, the BMR may be below normal in those whose thyroid gland is under-active (hypothyroidism) and above normal in hyperthyroidism. Serious diseases or malnutrition may also affect the BMR. Fever increases the human metabolic rate by about 10 percent per Celsius degree. Although mental work may leave you tired psychologically, it increases your metabolic rate only slightly. However, emotional stress may increase the metabolic rate by 5–20 percent, and heavy work or strenuous exercise may increase the rate by as much as 50 times.

As mentioned earlier, this page contains much food for thought, but you would not get much energy from eating it because you lack the

TABLE 12.1
Average Caloric Needs for Various Activities

Activity	Kilocalories/30 Minutes	
	Female, 55-kg (121 lb)	Male, 73-kg (161 lb)
Sleeping	28	38
Sitting	29	39
Eating	33	44
Standing, at ease	34	45
Standing, light activity	58	78
Driving a car	72	96
Playing volleyball	83	110
Slow walking, on level	84	112
Playing Ping-Pong	93	124
Slow pleasure swimming	104	139
Calisthenics	120	160
Pleasure bicycling, on level	120	160
Golfing	130	173
Gardening and weeding	141	188
Fast walking, on level	159	212
Walking downstairs	160	213
Playing tennis	166	221
Playing basketball	169	225
Fast swimming	213	286
Mountain climbing	241	321
Running, long distance	361	481
Walking upstairs	416	555
Sprinting	516	748

enzymes needed to digest the cellulose that it contains. Although burning this page would yield energy in such forms as heat and light, combustion and cellular respiration are not identical (Figure 12.6). In combustion, many chemical bonds break quickly, with a sudden release of energy, but during digestion enzymes control the rate at which chemical bonds are broken. Because of this, our body temperature remains fairly constant, as opposed to rising quickly after meals and returning to a lower level later.

The approximate total daily caloric needs for sedentary American males and females is 2500 and 2000 kcal, respectively. The comparable sets of figures for more active males and females, respectively, are 3000 and 2400 kcal (moderately active) and 3600 and 2800 kcal (very active). Pregnant and lactating females have caloric needs similar to those of males in the same activity class. However, these are average figures only, as is shown by the fact that a large person doing heavy work in very cold weather may need over 6000 kcal daily.

Although children need less total food than adults, their metabolic needs exceed those of adults in proportion to their body weight. For example, human babies require about twice as many calories per unit of body weight than adults. The explanation for this includes changing surface area-to-volume relationships of individual cells and of the entire multicellular animal body (Figures 2.10 and 2.11). The metabolic needs of a cell (and in a general way of an animal body) are related to the cell's volume. However, the ability to satisfy those needs, such as securing nutrients and eliminating wastes, depends upon the cell's surface area. As cells and entire animals grow, their volumes (metabolic needs) increase faster than their surface areas (ability to satisfy their needs). One common result is a decrease in metabolic rate with age.

Weight gain or loss depends upon caloric intake and caloric use. If you consume more calories than you use, then you will gain weight. Weight loss occurs when you use more calories than you consume. It is often said that the best form of exercise is to push yourself away from the table!

You may estimate your potential for gaining or losing weight by using references to find the caloric value of the quantity of each food you consume during a typical day. By recording all of your activities during the same period, and the time spent doing each, you can use Table 12.1 to estimate the kilocalories used for each activity.

VITAMINS AND MINERALS

Vitamins, minerals, and water are important nutrients that contain few calories. Water plays a role in many chemical reactions, and the average adult drinks about 1650 ml (1.7 qt) of water directly each day. An additional 1.1 L is obtained from food.

Vitamins

About two-thirds of Vasco da Gama's crew in his 1497 voyage around the Cape of Good Hope died of **scurvy**, a condition that plagued explorers and sea voyagers for centuries. During the 1700s the British Navy discovered that scurvy was reduced when citrus fruits were eaten, and British sailors have often been called "limeys" ever since. During the 1800s the Japanese Navy suffered from **beriberi**, a serious problem that affects the nervous, digestive, and circulatory systems. When the diet of mainly white rice was supplemented with vegetables, meat, and fish, the condition disappeared. Thus, it appeared that small things in the diet were highly important, although it was not until this century

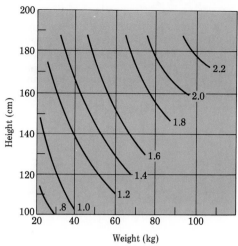

FIGURE 12.5
Graph for estimating the average surface area of the human body. The body's surface area is obviously in contact with its immediate environment. Among other things, this is very important in such functions as thermoregulation (Chapter 2).

TABLE 12.2

Heat Production per Hour per Square Meter of Surface Area for Average Adult Humans

	Kilocalories	
Age (years)	Males	Females
16–18	43.3	38.8
18–20	41.5	37.5
20–30	40.0	36.9
30–40	39.3	36.3
40–50	38.2	35.2
50–60	37.1	34.3

FIGURE 12.6
The ultimate source of our food is the process of photosynthesis. During cellular respiration, enzymes break the chemical bonds between carbon and other atoms, and energy is released in a controlled manner. During combustion, in contrast, many bonds are broken simultaneously, and there is a rapid release of energy.

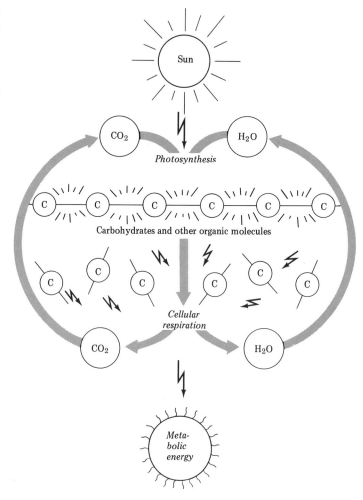

that scurvy, beriberi, and many other dietary problems were recognized as **vitamin deficiency diseases.** Earlier, it was thought that carbohydrates, proteins, lipids, and minerals were the only basic nutrients.

Vitamins are essential organic compounds found in minute amounts in foods. A Polish biochemist coined the term *vitamine* in 1911 for the antiberberi factor he was studying. The term was quickly accepted and used for all unknown dietary factors, and in 1919 the final *e* was dropped.

Some of the major vitamins are outlined in Table 12.3. Only deficiency symptoms are listed because we know relatively little yet about the effects of the excesses that lead to **hypervitaminosis.** However, as examples, headaches, vomiting, anorexia, and skin and bone problems have been reported for excesses of vitamin A. Excess vitamin D has led to vomiting, diarrhea, weight loss, and kidney damage. Hypervitaminosis is unlikely to develop from a normal diet, but it may result from the intentional ingestion of vitamins. Some people believe that large doses (megadoses) of vitamins will enhance a condition that results from a deficiency of a vitamin. For example, knowing that a deficiency of vitamin E interferes with reproductive structures and functions in some laboratory animals, some people take excessive vitamin E hoping that it will keep their reproductive systems in good order, or even improve them. In most cases this is false thinking, and toxic symptoms may occur. As with most body functions, there is an optimum, and either excesses or deficiencies may cause problems.

Although vitamins were first named by letters, some vitamins proved to be mixtures of several different vitamins (e.g., vitamin B). Now, many

TABLE 12.3
Vitamins in Human Nutrition

Vitamin	Common Sources	Major Body Functions	Possible Effects of Deficiency
Water-soluble			
B_1, thiamin	Liver, legumes, yeast, whole grains, pork	Coenzyme in cellular respiration	Beriberi, fatigue, loss of appetite
B_2, riboflavin	Green leafy vegetables, dairy foods, eggs, organ meats	Coenzyme in cellular respiration (in FAD)	Cracks at corner of mouth, lesions of eye, skin irritation
Niacin	Liver, lean meats, yeast, fowl, grains, legumes	Coenzyme in cellular respiration (in NAD and NADP)	Pellagra, skin and gastrointestinal tract lesions, nervous and mental disorders, diarrhea
B_6, pyridoxine	Liver, meats, vegetables, whole grains, dairy foods	Coenzyme in amino acid metabolism	Anemia, skin problems, convulsions, muscular twitching, slow growth, kidney stones
Pantothenic acid	Liver, eggs, yeast	In coenzyme A (energy metabolism)	Adrenal and reproductive problems, fatigue, nausea, sleep disturbances, impaired coordination
Biotin	Liver, yeast, intestinal bacteria, meat, eggs, milk, vegetables, legumes	In coenzymes for amino acid metabolism, fat synthesis, glycogen formation	Fatigue, nausea, muscular pain, dermatitis, depression, hair loss, skin problems
B_{12}, cyanocobalamin	Liver, meat, dairy foods, eggs, intestinal bacteria	Coenzyme in formation of nucleic acids and proteins, erythrocyte formation	Pernicious anemia, neurological disorders
Folic acid	Liver, eggs, vegetables, whole grains	Coenzyme in formation of heme and nucleotides; amino acid metabolism	Anemia, gastrointestinal tract disturbances
Choline	Liver, egg yolk, grains, legumes	Part of phospholipids; precursor of neurotransmitter acetylcholine	None reported
C, ascorbic acid	Citrus fruits, tomatoes, potatoes, green leafy vegetables, green peppers	Aids formation of connective tissue (bone, cartilage, dentin); collagen synthesis; prevents oxidation of cellular components	Scurvy, poor bone growth, slow healing
Fat-soluble			
A, retinol	Fruits, vegetables, liver, dairy foods	Constituent of visual pigments; maintenance of epithelial tissues	Night blindness, damage to mucous membranes, dry and flaky skin
D, calciferol	Fortified milk, fish oil, sunshine, eggs	Promotes bone growth and mineralization; absorption of calcium and phosphorus	Rickets in children, osteomalacia in adults
E, tocopherol	Meat, dairy foods, whole grains, green leafy vegetables	Muscle maintenance; prevents oxidation of cellular components	Possibly anemia, infertility (in rats)
K, menadione	Liver, intestinal bacteria, green leafy vegetables	Blood clotting	Blood clotting problems

biologists prefer to use the vitamin's chemical name. Vitamins are also often classified as **fat-soluble** or **water-soluble.** Fat-soluble vitamins (e.g., A, D, E, and K) are generally required by vertebrates but not by other animals. All animals require water-soluble vitamins (e.g., B complex and C). Although excess water-soluble vitamins are usually eliminated in the urine, excess fat-soluble vitamins tend to be excreted in the feces.

Within the body, vitamins are converted into **cofactors,** substances needed for some enzymes to function (Figure 12.7), or are substances from which enzymes are made. Because of these functions, only small quantities are needed, and vitamins make up a minute part of the diet. However, because of our limited ability to store vitamins, we must have a constant supply in our daily diet. Because some vitamins are made from precursors, it is important to discuss the vitamin value of food rather than simply its vitamin content. For example, carrots contain no vitamin A but are rich in the precursor carotene used by the body to make the vitamin.

Minerals

Unlike vitamins, **minerals** are inorganic substances. Although the body contains many different minerals (Chapter 4), most are trace elements or micronutrients, and only about 4 percent of body tissue is composed of minerals. Table 12.4 summarizes the major minerals, and three minerals of particular importance are described below.

Calcium.   The maintenance of a proper calcium balance in the body is another excellent example of homeostasis (Figure 12.8). The average adult contains more than 1 kg (2 lb) of calcium, mostly in bones and teeth. Small quantities of calcium elsewhere are important in blood clotting and in muscle and nerve functioning. To keep the latter functions operating properly when calcium is deficient in the diet, calcium is withdrawn automatically from the bones. Therefore, skeletal defects, such as **rickets** in growing children, are common early signs of calcium deficiency. If the blood calcium level becomes too low, the nerves and muscles become very irritable and the slightest stimulus may cause spasms or convulsions, a condition called **tetany.** Tetany may occur in pregnant and lactating women or in children with rickets.

Although many foods contain some calcium, an easy way to obtain the **recommended daily allowance** (RDA) is to consume dairy products.

Iron.   Iron is needed for hemoglobin, the oxygen-carrying compound in red blood cells. Therefore, an iron deficiency may cause one important

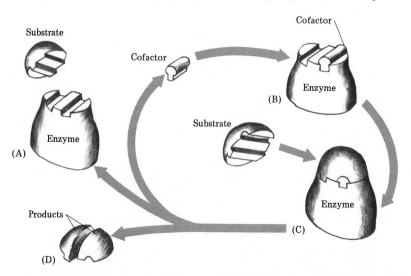

FIGURE 12.7
A cofactor acts as an ''adapter,'' allowing the substrate to attach to an enzyme.

TABLE 12.4
Minerals Required by Vertebrates

Element	Common Sources for Humans	Major Body Functions	Possible Effects of Deficiency
Macronutrients			
Calcium (Ca)	Dairy foods, eggs, whole grains, green leafy vegetables, legumes, nuts	Bone and tooth formation; blood clotting; nerve and muscle action; enzyme activation	Rickets, stunted growth, osteoporosis, irritable nerves and muscles
Chlorine (Cl)	Table salt	Water and acid–base balances; major negative ion in intercellular fluid; gastric juice (as HCl)	Muscle cramps, loss of appetite, mental apathy
Magnesium (Mg)	Green leafy vegetables, whole grains	Activates enzymes; part of bones and teeth; protein synthesis	Poor growth, weakness, spasms, behavioral changes
Phosphorus (P)	Dairy foods, meat, eggs, whole grains, nuts, legumes	In ATP, nucleic acids, and phospholipids; bone formation; metabolism of sugars; acid–base balance	Demineralization of bone, loss of calcium, weakness
Potassium (K)	Meat, milk, whole grains, vegetables, fruits, legumes	Water and acid–base balances; major positive ion in intercellular fluid; nerve function	Muscular weakness, paralysis
Sodium (Na)	Table salt, meat, dairy foods, eggs, vegetables	Nerve and muscle action; water and acid–base balances; major positive ion in intercellular fluid	Muscle cramps, reduced appetite, mental apathy
Sulfur (S)	Meat, dairy foods, eggs, legumes, nuts	In proteins and coenzymes	Unclear
Micronutrients			
Chromium (Cr)	Meat, dairy foods, yeast, whole grains, peanuts	In glucose metabolism	Unclear
Cobalt (Co)	Meat, milk, some tap water	Part of vitamin B_{12}; erythrocyte formation	Unclear
Copper (Cu)	Liver, meat, fish, shellfish, whole grains, nuts, tap water	Part of enzymes involved with iron metabolism; production of hemoglobin; bone formation	Anemia, bone changes
Fluorine (F)	Tap water, seafood, tea	Improves resistance to tooth decay; maintenance of bone	Higher frequency of tooth decay
Iodine (I)	Seafood, dairy foods, iodized salt	In thyroid hormone	Simple goiter
Iron (Fe)	Liver, lean meats, green leafy vegetables, eggs, whole grains, nuts, legumes	Part of hemoglobin and myoglobin; enzymes involved in cellular respiration	Iron-deficiency anemia (e.g., weakness, reduced resistance to infection)
Manganese (Mn)	Organ meats, whole grains, nuts, legumes, coffee, tea	Activates many enzymes	Unclear
Molybdenum (Mo)	Organ meats, dairy foods, green leafy vegetables, legumes	Required by some enzymes	Unclear
Selenium (Se)	Meat, milk, seafood, eggs, chicken, whole grains, garlic	In lipid metabolism	Unclear
Zinc (Zn)	Liver, fish, shellfish, many other foods	Required by digestive enzymes; involved in insulin physiology	Impaired growth; small sex glands

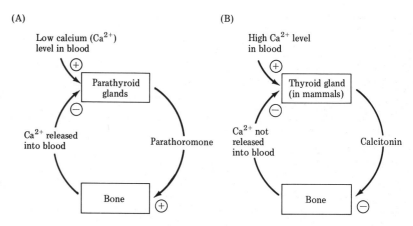

FIGURE 12.8
A negative feedback diagram of the control of body calcium. If blood calcium becomes too low, nerves and muscles become very irritable and the slightest stimulus may cause the spasms or convulsions of tetany. The effects of excess calcium are somewhat less clear, however.

form of **anemia,** and the lack of iron is one of the major dietary deficiencies in the United States. Some authorities estimate that about one-third of pregnant women, especially during the last trimester, develop some form of iron deficiency.

Iodine. Iodine forms part of the thyroid gland hormone thyroxin, which plays a major role in controlling the body's overall metabolic rate. The addition of iodine to table salt has eliminated most deficiency problems, such as goiters. Good natural sources of iodine include most seafoods, especially shellfish.

See the essay "Vegetarianism" for a discussion of how some humans secure their nutrients.

NUTRIENT ADDITIVES AND CONTAMINANTS

The growing, processing, and marketing of foods are among America's largest businesses, and food is one of the most important purchases that people make. Growers, intermediaries, and/or retailers add thousands of chemical compounds to various foods for numerous purposes.

Food Additives

As people become more dependent upon processed and convenience foods, the number and variety of **food additives** increase, and the average American may consume about 2 kg (5 lb) of these annually. Many humans who are extremely concerned about the effects on wildlife of the many substances that are being added to the environment think little about the effects of the substances that they add to their own bodies daily.

Some people believe that all chemicals added to food are bad and that only "natural" food should be eaten. Others believe that food additives serve many useful purposes, such as preserving and enriching food and making its color or flavor more appealing. The important questions are whether additives are needed and, if so, whether they are safe. Note the food additives contained in some of the foods that you consume by studying the labels, and try to categorize the additives into the classes shown in Table 12.5.

Food additives do not necessarily mean that a food is harmful, nor does the fact that a food is natural or "organic" ensure its safety. For example, poisoning from excessive vitamin A could result from eating the livers of halibut or polar bears. A known carcinogen (3,4-benzopyrene) found in automobile exhausts and cigarette smoke also occurs in small amounts in lettuce, spinach, tea, cabbage, and charcoal-broiled meat. Small amounts of hydrogen cyanide are released during the digestion of plums, apricots, cherries, yams, and lima beans. Goiter may

Vegetarianism

Some people, especially in developing nations, must live on a diet high in plant matter and low in foods of animal origin. **Vegetarian** diets, traditional among certain ethnic and religious groups, have become so popular that numerous restaurants include vegetarian menus. Some vegetarians object mainly to the killing of animals, although some who abstain from consuming red meat eat fish and/or fowl. Others also object to eating any animal products, such as milk, eggs, or cheese. Although some diets of these pure vegetarians can be nutritionally sound, others may not be. It is often more difficult for children and pregnant and lactating females to thrive on vegetarian diets than it is for others to do so.

Because no single food furnishes all nutrients, it is especially important for vegetarians to eat a wide variety of foods. Thus, vegetarians must know far more about nutrition than others do. It is easier to secure needed nutrients if there is no objection to eggs and/or dairy products.

Several nutrient problems faced by pure vegetarians are listed below.

Cyanocobalamin. This vitamin (B_{12}) is found only in foods of animal origin, and vegetarians may secure it from vitamins or from foods fortified with it (e.g., breakfast cereals). Because deficiency symptoms often take time to develop, "experimental" vegetarians may not experience them.

Riboflavin. Milk and meat are sources of vitamin B_2 for most people. Pure vegetarians may get riboflavin from yeast, vegetables (especially green ones), and some nuts and seeds.

Vitamin D. Large amounts of calciferol occur in only a few foods of plant origin. However, many common foods are enriched with vitamin D (e.g., milk, margarine, and breakfast cereals), and vitamin D may be ingested directly. A nonnutrient way many people get some calciferol is through exposure to sunlight, because the body can synthesize this vitamin under these conditions.

Calcium. Although milk is a good source of calcium, pure vegetarians may get calcium from certain dark-green leafy vegetables (e.g., collard, kale, dandelion, turnip, and mustard greens).

Protein. Although foods of both animal and plant origin contain protein, their amino acid content varies and, in general, animal protein more closely matches the body's needs. Soybeans and some other legumes are good sources of plant protein.

Vegetarians, especially purists, must be prepared to eat larger quantities of food than those eaten by omnivores. Vegetarian diets are simply not as "concentrated" as those that include meat and animal products, and they also contain far more indigestible material such as cellulose.

If you are not a vegetarian, try planning a vegetarian diet that you believe would provide you with all necessary nutrients. If you are a vegetarian, share your knowledge and experiences with others.

develop in susceptible people when they eat collard and mustard greens, cabbage, turnips, and cauliflower. Consuming bananas and certain cheeses, beers, and wines contribute to hypertension in some people. In general, a chemical is a chemical, whether it is synthetic or comes from food.

Few chemicals are safe for all humans under all circumstances and the effects of additives vary greatly from person to person, partly because of many synergistic interactions. **Synergism** refers to various "1 + 1 = 5" interactions, in which the combined effects of two or more agents exceed what we expect from knowing their independent effects. Thus, usually harmless chemicals sometimes interact synergistically with other usually harmless substances to produce hazardous products. Many chemicals ingested in small amounts over a period of time may take years to produce adverse effects. Obviously, testing for all possible effects is costly and time consuming.

Some chemicals are added to water as well as to food. Chlorine is added to many water systems to kill pathogens, and many Americans drink water with added fluorides. Although most dentists claim that fluorides prevent cavities or caries, some people object to all or most substances being added to their drinking water. See the essay "Fluoridation" for information about this issue.

TABLE 12.5
Some Classes of Food Additives

Class	Purpose	Examples	Typical Uses
Preservatives	To retard spoilage from microorganisms	Chemicals: salt, sodium nitrate, benzoic acid, sugar, citric acid, sodium and calcium propionate, sorbic acid, potassium sorbate, sulfur dioxide, sodium benzoate Processes: drying, dehydration, curing, smoking, canning, pasteurization, refrigeration, freezing	Meat, fruits, jellies, breads, vegetables, cheeses
Nutrient supplements	To increase the nutritive values of foods or to replace nutrients lost in processing	Vitamins, amino acids	Milk, flour and bread, rice, cereals, corn meal
Flavors and flavor enhancers	To flavor or enhance the flavor of foods	Monosodium glutamate, artificial sweeteners, oils from bananas, vanilla, cinnamon	Ice cream, soft drinks, fruit juices, low-calorie foods, candy, toppings, salad dressings, pickles, many convenience foods
Coloring agents	To color foods for esthetic or sales appeal (e.g., items that show a lack of freshness)	Natural and synthetic dyes	Cheese, ice cream, puddings, soft drinks, breakfast cereals, sausages, cake mixes
Antioxidants	To retard spoilage (rancidity) of fats by excluding oxygen	Chemicals: lecithin, propyl gallate, BHA, BHT Processes: refrigeration, sealed cans, wrapping	Shortenings, cooking oils, crackers, salted nuts, soups, cereals, artificial fruit drinks, some toppings
Stabilizers and thickeners	To give foods smooth texture, consistency, and "body"; to prevent separation of components	Vegetable gums (e.g., gum arabic), seaweed extracts (e.g., agar, algin), gelatin, dextrin, sodium carboxymethyl cellulose	Ice cream, sherbet, pie fillings, cheese spreads, soft drinks, beer, icings, salad dressings, diet foods, cake and dessert mixes, some whipped creams
Acidulants	To mask undesirable aftertastes or to add a tart taste	Citric acid, fumaric acid, phosphoric acid	Soft drinks, fruit juices, soups, gravies, desserts, cheeses, salad dressings
Alkalis	To reduce acidity	Sodium carbonate, sodium bicarbonate	Some wines, olives, canned peas, some pies and pastries
Emulsifiers	To disperse one liquid in another (e.g., oil in water)	Lecithin, mono- and diglycerides, propylene glycol, polysorbates	Ice cream, salad dressings, mayonnaise, shortening, icings, nondairy creamers, candy, margarine
Sequestrants (metal "scavengers" and chelating agents)	To prevent clouding in foods; to add color, texture, and flavor; and to "tie up" trace metal ions that catalyze spoilage of food	Citric acid, chlorophyll, ethylenediaminetetraacetic acid (EDTA), sodium phosphate	Artificial fruit drinks, soft drinks, beer, some canned and frozen foods, desserts, soups, salad dressings

Fluoridation

Most dentists and public health specialists believe that community water **fluoridation** is a safe, effective, and proven way to prevent dental decay. Fluorides occur naturally in most water but not always in the right amounts. The U.S. Public Health Service (USPHS) estimates that by raising or lowering the fluoride content of community water supplies to an optimal level, about two of three dental cavities can be prevented in children. USPHS research shows that the costs for children's dental care are about half as much in fluoridated communities as in fluoride-deficient communities. Trace quantities of fluoride are also needed for the proper mineralization of bone, in reproduction, and in the action of some enzymes. Some fluoride is found in almost all soft tissues in the body.

The above gives little hint as to the fluoride controversy, and some basic facts about fluorine will help set the stage for an analysis of the fluoride issue. Fluorine is the seventeenth most abundant element in Earth's crust (about four times more common than copper). It is so highly reactive chemically that it is found in its elemental gaseous form only in some industrial processes. Besides being present in water, fluorine (as fluoride ions) occurs in certain foods (e.g., tea, fish, eggs, and cheese) and in some medications, dust, and industrial contaminants.

Excess fluoride in the body is called **fluorosis**. Several million tons of fluoride-containing minerals (e.g., fluorospar) are used annually in steel making and in some other industries. The inhalation of fluoride dust during the mining and processing of these minerals has caused many cases of fluorosis. The effects of fluorides may result from a single massive dose (acute) or from large doses over a number of years (chronic). Acute fluoride poisoning is rare but may be fatal.

The most common chronic dental effects of high fluoride levels are brittle teeth and mottled enamel, especially on the permanent teeth. High fluoride levels also damage the bones and cause joint pains. Bone becomes dense, brittle, irregular, and less flexible. The typical amount of fluoride in artificially fluoridated water in the United States causes neither acute nor chronic fluorosis.

Absorption of fluorides occurs mainly in the lungs and in the digestive tract, and as long as the fluoride level in drinking water is within the normal range, the blood level of fluorides is maintained homeostatically. About half of ingested fluoride usually becomes part of bones and teeth. The average fluoride content of the dentin is higher than that of the enamel, and most enamel fluoride is on the outer surface. Fluorides may protect teeth against caries by reducing the solubility of enamel in acid and/or by inhibiting the bacteria that produce enamel-attacking acids.

Fluorides are excreted mainly in the feces and urine. Smaller amounts are lost when skin cells and hair are shed and in sweat, saliva, tears, and milk.

In areas of nonfluoridated or low-fluoride water, many dentists suggest the use of fluoride tablets or fluoridated toothpaste for children. Children retain far more absorbed fluoride than adults. Fluoride tablets appear to be more effective in preventing cavities in children's teeth than fluoridated toothpastes. Little is known about the long-term usefulness of fluoridated mouthwash, gum, or vitamin preparations.

Those who object to the fluoridation of community water systems usually agree that a certain level of fluoride prevents cavities in children's teeth. However, they suggest that fluoride can be provided specifically to children in other ways, and either believe (without evidence) that fluorides are harmful to older adults, object to governmental decisions about what they drink, and/or believe that the cost is too high for the benefits.

Do you believe that humans have a "right" to breathe natural air and to ingest natural food and water? Do you believe that communities should fluoridate their water supplies if most voters agree, or even if most disagree? How much credence do you give to information from groups such as the American Medical Association, American Dental Association, American Public Health Association, American Water Works Association, and the various antifluoridation associations?

Consumer Protection

The present U.S. Food and Drug Administration (FDA) had its beginnings in 1906. However, only since 1958 have laws required that the safety of food additives be tested by manufacturers. Because hundreds of food additives were already in use then, the FDA decided not to stop all usage and initiate expensive and time-consuming tests on all these chemicals. Instead, it drew up a list of food additives in use and circulated it among experts. Based on the professional opinions received, the FDA

FIGURE 12.9
These tiny bacteria (*Clostridium botulinum*) cause botulism, one of the most severe and usually fatal types of food poisoning. The bacterial spores may contaminate raw foods before harvest or slaughter, or foods being prepared or stored later. Properly cooked food is usually harmless, even if contaminated, because the neurotoxin produced by the bacteria is destroyed by sufficient heat.

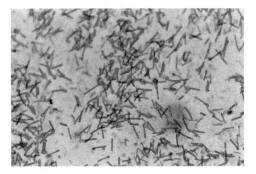

deleted a few items from the list and then published a list of about 600 chemicals "Generally Regarded As Safe." The chemicals on this GRAS (pronounced "grass") list were assumed to be safe because they had been in use for a number of years. Many additional chemicals are now on the GRAS list.

Today, after conducting what they consider adequate testing, manufacturers may use chemicals in food without consulting with the FDA. Although the FDA may challenge tests and/or interpretations, its staff and budget are limited. Because the FDA cannot test each substance independently, it usually accepts the manufacturer's tests as valid and sufficient. Most chemicals on the GRAS list are probably safe, although further testing has resulted in the ban of a few chemicals that were once allowed (e.g., cyclamates and some dyes and vegetable oils).

One part of the act that governs the FDA is a clause that prohibits the use in food of any chemical that causes cancer in animals or humans when fed at any level. However, the clause does not cover chemicals converted into carcinogenic substances in the body. For example, there is concern about nitrate and nitrite preservatives being converted into carcinogenic compounds in the digestive tract.

Food Contaminants

It is important to distinguish between food additives that are added to food intentionally and the **food contaminants** that result from some phase of production, processing, storage, or packaging. For example, pesticides may enter foods while they grow, and some chemicals from storage containers may contaminate foods. Also of concern are heavy metals (e.g., mercury and lead), animal wastes, insects, radioisotopes, and pathogens from poor hygiene. Some contaminants get into the body by "tagging along" with essential nutrients (e.g., radioactive strontium), and many accumulate in the body because they are not excreted readily.

Most cases of **food poisoning** are not caused by food itself or by infection from pathogens living in it but rather by toxins produced by microorganisms. Improperly refrigerated or overaged meats, cheeses, custards, cream pies, and food containing mayonnaise are especially hazardous. Unfortunately, such foods sometimes appear and smell normal. Digestive distress usually begins two to four hours after eating the food and may continue for three to eight hours. Common symptoms include nausea, vomiting, diarrhea, and cramps.

Botulism is a serious and often fatal form of food poisoning caused by toxins produced by a bacterium that is related to the species that causes tetanus (Figure 12.9). Most cases result from eating bacteria-contaminated food that was inadequately processed and sealed. One drop of the potent bacterial poison called botulin can kill tens of thousands of humans. Symptoms usually begin within 18–36 hours and often include dizziness, fatigue, and blurred and double vision. Breathing becomes difficult, and death may result from respiratory failure or heart paralysis. Exposing the food to boiling temperatures for 10 minutes may inactivate the toxin.

SUMMARY

1. **Metabolism,** the sum of all the chemical reactions in an organism, involves both the breaking down of large substances into simpler products and the building up of complex materials from simpler ingredients.

2. In contrast to autotrophs, heterotrophs cannot synthesize their own food. Heterotrophs depend upon autotrophs for both **nutrients** and

oxygen. Whereas the material substances of life cycle, energy does not and life depends upon a constant input of solar energy.

3. Because heat is the basic form of metabolic energy, the energy content of food is expressed in **calories** (kcal in humans). The metabolism when the body is at complete rest is called the **basal metabolic rate,** and activities require additional calories. Variables that affect the amount of body heat released include age, sex, surface area, environmental conditions, and body activity. Weight gain or loss depends upon caloric intake and caloric use.

4. **Vitamins** are organic and **minerals** are inorganic compounds. Vitamins are either **fat-soluble** (e.g., A, D, E, and K) or **water-soluble** (e.g., B and C), and they tend to be **cofactors** or precursors of enzymes. Minerals are usually needed in larger quantities than vitamins. Both deficiencies and excesses of vitamins and minerals may cause problems.

5. Many modern nutrients contain **additives** (e.g., vitamin D and fluorides) and sometimes **contaminants** (e.g., chemical residues). Many cases of **food poisoning** are caused by toxins produced by microorganisms in food. **Synergism** often operates as nutrients interact with one another and with a person's physiological state.

REVIEW QUESTIONS

1. Describe metabolism and state specific examples of catabolic and anabolic reactions.

2. Distinguish between an autotroph and a heterotroph, and explain how some substances are exchanged between the two categories of organisms.

3. Why is it said that the material substances of life cycle but that there is a one-way flow of energy?

4. Define the following categories of nutrients, and name specific examples of each: essential amino acid, trace element, micronutrient, and macronutrient.

5. Distinguish between a calorie and a Calorie, and explain how a calorimeter works.

6. What is the basal metabolic rate and what major variables affect it?

7. Distinguish between a vitamin and a mineral, and explain the general functions of the two categories of nutrients.

8. Name the major water-soluble and fat-soluble vitamins.

9. State examples of hypovitaminosis and hypervitaminosis problems. Do the same for deficiencies and excesses of minerals.

10. Distinguish between a food additive and a food contaminant.

11. Explain how most cases of food poisoning occur, and describe some methods of avoiding them.

12. Describe synergism and state at least one dietary example.

13. What are some of the main features of modern consumer protection laws pertaining to food? How would you improve them?

14. What nutritional problems must vegetarians be especially knowledgeable about?

CHAPTER 13

NUTRITION: THE DIGESTIVE PROCESS

In humans and the other higher animals, digestion occurs outside of the body, or extracellularly, following the release of digestive juices into an opening of a digestive organ. Foods are reduced enzymatically into their soluble, molecular units before they are absorbed selectively by epithelial cells, circulated, and assimilated into and used by body cells. Intracellular digestion is typical in the lower invertebrates. Food particles are phagocytized and digestive enzymes are added to the vesicles that surround them. Digestive products are then either absorbed into the cell's cytoplasm for direct use or transferred to other cells.

The great diversity of animals is partially explained by their various adaptations for securing and digesting food. In this chapter, you will study how digestion occurs; Chapter 12 dealt with the nature of nutrients.

While only small food particles can be processed by intracellular digestion, extracellular digestion allows animals to consume larger food. For example, various vertebrates, like this snake, swallow their prey whole, sometimes being aided by such structures as recurved teeth and distensible jaws and stomachs. In general, the digestive tracts of the higher animals are more specialized and efficient than those of the lower animals.

THE PATTERN OF NUTRITION

From the general pattern of nutrition shown below it is apparent that the digestive system works closely with the respiratory (Chapter 14), circulatory (Chapter 15), and excretory (Chapter 17) systems.

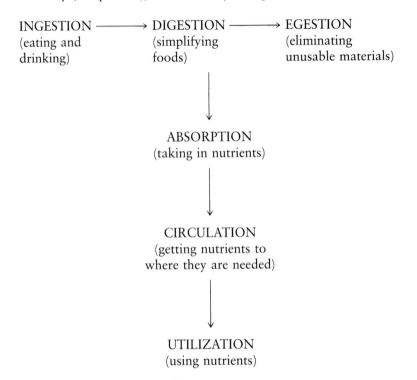

INGESTION ⟶ DIGESTION ⟶ EGESTION
(eating and   (simplifying   (eliminating
drinking)   foods)   unusable materials)

↓

ABSORPTION
(taking in nutrients)

↓

CIRCULATION
(getting nutrients to
where they are needed)

↓

UTILIZATION
(using nutrients)

Ingestion

Ingestion involves taking bulk nutrients into the body. How do we know when to eat and when to stop eating? In vertebrates, hunger is largely under chemical control and involves centers in the hypothalamus of the brain. Damage to these centers (or their control by other brain centers) may either reduce the desire to eat or lead to overeating. Although the level of blood glucose is of prime importance in explaining hunger, other nutrients (e.g., fatty acids) are also involved in this complicated phenomenon. Because the digestion and absorption of foods is barely under way when we stop eating, none of these factors alone controls the amount of food eaten. (Thirst is discussed in Chapters 2 and 17).

Digestion

Digestion includes all processes by which the condition of food is changed so that it can be absorbed. This includes both **mechanical digestion,** the physical breakdown of food, and **chemical digestion,** the reduction of nutrients to their molecular components. Physical digestion is accomplished by structures such as teeth, beaks, and gizzards.

Mechanical Digestion. Sound teeth, gums, and bony tooth sockets are important for good health, yet over half of Americans over age 50 have lost all of their teeth, more often to the gum and bone problems called **periodontal diseases** than to cavities or **dental caries.** Unfortunately, many people think of their teeth only when a problem develops, and some even believe that they can get along fine with only two teeth, providing they meet. Dentists often say that if we cared for the rest of

Some Teeth Get "Caried" Away

Three factors are generally needed for the development of what are popularly called cavities or, more precisely, **dental caries:** a susceptible tooth, carbohydrates (usually sugar), and certain bacteria. It is not known why some teeth are more susceptible to decay than others, but genetics appears to be one important factor. Perhaps because they are usually close together and have deep pits and furrows, large teeth are especially prone to develop caries.

Most teeth are covered by **dental plaque,** a combination of sticky mucus, food debris, and bacteria. Unfortunately, not all plaque is removed by most people while brushing and/or flossing. Most caries develop near or below the gum line, where plaque and associated materials are most likely to accumulate. Plaque itself does not cause caries, but it keeps the caries-producing chemicals next to the tooth.

Mouth bacteria (especially *Streptococcus mutans*) need little time to convert carbohydrates to destructive acids (Figure B13.1). Because most damage is done within 15–30 minutes after eating, immediate brushing, rinsing, and/or flossing is wise. Tooth enamel is destroyed first, but no pain is involved until the nerve endings in the dentin are reached (Figure B13.2). Untreated infections may break down the pulp and destroy the teeth. The amount of sugar or carbohydrate ingested is not as important as the frequency or time exposure of sugar ingestion. Thus, sucking on hard candy may cause more cavities than eating a far larger volume of cake quickly.

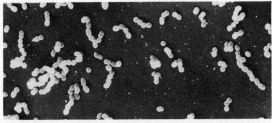

FIGURE B13.1
Although many species of bacteria live in the mouth, several types of *Streptococcus* are especially involved in producing dental caries.

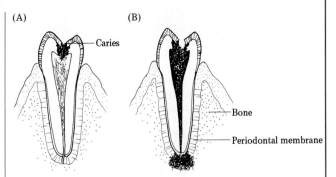

FIGURE B13.2
How dental decay occurs. (A) Caries has penetrated the enamel and dentin and caused inflamation of the pulp. (B) Untreated, the decay may spread to the periodontal membranes and bone at the base of the tooth.

More teeth are lost to **periodontal diseases,** diseases of the gums and bony sockets, than to dental caries. People affected by **gingivitis** have swollen, red gums that bleed easily and that force the teeth outward. This condition, which is usually not painful, is often caused by hard calcified deposits that form on the surfaces of teeth. Untreated, chronic gingivitis may lead to **periodontitis.** In this condition, the periodontal membrane becomes inflamed and eroded, pus may form, and the surrounding bone may be resorbed. Teeth may be lost when they loosen and when the gums recede. Good dental hygiene, regular visits to a dentist, and a proper diet help prevent periodontal diseases. Vitamins B and C and adequate protein are important for healthy gums and calcium for proper bone development. Eating foods that require thorough chewing seems to help prevent gingivitis.

Teeth sometimes come in at the wrong time or in the wrong order or angle, causing the bite to be out of line. **Orthodontia,** or the use of braces to bring teeth back into line, can correct many such problems. Cosmetically, this expensive, tedious, and sometimes painful and psychologically embarrassing procedure usually works. However, its value in dental health is less clear. It appears that those who wore braces as children may have as many gum problems later as matched individuals with bad bites who did not wear braces.

our bodies the way we care for our teeth, we might not survive past our 20s! See the essay "Some Teeth Get 'Caried' Away" for more information.

Figure 13.1 shows a cross-section of a typical tooth. The **enamel** that covers the part of the tooth that projects above the gum is composed primarily of a calcium compound and is the hardest substance produced by the body. In fact, teeth are so resistant that they have been preserved in sediments for millions of years, giving us important insights into

FIGURE 13.1
Teeth are the prime agents of mechanical digestion in mammals. Note how well the structure of a tooth is related to its function. The volume of food is not reduced significantly by chewing. However, by breaking up food into small particles, mechanical digestion greatly increases the surface area of food that is exposed to digestive juices. Birds lack teeth but swallow grains of sand, which grind ingested food in the muscular gizzard.

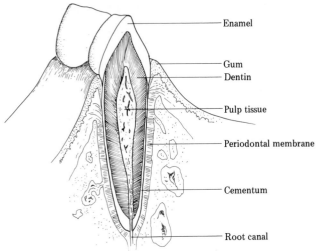

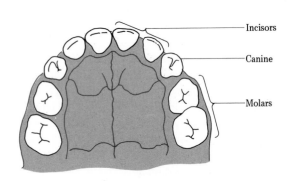

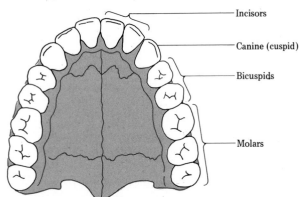

evolution (Chapter 24). Continuous with the enamel below the gum line is the **cementum,** and this is bound to the bony socket by a mass of tough fibers called the **periodontal membrane. Dentin** is also a calcium compound and is about as hard as bone. Inside it is the **pulp cavity** that contains the blood vessels and nerves.

Our first set of teeth (baby, milk, or primary teeth) are called **deciduous teeth** because, like the leaves of many trees in the fall, they are shed. They form before birth but usually do not appear or erupt until the teething period postnatally. At variable times before about 12 years the shallow roots of the deciduous teeth are resorbed and the teeth loosen, are lost, and are replaced by **permanent teeth** (Figure 13.2 and Table 13.1).

The deciduous teeth have a major influence on how adult dentition comes in. The third molars, or "wisdom teeth," do not always appear, and when they do they do not always erupt easily and cause many people problems. Tooth patterns and sizes vary widely, and "oral fingerprints" are nearly unique for each person. This is why dental records are often useful in identifying accident victims.

Chemical Digestion. Chemical digestion requires water and appropriate enzymes. Digestive enzymes hydrolyze large molecules into smaller components that can cross a plasma membrane and enter a cell. The three principal places where this occurs are the mouth, stomach, and small intestine. At each site, one or more digestive juices containing enzymes and other substances are released by secretory cells or glands. In the mouth and small intestine, secretions also often enter from accessory organs (Table 13.2). The body decomposes nutrients and synthesizes new tissues, and enzymes promote reactions in both directions.

FIGURE 13.2
The permanent teeth are shown here above the deciduous teeth, which they ultimately cause to shed.

| | Usual Time of Eruption | | TABLE 13.1 |
Type	Deciduous Teeth (months)	Permanent Teeth (years)	Human Teeth
Central incisors	6–8	6–8	
Lateral incisors	9–11	7–9	
Canines (cuspids)	18–20	11–14	
First premolars	—	9–12	
Second premolars	—	10–12	
First molars	14–17	6–7	
Second molars	24–26	12–16	
Third molars	—	17–21	

TABLE 13.2
Functions of Digestive Structures

Region and Structures	Secretion(s)	Major Functions
Mouth		Mechanical digestion; limited chemical digestion
Teeth	—	Mechanical breakdown of food
Salivary glands	Water	Moistens food
	Mucus	Lubricates and binds food particles
	Amylase	Starts breakdown of carbohydrates
	Bicarbonate	Neutralizes acidic food
Stomach		Stores, dissolves, mixes food; regulates emptying of food (chyme) into duodenum
Mucosa cells	Hydrochloric acid	Dissolves food particles; kills many ingested organisms
	Pepsinogen	As pepsin, splits protein chains
	Mucus	Lubricates and protects lining
	Gastrin	Stimulates HCl and pepsinogen secretions
Small intestine		Chemical digestion and absorption of most foods; mixes and propels chyme
Mucosa cells	Numerous enzymes	Break down all major food groups
	Mucus	Lubricates food (chyme)
	Gastrin	Stimulates HCl secretion
	Secretin	Stimulates pancreatic bicarbonate secretion
	Cholecystokinin	Stimulates gallbladder contraction, pancreatic enzyme secretions, stomach emptying
	Gastric-inhibitory peptide	Inhibits stomach acid secretion and motility
Pancreas	Numerous enzymes	Break down all major food groups
	Bicarbonate	Neutralizes HCl entering duodenum from stomach
Liver	Bile salts	Hydration of emulsified fat droplets
	Bicarbonate	Neutralizes HCl entering duodenum from stomach
Gallbladder	—	Stores and concentrates bile from liver
Large intestine		Concentrates undigested matter by absorbing water and salts; stores, mixes, and propels material
Mucosa cells	Mucus	Lubricates undigested matter
Rectum	—	Expulsion reflex egests undigested matter

Fortunately, the body recognizes the synthesized materials as "self," thus preventing attack by the immune system (Chapter 16).

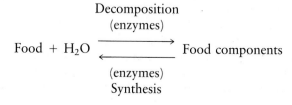

THE GASTROINTESTINAL TRACT

Basically, our **gastrointestinal tract** is a tube-within-a-tube that is a continuation of the body's outer covering (Figures 13.3 and 3.11). This means that food within the digestive tract is surrounded by but is not yet inside the body. Although most enzymes function within the cell in which they are produced, the actions of human digestive enzymes are extracellular. Thus, we release digestive secretions outside of the body (i.e., into the gastrointestinal tract), and absorb food into cells after it has been digested chemically. Many digestive enzymes are produced in an inactive or **zymogen** form within the cell, and they are activated by a change in pH or by another enzyme only when they are released.

FIGURE 13.3

(A) The main organs of the human digestive system. (B) Generalized cross-sectional view of the gastrointestinal tract walls. The inner mucosa cells often secrete diverse products that function in digestion. Epithelial cells in the intestinal mucosa also absorb simplified nutrients. Muscles in the tract walls mix and propel food and undigested matter. Although the tubular digestive tract of higher animals may become quite long, twisted, and specialized, its interior is actually outside the body, just as the hole in the donut (Figure 3.1) is outside of the donut. Technically, digestive juices are released to the outside of the body, just as in many bacteria and fungi, and chemical digestion occurs extracellularly.

(A)

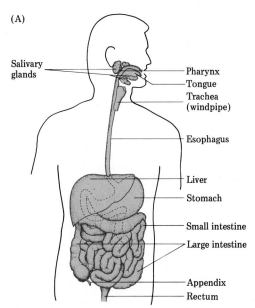

- Salivary glands
- Pharynx
- Tongue
- Trachea (windpipe)
- Esophagus
- Liver
- Stomach
- Small intestine
- Large intestine
- Appendix
- Rectum

(B)

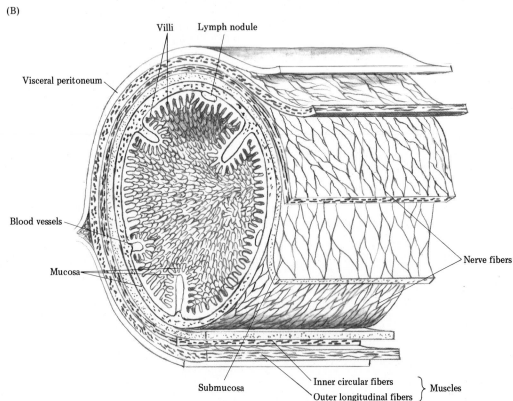

- Villi
- Lymph nodule
- Visceral peritoneum
- Blood vessels
- Nerve fibers
- Mucosa
- Submucosa
- Inner circular fibers
- Outer longitudinal fibers
- } Muscles

To learn more about the digestive tracts and processes of various other animals, see the essay "Comparative Digestive Systems and Feeding Strategies."

Mouth, Salivary Glands, and Esophagus

Major physical digestion but only limited chemical digestion occurs in the **mouth.** Many foods already contain a high water content, but up to about 1.5 L (3 pt) of liquid is produced by the **salivary glands** daily. Saliva contains bicarbonate ions (HCO_3^-), which keep its pH nearly neutral even when acidic food is in the mouth. An enzyme (amylase) in saliva hydrolyzes starch and glycogen into sugars, which you can confirm by chewing a cracker for about five minutes. (Use an unsalted cracker so the salt taste does not mask the results.) However, because of the short time food stays in the mouth, chemical digestion there is relatively unimportant. The mucus in saliva lubricates food and binds it into a softened ball, or **bolus.** Antimicrobial agents are also found in saliva. The tongue contains the taste buds (Chapter 20) and performs the nondigestive function of helping to shape our words.

The **pharynx,** or throat cavity behind the mouth, connects with one tube that leads to the lungs, the **trachea,** and with another that leads to the **esophagus** (Figure 13.4). Food and liquid in the lungs cause problems, and lots of air in the stomach does little good. Rings of cartilage surround the trachea to keep it open for breathing (Chapter 14), but the esophagus is soft and closed except when material is being swallowed.

Swallowing "the wrong way" is uncommon for several reasons. The **larynx,** or voicebox, rises against the base of the tongue and the vocal cords inside it come together, the tongue usually moves backward, and the flaplike **epiglottis** closes the space between the vocal cords (glottis). The swallowing reflex also usually inhibits breathing briefly while food is in the pharynx. Also, the flaplike **uvula** that extends from the soft palate rises during swallowing, and this closes the airways to the nose (Continues on p. 269.)

FIGURE 13.4

Events in the throat associated with (A) breathing and (B) swallowing. To get a feel, literally, of how air and nutrients get into the right tubes, place your hand on your throat, swallow, and then describe the movements that you feel. When you talk, air moves out of your body, which creates a problem if you try to talk and swallow at the same time. Can you swallow and breathe simultaneously?

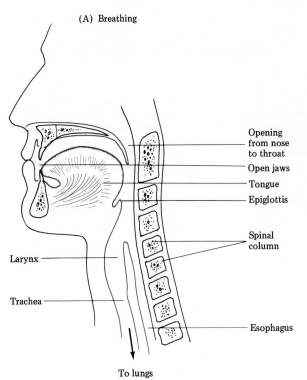

(A) Breathing

Opening from nose to throat
Open jaws
Tongue
Epiglottis
Spinal column
Larynx
Trachea
Esophagus

To lungs

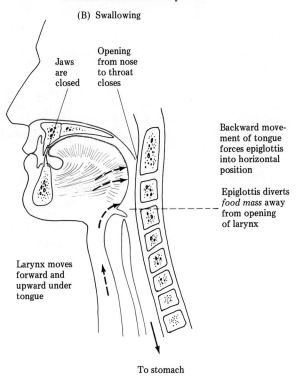

(B) Swallowing

Jaws are closed
Opening from nose to throat closes
Backward movement of tongue forces epiglottis into horizontal position
Epiglottis diverts *food mass* away from opening of larynx
Larynx moves forward and upward under tongue

To stomach

Comparative Digestive Systems and Feeding Strategies

Small or microscopic protistans and plants living in the ocean and in lakes accomplish much of Earth's photosynthesis. Most of these organisms, called **phytoplankton,** are digested easily by the animals that consume them, which includes some of the **filter feeders** (Figure B13.3). While some filter feeders move quickly (e.g., some fish, such as herring), most are slow moving (e.g., clams) or are sessile (e.g., fan worms). Many produce mucus to entrap nutrient particles found in water, and most also possess cilia that create currents to draw food particles into their digestive tracts. In many cases, filter feeders are nonselective, except for particle size, and they consume much of what they procure.

(A)

(B)   (C)

FIGURE B13.3
(A) Although some carnivorous whales eat large animals, many of the large species are filter feeders, using baleen to strain from water huge quantities of small organisms, many of which are too tiny for them to see. (B) Many other filter feeders, such as these sponges, are sessile. They extract food from water that enters through some pores and exits through others. (C) The familiar barnacles in tide pools have been said to stand on their heads and kick suspended food into their mouths.

Detritus includes the wastes of living organisms and the remains of the dead. Detritus often accumulates in the sediments of aquatic habitats as well as in the soil. This organic material provides an abundant food supply for many organisms (Figure B13.4). Some detritus feeders gain as much or more food from the bacteria that live on the organic material as they do from the detritus itself.

Numerous marine invertebrates absorb small organic molecules such as glucose and amino acids directly from the water. Not only do these chemicals provide energy but some also affect metamorphosis, act as antibiotics, or stimulate growth.

FIGURE B13.4
Earthworm castings are familiar sights in the spring and summer. Earthworms literally eat their way through soil, digesting the organic material that it contains. The soil that passes through their bodies is deposited occasionally on the soil surface as castings.

Many animals live in intimate and constant association with some of their nutrient sources. For example, some aquatic animals contain photosynthetic organisms in their tissues (Figure B13.5). Also, feeding either on or in animals are various **parasites** such as ticks (ectoparasites) and tapeworms (endoparasites). Because such fluid-feeding animals as mosquitoes and vampire bats do not remain attached to the **host** from which they secure their nutrients for long, they may be thought of as intermittent parasites. In their feeding activities, some fluid feeders serve as carriers or **vectors** of pathogens. For instance, various mosquitoes (only females feed on blood) carry pathogens that cause malaria, encephalitis, and yellow fever.

(A)   (B)

FIGURE B13.5
(A) The phrase "live happily ever after" may describe such associations as the autotrophic green algae that live in the tissues of heterotrophs like the freshwater *Hydra* and this marine giant clam. (B) The tick attached to the toad's skin is called an *ectoparasite* because it lives on the outside of an animal, whereas the tapeworm is called an *endoparasite* because it secures nutrients from inside its host.

Among the larger animals, the most familiar **predation** strategies are **herbivory** (eating plants) and **carnivory** (eating flesh) (Figure B13.6). Animals that eat both plant and animal food are called **omnivores.** Carnivores possess many means of finding and subduing their **prey,** just as prey use various methods to detect and elude their predators. Plant adaptations for avoiding predators include both structural features such as spines and chemical traits such as toxins.

The teeth and beaks of vertebrates are excellent examples of structure-and-function relationships (Figure B13.7). For example, while the canines of herbivores

FIGURE B13.6
Lurking in the vegetation upon which the herbivores are feeding are carnivores. Waiting unseen or circling overhead are the various scavengers who will consume the leftovers. The predation of herbivores on plants is certainly less dramatic but no less important than the hunting behavior of carnivores.

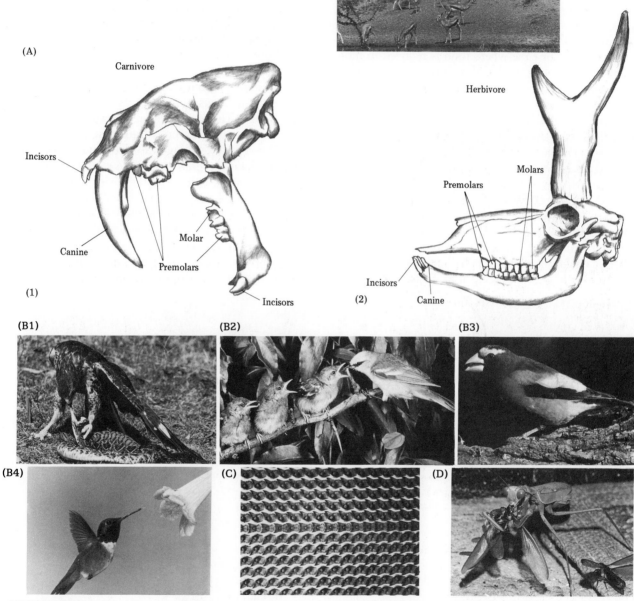

FIGURE B13.7
(A) The skulls and teeth of a mammalian (1) carnivore and (2) herbivore. Chisel-shaped incisors are used to bite, strip, or cut off chunks of food; the more cone-shaped incisors grasp, pierce, and tear food; and the flat-topped molars are used to crush and grind food. In some animals, teeth have become modified for nondigestive functions. Thus, the elephant's tusk, a modified upper incisor, is often used for attack and defense. (B) While most vertebrates have teeth, birds have beaks that serve some of the same functions: (1) Tearing beaks are common among carnivorous birds, such as this hawk that is eating a rattlesnake. (2) Many carnivorous birds that eat small animals such as worms or insects have chisellike beaks, like these warblers. (3) Seed-eating birds like this grosbeak have strong, crushing beaks. (4) Long, thin bills are useful for birds such as the hummingbird, which sucks nectar from flowers. (C) This filelike structure (a radula) is used by certain snails to scrape food from surfaces. (D) The intricate mouthparts of many insects, such as this praying mantis, aid them in securing, tasting, reducing, and swallowing food.

are relatively smaller than those in carnivores, the molars are well developed, with thick enamel ridges for grinding tough plant material. The incisors of rodents grow throughout life and must be worn away constantly by the self-sharpening gnawing process. Many toothlike devices in invertebrates aid in the capture, holding, and tasting of food and in reducing it to particles small enough to be swallowed.

The digestive systems of most animals include a digestive cavity or tube where food is reduced to molecules small enough to be absorbed. However, such simple and ancient animals as sponges have no digestive cavity, and digestion of small food particles occurs through intracellular processes. Animals with **incomplete digestive systems** have a digestive chamber with only one opening, through which both food and wastes pass (Figure B13.8A). Regions of the digestive cavity of these animals are not highly specialized and such animals generally process one meal at a time. Some food is digested extracellularly in the sac, and the remainder intracellularly by the cells that line the cavity. The evolutionary advent of a digestive cavity allowed animals to digest objects larger than individual cells.

A tube-within-a-tube body plan characterizes animals with **complete digestive systems.** Food usually enters at one end and is moved by either cilia or muscular movements; wastes exit through the other end. In between, various regions are specialized for food storage or processing. For example, in earthworms and birds (Figure B13.8B), **crops** store food and muscular **gizzards** help reduce the food to smaller particles, often with the aid of grit, sand, or pebbles that the animal has eaten. In most animals the main structure where food is both digested and absorbed is the intestine. Animals with a tubular gut can often process food continually.

Most animals do not feed continually, although herbivores such as cows and sheep seem to do little else (Figure B13.9). The food of these grazers contains much cellulose, a complex carbohydrate that most animals cannot digest easily. Along with deer and goats, these **ruminants** have several stomachlike chambers. The enormous numbers of bacteria that exist in the first two chambers produce enzymes that break down the cellulose and other nutrients. When they "chew their cud," ruminants have regurgitated the food from the first two chambers for further chewing. After being reswallowed, the food moves into two more stomachs and, later, into the intestines for final digestion and absorption.

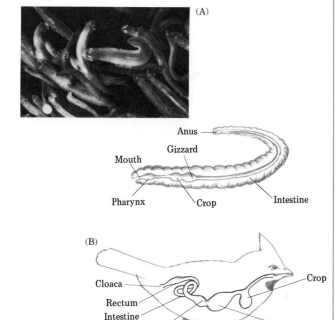

FIGURE B13.8

(A) The common flatworm *Planaria* feeds through a muscular tube (pharynx), and wastes exit through the same structure. Note that such small, flattened animals have very large surface areas (external and internal) relative to their volumes. (B) Humans have a tube-within-a-tube body plan, as do earthworms and birds.

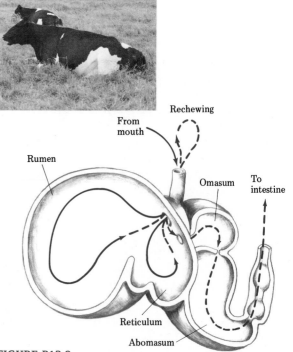

FIGURE B13.9

Ruminants subject to predation eat as much as possible and store the food temporarily for more complete processing later in comparative safety. Much of the digestion of cellulose is accomplished by bacteria. As with most animals, digestion by bacteria is extracellular.

and helps to stop breathing briefly. (Incidentally, vibrations of the uvula cause many of the snoring sounds people may make during sleep.)

Both mixing and propulsive muscular movements are involved in digestive processes. At both ends of the digestive system, striated muscles are most important, but from the upper esophagus through the large intestine, smooth muscles are primarily involved. The involuntary but well-regulated contraction of the smooth muscles that moves nutrients is one form of **peristalsis** (Figure 13.5).

Although swallowing may be voluntary, it is virtually impossible to swallow if the mouth is dry. Liquids pass through the long (25-cm, or 10-in.), muscular, and extensible esophagus into the stomach in about one second, although it takes five to ten seconds for a semisolid food mass to move down the esophagus by peristalsis. No digestive enzymes are secreted by the esophagus, but the mucus secreted by its cells provides a protective lubricant.

The acidic stomach fluids sometimes move upward to the lower esophagus, and the burning sensation that results is often called "heartburn" because it appears to be localized in the heart region.

Stomach

The fist-sized **stomach** is a distensible J-shaped sac with an average capacity of about 1–3 L (0.9–2.7 qt) (Figure 13.6). When empty it is relatively thin and tubelike. Obese people do not have unusually large stomachs, but fat deposits give them their "big belly" appearance. The main functions of the stomach are (1) to store and mix the food that it receives and (2) to help regulate the movement of food into the small intestine. The numerous longitudinal ridges of the inner wall facilitate stomach functions by greatly increasing the surface area exposed to food.

The most important function of the stomach is storage, and the volume of a full stomach is about 50 times larger than that of an empty stomach. Although a large meal may take 15–30 minutes to eat, parts of the meal may remain in the stomach for about three to five hours, during which time the food is mixed with **gastric juices** by peristalsis. However, food is not crushed physically by the stomach, as it is in the gizzards of some animals. The activities of the stomach are often noisy, as when foods slosh around and trapped gas rumbles. Sometimes excess gas moves upward into and out of the esophagus—a belch. Although belching is a compliment after meals in some societies, it is unacceptable in others. The vomiting of material from the stomach is an important reflex to protect us from potentially harmful substances. Although it is

FIGURE 13.5
In peristalsis, rings of circular smooth muscles contract behind a mass of food material and propel it. By contracting and relaxing repeatedly, the rings of smooth muscle create an oscillating movement of the contents. In both cases, the mixing movements bring the nutrients into close contact with digestive juices.

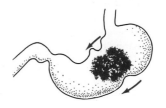
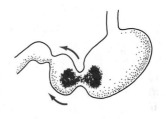
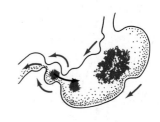

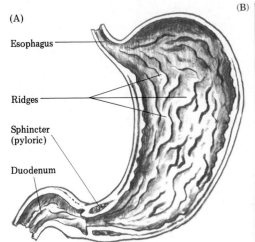

(A)

Esophagus

Ridges

Sphincter (pyloric)

Duodenum

(B)

FIGURE 13.6
Cross-section view of the stomach (A). The ridges provide additional surface area for the release of gastric juice and the limited absorption of some simplified nutrients (B).

generally a useful reflex, prolonged vomiting can lead to dehydration and the resulting change in pH (alkalosis) of body fluids may be fatal, especially in children.

A true insight into stomach function occurred in the last century when a guide was shot in the stomach and an inquiring army surgeon, William Beaumont, treated the wound. Although the guide recovered, the wound never healed completely. The stomach grew to the body wall so a hole led directly into the stomach, allowing the curious to peek in! Dr. Beaumont did just that over many years and conducted numerous interesting experiments to learn about stomach function. For example, he tied pieces of meat, vegetables, and bread to strings and lowered these into the man's stomach. By extracting the strings at various times, the surgeon learned how different foods were digested by gastric juices.

About 2 L of complex gastric juice are secreted daily in humans by tubular **mucosa cells** in the stomach wall. One type of mucosal cells (chief cells) secretes the zymogen pepsinogen. This molecule may be converted to the active enzyme pepsin that begins the hydrolysis of proteins. Other mucosa cells secrete a viscous substance (mucin) that coats and protects the mucosa, mixes with food, and serves a lubricating role. Still other mucosal cells (parietal) secrete hydrochloric acid (HCl), which gives gastric juice the lowest pH (1–3) of any body fluid. Although some gastric juices flow even during periods of starvation, the normal flow is intermittent. The sight or smell of food, the presence of food in the stomach, and stressful emotions stimulate the flow of gastric juices.

Mechanical receptors in the stomach wall react to its being distended with food, and this leads to nerve impulses that result in the release of HCl. A hormone (gastrin) that is produced by certain mucosa cells regulates the synthesis and release of both HCl and pepsinogen. HCl both initiates the conversion of pepsinogen to pepsin and provides the optimum pH for the action of this enzyme. Hydrochloric acid cannot break peptide bonds but it can "soften" proteins by changing their three-dimensional structures, thus facilitating their chemical digestion. HCl also kills many microorganisms that are ingested.

Because gastric juices are strong enough to dissolve some metals and because pepsin hydrolyzes proteins, it is logical to wonder why the stomach wall itself is not digested. In fact, it sometimes is. A peptic **ulcer** may penetrate the mucosa and break down small blood vessels. This resembles a raw, open sore, and it may ultimately perforate the stomach lining. Excessive acid may also produce ulcers (duodenal) in the upper small intestine. About 10 percent of Americans, including about four times as many males as females, develop ulcers. Excess HCl or bile acids, undersecretion of mucus, stress, and/or irritating nutrients or drugs are common causes of ulcers. If the stomach or small intestine is perforated, **peritonitis** may result. The latter is a bacteria-caused inflammation of the body cavity that was often fatal prior to the introduction of antibiotics.

Surgery can repair some ulcers. Nonsurgical treatment often includes reducing stress; increasing worry-free rest; eliminating alcohol, tobacco, caffeine, or other irritants; ingesting alkalis or buffers (e.g., milk or antacids); and/or using drugs that block the stimuli for acid secretion.

Food absorption through the stomach is limited, although some substances such as alcohol and various drugs and medications may be absorbed. For example, only about 20 percent of alcohol is absorbed in the stomach, but this accounts for some of its fast-acting effects. Food in the stomach slows but does not inhibit alcohol absorption.

The stomach empties gradually over a period of hours, thus allowing the small intestine to digest and absorb nutrients slowly over most of the time between meals. The rate of stomach emptying is affected by the amount and type of food eaten, substances from the small intestine, and one's emotional state.

Small Intestine

Most digestion and absorption of nutrients occur in the **small intestine.** The intestines are named for their diameters, not their lengths. For example, the small and large intestines are about 6 m (20 ft) and 1.5 m (5 ft) long, respectively. The intestines are highly coiled, filling most of the abdominal cavity. Most chemical digestion occurs in the **duodenum** or upper part of the small intestine (Figure 13.7). In the 3 L of **intestinal juices** produced daily are carbohydrases, lipases, proteases, and nucleases, as well as mucus and several hormones. Also entering the duodenum are products from various accessory glands (Table 13.2). Pancreatic juice from the outer (acinar) exocrine cells of the **pancreas** is a complex, active, and versatile digestive secretion. The pancreas also secretes bicarbonate that neutralizes the acidity of the partially digested food (chyme) entering from the stomach. "Intestinal gas" is largely CO_2 that results from the reaction of HCl and $NaHCO_3$.

Intestinal juice, pancreatic juice, and bile are all basic, in contrast to the strongly acid gastric juice. Foods differ widely in pH, from oranges at 3.5 and tomatoes at 4.2 to corn at 6.2 and shrimp at 6.9, to mention just a few examples. Because each enzyme has an optimum pH, not all digestive enzymes function when we consume different foods.

Each day the **liver** produces about 0.5–1 L of **bile,** and this is stored between meals in the **gallbladder.** When it reaches the duodenum through the bile duct, bile emulsifies and hydrates fats. In emulsification, large fat globules are broken down into small droplets, thus providing a large surface area exposed to the actions of fat-splitting lipases. Hydration helps transport fats to the absorptive sites in the intestinal mucosa. Bile contains no enzymes but consists mainly of water, inorganic salts, and pigments. The main brown-red pigment of bile (bilirubin) is derived from the breakdown of hemoglobin in old red blood cells by the liver, and this gives **feces** their dark color. Yellow bile pigments give blood plasma and urine much of their color. The presence of lipids and HCl in the duodenum lead to the release of a hormone that stimulates the

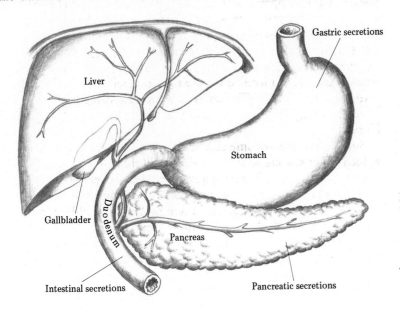

FIGURE 13.7
The main digestive structures in the region of the duodenum. Substances enter this upper portion of the small intestine from the stomach (chyme), pancreas (pancreatic juice), and the liver and gallbladder (bile). Intestinal juice is also released from the mucosa of the duodenum.

FIGURE 13.8
It is easy to see why gallstones may be painful, and why they often must be removed surgically. Stone-dissolving drugs and diet modification are other common forms of treatment.

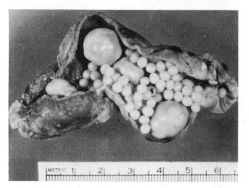

emptying of the gallbladder. Although the gallbladder adds nothing to bile, it changes bile's concentration by absorbing water from it.

Sometimes the bile pigments are not excreted properly but accumulate in the blood and give the skin and whites of the eyes a yellowish color, **jaundice**. Jaundice may be caused by the blockage of the bile duct, liver damage (e.g., cirrhosis), toxic chemicals (e.g., poisons or excessive alcohol), rapid red blood cell destruction (e.g., pernicious anemia), or disease (e.g., hepatitis or malaria). **Gallstones** sometimes form in the gallbladder or bile ducts (Figure 13.8). These may result from foreign substances in bile or from the precipitation of minerals or fatty materials.

About 90 percent of digested food and about 10 percent of water and minerals are absorbed by the small intestine. To facilitate absorption, the surface area of the small intestine is enormous (Figure 13.9), roughly the size of a tennis court. First, there are many shelflike folds that ensure the frequent contact of food with the surface. Next, there are millions of moving, fingerlike **villi**. Finally, the plasma membranes of the cells are folded into numerous still smaller microvilli that further increase the secretory and absorptive surface area. The cells at the top of the villi are sloughed off continually and are replaced by mitotic activity of the basal cells. The entire intestinal epithelium is replaced about every 36 hours, and about 25 percent (by weight) of each stool is made up of dead intestinal cells.

Both active transport and diffusion are important in the absorption of simplified nutrients during the three to six hours that food remains in the small intestine. Most absorption has occurred by the time food reaches the middle small intestine (jejunum), and the lower small intestine (ilium) is not involved in most absorption unless other parts fail to perform. However, the terminal portion is very important in the absorption of vitamin B_{12} and bile salts.

Amino acids, monosaccharides, some vitamins and minerals, and water enter the capillaries of the villi. However, most fatty acids and glycerol are resynthesized as fats (triglycerides) after absorption and then enter the lymph as tiny droplets (Figure 13.10). Lymph eventually enters the bloodstream in the upper thorax (Chapter 15).

FIGURE 13.9
(A) The structure of the small intestine is such that an enormous surface area is exposed to the nutrients that are being digested and absorbed. Both the movements of the villi and the twisting of the entire small intestine also assist absorption. (B) Other animals have adaptations for a similar function. Earthworms have a simple infolding of the intestinal wall, and sharks force food past the large surface area of the spiral valves.

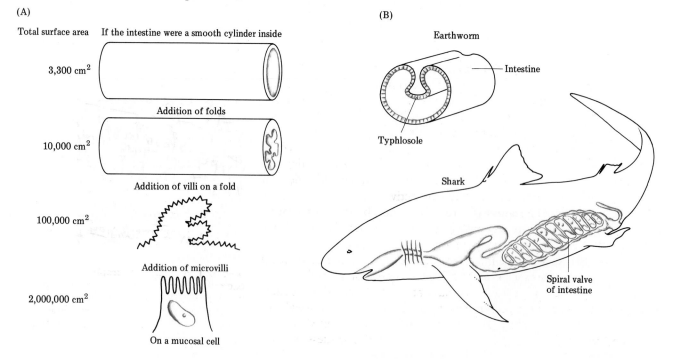

(A)

Total surface area   If the intestine were a smooth cylinder inside

3,300 cm^2

Addition of folds

10,000 cm^2

Addition of villi on a fold

100,000 cm^2

Addition of microvilli

2,000,000 cm^2

On a mucosal cell

(B)

Earthworm

Intestine

Typhlosole

Shark

Spiral valve of intestine

(A)

Villi

Lacteal

Capillary network

Glandular epithelium

Muscle

Vein

Lymph vessel

Artery

(B)

(C)

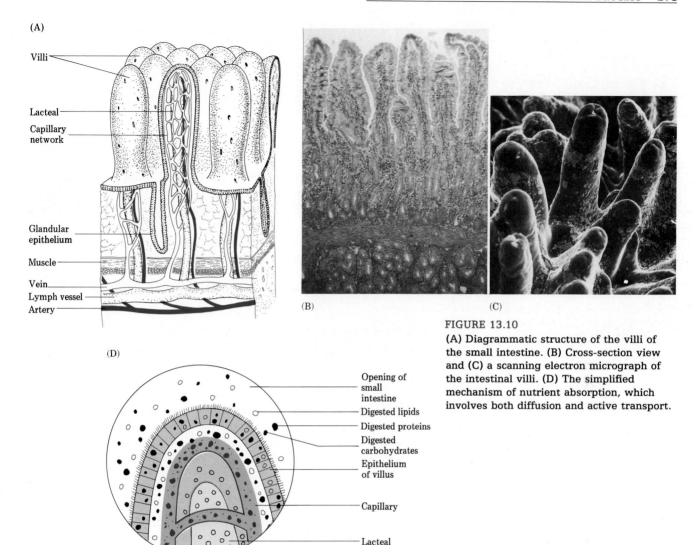

(D)

Opening of small intestine

Digested lipids

Digested proteins

Digested carbohydrates

Epithelium of villus

Capillary

Lacteal

FIGURE 13.10

(A) Diagrammatic structure of the villi of the small intestine. (B) Cross-section view and (C) a scanning electron micrograph of the intestinal villi. (D) The simplified mechanism of nutrient absorption, which involves both diffusion and active transport.

Large Intestine

Material that takes only a few hours to pass through the small intestine is moved by peristalsis into the **large intestine** or **colon,** where it may spend a day or so. Yet, because few nutrients are left to be absorbed, the colon is less involved with usable than with unusable materials. Part of the liquified material (chyle) that enters the colon leaves the body via the **anus** as semisolid feces that were stored in the **rectum.** The colon recovers much of the remaining water (about 300–400 ml daily) and returns it to the bloodstream. Some minerals (e.g., sodium, chloride, and potassium) and bile salts are absorbed by active transport, but excess calcium is excreted. **Defecation** is a reflex initiated by stretch receptors that are activated when the rectum fills with fecal matter. The feces leave the anus when the anal sphincters relax.

The colon houses immense numbers of bacteria, including *Escherichia coli,* which is so important in research in molecular biology and genetics (Figure 13.11). These and other bacteria that subsist on materials largely indigestible by humans make up about one-third of the dry weight of the feces. These gut bacteria are able to synthesize vitamin K, biotin, and some amino acids. These bacteria are normally harmless because

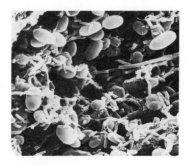

FIGURE 13.11

Scanning electron micrograph of intestinal bacteria. Living on what we cannot digest, these bacteria produce metabolic wastes that are essential to us. Such mutualistic associations, where both species benefit, are quite common.

No Swimming

"Beach Closed" and "No Swimming" signs are not welcome sights to people wanting to swim. Public water supplies, whether for domestic or recreational use, are tested regularly for contaminants, and finding *Escherichia coli* in such waters is one reason for concern. Why are these coliform bacteria that are so common inside the body of such concern when they are found outside the body? The reason is simply that they indicate fecal contamination of the water. Tests for *E. coli* are relatively quick, easy, and inexpensive, whereas tests for many pathogens are often expensive and time consuming. Therefore, while a few *E. coli* in water may be acceptable for some water uses, high numbers prompt health officials to search for various pathogens, and beaches may be closed in the process.

Finding many coliform and other bacteria may also indicate low oxygen conditions in water. Dissolved oxygen (DO) is important to aquatic life, and finding low DO in a lake or river may indicate unhealthy conditions. Fish may die and certain cyanobacteria and algae may take over, diminishing or destroying the recreational or esthetic values of the waters. DO results mainly from photosynthesis by aquatic algae and plants and by diffusion from the atmosphere.

they usually do not cross the colon mucosa and enter the bloodstream. However, a break or wound in the colon wall may allow the bacteria to enter the blood (septicemia) or the abdominal cavity. See the essay "No Swimming."

One area where infections sometimes originate is the short, twisted, wormlike **appendix** that projects from the blind pouch below where the small intestine joins the colon. This organ has no digestive function in humans, although its lymphatic tissue may play a role in body defense. If its opening is blocked, the accumulation of food and bacteria may cause it to become infected and inflamed, **appendicitis.** Unless treated promptly with antibiotics or removed surgically, the appendix may rupture and cause serious or fatal peritonitis.

There are no peristaltic movements in the colon about 99 percent of the time. However, when they occur they are strong and lead to mass movements of large amounts of material for long distances. Defecation typically occurs two to four times daily and is often initiated by reflexes when food enters the stomach and small intestine. However, the frequency of defecation varies from person to person, and some people defecate only once every several days. Despite some advertisments for laxatives, there is little chance of "poisons" accumulating in the body if defecation is infrequent. Instead, the headaches and other discomfort from **constipation** result from distention of the colon or rectum. Because of excessive water absorption, the passage of hard, dry feces is usually painful. An estimated 20 percent of Americans are often or usually constipated, and about half of these people take laxatives regularly. Laxatives often work by irritating the colon mucosa or by causing more fluid to be released into the large intestine. However, long-term laxative use decreases motility of the colon and may cause psychological dependence.

Constipation, abdominal pressure (as during pregnancy or obesity), or hereditary predisposition sometimes causes the blood vessels in the walls of the rectum to enlarge. These **hemorrhoids** may cause pain and bleeding.

One easy way to reduce constipation is to include sufficient **roughage** in the diet. Roughage is indigestible plant fibers (bulk), and many people do not eat enough of it. Roughage seems to help reduce colon cancer, inhibit the formation of gallstones, and control obesity, although it is not completely clear how it does this. However, roughage activates mucus secretion and peristalsis, speeds the passage of wastes through the intestines, increases stool size, and reduces blood cholesterol. Cancer of the colon and appendicitis are rare in rural areas of Africa and India where the diet is high in roughage. The fact that these problems increase in individuals who move from those areas into affluent urban centers gives additional evidence for the key role nutrition plays in many human ailments.

Diarrhea, usually a far more serious homeostatic problem than constipation, involves the frequent evacuation of watery, unformed feces. The proper absorption of water may be hindered by pathogens (e.g., those that cause dysentery), poisons, vitamin deficiencies, certain foods (e.g., excessive fruits), and/or emotional stress. Besides causing excessive water and mineral loss, severe or prolonged diarrhea may cause ulcers in the colon because of the rapid flow of digestive juices into the large intestine. **Acidosis** or lowered pH of the blood is another potentially serious result of uncontrolled diarrhea.

FOOD PROCESSING AND DISTRIBUTION

Absorbed nutrients may provide energy for body activities, be changed to a storage form (e.g., fat), or be used in making secretions or new cellular substances. The liver plays a central role in the interconversions associated with the metabolism of nutrients. The liver also performs a wide variety of other functions, such as detoxifying the blood, degrading worn-out red blood cells, inactivating hormones, and sending various products to the kidneys for excretion.

Figure 13.12 shows the overall pattern of nutrient metabolism. The carbohydrates are absorbed into capillaries across the intestine wall,

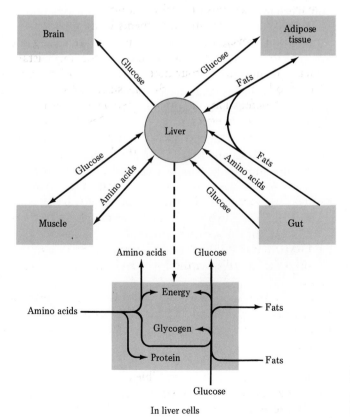

FIGURE 13.12
The homeostatic patterns of carbohydrate, protein, and lipid metabolism. Notice that many of the arrows are two-headed.

mostly in the form of glucose, and reach the liver via the hepatic portal vein (Chapter 15). The blood glucose level must be kept within very narrow limits, and excess glucose may be converted to **glycogen** by the liver and stored there. Many tissues can also store glycogen as well as oxidize it for energy. Due to the size of the body's muscle mass, the total amount of muscle glycogen usually exceeds that of the liver. The liver can also change glucose to fat, and more people "get fat" from consuming excess carbohydrates than from excess lipids.

Although some glycolysis occurs in all cells, the most active cells in the process are those in the skeletal muscles, but brain cells are most dependent upon it. Because they cannot store glycogen, neurons need a continuous supply of glucose.

Proteins reach the liver in the form of amino acids. Because the liver cannot store many amino acids, these must be a regular part of the diet. Thus, amino acids are either synthesized into proteins or carbohydrates (e.g., glycogen), used as an alternative energy source, or excreted.

Upon absorption, fats in the blood or lymph go either to the liver or to depot sites where most are either stored or oxidized for energy. Others are excreted, some are present in secretions such as milk or oil from skin glands, and some become part of the phospholipids that are important components of cell and organelle membranes. Much depot fat represents the intake of food energy in excess of that expended. Although fat storage is in adipose tissue found throughout the body, about half is in subcutaneous tissue, where it serves as insulation.

NUTRITIONAL PROBLEMS

Although few people reading this book are likely to starve, the prime nutritional problem for humans today, as in the past, is getting enough food. Perhaps one-fourth of humans suffer from undernutrition, and untold millions die from starvation (Figure 13.13).

Undernutrition is especially evident in children, where it takes several forms. For example, **kwashiorkor** results mainly from protein deficiency and **marasmus** from a diet low in both calories and protein. Children with these problems often have access to their mother's milk, if at all, for only about one year, and thereafter their diets usually consist mainly of carbohydrates. Symptoms often include a bloated body due to edema and fluid imbalance, anemia, severe diarrhea, susceptibility to infections, physical weakness, and psychological lethargy. Both physical and mental growth are impaired, usually irreversibly. Over half of infant deaths

(A)   (B)

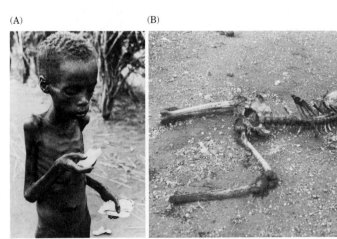

FIGURE 13.13

(A) A child with the typical symptoms of kwashiorkor. (B) The landscape in some areas is littered with victims of starvation, all too often caused by overpopulation and reduced capability to grow crops due to human-caused changes and to human-influenced climatic changes.

(ages 1–4 years) in many developing nations typically result from kwashiorkor and marasmus.

Actually, the body has a remarkable ability to survive without food for long periods. Under medical supervision, healthy adults have gone without food for several months, although regular water is needed. Much of the early weight loss during fasting or starvation is due to water loss and use of adipose stores.

While most people eat to live, others seem to live to eat, and some of the latter suffer from **obesity**, another serious nutritional problem. Defined roughly as 120 percent of the appropriate weight, obesity affects 15–30 percent of Americans, especially in terms of the increased incidence of diabetes and coronary heart diseases. Many Americans also consume too much salt (about 20 times the needed quantity on average) and this is implicated in hypertension. Excess cholesterol is correlated with atherosclerosis and heart attacks (Chapter 15). About 30 percent of surgeries performed in the United States are to correct problems associated with the gastrointestinal tract.

SUMMARY

1. **Nutrition** includes **ingestion, digestion,** and **egestion.** Hunger is controlled by the hypothalamus and, to a large degree, by the blood glucose level.

2. Food is broken down physically by **mechanical digestion,** and the surface area of food exposed to digestive substances is increased in the process. The reduction of food to its molecular components is accomplished by **chemical digestion.**

3. **Teeth** are covered by hard **enamel** on the outside and by somewhat softer **dentin** just below. Blood vessels and nerve endings are located in the **pulp cavity. Cementum** helps bind teeth to the bony socket. **Deciduous teeth** appear first and the more numerous **permanent teeth** erupt later.

4. More teeth are lost to **periodontal problems** than to **dental caries. Dental plaque** and various bacteria are involved in caries.

5. The human **gastrointestinal tract** is basically a tube within a tube. Major mechanical but only minor chemical digestion occurs in the **mouth.** The **stomach** is mainly a storage structure, but **gastric juices** begin numerous processes there and limited absorption occurs. **Ulcers** may result from excessive HCl.

6. The major digestive and absorptive region is the **duodenum,** where **intestinal juices** are basic. Other inputs to the duodenum include **pancreatic juice** and **bile** from the **liver** via the **gallbladder.** Absorption takes place through millions of **villi.**

7. Major water absorption occurs in the **colon,** where nutrients may remain for about a day. Intestinal bacteria digest some things we cannot and contribute useful nutrients, such as certain vitamins, to us in the process. The **feces** are stored in the **rectum** and expelled through the **anus. Constipation** and **diarrhea** are common problems of **defecation.**

8. The liver coordinates the distribution and use of most nutrients. Excess glucose may be stored as **glycogen** in the liver or in other tissues. Limited amino acid storage occurs in the liver, and the liver sends most lipid molecules to storage sites.

9. Malnutrition includes both too little to eat (**marasmus, kwashiorkor,** and vitamin and mineral deficiencies) and too much (**obesity**).

10. Most **filter feeders** are sessile or slow moving, and many use mucus and cilia to entrap and move small organisms and/or food particles. **Detritus** feeders secure nutrients from wastes and dead organisms and from the bacteria that live on the organic matter. Some small invertebrates absorb nutrients from their aquatic environment. **Parasites** and other fluid feeders secure predigested nutrients from a **host. Predation** patterns include **herbivory, carnivory,** and **omnivory.**

11. Many simple animals have **incomplete digestive systems,** with one opening for both food and wastes. Higher animals have a **complete digestive system** or a tube-within-a-tube body plan.

REVIEW QUESTIONS

1. Distinguish among ingestion, digestion, and egestion.

2. Describe some ways that the digestive and circulatory systems work together.

3. What factors control when, what, and how much we eat?

4. What is the function of mechanical digestion? How does it differ from the role of chemical digestion?

5. Distinguish between the causes and effects of dental caries and periodontal diseases.

6. Use our teeth as examples of structure-and-function relationships.

7. Distinguish between permanent and deciduous teeth.

8. Describe some major differences in nutrition among filter feeders, animals with incomplete digestive systems, and animals with complete digestive systems.

9. Follow a bite of hamburger (in a bun with lettuce and tomato; hold the onions if you wish) through our gastrointestinal tract, describing where and how various components are digested and absorbed.

10. What mechanisms and structures do we have to guide food and liquids into the esophagus and air into the trachea?

11. Why doesn't the stomach digest itself regularly? Why do ulcers occasionally develop in some people?

12. Compare and contrast the sizes, lengths, locations, and functions of the small and large intestines.

13. What changes in pH occur in various locations within our digestive tract?

14. Describe the role of the liver in chemical digestion and in processing absorbed nutrients.

15. Where are the gallbladder and pancreas located, and what are their functions in digestion?

16. Distinguish between constipation and diarrhea.

17. Distinguish between marasmus and kwashiorkor. How would you reduce or eliminate the two problems from the world?

18. Why is there public health concern when *E. coli* are found in drinking or recreational waters?

CHAPTER 14

RESPIRATION: GAS EXCHANGE

"Man does not live by bread alone" is a familiar saying. (To be perfectly fair about it, neither does Woman.) In this context, oxygen fuels the "fire of life," allowing us to secure energy through oxidizing the nutrients we absorb. The respiratory system includes all structures involved in exchanging gases with our environment. Besides obtaining oxygen, the respiratory system also eliminates carbon dioxide, helps maintain blood pH, and aids speech.

All organ systems are important in maintaining homeostasis, but in different ways and with some being more active than others. For example, the digestive system may become relatively inactive when food is unavailable, muscles may be paralyzed, bones may break, and reproduction may not occur. These conditions may be important and even grave but they need not be fatal. However, death may follow quickly if the lungs, heart, kidneys, or brain stop functioning. The brain in particular requires an abundance of oxygen. Obviously, breathing is one of our most repetitive and important functions, although we usually take it for granted. The time that we can go without breathing is measured in minutes, as opposed to days for drinking and eating.

The mammalian respiratory system is relatively compact compared with many other organ systems. Air enters and leaves the system through a common set of tubes, and gas exchange occurs in bubblelike structures in the lungs.

THE BREATH OF LIFE

Because air is normally colorless, odorless, and tasteless, most people pay little attention to it. However, we take in over five times as much air daily (about 16 kg, or 35 lb) as we do food (1.2 kg) and fluids (1.8 kg) combined. As a neonate, you breathed about 4–5 times faster than you do now, another good example of changing metabolic rates with age.

No one wants to drink polluted water or eat spoiled or contaminated food. Fortunately, we can choose not to ingest such materials, at least for a time, but we must breathe regularly whether or not air is polluted. Therefore, we should be as concerned about air quality as we are about the hazards of food poisoning and water pollution. Particular air pollutants and their effects are discussed later in this chapter.

THE INS AND OUTS OF BREATHING

The breathing system is basically a series of tubes and sacs that is constructed to provide an immense surface area for gas exchange (Figure 14.1). See the essay "Comparative Methods of Gas Exchange" for a discussion of breathing in other animals.

Most people breathe through their noses most of the time, for good reasons. The nasal passages are narrow and winding, with many grooves and several shelflike **turbinates.** Coarse hairs just inside these entrances screen out large particles, sometimes even insects. Some nasal cells are ciliated, and others secrete mucus. Adults inhale over 16,000 L (4200 gal) of air daily, and even seemingly clean air usually contains hundreds of thousands of pathogens and small particles. Many of these materials stick to the mucus and are swept to the mouth by beating cilia. Unless air is very dirty, you usually swallow these substances unconsciously. The mucus has some antibiotic qualities and gastric juice kills many or all of the remaining pathogens. Very small particles may reach the lung sacs or **alveoli,** and fibrous tissue may form near any particles that remain in the lung, such as in miners who suffer from **black lung disease** (Figure 14.2). Some particles are consumed by scavenger cells, while others are trapped and destroyed in lymph nodes. Triggered by a variety of nasal irritants, our sneeze reflex also helps keep out contaminants.

FIGURE 14.1
Air enters the human respiratory system through either the nose or the mouth. After passing through the pharynx and larynx, air enters the cartilage-supported trachea. The bronchioles that branch from the two bronchi support the grapelike clusters of alveoli at their thin tips. Gas exchange occurs in the alveoli.

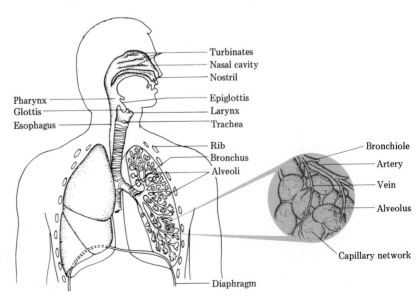

Turbinates
Nasal cavity
Nostril
Pharynx
Glottis
Esophagus
Epiglottis
Larynx
Trachea
Rib
Bronchus
Alveoli
Bronchiole
Artery
Vein
Alveolus
Capillary network
Diaphragm

The skull bones contain several hollows or **sinuses** that connect with the nasal passages. Other openings into the nasal passages include the **eustachian tubes,** which help keep air pressures equal on both sides of the eardrum despite changes in altitude. Another pair of openings comes from the tear ducts, and the lymphlike fluid from the **tear glands** keeps the cornea moist. This fluid and the mucus secretions help humidify inhaled air for better gas exchange in the lungs.

The redundancy of being able to breathe through both the nose and the mouth is fortunate, for when we have a cold or the flu or are reacting to an allergen, the narrow nasal and sinus passages may become inflamed and filled with liquid. This approaches positive feedback, because at the time when many pathogens are present the efficient filtering action of the nasal passages is bypassed.

Pathogens that enter the nose may reach the middle and inner ear through the eustachian tubes. Because the eustachian tubes in children are wider and shorter than those in adults, children are especially prone to these ear infections. A major function of lymphatic tissue (Chapters 15–16) is to destroy pathogens, and the nasal passages are a major route by which pathogens enter the body. Therefore, the work of the several lymphatic nodules called **tonsils** are important. When one tonsil (the adenoid or pharyngeal tonsil) enlarges as it "fights" pathogens, as is common in children, it may block the nasal passages and interfere with breathing. **Tonsillitis** occurs when another tonsil (palatine) is swollen. Because lymphatic tissue is protective, these tonsils are not removed as often now as in the past.

The Windpipe

From the nasal passages or the mouth, air moves through the **pharynx** into the **larynx.** Also called the voicebox or Adam's apple, the larynx is larger and more prominent in men than in women. Vibration of the fibrous **vocal cords** within the larynx produces sounds as air is expelled (Figure 14.3). Pitch is determined by the tension on and the length of the cords, and males generally have lower-pitched voices than females because their vocal cords are larger. Basically, the vocal cords operate like reed instruments such as clarinets and bassoons. Growth of the male larynx may be rapid during adolescence, and a young man may experience temporary difficulty in controlling this strange instrument. Inflammation of the larynx, **laryngitis,** may swell the vocal cords, making speaking and breathing difficult. If a cancerous larynx is removed, esophageal speech may be learned. This involves swallowing air and forcing it out through the esophagus. *(Continues on p. 288.)*

FIGURE 14.2
Breathing coal dust particles over time may lead to the condition known as black lung disease, as is common in the lungs of coal miners. People who work in dusty environments may develop equally debilitating brown lung disease. These photos compare the lung tissue of smokers and nonsmokers.

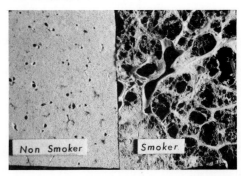

FIGURE 14.3
(A) A front view of the cartilaginous larynx shows the location of the vocal cords relative to the epiglottis above and the trachea below. The opening into the trachea is called the glottis. We speak when air moves out of the glottis, meaning that the epiglottis is open, and this is not a good time to swallow food or fluid. (B) The vocal cords shown closed and open. The inner wall of the cartilage-ringed trachea can be seen in the latter view. Our teeth and tongue are also involved in speech.

(A)

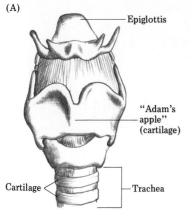

Larynx, frontal view

(B)

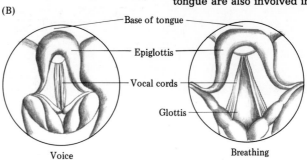

Larynx viewed from above

Comparative Methods of Gas Exchange

Whether the organism is a microscopic protistan like an amoeba or a huge animal like an elephant, gas exchange takes place by diffusion. The fact that gases must be dissolved in water in order to be used by organisms creates little difficulty for most small aquatic animals (Figure B14.1). These animals generally have low oxygen needs and a large surface area exposed to their water environment.

Oxygen is far less available in water than it is in air. Oxygen also dissolves more readily in fresh water than it does in salt water at the same temperature. Also, oxygen dissolves more easily in cold water than it does in hot water. Thus, when hot water speeds the metabolic rate of ectotherms, and increases their need for oxygen, their environment contains only small amounts of oxygen (Chapter 2). In many cases the hot water influent also contains sewage and/or other organic materials. The bacteria that decompose (oxidize) these wastes also require oxygen, and oxygen depletion caused by their respiratory activities may lead to massive fish kills.

Above a body size of about 1 mm in diameter (approximately twice the size of the period in this sentence), diffusion across the body surface is neither adequate to provide the needed oxygen nor sufficient to remove waste gases. This is especially true if animals have evolved a waterproofed covering. Whether aquatic or terrestrial, large animals must possess a system by which gases can come close to the individual cells. Therefore, all but the smallest animals must also possess a circulatory system to both distribute oxygen and other materials after they have entered cells and to assist in ridding the body of wastes.

Some of the special respiratory adaptations of larger animals are summarized in Figure B14.2 and described below.

Gills

Gills, the branching external or internal extensions of the body surface that provide a large surface area for gas exchange with water, are found in many crustaceans, mollusks, amphibians, and fish. Gilled animals often actively move water past their gills, especially if the gills are covered and protected by a shell or flap. For example, millions of beating cilia in clams move water over their enclosed gills.

The gills of most fish are elaborate, highly branched, and blood-filled structures. Fish gills are protected by a flap or operculum on the outside and the gill slit is connected to the mouth cavity on the inside. Water that passes into the mouth exits over the gills where gas exchange occurs. Many fish actively and continuously pump water through the gill slits by muscular movements of the mouth cavity and the operculum. Many others rely on their movements through water to maintain water flow over the gills; if they stop swimming they die. However, some fish such as eels may secure half of their oxygen through their highly vascular skin.

Countercurrent circulation patterns are common in animals. The role of countercurrent flow in thermoregulation is illustrated in Figure 2.22. Gas exchange in gills is facilitated by having blood flow in the opposite direction to that of water (Figure B14.3).

Lungs

An extensive wet surface area such as a gill exposed to air could lead to excessive water loss. **Lungs,** extensive wet surface areas inside the body, lessen that problem. For gas exchange, evaginations of the body surface are best for aquatic animals whereas invaginations are more suitable for air breathers. Simple lunglike structures are found in such diverse animals as certain snails,

(A)   (B)   (C)

FIGURE B14.1
The concentration of oxygen dissolved in water is at least 20 times less than it is in air. Oxygen also diffuses more slowly in water than in air. This normally creates few problems for a freshwater *Hydra* (A) or a saltwater sponge (B). Diffusion across their surfaces is adequate to satisfy their low metabolic rates. However, when the dissolved oxygen level becomes too low, many animals die (C). This problem often occurs when water warms and especially when it also receives quantities of organic matter.

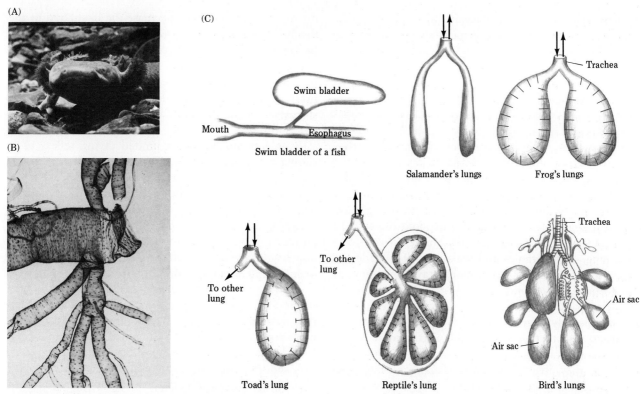

FIGURE B14.2

In all animals, a wet membrane is necessary for the diffusion of gases. Although free diffusion of gases across the whole body surface is adequate for small animals, large animals have evolved some of the special respiratory mechanisms shown here. (A) The branched gills that are common in aquatic animals provide a large surface area for gas exchange. Water loss is reduced in terrestrial animals by having wet gas exchange surfaces located inside the body. (B) The branched tubes called tracheae are found in many arthropods such as insects. (C) Saclike lungs are common in vertebrates, where the general evolutionary trend has been from simple sacs with relatively little surface area for gas exchange to extensive, complex, and highly vascularized internal surfaces.

FIGURE B14.3

(A) The surface area of fish gills may be 10–60 times that of the rest of the animal's body surface. Active fish have relatively more gill surface area than sluggish fish. (B) Gases can be exchanged with water all along the thin-walled and highly vascularized gills, and fish may extract up to 80–85 percent of available oxygen. Blood flow is toward the mouth cavity, opposite that of water flow. Such countercurrent flow systems are important in the functioning of numerous body systems in many animals.

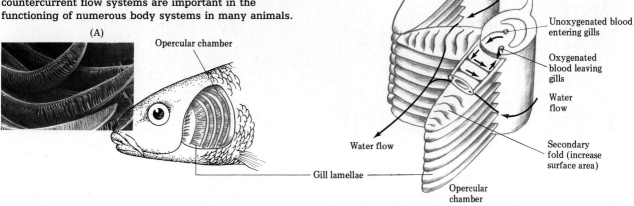

spiders, scorpions, and crustaceans. In amphibians, the skin and mouth cavity supplement the gas exchange that occurs in the lungs. In a few aquatic reptiles, gas exchange in the lungs is supplemented by exchanges through the skin and across the wall of the terminal chamber of the digestive tract (cloaca).

Why whales become stranded on beaches, sometimes in large numbers, is still a mystery (Figure B14.4). What is known is that these air-breathers die quickly if they cannot be coaxed back into the sea. Humans, who have about the same oxygen requirements and respiratory structures as whales and other mammals, usually die quickly if they are submerged without breathing gases for more than a few minutes. About 1 out of every 14 accidental deaths in the United States results from drowning. Some whales and seals may dive to depths of 1000 m (about 3000 ft) or more, stay submerged for more than an hour, and swim several kilometers before surfacing for air. Yet the lungs of diving mammals are not significantly larger than the lungs of other mammals in proportion to their body sizes. Furthermore, while we generally take a deep breath before diving, seals exhale as they dive. Clearly, diving mammals have special adaptations, including automatic responses known collectively as the **diving reflex.**

During a dive, the seal's or whale's heart rate slows dramatically and the blood flow is reduced drastically to tissues that can tolerate oxygen deprivation, such as the skin, muscles, and digestive organs. Most oxygen is shunted to the brain and heart cells, which usually begin to die within minutes without oxygen. By exhaling while diving, marine mammals decrease the likelihood of the bends developing. Underwater survival is also increased by adaptations of their circulatory systems. Diving mammals have 1.5–2 times as much blood as most land mammals of the same size. They also have relatively more oxygen-carrying red blood cells and the oxygen-storing pigment myoglobin in their muscles is more concentrated.

Even humans exhibit a limited diving reflex, as is shown by the reduced heart rate when the face is immersed in cold water. This and the fact that cold water reduces metabolism, and therefore the need for oxygen, help explain why humans sometimes survive longer in water below 21°C (72°F) than they do in warmer water. However, the main importance of this reflex in humans may be during birth, when the oxygen supply is reduced or interrupted temporarily.

Because of the high energy needs of flight, the bird respiratory system is more extensive and efficient than that in mammals (Figure B14.5). In addition to lungs, birds possess an extensive system of **air sacs** that extend into the wings and legs, where they replace bone marrow and thus assist flight by reducing the bird's weight. Air sacs have no capillary networks and they are not involved directly in gas exchange. Rather, they serve as air reservoirs and bellows. During inhalation, most incoming air passes through a tube (bronchus) to the posterior air sacs and that forces air already in the lungs to move into the anterior air sacs. During exhalation, the posterior air sacs empty into the lungs and the anterior air sacs empty to the outside. Thus, the lungs

FIGURE B14.4
Although whales are air-breathing mammals, their massive weight makes it impossible for them when stranded to use their breathing muscles and survive for long.

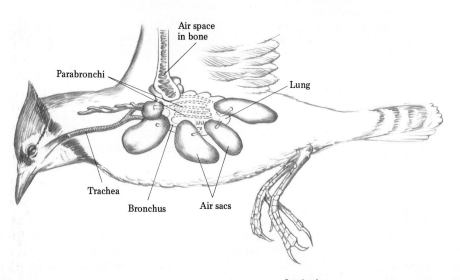

FIGURE B14.5
Bird lungs are small but highly efficient, a trait important to supporting the very high metabolic requirements of flight. The several air sacs that are attached to each lung act as bellows, flushing the lungs with fresh air with each breath. The tiny air tubes (parabronchi) traverse the spongelike lung tissue, with its enormous surface area and capillary network for gas exchange.

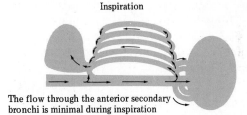

Inspiration

The flow through the anterior secondary bronchi is minimal during inspiration

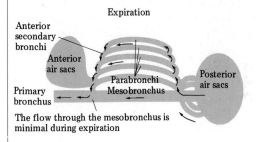

Expiration

The flow through the mesobronchus is minimal during expiration

are completely filled with fresh air with each breathing cycle, and there are not large amounts of "dead air" remaining in the lungs during each inhalation-exhalation cycle as in mammals. Gas exchange in birds is also more efficient than in mammals because air flows through the lungs in one direction only (front to back) rather than back and forth. Further, blood in the pulmonary vessels in the bird lung flows in a direction opposite that of air, providing another countercurrent-type system to increase the efficiency of gas exchange.

Tracheae

An alternative to lungs for air-breathing animals is to channel air directly to the tissues. Such a system of branching chitin-lined tubes called **tracheae** is common in insects and spiders. The tracheae open to the outside through holes called **spiracles** on the animal's surface. Valves control the opening and closing of the spiracles, which are bristle-guarded to keep out particles and enemies. Gas exchange occurs near the thin, moist tips of the finest branches (tracheoles). However, the large surface area that is advantageous for gas exchange is dis-

advantageous in terms of evaporative water loss, and desiccation is a major cause of death in adult insects. Simple tracheae are not efficient enough for such large active insects as grasshoppers. These insects have some tracheae expanded into air sacs that fill and empty by body movements. Even so, the relatively inefficient gas exchange capabilities of a tracheal system place a severe limitation on the sizes of insects, despite the giant insects described in some science fiction books.

FIGURE 14.4
Normal alveoli (A) compared with those in humans who are affected by emphysema (B). The combination of abnormal bronchioles and a major reduction in the surface area of the alveoli available for gas exchange makes breathing difficult for people with emphysema. The increasing incidence of this disease can be correlated with smoking and exposure to air pollutants.

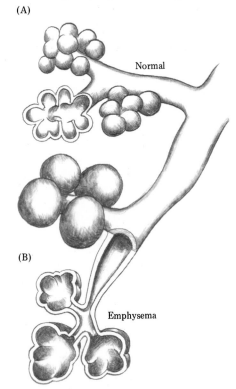

(A)

Normal

(B)

Emphysema

Because our breathing system depends upon differences in air pressure inside and outside of the body, it is important for the windpipe or **trachea** to be rigid so that it does not collapse as the empty esophagus does. The trachea is lined with C- and Y-shaped rings of supportive cartilage. Scattered plates of cartilage also occur in the **bronchi** and larger **bronchioles,** but there is no cartilage in the smaller bronchioles. As the amount of cartilage decreases in the smaller bronchioles, the quantity of smooth muscle increases to support the tube walls. Spasms of these muscles lead to the labored breathing of those with **asthma.** One cause of **emphysema** is for the small bronchioles to narrow, twist, or collapse (Figure 14.4). This greatly reduces the area for gas exchange in the 1.5 million Americans who suffer from emphysema.

The Lungs

The space between the **lungs** contains the heart, large blood vessels, and the esophagus. A membrane called the **pleura** covers the outer surface of the lungs, the inner wall of the **thorax** or chest cavity, and the top of the muscular **diaphragm** that separates the thoracic and abdominal cavities. A thin film of liquid normally lubricates the pleural surfaces so they slide past one another easily as the lungs change size. The inflammation of these membranes is called **pleurisy** and this often leads to excess fluid (edema) which collects in the thoracic cavity. Insufficient fluid between the two pleural layers causes pain during breathing as the dry membranes rub against one other. The thin film of liquid between the thorax wall and the lungs allows the lungs to cohere as the chest cavity enlarges during inhalation.

Although the lungs weigh a little over 1 kg (2 lb), their density is relatively low because of the several hundreds of millions of alveoli per lung. Alveoli form grapelike clusters around the ends of small bronchioles. Their total surface area is enormous, many times that of the body's skin surface area. Only a thin membrane separates the alveoli from the surrounding capillaries, and both of these structures consist of only a single layer of flattened epithelial cells. Gas exchange occurs between the alveoli and the adjacent capillaries, the largest capillary network of any organ in the body (Figure 14.5). The alveoli are kept open by a secretion called a surfactant which reduces the surface tension of the inner fluids and thereby prevents the lung walls from collapsing and sticking together. Children, especially low-birth-weight infants, are sometimes born with a disease (hyaline membrane disease) in which there is inadequate production of the surfactant secretion, a condition which may be fatal because there is insufficient surface area for gas exchange.

(A)   (B)

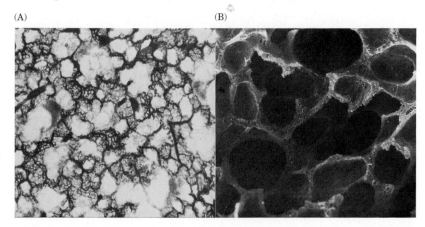

FIGURE 14.5
Light (A) and electron micrographs (B) of some of the estimated 300 million alveoli in the human lungs. Note the very large surface area created by the bubblelike structures.

THE BREATHING MECHANISM·

Functionally, lungs are elastic bags not unlike rubber balloons. The absence of muscles in lungs no more allows them to change size by themselves than balloons can. Instead, the lungs react rather passively to pressure changes within the thoracic cavity. During contraction, the muscular, dome-shaped diaphragm flattens and thereby enlarges the thoracic cavity from top to bottom. Contraction of a neck muscle and the intercostal muscles between the ribs pull the rib cage upward and outward, enlarging the thoracic cavity from front to back and from side to side (Figure 14.6). Inhalation occurs because these actions create a pressure within the lungs that is lower than the atmospheric pressure is outside. Exhalation occurs when relaxation of the diaphragm and the intercostal muscles reduces the volume of the thoracic cavity, creating a pressure there that exceeds atmospheric pressure.

In normal breathing, the diaphragm accounts for about 80 percent of the air movement and the intercostal muscles for the remainder. The elastic recoil of the lungs, thoracic wall, and abdominal organs are also important in exhalation. If there is a loss in lung elasticity, such as in those with emphysema or asthma, exhalation must be a more active and conscious process.

The fact that breathing is caused by pressure differences is fortunate in several cases. For example, it makes possible **artificial resuscitation** of drowning or shock victims. The older, back-pressure arm-lift method is now largely replaced by the mouth-to-mouth method (Figure 14.7). The latter is more efficient but suffers somewhat because many people practice only on models rather than on live "victims" in first-aid classes.

First aid is also helpful for a person who has the trachea blocked by food. The cough reflex is triggered by irritation of the throat or trachea, but this may not be sufficient to dislodge a large chunk of food. Thousands of people owe their lives to an alert friend or bystander who placed their arms around the choking person's waist and made a quick

FIGURE 14.6
The breathing mechanism. As in most air-breathing vertebrates, the filling and emptying of our lungs depends upon pressure differences.

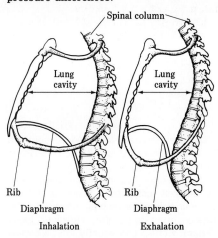

FIGURE 14.7
Artificial resuscitation by the mouth-to-mouth method.

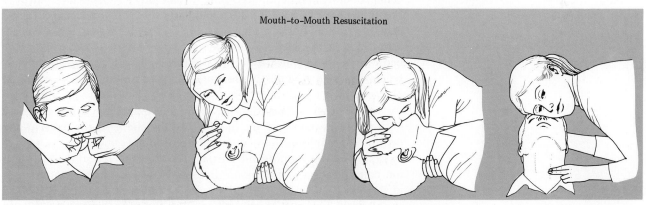

Mouth-to-Mouth Resuscitation

(A) Wipe any foreign matter, blood, vomit, or broken teeth from the victim's mouth quickly.

(B) Tip the victim's head back as far as you can (except in the case of neck or back injuries). Look at the chest and listen and feel at the mouth to detect breathing.

(C) If the person is not breathing, give four quick, full breaths mouth-to-mouth right away. Take a deep breath for each of the four breaths, open your mouth wide, and pinch the victim's nose.

(D) Next, check the pulse and breathing. If the person is not breathing but has a pulse, give mouth-to-mouth breathing once every 5 seconds. If there is no pulse, cardiopulmonary resuscitation (CPR) is needed. However, if you can not give CPR, continue mouth-to-mouth breathing because there may be a heartbeat that you cannot feel. If the victim is a baby, blow into the mouth *and* the nose, giving small breaths every 3 seconds.

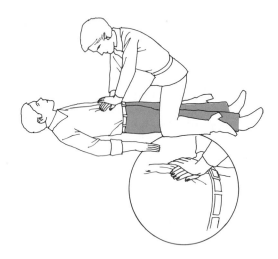

FIGURE 14.8
Despite increased knowledge about the Heimlich maneuver, about 3,000 Americans die annually from food strangulation. Death may occur in only 4–5 minutes if the airway is blocked completely. Thinking that the victim is experiencing a heart attack, onlookers often do nothing. However, heart attack victims can usually talk, whereas those whose trachea is blocked obviously cannot. The procedure shown, which is well worth practicing, is basically similar whether the victim is lying, sitting, or standing.

upward thrust while their fists were against the victim's abdomen (Figure 14.8). The increased pressure of this **Heimlich maneuver** elevates the diaphragm, compresses the lungs, and usually causes the trapped food to pop out of the trachea of the breathless and blue-lipped victim.

Some victims of problems such as polio may be assisted in breathing by mechanical devices ("artificial lungs") that simulate the normal rhythmical pressure changes. Also, injuries such as a gunshot wound in the chest may collapse one lung because the pressure inside and outside the lung are then identical. Because the pleural membranes of the two lungs are separate, the collapse of one lung does not cause the collapse of the other.

CONTROL OF BREATHING

How does breathing continue unconsciously, even during sleep, and change its rate and depth according to the body's needs? The control of breathing involves several factors and body structures and is an excellent example of homeostasis (Figure 14.9). The neurons that control respiration are in the medulla oblongata of the brain (Chapter 19). The neurons in the **respiratory centers** control normal quiet breathing by regularly stimulating the diaphragm and intercostal muscles to contract. However, it is the carbon dioxide level in the blood that precisely controls the rhythm and depth of breathing. For example, vigorous exercise both increases manyfold the blood CO_2 and the need for oxygen, and breathing becomes faster and deeper accordingly.

Brain neurons monitor the partial pressure of CO_2 and the hydrogen ion concentration of the blood directly. However, chemoreceptor cells in the aorta and in the carotid arteries of the neck that supply oxygen to the brain also do this, as well as sense the partial pressure of blood oxygen. The medulla cells also receive inputs from stretch receptors in the walls of the lungs.

Because we can voluntarily control breathing somewhat for talking, eating, coughing, and many other activities, the higher brain centers are obviously also involved. However, parents need not worry about a child's threat to hold his or her breath until death because it is impossible to commit suicide by holding one's breath. When one becomes unconscious or when the partial pressures of blood CO_2 rise and O_2 fall sufficiently, and the H^+ concentration changes accordingly, no amount of willpower can prevent the medulla's breathing center from "taking over" the cerebral control. Unfortunately, this is why some people drown

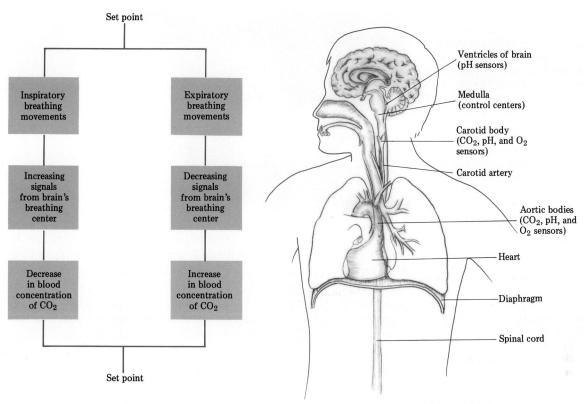

FIGURE 14.9
Feedback diagram summarizing the control of breathing.

even though they know that they might reach the surface if they could only hold their breaths a little longer.

It seems strange to many people to learn that their breathing is regulated by CO_2, the waste product of their metabolism, rather than by blood oxygen levels. In lung breathers, however, arterial blood is usually nearly saturated with oxygen, even during exercise. Although venous oxygen becomes depleted during exercise or work, the respiratory centers in the brain receive and respond to arterial blood. However, under low environmental oxygen conditions, such as at very high altitudes, messages from the oxygen-sensitive chemoreceptors assist in stimulating increased lung ventilation. Unlike lung breathers, aquatic animals such as fish and crustaceans depend upon blood oxygen levels for respiratory control. This is because CO_2 levels in most natural waters are low and CO_2 is removed from the blood about as quickly as it is formed.

Trying to remain submerged longer, some swimmers have drowned by **hyperventilation,** or breathing deeply and rapidly, before they enter the water. Although this removes much CO_2 from the blood, and thus delays the urge to surface and breathe, blood oxygen is depleted just as rapidly as without hyperventilation. This may cause fatal unconsciousness when the oxygen supply to the brain falls below a critical level.

THE UPS AND DOWNS OF BREATHING

Breathing is complicated when people explore high altitudes or great depths (Figure 14.10). Nevertheless, humans have climbed the highest mountain and descended to the deepest ocean trench, and although breathing problems at these extremes can be solved, they still affect many people.

Although the percentage of gases remains about the same at different altitudes, the total pressure of the atmosphere drops as altitude increases.

FIGURE 14.10
This physician-scientist is measuring his breathing capacity on top of Mount Everest. The low pressure at high altitudes creates many problems, although oxygen still makes up about 21 percent of oxygen at all altitudes. People not adapted to high altitudes may lose consciousness and die at about 7,000 m (over 21,000 ft). Well below that altitude the oxygen deficiency may cause headaches, nausea, and lethargy. In response to lower oxygen pressures, both the heartrate and breathing rate generally increase.

Thus, above about 5500 m (18,000 ft), the approximate altitude of the highest permanent human habitations, there is only about one-half as much oxygen as at sea level. The brain is particularly sensitive to oxygen deprivation, and symptoms such as blurred vision and mental confusion are common. Although people might get enough oxygen if they breathed deeply and rapidly, the conscious control of breathing is dulled by a lack of oxygen. Also, CO_2 production, which would stimulate breathing, is reduced by the slow physical activity. The cabins of most commercial aircraft are pressurized at a pressure equivalent to about 1500 m (5000 ft).

Pressure increases by about 1 atmosphere (normal air pressure at sea level) for each 10 m (30 ft) when one descends into water. Most divers may breathe compressed air without problems to a depth of about 60 m (Figure 14.11). Below that depth, oxygen and nitrogen are forced into the blood in above-normal quantities. The brain is very sensitive to these changes, and mental confusion, euphoria (sometimes called "rapture of the deep"), nervous twitching, convulsions, and/or coma may result. Nitrogen gas is inert in the body but creates problems if a rapid ascent is made. Nitrogen bubbles may form in the blood when the body is exposed to less pressure, much as dissolved CO_2 escapes as bubbles when the pressure is reduced when the cap of a carbonated drink is removed. These nitrogen bubbles in the blood, spinal fluid, and cells cause the **bends,** or nitrogen narcosis. The bubbles may deprive tissues

FIGURE 14.11
These divers are exploring a shallow coral reef and are using scuba tanks of compressed air for breathing. At greater depths the increased pressure may force nitrogen bubbles into the blood and divers must plan ascents carefully in order to avoid problems such as the bends.

of oxygen by blocking blood vessels, much as blood clots do. Other problems include damaged nerve fibers, fluid accumulation in the lungs, painful joints and muscles, and sometimes paralysis and death. A slow ascent or entrance into a decompression chamber for controlled pressure changes eliminates or reduces such damage. For deep diving, breathing mixtures with less oxygen and with helium to replace nitrogen are useful, although helium affects the vocal cords so that divers sound like Donald Duck.

LUNG CAPACITIES

Several different lung capacities are commonly identified and measured. The 500 ml (1 pt) of air that the average person moves into and out of the lungs during a normal resting breath is called **tidal air** (Figure 14.12). At the end of a normal resting breath, one can forcefully exhale about 3 L of air, a volume called the **inspiratory reserve. Expiratory reserve** is the somewhat smaller volume of air that can be expired forcibly after a normal exhalation. **Vital capacity** is the maximum volume of air that one can move into and out of the lungs. Most males have a larger vital capacity than equal-sized, equal-aged females. During normal breathing, only about 10 percent of air is exchanged with each breath but deliberate deep breathing may exchange as much as 80 percent of air.

Even after a maximum exhalation about 1.5 L of **residual air** remains in the lungs. This air, which is not under our conscious control, seems to be important in conserving water that results from the partial stagnation of air in the alveoli. Residual air is also involved in some legal cases, such as when it is important to know whether a child was stillborn or born alive. Because the lungs are shriveled until a child takes its first breath, the lungs of a stillborn child sink during an autopsy, whereas those of a child born alive float.

Dead air in the breathing structures is also not available for gas exchange. The volume of this air in healthy adults averages about 100–150 ml but may equal 1–2 L in those affected by emphysema or asthma.

GAS EXCHANGE AND TRANSPORT

Like food in the digestive tract, air in the lungs is outside of but is surrounded by the body (Figure 3.11). Gas exchange occurs in the alveoli (Figure 14.13), one of the few places in the body where diffusion is far more important than active transport in exchange processes. About one-

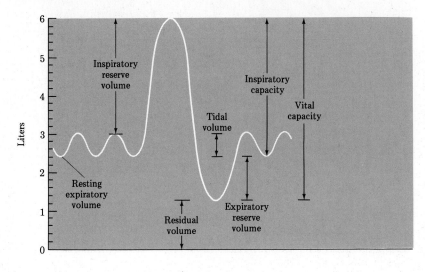

FIGURE 14.12
The maximum amount of air that one can move into and out of the lungs voluntarily is called the vital capacity. Because of the importance of oxygen in supporting high metabolic rates, vital capacity may be measured to indicate improvement during a physical conditioning program. About 1.5 L of so-called residual air that is not under our conscious control remain in the lungs even after a maximum exhalation. That plus the so-called dead air that remains in other parts of the respiratory system provide one reason that gas exchange in mammals is less efficient than it is in birds.

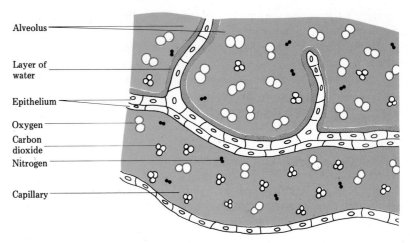

FIGURE 14.13
The pulmonary artery brings low-oxygen blood to the capillaries that surround the alveoli, and oxygen-rich blood returns to the heart in the pulmonary vein, to be pumped to the rest of the body. As with most gas exchange, a wet surface is required and some water leaves the body with each exhaled breath. Gases are exchanged by diffusion, dependent upon the relative partial pressures of oxygen and carbon dioxide in the alveoli and alveolar capillaries.

Alveolus

Layer of water

Epithelium

Oxygen

Carbon dioxide

Nitrogen

Capillary

fourth of our blood supply is concentrated in the fine capillary nets that surround the alveoli. Most of the liquid part of blood, the plasma, is water, and oxygen is relatively insoluble in water. Therefore, oxygen combines with the iron-containing compound **hemoglobin** in the red blood cells. Without hemoglobin, the blood would be able to carry only a small percentage of the oxygen it transports. The hemoglobin molecule consists of four heme groups with their polypeptide units intertwined (Figure 14.14). Because the iron in each heme group can combine with one molecule of oxygen, the hemoglobin molecule can hold up to four oxygen molecules.

The **oxygen dissociation curves** shown in Figure 14.15A for several mammals illustrate some important points. Whether the oxygen combines with or is released from hemoglobin depends upon the partial pressure of oxygen in the surrounding blood plasma. (Partial pressures of gases are commonly expressed in millimeters of mercury.) As the partial pressure of oxygen rises, hemoglobin picks up oxygen; as it drops, oxygen and hemoglobin dissociate. Thus, the attachment of oxygen to hemoglobin is loose and easily reversible. Because the partial pressure of oxygen is high in the alveolar capillaries due to the oxygen that has diffused into them, most hemoglobin becomes saturated with oxygen. Elsewhere, however, where the partial pressure of oxygen is low, oxygen

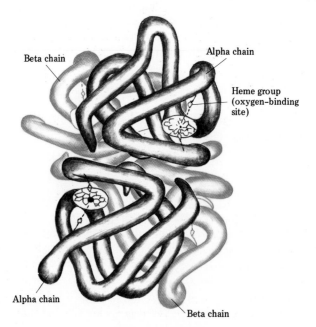

Beta chain

Alpha chain

Heme group (oxygen-binding site)

FIGURE 14.14
The hemoglobin molecule. In heme, an iron atom is held by the nitrogen atoms that are part of a porphyrin ring. Each heme group consists of 150 amino acids, or 600 in all per hemoglobin molecule. Normal mature red blood cells contain about 265 million hemoglobin molecules. Hemoglobin allows our bloodstream to carry about 60 times more oxygen than the plasma could alone.

Alpha chain

Beta chain

dissociates from hemoglobin and moves into the tissues. Even then, the hemoglobin retains most of its oxygen, a useful reserve for work, exercise, or reaction to emergencies. Figure 14.15B shows the curves for the fetus and its mother. The fetus derives all of its oxygen from the maternal blood and its hemoglobin has a higher affinity for oxygen than the hemoglobin of its mother does.

Hemoglobin is only one of several oxygen-carrying protein molecules known as **respiratory pigments** that are found in most active animals. The pigments are carried in red blood cells in vertebrates and echinoderms (e.g., starfish) but are simply dissolved in the blood plasma in most invertebrates. While iron is important in hemoglobin, a different metal may be involved in other respiratory pigments. For example, the most common respiratory pigment (hemocyanin) in mollusks (e.g., snails) and arthropods (e.g., insects) contains copper.

No special pigments are used to transport CO_2 from the tissues to the lungs. Because CO_2 has a higher partial pressure in the tissues where it is produced by cellular respiration than it does in the blood, it diffuses from the area of its higher concentration. Only about one-fourth of CO_2 is carried in hemoglobin and only a little dissolves in the plasma. Most CO_2 is carried in blood as bicarbonate ions (HCO_3^-), which are produced in the following two-stage reaction:

$$CO_2 + H_2O \underset{\text{enzyme in red blood cells}}{\overset{\text{carbonic anhydrase}}{\rightleftharpoons}} \underset{\text{carbonic acid}}{H_2CO_3} \rightleftharpoons \underset{\text{bicarbonate ion}}{HCO_3^-} + \underset{\text{hydrogen ion}}{H^+}$$

Any increase in CO_2 drives this reaction to the right, releasing more hydrogen ions. In the tissues where oxygen is needed, the decrease in blood pH promotes the rate at which hemoglobin unloads oxygen. The uptake of oxygen in the lungs is facilitated by the increase in pH of the blood there. Because the partial pressure of CO_2 is higher in the pulmonary capillaries than in the alveoli, it diffuses to the area of its lower concentration.

Although we need nitrogen for such compounds as proteins, we get no significant nitrogen from our very nitrogen-rich gaseous environment. Essentially, the nitrogen that we inhale (over three-fourths of air) we also exhale. It is mainly through the activities of nitrogen-fixing monerans, especially the bacteria that live in root nodules of legumes, that usable nitrogen becomes available for use by most animals (Figure 14.16). Through photosynthesis, green plants are also the ultimate source of oxygen for animals (Chapter 12). While inhaled air is about 21 percent oxygen, exhaled air is about 16 percent O_2. On the other hand, exhaled air contains about 4 percent CO_2, an enormous increase over the tiny percentage of CO_2 that we inhale (0.03–0.04 percent).

SMOKING VERSUS THE BREATHING SYSTEM

The Surgeon General's warning on cigarette packages is no idle public service message. Of all the self-destructive things that humans do, few are as common or pervasive in their adverse effects on the body as smoking. Tobacco smoke contains over 600 chemicals, including those shown in Table 14.1.

Although people have smoked for centuries, the detailed scientific evidence about the harmful effects of these substances has come mainly

FIGURE 14.15

(A) These oxygen dissociation curves show that because small mammals have a higher metabolic rate than large mammals, oxygen is given up more readily at any given pressure. Llama hemoglobin allows it to take up oxygen at low atmospheric pressures, a useful adaptation to its high-altitude habitat in the Andes. (B) Fetal blood becomes saturated at lower partial pressures of oxygen than maternal blood does. Because of this, oxygen tends to leave the maternal blood, the fetus's only oxygen source, and enter the fetal blood.

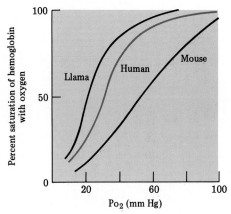
(A)

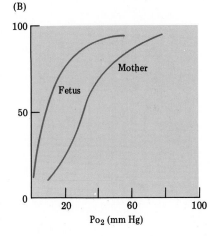
(B)

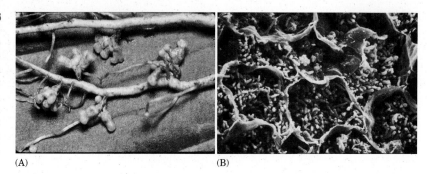

FIGURE 14.16
Nodules such as these on legume roots (A) are filled with bacteria (B) that are able to utilize atmospheric nitrogen and in the process to create nitrogen compounds useful for the plants, herbivores, and ultimately carnivores and various decomposers. Although we live in a very nitrogen-rich environment (air is about 79 percent nitrogen), we cannot secure the nitrogen we need for proteins from that source.

(A)   (B)

during the past few decades. While medical treatment can cure many ailments, the death rates for several problems of the respiratory tract are increasing. Cigarette smoking is the major cause of **lung cancer,** and the more one smokes, and inhales, the greater the risk (Figure 14.17). In the past, death rates for lung cancer were far higher for males than for females, but female death rates are rising rapidly as more females smoke, especially young females. Smokers have 4–25 times more risk than nonsmokers of dying from chronic bronchitis and emphysema. A smoker's risk of dying from cancer of the esophagus, the mouth, or the larynx is 2–9 times greater, 3–10 times greater, and 3–18 times greater, respectively, compared with nonsmokers.

Nor are the harmful effects of smoking limited to the respiratory system. Cancers of the bladder and of the pancreas occur 7–10 times and 2–15 times more often, respectively, among smokers than among nonsmokers. Smoking increases allergic responses and destroys some of the beneficial scavenger cells in the respiratory tract. It is also a major contributing factor to heart disease. Stillbirths and vulnerable low-birth-weight infants occur more often in smoking than in nonsmoking pregnant females. In part, this is because of the carbon monoxide in cigarette smoke and the fact that CO has a much higher affinity for hemoglobin than oxygen does. Overall, smokers in their mid-twenties who smoke two packs of cigarettes per day have a life expectancy of over eight years less than nonsmokers of the same age.

Why the ranges listed above? Humans are highly variable and the conditions surrounding smoking vary. Therefore, most studies are statistical and the rates derived apply to populations, not necessarily to all individuals.

Does it do any good to stop smoking? One reason for a qualified "yes" answer is evident in the series of changes in tracheal tissue that

TABLE 14.1
Some Chemicals in Tobacco Smoke

Chemical	Characteristics and Comments
Ammonia	Cleaning fluid; part of smelling salts; irritates nasal passages
Arsenic	Very poisonous element
Carbolic acid	Germicide; powerful antiseptic; disinfectant
Carbon monoxide	Deadly gas; colorless, odorless; hinders oxygen-carrying ability of red blood cells
Carcinogens	Over two dozen cancer-producing substances
Formaldehyde	Colorless; disinfectant; a preservative for biological specimens
Hydrogen cyanide	Poisonous gas
Hydrogen sulfide	Poisonous gas
Methanol	May cause blindness, intoxication, or death
Nicotine	Deadly poison

(A)

(B)

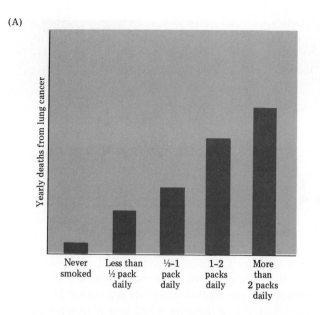

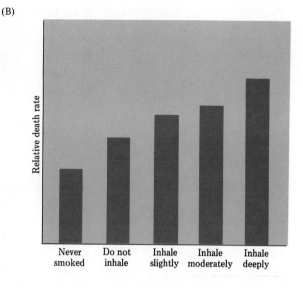

FIGURE 14.17

The relationships between (A) relative lung cancer death rates and the degree of smoking and (B) death rates related to how one smokes.

result from smoking, shown in Figure 14.18. Repair of tissue may occur if a sufficient population of healthy cells remains. The fewer such cells, then the less chance there is for complete recovery (Figure 14.19). While it helps to stop smoking, it is far better not to start. Convincing others of this is also important in order to reduce the harmful effects of "second-hand" smoke. Fortunately, thousands of people stop smoking daily and regulations increasingly restrict smoking in public places. From the bi-

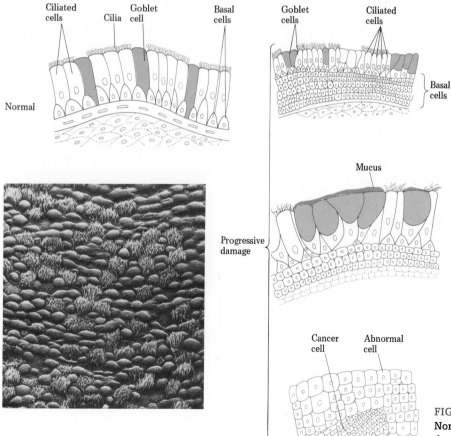

FIGURE 14.18

Normal tracheal tissue and tissue damaged by smoking.

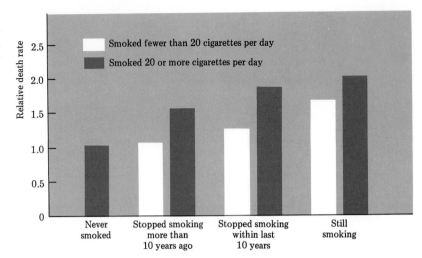

FIGURE 14.19
Comparative death rates of those who have never smoked, those who have stopped smoking for various time spans, and those who still smoke.

ological standpoint, it is surprising that so many people still smoke. However, the ambiguity of the situation is that governments provide economic support for the tobacco industry while they also warn against the health hazards of smoking.

AIR POLLUTANTS AND THEIR EFFECTS

Air pollution is caused by the release into the atmosphere of any unnatural material or by the abnormal increase of natural substances such as pollen or dust. Air pollution includes gases, aerosols, and particles, but most are invisible, odorless, and tasteless except at high concentrations. Most air pollution results from combustion, and even the burning of a common match illustrates that combustion is chemically complex (Figure 14.20).

Although air pollution is an old problem, it has taken several incidents to alert people to the problem. A Belgium **episode,** as public health officials refer to major air pollution problems, occurred on three December days in a densely populated valley. Over 60,000 people became violently ill and about 60 elderly people died from acute heart failure. The episode in Donora, Pennsylvania, occurred in late October, and nearly half of the people in the area suffered symptoms. Six thousand people became very sick, and about 20 died. A five-day winter episode

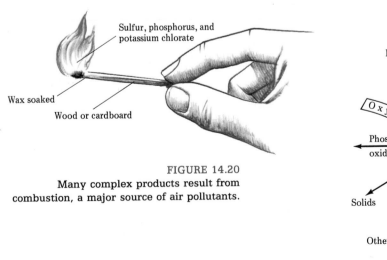

FIGURE 14.20
Many complex products result from combustion, a major source of air pollutants.

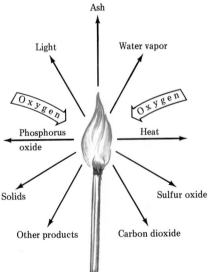

in London killed about 6000 people, and another 18-hour episode in London killed over 1000 people. Today, air pollution damage is no longer limited to isolated episodes but kills untold thousands annually, especially in urban and industrialized areas. Public health statisticians now express pollution-related deaths as "excess deaths."

Table 14.2 shows the main sources and types of air pollutants in the United States. The total is perhaps a metric ton for every American. Are you breathing your "share"? Although comedians in Los Angeles joke about waking up to the sound of birds coughing, air pollution is no joke. Most people are exposed to chronic rather than acute air pollution levels, and with every breath come closer to one or more of the four major types of problems described below.

The person affected by **chronic bronchitis** has a reduction or failure of ciliary action and an increase in mucus production. As the mucus accumulates and reduces the diameter of the bronchial tubes, the person develops a chronic cough and a shortness of breath, much like that of the heavy smoker.

The fine bronchioles become constricted in the person with emphysema, and this results in more air being retained in the alveoli during exhalation than is optimal. When air is inhaled and the alveoli enlarge still more, they may burst and destroy the surrounding capillaries. The loss of oxygen exchange capacity in the lungs causes chronic shortness of breath and slow oxygen starvation of the whole body.

Bronchial asthma has many causes but it is often an allergic reaction of the bronchial membranes to pollutants. The swelling membranes cause difficulty in exhaling, wheezing, and shortness of breath.

Among the many causes of lung cancer are several carcinogenic air pollutants, such as benzopyrene from coal smoke and tars in cigarette smoke.

Although there is a lot of air to dilute pollutants, most contaminants stay in the lower atmosphere. There, concentrations in the 10–50 parts per million (ppm) range are common, and concentrations of 50 to over 100 ppm are known. To give some perspective to ppm's, the following everyday comparisons are made: length—1 in. to 16 mi; money—1¢ to $10,000; area—1 ft² to 23 acres; time—1 min to 2 yr; and weight—1 oz to 31 tons. Even these extremely small amounts are measurable, are important biologically, and are the basis for many laws and regulations.

TABLE 14.2
Air Pollution in the United States

	Relative Health Effect		Emissions	
	Percentage of Total	Rank	Percentage of Total	Rank
Pollutant				
Sulfur oxides	34	1	15	2
Particulates	28	2	5	5
Nitrogen oxides	19	3	13	4
Hydrocarbons and volatile organic compounds	18	4	14	3
Carbon monoxide	1	5	53	1
Source				
Stationary fuel combustion	44	1	21	2
Industry	27	2	16	3
Transportation	24	3	55	1
Solid waste disposal	3	4	2	5
Miscellaneous	2	5	6	4

Carbon Monoxide

Because carbon monoxide is a colorless, odorless, and tasteless gas, we cannot sense it. CO forms a stronger and more permanent bond with hemoglobin than oxygen does. Figure 14.21 shows that the effects of CO are both time and concentration related.

Above a concentration of about 20–30 ppm, CO causes headaches, fatigue, dizziness, and slowed reactions. These concentrations are common at busy intersections in urban areas, and higher concentrations are typical behind a car stopped for a traffic light. The danger level of CO for any extended exposure is about 100 ppm. Such levels may accumulate in a closed garage when a car's engine is running and in tunnels, public parking garages, and heavy traffic. Cigarette smoke may contain over 40,000 ppm of CO, so even nonsmokers in smoke-filled rooms or vehicles may breathe harmful levels of CO.

Cigarette smoking by people in air-polluted cities is an excellent example of a synergistic effect. Moderate smokers may inhale 20–30 ppm of CO from smoking and 20–30 ppm of CO from the surrounding air. The synergistic effectiveness of these two sources of CO may be at or above the danger level of about 100 ppm.

Hydrocarbons

Hydrocarbons are complex organic compounds that contain hydrogen and carbon. Gasoline, methane, toluene, butane, pentane, benzene, propane, and ethylene are common examples. However, tens of thousands of other hydrocarbons are known and those most important in air pollution are gases within normal temperature ranges. They enter air from liquids that vaporize quickly, from tire wear, and from incomplete combustion. Some have bad odors, many irritate mucous membranes and the respiratory system, and some are carcinogenic. In the presence of sunlight, hydrocarbons combine with some other air pollutants to produce the familiar hazy, **photochemical smog** (Figure 14.22).

Nitrogen Oxides

The several **oxides of nitrogen** are abbreviated NO_x. Nitric oxide (NO) is colorless and odorless, but nitrogen dioxide (NO_2) has a reddish-brown color and a pungent odor. Nitrogen and oxygen do not react readily at room temperatures but do so at the higher temperatures of many industrial processes and when many fuels are burned, such as in internal combustion engines.

Urban atmospheres often contain 10–100 times more NO_x than rural atmospheres. Many of the harmful effects of NO_x result after it forms photochemical oxidants. NO_2 is about four times more toxic to humans

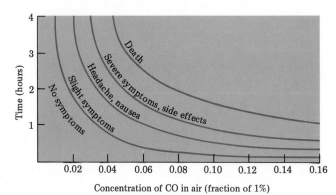

FIGURE 14.21
The time and concentration effects of carbon monoxide upon health.

than NO, with most of the damage being done to the breathing system. Part of this damage results when NO_x combines with moisture to form nitric acid. NO_x may also be carcinogenic, and levels of 5 ppm, common in heavy smog, are dangerous. Cigarette smoke contains 250 ppm.

Sulfur Oxides

Sulfur oxides (SO_x) usually form when fuels containing sulfur are burned. Sulfur dioxide (SO_2) and sulfur trioxide (SO_3) are mainly involved in air pollution. SO_x are not particularly toxic by themselves but they combine readily with water to form sulfuric acid, which irritates the eyes and damages mucous membranes and the breathing system. Many physicians believe that SO_x are the most damaging air pollutants overall. At high concentrations the gases have a pungent odor and can be tasted. See the essay "Acid Rain" for information on other biological effects of NO_x and SO_x.

Particulates

Common **particulates** include dust, pollen, soot, ashes, and smoke. Most particles are one-hundred-thousandth to one-half the size of the period at the end of this sentence. Despite this, dust falls amounting to hundreds of kilograms per square kilometer per month are common in many cities (Figure 14.23). Like most gaseous air pollutants, many particles result from combustion, evaporation, and friction. Because of the large variation in the size and chemical properties of particles, a general discussion is not easy. Particles damage the breathing system directly or indirectly because they may carry toxic materials such as trace metals. The black lung disease of miners is a well-known form of particulate damage.

Other Air Pollutants

Some serious air pollutants not listed in Table 14.2 are described briefly below.

Ozone. Ozone is O_3, in contrast to the O_2 we normally breathe. In a layer about 25–45 km (15–28 mi) above Earth, ozone filters out much dangerous cosmic and ultraviolet radiation. Without this protective layer, human and much other life would be in danger, and we would see a dramatic rise in cancer, especially skin cancer. Some constituents used as propellants in spray cans are suspected of leading to the depletion of the ozone layer, a process that could continue for years even if all usage of the materials stopped today.

Although ozone is helpful high in the atmosphere, it is harmful near Earth's surface. Ozone is colorless but has a pungent odor, and perhaps you have smelled O_3 during a major thunderstorm. Ozone causes headaches, coughing, and severe fatigue, and it impairs vision and lowers the body temperature.

Asbestos. Asbestos was used in many products, such as insulation, roofing, ceiling and floor tiles, and brake linings. Asbestos fibers are still found in some environments even though asbestos usage is reduced and major efforts have been made to remove or cover asbestos-containing objects. Asbestosis has been especially common among construction and many industrial workers. The fibers that collect in the lungs, such as on tar deposits in the lungs of smokers, cause many problems, including cancer.

FIGURE 14.22
Photochemical air pollution is often called smog in the western United States but, more politely, simply haze in the East. This type of air pollution is especially common in areas with much sunshine.

FIGURE 14.23
Particulate air pollution is especially common in industrial areas. As this old woodcut shows, the problem is not new.

Acid Rain

Both sulfur and nitrogen oxides mix readily with water in air to form sulfuric and nitric acids, respectively. Breathing acidic air damages the respiratory system, and this contributes to bronchitis, asthma, and emphysema. However, some effects of acid rain affect humans more indirectly. Fish catches are declining in some areas because of the low pH of water, and thousands of lakes that are otherwise still beautiful esthetically are largely "dead" biologically due to **acid rain** (Figure B14.6). Acid precipitation also damages many plants, including crops, and usually reduces soil fertility.

Ironically, the northeastern United States, which has been especially damaged by acid rain, is not the main source of the pollutants that cause the problem (Figure B14.7). Most of these pollutants originate in industrial and coal-burning regions to the south and west. Taller stacks used there to reduce local air pollution are allowing pollutants to travel more widely. The dying lakes and forests of New England may result from power plants in Indiana or Ohio, for example.

Rivers have fairly definite boundaries, watersheds can be recognized easily, and the pathway of aquatic pollutants can often be traced. However, we are just beginning to think in terms of the more indistinct "airsheds," and this raises many questions. Should a resort owner next to a dead lake in upper New York, or a timber company owning a dying forest in Vermont, or a victim of air-pollution-induced emphysema in Massachusetts be able to sue a West Virginia industry, for example, for damages due to its role in acid rain and air pollution?

(A)

(B)

FIGURE B14.6
Acid rain affects many materials, such as this statue (A), as well as organisms, such as these dead trees on a New England mountain (B). Acid rain-affected lakes may remain beautiful as well as suitable for many forms of recreation, but they are often largely dead biologically.

FIGURE B14.7
Partly because many air masses converge on the northeastern United States, it is especially susceptible to acid rain and air pollution damage. Tall stacks often carry pollutants away from the areas where they are produced and lead to effects long distances away.

Metals. Metals include mercury, beryllium, cadmium, nickel, lead, boron, germanium, arsenic, selenium, yttrium, vanadium, and antimony. They come from many industrial processes and the burning of leaded gasoline. "Get the lead out" is not only a common phrase but also a public health necessity.

More is known about the effects of lead than about most of the other metals. Like most other trace metals, lead may occur in food and water as well as in the air. However, about half of inhaled lead enters the body, compared with only about 10 percent of the lead in food and water. Lead is especially harmful to children, and hundreds of thousands of American children are treated for lead poisoning annually. Thousands of children suffer moderate to serious brain damage, hundreds develop other major problems, and some die.

Because most metals are not needed for normal metabolism and are not excreted easily, many accumulate in the body. Although many organs may be harmed, the bloodforming (hematopoietic) and nervous tissues are especially damaged.

SUMMARY

1. **Respiratory systems** include all structures involved in exchanging gases with the environment. In most animals, **gas exchange** is one of the most regular body functions. Metabolic rates are often calculated by measuring O_2 consumption or CO_2 production. Partly because of changing surface area-to-volume relationships, **breathing rates** tend to decrease with age.

2. Gas exchange occurs by diffusion in all animals and a wet surface must be present. This creates fewer problems for aquatic than for terrestrial animals. However, warm and salt water hold less oxygen than cold and fresh water, respectively.

3. Many tiny animals exchange gases directly with their environments without special respiratory surfaces. A flat shape facilitates gas exchange in somewhat larger animals. **Gills** are branching extensions of the body surface for gas exchange with water. Countercurrent flow of blood and water facilitates gas exchange in some gilled animals. **Tracheae** are systems of branching tubes that are found in some arthropods. Air enters and exits the body through **spiracles** on the body surface.

4. **Lungs** are thin and wet internal gas exchange surfaces that are surrounded by capillaries. Lungs vary from simple sacs with little surface area for gas exchange to extensive, complex, and highly vascularized surfaces. Birds possess **air sacs** in addition to lungs and their respiratory systems are more efficient than those found in mammals. Even after oxygen enters air sacs and lungs it remains outside of body cells, and it must still be absorbed. All but the simplest of animals possess a circulatory system to distribute oxygen and pick up CO_2.

5. Air may enter the human respiratory system through the mouth or nose; nose breathing is healthier because air is better cleansed, warmed, and humidified. Air then passes through the **pharynx** into the **larynx.** Sound is produced by vibrating **vocal cords** as air leaves the body.

6. The **trachea** is rigid due to cartilage rings that support it and keep it open. The two **bronchi** that branch off of the trachea divide repeatedly and form the **bronchioles.** The **alveoli,** the sites of gas exchange, are found in clusters at the tips of bronchioles.

7. The **pleural membrane** lines the outer surface of the lungs, the inner wall of the **thoracic cavity,** and the top of the muscular **diaphragm.** Inflammation of the pleura is called **pleurisy.**

8. Breathing depends upon differences in air pressure inside and outside of the body. Contraction of the diaphragm and the chest muscles

leads to **inhalation** by creating a pressure within the lungs that is less than outside air pressure. **Exhalation** occurs when the diaphragm and chest muscles relax.

9. The **Heimlich maneuver** is useful in clearing a food-clogged trachea.

10. The iron-containing **hemoglobin** molecule found in the red blood cells has a high affinity for oxygen and it allows the blood to carry far more oxygen than plasma alone could. Hemoglobin is only one of several **respiratory pigments** found in animals.

11. **Oxygen dissociation curves** relate the saturation of hemoglobin with oxygen to the partial pressures of gases such as oxygen. The attachment of oxygen to hemoglobin is loose and easily reversible.

12. Most CO_2 is carried in the blood plasma as bicarbonate ions, although a little is dissolved in the plasma and some is carried on the red blood cells.

13. The body's **respiratory center** is located in the medulla of the brain. The precise regulation of the rate and depth of breathing is by means of the blood CO_2 concentration. Chemoreceptors in some arteries and stretch receptors in the lungs are also involved.

14. Although the percentage of O_2 in air is relatively constant at all altitudes, the air pressure at high altitudes is too low to support efficient breathing. Water pressure forces gases into the lungs. During a rapid ascent, nitrogen bubbles may form and cause the **bends**.

REVIEW QUESTIONS

1. How and why do breathing rates change with age?

2. Why must respiratory surfaces be wet for gas exchange to occur?

3. How do the temperature and salinity of water affect its ability to dissolve oxygen?

4. Describe the interrelationships between the respiratory and circulatory systems.

5. Describe examples of surface area-to-volume relationships within the respiratory system.

6. Describe examples of countercurrent circulation patterns in animal respiratory systems.

7. Describe and explain the general trend in the evolution of lungs in vertebrates.

8. How does the respiratory system in birds differ from that in mammals?

9. What are tracheae and how do they function in gas exchange?

10. Describe the causes and effects of asthma, emphysema, black lung disease, lung cancer, bronchitis, and pleurisy.

11. Why is the esophagus a soft tube and the trachea a rigid tube?

12. Describe and explain inhalation and exhalation in human breathing.

13. Describe the Heimlich maneuver and explain its function. How does it differ from artificial resuscitation?

14. Describe the nature and function of respiratory pigments such as hemoglobin.

15. Draw oxygen dissociation curves for a small and a large mammal. Why are the oxygen dissociation curves for the fetus and the mother different?

16. Compare and contrast the transport of oxygen and CO_2 in the blood.

17. Describe the neural and chemical controls of breathing rate and depth.

18. Explain how and why the bends occurs. How can the problem be avoided?

19. What are the main sources and biological effects of hydrocarbon, nitrogen oxide, sulfur oxide, CO, and particulate air pollutants?

20. What causes acid rain, what are some of its effects, and what are some possible solutions?

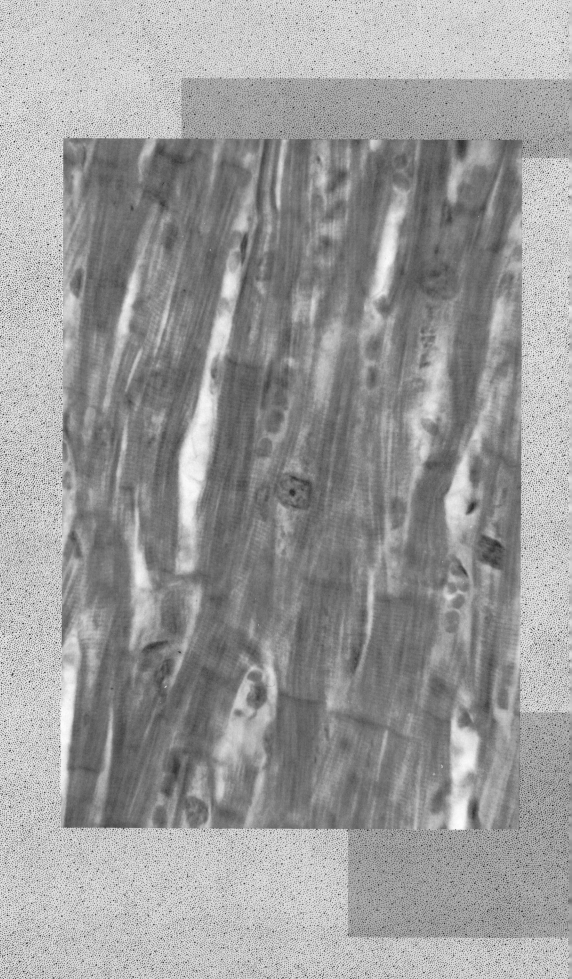

CHAPTER 15

CIRCULATION: INTERNAL TRANSPORT

Somehow blood and personality became intertwined in human thinking. Some ancient philosophers believed that various human personalities resulted from varying mixtures of four body fluids or "humors"— phlegm, yellow and black bile, and blood. Phrases such as "hot blooded," "cold blooded," "blue blooded," "red blooded," and "bad blood" are common. Some people also speak of "blood relatives" or of having a certain percentage of "Indian blood" or blood from some other racial or national group. Today, we know that genes rather than blood control inheritance. The blood of royal family members is as red as that of "commoners," and blood temperature of cruel, kind, criminal, law-abiding, or other humans varies little from person to person.

Only some relatively simple, small, and sluggish organisms lack a circulatory system to transport materials from cell to cell. In humans and other multicellular animals, blood and lymph bathe and interconnect cells; transport nutrients, wastes, many enzymes, and hormones; help protect the body against diseases; and play an important role in thermoregulation.

Cardiac muscle is unique in many respects, such as its ability to contract spontaneously. It is fast acting, somewhat like striated muscle, but it also resembles smooth muscle in that it does not tire easily and is not under significant conscious control. While smooth and striated muscles are distributed widely in the body, cardiac muscle is restricted to the heart. To appreciate the difference between cardiac and typical striated muscle, see how long you can squeeze and relax your fist, which is about the size of your heart, at a rate of about 70 times per minute.

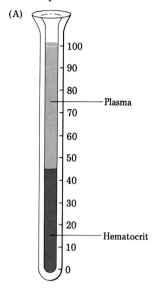

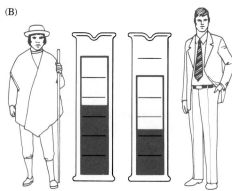

BLOOD

The pale, straw-colored fluid part of blood is called **plasma,** and the portion that is occupied by cells or so-called formed elements is called the **hematocrit.** Centrifuging is a common way of separating blood into its plasma and cellular components (Figure 15.1). In humans living near sea level, the plasma constitutes over half of the blood.

Blood makes up about 6–8 percent of the body weight and its volume equals about 5–6 L in average adults. Over 90 percent of plasma is water, and about 7 percent consists of proteins of about 70 different kinds. The proteins in this slightly alkaline fluid (average pH of 7.4) make plasma thicker than we might expect from such a watery substance. Sometimes called our "inner sea," blood contains many of the same inorganic salts as ocean water, although in different proportions. If you have tasted blood, such as when sucking a cut finger, you know its salty taste. The main components of blood plasma are shown in Table 15.1 and the cells are illustrated in Figure 15.2.

BLOOD CELLS

Some of the traits and functions of the three types of blood cells are summarized in Table 15.2.

Red Blood Cells or Erythrocytes

Because of their large number and the presence of the iron-containing protein **hemoglobin, red blood cells** (RBC) or **erythrocytes** give our blood its red color. As they mature, these biconcave disks lose their nuclei and most other organelles. The large oxygen-carrying ability of blood is due mainly to hemoglobin in the RBCs, and each of these floating sacs of hemoglobin contains about 265 million molecules of this respiratory pigment. Oxygen binding to hemoglobin is reversible and is facilitated by the large surface area of the RBCs. The oxygen content in the lungs is high, and hemoglobin "loads up" and becomes bright red there (Chapter 14). As oxygen is unloaded to body cells where the concentration is

TABLE 15.1

Components and Functions of Blood Plasma

Component	Functions
Water	Maintains blood volume and pressure; forms lymph; carries other materials
Inorganic salts (electrolytes)	Help maintain osmotic and pH balance; buffer chemical reactions; have many effects on cells
Plasma proteins	Help maintain osmotic and pH balance; participate in blood clotting; some act as enzymes; act as antibodies; help maintain blood pressure and viscosity
Nutrients (e.g., glucose, vitamins)	
Gases (e.g., CO_2, O_2)	Transport to and from cells
Hormones	
Metabolic wastes (e.g., urea)	

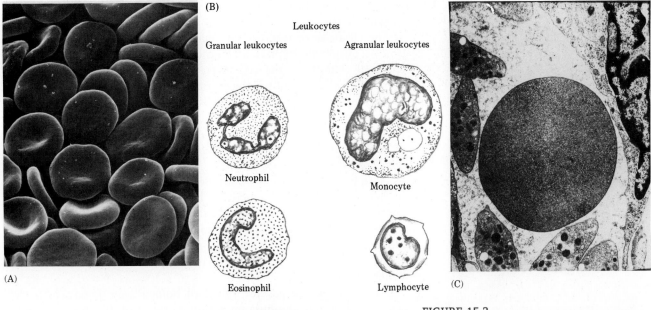

(A)

(B)

Leukocytes

Granular leukocytes   Agranular leukocytes

Neutrophil

Monocyte

Eosinophil

Lymphocyte   (C)

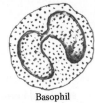

Basophil

FIGURE 15.2
The three major types of blood cells. (A)
Blood is red because of the large number of
erythrocytes and the hemoglobin they
contain. Each molecule of hemoglobin may
hold up to 4 molecules of oxygen. A human
erythrocyte loses its nucleus before it enters
the circulation from the bone marrow. The
mature erythrocytes of most nonmammalian
vertebrates possess a nucleus, and most
have different cell shapes. (B) The various
types of leukocytes, some almost twice the
size of others, are recognized by their
staining qualities and the shape of their
nuclei. (C) Thrombocytes or platelets, so-
called because of their shape, are
noncellular, colorless disks that are smaller
than erythrocytes.

low, blood becomes maroon in color. It is estimated that without hemoglobin the body would need about 65 times more blood to supply the
tissues with oxygen!

Erythrocytes form from specialized nucleated cells (erythroblasts) in
the red bone marrow at a rate of over 200 million per second. In children
there is blood-forming or **hematopoietic** tissue in many bones. However,
during development the red bone marrow in most bones is replaced by
fat and becomes yellow marrow, and most hematopoietic tissue in adults
is in the ribs, sternum (breastbone), and vertebrae. In the embryo, before
bones mature, the liver and the **spleen** produce RBCs. The adult spleen
stores blood and squeezes it into the general circulation when it is

TABLE 15.2
Blood Cells

Cells	Typical Diameter (micrometers)	Number per Microliter	Main Functions
Erythrocytes	7	4.2–5.4 million (males); 3.6–5.0 million (females)	Oxygen transport; some CO_2 transport
Leukocytes	8–20	5,000–10,000	Defense against pathogens and foreign substances
Thrombocytes	1–2	150,000–400,000	Blood clotting

needed, such as during an accident when extensive bleeding or **hemorrhaging** occurs.

The elasticity of the RBC plasma membrane is gradually lost. This process during the typical lifetime of 4 months occurs as a result of chemical changes in the cell surface and because of being squeezed, bumped, and bent during each trip through the smallest capillaries. During its existence, an average RBC travels about 1100 km (685 mi). Old erythrocytes are destroyed in the liver by phagocytic cells and much of their iron is recycled to the bone marrow. Some breakdown products of hemoglobin (e.g., bilirubin) are expelled with the feces, giving them their reddish-brown color.

Note in Figure 15.1 how altitude affects the quantity of blood and its hematocrit. The hematocrit in high-altitude-adapted humans may approach 60 percent, whereas it is about 45 percent in humans adapted to sea-level conditions. The high-altitude humans also produce more blood to assist them in surviving in the rarefied atmosphere of their environment. Similar trends occur in many other mammals and other vertebrates.

According to the general model of homeostasis (Figure 2.5), each body function has an optimum within its range of tolerance. Thus, too little or too much of a substance or too slow or too fast a rate of metabolism may have serious effects. This concept applies very clearly to our circulatory system, and especially to the numbers of erythrocytes and leukocytes. For example, under the low-oxygen conditions of high altitudes, the marrow produces blood cells more quickly and the liver destroys them more slowly than under the high-oxygen conditions of low altitudes.

Millions of people donate blood annually and the loss of the usual pint of blood does no harm to most people. The plasma is made up within the body in about two days and the RBCs within about seven weeks. American Red Cross donors are typically allowed to donate blood once every eight weeks but no more than five times in a year. Some people have donated over 50 L of blood in their lifetimes.

As with every homeostatic process, RBCs sometimes become too concentrated. This usually results from overproduction or from loss of plasma, and in both cases the blood becomes viscous and this may clog small blood vessels. It may also result from "blood doping," such as when athletes store some of their blood and inject it prior to competition.

Anemias. **Anemia** is a general term for a variety of conditions that result from a lack of hemoglobin or from abnormal numbers of normal erythrocytes. Insufficient numbers of erythrocytes may be produced, such as when too little iron is available or when marrow degenerates. In addition, erythrocytes may be destroyed too quickly, their hemoglobin may be abnormal (as in sickle-cell anemia), or loss of erythrocytes by hemorrhage may occur. Several of the most common forms of anemia are described below.

Iron-Deficiency Anemia. Inadequate dietary iron or the inability to absorb iron properly may inhibit normal hemoglobin production. If chronic, especially if coupled with a bleeding ulcer or excessive menstrual bleeding, this anemia can be serious.

Giant-Cell Anemia. The rate of cell division during the production of RBCs is one of the most rapid such rates in the body. When certain substances needed for RBC production are deficient, the RBCs formed may be abnormally large and fragile. This form of anemia (megaloblastic) results from a deficiency of either folic acid or cyanocobalamin (vitamin B$_{12}$). The latter problem (once called pernicious anemia) is usually due to the inability to absorb cyanocobalamin rather than to a lack of the vitamin in the diet. Patients are now usually treated easily by injections of vitamin B$_{12}$. Cures for folic acid deficiencies, which are most common in infants and in pregnant women during the last trimester, usually take just a few weeks.

Sickle-Cell Anemia. Sickle-cell anemia and sickle-cell trait are due to an abnormal form of hemoglobin (Chapter 9). The RBCs are fragile, misshapen, and have their oxygen-carrying capacity reduced. The incidence of this form of anemia is higher in blacks than in members of other races.

Hemolytic Anemia. In this condition the RBCs survive for only a short time. Causes include poisons, immune disorders, inherited defects, and Rh incompatibility of the mother and fetus (Chapter 16).

Aplastic Anemia. This anemia results when the marrow degenerates so that few or no RBCs form. It is usually caused by overexposure to radiation (e.g., X-rays, radioisotopes, or nuclear fallout) or by certain drugs.

Blood-Loss Anemia. During hemorrhaging, the blood flow to the heart is reduced, the cardiac output declines, and blood pressure drops. Contraction of the spleen, a reduction in the amount of urine formed, and movement of the lymph into the blood vessels help restore blood pressure. The latter is why those who suffer a large blood loss are so thirsty. Various reflexes work to keep the blood flow to the brain and lungs normal. Thus, blood flow to many tissues is reduced so an adequate flow to the vital organs is maintained.

Some Environmental Villains.   Remember the nonsense sentences of the Mad Hatter in *Alice in Wonderland*? At one time, hatters often suffered brain damage from breathing the mercury fumes that were used to help shape felt. Mercury and other heavy metals, especially lead, also tend to accumulate in the bone marrow and interfere with the formation of RBCs.

Another environmental contaminant that seriously affects oxygen transport is carbon monoxide (CO). The affinity of CO for hemoglobin is over 200 times as great as oxygen's affinity (Chapter 14).

White Blood Cells or Leukocytes

The **leukocytes** are larger but far less numerous than the erythrocytes. Five main types of **white blood cells** (WBCs) are recognized by their staining qualities and the shape of the nucleus. Despite their name, WBCs are largely colorless and transparent, although some contain cytoplasmic granules (granulocytes; Chapter 16).

WBCs have flexible membranes and some can move like amoebas (Figure 15.3). This allows them to squeeze between the cells of thin-walled capillary and lymphatic vessels and, like bloodhounds on a trail, follow traces of chemicals into regions containing pathogens, other foreign substances, or damaged cells. Over 60 percent of WBCs are the **neutrophils** that function mainly in protecting us against infections by engulfing pathogens and cellular debris. **Monocyte** WBCs also move to infection sites, but it takes several days for them to accumulate in large numbers. Thus, monocytes are usually associated more with chronic infections rather than acute ones (Chapter 16).

A rapid rise in the white blood cell count is usually an indication of an infection. Often the infection is internal, such as a bruise. Neutrophils

FIGURE 15.3
(A) Sensing foreign substances or injured cells, this leukocyte is leaving a capillary. (B) At the site of the infection or tissue damage, specialized leukocytes engulf particles as amoebas do. Thus, they constitute one important element in our various defense mechanisms.

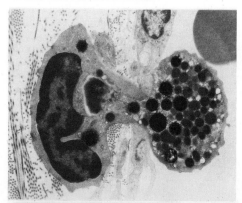

(A)

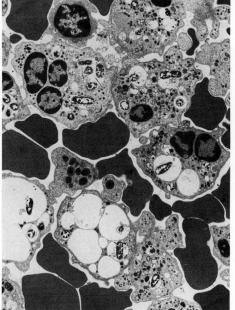

(B)

FIGURE 15.4

(A) Erythrocytes are seen here enmeshed in a network of fibrin. The complex process of blood clotting involves numerous factors, including vitamin K and calcium. A missing or faulty factor may prevent or slow clotting, as in hemophilia. Obviously, clotting must be prevented if whole blood is being stored for transfusions. This is often done by adding agents (anticoagulants) that sequester calcium. (B) The simplified sequence of events in blood clotting.

(A)

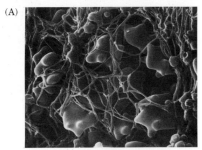

(B)   Platelets in blood

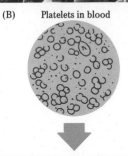

Thromboplastin

Prothrombin

Thrombin

Fibrinogen

Fibrin

Fibrin net

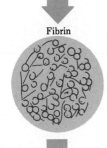

Trapped red blood corpuscles

move to the area to engulf dead cells, and in the process the color of the bruise changes from blue to yellow to normal due to the breakdown of blood pigments.

An interesting trait found in about 3 percent of neutrophils is a drumstick-shaped extension of the chromatin of the XX chromosome pair, and this trait can be used to identify females and several genetic syndromes (Chapter 9).

About 2 percent of WBCs are **eosinophils,** cells that increase during allergic reactions or infections by parasites. The role of the **basophils,** the least common of the WBCs, is poorly known, but they may produce an anti-clotting substance. **Lymphocytes** play a central role in immunity and are more common in lymph than in blood.

In general, blood diseases may be caused by problems in the blood itself or they may reflect diseased nonvascular tissue. For example, both blood cancer, or **leukemia,** and **leukocytosis** may lead to high counts of WBCs. In the former case, large numbers of immature WBCs are produced and released into the blood. The latter disease is caused by an infection, such as an inflamed appendix. Unlike most cancers that form solid tumors, leukemic cells are spread throughout the bloodstream. There are several forms of leukemia depending upon the sites of origin of WBCs and upon which WBC is transformed.

Like most diseases, leukemias may be chronic or acute. People with chronic leukemia may produce large numbers of fairly normal WBCs for many years before serious damage results. The rapidly dividing WBCs of acute leukemia tend to spread or metastasize to other organs, and death may occur within months. Leukemias eventually invade surrounding tissues and, for example, bone is weakened and fractures easily when cancerous cells spread to it from the bone marrow. Anemia and poor blood clotting are also common symptoms when leukemic cells crowd out developing erythrocytes and thrombocytes in the bone marrow. Normal tissues everywhere deteriorate because of the high metabolic demands of the growing cancer cells. Finally, the many abnormal WBCs do not function normally to combat infections, engulf debris, or produce protective antibodies. Thus, those with leukemia are affected by many ailments that they would usually be able to control otherwise. Historically, the development of leukemias among radiologists and dental technicians was an early indication that radiation caused some kinds of leukemias.

A deficiency of WBCs often happens in people with tuberculosis or viral pneumonia, and the low WBC count is an aid in diagnosing these diseases. An abnormally large number of monocytes is typical of **mononucleosis,** the cause of which is probably viral. Symptoms of this relatively common disease of young adults include swollen lymph glands, fever, and fatigue. Rarely fatal, this disease may last from one to several weeks.

BLOOD CLOTTING

Platelets or **thrombocytes** are cell fragments that break off of very large bone marrow cells (megakaryocytes). Although they have a plasma membrane, these small structures lack a nucleus. Their crucial function is in the formation of blood clots (Figure 15.4). The chain of chemical events that results in most clots forming within four to ten minutes after an injury typically begins when thrombocytes encounter rough surfaces such as damaged tissue. They and the damaged cells release substances (thromboplastins) that convert a blood protein (prothrombin) into an active form (thrombin). That, in turn, converts a soluble plasma protein (fibrinogen) into an insoluble protein (fibrin) which forms the fibrous network that traps erythrocytes and thrombocytes to form the clot. Fibrin also shrinks as it forms, thus pulling together the rough edges of a wound.

Vitamin K is necessary for the production of blood-clotting agents in the liver. We cannot make our own vitamin K but we usually get enough from the bacteria that live in our colon (Chapter 13). However, prolonged antibiotic treatment or chronic alcoholism may kill these useful bacteria and thereby interfere with blood clotting.

Clotting prevents excessive external blood loss but may cause problems if it occurs within blood vessels. Normally, the smooth lining of blood vessels prevents platelets from adhering and breaking down. Also, blood contains some **anticoagulants** or anticlotting agents. However, roughened surfaces of blood vessels may be caused by **atherosclerosis**, or hardening of the arteries, or by injury or inflammation, and internal blood clots may form. The type of blood clot called a **thrombus** sometimes enlarges to block the blood flow to a tissue. This may lead to certain kinds of heart attacks and strokes (e.g., coronary thrombosis and cerebral thrombosis, respectively). A clot that breaks free and floats away to clog a blood vessel elsewhere is called an **embolus**. (An embolus may also be caused by a gas bubble or a foreign body such as a mass of bacteria.)

Obviously, clotting must be prevented when donated blood is stored. Although whole blood to be used directly for transfusions is usually stored for only a few weeks, many useful chemicals can be extracted from blood that is too old to be used for transfusions.

THE BASIC DESIGN OF THE CIRCULATORY SYSTEM

The circulatory system consists of the heart, various vessels, and blood and lymph. The system is often likened to a plumbing system, but although there are similarities, the former is far more complicated. However, the basic design of the circulatory system is relatively simple (Figure 15.5). Each of the four heart chambers has a large vessel attached to it. Oxygen-poor blood is pumped to the surrounding lungs, where it picks up oxygen and releases CO_2. After returning to the heart, this blood flows to the body and slowly releases its O_2 to and receives CO_2 from the cells.

Although today we are all aware that blood circulates, this was not always common knowledge. Prior to 1628, when William Harvey (1578–1657) published his book on hearts and the circulation of blood, it was generally believed that blood was immobile or that it moved only sluggishly and erratically. Aristotle believed that blood formed in the liver and moved to the heart, but Harvey doubted that livers could produce the quantities of blood that he found while dissecting many kinds of animals. Many once believed that the beating of the heart and the pulses resulted from the "vital spirits" that they contained. It was also generally thought that there were two kinds of blood, one corresponding to what we know now to be the pulmonary circulation and the other to the systemic circulation. However, Harvey's study of the valves in the heart, in the arteries that left the heart, and in the veins convinced him that we have one supply of blood that is driven by the heart into the arteries and that returns in the veins. Some of Harvey's simple but elegant observations on the functions of valves are shown in Figure 15.6.

FIGURE 15.5
The human circulatory system is typical of that in most endotherms. Deoxygenated blood (lighter color) from the body returns to the right atrium through the vena cava. After passing into the right ventricle, it is pumped through the pulmonary loop for gas exchange in the lungs. Oxygenated blood (darker color) returns to the heart's left atrium, passes into the left ventricle, and is pumped through the extensive systemic loops. In each tissue and organ, the arteries branch into smaller arterioles and then into capillary networks where gas, nutrient, waste, and other exchanges occur. From the capillaries, oxygenated blood flows back toward the heart in venules and then in veins, including the vena cava, thus completing the circuit.

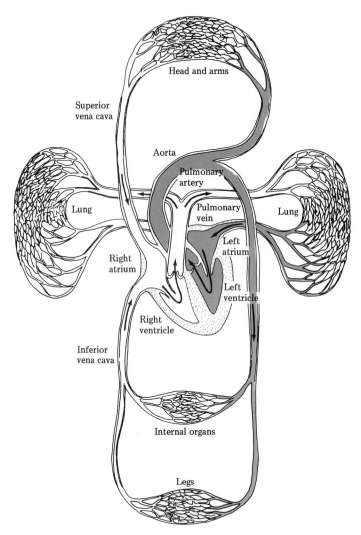

While Harvey established the concept of blood circulation, he did not live to see the physical connection between arteries and veins. Capillaries were discovered and described (by Marcello Malpighi) four years after Harvey's death. See the essay "Comparative Circulatory Systems" for a discussion of circulation in other animals.

BLOOD VESSELS

If our blood vessels were laid end to end, they would stretch about 100,000 km (62,000 mi), nearly one-fourth of the way from Earth to the moon. The structures of the three main types of blood vessels are shown in Figure 15.7.

The heart and large vessels are important largely in terms of the thin-walled **capillaries,** where the major exchange functions of the system

FIGURE 15.6
An example of William Harvey's observations. (A) Harvey tied a cloth around a subject's upper arm and the subject gripped an object to make the veins in the forearm bulge. When a finger was pressed on a vein, no blood was visible between that point and the next bulge, the location of a valve. (B) As long as the first finger continued to apply pressure, no blood could be pushed back into the emptied vein from beyond the next valve. (C) Similarly, blood did not return to the emptied vein when pressure was applied in the opposite direction.

(A)   (B)   (C)

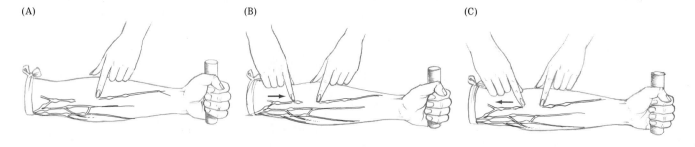

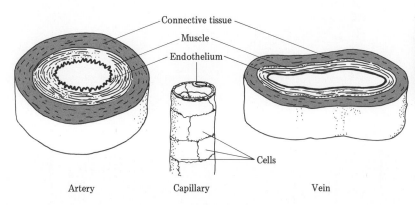

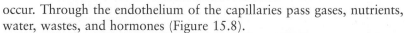

Artery   Capillary   Vein

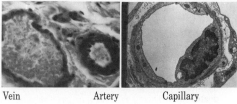

Vein   Artery   Capillary

FIGURE 15.7
The structure of blood vessels. The exchange of gases, nutrients, and wastes takes place across the thin walls of the capillaries. Veins are usually larger than arteries. For example, the vena cava, the largest vein, has a diameter of about 3 cm, compared with the 2.5-cm diameter of the aorta, the largest artery. Small capillaries may be only 8–10 micrometers in diameter, just wide enough for the erythrocytes to squeeze through, usually in single file.

occur. Through the endothelium of the capillaries pass gases, nutrients, water, wastes, and hormones (Figure 15.8).

An important part of homeostasis is **vasomotion,** the ability to respond to the constantly changing vascular needs of various parts of the body. Tiny rings of smooth muscle called **precapillary sphincters** control the flow of blood into parts of capillary beds depending upon local metabolic needs (Figure 15.9). For example, the blood flow into an active skeletal muscle may increase 20 times or so over the flow into a resting muscle. The phrase "cold feet" commonly describes the circumstance when we are frightened and when much blood flows to the heart, lungs, and skeletal muscles to prepare us for the flight-or-fight situation. As a result, the hands and feet may cool due to the reduced blood flow (vasoconstriction) to the skin.

The body's blood volume is much less than the total capacity of the blood vessels if all vessels were dilated at a given moment. If all capillaries were open at once there would be insufficient blood flow to the brain and death could result. This is what happens during **shock,** and is why the condition is so serious.

The walls of the **arteries** that carry blood away from the heart and the **veins** that return blood to the heart are thicker than those of capillaries. In addition to the thin endothelium that lines their inner surfaces,

FIGURE 15.8
(A) Some so-called thoroughfare capillary beds exchange little material with the surrounding extracellular fluid, but they allow blood to flow quickly through relatively inactive tissues. (B) Other capillary beds provide very large surface areas for the exchange of many substances with the slow-moving blood. (C) Erythrocytes can be seen lined up in this small capillary.

(A)

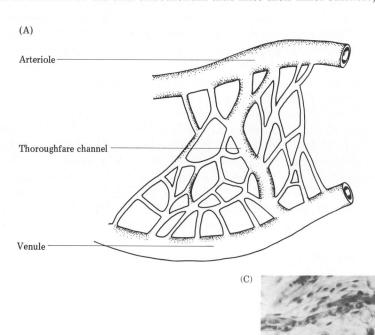

(B)

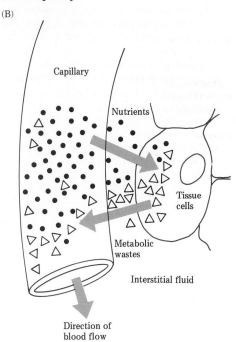

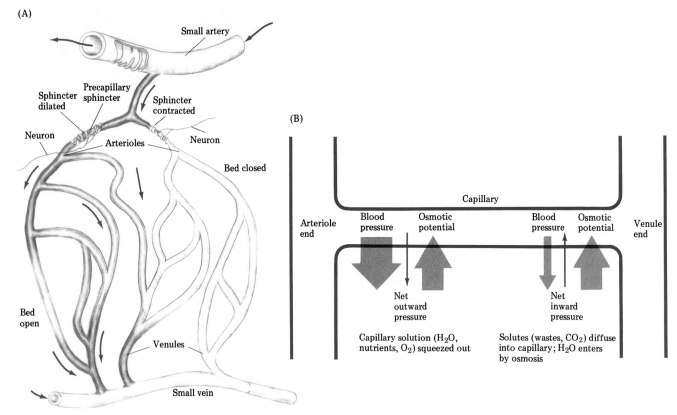

(A)

Small artery

Precapillary sphincter

Sphincter dilated

Sphincter contracted

Neuron

Arterioles

Neuron

Bed closed

Bed open

Venules

Small vein

(B)

Capillary

Arteriole end

Blood pressure

Osmotic potential

Blood pressure

Osmotic potential

Venule end

Net outward pressure

Net inward pressure

Capillary solution (H_2O, nutrients, O_2) squeezed out

Solutes (wastes, CO_2) diffuse into capillary; H_2O enters by osmosis

FIGURE 15.9

(A) Contraction of precapillary sphincters at the junctions where capillaries branch into arterioles slows or blocks the blood flow into regions fed by those capillaries. This may occur because other parts of the body are more active metabolically. The precapillary sphincters in the latter locations relax to increase the blood flow into the capillaries. Both nerves and hormones are important in the control of precapillary sphincters. (B) Due to differences in blood pressure and osmotic potential, nutrients, oxygen, and some water tend to move out of the capillaries close to the arteriole end, and wastes, CO_2, and other water molecules tend to enter near their venule end.

arteries and veins are also surrounded by a middle layer of smooth muscle and elastic fibers and an outer layer of elastic connective tissue that contains collagen and other supportive compounds. To withstand initial high pressure of surging blood leaving the heart, the arteries have thick but resilient walls. Most arteries are also buried deeply in the tissues for protection. The thinner, more flexible walls of the veins reduce resistance to the flow of blood returning to the heart. The smaller arteries and veins are called **arterioles** and **venules,** respectively.

A weakening of the arterial wall so that it bulges is an **aneurysm,** and if the wall bursts the resulting hemorrhage may be fatal. Alternatively, the blood that flows into adjacent tissues may create pressure that injures nerves and causes paralysis. The most common site of wall weakness is in the ascending **aorta,** the body's largest artery, but cerebral vessels are also prone to this problem.

Although most blood that leaves capillaries flows into larger and larger veins on its return to the heart, there are also **portal systems** (Figure 15.10). In these areas the blood leaving the capillaries flows into veins that terminate in other capillaries before it returns via still other veins to the heart. For example, much of the blood that flows to the liver arrives via the hepatic portal vein, carrying with it freshly absorbed nutrients from the digestive tract (Chapter 13). The hepatic artery provides the liver with the oxygen it needs for its numerous metabolic activities.

When blood has reached the veins its pressure is too low to return it to the heart without help, a special problem for animals like humans who stand erect. Muscular activity, including breathing movements, must squeeze venous blood to keep it flowing (Figure 15.11). For example, the negative pressure that is created in the thoracic cavity as the diaphragm contracts, and that allows us to inhale, also creates a positive pressure in the abdominal cavity and that pushes against the veins.

Valves at regular intervals usually prevent back flow and keep the blood moving in one direction. However, the valves sometimes become weak and blood stagnates in the dilated veins. These **varicose veins** are common in the legs, especially near the skin where there is less squeezing action by the surrounding muscles. The condition is unsightly, and the pressure on adjacent tissues may be painful. Dilated veins in the rectum walls are called **hemorrhoids,** and inflammation of the inner wall of a vein is called **phlebitis.** A danger of phlebitis is that the walls may become "sticky" or rough and cause a thrombus to form within the vessel.

THE HEART

To primitive humans the heart was a source of bravery. In some societies a person was assured of courage by eating the heart of an admired animal, such as a lion, or that of an enemy. Like the blood, the heart has many personality-related phrases associated with it, such as "good hearted" or "hard hearted," and it is obvious that we do not attach similar significance to most organs. For example, imagine telling someone to have a nice trip and that your liver will go with them, or that you are broken-kidneyed over their problem! Yet, if the fist-sized, 0.4-kg (1-lb) heart is only a pump, it is a very impressive "bag" of muscles that typically moves 9500 L (2600 gal) of blood daily or, in unscientific units, about 18 million barrels of blood in an average lifetime.

The heart is located near the middle of the thoracic cavity, meaning that those holding their hand or hat over their chest during the Pledge of Allegiance or similar patriotic activity are really covering their left

FIGURE 15.10

The hepatic portal system. In addition to absorbed nutrients, substances such as some drugs, alcohol, and pesticide residues on food also travel to the liver, where they are detoxified. Another portal system is associated with the pituitary gland.

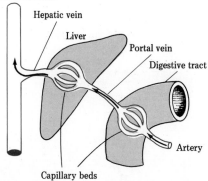

Posterior vena cava

Hepatic vein

Liver

Portal vein

Digestive tract

Artery

Capillary beds

(A)

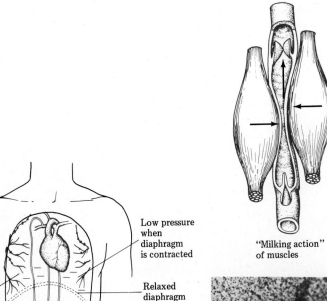

(B)

Chest cavity

Low pressure when diaphragm is contracted

Relaxed diaphragm

Contracted diaphragm

Abdominal cavity

High pressure when diaphragm is contracted

Veins and lymphatic vessels

Breathing

"Milking action" of muscles

FIGURE 15.11

Methods by which the flow of blood and lymph is facilitated. (A) When blood in veins or lymph in lymphatic vessels is squeezed, as when nearby muscles contract, it moves through open valves. (B) Breathing movements also help propel body fluids through vessels with valves. As in the case of valves elsewhere in the circulatory system, pressure on valves in the opposite direction generally closes them and prevents the back flow of blood. (C) Microscopic view of a valve within a lymphatic vessel.

(C)

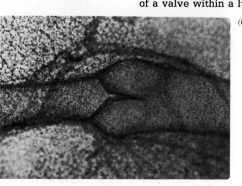

Comparative Circulatory Systems

The primitive multicellular jellyfish shown in Figure B15.1, as well as the flatworm illustrated in Figure 13.8A, have simple structures and low rates of metabolism. Like the smaller protistans, they can secure nutrients and gases and rid themselves of wastes directly across their body surfaces.

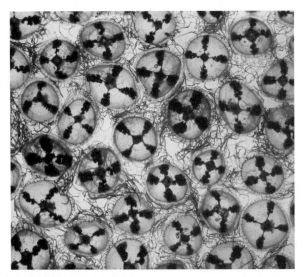

FIGURE B15.1
Although jellyfish and some of their relatives move, they have little internal complexity. Their large external and internal surfaces exposed to their environments allow them to live without a circulatory system, and their simple body fluids are not generally called blood.

Many invertebrates have **open transport systems** (Figure B15.2). In the simplest form of such systems, the blood that bathes the internal structures only moves as the animal moves. In animals with somewhat more advanced open systems, blood that forms pools around internal organs is stirred by the rhythmical filling and emptying of one or more tubular hearts. However, the increased efficiency of the circulatory system alone does not support the high metabolic rate of most insects. For oxygen support, insects generally rely more on their tracheal respiratory system than on their blood (Chapter 14). Insects also use active transport mechanisms to extract nutrients from blood. Crustaceans, which possess gills rather than a tracheal system and do not transport oxygen in their blood, are generally far less active than insects.

In animals with **closed transport systems,** blood is confined to a set of branching vessels, through which it is propelled by one or more hearts (Figure B15.3). A major advantage of closed systems is that blood can be moved quickly when conditions require it to regions that must support a high metabolic rate. Closed systems are found in some worms (earthworms), certain mollusks (octopus and squid), and vertebrates.

Pressure is important in a closed transport system (Figure B15.4). In theory, having a heart that pumps blood to gills or lungs, then to the other tissues, and then back to the heart sounds inefficient because of the drop in pressure that would occur along such a long pathway. Yet this is the design in fishes, the most numerous group of vertebrates. The large and active invertebrate squid has extra booster hearts to overcome the problem of pressure drop.

(A)

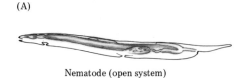

Nematode (open system)

(B)

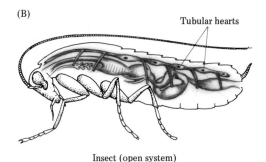

Tubular hearts

Insect (open system)

FIGURE B15.2
Open transport systems. (A) The nematode worm's muscular movements cause its blood to slosh around rather haphazardly, thus limiting the metabolic rate of these abundant and often destructive animals, some of which are endoparasites in humans. (B) The beating of the tubular heart in arthropods, such as this insect, stirs blood more efficiently.

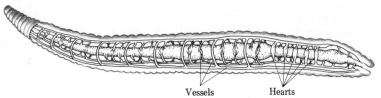

FIGURE B15.3
Closed transport systems are quite diverse and probably evolved independently in various animal groups. Earthworms have a set of top (dorsal) and bottom (ventral) longitudinal vessels that are connected by ringlike vessels, some of which are enlarged to serve as hearts. The larger vessels branch within the tissues to form networks of thin capillaries.

In most amphibians, reptiles, birds, and mammals, there are two circulatory pathways, the pulmonary and systemic loops, and a single, multichambered heart pumps the blood through the two circuits at once. Because the amphibian heart has two atria but only one ventricle, there is mixing of the oxygenated blood returning from the pulmonary loop and the deoxygenated blood returning from the systemic loop. However, ridges in the heart direct the blood somewhat so about three-fourths of the systemic blood is freshly oxygenated. Most reptile hearts are also three-chambered, but crocodile, bird, and mammal hearts have four chambers. Because the two ventricles are partitioned in the latter hearts, there is no mixing of oxygenated and deoxygenated blood.

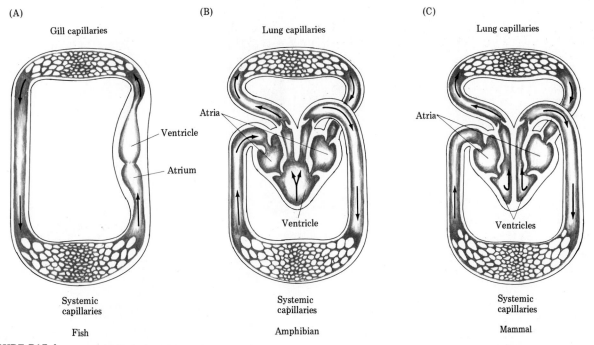

FIGURE B15.4
Comparison of three closed transport systems. (A) In fish, the two capillary networks are arranged in series, and the heart consists of only one atrium and one ventricle. (B) In other vertebrates, there are generally two circulatory loops, pulmonary and systemic, connected in parallel. In amphibians, some mixing of oxygenated and deoxygenated blood occurs in the single ventricle. (C) A partition creates two separated ventricles, and thus a four-chambered heart, in crocodilians, birds, and mammals. In these animals, there is no mixing of oxygenated blood returning from the lungs with the deoxygenated blood returning from the systemic loop.

FIGURE 15.12
The wrist and carotid pulses (A), and other sites where a large surface artery can be pressed against a bone or firm tissue to measure the pulse (B).

(A)

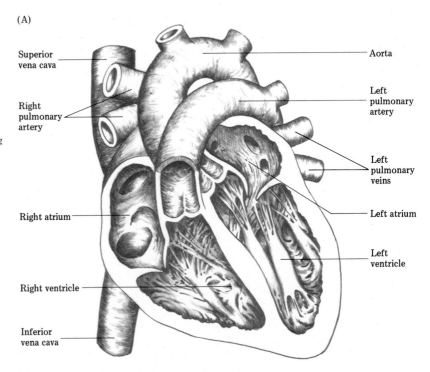

Wrist pulse

(B)

Carotid pulse

Other possible pulse–measuring sites

lung rather than their heart. The heart is surrounded by a double-walled sac (pericardium) and fluid in the sac reduces friction between tissues as the heart beats.

The heart begins beating about eight months before birth, and in some cases it can continue to beat for years after its "owner" is brain dead. The throbbing is fast in the neonate, about 135–140 beats per minute, but it slows to about 65–75 in the age range of 15–30 years, after which it increases slightly. You can measure your heartbeat rate directly or measure **pulses** in the wrist, carotid artery, or other sites (Figure 15.12). Do not use your thumb because its pulse may become confused with the other pulses.

Like the blood vessels, the heart's four chambers are lined with endothelium. The two sides of this double pump are separated by a thick muscular wall (Figure 15.13). Like every other organ, the heart must be

(A)

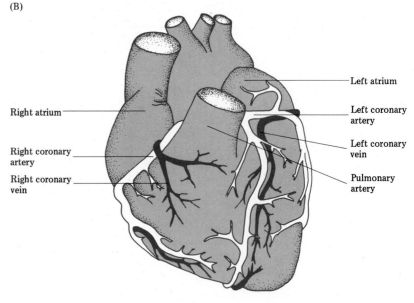

Superior vena cava

Right pulmonary artery

Right atrium

Right ventricle

Inferior vena cava

Aorta

Left pulmonary artery

Left pulmonary veins

Left atrium

Left ventricle

(B)

Right atrium

Right coronary artery

Right coronary vein

Left atrium

Left coronary artery

Left coronary vein

Pulmonary artery

FIGURE 15.13
The human heart. (A) In this cross-sectional view, note that the chambers are labeled as if the heart were in a body. Thus, the right atrium, for example, that is on the left side of the diagram would be on the person's right side. Also note the valves between the atria and ventricles and, with their ends opening in the opposite direction, in the aorta and pulmonary artery that leave the heart. (B) Part of the heart's own circulatory system is seen in this external view.

supplied continuously with oxygen and nutrients, yet the heart receives no nutrition from the 280 or so liters of blood that flow through it each hour. However, although most tissues use only about 25 percent of the oxygen supplied to them by blood, the heart uses about 80 percent of its blood-oxygen supply. This oxygen, which is supplied in the coronary arteries, provides the heart with only a minimal margin of safety. The bulk of the heart is made up of **cardiac muscle,** or myocardium, and it is supplied with more capillaries than any other tissue.

Because the function of the **atria** is simply to fill the **ventricles** below them, they are relatively thin-walled. The thick-walled ventricles provide the larger forces needed to propel blood away from the heart. However, the right ventricle pumps blood only through the pulmonary artery to the lungs, which are nearby and at about the same level. The left ventricle pumps blood through the aorta to the rest of the body and it is the most powerfully muscled of the heart's chambers.

The direction of blood flow through the heart is controlled by one-way **atrioventricular valves.** Whether a valve is open or closed depends upon the relative pressure applied to the two sides. Normally, each valve closes at the proper time to prevent the backward flow of blood, and small muscles ("heart strings") prevent the flaps of the valves from moving freely in all directions. Although defective valves may allow some back flow or reduce the normal flow, thus reducing cardiac efficiency, not all such cases are serious threats to health (Figure 15.14). These **heart murmurs** are so named because of the abnormalities in the heart sounds that can be detected by a physician. To compensate for the decrease in blood being pumped out, the heart may pump harder and faster and this strain sometimes weakens the heart. The removal of the faulty tissue and the implantation of artificial valves has reduced the problem for many thousands of people.

MAJOR HEART PROBLEMS

About 700,000 people in the United States die annually of **heart attacks,** by far America's number one killer. About two-thirds of victims die within hours, and often minutes, after the onset of the problem, which

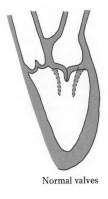

Normal valves

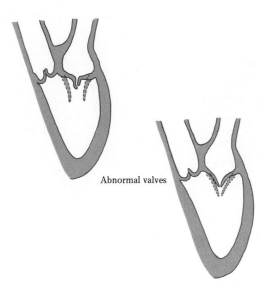

Abnormal valves

FIGURE 15.14
Some causes of heart murmurs. Because valves are sometimes of unequal sizes or are too short, some blood flows back into the atria when the ventricles contract, reducing the efficiency of the system.

usually involves blockage of a coronary artery (Figure 15.15). The cardiac muscle served by the artery dies, and ventricular muscle flutter or fibrillation is the most common cause of death following this. Many people who survive their first heart attack succumb to a later one. About 30 million Americans have some form of heart disease, and the incidence of such problems has increased dramatically this century.

The blockage of a coronary artery is often affected by atherosclerosis, which commonly involves the deposition of cholesterol and the growth of fibrous tissue or **plaques** containing calcium in the vessel wall. In time, this may reduce the vessel's diameter and increase resistance to blood flow. Also, blood clots may form on the rough inner surface of the artery. A person with a gradual closing of a coronary artery often feels intense pain or a feeling of squeezing or crushing in the chest and left shoulder and arm, a problem known as **angina pectoris**. However, damage from oxygen starvation may occur so slowly and be so localized that there are few if any symptoms. Past the age of 30 years most people have a few dead areas and patches of scar tissue in their hearts. Heart attacks come in all sizes, and small coronary blockages may be mistaken for "indigestion" pains. Oral medications such as nitroglycerin relax coronary vessels and allow oxygenated blood to reach the heart tissue and relieve the pain.

Atherosclerosis is one form of **arteriosclerosis**, a general term for hardening of the arteries or degenerative diseases of arteries. With aging, the elastic tissue within the artery wall is replaced with other less flexible connective tissue. About 80 percent of all deaths (850,000 per year) caused by cardiovascular problems in the United States are due to atherosclerosis, and about 1 in 20 Americans has heart disease due to this condition. The fact that autopsies on very young children killed in accidents have shown fatty streaks in the blood vessels indicates that atherosclerosis begins early in life and is a progressive problem. Autopsies of American casualties in Vietnam showed a 45 percent incidence of atherosclerosis. Although the average age of these men was only 22 years, 5 percent had severe cases of atherosclerosis.

In recent years many tens of thousands of American victims of coronary artery problems have undergone **bypass operations**. To avoid rejection of foreign tissue, a length of vein from the person's own leg is sewn onto the heart to bypass a blocked artery. Pain is relieved by this operation, but many of those undergoing sham operations also report less angina pain, which suggests that there may be a psychosomatic or

FIGURE 15.15

The coronary arteries that branch from the aorta and supply the heart with oxygen and nutrients are common sites for the blood clots that lead to heart attacks, such as a coronary thrombosis.

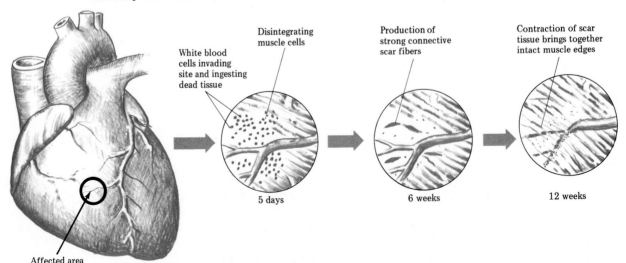

White blood cells invading site and ingesting dead tissue

Disintegrating muscle cells

Production of strong connective scar fibers

Contraction of scar tissue brings together intact muscle edges

5 days

6 weeks

12 weeks

Affected area

placebo component to this procedure, as there is to some other forms of treatment. The degree to which long-term survival is affected by this surgery alone remains to be seen.

The risk of heart attack is affected by heredity, diabetes, hypertension, age, sex (males more prone), race (American blacks more prone than nonblacks), and diet. Because cholesterol is implicated in atherosclerosis, it is commonly suggested that people eliminate or reduce foods that contain cholesterol from their diets (e.g., eggs). However, some forms of cholesterol are synthesized by the body, and these may also be deposited in the coronary arteries. The matter is not simple, and sugar, vitamin B_6, vitamin C, some trace elements, and other dietary elements have also been implicated. To further complicate the situation, there is apparently no link between cholesterol intake and the incidence of heart attacks in several human populations. However, the offspring of American victims of coronary heart disease run a risk of developing heart problems that is about five times higher than that for others. Whether this represents a genetically determined difference in body chemistry or results from learned habits of exercise, handling stress, weight control, and diet remains to be determined. Finally, smokers are more likely to develop atherosclerosis than nonsmokers.

According to some researchers, the behavior pattern known as **Type A behavior** increases the chance of developing heart attacks. Type A people are often described as perfectionists, tense, impatient, achievement oriented, energetic, and unable to relax. Conversely, **Type B** people are generally said to feel no habitual sense of urgency. Some researchers report that healthy Type A people are about twice as likely as Type B people to have a **myocardial infarction,** damaged heart muscle due to a reduction in its blood supply. The same research suggests that Type A people are about five times more likely to experience a second myocardial infarction than those identified initially as Type B people. Far more research is needed to confirm these theories and, if true, to determine how a person becomes Type A or B and why Type A behavior may predispose some to heart attacks.

CARDIAC CYCLES AND CONTROLS

When placed into an appropriate solution, an isolated frog or turtle heart may beat for hours, days, or years. Human hearts being flown from a donor in one city to a potential recipient in another city do much the same. In vertebrate hearts, the heartbeat is initiated within the cardiac muscle itself (myogenic) rather than by an outside stimulus carried by a nerve, such as in insects.

The **sinoatrial (SA) node** in the rear wall of the right atrium triggers the heartbeat (Figure 15.16). Because it can contract and initiate impulses, nodal tissue has traits common to both muscle and nerve. The SA node is commonly called the **pacemaker** because it sets the tempo for the whole heart, which beats about 70 times per minute in normal resting adults. The waves of contraction spread over the walls of both atria, causing them to contract. When the signals reach the **atrioventricular (AV) node** embedded in the partition between the atria, they are relayed to the ventricles, which contract in unison. If the SA node is damaged by disease or aging, artificial pacemakers can be used to stimulate cardiac muscle electrically (Figure 15.17).

The **cardiac cycle** consists of the contraction and relaxation of the heart and is often divided into three phases. During **diastole** the atria and ventricles are relaxed. After diastole, the atria contract in unison

(A)

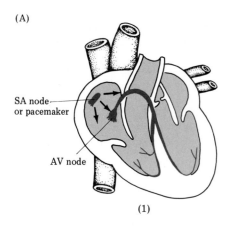

(1)

FIGURE 15.16

(A) The heart's pacemaker, the sinoatrial node, not only causes both atria to contract but also relays a signal to the ventricles by way of the AV node and other nodal tissue (bundle of His). Fibers from the nodal tissue extend to the bottom of the heart and branch to form fibers (Purkinje) that penetrate all parts of the ventricle walls. (B) Although cardiac muscle has an inherent ability to contract, many factors influence the exact pace of the heart's activity.

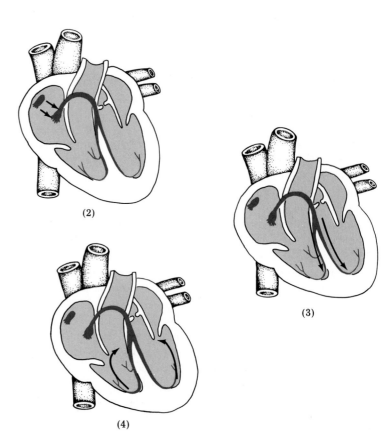

(2)

(3)

(4)

(B)

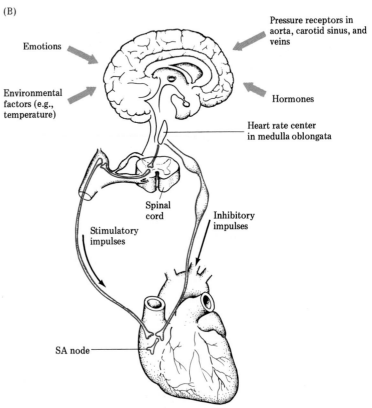

and propel additional blood into the ventricles. Heart contraction is known as **systole** and consists of atrial and ventricular phases. Atrial systole takes about 0.1 second, ventricular systole about 0.3 second, and diastole about 0.4 second. Thus, the typical resting cardiac cycle takes less than a second, with the heart at rest for about half of that time.

Closing valves make the familiar lub-dub sounds that can be heard with a stethoscope during each cardiac cycle. Other details about heart actions can be learned by means of an **electrocardiogram** (ECG), or a recording of the electrical activity of the heart (Figure 15.18).

Nervous and hormonal controls affect the heartbeat rate (Chapters 18 and 19). For example, exercise, excitement, and stress accelerate the heartbeat rate through direct stimulation by nerves or through the action of hormones such as adrenaline which reach the heart via the bloodstream. Thus, although the **heartrate center** is in the medulla of the brain, the cerebrum is also important in detecting and interpreting various emotional events.

The pressure and chemical composition of the blood also affect the heart rate. Pressure receptors in the walls of the aorta and the carotid arteries of the neck are very sensitive to blood pressure changes. High blood pressure stretches the walls of those vessels and the stimulation of the pressure receptors leads to nerve impulses being sent to the cardioinhibitory center in the medulla. Also within the aorta and the

FIGURE 15.17
An artificial pacemaker compensates for the heart's sinoatrial node failing to excite the atria and ventricles to contract. Wires are often threaded from the power supply down a neck vein, through the vena cava, right atrium, and valve, and into the right ventricle. Open-heart surgery is not needed. Some devices allow the heart to adjust to changing body activities, and some units can be worn externally.

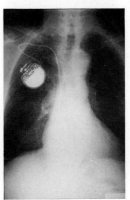

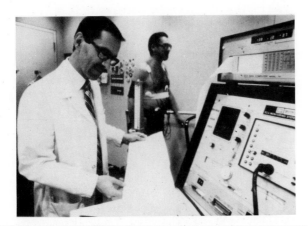

FIGURE 15.18
An electrocardiogram. Many cardiac problems can be detected by ECGs and treated before they become serious.

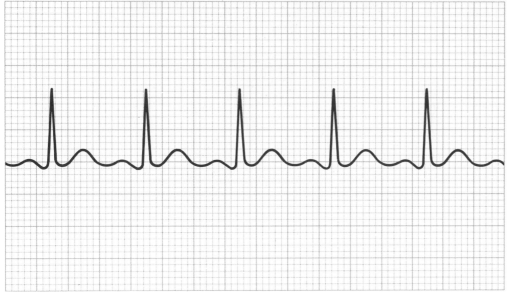

Normal electrocardiogram
(ECG)

(A)

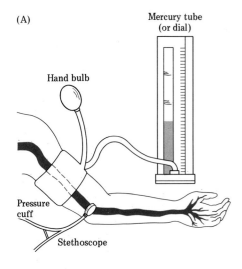

Hand bulb

Mercury tube (or dial)

Pressure cuff

Stethoscope

Artery (upper arm)

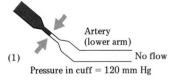

(1) Artery (lower arm) — No flow
Pressure in cuff = 120 mm Hg

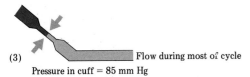

(3) Flow during most of cycle
Pressure in cuff = 85 mm Hg

FIGURE 15.19

(A) The blood pressure changes during the cardiac cycle occur as the heart contracts (systole) and relaxes (diastole). When a sphygmomanometer is used to measure blood pressure, a cuff is placed around an upper arm and inflated to prevent arterial blood flow. While listening with a stethoscope, the person measuring the blood pressure slowly releases air from the cuff. The pressure is noted both when the sounds of the pulsing blood flow appear (systolic pressure) and when they disappear (diastolic pressure). **(B)** The pattern of blood pressure and flow rates in different parts of the circulatory system. Notice the major reduction in both pressure and flow rate when the blood reaches the small arterioles and capillaries.

carotid arteries are several chemoreceptors that respond to changes in pH and to the partial pressures of oxygen and carbon dioxide in the blood (Chapter 14). High CO_2 and low O_2 levels lead to impulses to the cardioaccelerator center. Stretch receptors are also found in the walls of the major veins.

Like most other muscles, the greater the heart stretches the stronger it contracts. Thus, during sleep or rest when the heart fills slowly and receives little blood, its contraction is relatively weak. However, the heart fills quickly during exercise, the contraction is powerful, and the rate is fast.

BLOOD PRESSURE

Arterial blood pressure (expressed in millimeters of mercury, abbreviated mm Hg) reaches its peak during systole and its minimum during diastole (Figure 15.19). Blood pressure is read as systolic over diastolic, such as

(2) Flow near end of systole
Pressure in cuff = 115 mm Hg

(4) Flow is continuous
Pressure in cuff = under 80 mm Hg

(B)

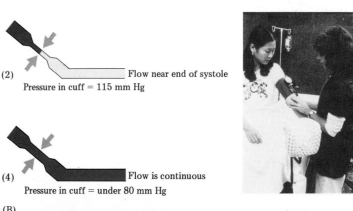

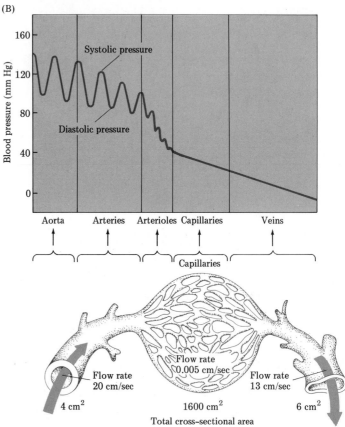

Systolic pressure

Diastolic pressure

Aorta | Arteries | Arterioles | Capillaries | Veins

Capillaries

Flow rate 0.005 cm/sec

Flow rate 20 cm/sec

Flow rate 13 cm/sec

4 cm² | 1600 cm² | 6 cm²

Total cross-sectional area

110/85 mm Hg. The diastolic pressure is not zero, as might be expected, because the thick, elastic arterial walls recoil slowly after being stretched. The highest blood pressures are found in the large arteries near the heart, and blood pressure is somewhat lower in the arm where it is commonly measured with a **sphygmomanometer** (from the Greek for "pulse"). The friction of blood within the vessel walls reduces the pressure steadily with increasing distances from the heart.

Neonates have blood pressures of about 80/45, and the blood pressures of children at the ages of 5 and 10 years average about 94/55 and 109/58, respectively. Blood pressure for adults in their twenties averages about 115/75 mm Hg for females and 120/80 for males. With age, the arteries become somewhat less elastic and blood pressure rises.

High blood pressure or **hypertension** is excessively high systolic pressure (usually above 145–160 mm Hg) or diastolic pressure (often above 90–100 mm Hg). Some physical causes of hypertension include obesity, inactivity, heavy smoking, constriction of a kidney artery, a tumor of the adrenal medulla, and an overactive adrenal cortex (Chapter 18). However, most cases of hypertension are not associated with any known physical abnormality and are called essential hypertension. The use of the word *tension* in this context has little to do with stress, although some emotional factors increase blood pressure, and the prefix *essential* in medical terminology means unexplained rather than necessary. Perhaps 20 percent of Americans have hypertension, although many are unaware of their condition.

One danger of untreated hypertension is the chronic strain it places on the left ventricle. As those muscle fibers increase in size or hypertrophy, the coronary arteries must usually supply them with more oxygen. Therefore, any decrease in coronary blood flow affects hypertensive people more than others, and a small narrowing of a coronary artery that might not seriously affect a normal person may cause a heart attack in a hypertensive individual. Hypertension also increases the chances of strokes. Numerous drugs are available to treat hypertension and many patients are also placed on a low-salt diet because salt promotes water retention by the tissues (Chapter 17).

LYMPH AND LYMPHATIC CIRCULATION

Blood pressure forces many substances out through the walls of capillaries. This fluid that bathes and nourishes cells is similar to plasma but it contains no erythrocytes or thrombocytes and much less protein. As is shown in Figure 15.20, much of this fluid is reabsorbed into capillaries before they merge into venules. When the remaining almost colorless fluid enters lymphatic vessels, it is called **lymph.** The lymphatic system is a one-way system that returns lymph to the heart via a vein (subclavian) that empties into the **vena cava.** The small, thin-walled, and highly permeable lymphatic vessels extend into most tissues and tissue fluid flows into their open ends. Like veins, lymphatic vessels have valves and lymph is also moved by muscular contractions (Figure 15.11).

Edema is an accumulation of fluid in the tissues that occurs either because of the excessive flow of fluid into the tissues or because something interferes with the drainage of fluid into the lymphatic vessels. **Elephantiasis** is an example of the latter (Figure 15.21). After mosquitoes introduce a parasitic worm into the blood, the young worms migrate to the **lymph nodes** where they mature and block lymph circulation. As tissue fluids accumulate, body parts enlarge (i.e., assume elephantine proportions), sometimes to several times their normal size. Burns often

FIGURE 15.20
(A) The extensive lymphatic system consists of a network of veinlike vessels, with valves, that return lymph to the blood. (B) Tissue fluid moves into lymphatic vessels whenever its pressure exceeds that of lymph in the vessels. (C) Lymph nodes contain sinuses within which are many phagocytic cells and leukocytes important in our defense system (Chapter 16). Everyone is familiar with the swollen, painful lymph nodes during infections.

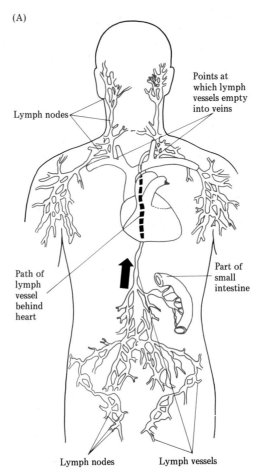

(A)

Lymph nodes

Points at which lymph vessels empty into veins

Path of lymph vessel behind heart

Part of small intestine

Lymph nodes   Lymph vessels

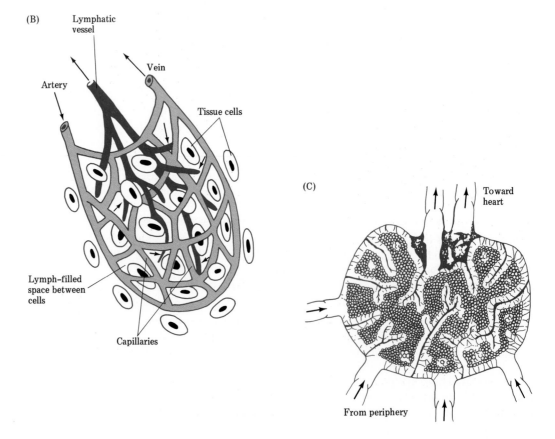

(B)

Lymphatic vessel

Artery

Vein

Tissue cells

Lymph–filled space between cells

Capillaries

(C)

Toward heart

From periphery

cause local edema, as does anything that blocks lymphatic flow, such as tight undergarments or the pressure of a growing fetus in a pregnant female. Kidney diseases may also disrupt the body's water balance and cause edema (Chapter 17). The swollen abdomen of a child with kwashiorkor is due partly to edema (Chapter 13).

The lymphatic system is extremely important in our defense system and that topic is discussed in Chapter 16.

FIGURE 15.21
Elephantiasis due to filiarial worms transmitted by mosquitoes is common in many tropical regions of Earth.

SUMMARY

1. Although **blood** is associated popularly with personality and heredity, its function is to transport gases, nutrients, hormones, many enzymes, antibodies, wastes, and to assist in thermoregulation.

2. **Hemoglobin** is the respiratory pigment that carries oxygen. The **hematocrit** is the total of all formed elements, or **blood cells,** in the blood.

3. In **open transport systems,** blood moves only as the animal moves or as it is stirred by the beating of simple **hearts.** In **closed transport systems,** as in humans, blood is confined within a set of blood vessels, through which it is moved by one or more hearts.

4. Blood vessels that transport blood from the heart are **arteries,** and those that return blood to the heart are **veins. Capillaries** are microscopic vessels that connect arteries and veins. The largest artery is the **aorta** and the largest vein is the **vena cava.**

5. Humans have **four-chambered hearts,** with the right half pumping blood to and from the lungs (pulmonary loop) and the left half circulating blood to and from the rest of the body (systemic loop).

6. The volume of blood in average humans is about 5–6 L. Mammals adapted to high altitudes have more blood and a higher hematocrit than mammals adapted to low-altitude conditions.

7. The major blood cells are the **erythrocytes** (oxygen and some CO_2 transport), **leukocytes** (defense against pathogens and foreign substances and destruction of cellular debris), and **thrombocytes** (blood clotting).

8. Many blood cells form in the red bone marrow. The liver destroys worn-out erythrocytes and recycles much of their iron to the marrow. The **spleen** stores blood and releases it into the circulation in emergencies.

9. **Blood clots** occur as a result of a series of chemical events that follow an injury. An insoluble fibrous protein traps erythrocytes and thrombocytes and prevents excessive **hemorrhaging.** Internal blood clots may clog vessels and cause damage, such as heart attacks and strokes. A **thrombus** is such a clot that stays where it forms, whereas an **embolus** is a clot that forms in one location and floats to settle elsewhere.

10. **Portal systems,** such as the hepatic portal system, occur where blood flowing in a vein enters capillary networks before it continues its return flow to the heart in another vein.

11. **Blood pressure** in veins and **lymph** pressure in **lymphatic vessels** are low, and the presence of **valves** and muscular movement are important in the return of venous blood and lymphatic fluid to the heart.

12. **Precapillary sphincters** relax or contract, thereby affecting the blood flow into capillary networks, an important aspect of homeostasis. Thus, blood can be shunted from less active areas to areas with high metabolic needs.

13. **Cardiac muscle** is active, somewhat like striated muscle, but it does not tire easily nor is it under significant conscious control, thus resembling smooth muscle.

14. Because the **atria** must fill only the **ventricles** below them with blood, they are relatively thin walled. The ventricles are thick walled, with the left ventricle being the most muscular chamber. Because of valves between the atria and the ventricles and near the openings of the pulmonary artery and the aorta, blood normally flows through the heart in one direction only. **Heart murmurs** are caused by faulty valves.

15. The **heartbeat** is initiated within cardiac muscle itself rather than being stimulated by a nerve. The **sinoatrial node** in the wall of the right atrium is the heart's **pacemaker**. The waves of contraction that start there not only affect the atria but they also stimulate the **atrioventricular node**. The AV node, in turn, leads to contractions of the ventricles.

16. The relaxed heart is said to be in **diastole** and the contracted heart is in **systole**. The high systolic and the low diastolic blood pressures that result are expressed as two numbers, such as 125/85 mm Hg. **Hypertension** is high blood pressure, usually at least 145/90 and higher.

17. The lymphatic system consists of valved, veinlike vessels that collect and return excess tissue fluid to the blood. Fluid tissue bathes cells and results from blood pressure forcing **plasma** out through permeable capillaries. Intercellular fluid is called lymph when it enters lymphatic vessels. **Edema,** the accumulation of fluid in tissues, results from excessive flow of fluid into the tissues or because the lymphatic vessels are blocked, as in **elephantiasis.**

REVIEW QUESTIONS

1. What are the major functions of a circulatory system, and what animals can exist without such a system?

2. Compare and contrast the structures of open and of closed circulatory systems.

3. Compare and contrast the functions of valves in the heart, large arteries, veins, and lymphatic vessels.

4. Relate the structures of erythrocytes, leukocytes, and thrombocytes to their functions.

5. Compare and contrast the structures of and the work performed by the atria and the ventricles.

6. Compare the work performed by the heart in moving blood through the pulmonary and systemic circulatory loops.

7. What are some adaptations of the circulatory system to mammalian life at high altitudes?

8. Describe the general features of the blood-clotting mechanism.

9. Compare and contrast the structures and functions of arteries, veins, and capillaries.

10. Distinguish between a thrombus and an embolus.

11. What is a portal system?

12. Of what importance are precapillary sphincters to homeostasis?

13. Compare and contrast cardiac, smooth, and striated muscle.

14. Distinguish between the functions of the atrioventricular and sinoatrial nodes.

15. Explain diastolic and systolic blood pressures. What levels of each generally signify hypertension?

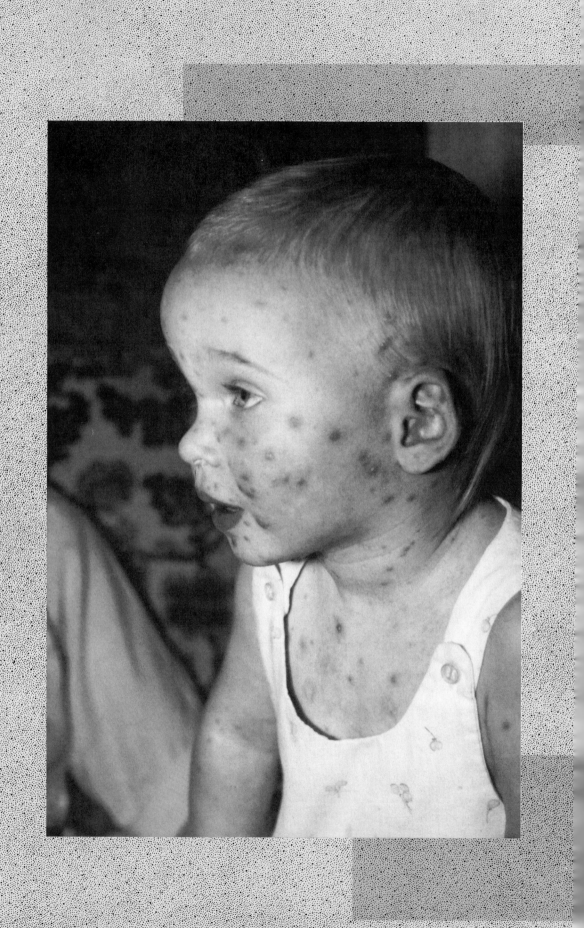

CHAPTER 16

DISEASE, CANCER, AND DEFENSE

Disease *and* cancer *are familiar but feared terms. Some diseases may be unique to certain species but others can be transmitted from one species to several others. Everyone gets sick sometimes, many recover from most such occurrences, but many others die. How do we maintain our various homeostatic balances when we are diseased? What is cancer? How do normal cells recognize themselves as different from invading organisms and from diseased or cancerous cells? What are our major defenses against cancer and disease-causing organisms? Why don't these defenses always work? Why are tissue or organ transplants sometimes rejected? This chapter attempts to answer these and many related questions.*

Chicken pox, a virus-caused disease, is not a pleasant experience for children. Fortunately, serious and lasting damage to the body from this disease is uncommon, which is not always the case with other pathogens or disease-causing agents and organisms. This chapter deals with some of the ways that pathogens enter and affect the body, and how we protect ourselves from their effects.

FIGURE 16.1
A variety of common pathogens. (A) This bacterium causes syphilis, one of the more serious sexually transmitted diseases. Another species of bacterium causes gonorrhea, which is more common but usually less serious (Chapter 8). (B) Viruses cause innumerable diseases, such as these Epstein-Barr viruses (magnified 80,000 times) that are associated with infectious mononucleosis. (C) Hundreds of millions of humans, especially in tropical developing nations, are infected with these flukes (flatworms) and suffer from schistosomiasis. (D) The tiny protozoans in some of these blood cells cause malaria, a disease that claims even more human victims than schistosomiasis. (E) The familiar "athlete's foot" problem is caused by a fungus. Many other fungal diseases harm humans in a somewhat less direct but far more extensive way—by destroying food plants, as these corn smut fungi are doing.

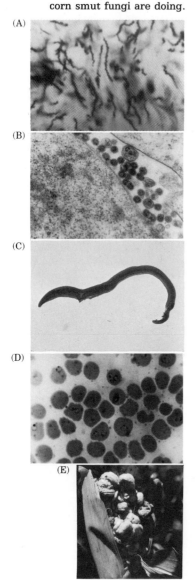

(A)

(B)

(C)

(D)

(E)

AN OVERVIEW OF DEFENSE SYSTEMS

The sentence "It's a jungle out there" is used in many contexts. Biologically, it could mean that humans, and all other species, are surrounded by potential disease-causing organisms or **pathogens.** Many pathogens are microorganisms, such as certain bacteria, viruses, protozoans, and fungi; others are macroscopic, such as parasitic worms (Figure 16.1). These pathogens reach us in various ways—in food and water; in the droplets of a sneeze; in the bite of an animal such as a mosquito, tick, or dog; from contact with a contaminated object in our environment; or by direct contact with an infected person. Many pathogens are themselves potential hosts to other pathogens. A **host** is the organism on or in which pathogens live.

The fact that pathogens are so common and diverse shows that this "life-style" has become quite successful in evolution. During evolution, the hosts of pathogens have also developed methods of defense against pathogens. Two main categories of defense are often recognized in the vertebrates. In **nonspecific defense systems** the organism has adaptations to prevent the pathogen from entering the body in the first place. It may also have generalized means of protecting itself from pathogens that manage to partially penetrate the body, such as in a surface wound. **Specific defense systems** are those that differ from one pathogen or foreign material to another. In these systems the host recognizes, eliminates, and remembers each specific pathogen or foreign or abnormal substance.

NONSPECIFIC DEFENSE SYSTEMS

Initial Defenses

Just as products are protected by their packaging, the skin and mucous membranes represent the "outer and inner wrappers" of animals. Healthy skin, the body's first line of defense, contains a tough layer (keratin) that keeps out most pathogens as long as it is unbroken (Figure 16.2). The epithelial cells in the upper epidermis divide frequently to replace the dead cells and to repair injuries. The more fragile membranes that line many internal structures are protected by mucus, which often has antibiotic properties. As is illustrated in Figure 3.11, mucous membranes are continuous with the skin and are still in contact with the body's external environment. Some mucous membranes are also ciliated, and the beating cilia help to sweep away entrapped microorganisms and particles. Despite these adaptations, mucous membranes are common sites of entry for pathogens or their toxins.

Enormous numbers of normally harmless bacteria and fungi grow on many body surfaces. These so-called normal flora occupy sites on which pathogens might otherwise grow. They also compete with any pathogens for available nutrients, and some produce compounds toxic to pathogens. Many secretions of mucous membranes also engage in chemical warfare with pathogens. For example, saliva, sweat, nasal secretions, and tears contain an enzyme (lysozyme) that attacks the cell walls of bacteria. Cells in both the stomach and vagina produce acidic secretions that kill or inhibit many pathogens, as well as foreign substances. The bile salts that are released into the small intestine do much the same (Chapter 13).

Pathogens that penetrate the surface cells are hardly "home free." Several varieties of white blood cells or leukocytes ingest pathogens by phagocytosis, much as amoebae ingest food particles (Figure 15.3). **Neutrophils,** making up about 60 percent of leukocytes, move quickly to infection sites. **Macrophages** ("big eaters") are giant scavenger cells that develop in the tissues from **monocytes.** Because they often take several days to accumulate in large numbers, increases in macrophages are more often associated with long-term or chronic infections than with short-term or acute infections. Potent digestive enzymes released by lysosomes, the cytoplasmic granules which give the granulocytes their name, may destroy the engulfed pathogens, other foreign substances, or cellular remains. These phagocytic leukocytes sometimes consume scores of bacteria before they succumb to the accumulating toxic breakdown products. However, some bacteria produce enzymes that destroy the lysosome membranes, causing, in turn, the destruction of the phagocyte. Other bacteria, such as those that cause tuberculosis, have cell walls or capsules that are very resistant to the lysosomal enzymes.

The Inflammatory Response

The **inflammatory response,** another second line of defense to an infection or injury, involves the accumulation of neutrophils and fluid released from dilated blood vessels at the site of infection or injury. The release of **histamine** from injured cells makes nearby capillaries more porous to neutrophils and monocytes. Blood in dilated capillaries causes redness, heat, and swelling (edema). Pain and tenderness result from the distended tissues and from substances released by the injured cells.

When most pathogens are engulfed by phagocytes, the healing process begins and the cavity that often forms contains fragments of damaged cells, dead neutrophils, and pathogens. Collectively, this mixture of pus remains until a scab forms or the wound heals. If the pus cavity should break into a body cavity, the debris is usually removed gradually by the lymphatic system. The fever that commonly accompanies an infection is caused by proteins that are released by host cells or pathogens. The higher body temperature may inhibit the pathogen's reproduction or interfere with its iron metabolism.

Interferons

While infected with one virus, vertebrates are unlikely to become infected with another virus. Ultimately, this observation led to the discovery of **interferons,** the polypeptides produced in small quantities by cells of the virus-infected animal. The interferon proteins prevent the cell from synthesizing macromolecules required by the virus, and they stimulate other cells to produce antiviral proteins. Not only do the types of interferons produced differ from species to species but one species may also produce several different interferons. Whether interferons will become "miracle cures" for viral diseases and some forms of cancer remains to be seen. However, the chemical structures of some human interferons are known, and genetic engineering makes it possible to utilize bacteria to make large quantities of this, just as *Escherichia coli* are used to produce insulin.

SPECIFIC DEFENSE SYSTEMS

Animals can distinguish between their own macromolecules and cells, **self,** and matter foreign to the body, **nonself.** An animal's **immune system** (from the Latin for "safe") is responsible for the recognition, selective destruction, and memory of nonself substances. Although cancer cells

FIGURE 16.2
The outer layer in this cross-sectional view through mammal skin consists of dead cells. The underlying cells divide rapidly to replace the cells that die.

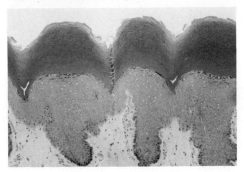

are not foreign, as pathogens are, their altered chemical composition and behavior may cause the immune system to recognize them as nonself and to selectively attack them. An individual with a faulty immune system may succumb even to normal flora, especially when its body is stressed in some way.

Nonself substances are constantly entering the body, and the unique chemical traits of each substance that differ from those of the body act as antibody generators or **antigens.** Any substance capable of stimulating an immune response is an antigen. Common foreign antigens include pathogens, pollen, spores, and many large molecules. The response of the body to these invading substances is to produce highly specific protein molecules called **antibodies** that combine with the antigen to destroy or inactivate it or to identify it for destruction. The human body can distinguish millions of different antigens, and those antibodies effective against one antigen generally have no effect upon another antigen. No prior exposure to an antigen is necessary to initiate one of these tailor-made immune responses.

Typical antibodies are Y-shaped molecules that contain two binding sites (Figure 16.3). When they meet, antigen and antibody fit together, somewhat like a lock and key in that they must bind in just the right way for the antibody to be effective. Each antibody recognizes and binds to a particular site, called an **antigenic determinant,** of the foreign or abnormal cell or biological macromolecule. It is the specific sequence of amino acids that constitutes an antigenic determinant and that gives that part of the protein molecule its particular configuration. Even small antigens have several antigenic determinants, and large antigens, such as entire cells, may have several hundred such sites on their surfaces, to each of which a specific antibody may bind. Some molecules (called haptens) found in substances such as dust or certain drugs are too small to be antigenic alone but they stimulate an immune response by combining with a larger, usually protein, molecule.

Although two forms of immune responses are often recognized, they have overlapping functions. The **humoral immune responses** (antibody-mediated immunity) that involve antibodies present in the blood plasma act primarily against viruses and bacteria that have not yet entered cells.

FIGURE 16.3
(A) Antigens, (B) antibodies, and (C) antigen-antibody complexes. The three-dimensional shape of the two binding sites is what makes them the "key" that is unique for each specific antigen or "lock." Antibodies are also known as *immunoglobulins* (abbreviated Ig) and exist in several varieties (e.g., IgG, IgM, and IgA).

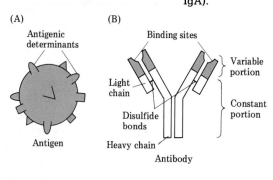

(A)

Antigenic determinants

Antigen

(B)

Binding sites

Light chain

Disulfide bonds

Heavy chain

Variable portion

Constant portion

Antibody

(C)

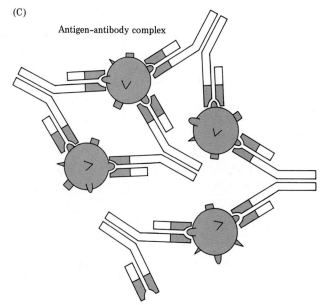

Antigen-antibody complex

Cellular immune responses (cell-mediated immunity) are based on cells sensitive to nonself substances. These responses occur most often against established viral infections, cancer cells, foreign tissue, fungi, and multicellular parasites.

Immune System Cells

The major features of blood and lymph are described in Chapter 15. The red blood cells, or erythrocytes, are involved primarily with oxygen transport. However, on their surfaces are inherited, as opposed to foreign, antigens that are important in the typing of blood, described later in this chapter and in Chapter 9. The white blood cells, or leukocytes, are larger than erythrocytes but far less numerous. Unlike the erythrocytes, the leukocytes are capable of independent locomotion, which is very important in their immune system functions.

The five major types of leukocytes are illustrated in Figure 16.4. Lysosomes are among the cytoplasmic granules that give the granulocyte

FIGURE 16.4

(A) The five types of leukocytes and their relative percentages (Chapter 15). (B) Several leukocytes ingest pathogens and foreign, dead, dying, or abnormal cells by phagocytosis. The vaginal macrophage shown here contains several sperm that it engulfed by flowing around them and enclosing them within endocytic vesicles. Even when these cells (often called *phagocytes* in general) do not ingest and kill all pathogens, they often reduce the number of pathogens to the point where other defenses can complete the task. While some macrophages wander through the tissues, others remain in an organ, such as the tanklike form shown in (C) that wanders through the lungs and destroys foreign matter in inhaled air.

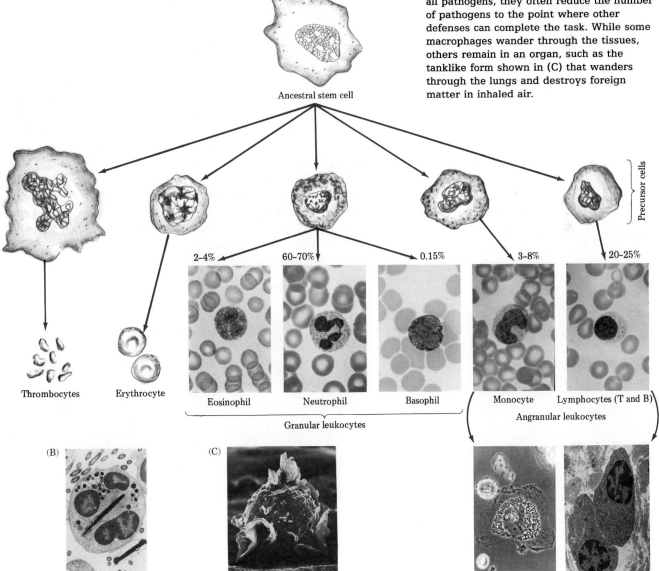

(A)

Ancestral stem cell

Precursor cells

2–4% 60–70% 0.15% 3–8% 20–25%

Thrombocytes   Erythrocyte   Eosinophil   Neutrophil   Basophil   Monocyte   Lymphocytes (T and B)

Granular leukocytes   Angranular leukocytes

(B)   (C)

Macrophage   Plasma cell

group their name. The granulocytes form in the red bone marrow and, among other functions, are the principal phagocytes in the blood and other tissues. **Eosinophils** participate especially in allergic reactions and in parasitic infections (e.g., tapeworms). In addition to their involvement in allergic reactions, **basophils** contain an anticlotting compound (heparin) that probably reduces the formation of internal blood clots (emboli and thrombi). Neutrophils were described previously, as were the monocytes among the agranulocyte group. The agranulocytes, including the very important **lymphocytes** which are specialized to form antibodies, develop mainly in lymphatic tissue such as the lymph nodes, thymus, spleen, and tonsils. The billions of lymphocytes that are the main chemical warriors in the specific immune responses are more common in lymph than in blood.

The Immune Response

Foreign organisms and materials that escape the body's nonspecific defense systems encounter the immune system. The lymphocytes are of particular importance in the immune response (Figure 16.5). There are two varieties of lymphocytes, T and B cells, that originate from ancestral cells (stem cells) in the embryonic bone marrow. During the fetal stage of development, some of these cells travel to the thymus gland in the chest region and multiply rapidly. Some of these thymus-dependent or

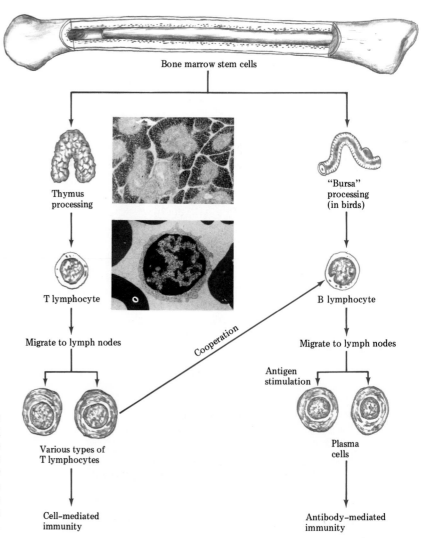

FIGURE 16.5
The formation, movements, and general functions of the B and T cells. B cells are named after a lymphatic structure in birds (bursa of Fabricus in the cloaca) that other vertebrates lack. The variety of a lymphocyte called a *plasma cell* produces antibodies, a group of substances that can react with any foreign material.

T cells remain in the thymus, others enter the bloodstream, and some travel to the spleen and lymph nodes. T cells become specialized for phagocytosis. The **B cells** differentiate in the bone marrow or in the fetal liver or spleen and then travel to the lymph nodes.

The B lymphocytes are responsible for the humoral immune response. A B cell activated by an antigen grows and divides repeatedly by mitosis to form a large number of identical cells called a **clone** of cells (from the Latin word for "twig") (Figure 16.6). Most of these cells differentiate

FIGURE 16.6
The humoral immune response. Each foreign body has antigenic determinants on its surface that are recognized by specific antibodies. There are innumerable varieties of B lymphocytes, each capable of responding to a specific antigen. Usually, some B cells continue to secrete small amounts of antibody for years after an infection. Testing for specific antibodies in a person's body, such as those for the AIDS virus, is commonly done to identify people who have come into contact with a particular pathogen.

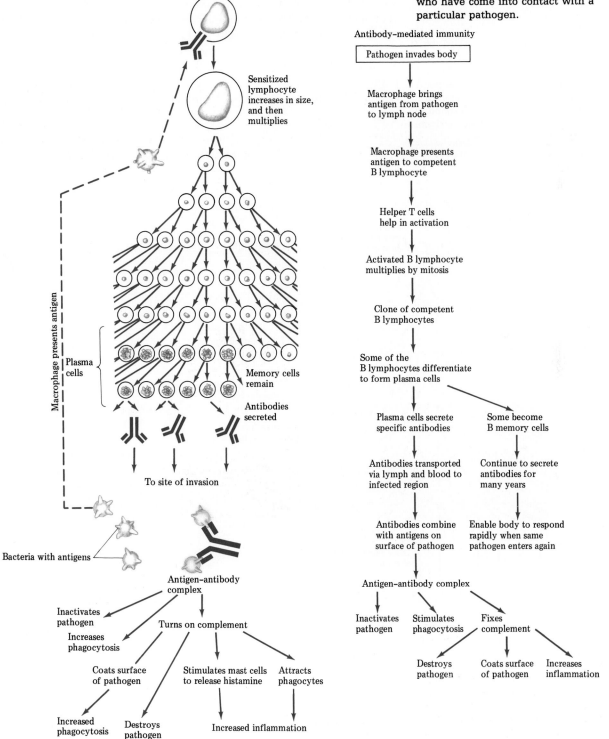

into **plasma cells** in the lymph nodes, within which they synthesize appropriate antibodies. Most antibodies move through the lymph and blood to the location of nonself substances, but some remain bound to the lymphocytes.

In contrast to the humoral system, the cellular immune system depends upon cellular warriors, macrophages, and T cells, rather than on the production of antibodies. Most lymphocytes are small and inactive. However, they grow and divide after being activated by antigens, which are often carried to them by macrophages (Figure 16.7). T cells (especially so-called killer T lymphocytes) have specific surface molecules that recognize and react to antigenic determinants on the surfaces of nonself cells, including the body's own cells that have become changed by cancer, aging, or viral infections. T cells also play a major role in the rejection or destruction of transplanted tissues or organs. The killer T cells release a powerful group of substances called **lymphokines** that either destroy pathogens directly or prevent or inhibit them from reproducing. The lymphokines also activate other lymphocytes, attract and stimulate mac-

FIGURE 16.7
The cellular immune response. There are other types of T cells than the killer T lymphocytes. For instance, specific helper T cells must often bind an antigen before a B cell can become activated and respond to the antigen.

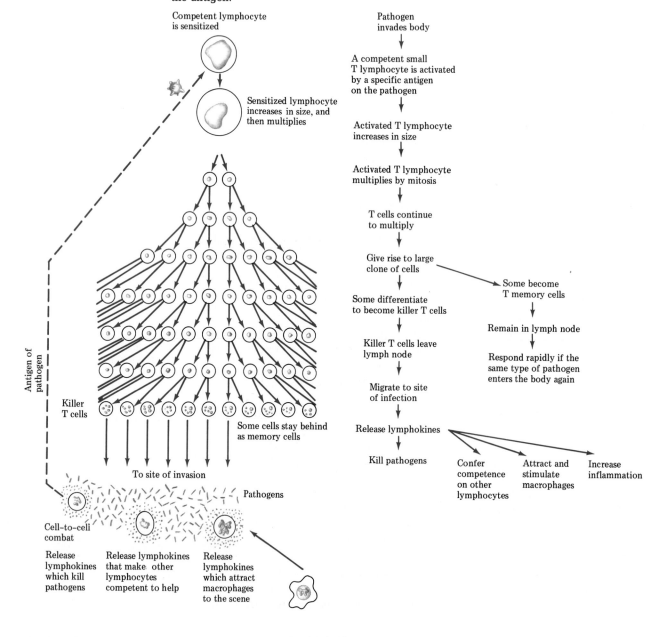

rophages (make them "angry"), and enhance the inflammatory response.

There is typically a time lag of several days between an animal's first exposure to a particular antigen and the production of adequate numbers of antibodies and activated lymphocytes to attack the antigen (Figure 16.8). **Immunological memory** means that the immune system "remembers" that particular antigen, sometimes for life. Thus, even years later, it may remain capable of responding quickly and massively to the particular antigen. This explains why we usually do not get the same disease again. Getting many colds or suffering from the flu repeatedly constitutes no exception to this principle. It simply means that the terms *cold* and *flu* are generic and encompass a large number of viruses, each with slightly different antigens that may prevent recognition by memory cells.

Active and Passive Immunity

Infectious diseases such as chicken pox, measles, and whooping cough are generally associated with childhood. Through the immune response, those who survive their first exposure to a disease acquire a **natural active immunity,** so named because the person's body made specific antibodies upon exposure to each pathogen. In many cases today, however, it is not necessary to have a disease in order to acquire immunity against it because **vaccines** may induce antibody formation. Some vaccines contain dead or inactivated pathogens or live pathogens with little chance (low virulence) of causing the disease that one is attempting to protect oneself against. As long as the vaccines still have the antigens necessary to stimulate the immune response, they have the same effect as having had the disease itself. Because the body makes the antibodies against the disease, this is called **active immunity by vaccination.** The immunity conferred by active immunity processes is usually long lasting, although in some cases booster vaccines are needed at intervals.

Immunization has been so effective against former killers and cripplers such as diphtheria and polio that these diseases are now very uncommon in the developed nations. In fact, the only known smallpox viruses are those maintained in a few laboratories; smallpox appears to be extinct. See the essay "Monoclonal Antibodies" for information about a new type of vaccine.

Sometimes it is risky to use a live pathogen in a vaccine. However, domestic animals such as horses or goats often can be injected with the pathogen without harmful effects to them. The animal produces antibodies against the antigen and the animal's blood is used to prepare a vaccine for humans. Alternatively, vaccines can be prepared from the blood of humans who have been exposed to the antigen. Because the recipients of these vaccines do not produce the antibodies themselves, this borrowed immunity is called **passive immunity** (Figure 16.9). Human sera are usually superior to animal sera because the nonhuman proteins can act as antigens, stimulating an immune response known clinically as *serum sickness*. The duration of immunity achieved by passive immunity is usually just a few months and no memory cells develop.

Natural passive immunity is conferred to a fetus by the mother's antibodies that pass to it through the placenta. This provides some defenses for a newborn infant before its own immune system matures. A breast-fed baby continues to receive some antibodies in its mother's milk.

The harmful effects of many bacteria are due to **toxins,** poisonous substances they release as waste products. Protein toxins are antigens and the specific antibodies produced against them are called **antitoxins.**

FIGURE 16.8

It often takes a few days for the body to respond after initial exposure to a particular antigen. Another exposure, even years later in many cases, stimulates a faster and larger production of antibodies. So-called memory cells, which are activated B and T cells that revert back to small lymphocytes, are responsible for this phenomenon.

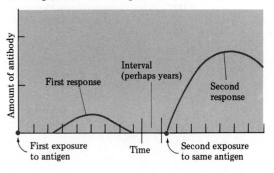

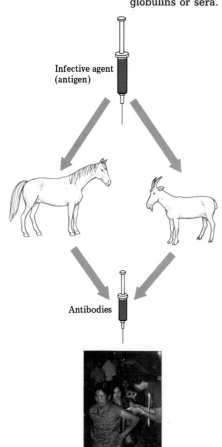

Infective agent (antigen)

Antibodies

Some toxins can be changed chemically into **toxoids,** and because toxoids retain their antigenic traits but lose their poisonous ones they can be used in vaccines (e.g., whooping cough, tetanus, and diphtheria).

TRANSFUSIONS AND TRANSPLANTS

We often hear the sentence "The operation was a success but the patient died." Such was the case in the 1800s when blood **transfusions** sometimes worked but often killed the patient, with no apparent pattern to the successes and failures. It was not until around the turn of the century that it was discovered that only certain kinds of blood were compatible. Individuals who received incompatible blood had their erythrocytes clump, which often clogged smaller blood vessels.

In the familiar **ABO blood group,** two kinds of antigens may be found on the erythrocytes, A and B. In the plasma, there may be the corresponding antibodies, called *anti-A* and *anti-B*. Obviously, it is not good to have an antibody in one's blood plasma that could attack an antigen on one's red blood cells. Normally, as shown in Table 16.1, we have plasma antibodies that could attack the cellular antigens that we do not have. For example, a person with antigen A has anti-B antibodies. Type O people, so named because their erythrocytes have neither antigen, have low amounts of both antibodies. For that reason, type O persons are often called *universal donors* because small quantities of their blood can be transfused to anyone without adverse effects. Because their blood contains neither anti-A nor anti-B antibodies, persons with AB blood are called *universal acceptors*. Although short-term emergency transfusions may lead to few problems in most cases, it is best to have properly matched blood in order to avoid the clumping and bursting of erythrocytes, fever, jaundice, tissue damage, or death. The genetics of blood groups is discussed in Chapter 9.

Another important blood antigen is called the **Rh factor** because it was observed first in the familiar <u>rh</u>esus monkey. About 83 percent of whites, about 92 percent of blacks, and almost all Orientals have the Rh antigen on the surfaces of their erythrocytes. These individuals are called Rh$^+$ (Rh positive), in contrast to the Rh$^-$ (Rh negative) individuals who lack the antigen. Unlike the ABO blood group situation, the Rh$^-$ individual has no plasma antibody to Rh. Upon exposure to Rh$^+$ blood, however, the Rh$^-$ individual synthesizes the Rh antibody, sometimes with serious implications for an Rh$^+$ fetus being carried by an Rh$^-$ female (Figure 16.10). Although maternal and fetal blood usually do not mix during pregnancy, a little of the fetus's blood sometimes enters the mother's blood, especially near the time of birth as the uterine and placental tissues separate. The mother's body recognizes the Rh antigen as nonself and makes antibodies against it. The woman's first child is not usually affected seriously if the blood mixes only late during the birth process but the presence of anti-Rh antibodies in the maternal blood may threaten subsequent children. In the past, millions of infants

TABLE 16.1
The ABO Blood Group

Blood Type	Antigens Present	Antibodies Present	Can Act as a Donor to	Can Receive Blood from
A	A	Anti-B	A, AB	A, O
B	B	Anti-A	B, AB	B, O
AB	AB	Neither	AB	A, B, AB, O
O	Neither	Anti-A, anti-B	A, B, AB, O	O

Monoclonal Antibodies

Until the mid-1970s, research on antibodies had been complicated by the fact that most antigens carry many antigenic determinants. Although a single large lymphocyte produces only one kind of antibody, it was not possible to clone them (produce large numbers of identical cells) in pure cultures. However, cells capable of producing antibodies could be obtained from cancerous tumors (myelomas) of the immune system that would proliferate quickly and continuously. It is now possible to produce hybrid cells to take advantage of the useful traits of both types of cells.

An animal is first inoculated with an antigen to elicit antibody production. This takes time, usually several weeks, and often requires multiple injections of antigens to maximize the response. Lymphocytes are then isolated from the spleen or other lymphatic tissue and mixed with isolated myeloma tumor cells. Cell fusion is induced chemically between the lymphocytes and the myeloma cells and results in the formation of hybrids containing both genomes, which are spread onto a proper medium so that clones form from each hybridoma cell. Individual clones are tested to determine which specific antibody they produce and the cells producing the desired antibody are grown separately in large numbers. Because antibodies that form from a single clone of cells are uniform, they are called **monoclonal antibodies.**

Monoclonal antibodies are very useful in studies of membrane specificity, in tissue typing, and in many developing medical procedures. For example, instead of inoculating someone with an antigen in order to cause the person to develop his or her own antibody, it is now possible to inoculate people with specific antibodies produced using the monoclonal technique.

died or developed such problems as anemia, mental retardation, jaundice, brain and heart damage, and deafness due to Rh incompatibility. Today, protective sera are available that prevent the buildup of antibodies within the Rh$^-$ woman if one is given within 72 hours after the miscarriage or birth of an Rh$^+$ child. These sera are interesting because they save one potential life by being given to another person. Because Rh incompatibility can also cause problems during blood transfusions, the donor and recipient must be matched for Rh as well as for ABO antigens.

Many **transplants** of organs or tissues fail because the transplant is recognized as nonself and attacked by the immune response. Success is improved if the transplant is done immediately after birth or if the donor

(A)

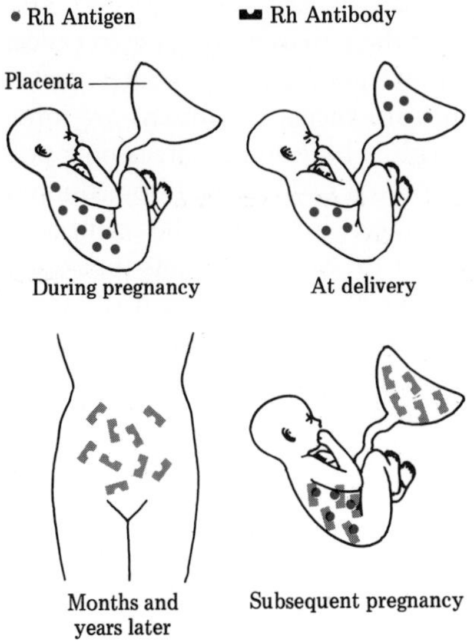

● Rh Antigen   ◣ Rh Antibody

Placenta

During pregnancy

At delivery

Months and years later

Subsequent pregnancy

(B)

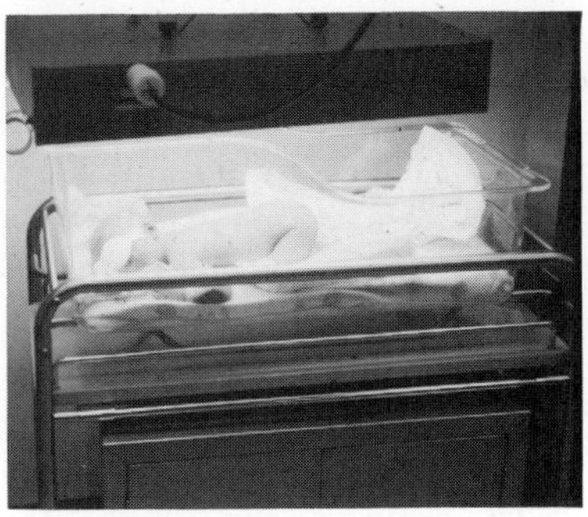

FIGURE 16.10

(A) The cause of the Rh problem, which is no monkey business for the affected pregnant Rh$^-$ females who are carrying Rh$^+$ fetuses. The problem can now be alleviated by injection of Rh antibodies into the Rh$^-$ female within about 72 hours of the miscarriage or birth of an Rh$^+$ child. The anti-Rh antibodies destroy any Rh$^+$ fetal red blood cells that enter the maternal blood before the woman's immune response occurs. (B) Even first-born children may be affected if they are not kept under ultraviolet light for several days after birth to prevent jaundice. This infant has her eyes covered to avoid damage to them by exposure to the UV light.

FIGURE 16.11
Barney Clark received the first artificial heart in 1982. Artificial organs are constructed of materials, such as plastic, that do not elicit an immune response. With a reduction in internal rough edges that led to blood clots and caused strokes in some humans using the early devices, such artificial organs should become more common and more successful.

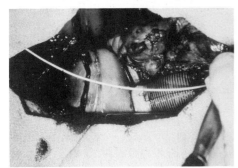

is an identical twin. (Nontwin siblings are more closely related than parents and, therefore, make better alternative donors.) The use of drugs (immunosuppressants) to prevent rejection is possible but this leaves the recipient vulnerable to pathogens. Artificial organs and tissues, including blood, are of growing interest (Figure 16.11).

There are two common exceptions to rejection. First, corneal transplants are highly successful because the cornea has almost no blood or lymphatic vessels associated with it. Second, pregnancy is so common that we usually forget that the sperm that fertilized the egg was nonself. However, the embryo normally develops its own biochemical identity safely in the uterus without initiating an immune response.

CANCER

Although **cancer** is a group of diseases rather than a single condition, in all cancers there is a largely unrestrained proliferation of cells. Normally when body growth stops, mitosis occurs at a rate sufficient only to replace worn-out or dead cells. Cancerous cells, however, divide again and again and again. The mass of new cells becomes a **tumor** (neoplasm) that may compress, invade, and destroy normal tissues. Small, relatively localized and harmless tumors are called **benign.** On the other hand, untreated **malignant** tumors may cause serious harm or death, although the exact effect depends upon their site and upon the age, sex, race, and health of the victim. Often, the term *cancer* is applied only to malignant tumors. Although some malignant cells grow into a solid tumor, others **metastasize** or disperse into the bloodstream or the lymphatic system and then into distant tissues. Thus, surgical removal of a tumor may not remove all malignant cells from the body.

Normal cells have surface recognition factors that allow cells of like type to recognize them as self and interact normally with them. Malignant cells somehow lose these surface traits and become nonself. Cancer cells look more like immature than mature cells, and they show various abnormal genetic and chemical traits. They tend to adhere indiscriminately to any other cell types that they encounter. That, plus the lack of proper surface markers for intercellular recognition, helps explain the ability of malignant cells to migrate and establish themselves throughout the body.

In 1900, 4 percent of American deaths were attributed to cancer, by 1950 the figure had risen to 15 percent, and now over 20 percent of Americans can expect to die from cancer. These percentages are deceptive because better diagnosis is one reason for the apparent increase in cancer incidence. However, the World Health Organization estimates that a very large number (perhaps 80 percent) of cancers in humans are influenced by environmental agents, and Earth is becoming increasingly polluted. The fact that a London physician noted in 1775 that boy chimney sweeps were unusually liable to scrotal cancer shows that the connection between environmental factors and cancer is not new. Agents that cause cancer are called **carcinogens,** and soot was apparently the first carcinogen recognized. Today, known carcinogens include numerous chemical compounds, radiation (including excessive exposure to sunlight), and certain cancer-producing or **oncogenic viruses.**

The nature of the cellular transformation from normalcy to the cancerous state is unknown. Three possible causes include accumulated mutations, the activation of one or more genes that are not normally expressed, or the effect of an oncogenic virus.

Because the surface recognition factors of cancer cells have changed, they often or usually elicit an immune response. Perhaps most cancers are controlled by the body's immune system and maybe only the occasional cancer that escapes this immune surveillance net comes to the attention of the patient and physician. It may be, in fact, that the vertebrate immune system evolved as much to protect the organism against its own wayward cells as it did to neutralize foreign invaders.

Once cancers have been diagnosed, there are several lines of treatment. First is surgery if the affected organ or tissue is not irreplaceable, if the cancerous cells are localized, and if metastasis has not occurred. **Radiotherapy,** the exposure of the cancerous tissues to X- or gamma rays, is also common, especially if the affected structure cannot be removed by surgery (Figure 16.12). However, radiation also affects normal cells, although the rapidly dividing cancerous cells are usually more susceptible to radiation damage. One major concern is that radiation may affect the immune system itself, rendering the victim defenseless against pathogens, or induce the development of other cancers in the attempt to eliminate an existing cancer. In contrast to both of these methods, where the exact location of the cancer is critical knowledge, is **chemotherapy** or treatment with drugs that preferentially affect cancer cells. Unfortunately, side effects of chemotherapy are common and sometimes severe, often because other cells that also proliferate quickly (e.g., epithelial cells of hair, teeth, and the intestine, as well as the cells of the immune system) are affected adversely.

Immune therapy holds promise and is the subject of much current research. For example, some agents may trigger the general immune alarm and mobilize the body's immune system against the cancer. The use of interferons against specific oncogenic viruses is another possibility, as is the use of monoclonal antibodies. (See the essay on this topic.) Radioactively labeled monoclonal antibodies that are specific for certain kinds of cancer can be used to precisely locate the cancers and determine their sizes. If drugs can be bound to the monoclonal antibodies (so-called targeted drug therapy), then only the cancerous and not normal cells would be destroyed by the chemotherapy. See the essay "Training Your Immune System" for yet another possibility.

DISORDERS OF THE IMMUNE SYSTEM

In the **autoimmune diseases** (such as rheumatic fever, rheumatoid arthritis, multiple sclerosis, myasthenia gravis, systemic lupus erythematosus, and ulcerative colitis), parts of the body are destroyed by its own immune system (Figure 16.13). Why the body fails to recognize some of its cells as self and attacks them is unknown. Perhaps "silent" genes become activated, as is suspected in some cases of cancer. Or perhaps a previous viral infection in the affected tissue is implicated. That is, even though a virus was destroyed, the antibodies produced against it may continue to attack the noninfected tissue. In general, autoimmune diseases increase with age and are more common in women than in men.

Another common disorder of the immune system is **hypersensitivity,** or abnormally increased sensitivity to mildly antigenic agents called **allergens** that do not elicit an immune response in others. Perhaps 15 percent of Americans suffer from serious allergic reactions, with larger numbers of individuals experiencing more minor irritation from any number of environmental agents.

Hives, hay fever, and allergic asthma are common examples of immediate-type hypersensitivity. When an allergic person inhales ragweed pollen, for example, antibodies are released by sensitized plasma cells in

FIGURE 16.12
Radiotherapy is especially useful for localized cancers in critical organs that cannot be removed by surgery or are difficult to reach. Radiation, however, kills some noncancerous cells and is also a powerful mutagen that may induce genetic changes in cells. Modern cancer treatment is often not a choice between the sole use of one method or another but, rather, usually involves several methods sequentially or simultaneously.

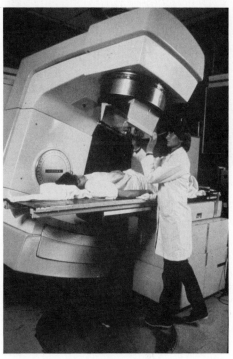

FIGURE 16.13
The person whose twisted hands are shown
in this X-ray suffers from rheumatoid
arthritis, one common autoimmune disease.
In this case the cartilage tissue in joints is
attacked. In myasthenia gravis, the
neuromuscular junctions are impaired
(Chapter 19), leading to muscular weakness.

the nasal passages (Figure 16.14). The straight ends of these antibodies attach to the membranes of certain (mast) cells, leaving the Y ends free to attach to allergens. When that attachment occurs, the mast cells release various substances (e.g., histamine) that cause inflammation. The resulting edema, runny nose, sneezing, watery eyes, redness, and constriction of the respiratory passages occur as blood vessels dilate and capillaries become more permeable. Similar reactions may occur in some people who are allergic to foods. The red welts known as hives occur when the allergen-antibody reaction takes place in the skin. Once sensitized to an allergen, a person may suffer symptoms very quickly upon subsequent exposure.

Most allergic reactions are localized near where the allergen contacts the body, with the specific symptoms depending upon the location of the affected tissue. In addition to the above cases, for example, asphyxiation may result from an affected larynx, and diarrhea and cramps may occur if the digestive tract is involved. A generalized reaction (systemic anaphylaxis) is far more serious and may be fatal. This may happen when a person develops an allergy to a specific drug such as penicillin or to certain substances in the venom injected by a stinging insect. So much histamine is released so quickly into the bloodstream from many parts of the body that the widespread plasma loss from the dilated and permeable blood vessels may lead to shock and death within minutes.

Drugs called *antihistamines* are often used in the treatment of allergic symptoms. These substances combine with the same receptor sites on cells that histamines would and thus they prevent histamines from doing so. Antihistamines are not totally effective, however, because some cells

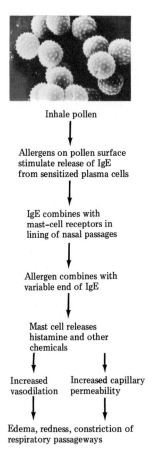

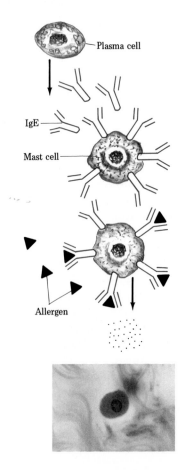

FIGURE 16.14
A summary of the hay fever allergic response. Substances as diverse as fungal spores, hair, plant chemicals (e.g., poison ivy), cosmetics, feathers, and dust may act as allergens and elicit hypersensitivity responses. The tendency to develop some allergies is apparently inherited. Also, people who have one type of autoimmune response are likely to develop others.

Training Your Immune System

When told that they have cancer or another serious disease, some people deny it. Others accept the fact stoically, and still others feel intense hopelessness and helplessness. But some are fighters, saying through their actions and attitudes, "I'm going to conquer this thing." Some of the former three types survive but many die; many of the fighters survive although some do not. It is not a new idea that one's emotions are linked to getting and surviving diseases, nor is it a proven idea. But, do we just remember the fighters who win and forget those who lose? Many think not, and research in this field is increasing.

It is generally agreed that stress can affect the incidence of certain disorders, such as when some air traffic controllers, urban cab or bus drivers, or hard-driving executives develop ulcers, asthma, or hypertension. We even have a common name for such a relationship—*psychosomatic,* meaning literally that psychological traits affect somatic or bodily functions (Chapter 11). Medical schools now often offer training in psychosomatic medicine, including the treatment of hypochondriacs, or charter members of the disease-of-the-month club, with verbal or substantial placebos.

Actually, the opposite condition is also common, although the term for it, *somatopsychological,* is not so well known. People's self-images and the ways others react to them can obviously be affected by deviations from physical or societal norms, such as being very thin or very fat, very short or very tall, very beautiful or very ugly—or very anything. Also, being deaf or blind, lacking an arm or a leg, having a chronic disease, or feeling constantly fatigued and listless, for example, usually affect one's psychological state. In other words, our physical traits can also affect our emotions.

Some evidence, often based upon studies of small numbers of people, suggests that anxiety, apathy, and depression may suppress the immune response, whereas anger and distress about one's condition may lead to increases in the numbers of lymphocytes and decreases in the rates of mitosis in cancerous tissue. Other studies suggest that the stress of major academic examinations reduces the immune response, including the production of interferons and, thus, increasing the incidence of viral colds and flu at those times. For example, blood T cell levels decline when those who carried the virus for genital herpes felt depressed.

Proof of a direct link between the brain (especially the limbic system, discussed in Chapter 19) and the immune system is not at hand, but diverse research efforts are underway and suspicions about such connections are high. Keep a stiff upper lip!

(e.g., mast cells) secrete substances other than histamine upon being exposed to allergens. So-called desensitization therapy is used in some cases of very serious allergic reactions. By regularly exposing the person to very small amounts of the allergen over a long period of time, sufficient antibodies of one type (IgG) are produced so that they combine with the allergen, thus preventing the antibody that would result in the allergic reaction (IgE) from doing so.

Victims of **acquired immune deficiency syndrome** (AIDS) have a deficiency of T cells. As a result, their ability to resist antigens is severely reduced, and AIDS victims often die from diseases such as pneumonia, tuberculosis, or cancer. One cause of AIDS is a mutant form of a virus that causes a rare form of leukemia (blood cancer). AIDS is transmitted primarily by sexual contact, by contaminated blood transfusions, or by the use of unsterile needles, such as those used by many addicts to inject drugs into the body. Massive research efforts are underway to understand AIDS but no cures are yet known (Chapter 8).

SUMMARY

1. **Pathogens** are disease-causing organisms, such as certain bacteria, viruses, protozoans, fungi, and worms. A **host** is the organism that supports the pathogen. Pathogens often reach the host in air, food, and/ or water; by the bite of an animal; or by direct contact with an infected person or a contaminated object.

2. In vertebrates, two main classes of defenses are often recognized. **Nonspecific defense systems** help prevent the pathogens from entering the body in the first place. In **specific defense systems,** the host recognizes, eliminates, and remembers each specific pathogen or foreign or abnormal substance.

3. The unbroken skin is the body's first line of defense against foreign materials. Mucus helps protect the body's more fragile internal membranes. Some mucous membranes also have cilia to sweep away harmful substances. Normal flora also occupy sites on which pathogens might otherwise grow, compete with pathogens for available resources, or produce compounds toxic to pathogens.

4. Foreign materials that enter the body, as well as damaged cells, may be consumed through phagocytosis by specialized leukocytes such as **macrophages** that multiply and move to infection sites.

5. A second line of defense to an infection or injury involves the **inflammatory response,** the accumulation of **neutrophils** and fluid released from dilated blood vessels at the site of infection or injury. Pus is the mixture of damaged cells, dead neutrophils, and pathogens that remains until a scab forms or the wound heals.

6. **Interferons** are proteins produced in small quantities by cells of virus-infected animals. Specific interferons prevent cells from synthesizing the macromolecules required by each virus, and they stimulate other cells to produce antiviral compounds.

7. Animals can distinguish between their own cells and macromolecules—**self**—and substances foreign to the body—**nonself.** An animal's **immune system** is responsible for the recognition, selective destruction, and memory of nonself matter.

8. Any substance (e.g., pathogens, pollen, spores, and many macromolecules) capable of stimulating an immune responce is an **antigen.** The body's response to these invading substances is to produce highly specific **antibodies** that combine with the antigen to destroy or inactivate it. The human body can distinguish millions of different antigens.

9. Typical antibodies are Y-shaped molecules that contain two binding sites. Each antibody recognizes and binds to a particular site or **antigenic determinant** on the antigen.

10. **Humoral immune responses** involve antibodies present in the blood plasma that act primarily against viruses and bacteria that have not yet entered cells. **Cellular immune responses** involve cells sensitive to nonself substances. These responses usually occur against established viral infections, cancer cells, foreign tissue, fungi, and multicellular parasites.

11. Some antigens are inherited, such as those found on the surfaces of erythrocytes that are important in blood typing.

12. Leukocytes are very important in the immune response. **Eosinophils** participate in allergic reactions and in parasitic infections. **Basophils** contain an important anticlotting compound and are also involved in allergic reactions. Neutrophils are important phagocytes, as are the macrophages produced from monocytes. **Lymphocytes** are specialized to produce antibodies.

13. Both B and T lymphocytes originate from ancestral cells in the embryonic bone marrow. The cells that will become **T cells** travel to the thymus gland and multiply. T cells become specialized for phagocytosis and are important in the cellular immune response. The **B cells** are responsible for the humoral immune response. Most B cells differentiate into **plasma cells** in the lymph nodes and synthesize antibodies.

14. Through **immunological memory,** the immune system remembers antigens, sometimes for life, and often mounts a rapid and massive attack on the antigen when it is encountered again.

15. In gaining **active immunity,** which is usually long lasting, a person's body makes antibodies against a pathogen. Such immunity may develop from exposure to a pathogen naturally or to a live, dead, or inactivated pathogen in a **vaccine. Passive immunity,** which is usually short term, is gained when one receives antibodies made by a domestic animal or another human who has been exposed to a pathogen.

16. The harmful effects of many bacteria are caused by **toxins,** which they release as waste products. Antibodies produced against these toxins are called **antitoxins.** Because **toxoids** retain the antigenic determinants of toxins but are nonpoisonous, they can often be used in vaccines.

17. Successful **blood transfusions** depend upon a proper match between the blood of the donor and that of the recipient. Normally, we have plasma antibodies that could interact only with red blood cell antigens that we do not have. Thus, a type A person in the **ABO series** has antigen A on the erythrocytes but anti-B antibodies in the plasma. The mixing of incompatible blood leads to clumping of the erythrocytes.

18. Most humans possess the **Rh antigen** and are called Rh$^+$. Problems may develop in the Rh$^+$ fetus or child of an Rh$^-$ woman. Protective serums are available today to eliminate such problems. In blood transfusions, blood must be matched for both ABO and Rh factors.

19. Many **transplants** fail because the organ or tissue is recognized as nonself and antibodies are produced against it. Artificial organs and tissues may reduce this problem.

20. **Cancer** is a group of diseases characterized by rapid cell proliferation. The **tumor** that results may be localized and **benign** or it may become **malignant** and dangerous, especially if it **metastasizes** or spreads. Malignant cells lose their surface recognition traits and become abnormal in appearance, behavior, and chemical and genetic traits. The body usually recognizes cancer cells as nonself.

21. Cancer death rates are rising in part because of exposure of more and more humans to diverse chemical **carcinogens.** Radiation and **oncogenic viruses** are also common causes of cancer.

22. Treatment for cancer often involves surgery, especially if the tumor is localized in an organ that is not irreplaceable and if metastasis has not occurred. **Radiotherapy** is also common, although radiation may adversely affect normal cells, including those of the immune system, as well as cancerous ones. **Chemotherapy** involves the use of drugs that preferentially affect cancer cells. Unfortunately, the drugs also often adversely affect other rapidly dividing cells, including those of the immune system. **Immune therapy** holds both the promise of better treatment and fewer side effects.

23. In the **autoimmune diseases,** parts of the body are attacked by its own immmune system for unknown reasons. **Hypersensitivity** is another common disorder of the immune system. **Allergens** that have little or no effect on most people cause mild to major reactions in others. **Acquired immune deficiency syndrome** is a serious, usually fatal, problem.

REVIEW QUESTIONS

1. Distinguish between the general features of specific and nonspecific defense systems.

2. What traits protect mucous membranes from pathogens?

3. Describe the role of the normal flora in preventing the invasion of pathogens.

4. Distinguish between the traits and functions of the granulocytes and the agranulocytes.

5. Describe the function and characteristics of the inflammatory response.

6. Where are interferons produced and what are their functions?

7. How do body cells distinguish between self and nonself materials?

8. What is the difference between a "native" and a "foreign" antigen?

9. How do antigens elicit the production of specific antibodies?

10. Distinguish between the humoral and the cellular immune responses.

11. Describe the formation and function of B, T, and plasma cells.

12. How does immunological memory occur?

13. Distinguish between active and passive immunity and between natural active immunity and active immunity by vaccination.

14. Describe at least three methods by which vaccines are produced, including antitoxins.

15. To whom can a person with B⁺ blood give blood for use in a transfusion? Receive blood from?

16. How does the Rh factor differ from the ABO blood types?

17. Distinguish between a malignant and a benign tumor.

18. How does metastasis complicate the treatment of cancer?

19. Discuss the pros and cons of surgery, radiotherapy, chemotherapy, and immune therapy in the treatment of cancer.

20. What is an allergen and what does it do in the body?

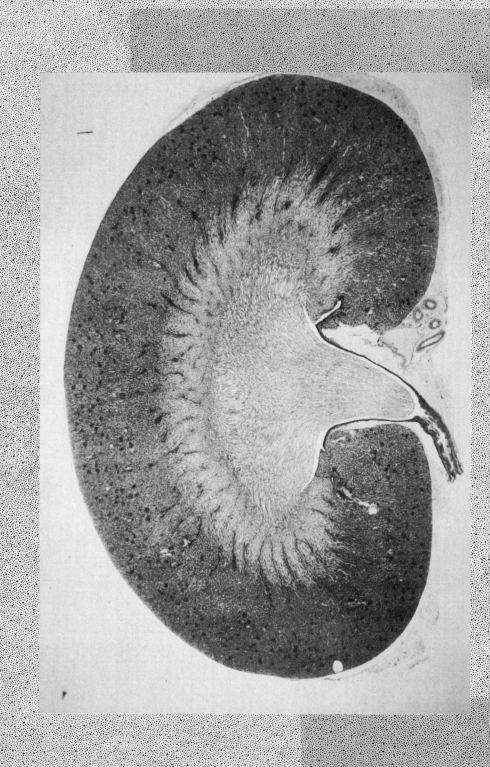

CHAPTER 17

EXCRETION AND OSMOREGULATION

Most animals are mostly water, and water is the medium in which most metabolic reactions typically occur. Water is also the home for many animals, and some forms secure food and water from and release wastes directly into their environment. Larger aquatic and terrestrial animals generally maintain their own "internal sea" to bathe cells and to transport and exchange nutrients, gases, and wastes. In this chapter, you will explore the diverse ways that animals secure water, maintain the volume and chemical composition of their body fluids, and excrete their "metabolic ashes." Special attention is paid to the kidneys because they are such prime organs of homeostasis.

As with several other important organs, our kidneys are bilateral, and we can usually live with only one if necessary. While other organs contribute in various ways to ridding the body of wastes, our kidneys, seen here in cross-sectional view, are our prime organs of excretion. Although the kidneys are relatively large organs, their essential work is performed mainly by microscopic components.

EXCRETORY SYSTEMS AND WASTES

The homeostatic regulation of the volume and chemical composition of body fluids is called **osmoregulation.** Osmoregulation is relatively simple for many small aquatic animals whose cells are often in direct contact with their environment. However, in larger animals an **excretory system** is needed to collect excess water, ions, and metabolic wastes, reabsorb any materials that are still useful, and expel the remainder from the body. Thus, excretory systems are not merely passive "plumbing" systems but, rather, sets of structures which actively sort out the substances to be retained from those to be excreted. Obviously, for materials to be excreted from the body before they reach toxic levels that could threaten homeostasis, they must once have been in the body. Thus, the undigested food materials that are egested from the digestive tract are not metabolic wastes because they were never in the body; they were only surrounded temporarily by the body (Figure 3.11).

The three principal metabolic wastes in animals are carbon dioxide, water, and various nitrogenous (nitrogen-containing) compounds. Because the excretion of carbon dioxide primarily involves respiratory surfaces such as gills, tracheae, and lungs, the details of those functions are found in Chapter 14. This chapter treats the excretion of water and nitrogenous wastes.

The main nitrogenous wastes that form from the metabolism of amino acids and nucleic acids are ammonia, urea, and uric acid (Figure 17.1). Note that there is a close correlation between the type of nitrogenous compound excreted and the environment and life-style of the animal. **Ammonia** results when the amino group is removed from amino acids, a process called *deamination*. Ammonia is highly toxic and cannot be allowed to accumulate in an animal's body or in its immediate environment. Aquatic invertebrates such as sponges and jellyfishes and freshwater fishes excrete nitrogen mainly as ammonia. Ammonia is highly soluble and the surrounding water usually dilutes it quickly. However, many aquatic and terrestrial animals must convert ammonia to a less toxic waste product such as urea or uric acid.

The **urea** that is produced in the liver is the main nitrogenous waste of marine fishes, amphibians, and mammals. Although much metabolic energy is required to form urea through a sequence of enzymatic reactions, less water is lost in the excretion process compared with ammonia. Also, being less toxic than ammonia, but more so than uric acid, urea can accumulate in the body in higher concentrations and be excreted in more condensed form.

Uric acid may be produced from ammonia or when nucleotides from nucleic acids are metabolized. It is an especially suitable form of nitrogenous waste for terrestrial animals with a high need to conserve water. Uric acid can be excreted as a paste with little water loss because it is poorly soluble and precipitates as crystals. This least toxic of the nitrogenous wastes is excreted by insects, certain snails, many reptiles, and birds. You may have become aware of the latter when you sat on a park bench with pigeons roosting on the branches above. More metabolic energy is required to produce uric acid than to form urea but less water is lost in the excretion process.

Actually, most animals produce more than one of these nitrogenous wastes. For example, although humans produce mostly urea, we also excrete some ammonia and uric acid. (Part of the ammonia that one smells while changing a baby's diaper results from the breakdown of urea by bacteria.) Also, the nature of the nitrogenous waste may change

Ammonia

$$H_2N-H$$ with structure showing N bonded to three H

Urea
$$H_2N-C-NH_2$$ with O double bonded to C

Uric acid

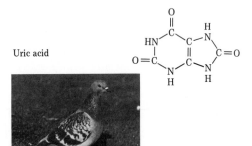

FIGURE 17.1

The general features of the three main nitrogenous wastes. Some animals secrete ammonia, the first metabolic product of deamination, directly into their aquatic environment. Other animals may convert ammonia to urea or uric acid. The latter waste can also be formed from the metabolism of nucleic acids. Little metabolic energy is required to form ammonia but abundant water is needed to excrete it. Thus, the more metabolically expensive processes of forming urea or uric acid are more typical of terrestrial animals.

during an animal's life. Thus, aquatic tadpoles excrete mainly ammonia, whereas terrestrial adult frogs and toads excrete mainly urea.

The environments of animals vary from those in which salts are highly concentrated (e.g., salt lakes and oceans) to those in which salts are scarce (e.g., most terrestrial habitats). As when responding to any other physical factor in the environment, such as heat or light, some animals have a wide range of tolerance while others can tolerate little variation. The prefix **eury-** is used for species with wide tolerances, and **steno-** applies to those with narrow ranges of tolerance. A suffix describes the appropriate physical factor. Thus, an animal able to survive exposure to a wide range of temperatures is called *eurythermal;* one with the ability to tolerate a minimal range of salinity is called *stenohaline.* Within both eury- and steno- groups are some **osmoregulator** species, such as humans, that are able to maintain their internal conditions at a relatively constant level despite changes in their environments (Figure 17.2). **Osmoconformer** species lack this ability, and their internal conditions change as their environment changes. Chapter 2 describes a similar situation in terms of temperature regulation. See the essay "The Best of Both Worlds" for information about two very euryhaline animals.

Some aquatic environments are called **hypotonic** because they have salt concentrations less than those found in an animal's fluids (Chapter 3). Other waters are more concentrated or **hypertonic,** or are equal or **isotonic** in salt concentration compared with the salt concentrations in the animal's habitats. These terms are relative and are always used when one is comparing two conditions. Thus, while the environment of the freshwater fish is hypotonic to the animal, the fish is hypertonic relative to its environment. In addition to these aquatic habitats are terrestrial environments wherein organisms live in relatively dry air or soil that is not usually described in these terms. See the essay "Comparative Excretion and Osmoregulation" for more information.

THE MAMMALIAN URINARY SYSTEM

Fairly typical of mammals, the human body is about two-thirds water by weight, depending partly upon the amount of body fat (Figure 17.3). Far more body fluid is intracellular than extracellular. Naturally, if the body's fluid balance is to remain constant, then intake must equal output (Figure 17.4). This homeostatic balancing act is especially remarkable

FIGURE 17.2

(A) These marine shore crabs have euryhaline tolerances and scurry through many microclimates in the intertidal zone. Sessile intertidal animals also usually have euryhaline tolerances, as well as wide tolerances for most other physical factors in their environments. They spend roughly half of their lives submerged at high tides and half exposed to desiccating air at low tides, leading some biologists to call them "Dr. Jekyll and Mr. Hyde" animals. **(B)** The freshwater crayfish, however, is stenohaline, and lives in a relatively constant osmotic environment.

(A)

(B)

The Best of Both Worlds

Everyone watching sockeye salmon attempting to leap powerful falls on their way upstream to breeding sites marvels at their efforts, and persistence. However, many of the marvels involved are not so visible, such as the osmotic challenges that these animals face during their lifetimes.

Salmon and other anadromous fishes hatch in fresh water, migrate to the ocean to live most of their lives, and then return to fresh water to spawn. Thus, twice in their lives, these very euryhaline fish must reverse their osmoregulatory mechanisms. They normally make

these physiological adjustments during the time when they are in estuaries, the waters of intermediate salinity where freshwater rivers meet the sea. However, the euryhaline tolerances of many anadromous fish are such that they can survive being placed in water of almost any normal salinity range at any time.

Catadromous fish, such as freshwater eels, must have similar osmotic adaptations, although they live lives opposite those of anadromous species. Catadromous species hatch in salt water, migrate to fresh water to live, and return to the ocean to breed.

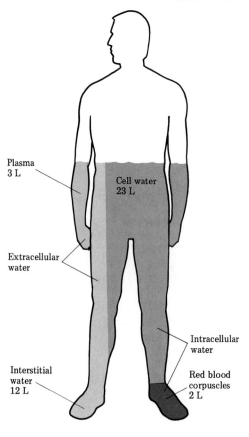

FIGURE 17.3

The average distribution of body fluids in humans. Intracellular fluids are those found within cells, whereas extracellular fluids include blood plasma and lymphatic and tissue fluids.

Plasma
3 L

Cell water
23 L

Extracellular water

Intracellular water

Interstitial water
12 L

Red blood corpuscles
2 L

when we realize all of the variables involved—summer heat vs. winter cold, work or exercise vs. rest, changing diets, and sickness vs. health. While moist foods contain free water, so-called metabolic water is released only during the oxidation of food being digested. For example, we secure about 0.6 g of water from oxidizing 1 g of glucose. Comparable figures for oxidizing 1 g each of protein or lipid are about 0.3 g and 1.1 g, respectively. Some mammals, such as the desert kangaroo rats, may never drink, securing all of their water from oxidizing the seeds they eat. The brain centers that control both fluid and food intake are located in the hypothalamus and in nearby structures (Chapter 19).

Kidneys are found only in vertebrates and they are the principal organs of excretion (Figure 17.5). Their working units, the numerous **nephrons,** are more elaborate than the nephridial organs of invertebrates, although their basic function is similar. The kidneys that are such indispensable organs of homeostasis perform three main functions: (1) They filter the blood of water, salts, wastes, and other materials; (2) they secrete several substances; and (3) they selectively reabsorb needed materials and excrete urine. In this way, they help to maintain the volume and chemical composition of the body fluids. The deep reddish-brown color of human kidneys reflects the very large blood supply of these bean-shaped (kidney beans, of course!), fist-sized organs. The volume of blood passing through the human kidneys is about 1200 ml per minute, or about 25 percent of the entire cardiac output. The paired kidneys that make up only about 0.4 percent of body weight (300 g, or 11 oz) are located just below the diaphragm in the rear of our abdominal cavity, and they are protected by shock-absorbing deposits of fat.

The urine that is produced in a constant trickle collects first in the funnel-shaped medial cavity of each kidney, the **renal pelvis.** It then

(Continues on p. 362.)

Daily Water Balance

Sources		Losses	
Fluid intake	1.2 L	Urine	1.5 L
Moist food	1.0 L	Skin	0.6 L
Metabolic water	0.35 L	Gut	0.1 L
		Lungs	0.3 L
Total intake	About 2.5 L	Total loss	About 2.5 L

40 L

FIGURE 17.4

The average daily human intake and output of fluids. The fact that these are only average figures is clear when we think about perspiration. On a cool day we may lose less than 500 ml of fluid through the skin, although on a hot day we may perspire more than 2 or 3 L. While exercising or working hard under hot conditions, we may lose 3 or 4 L of sweat per hour for short periods of time. Also, diarrhea increases and constipation reduces the amount of water lost in the feces. Normally, we become aware of the moisture in exhaled air only when we "see" our breath on cold days.

(A) Urinary System

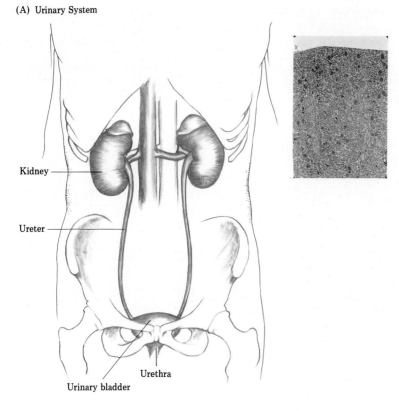

Kidney

Ureter

Urethra

Urinary bladder

(B) Kidney

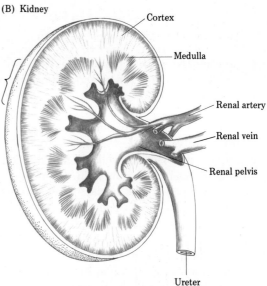

Cortex

Medulla

Renal artery

Renal vein

Renal pelvis

Ureter

FIGURE 17.5
(A) The kidneys and their relationship to other structures of the urinary system. Urine from the kidneys reaches the urinary bladder through the paired ureters and is then released from the body through the urethra. **(B)** This cross-sectional view of the kidney shows the abundant blood supply. The functional units, the nephrons **(C)**, are found partly in the outer cortex and partly in the roughly pyramidal inner medulla regions.

(C)   Nephron

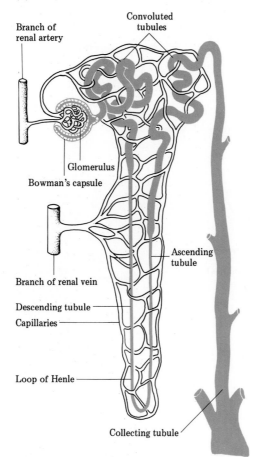

Branch of renal artery

Convoluted tubules

Glomerulus

Bowman's capsule

Ascending tubule

Branch of renal vein

Descending tubule

Capillaries

Loop of Henle

Collecting tubule

Comparative Excretion and Osmoregulation

Sponges and such cnidarians as jellyfish, most of which are marine, have no specialized excretory structures. Water and wastes are generally exchanged by diffusion and osmosis between their cells and their typically isotonic environment. Flatworms and some relatives are the simplest animals with specialized excretory organs. The **nephridial organs** that are common among invertebrates consist of simple or branching tubes that open to the outside through pores (Figure B17.1). The flatworm's **flame cells,** so named because of the apparent "flickering" of the moving cilia, lie in the fluid that bathes the animal's cells. The wastes that diffuse into these blind excretory cells are swept into the excretory ducts by the beating cilia. Another major role of the flame cells is to regulate the body's fluid content. Flame cells and related structures are also found in rotifers and in the larvae of some mollusks.

Earthworms have nephridial organs in each body segment, and these organs are open at each end (Figure B17.1B). The ciliated and funnellike inner end opens into the body cavity, and the associated tubule is surrounded by a network of capillaries that permit the removal of wastes from the blood. The body fluid that also passes into the long, looped tubule has its composition adjusted along the way. For instance, much water, glucose, and salts are reabsorbed and returned to the blood, just as wastes are concentrated. Earthworm urine is dilute and copious, being excreted at the rate of about 60 percent of the animal's body weight per day. (Imagine losing 60 to 100 or so pounds of urine daily instead of the usual 2 percent of your body weight!) The excretory structures of molluskans such as clams and snails are basically like those of earthworms.

The main excretory organs of crustaceans such as crabs and lobsters are paired structures called **green glands** (Figure B17.2). These greenish chambers with convoluted interiors are connected by tubules to bladders that open to the outside through pores. Each gland is bathed in blood, and the high pressure of that fluid forces out many materials except macromolecules and

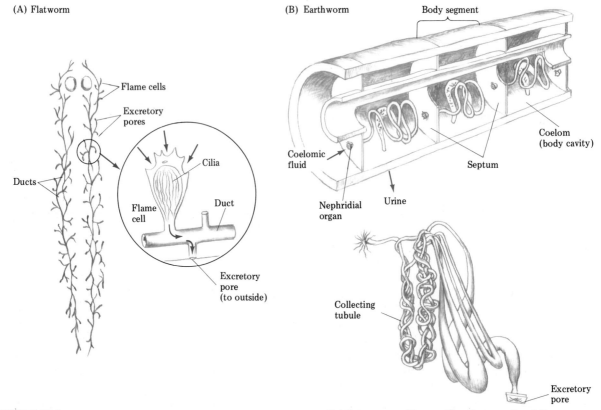

(A) Flatworm

Flame cells
Excretory pores
Cilia
Ducts
Flame cell
Duct
Excretory pore (to outside)

(B) Earthworm

Body segment
Coelom (body cavity)
Coelomic fluid
Septum
Nephridial organ
Urine
Collecting tubule
Excretory pore

FIGURE B17.1
(A) The flat shape of flatworms facilitates the loss of some wastes to the environment by diffusion. However, like many other invertebrates, flatworms also possess nephridial organs. The saclike flame cells of flatworms release wastes through pores that open to the outside of the body. (B) Paired nephridial organs are found in the body segments of earthworms.

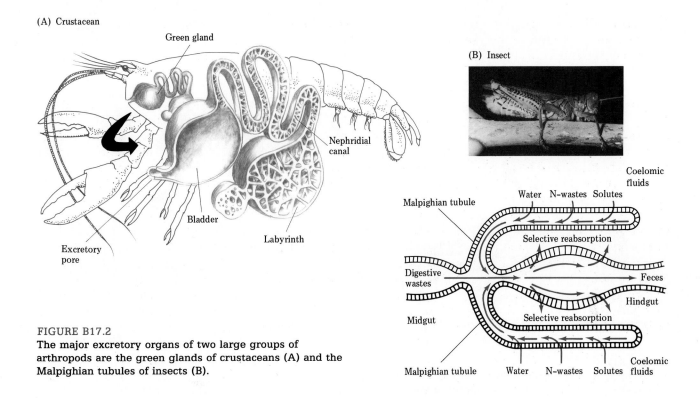

(A) Crustacean

Green gland

Nephridial canal

Bladder

Labyrinth

Excretory pore

(B) Insect

Coelomic fluids

Malpighian tubule

Water   N-wastes   Solutes

Selective reabsorption

Digestive wastes

Feces

Hindgut

Midgut

Selective reabsorption

Malpighian tubule

Water   N-wastes   Solutes

Coelomic fluids

FIGURE B17.2
The major excretory organs of two large groups of arthropods are the green glands of crustaceans (A) and the Malpighian tubules of insects (B).

cells. The fluid that passes from the blood has its volume and composition adjusted, with useful materials being reabsorbed and wastes concentrated for excretion.

The number of **Malpighian tubules** found in insects and other terrestrial arthropods varies from two to hundreds (Figure B17.2B). The cells of these thin-walled blind sacs with muscular walls, which float in the body's pool of fluids, receive wastes by diffusion or extract them by active transport. Water and several ions are reabsorbed into the blood both from the tubules and from the digestive tract to which the tubules are attached and into which they empty. Because terrestrial insects must conserve precious water to avoid desiccation (Chapter 14), the major waste product, uric acid, is excreted as fairly dry pellets.

To some extent, the metabolic wastes of vertebrates are excreted by the skin, the respiratory structures, and the digestive tract. For example, the lungs excrete water and carbon dioxide, the liver excretes bile salts, and the skin may excrete salts, urea, and water. Some animals also have specialized **salt glands** to excrete excess salts (Figure B17.3). However, the prime excretory organs of vertebrates are the kidneys, which with the urinary bladder and various tubes and ducts form the urinary system.

Like Coleridge's Ancient Mariner, marine bony fishes have a serious osmoregulation problem in that their body fluids are hypotonic to seawater (Figure B17.4). Even though surrounded by water ("Water, water, everywhere, nor any drop to drink"), they are in danger of dehydration because water is lost from their cells by osmosis. As a result, many drink salt water constantly and excrete the excess salt that it contains by means of specialized cells in their gills. Only a little isotonic urine is produced by their small kidneys, as most water must be retained.

Marine sharks and rays, the cartilaginous fishes, and a few other vertebrates solve the problem differently (Figure B17.4B). Their cells can tolerate concentrations of urea that would be toxic to most other animals. Because the urea renders their body fluids slightly hypertonic to seawater, some water enters their gills osmotically, and some additional water is ingested with food. The large and complex kidneys produce copious amounts of hypotonic urine. Excess salt is excreted both by the kidneys and by a specialized rectal salt gland.

The opposite problem is faced by freshwater fish, which are in constant danger of drowning because their body fluids are hypertonic to their surroundings (Figure B17.4C). Although they drink virtually no water and

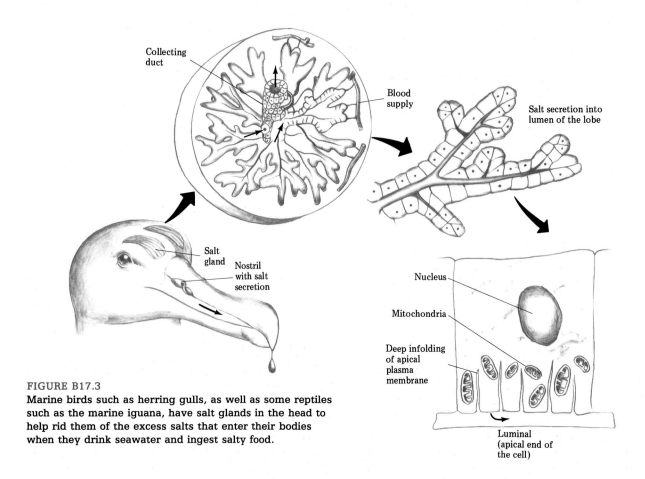

FIGURE B17.3

Marine birds such as herring gulls, as well as some reptiles such as the marine iguana, have salt glands in the head to help rid them of the excess salts that enter their bodies when they drink seawater and ingest salty food.

have mucus-covered scales generally impermeable to the passage of water, there is a large osmotic inflow of water from their environment to their hypertonic gill cells. The kidneys produce large amounts of hypotonic urine. To compensate for the potential salt loss to the environment, special cells in the gills actively transport salt into the body. Some salt also enters the body in the food that is ingested. Many freshwater fishes look as if they are drinking constantly but the water is simply entering their mouths, and not their digestive tracts, for passage past the gills for gas exchange and salt intake.

The osmoregulatory adaptations of most amphibians are similar to those of freshwater fish in that amphibians produce large amounts of dilute urine. However, the majority of amphibians, which are only semiaquatic, lose water and some salt through their skins by evaporation.

To compensate for the salt water that is ingested while they feed, the kidneys of marine mammals such as whales and dolphins produce very concentrated urine. The high protein diet of marine carnivores results in the production of large amounts of urea, which must be excreted in the urine.

Terrestrial environments create serious water conservation problems for many animals because of contact with relatively dry air. Of course, choosing moist and often cool microenvironments to live in, or being nocturnal when humidities are usually higher and temperatures are generally lower than during the day, helps many small terrestrial animals adapt to osmotic challenges. Becoming dormant during an especially harsh time is as much an adaptation to osmotic stresses as it is to adverse temperatures (Chapter 2). Because of their relatively high metabolic rates and their large surface areas relative to their volumes, small terrestrial animals often face more serious water balance problems than larger animals.

Two obvious adaptations to living on dry land are to have internal gas exchange structures such as lungs and to possess body surfaces relatively impermeable to water. Thus, the exoskeleton of many terrestrial arthropods such as insects and spiders is covered with a waxy

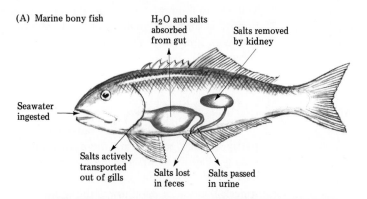

(A) Marine bony fish

H₂O and salts absorbed from gut

Salts removed by kidney

Seawater ingested

Salts actively transported out of gills

Salts lost in feces

Salts passed in urine

(B)

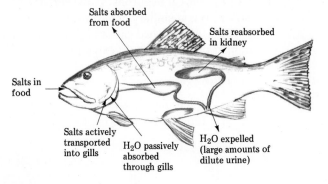

(C) Freshwater fish

Salts absorbed from food

Salts reabsorbed in kidney

Salts in food

Salts actively transported into gills

H₂O passively absorbed through gills

H₂O expelled (large amounts of dilute urine)

FIGURE B17.4
Osmoregulation in marine and freshwater fish. (A) Bony marine fish compensate for the water loss across the gills to their hypertonic environment by drinking seawater. Special cells in the gills excrete the excess salt, and the animals produce little urine. (B) Marine sharks become hypertonic to their environment by tolerating high concentrations of urea. Water is gained by osmosis and while consuming food. The kidneys produce abundant hypotonic or isotonic urine, and rectal salt glands excrete excess salt. (C) Because the osmotic concentration inside their cells is hypertonic to their environment, freshwater fish usually do not drink. Their kidneys produce large quantities of hypotonic urine from the water that is gained constantly by osmosis across their gills.

surface layer. The scales of fish and reptiles, the feathers of birds, and the hair of mammals also reduce surface water losses. Even so, other adaptations are needed for survival in very dry environments. Thus, some desert insects are able to secure moisture directly from air after it condenses on their bodies during times of high humidity. Reproduction in aquatic organisms often involves the release of large numbers of sperm and eggs into the environment, where they may meet by chance. Terrestrial animals must have mechanisms for reproduction, such as internal fertilization, without such a dependence upon external water.

passes through one of the paired **ureters,** which help to move it along by peristaltic contractions, and then enters the **urinary bladder.** This hollow, muscular organ can stretch to accommodate urine (up to about 800 ml in humans) until **urination** occurs through the **urethra.** The urethra is short in females and transports only urine, whereas it is longer in males and may also transport semen, although at different times (Chapter 5). Partly because of the short urethra in females, bladder infections are more common for them than for males.

The Nephron

The kidney consists of an outer **cortex** and an inner **medulla.** Found partly in each portion are the approximately one million nephrons per kidney (Figure 17.6). The renal artery branches repeatedly into an arteriole for each nephron. Each arteriole divides several times to form a spherical mass of capillaries, the **glomerulus,** which is surrounded by a cuplike, double-walled **Bowman's capsule,** or renal corpuscle. The glomerular capillaries merge to form another arteriole that leads out of Bowman's capsule. It branches again to form a network of capillaries that wrap around the tubular portions of the nephron. The venules that form from these capillaries merge into the renal vein, which returns the blood from the kidney to the heart.

The tubular part of the nephron consists of **proximal** (near) and **distal** (far) **convoluted tubules** that are connected by the U-shaped **loop of Henle.** The ends of the tubules connect to **collecting ducts** that carry urine into the renal pelvis. The capsular portions of nephrons are located in the cortex, as is most of the tubular portion in about 80 percent of human **cortical nephrons.** However, in about one-fifth of our nephrons, called **juxtamedullary nephrons,** the loops of Henle extend deep into the medulla. If stretched out and placed end to end, the two million or so human nephrons would extend over 60 km.

Urine Formation

The glomerulus and Bowman's capsule act as a blood-filtering unit. Because blood pressure is the force for filtration, excessively high or low

FIGURE 17.6

The major features of the two types of human nephrons (A), and micrographs of entire renal corpuscles (B) and of a cross-section of a single renal corpuscle (C).

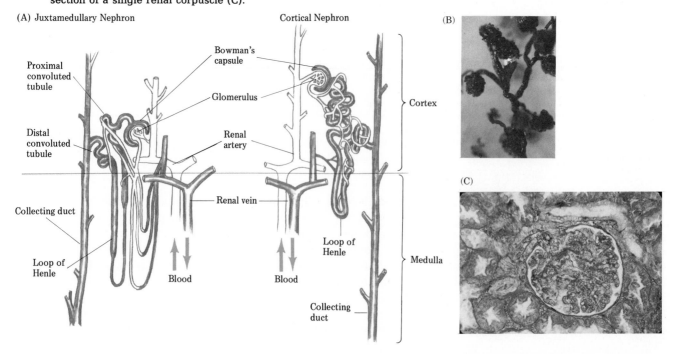

(A) Juxtamedullary Nephron

Cortical Nephron

(B)

Proximal convoluted tubule

Bowman's capsule

Glomerulus

Cortex

Distal convoluted tubule

Renal artery

(C)

Renal vein

Collecting duct

Loop of Henle

Loop of Henle

Medulla

Blood

Blood

Collecting duct

blood pressure may create problems. The entering arteriole is larger than the exiting arteriole, thus creating a higher blood pressure than that found in most capillaries. Glomerular capillaries also have thin walls. The linings of the filtering units contain pores that account for the rapid movement of water and many solutes of low to medium molecular weight from the blood. However, the openings are normally too small for the passage of blood cells, the abundant plasma proteins, and other macromolecules such as fat globules. The plasma passing through the glomerulus loses about one-fifth of its volume to the **glomerular filtrate.** The latter is not usually called urine yet because filtration is a nonselective process and the filtrate is still very similar to blood plasma (Table 17.1).

The glomerular filtration rate is about 180 L per day, equal to about 60 times the body's blood plasma content! At this rate, dehydration would quickly be life threatening unless **reabsorption** began immediately. The kidney's function of retention is as vital as its role in excretion. About 99 percent of the filtrate is reabsorbed into the blood through the nephron tubules, including about 178 L of water, 1200 g of salts, 250 g of glucose, plus amino acids, vitamins, and other nutrients and useful materials daily. About two-thirds of the filtrate is reabsorbed as it passes through the proximal convoluted tubule. The epithelial cells that line the tubule have numerous microvilli to provide an enormous surface area for reabsorption (Figure 17.7). Some materials move by diffusion, whereas others are actively transported. The many mitochondria in the tubule cells provide the ATP necessary for the active transport processes. Reabsorption continues across the loops of Henle, the distal convoluted tubules, and the collecting ducts. From an engineering standpoint it may sound inefficient to push out almost everything and then to sort out and reabsorb the materials that are still useful, but it works. In this sense, urine is very highly modified blood.

Secretion of some substances from the blood into the filtrate also occurs (Figure 17.8). For example, potassium, hydrogen and ammonium ions, as well as foreign substances such as certain dyes and some drugs such as penicillin, are removed from the blood in this manner. Among other functions, secretion helps to maintain the pH of blood.

Our two types of nephrons perform somewhat differently. In the cortical nephrons, sodium (Na^+) is transported to the interstitial fluid from the filtrate. Because this makes the surrounding fluid hypertonic to the filtrate, water flows out of the tubule into the interstitial fluid by osmosis. Salts, water, and other materials then enter the capillary network that surrounds the tubule.

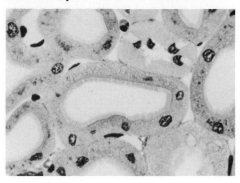

FIGURE 17.7
A micrograph of kidney tubule cells. Note the large surface area for reabsorption that is created by the numerous microvilli.

Substance	In Blood Plasma	In Urine
Inorganic ions		
Sodium	0.30	0.30
Chloride	0.40	0.60
Potassium	0.015	0.15
Phosphate	0.003	0.01
Nitrogenous wastes		
Ammonium ions	0.001	0.04
Urea	0.03	2.0
Uric acid	0.004	0.05
Conserved materials		
Glucose	1.0	0.0
Protein	7.0	0.0
Amino acids	0.03	0.0

TABLE 17.1
Concentrations of Some Materials in Blood Plasma and in Urine (grams per 100 milliliter of fluid)

FIGURE 17.8
The general mechanism by which glomerular filtrate is concentrated in a nephron. Osmosis occurs along much of the tubule and across the wall of the collecting duct. The sodium that is actively transported out of the ascending tubule into the surrounding interstitial fluid diffuses into the descending tubule. This helps create a countercurrent exchange that conserves water. Sodium is also actively transported out of the proximal and distal convoluted tubules.

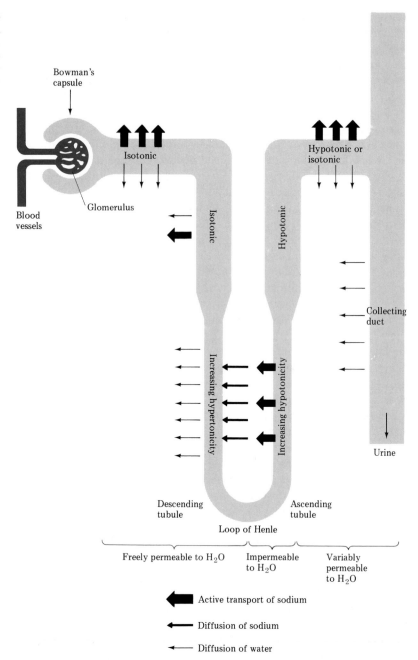

In the more internal juxtamedullary nephrons, the relatively long loop of Henle is specialized to produce a highly hypertonic interstitial fluid. The ascending loop allows chloride and sodium ions to pass through but it is impermeable to water. When the concentration of Cl^- and Na^+ ions becomes greater outside the tubule than inside, water moves passively out of the filtrate in the collecting ducts by osmosis. Some of the inorganic ions that leave the ascending loop move into the descending loop and recirculate back to the ascending loop. This **countercurrent exchange** results in the filtrate that flows through the distal convoluted tubule becoming isotonic, or even hypotonic, to blood. Because the interstitial fluid near the renal pelvis is very hypertonic to the filtrate, water is drawn osmotically from the filtrate in the collecting ducts and the fluid is then generally called **urine**. By creating hypertonic urine, this process conserves water.

Mammals that live in environments where water is scarce have a higher percentage of juxtamedullary nephrons than humans do, and the loop of Henle in each tends to be longer. Thus, some desert mammals produce urine that is over 20 times as concentrated as blood plasma. Some of these animals can actually drink saline water, something human sailors know is potentially lethal. That is, our kidneys cannot produce urine with a salt concentration greater than about 2 percent, whereas the salt concentration in seawater is about 3.5 percent. Therefore, if we drink seawater we lose more water than we consume, over 1300 ml of urine per liter of water ingested. This is because to rid ourselves of salt, we eliminate much water in order to dilute the salt to 2 percent. Furthermore, the magnesium and sulphate ions in seawater act as laxatives, causing diarrhea and further dehydration. Also, humans cannot get the water they need by eating fish. This is not because fish fluids contain too much salt for the kidney to handle but because fish are high in protein and, like salt, the excretion of protein requires much water.

Urine reaches the thin-walled bladder by gravitational flow and occasional peristaltic movements of the ureters. Sphincter muscles in the ureters help prevent the backflow of urine, as does the fact that the ureters enter the bladder at an oblique angle to form a flaplike valve. Pressure in the bladder tends to close the ureter openings, and as urine accumulates in the bladder, stretch receptors in the walls are stimulated (Figure 17.9). A smooth muscle layer allows for the rapid elimination of urine, and in most people the desire to urinate begins when the bladder holds about 300 ml of urine. Most people can suppress the urge to urinate until about 700–800 ml of urine have accumulated, but thereafter it is often painful or impossible to suppress urination.

Surrounding the part of the urethra closest to the bladder are two sphincter muscles controlled by the nervous system. The external sphincter is under conscious control, but the internal sphincter is not, except during urination when both sphincters are usually closed. Stretch receptors may cause the bladder to contract and the internal sphincters to relax. For urination to occur, either conscious activity of the brain or the passage of urine into the urethra must cause the external sphincter to relax. One may urinate before the natural reflex occurs by contracting muscles in the abdominal wall, stimulating the bladder's stretch receptors.

Neonates have no control over the external sphincter and urination occurs whenever the urination reflex begins. Bedwetting or **enuresis,** which often occurs for a number of years, may be an emotional, neurological, and/or structural problem. **Incontinence,** the loss of control over urination, occurs when the spinal cord or certain spinal nerves have been damaged. Affected people may have a tube (catheter) inserted into the urethra so that urine can flow into a container. Bladder and emotional disorders may cause temporary incontinence.

The 1–2 L of urine that is voided daily is about 96 percent water, 2.5 percent nitrogenous waste (mostly urea), and 1.5 percent salts, plus traces of other substances (Table 17.1). The yellow color of urine comes from the breakdown of bile pigments. The pH of urine is generally about neutral or slightly acid (4.5 to 8). The fact that no bacteria are usually present has allowed urine to be used to wash wounds under field conditions when pure water was unavailable!

Because the composition of urine reflects the status of so many body functions, the examination of urine chemically, physically, and microscopically, a **urinalysis,** is a powerful diagnostic tool. For example, the presence of blood cells, plasma proteins, or quantities of materials that are usually reabsorbed (e.g., glucose) indicates problems.

FIGURE 17.9
The structure of the bladder.

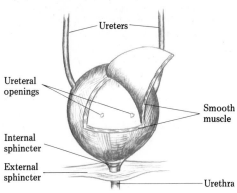

Ureters

Ureteral openings

Smooth muscle

Internal sphincter

External sphincter

Urethra

Ailing and Failing Kidneys

Among the major diseases, kidney problems rank about fifth in the United States, killing over 100,000 Americans annually. Many more are incapacitated and several million Americans have undiagnosed kidney problems. The kidneys are susceptible to various pathogens, poisons (e.g., carbon tetrachloride and mercury), shock, many circulatory system problems, and some of the other problems described below.

Some metabolic wastes are insoluble and are transported as small crystals. Occasionally these crystals aggregate to form **kidney stones,** which may alter or block urine release. The passage of such a stone through the kidney and ureter is extremely painful. Some kidney stones can be dissolved but others must be removed surgically or treated with ultrasound. **Gout** results when uric acid crystals form in the joints, such as those of the foot (Chapter 21).

Nephritis is the inflammation of the glomerulus or nephron tubules and this may allow blood cells and plasma proteins to enter the glomerular filtrate. Hypertension also results, and kidney failure may occur if nephritis is not treated successfully. Elevated arterial blood pressure not only damages the kidney and cardiovascular system, but may also cause liver failure.

Like most problems, the various kidney diseases are treated most easily and successfully if they are diagnosed early. No symptom is a sure indication of kidney problems, but the following are common early warning signs. If any of these symptoms persist, consult a physician.

- Puffiness around the eyes
- Edema of various body parts
- Flank or lower back pain or tenderness
- Bloody or coffee-colored urine
- Changes in the normal frequency of urination
- Burning or other abnormal sensation during urination

If one kidney fails it is commonly removed surgically, and otherwise healthy people can usually live normal lives with one good kidney. However, the most dangerous circumstance is obviously the failure of both kidneys. When the nephrons cease to function properly, edema occurs, acidosis results from the increase in hydrogen ions, and nitrogenous wastes accumulate, all of which may lead to nausea, fatigue, mental changes, and, in advanced cases, coma and eventual death. Kidney dialysis and kidney transplants are common forms of treatment.

The term **dialysis** means the separation of substances across a differentially permeable membrane between solutions of differing concentrations. Even modern equipment, often called an artificial kidney, is expensive to construct, maintain, and use, and it is not available to all who need it (Figure 17.10). The selection of those who will receive care from the larger number of those who would benefit from dialysis is not easy, is fraught with moral and ethical considerations (such as "worthiness" or financial status), and is often a life or death decision. (For example, should a highly educated person, such as a physician or lawyer, have precedence over a relatively uneducated laborer, or a young person over an elderly individual, or a mother of four over a childless female? Should individual physicians make these crucial decisions or should a committee be involved? If the latter, then how should committee members be selected?) For temporary disorders, the use of kidney dialysis may be short term, but for serious or chronic kidney disease its use may be for life, or until a kidney transplant is attempted.

Continuous ambulatory peritoneal dialysis (CAPD) is a different di-

(A)

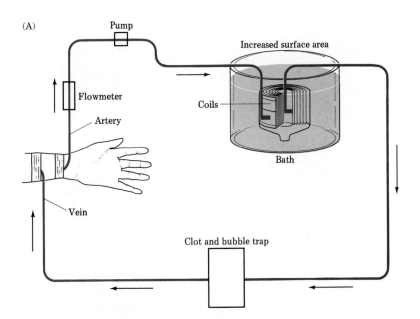

(B)

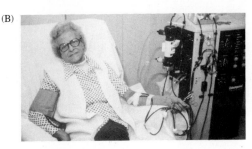

FIGURE 17.10
The artificial kidney is a device through which the patient's arterial blood flows (A). Pores in the dialysis equipment sheets allow waste molecules to pass through, and the cleansed blood returns to the patient through a venous catheter. Anticoagulants are used to keep the blood from clotting. The frequency of treatment depends upon the severity of the problem but it is often several times a week, with each treatment lasting several hours, for the life of the patient (B). Although some dialysis machines, including portable devices, are now available for home use, most treatment occurs in hospitals or clinics.

alysis technique. CAPD depends upon the fact that the lining of the abdominal cavity, the peritoneum, is a natural differentially permeable membrane. The kidney patient may move around while the dialysis fluid in an attached bag circulates through the abdominal cavity. Bags with fresh fluid periodically replace those with old fluid, which are then discarded. In all dialysis, precautions must be taken to avoid bacterial infections, internal clotting of blood, and the introduction of gas bubbles into the blood.

Interest in transplants is high because a real kidney is usually more desirable than the use of dialysis. Also, although the initial surgical expenses are high, the long-term treatment costs are often lower than those for dialysis. Successful transplants usually allow the recipient to live a normal life for at least a few, and often for many, years. Kidney transplants are now one of the most common and successful organ transplant procedures, partly due to drug treatment that suppresses rejection of the foreign organ. Transplants from twins or other siblings are more successful than those from parents, who are less related genetically, or from totally unrelated donors. Unfortunately, the supply of healthy kidneys for transplants is far below the number needed. Perhaps all driver's licenses will soon indicate whether the driver, in case of a fatal accident, has consented to have his or her kidneys or other organs donated as potential transplants.

Regulation of Urine Production

Blood is circulating constantly and the rate of its filtration is fairly constant, as is the reabsorption rate in the proximal convoluted tubules. However, the exact rate of urine production depends upon such variables as fluid and salt intake, the climate, and one's activities and emotions. These factors have their varying effects because the rate of reabsorption changes in the collecting ducts. The hormone that controls the permeability of the collecting ducts is **antidiuretic hormone** (ADH), which is synthesized in the hypothalamus and released by the posterior pituitary. ADH increases the permeability of the collecting ducts so that more water is reabsorbed (Figure 17.11).

When the body is dehydrated, as when fluid intake is low or during prolonged diarrhea, the concentration of dissolved salts in the blood

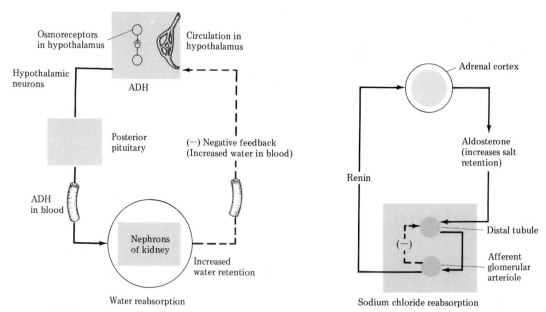

FIGURE 17.11

The major features of the regulation of urine output. Antidiuretic hormone (ADH) that is released from the pituitary gland aids in conserving water by increasing the reabsorption of water from the collecting ducts. The stress of cold and alcohol inhibit the release of ADH, whereas dehydration, hemorrhage, pain, and emotional stress stimulate ADH secretion. Reabsorption is regulated so precisely that the fluid content of the body tissues remains virtually constant.

increases. The increased osmotic pressure of the blood is sensed by specialized osmoreceptors in the brain and in some large blood vessels. This leads to increased release of ADH, with the result that more water is reabsorbed from the filtrate passing through the collecting ducts, whose walls are then more permeable to water. The ADH causes water to be conserved, the blood volume to increase, the production of only a small volume of hypertonic urine, and, in general, the restoration of homeostatic normality. ADH release is also increased during sleep.

Thirst and fluid balance in general are linked closely to sodium balance because Na^+ makes up about 90 percent of extracellular positive ions. Where salt concentration increases, water is drawn to the region osmotically. The increased ingestion of salt leads to an increased osmotic pressure of the blood. This is what the thirst center in the hypothalamus senses that leads it to cause the sensation of thirst. Drinking water restores the osmotic balance of the blood but it may also increase blood volume. That, in turn, leads to a decrease in ADH secretion, which results in increased excretion of water. This is an excellent example of the fine tuning of body functions by the interaction of many body parts, the negative feedback processes of homeostasis. (If the sodium concentration in the blood is low, its reabsorption is also controlled by another hormone, aldosterone, that is described in Chapter 18.)

Agents that increase the flow of urine are called **diuretics.** Caffeine, a common diuretic in coffee, tea, and many soft drinks, decreases the amount of water reabsorbed. Drinking large amounts of fluid results in dilution of the blood and a fall in its osmotic pressure. That reduces the release of ADH, lessening the amount of water reabsorbed, and leads to the production of copious, hypotonic urine. Although the diuretic effect of beer (which is about 4 percent alcohol) is due largely to its water content, concentrated alcohol is a powerful diuretic that results in several milliliters of water being excreted for every milliliter of liquor ingested. Straight whiskey and vodka are about 40–50 percent alcohol, and wine is about 12 percent alcohol.

The several forms of diabetes affect the volume and chemistry of urine (Chapter 18). People with relatively rare **diabetes insipidus** may produce up to 25 L (7 gal) of urine daily, have an incessant thirst, and suffer from a salt imbalance in the body fluids. Because ADH is deficient, resorption of water is low and large amounts of hypotonic urine are produced. **Diabetes mellitus** is far more common and involves the presence of glucose in the urine. The excess glucose in the tubule increases the osmotic pressure of the filtrate and thus reduces resorption of water.

SUMMARY

1. **Osmoregulation** is the homeostatic regulation of the volume and chemical composition of body fluids.

2. Larger animals have **excretory systems** to collect excess water, ions, and metabolic wastes, reabsorb useful materials, and expel the remainder. Small and simple animals may exchange water and wastes directly with their environment, without specialized structures.

3. Water, carbon dioxide, and nitrogenous compounds are the principal metabolic wastes. The latter exist in three common forms—**ammonia, urea,** and **uric acid.**

4. Ammonia is highly toxic but also highly soluble and is generally released by such animals as aquatic invertebrates and freshwater fishes directly into their environments, which dilute it quickly.

5. Marine fishes, amphibians, and mammals release nitrogenous wastes mainly in the form of urea. Compared with ammonia, urea requires more metabolic energy to form but it is less toxic and less water is lost during its excretion.

6. Many terrestrial animals excrete uric acid, which helps them conserve water. Uric acid requires much metabolic energy to synthesize but it is the least toxic of the nitrogenous wastes.

7. The prefix **eury-** is used with an appropriate suffix (e.g., euryhaline) for animals with a wide range of tolerance; **steno-** describes those with narrow ranges of tolerance.

8. Aquatic environments with salt concentrations less than those found in the body fluids of their animal inhabitants are called **hypotonic.** Other waters that have salt concentrations equal to or greater than those of aquatic animals are called **isotonic** or **hypertonic,** respectively.

9. The **nephridial organs** that are common among invertebrates consist of simple or branching tubes that open to the outside through pores. The **flame cells** of flatworms, the **green glands** of crustaceans, and the **Malpighian tubules** of insects are common examples. As in kidneys, nephridial organs adjust the volume and concentration of the body fluids that pass through them.

10. Vertebrates lose some wastes through their skins and their respiratory and digestive systems. Some also have specialized **salt glands.** However, the prime organ of excretion, found only in this animal group, is the **kidney.** Together with the **urinary bladder** and various tubes and ducts, these organs of major importance in homeostasis constitute the urinary system.

11. Because their body fluids are hypotonic to seawater, marine bony fishes must drink constantly to prevent dehydration. Specialized gill cells excrete excess salts, and a fish's small kidneys produce little isotonic urine. In contrast, marine cartilaginous fishes tolerate high levels of urea, which makes their body fluids somewhat hypertonic to seawater. Because water enters their gills osmotically, and their bodies when they eat, these

fishes produce large amounts of hypotonic urine. Rectal salt glands excrete excess salts.

12. Because their body fluids are hypertonic to their environment, freshwater fish face the problem of excess water inflow through their gills. Even though these fishes do not drink, their kidneys produce copious amounts of hypotonic urine.

13. Amphibians generally have osmoregulatory adaptations similar to those of freshwater fish. In contrast, most marine mammals produce very hypertonic urine because of their high-protein diet and because of the salt that is ingested when they feed.

14. Terrestrial environments create serious water conservation problems for many animals. Some animals lessen the impact of contact with relatively dry air by their behavior and through their choice of habitats. In general, small animals face more serious osmoregulatory problems than large animals. Most terrestrial animals have body surfaces that are relatively impermeable to water, and reproductive patterns that do not depend heavily upon an aquatic environment.

15. Many mammals are about two-thirds water by weight, with most water occurring intracellularly. For a constant fluid balance, intake (mainly drinking and eating) must equal fluid output (mainly urine, sweat, and losses through digestive and respiratory processes). The brain centers that control fluid and food intake are located in the hypothalamus.

16. To facilitate the filtration of blood, there is a very large blood supply to the paired kidneys, which are located in the abdominal cavity. The urine that is produced in a constant trickle in the kidneys passes through the **ureters** to the hollow, muscular, and expandable urinary bladder, from which it is expelled periodically through the **urethra** during **urination.**

17. The approximately two million filtering units in the human kidneys, the **nephrons,** are found partly in the outer **cortex** and partly in the inner **medulla.** The arterioles that form the spherical **glomerulus** are surrounded by the cuplike, double-walled **Bowman's capsule.** The exiting arteriole divides to form the capillary network that surrounds the **proximal** and **distal convoluted tubules** and the U-shaped **loop of Henle.** The venules that form from those capillaries merge into the renal vein.

18. Blood pressure forces many substances from the blood plasma through Bowman's capsule to form the **glomerular filtrate.** Normally, blood cells, plasma proteins, and other macromolecules remain in the blood plasma.

19. **Reabsorption** begins immediately in the proximal convoluted tubule, and eventually about 99 percent of the filtrate is reabsorbed into the blood. Some materials move by diffusion and others are actively transported. Numerous microvilli and abundant mitochondria in the tubule cells facilitate reabsorption, as does **countercurrent exchange.** Some materials are also **secreted** from the blood into the filtrate.

20. **Urine** is mostly water but also contains urea and various salts. The yellow color results from the breakdown products of bile salts. Urine is generally neutral or slightly acidic and is usually free of bacteria. Because urine reflects many body functions, a **urinalysis** yields an abundance of useful diagnostic information.

21. The rates of blood filtration and reabsorption in the proximal convoluted tubules are fairly constant. However, the rates of reabsorption in the collecting ducts vary under the control of **antidiuretic hormone**. ADH increases the permeability of the collecting ducts. Agents that increase urine production are called **diuretics** (e.g., alcohol and caffeine).

REVIEW QUESTIONS

1. How does excretion differ from secretion and from egestion or elimination?

2. Describe how ammonia, urea, and uric acid are formed.

3. What is the relationship between an animal's habitat and life-style and the form of the nitrogenous wastes that it excretes?

4. What is the difference between a euryhaline and a stenohaline animal?

5. Use the terms *hypotonic, isotonic,* and *hypertonic* to describe the osmotic relationships between aquatic animals and their environments.

6. Describe the structures and methods by which the nephridial organs called *flame cells, green glands,* and *Malpighian tubules* adjust the volume and composition of the body fluids that pass through them.

7. Compare and contrast osmoregulation and waste disposal in marine and in freshwater fishes.

8. How do marine cartilaginous fishes excrete wastes and osmoregulate differently from marine bony fishes?

9. Describe some of the major water conservation adaptations of terrestrial animals.

10. State the average quantity of body fluids, identify their general locations in the body, and describe how fluid intake relates to fluid output.

11. Describe the structure of a typical nephron and explain the filtration of blood, the selective reabsorption of useful materials, and the production of urine.

12. Compare the volume and composition of blood plasma, glomerular filtrate, and urine.

13. Distinguish among the location, structures, and functions of the two main types of nephrons (cortical and juxtamedullary).

14. Describe the source and function of antidiuretic hormone in maintaining the homeostatic control of the body's fluid balance.

15. Name at least two diuretics and explain how some such substances exert their effects.

16. Would you donate a kidney to a close relative, or to a total stranger, while you are alive? Allow your kidneys to be used after you die? Accept a donated kidney?

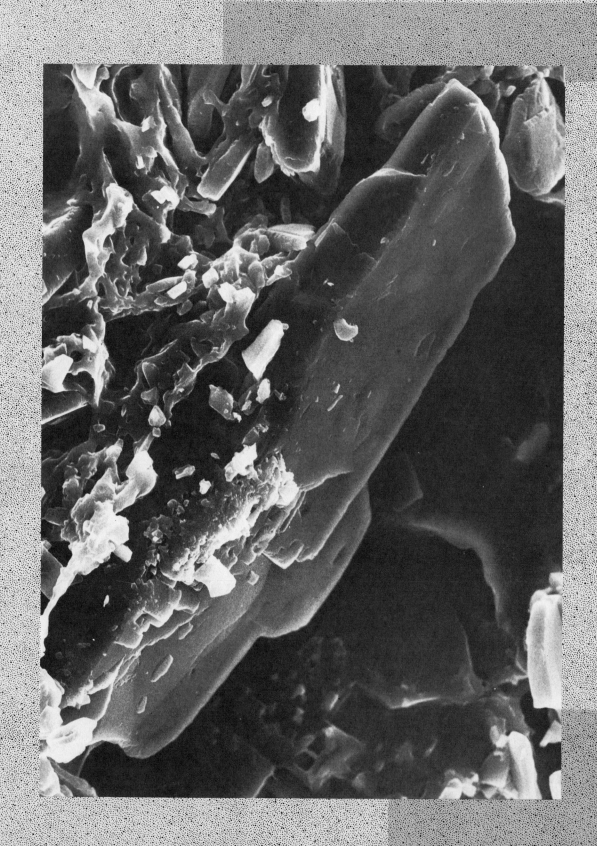

CHAPTER 18

REGULATION: THE ENDOCRINE SYSTEM

Homeostasis depends upon the continuous regulation of body processes. The division of labor that one finds among the body parts of animals is possible only if the parts are integrated and coordinated by a communications network. Animals have nervous and chemical networks, which work in concert. Nervous regulation, which is a rapid method of ongoing control, is treated in Chapters 19 and 20, and control by chemicals, often more long term, is treated here. Both systems play crucial homeostatic roles in directing appropriate responses to changing external and internal conditions. In this chapter, you will study the chemical controls that are especially important in reproduction, growth, metabolism, the maintenance of blood and body fluid volume and composition, digestion, responding to stress, and behavior.

Even though it has been magnified 3000 times, the hormone crystal does not look like much to most people. Hormones are produced by many tissues throughout the body, and, like the old saying, "a little goes a long way." In this chapter, you will learn about the major endocrine glands and about the variety of their hormones and effects.

FIGURE 18.1
(A) The major mammalian endocrine glands are located throughout the body. A connection with the bloodstream is more important than their specific location. (B) Endocrine glands are ''powerful'' for their sizes, and most human endocrine tissue will fit in the palm of one hand.

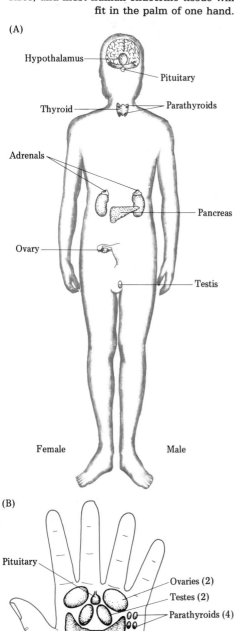

(A)

Hypothalamus

Pituitary

Thyroid

Parathyroids

Adrenals

Pancreas

Ovary

Testis

Female   Male

(B)

Pituitary

Ovaries (2)

Testes (2)

Parathyroids (4)

Thyroid

Adrenals (2)

Pineal

Approximate quantity of islet cells in pancreas

GLANDS AND HORMONES

An interesting experiment was done in 1849 by the German physician A. A. Berthold using six young male chickens. In two castrated birds, the combs, wattles, plumages, and behaviors of normal roosters failed to develop. Two other birds were also castrated, but the testes were reimplanted into distant sites where the testes reestablished access to blood vessels but not to nerves. These birds developed normally, as did two control birds that were not operated upon. The results showed that substances made by one part of the body could affect distant tissues. This was the first successful and clear-cut experiment in **endocrinology,** the study of the endocrine glands, similar specialized cells or cell groups, and of chemical messengers in general. **Endocrine glands,** the so-called glands of internal secretion, are ductless glands that release one or more **hormones** into the bloodstream or interstitial fluids, which carry the substances to other parts of the body. Other familiar glands, such as sweat, salivary, and digestive glands, are called **exocrine** because they secrete through a duct to some body surface or cavity. Hormones, from the Greek word meaning "to excite," do just that to the tissues that are sensitive to them, the so-called **target tissues.** That is, most cells are in constant contact with blood, and, therefore, with most hormones, but only target cells respond biochemically to a particular hormone. While the target tissue is often another endocrine gland, it may also be another body structure or, in some cases, most body cells.

The major endocrine glands of mammals, shown in Figure 18.1A and summarized in Table 18.1, are scattered throughout the body. The fact that hormones are released into the blood means that the location of endocrine glands is relatively unimportant. The thyroid, for example, might be as effective in your big toe as it is in your neck, provided it had access to the bloodstream.

Note in Figure 18.1B that size is not always a good criterion of how important body parts are. Hormones are effective in very tiny amounts, even as low as about 1 part per million parts of blood. A woman produces only about a teaspoon of estrogen in her lifetime!

Although they are extremely potent compounds, hormones act quite slowly in comparison with responses to nerve impulses. Even our endocrine reaction to stress (our "fight-or-flight" response) is sluggish compared with most related responses controlled by the nervous system. However, hormones or hormonelike substances are also secreted by nerve cells or neurons (Figure 18.2). Logically enough, the cells that secrete these **neurohormones** in both invertebrates and vertebrates are called **neurosecretory cells.** Because they are chemical messengers, the **neurotransmitters** secreted by one neuron to communicate with a neighboring neuron or with a muscle cell are studied by endocrinologists as well as by neurologists (Chapter 19).

Most individual cells and tissues also release chemical messengers by which they communicate with other cells. The histamine that is released during inflammation and allergic reactions is one example (Chapter 16). Some of these chemical messengers flow in the blood but others simply diffuse through interstitial fluids to reach their target tissues. Collectively, all of these structures that secrete chemical messengers constitute the **endocrine system.**

THE CHEMISTRY AND ACTIONS OF HORMONES

Target cells detect and respond to a hormone because they have receptor molecules to which the hormone can bind (Figure 18.3). As are some

TABLE 18.1
Principal Vertebrate Endocrine Glands, Their Hormones, and Major Actions[a]

Gland	Hormone(s)	Major Actions
Hypothalamus	Releasing and inhibiting hormones	Stimulate or inhibit secretions of hormones by the anterior pituitary
	Oxytocin and antidiuretic hormone	Stored and released by posterior pituitary (see below)
Posterior pituitary	Oxytocin	Stimulates uterine contractions and release of milk from mammary glands
	Antidiuretic hormone	Stimulates reabsorption of water in kidneys; contractions of smooth muscles in walls of small arteries
Anterior pituitary	Adrenocorticotropic hormone	Stimulates release of hormones from adrenal cortex
	Thyrotropic hormone	Stimulates thyroid to release hormones and to grow
	Follicle-stimulating hormone	Stimulates growth of ovarian follicles in females and sperm development in males
	Luteinizing hormone	Stimulates conversion of ovarian follicle into corpus luteum; stimulates secretion of sex hormones by gonads
	Prolactin	Stimulates milk production in mammary glands
	Growth hormone	Stimulates growth
Thyroid	Thyroxin	Controls metabolic rate; essential for normal growth and development
	Calcitonin	Lowers blood calcium level
Parathyroids	Parathormone	Increases blood calcium level
Adrenal cortex	Mineralocorticoids	Control fluid balance; regulate metabolism of salts
	Glucocorticoids	Control metabolism of nutrients; reduce tissue inflammation; adaptation to long-term stress
Adrenal medulla	Epinephrine	Mobilizes reactions to stress
	Norepinephrine	Helps regulate reactions to stress; sustains blood pressure
Pancreas	Insulin	Facilitates glucose uptake and utilization by cells; stimulates glycogen formation, fat storage, and protein synthesis
	Glucagon	Increases blood glucose level
Pineal	Melatonin	Onset of puberty in humans; aspects of reproduction and skin pigmentation in some animals

[a] The structures and functions of some endocrine glands are treated in other chapters (e.g., thymus in Chapter 16 and the gonads in Chapter 5). See text for details about prostaglandins, endorphins, and pheromones.

other biochemical reactions, this is often likened to a lock-and-key mechanism. Some target cells can respond differently and specifically to more than one hormone because they possess more than one type of hormone receptor. However, not all target tissues respond to the same hormone in the same way at all times. In part, this is due to synergistic interactions, such as when the presence of one hormone enhances the effect of another. Cells that lack appropriate hormone receptors are in contact with the hormones that flow in adjacent blood vessels and tissue fluids but they are not affected directly by the hormones.

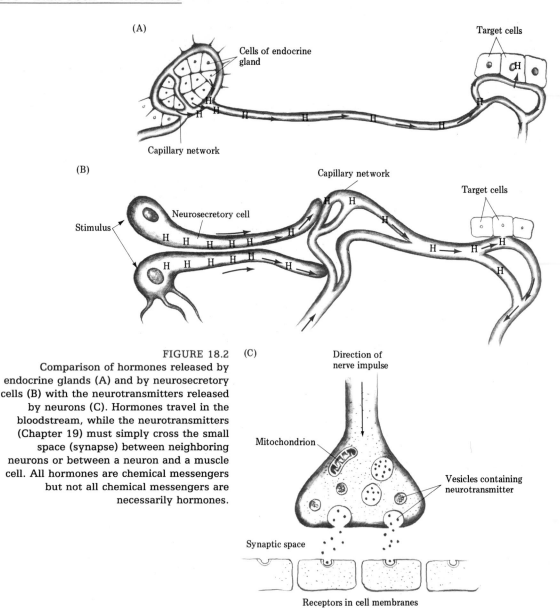

FIGURE 18.2
Comparison of hormones released by endocrine glands (A) and by neurosecretory cells (B) with the neurotransmitters released by neurons (C). Hormones travel in the bloodstream, while the neurotransmitters (Chapter 19) must simply cross the small space (synapse) between neighboring neurons or between a neuron and a muscle cell. All hormones are chemical messengers but not all chemical messengers are necessarily hormones.

Most hormones are medium-sized molecules that do not belong to any one chemical group (Figure 18.4). Some are relatively small polypeptides or proteins, and others are steroids, amino acids, or amino acid derivatives. Because of their size, hormones are large enough to have specific characteristics but small enough to circulate easily. Human protein hormones, for example, vary from three amino acids (thyrotropin-releasing hormone) to 191 amino acid units (growth hormone). During evolution, various organic substances have come to serve as chemical messengers.

Although the chemical structure of hormones generally varies little from species to species, the hormone's functions may. Thus, while thyroxin from the thyroid gland regulates the metabolic rate in mammals, it controls the metamorphosis of an amphibian tadpole into an adult. Metamorphosis is a complex process, just as control of multifaceted mammalian metabolism is. For example, the tadpole's tail, gills, and teeth must be resorbed while concurrent constructive changes are causing limbs to develop, as well as producing changes in the digestive tract, tongue, middle ear, and retina. Thus, the specific responses to a hormone

(A)

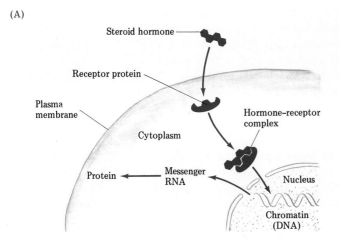

Steroid hormone

Receptor protein

Plasma membrane

Cytoplasm

Hormone-receptor complex

Protein ← Messenger RNA

Nucleus

Chromatin (DNA)

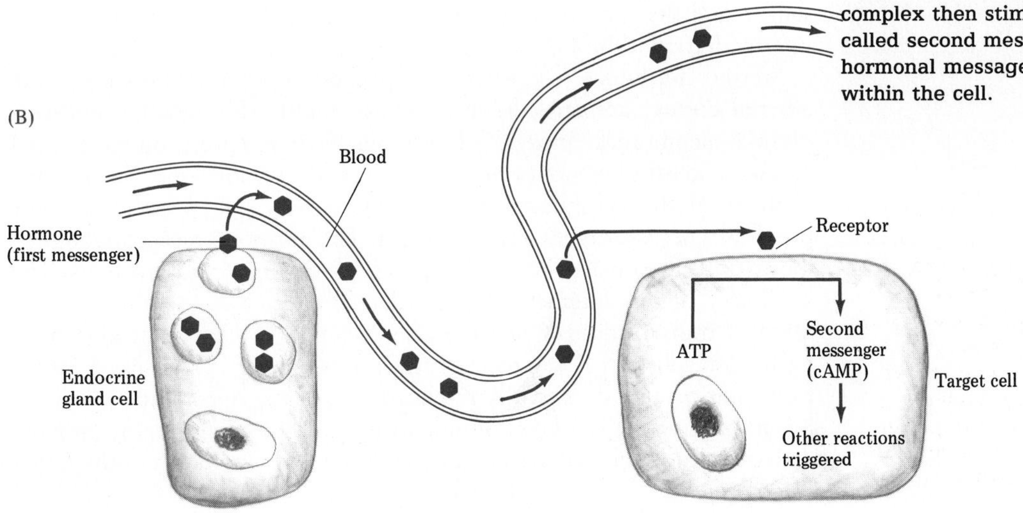

FIGURE 18.3
(A) The general mode of action of steroid hormones. Further details of how particular proteins are synthesized on the ribosomes are found in Chapter 10. In brief, however, these and some other types of fat-soluble hormones pass easily through the plasma membrane of a target cell and attach to specific protein receptors. In the nucleus, this hormone-receptor complex combines with another protein receptor that is associated with DNA. This combination activates certain genes and leads to the synthesis of the messenger RNAs coding for specific proteins. **(B)** In most cases, especially for larger protein hormones, the hormone attaches to receptors in the target cell's surface. That hormone-receptor complex then stimulates enzymatically a so-called second messenger to relay the hormonal message to the appropriate site within the cell.

(B)

Blood

Hormone (first messenger)

Receptor

Endocrine gland cell

ATP

Second messenger (cAMP)

Target cell

Other reactions triggered

$$S \text{———} S$$
cys — tyr — ile — gln — asn — cys — pro — leu — gly(NH$_2$)

Oxytocin (mammals, birds)

SMALL POLYPEPTIDE

his — ser — gln — gly — thr — phe — thr — ser — asp — tyr — ser — lys — tyr — leu — asp —

ser — arg — arg — ala — gln — asp — phe — val — gln — trp — leu — met — asn — thr

Glucogon

SMALL PROTEIN

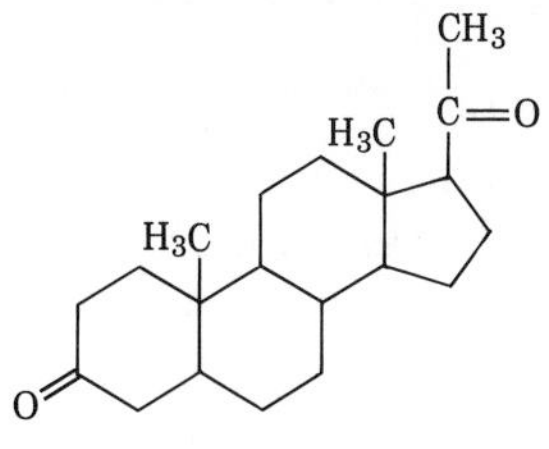

Progesterone

STEROID

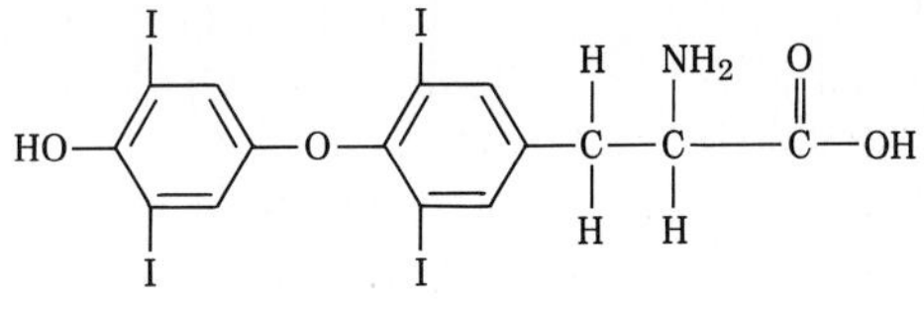

Thyroxin

AMINO ACID DERIVATIVE

FIGURE 18.4
The chemical structures of some representative mammalian hormones. Note that hormones belong to no one particular chemical group. The sources and functions of these hormones are summarized in Table 18.1 and described in the text. Sometimes, seemingly small differences, such as the nature of the amino acids present in oxytocin and antidiuretic hormone, lead to quite large differences in function. Certain hormones contain some relatively rare elements, such as iodine in thyroxin.

depend upon the target cell, not on the hormone itself. Each cell is programmed genetically to respond to a specific hormone in a particular way. That response is also affected by the animal's developmental stage and its physiological state at any given time.

In vertebrate blood, 30 to 40 different hormones may be circulating at any time. Most of these hormones are transported bound to plasma proteins. The free hormones in the blood plasma are removed continuously from circulation by target tissues. The liver also inactivates some hormones, and the kidneys excrete others.

In the case of many protein, polypeptide, and amine-type hormones, the hormone forms a hormone-receptor complex in the target cell's plasma membrane. The hormone, or "first messenger," stimulates an enzyme (adenylate cyclase) to convert ATP to a so-called **second messenger** (cyclic AMP), which then relays the hormone's message to the appropriate site inside the cell. The hormone itself is an external messenger and the second messenger may be thought of as the hormone's intermediary within a cell.

Steroid hormones (e.g., estrogen, progesterone, and those from the adrenal cortex) are fat soluble and pass readily through the lipid-rich plasma membranes. Instead of reacting with receptors on target cell surfaces, steroid hormones become bound to hormone-specific receptor proteins in the cytoplasm. Nonresponsive cells lack specific receptor proteins whereas target cells have them. The hormone-receptor complex moves into the nucleus and associates with chromosomal proteins. The specificity of this interaction with selected regions of the chromosome initiates transcription. When the messenger RNA that is produced by particular genes in DNA reaches the cytoplasm, it is translated into a particular protein dictated by the hormone. Thus, steroid hormones act by stimulating the synthesis of new proteins rather than by affecting the activity of proteins that are already present in the cytoplasm of the target cell.

Negative feedback mechanisms explain how endocrine structures "know" how much of a hormone to secrete at any moment (Figure 18.5). For example, through the action of their hormone (parathormone) working in conjunction with other endocrine glands, the parathyroid glands that are found in the neck region of mammals regulate the calcium level of blood (see also Figure 18.11). A drop in blood calcium concentration stimulates the parathyroid glands to release more parathormone. Parathormone causes the kidney tubules to reabsorb more calcium and it causes the bones to release calcium ions into the blood, thus restoring

FIGURE 18.5
Negative feedback mechanisms are +,− or −,+, so designated because the response is opposite that of the stimulus, and thus tends to restore imbalances. For instance, above-normal or below-normal levels of various blood substances, such as calcium in this example, either stimulate or inhibit the release of one or more hormones. The substance is then either withdrawn from or released into the blood to maintain the level of the substance within an appropriate range of tolerance. This example is greatly simplified because, as with many other endocrine instances, any hormone interacts with one or more other hormones.

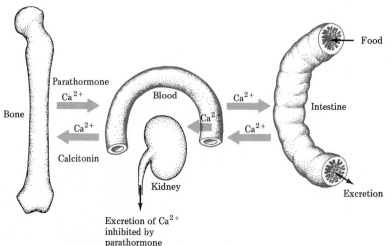

the homeostatic balance. Conversely, an above-normal rise in blood calcium inhibits the release of parathormone, with generally opposite effects (see later discussion).

To learn about the endocrine system in several invertebrates, see the essay "Invertebrate Hormones."

VERTEBRATE HORMONES

The positions of the endocrine glands have not changed much in the course of vertebrate evolution. Can such scattered, generally small masses of cells, which secrete such a variety of chemical substances that often have such distant effects, constitute a system in the same sense that we use the term *system* in discussing digestive or respiratory structures or functions? As is often the case in biology, ancient origins may offer a clue. Suppose, for example, that the genetic program of some animal was changed slightly (e.g., by a mutation; Chapter 10) in terms of its transmitter substances. Also suppose that the transmitter substance reached the bloodstream and that it had a surface configuration that allowed it to attach to or penetrate cells elsewhere. If the cell's metabolic activity were changed in a way that enhanced the ability of the animal to survive and reproduce, then such a beneficial mutation might be perpetuated. While these constitute small steps and chance events, this is one way that evolution often works. Today, neurosecretory cells are the main sources of hormones in many invertebrate groups.

Even though endocrine structures are diverse and dispersed, they truly work together. Often their hormones simply influence the same structure, whereas at other times their hormones interact more directly, sometimes in an antagonistic and other times in a synergistic ("1 + 1 = 5") manner. One example of the intricate negative feedback mechanisms involved is shown in Figure 18.6, the general relationships between the pituitary and thyroid glands. Thyroxin, a major thyroid hormone, speeds the

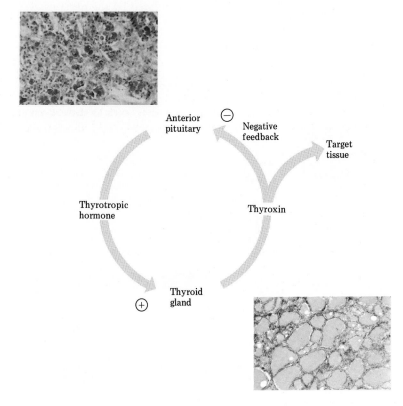

Anterior pituitary

⊖ Negative feedback

Target tissue

Thyrotropic hormone

Thyroxin

Thyroid gland

⊕

FIGURE 18.6
Negative feedback relationships between the pituitary and thyroid glands. Thyroid-stimulating hormone from the pituitary does exactly that, causing the thyroid to secrete its hormone thyroxin. High levels of thyroxin, in turn, inhibit the pituitary, which slows its release of thyrotropic hormone. The overall effect is the precise control of the rate of cellular respiratory metabolism.

Invertebrate Hormones

Most hormones in invertebrates are neurohormones rather than hormones released by endocrine glands. Their functions are numerous and diverse. They regulate the processes of regeneration in *Hydra* and various worms, control color changes in crustaceans, affect metabolism in many groups, control molting and metamorphosis in insects, and influence gamete production and reproductive behaviors in various species. Two specific examples are described below to illustrate the variety of these hormonal regulatory processes.

Color Change in Crustaceans

Crustaceans such as crabs, lobsters, and crayfish possess endocrine glands as well as neurosecretory cells. Color is important in many behaviors as well as in protective coloration, and color changes in crustaceans are under hormonal control. Various pigments are found in cells located beneath the external skeleton (exoskeleton). When the white, yellow, red, black, or even blue pigment granules are concentrated near the center of cells, their colors are barely visible; when dispersed throughout the cell, the colors show clearly. Under the influence of neurohormones, protective coloration is achieved by appropriate condensation or dispersal of cellular pigments. Other crustacean hormones control reproduction, heartbeat rate and other aspects of metabolism, changes in retinal pigments, and molting.

Insect Development

The various hormones from insect endocrine glands and neurosecretory cells regulate many body functions. For example, development in insects is often complex (Figure B18.1), and hormones and environmental factors interact to control it. Frequently, changes in an environmental factor such as light or temperature affect the neurosecretory cells in the brain. The hormone produced by those cells (brain hormone) is secreted by extensions of the neurons into a nearby structure for storage (corpora cardiaca). When released, the hormone stimulates two main target tissues. One target is an endocrine gland in the insect's thorax (prothoracic gland) that produces a hormone (ecdysone) which stimulates growth and molting. Another target is an endocrine gland (corpora allata) in immature insects that secretes a hormone (juvenile hormone) that keeps larval insects in that developmental stage by suppressing metamorphosis after each larval molt. When the juvenile hormone's concentration declines, the insect transforms into a pupa; in its absence, the pupa metamorphoses into an adult. Research continues on the promise of using juvenile hormone to control some pest species of insects. For example, the application of juvenile hormone may keep insects in the larval stage until they freeze in the fall, not having matured or reproduced as usual. These pesticide compounds would be useful because they are specific to a particular pest, unlike many chemical pesticides that sometimes kill rather indiscriminately numerous kinds of insects, and often other kinds of animals as well.

Thus, in insects as in some vertebrates, certain endocrine structures are regulated by the nervous system, and the concentrations of some hormones may change as animals age.

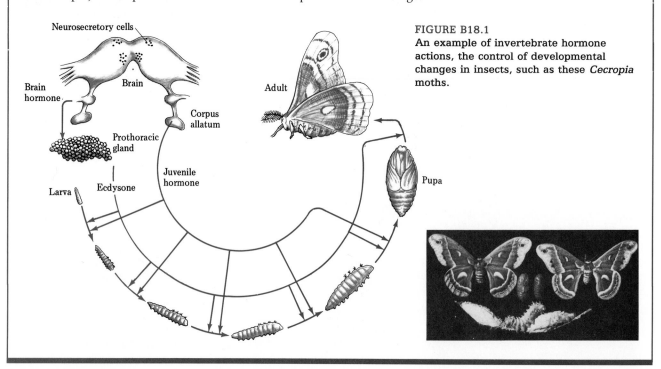

FIGURE B18.1
An example of invertebrate hormone actions, the control of developmental changes in insects, such as these *Cecropia* moths.

body's general metabolic rate. The production of thyroxin increases when a stimulatory thyrotropic hormone secreted by the pituitary reaches the thyroid in the blood. The rising level of thyroxin in the blood that flows through the pituitary reduces the release of the thyroid-stimulating hormone by the pituitary. The result of this finely balanced interaction is the homeostatic maintenance of a generally steady rate of cellular respiratory metabolism.

Because it is impossible to treat all vertebrate hormones extensively in a single chapter, only the principal functions of the major mammalian endocrine structures are described below. As with many other structures, endocrine glands may become abnormal, producing either too little hormone, **hyposecretion,** or too much, **hypersecretion.** Target cells may be deprived of stimulation in the first instance, and overstimulated in the latter case. Despite the secretion of normal amounts of hormone, abnormal function may also occur if the target cells have too few or abnormal hormone receptors.

The Hypothalamus and Pituitary Relationship

The small **hypothalamus** portion of the brain works closely with the pea-sized **pituitary gland** that is located below it, roughly in the center of the head. The maintenance of homeostasis in this manner is a prime example of the close relationship between nervous and endocrine functions (Figure 18.7). For example, two important hormones, oxytocin and antidiuretic hormone (ADH), that are produced by neurosecretory cells in the hypothalamus travel down neurons in a short stalk to the posterior pituitary gland, where they are stored prior to being released into the blood (see below).

Because the brain constantly senses both the external and internal environments, it is logical to expect that the monitoring of blood levels

FIGURE 18.7
(A) The two-lobed mammalian pituitary gland extends from the hypothalamus of the brain. The relationship between the posterior pituitary and the hypothalamus is especially close because of connecting nerves from the hypothalamus. However, the anterior pituitary is also in intimate contact with the hypothalamus via a circulatory connection, including a portal system (Chapter 15). The hormones released into the blood by the posterior pituitary are actually produced by the hypothalamus. (B) Releasing factors and inhibitory factors produced by the hypothalamus either stimulate or inhibit, respectively, the quantity of particular hormones released by the anterior pituitary. In this example, the thyroptropic-releasing hormone from the hypothalamus controls, via the pituitary and the thyroid, the rate of cellular respiratory metabolism.

(A)

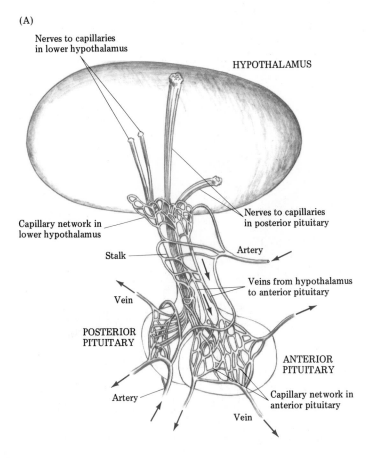

(B)

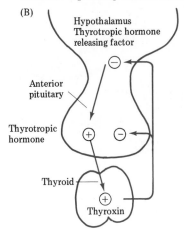

of hormones is among its long list of functions. Perhaps no other tissue the size of the hypothalamus controls as many functions. Through its **releasing** and **inhibitory hormones,** the hypothalamus either stimulates (releases) or inhibits the secretion of hormones by the anterior pituitary (Figure 18.8). Thus, the blood concentration of sex hormones in many vertebrates may change from season to season based upon changes in such environmental factors as light (photoperiod) and temperature. A releasing hormone, such as thyrotropin-releasing hormone (TRH), triggers another gland to secrete the hormone that it produces, such as thyroxin in this example. In turn—and this is typical of negative feedback control (Figure 18.5)—high levels of thyroxin reduce the output of TRH by the hypothalamus, via their more direct effect upon the pituitary. Thus, although the pituitary is often called the master gland because of the wide range of functions it affects, it has its own master, the hypothalamus. See the essay "Of Hormones and History" for an example of how a malfunctioning hypothalamus can produce a multitude of problems.

The pituitary glands of some vertebrates contain an intermediate lobe, although that structure is poorly developed in humans. Through forms of a hormone (melanocyte-stimulating hormone) produced there, skin pigmentation is controlled in some animals.

Posterior Pituitary Hormones

As described earlier, the structure and function of the posterior pituitary are an extension of the hypothalamus, and the hormones described below are produced in the hypothalamus and stored in and released by the posterior pituitary.

FIGURE 18.8
Summary of some of the major hormones secreted by the pituitary gland. Although hormones released by the pituitary affect numerous body functions, some of the hormones are produced by the hypothalamus. The hypothalamus also secretes several releasing and inhibiting factors that influence the pituitary and, in turn, other glands.

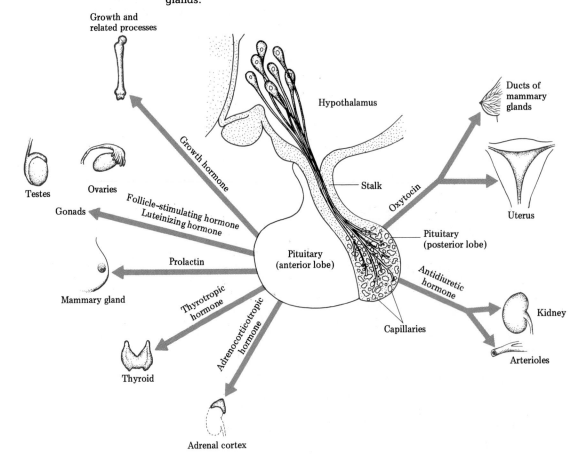

Of Hormones and History

Malfunctions of the endocrine glands can affect an individual's life in major ways, and by affecting certain people, endocrine malfunctions may also alter history. President Kennedy's Addison's disease was controlled so that apparently it did not affect his decision-making abilities. This was not true in the case of Napoleon Bonaparte (1769–1821) (Figure B18.2). As a young man, he was thin and dynamic, and because he considered sleep a waste of time he napped only 3 or 4 hours a day. By the age of 40, however, Napolean had changed considerably. His face became rounded and his body fat was distributed as it is in a female. His body hair almost disappeared, rounded breasts developed, and his penis and testes were small. His personality changed, he needed more sleep, he lacked a sex drive, and decisions were seemingly difficult to make. He was asleep in a chair for most of the 6 hours before the Battle of Waterloo. By the age of 45 he was physically and mentally worn out due to a diseased hypothalamus, which in turn led to abnormal functioning of his pituitary, thyroid, adrenals, and testes. What course would European and world history have taken if treatment had been available at that time for hypothalamic malfunction?

FIGURE B18.2
Napoleon in his 20s (left) and in his 40s (right).

Antidiuretic Hormone. Antidiuretic hormone (ADH) acts on the kidneys to cause the retention of water and the formation of concentrated urine (Chapter 17). The posterior pituitary releases ADH when the body must conserve water, as when there is a drop in blood volume or an increase in the concentration of various blood substances. Changes in these blood properties are sensed by receptors in the hypothalamus.

Diabetes insipidus occurs when the posterior pituitary does not release enough ADH. Because the kidneys cannot retain water normally, up to about 25 L (7 gal) of dilute urine may be produced daily. Obviously, the affected person must drink an equivalent amount of water to keep the level of body fluids constant. (Note: This disease is not the same as the more common and familiar diabetes mellitus, discussed later in this chapter.)

Oxytocin. One action of **oxytocin** is to stimulate uterine contractions during labor and delivery. Oxytocin is also released in response to an infant's suckling of the breast. It stimulates contraction of muscles associated with the ducts within the breast and thus initiates the letdown reflex, or the ejection of milk (Chapter 6).

FIGURE 18.9
(A) General Tom Thumb, a little person whom P. T. Barnum exhibited around the world. The effects of abnormal levels of growth hormone (GH) are quite different in children and adults. Hypo- and hypersecretion of GH before the long bones of the body cease growing may lead to dwarfism and giantism, respectively. (B) In adults, a reduction of GH does not lead to shrinkage and an increase does not result in further growth in height. However, the hypersecretion of GH in adults may lead to acromegaly, a condition caused by enlargement of cartilage and some bones.

(A)

(B)

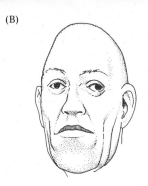

Acromegaly

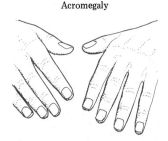

Anterior Pituitary Hormones

The anterior pituitary is often described popularly as the "maestro of the orchestra" (Figure 18.8). The maestro calls each orchestral section into action at the proper time so that harmony (negative feedback) is produced. Otherwise, the musicians might play what and when they wished, and ever-increasing noise (positive feedback) could result, as when musicians tune up. In addition to coordinating many body functions directly, the pituitary also secretes several **tropic hormones** that regulate other endocrine glands. Some of the major pituitary functions that are described below are expanded in later sections that treat other endocrine glands.

Adrenocorticotropic Hormone (ACTH).   Adrenocorticotropic hormone stimulates the adrenal cortex to release a group of hormones.

Gonadotropic Hormones.   The pituitary secretes several **gonadotropic hormones** that stimulate the growth and functions of the gonads. For example, **follicle-stimulating hormone** influences the development of eggs and sperm, and its absence leads to sterility. **Luteinizing hormone** (LH) causes cells in the testes to secrete sex hormones such as testosterone. In females, LH functions in ovulation and in stimulating the growth of a hormone-secreting structure (corpus luteum) in the ovary that develops after ovulation. Details about the roles of the gonadotropic hormones in regulating the menstrual cycle and in reproduction are found in Chapter 5; Chapter 6 treats their effects on prenatal and postnatal growth and development.

Prolactin.   **Prolactin** stimulates the breasts to produce milk after they have developed under the influence of other hormones during pregnancy (Chapter 6).

Thyrotropic Hormone (TH).   **Thyrotropic hormone** stimulates all the basic activities of the thyroid, and without TH the thyroid atrophies and ceases to release thyroxin. Many cases of thyroid disease are actually caused by excessive or deficient TH.

Growth Hormone (GH).   General Tom Thumb (Charles Stratton) and his wife, Lavinia, were only about 1 m (3 ft) tall, and these famous midgets were shown around the world by Phineas T. Barnum, the master showman (Figure 18.9A). One common type of **dwarfism** results when there is too little **growth hormone** in growing children. GH interacts with many body structures and substances in controlling growth and development. GH works primarily by stimulating protein synthesis, increasing the uptake of amino acids by cells, and assisting with the body's use and storage of nutrients. Secretion of growth hormone is controlled by GH-releasing and GH-inhibiting hormones from the hypothalamus. GH secretion is also affected by the glucose and amino acid concentrations in the blood, level of body activity, stress, sleep patterns, and emotional state.

Although the adult size of some pituitary dwarfs is only about twice their size at birth, the mental development of most "little people" is normal. In the past only a few of the thousands of pituitary dwarfs in the United States were treated with necessary human growth hormone. GH from other mammals is distinct enough to make it ineffective in humans, and the main but very limited source of human GH was from deceased human donors. Today, through recombinant DNA technology

(Chapter 10), human GH can be produced commercially. Providing that those in need receive GH early in life, before the body's long bones have completed their development, pituitary dwarfism may soon be eliminated.

Biotechnology raises many legal, ethical, and moral issues, among which is the range of appropriate uses for GH. Very few people object to using synthetic GH to aid normal growth in children who would otherwise become dwarfs. However, suppose a couple wanted their normal-sized child to be far taller in order to play professional basketball. Should parents be allowed to have synthetic GH used to alter the height of children otherwise destined to be normal in size?

Oversecretion of GH in children leads to pituitary **giantism,** a condition that may result from a tumor of GH-secreting cells. The excessive GH also promotes other aspects of growth, and "giants" have large hands, feet, jaws, and internal organs. A well-known American example was Robert Wadlow, who was about 2.7 m (8 ft 11 in.) tall, wore size 37 shoes, and weighed 224 kg (495 lb).

Hypo- and hypersecretion of GH after adolescence have different effects. For example, because the body's long bones have then completed their growth, hyposecretion of GH in an adult does not cause them to shrink. However, although hypersecretion of GH in an adult, **acromegaly,** does not lead to a lengthening of the long bones, those bones may thicken and cartilage in the hands, feet, and face may grow. The person acquires coarse features (Figure 18.9B), and although intelligence is not affected, the person may look "dull" according to a common social stereotype. As in other somatopsychological cases, the affected person's self-image may change and interpersonal interactions may be influenced (Chapter 11).

Because one effect of GH is to block entry of glucose into many types of cells, giants and acromegalic persons tend to have high blood glucose and diabetes mellitus. Treatment of giantism and acromegaly is difficult because of the relatively inaccessible location of the pituitary. X-ray treatment or removal of a pituitary tumor is sometimes possible.

Thyroid and Parathyroid Hormones

The butterfly-shaped **thyroid gland** is found in the neck just below the larynx (Figure 18.10). **Thyroxin,** or its metabolic product, increases the rate of cellular respiration in most tissues. **Cretinism** develops if a child's thyroid does not develop properly. Cretins are physically small, sexually immature, and mentally retarded, and most of their body functions occur at a subnormal rate. The early diagnosis and clinical administration of synthetic thyroxin may prevent cretinism. Insufficient thyroxin in an adult causes **myxedema,** the symptoms of which include fatigue, reduced body temperature, anemia, dry skin, mental deficiency, and reduced resistance to infection. The lack of dietary iodine may contribute to both types of hypothyroidism.

Among the common symptoms of hyperthyroidism, or excessive secretion of thyroxin, are a hearty appetite, heavy perspiration, hot and flushed skin, frequent diarrhea, insomnia, rapid heart rate, increased blood pressure, and nervousness and irritability.

Worldwide, enlarged thyroid glands or simple **goiters** affect several hundred million people, especially in developing nations. Sometimes the enlarged thyroid glands weigh a kilogram or more. The water, soil, and crops grown in glaciated areas (e.g., the Alps region of Europe and the Great Lakes area of the United States) often lack iodine. Deficient iodine means inadequate thyroxin in the blood to "turn off" the secretion of

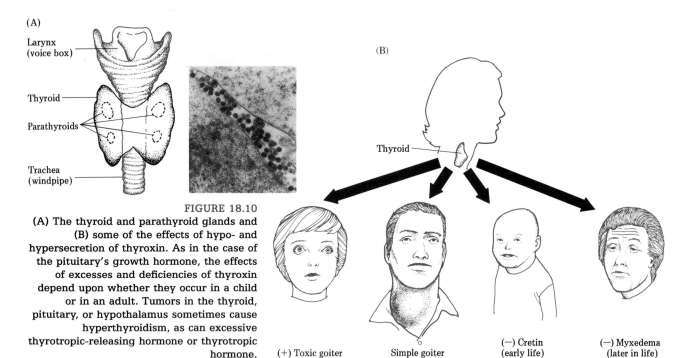

FIGURE 18.10

(A) The thyroid and parathyroid glands and (B) some of the effects of hypo- and hypersecretion of thyroxin. As in the case of the pituitary's growth hormone, the effects of excesses and deficiencies of thyroxin depend upon whether they occur in a child or in an adult. Tumors in the thyroid, pituitary, or hypothalamus sometimes cause hyperthyroidism, as can excessive thyrotropic-releasing hormone or thyrotropic hormone.

thyrotropic hormone by the pituitary or thyrotropin-releasing hormone by the hypothalamus. As a result, the thyroid grows, even though it produces little thyroxin. The addition of iodine to water or food (e.g., iodized salt) normally eliminates the problem. So-called toxic goiters may also occur, a condition in which excessive thyroxin is produced.

Calcium plays many important metabolic roles, such as in bone formation, in the functioning of neurons, in muscle contraction, as a coenzyme, in cell adhesion, in fertilization, and in blood clotting. The thyroid produces another hormone called **calcitonin** that reduces the calcium ion concentration of the blood (Figures 18.5 and 18.11).

The pea-sized **parathyroid glands,** which are located in or near the thyroid, secrete the hormone **parathormone,** which acts in antagonism to calcitonin. Parathormone increases the blood calcium concentration by causing calcium to be removed from bone (where about 99 percent of the body's supply is found), increasing the rate of calcium absorption from the digestive tract, and decreasing the amount of calcium excreted in the urine. Low blood calcium causes muscular spasms (tetany) and more excitable nerves. Serious muscle spasms may be fatal if they involve the muscular diaphragm that is important in breathing (Chapter 14). Hypersecretion of parathormone causes weak and brittle bones and increases the incidence of kidney stones. The negative feedback interactions of the thyroid and parathyroid glands and their target tissues normally maintain the body's calcium ion concentration within normal levels.

Adrenal Gland Hormones

The small, triangular **adrenal glands** perched on top of each kidney consist of two parts that each produce different hormones (Figure 18.12). The relatively large **adrenal cortex** secretes a variety of slow-acting hormones, whereas the smaller, central **adrenal medulla** secretes relatively fast-acting hormones.

The major effects of the steroid cortical hormones are to provide sustained control of overall metabolism, especially kidney function, salt

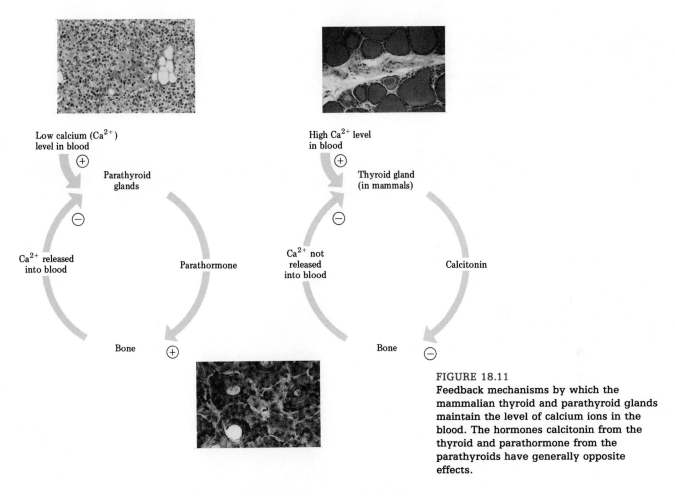

Low calcium (Ca^{2+})
level in blood

$\oplus$

Parathyroid
glands

$\ominus$

Ca^{2+} released
into blood

Parathormone

Bone $\oplus$

High Ca^{2+} level
in blood

$\oplus$

Thyroid gland
(in mammals)

$\ominus$

Ca^{2+} not
released
into blood

Calcitonin

Bone $\ominus$

FIGURE 18.11
Feedback mechanisms by which the mammalian thyroid and parathyroid glands maintain the level of calcium ions in the blood. The hormones calcitonin from the thyroid and parathormone from the parathyroids have generally opposite effects.

balance, tissue inflammation, and blood pressure. For example, **aldosterone,** the major hormone in the **mineralocorticoid** group, conserves sodium and water and increases potassium excretion through its actions on such target organs as the kidneys, salivary and sweat glands, and intestine. Hypersecretion causes edema (retention of water in the tissues) and hypertension (high blood pressure due to an increase in blood volume).

The **glucocorticoids** (e.g., cortisol) help to maintain normal blood sugar and normal glycogen stores in the liver, and they also aid in regulating lipid and protein use and storage. The hormones reduce the inflammation of tissues by decreasing the permeability of capillary walls, which thereby reduces swelling. They may also reduce the release of potent lysosome enzymes that could damage tissues.

Hyposecretion of cortical hormones causes **Addison's disease,** whose symptoms include digestive disturbances, lowered blood sugar, reduced blood pressure and body temperature, kidney malfunction, increased sodium loss in the urine (which may lead to shock), muscular weakness, and a bronzed skin. Collectively, such changes combine to make coping with stress and infections difficult. However, the symptoms can often be alleviated by treatment with cortical hormones, as they were in President John F. Kennedy's mild form of the disease. Hypersecretion leads to edema, rise in blood glucose, fat deposition, and a depressed immune response.

The direct nervous control of the adrenal medulla is both atypical of most endocrine glands and extremely useful in responding quickly to stress. (Details about the functioning of the autonomic nervous system

FIGURE 18.12 (A)

(A) Like the pituitary, the adrenal glands consist of two portions, each of which secretes distinct hormones, and the glands are influenced by a hypothalamic hormone, ACTH. (B) Hormones from both portions act in concert with numerous structures and processes in the body's response to stress. While short-term reaction to stress is helpful, chronic stress may contribute to heart disease, hypertension, atherosclerosis, ulcers, and possibly diabetes mellitus.

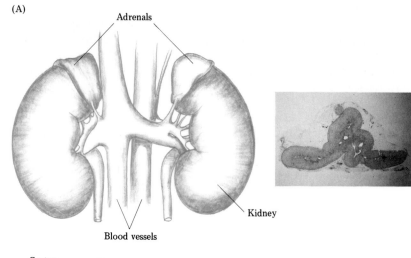

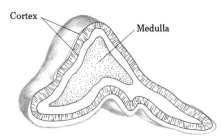

(B)

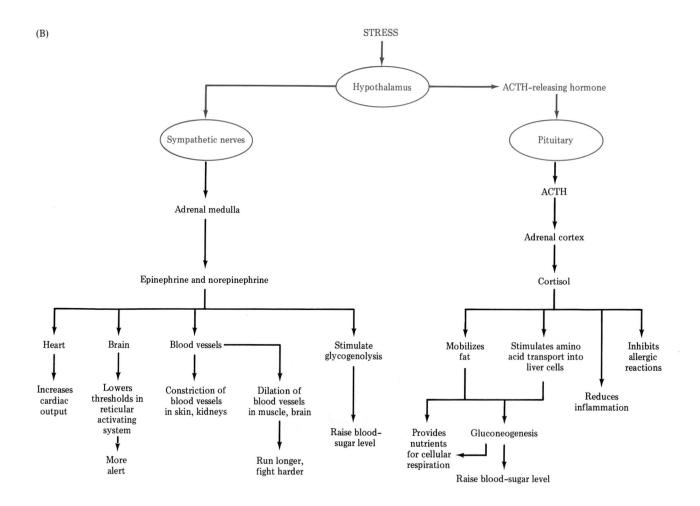

that is involved in this control are found in Chapter 19.) **Epinephrine** (adrenalin) and **norepinephrine** are the major medulla hormones. Norepinephrine is also a neurotransmitter released by many neurons. The combined effects of epinephrine and norepinephrine, in conjunction with the actions of the nervous system, prepare the body for a quick response to stress or to a real or fancied physical or psychological danger or peril. In the so-called fight-or-flight response, the heart rate and blood pressure increase, breathing is faster, levels of blood glucose and other nutrients increase, blood clots faster, the spleen releases stored blood cells into the bloodstream, some pain is ignored, and reaction time decreases. The greater blood flow (increased by as much as 100 percent) to the skeletal muscles, liver, lungs, heart, and brain is balanced by decreased circulation to support functions relatively unimportant in emergencies, such as digestion, reproduction, and excretion. The decreased blood flow to the skin not only causes the paling that is common in an alarm reaction but it also helps decrease blood loss in case of hemorrhage. Stress stimulates the hypothalamus to secrete ACTH-releasing hormone. Epinephrine also leads to the release of ACTH and some other hormones from the anterior pituitary. ACTH, in turn, promotes the release of adrenal cortical hormones which aid the body in adjusting to prolonged stress. ACTH may cause cortisol secretion to increase by up to 20 times within minutes. The norephinephrine that is released during stressful situations acts mainly to sustain blood pressure. Obviously, because the emergency hormones have such diverse effects, the target organs have different receptors, some being stimulated and others being inhibited.

Pancreas Hormones

The **pancreas,** a relatively large organ found below the stomach and near the small intestine, has both exocrine and endocrine functions (Figure 18.13). (The exocrine functions, important in digestion, are discussed in Chapter 13.) Scattered throughout the pancreas are more than a million small clusters of cells known as **islets of Langerhans.** Most of the islet cells produce the protein hormone **insulin** but some cells secrete the hormone **glucagon,** also a protein. Together, these two hormones normally maintain the blood glucose concentration homeostatically within a narrow range of tolerance. Because brain cells are normally

FIGURE 18.13
(A) Location of the pancreas and (B) light microscopic view of the islet of Langerhans cells and of the surrounding exocrine cells. About 70 percent of the islet cells (beta cells) produce insulin, and the remainder (alpha cells) produce glucagon.

(A)

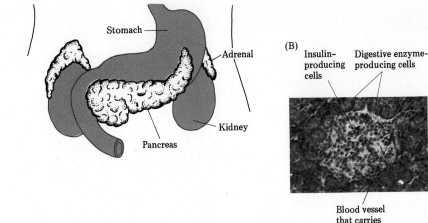

(B) Insulin-producing cells   Digestive enzyme-producing cells

Blood vessel that carries insulin

unable to use other nutrients for energy, they are especially dependent upon an adequate level of blood glucose.

Insulin has diverse effects on metabolism and is one of the body's most important regulators of the use of nutrients. The main action of insulin is to stimulate the uptake of glucose from the blood by most cells, but especially muscle and fat cells (Figure 18.14). Although liver cells do not require insulin to take up glucose from the blood, most cells are impermeable to glucose in the absence of insulin. Insulin affects the liver, however, by increasing the level of an enzyme (glucokinase) that alters (phosphorylates) glucose to prevent it from diffusing out of the liver cells. Insulin also promotes the formation of glycogen (glycogenesis) for storage in the liver and skeletal muscles, stimulates protein synthesis, and promotes fat storage.

Glucagon acts antagonistically to insulin by increasing the glucose concentration of blood. The main target organ of glucagon is the liver, which it stimulates to break down glycogen to form glucose (glycogenolysis) and to produce glucose from other nutrients (gluconeogenesis) (Chapter 13). Glucagon also increases the blood concentration of amino acids and fatty acids and it stimulates the release of epinephrine by the adrenals.

Blood glucose is the main stimulus that controls the release of pancreas hormones. Insulin is normally released when blood glucose increases, as after a large meal is digested and absorbed. As cells remove glucose from the blood, insulin secretion declines. A sufficiently reduced blood glucose level stimulates certain islet cells to release glucagon, which mobilizes the release of glucose into the blood. The changing

FIGURE 18.14

Some homeostatic mechanisms in the control of blood glucose by insulin and glucagon. These two protein hormones act in basically antagonistic ways.

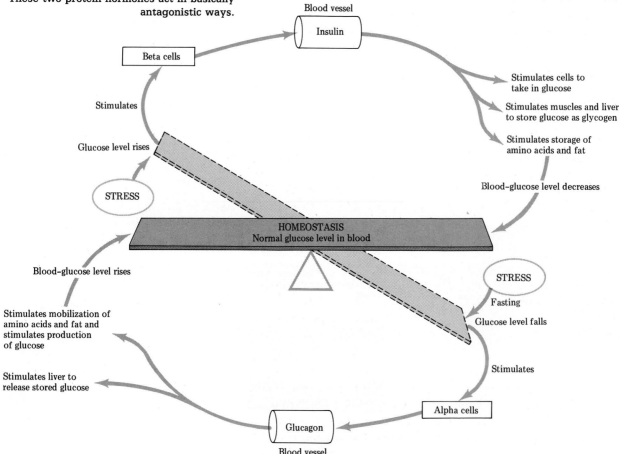

Blood vessel
Insulin
Beta cells
Stimulates cells to take in glucose
Stimulates muscles and liver to store glucose as glycogen
Stimulates storage of amino acids and fat
Stimulates
Blood-glucose level decreases
Glucose level rises
STRESS
HOMEOSTASIS
Normal glucose level in blood
STRESS
Fasting
Glucose level falls
Blood-glucose level rises
Stimulates
Stimulates mobilization of amino acids and fat and stimulates production of glucose
Stimulates liver to release stored glucose
Alpha cells
Glucagon
Blood vessel

levels of blood glucose are also important regulators of the sensation of hunger (Chapter 13).

Hippocrates described **diabetes mellitus** over 2000 years ago. Diabetes, a serious disorder of carbohydrate metabolism for millions of Americans, occurs when the pancreas cannot secrete enough insulin, or produces an ineffective form of insulin, or when target cells cannot take up and utilize insulin. Although blood glucose levels become very high (hyperglycemia), body cells cannot utilize blood glucose properly and the unused portion is eliminated in the urine. The excess glucose in the kidney reduces the reabsorption of water, and this results in constant thirst and an excessive output of dilute urine. Simple screening tests for detecting glucose in the urine provide an inexpensive diagnosis for diabetes, as well as aiding known diabetics to plan proper control techniques. (See the essay "Serendipity in Endocrinology.")

Because the cells of diabetics must often turn to fat stores as a fuel source, the blood concentration of fatty acids rises. High blood lipid levels, up to five times the normal concentration, may lead to atherosclerosis (Chapter 15). Also, some breakdown products of fatty acid metabolism (ketone bodies) accumulate in the blood and the pH of blood may become so low (acidosis) that coma or death occurs. The ketone bodies complicate the fluid balance problem but their presence in the urine provides another useful clinical indication of diabetes. However, the associated smell of acetone on the breath has sometimes been mistaken for alcohol. Some diabetics have passed out and died because bystanders thought they were drunk, a problem easily combated by the wearing of medical alert bracelets.

Although untreated diabetics may have good appetites, they often become thin and emaciated because lack of insulin causes protein wasting. In such people, proteins are not synthesized normally and the liver converts (deaminates) amino acids to glucose, thus aggravating the problem of excessive blood glucose. These events also adversely affect growth, slow the repair of tissues, and decrease the body's resistance to infection.

More than 90 percent of diabetics develop the disorder gradually, often when they are overweight and over 40 years of age. Insulin secretion may be normal but target cells may not be able to utilize it normally. In juvenile-type diabetes, the condition develops before age 20. Because insulin secretion is usually inadequate, daily insulin injections may be necessary. Even with treatment, diabetes kills tens of thousands of Americans annually and is a major cause of death, with more females than males being afflicted. Diabetes is currently a leading cause of blindness and is a major contributor to strokes and heart attacks. Above-average rates of birth defects and infant mortality occur in the offspring of diabetic mothers. In general, juvenile diabetes is far more serious than the adult-onset form of the disease. Insulin treatment eliminates many of the symptoms of diabetes but it does not cure the condition. The fact that its use allows most pregnancies by diabetic women to go full term also means that the incidence of the disease may increase because of the survival of those carrying genes for the disease.

Thymus Gland

The **thymus** secretes hormones (e.g., thymosin) that help certain white blood cells (T lymphocytes) mature and function, a role very important in the body's defense arsenal (Chapter 16). The thymus is large in children but it atrophies after puberty although its function in adults continues.

Serendipity in Endocrinology

Serendipity, the ability to find valuable things not sought, has always been important in scientific research (Chapter 1). A good example was the work of two German scientists in the late 1800s on the role of the pancreas in canine digestion. Their work led eventually to our ability to control the symptoms of diabetes mellitus. These researchers had removed the pancreas glands from some dogs but not from others. Seeing insects attracted to the urine of the former dogs but not of the latter led them to the "aha" experience—the pancreatectomized dogs had a disease equivalent to human diabetes.

At that time it was known that the symptoms of hyposecretion of the thyroid could be relieved by ingesting thyroid gland extracts. However, all early attempts to treat diabetics (canine and human) in this way failed. One reason was that the pancreas is a double-duty gland. When entire pancreas glands were homogenized during the preparation of the extract, the digestive enzymes released from the exocrine portion digested

the hormone insulin produced by the endocrine portion. The same thing happened in the diabetic's body when it became possible to prepare pure extracts of insulin. While the amine-type thyroid extract could be ingested, the protein insulin was digested before it could be absorbed into the bloodstream. However, injected into muscles and released gradually into the bloodstream, insulin relieves the diabetic's symptoms. The frequency of injections needed depends upon the severity of the disease.

That lactation (milk production) is under hormonal control was shown by conjoined (so-called Siamese) twins. One twin became pregnant and bore a child, but both twins produced milk. Because no nerve connection between the connected twins was found, hormonal control of lactation was indicated. This was another case of serendipity, since there are not that many conjoined twins and those who exist do not always do such interesting things that come to the attention of endocrinologists.

Pineal Gland

The tiny **pineal gland,** which lies on the surface of the brain, secretes the hormone **melatonin,** which seems to be involved in the onset of puberty. Melatonin secretion is stimulated by light falling on the retina, and this appears to inhibit the functions of the gonads, possibly working through the pituitary. For example, females blind from birth become sexually mature earlier than normally sighted females. Puberty is delayed in both males and females with pineal tumors that lead to hypersecretion.

Gonads and Placenta

The gonads produce a variety of hormones, as does the placenta in females. These hormones are involved extensively in reproduction and are discussed in Chapters 5 and 6.

Prostaglandins, Endorphins, and Pheromones

Prostaglandins (PGs) are a group of specialized lipids that were named because of their abundance in semen in the prostate gland. However, it is now known that the numerous PGs are produced by many tissues. These versatile compounds affect such diverse functions as blood pressure, blood clotting, pain, uterine contractions, menstruation, hormone secretion, muscle tone, and the inflammation response. Although prostaglandins travel in the bloodstream, their main effects appear to be in the tissues that produce them and in neighboring tissues. For example, PGs in semen help move sperm through the female reproductive tract by promoting uterine contractions.

Substances with such diverse effects as prostaglandins are bound to stimulate the search for clinical uses, and research on prostaglandins is active. PGs are used currently in treating hypertension, stomach ulcers, and asthma (by relaxing bronchial passages), and also during abortions, in reducing pain, and in contraceptive preparations.

One of the most effective pain-relieving compounds is morphine, an opiate or opium-derived substance, which works by binding to specific receptors in the brain. This raises the question as to why the human brain should have receptors for compounds occurring in poppies and certain other plants. It turns out that there are natural or endogenous brain chemicals called **endorphins** ("the morphine within") that have a high affinity for the opiate receptors. Endorphins are small polypeptides that are intermediate between hormones and neurotransmitters. Research continues to determine whether endorphins and related substances (e.g., enkephalins) can be used to reduce pain in a nonaddictive way. Acupuncture is a medical practice, developed in the Orient, that involves inserting thin, sterile needles through the skin into specific spots on the body to reduce pain. This approach may work because it promotes the release of more endorphins—but so, in fact, may the administration of placebos, substances inert in the process being studied or treated. For example, sugar pills and saline injections sometimes relieve pain in some people almost as well as morphine-based compounds. Perhaps the belief and hope that someone cares and is trying to be of assistance also promotes the production of endorphins that, in fact, reduce pain.

Zoologists studying hibernation have become interested in endorphins. One type of endorphin induces some of the hibernation-like changes that are found in certain mammals, such as reduced heart and breathing rates, decreased thyroid activity, reduced body temperature, conservation of water through the production of concentrated urine, and some reduction in response to pain. The injection of a chemical antagonist to this endorphin into hibernating hamsters caused a reversal of these changes but it had no such effects on nonhibernating mammals. Excessive levels of the same endorphin also cause overeating. Whether this is merely an effect useful to certain mammals in building fat stores prior to hibernation or whether it might also explain some forms of obesity remains to be seen.

The chemical messengers that are exchanged *between* organisms, as opposed to *within* organisms, are called **pheromones**. Pheromones are very important in numerous behaviors, especially reproductive, and are widespread among animals, as well as in microorganisms, protists, and many algae. Perhaps pheromones became modified into internal chemical messengers as multicellular animals closed off their body cavities and created their "inner seas."

SUMMARY

1. By showing that a substance produced in one part of the body could influence a distant part, **A. Berthold** began the experimental study of **endocrinology,** the study of chemical messengers and the structures which produce them.

2. **Endocrine glands** secrete **hormones** directly into the bloodstream or interstitial fluids. The secretions of **exocrine glands** are released through ducts to a body surface or cavity. Collectively, all structures that secrete chemical messengers constitute the **endocrine system.** Endocrine structures are diverse and widely dispersed in animal bodies.

3. Hormones and other chemical messengers are potent in very small amounts. However, their actions are relatively slow compared with those of the nerves, with which they frequently work in concert. Most hormones are medium-sized molecules that belong to various chemical groups (e.g., relatively small polypeptides, proteins, steroids, amino acids, or amino acid derivatives). Both **hyposecretion** and **hypersecretion** of hormones may cause problems.

4. **Target cells** have receptor molecules on their surfaces or in their cytoplasms that bind to chemical messengers, roughly like a lock-and-key mechanism. Some target cells can respond differently and specifically to more than one hormone because they possess more than one type of hormone receptor. However, not all target tissues respond in the same way at all times to the same hormone, sometimes because of synergistic interactions with other hormones.

5. Although the chemical structure of hormones generally varies little from species to species, a hormone's functions may. The specific responses to a hormone depend upon the target cell, not on the hormone itself.

6. After forming a hormone-receptor complex in the target cell's plasma membrane, many protein, polypeptide, and amine-type hormones work through a **second messenger** that relays the hormone's message to an appropriate site inside the cell. Fat-soluble steroid hormones pass readily through plasma membranes and bind to hormone-specific receptor proteins in the cytoplasm. Steroid hormones act by stimulating the synthesis of new proteins rather than by affecting the activity of existing proteins in the target cell's cytoplasm.

7. Hormones usually interact with structures and processes in a negative feedback manner. For example, a drop in blood calcium stimulates the **parathyroid** gland to release **parathormone.** This leads to events that increase the blood calcium level until above-normal levels inhibit the release of parathormone.

8. The hormones from different endocrine glands also usually control one another through negative feedback mechanisms. Thus, high levels of **thyroxin** inhibit the release of **thyrotropic hormone,** and low levels do the opposite.

9. Most hormones in invertebrates are **neurohormones,** and they regulate such diverse processes as regeneration, color change, molting and metamorphosis, and various aspects of reproduction.

10. The origin of hormones is unknown. One possibility is that altered **neurotransmitters** that entered the bloodstream somehow enhanced the ability of an animal to survive and reproduce. Altered **pheromones** present another possibility, with their effects being within an animal rather than between one animal and another.

11. **The hypothalamus** in the brain works closely with the small **pituitary gland** below it, a good example of the close relationship between the nervous and endocrine systems. The hormones released by the **posterior lobe** of the pituitary are produced by the hypothalamus (e.g., **oxytocin** and **antidiuretic hormone**). Through its **releasing** and **inhibiting hormones,** the hypothalamus either stimulates or inhibits the secretions of hormones by the **anterior pituitary.**

12.   Through its **tropic hormones,** the anterior pituitary gland controls several other endocrine glands (e.g., **adrenocorticotropic, gonadotropic,** and thyrotropic hormones). Gonadotropic hormones include **follicle-stimulating hormone** and **luteinizing hormone. Prolactin** (which stimulates the prepared breast to produce milk) and **growth hormone** (GH) are other important hormones secreted by the anterior pituitary. Excessive and deficient GH early in life leads to **giantism** and **dwarfism,** respectively. **Acromegaly** results from hypersecretion of GH in an adult.

13.   Thyroxin, a major hormone secreted by the thyroid, increases the rate of cellular respiration in most tissues. **Cretinism** in children and **myxedema** in adults are examples of hyposecretion of thyroxin. Simple and toxic **goiters** affect millions worldwide.

14.   **Calcitonin** from the thyroid and parathormone from the parathyroid, whose effects are antagonistic, interact with such target tissues as the bones, kidneys, and intestines to control the body's important calcium balance.

15.   Different hormones are released by the **adrenal cortex** and the **adrenal medulla.** The cortical hormones, which include the **mineralocorticoids** (e.g., aldosterone) and the **glucocorticoids** (e.g., cortisol), tend to be relatively slow acting compared with the fast-acting medulla hormones such as **epinephrine** and **norepinephrine.** Hyposecretion of cortical hormones causes **Addison's disease,** and those afflicted, among other traits, handle stress poorly. Hormones from both portions of the adrenals mobilize the body to combat stress and danger.

16.   The **pancreas** has both exocrine and endocrine functions. Most **islet of Langerhans** cells secrete **insulin,** but some secrete **glucagon.** Together, the two hormones, which have generally opposite effects, control the blood glucose concentration homeostatically within a narrow range. Brain cells are especially dependent upon and sensitive to blood glucose levels.

17.   The major action of insulin is to stimulate the uptake of glucose by most cells. Insulin also promotes the formation of glycogen for storage in the liver and skeletal muscles, stimulates an enzyme to alter glucose so it cannot diffuse out of the liver, stimulates protein synthesis, and promotes fat storage.

18.   **Diabetes mellitus** is a major disorder of carbohydrate metabolism that affects millions of people and is a leading cause of death and disability. It results from the inability of the pancreas to secrete sufficient quantities of effective insulin or from faulty target cells that cannot take up and utilize insulin. High blood glucose levels complicate fluid balance and excess glucose is excreted in the urine. The cells of the diabetic turn to fat stores as a fuel, and the high blood lipid levels further aggravate the maintenance of a normal fluid balance. Because proteins are not synthesized normally, growth is affected, repair of tissues is slowed, and resistance to infection is poor. Treatment with insulin alleviates the symptoms but does not cure the disease.

19.   Through its hormone **melatonin,** the **pineal gland** seems to be involved with the onset of puberty.

20.   The specialized lipids called **prostaglandins** that are secreted by many parts of the body help to regulate numerous functions (e.g., blood

clotting, muscular contractions, the sensation of pain, blood pressure, reproduction, and the inflammation reaction). Prostaglandins exert their effects primarily in the tissues that produce them or in nearby tissues.

21. **Endorphins** are small polypeptide chemical messengers that are produced in the brain. Endorphins, which have a high affinity for the opiate receptors in the brain, are involved in aspects of pain. They also induce hibernation-like changes in some mammals.

22. **Pheromones** are diverse hormonelike compounds that are important chemical messengers between individuals of numerous species of animals. It is possible that pheromones became modified long ago in animals to also serve as internal chemical messengers.

REVIEW QUESTIONS

1. In Berthold's experiment, why was it unimportant for him to reimplant the testes into their original sites in the chickens that he had castrated? Was his experiment controlled?

2. Distinguish between exocrine and endocrine glands.

3. Distinguish among hormones, neurohormones, and neurotransmitters.

4. How does the size of hormone and other chemical messenger molecules facilitate their actions?

5. What are the traits of target tissues that allow them to respond to chemical messengers?

6. Explain how the effects of hormones may differ from one species to another even though the chemical structures of specific hormones generally show little variation.

7. What is a second messenger and how does it function?

8. Distinguish among the typical modes of action of steroid hormones and those of many protein, polypeptide, and amine-type hormones.

9. Describe at least two examples of negative feedback mechanisms between two or more hormones. Do the same for interactions between two or more hormones and a target structure that they both affect.

10. Describe the general pattern of hormonal control in invertebrates, and state some examples of processes that are regulated hormonally.

11. State some possible evolutionary origins for hormones and related chemical messengers.

12. How does the connection between the hypothalamus and the anterior pituitary differ from that with the posterior pituitary?

13. How do tropic hormones differ in their functions and origins from releasing and inhibiting hormones?

14. Describe at least two examples of how hyposecretion and hypersecretion of hormones differ in their effects depending upon whether they occur in a child or in an adult.

15. Why is calcium metabolism important in mammalian metabolism, and how do thyroid and parathyroid hormones interact with various target tissues to regulate it?

16. How do the secretions of the adrenal cortex differ chemically and functionally from those of the adrenal medulla?

17. Describe the fight-or-flight response, and explain how adrenal hormones interact with other endocrine glands and various target tissues to affect the metabolic response to stress.

18. Why are brain cells so dependent upon and sensitive to blood glucose levels?

19. Describe the antagonistic interactions between insulin and glucagon.

20. Describe the symptoms and forms of diabetes mellitus, and explain their causes.

21. What is the apparent function of the hormone from the pineal gland?

22. Where do prostaglandins come from, what functions do they affect, and how do their actions differ from those of most hormones?

23. What are endorphins and what do they do?

24. What is the function of pheromones?

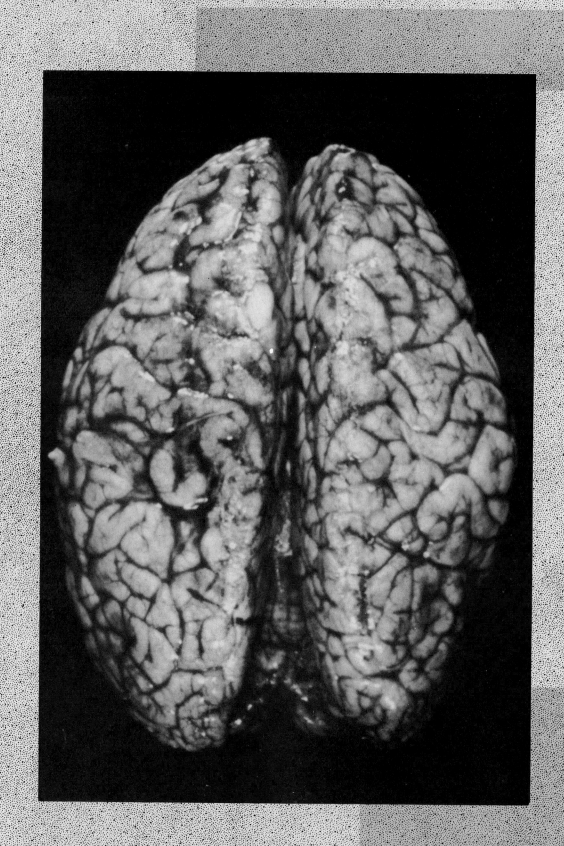

CHAPTER 19

REGULATION: NEURONS AND NERVOUS SYSTEMS

"Slow and easy" could describe regulation by chemicals in animals, because a time lag of seconds to days may occur between the release of a chemical messenger and a response by cells sensitive to it (Chapter 18). Nervous systems act far more rapidly and often even more specifically, usually transmitting information about events in one part of an animal almost instantaneously to other body parts. While chemical control of some kind operates in all organisms, nervous systems are found only in animals. In both systems of communication and regulation, the structure or process being controlled may be distant from a control center. In this chapter, you will explore the nature of nervous systems and their basic units. Survival and the maintenance of homeostasis depend heavily upon the ability to sense and react quickly to changes in the external and internal environments.

Billions of neurons make up this human brain, and the obvious and numerous convolutions greatly increase the surface area. Such an active organ requires a large blood supply and abundant nutrients and oxygen, more than any other organ. Some animals have larger brains but none can utilize this amazing organ to the same degree as humans—for example, to analyze its own structures and functions, as you will do in this chapter.

INFORMATION AND NERVOUS SYSTEMS

Detectable changes within organisms and in their external environments are called **stimuli.** From within and from without, the countless stimuli that bombard organisms daily are often called "information." Even simple homeostatic responses, including decisions to ignore or store information rather than act upon it immediately, involve the steps outlined in Figure 19.1. Although cytoplasm itself is irritable, or sensitive to changes in its environment, many animals have specialized **receptors.** Receptors are structures able to sense stimuli and to initiate nerve impulses (Chapter 20). Transmission to an **integrator** such as the brain is necessary in order to interpret the message and to initiate a response. A response requires transmission of the impulse to an **effector** such as a gland or muscle.

In vertebrates, the brain and spinal cord are the major structures that receive and integrate incoming information and determine appropriate responses. These structures constitute the **central nervous system** (CNS) (Figure 19.2). The sensory receptors and the nerves that lie outside of the CNS make up the **peripheral nervous system** (PNS). The vertebrate PNS includes several pairs of cranial and spinal nerves that link the CNS

FIGURE 19.1
The general flow of information through a nervous system. Many stimuli originate internally, informing us about our blood chemistry or pressure, for example. Other stimuli alert the animal to changes in its external environment (e.g., light and temperature). From the receptor, the nerve impulse travels to an integrator, and if the integrator "decides" that a response is called for, an impulse travels to an effector such as a muscle or gland.

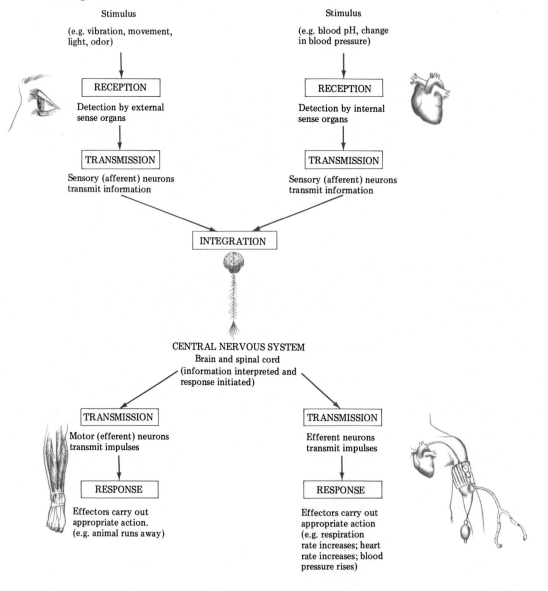

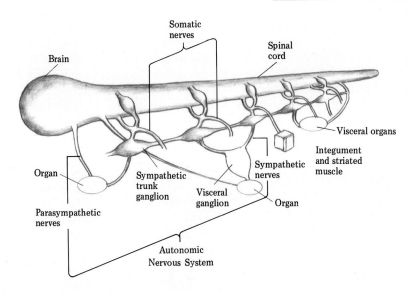

FIGURE 19.2
Generalized scheme of vertebrate nervous systems. The central nervous system (CNS) consists of a complex brain and a single spinal cord. The cranial and spinal nerves make up the peripheral nervous system (PNS), whose function it is to link the CNS with the remainder of the body. Some parts of the PNS are concerned mainly with the external environment (somatic system), and other parts largely regulate the internal environment (autonomic nervous system).

with the receptors and effectors. For convenience, the PNS is often divided further into two subsystems. The **somatic system** or voluntary motor system is concerned with receiving and transmitting information about the external environment, whereas the **autonomic nervous system** (ANS) or involuntary nervous system regulates the internal environment. The ANS subdivisions (sympathetic and parasympathetic) are described in a later section. For information on far simpler nervous systems, see the essay "Invertebrate Nervous Systems."

NEURONS AND ASSOCIATED CELLS

The anatomical and functional unit of the nervous system is the **neuron,** or nerve cell (Figure 19.3). Although neurons exist in many sizes and shapes, most share some traits (Figure 19.4). Most of the cytoplasm is in the cell body, which contains the nucleus and most other organelles.

FIGURE 19.3
The large variety of neurons in vertebrate nervous systems are often categorized into three basic groups according to the number of extensions. (A) Motor neurons are generally multipolar, with many short dendrites and one long axon. (B) Bipolar neurons, with one dendrite and one axon, are typically found in the retina, in nerves coming from the middle ear, and in the olfactory (smell) nerve. (C) Sensory neurons tend to be unipolar, with long processes that branch from a short process extending from the cell body. The branches are labeled here according to their functions.

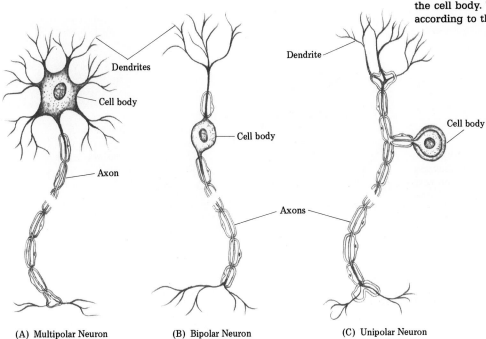

(A) Multipolar Neuron   (B) Bipolar Neuron   (C) Unipolar Neuron

FIGURE 19.4

FIGURE 19.4
(A) Light micrograph of several multipolar neurons and the nuclei of nearby glial cells. **(B)** Diagrammatic view of the major features of neurons, using a multipolar neuron as an example. Dendrites carry impulses toward the cell body and axons carry them away, toward another neuron or to an effector.

(A)

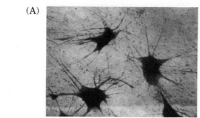

(B)

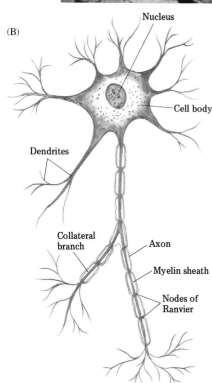

Nucleus

Cell body

Dendrites

Collateral branch

Axon

Myelin sheath

Nodes of Ranvier

Multipolar neurons contain numerous short extensions called **dendrites** (from the Greek for "tree"); bipolar neurons contain only one. Dendrites are specialized to receive stimuli and to carry nerve impulses to the cell body. The one or more extensions specialized to carry impulses from the cell body to another neuron or effector cell are called **axons.** Although usually only microscopic in diameter, some axons are very long. For example, the axon of a sensory neuron in our foot extends all the way up to the spinal cord, a distance of about a meter (3 ft). Because of their length, axons are often called *nerve fibers.* After branching near the target cells, axons end in specialized terminal structures. Branches or **collaterals** may also arise along the length of an axon, allowing for extensive interconnections among neurons.

Nerves are not the same as neurons (Figure 19.5). **Nerves** are bundles of parallel axons in the PNS that are held together by connective tissue, somewhat analogous to the wires within a telephone cable. The cell bodies of the axons are clustered in **ganglia.** Nerves can carry information about many things simultaneously because the axons are normally isolated from one another. Some nerves are primarily **sensory** (afferent), carrying information to the CNS; others are primarily **motor** (efferent), transmitting impulses to effectors; and still others are mixed, carrying both sensory and motor information. In the CNS, the masses of cell bodies are generally called **nuclei** rather than ganglia, and the bundles of axons are called **tracts** rather than nerves.

Various kinds of **glial** cells are associated with neurons (Figure 19.6). Glial cells support and nourish neurons and help to maintain a proper ionic environment for them. Although they do not generate or conduct nerve impulses, they assist neurons in the transmission process. For example, the axons of PNS neurons are enveloped by **Schwann cells** that have their cell membranes wrapped around the axons several times. On some axons the Schwann cells produce an inner **myelin sheath** of whitish, fatty material that electrically insulates the axon, speeds the nerve impulse, and aids in the regeneration of injured neurons. Most large axons are myelinated, and most smaller ones are not. Between adjacent Schwann cells, and occurring every millimeter or so, are unmyelinated sections called **nodes of Ranvier.**

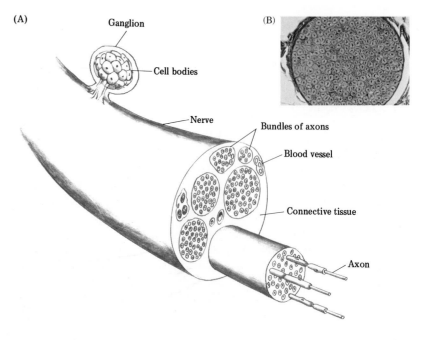

(A)

Ganglion

Cell bodies

Nerve

(B)

Bundles of axons

Blood vessel

Connective tissue

Axon

FIGURE 19.5
(A) Diagrammatic view of a nerve, showing the bundles of axons and a ganglion that contains their cell bodies. Some of the connective tissue consists of fat (adipose tissue) that helps to protect and insulate the nerve. Note also the blood vessels that provide nutrients and oxygen, and carry away wastes from active nervous tissue. **(B)** This micrograph shows a cross-section of a mammal nerve. Many vertebrate nerves contain hundreds or thousands of individual axons.

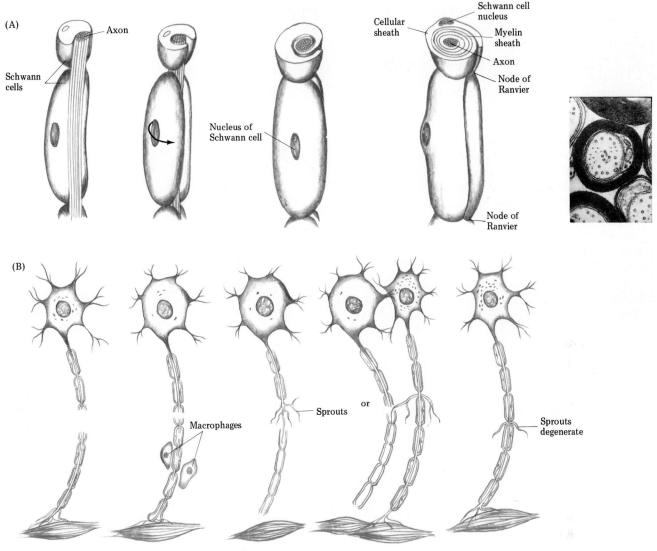

FIGURE 19.6
(A) Glial cells include the Schwann cells, which produce the myelin sheath around some axons of PNS neurons. Myelin acts as an electrical insulator and speeds the conduction of nerve impulses. (B) Some steps in the regeneration or replacement of injured axons in the PNS.

The myelin sheaths found around the axons of CNS neurons are produced by other types of cells (oligodendrocytes). Because of myelin's color, masses of myelinated axons in parts of the brain and spinal cord are called **white matter.** Their unmyelinated cell bodies constitute the **gray matter.** In the vertebrate CNS, there are far more glial cells (about nine times as many) than neurons.

Nerve function is impaired when myelin is destroyed or replaced by scar tissue. For example, the patchy deterioration of myelin that occurs at irregular intervals along neurons in the CNS causes **multiple sclerosis** (MS). Common symptoms of MS include paralysis of affected body parts, poor coordination, slurred speech, blurred vision, and tremor.

Axons themselves may also be injured, although regeneration is sometimes possible if a tube of Schwann cells, or another appropriate pathway, and an uninjured cell body remain. Function is also sometimes restored if a collateral from a neighboring neuron grows into the intact but empty cellular sheath of the injured neuron. Regeneration is far less likely to be successful in the CNS than in the PNS. Although regeneration is sometimes possible, neurons, unlike most other cells, lack the ability to divide after they have matured.

Invertebrate Nervous Systems

Sponges have no nervous system and respond to environmental changes only at the cellular level. The simplest organized nervous systems are the loose **nerve nets** found in *Hydra* and its relatives (Figure B19.1). Some nerve cells are specialized for reception and others for neurosecretion but there is no central integrator. Impulses travel in both directions in nerve cells and transmission diffuses outward in all directions from the point of stimulation. For such relatively slow and **radially symmetrical** animals, however, this system is generally adequate. In addition to having a nerve net, starfish, which are also radially symmetrical, have a nerve ring around the mouth and radial nerves that extend into each arm.

Compared with radially symmetrical animals, **bilaterally symmetrical** animals tend to be more active, and their nervous systems reflect this (Figure B19.2). As invertebrates become more complex, they usually have more nerve cells and more specialized neural structures, such as nerves, nerve cords, ganglia, and brains. The number of association neurons increases, as does the complexity of synaptic types. This provides more opportunity for control and integration and allows for a wider range of responses and behaviors.

Because bilaterally symmetrical animals tend to move forward, the ability to detect environmental changes, including the presence of enemies, is important. Such animals tend to have sense organs and other neural structures concentrated near the head end, part of the trend toward **cephalization.** They also usually have integrators such as ganglia or brains to organize sensory information and coordinate the responses accordingly.

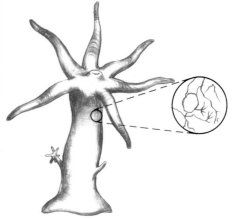

(A) Hydra—netlike arrangement

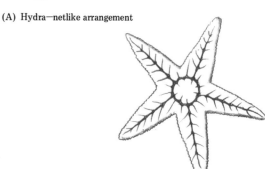

(B) Sea star—radial system

FIGURE B19.1
Some simple invertebrate nervous systems. (A) The nerve net of *Hydra* (cnidarian). (B) Starfish (echinoderms) have both a nerve net and a circumoral nerve ring from which radial nerves extend into the arms. Animals in both of these groups are primarily radially symmetrical.

THE NERVE IMPULSE

Most animals have a nervous system that aids them in maintaining homeostasis quickly and precisely. For example, the nerve impulse allows you to lift your foot almost instantly when you step on a sharp object. Favorite experimental structures for the study of nerve impulses are the giant axons (up to 1 mm in diameter) of invertebrates such as squid and annelid worms.

Most cells can maintain a slight difference in electrical charge, or **resting potential,** between the inside and the outside of the plasma membrane (Figure 19.7). Neurons are called excitable because they can use the potential difference to convey information from one part of a cell to another, and often to another cell.

Theoretically, the numbers of diffusible particles on one side of a plasma membrane should be balanced by an equal number of particles on the other side. However, because plasma membranes are not equally

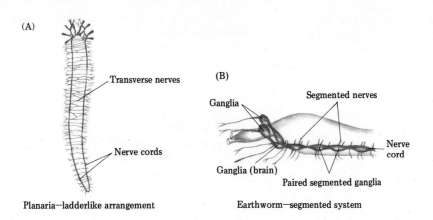

(A)

Transverse nerves

Nerve cords

Planaria—ladderlike arrangement

(B)

Ganglia

Segmented nerves

Nerve cord

Ganglia (brain)

Paired segmented ganglia

Earthworm—segmented system

(C)

Ganglia

Ganglia

Nerves

Octopus—centralized ganglia

FIGURE B19.2

(A) Planarians (flatworms) have a ladderlike nervous system, with considerable concentration of sense organs and a primitive brain (cerebral ganglion) in the head. (B) The longitudinal nerve cords in annelids such as earthworms and arthropods such as grasshoppers include numerous small ganglia along their length as well as larger brainlike ganglia in the head. (C) Although there is a general trend toward cephalization as one compares lower and higher invertebrates, the advanced cephalopods such as the octopus have their ganglia massed near the central region of the body. With their hundreds of millions of neurons, octopuses have considerable learning ability.

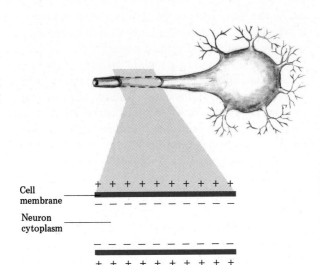

Cell membrane

Neuron cytoplasm

+ + + + + + + + + +

− − − − − − − − − −

− − − − − − − − − −

+ + + + + + + + + +

FIGURE 19.7

In the axon of a resting neuron, positive potassium ions tend to diffuse out, leaving the inside of the plasma membrane negative (−) with respect to the positive (+) outside. The membrane or electrical potential that results from the separation of opposite charges by the plasma membrane is measured by means of tiny electrodes.

FIGURE 19.8
When a stimulus is applied to the dendrites or the cell body of a neuron, the plasma membrane has a resting potential. A nerve impulse is a wave of depolarization that travels along the neuron. Sodium ions diffuse into the cell in the region of depolarization and momentarily cause the inside of the membrane to become positive relative to the outside. As the impulse moves rapidly along, the resting conditions are reestablished quickly behind it.

Plasma membrane

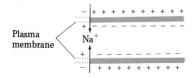

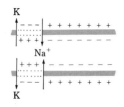

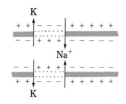

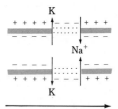

Direction of impulse

permeable to all substances, particles of the same kind are not distributed equally on the two sides of the membrane. Also, the large negatively charged protein ions inside the membrane do not diffuse out. The most important ions in the nerve impulse are potassium (K^+) and sodium (Na^+), although chloride (Cl^-), calcium (Ca^{2+}), and negatively charged organic ions are also involved.

In resting neurons, the potassium ion concentration inside the plasma membrane is much higher than it is in the tissue fluid outside. Conversely, the sodium ion concentration is much higher outside the membrane than it is inside. Even when an impulse is not being conducted, Na^+ and K^+ ions are being transported actively across the neuron membrane, with ATP providing the necessary energy for the so-called sodium-potassium pump. The K^+ ions that are transported in do not equalize the many negative ions inside of the cell. Thus, the outside of the resting plasma membrane is electrically positive and the inside is negative.

Events that cause depolarization of the excitable plasma membrane lead to a **nerve impulse,** or **action potential** (Figure 19.8). Somewhat like the traveling spark on a burning fuse, what moves in a nerve impulse is the site of the chemical reaction. The depolarizing electrical stimulus increases the permeability of the plasma membrane to sodium and the influx of Na^+ reverses the polarity. The inside of the membrane becomes positive relative to the outside because more sodium ions are entering than are leaving. Then the channels for the diffusion of sodium inward close and Na^+ ions are actively pumped out. The outward diffusion of K^+ ions quickly restores the resting potential.

Not just any stimulus will do this; stimuli must reach the **threshold,** or minimum level, that can elicit a nerve impulse. A stimulus above the threshold level does not result in a response greater than that caused by a threshold-level stimulus. Neurons either fire completely or not at all, a principle known as the **all-or-none law.** This has sometimes been compared to the setting of a common mousetrap because once the trap is set off, the force of the trap is built in and is not increased by stronger stimulation.

The all-or-none principle seems to contradict everyday experience, because everyone has noted that the greater the stimulus the greater the response. For example, when you touch a stove you can tell whether it is just warm or very hot. However, the all-or-none principle applies to single neurons, whereas many behaviors involve the larger nerves. The graduated response of a muscle is caused by the stimulation of a variable number of neurons affecting it. The greater the stimulus applied to a nerve, the more neuron fibers are excited and the more muscle fibers are stimulated. Also, a more intense stimulus may cause a greater frequency of nerve impulses per unit of time.

For a millisecond or so while the membrane is depolarized and while resting conditions are being reestablished, it is in a **refractory period.** Early in this short time span, no nerve impulse is possible; slightly later, only an above-threshold stimulus can elicit an action potential. Despite this, neurons can transmit hundreds of impulses per second. Do not blame mental fatigue on "tired" neurons because neurons can fire indefinitely providing they have sufficient oxygen.

Nerve impulses travel faster in thick than in thin axons. Conduction is fastest in large and myelinated axons, reaching rates of over 120 m (about the length of a football field) per second in humans. This occurs because the multimembraned covering prevents the formation and propagation of action potentials. Instead, action potentials leap from one uncovered node of Ranvier to another (saltatory conduction), moving far faster than if myelin were absent and conduction were continuous (Figure 19.9). Even though the word *fire*, commonly used to discuss the nerve impulse, conveys the notion of speed, the electrochemical action potential travels far slower than the flow of electricity through a wire.

Various substances increase or decrease the excitability of neurons, often by affecting the permeability of the membrane to calcium, which is so important for normal neural function. When calcium ions are deficient, some sodium ions leak into the cell between action potentials through incompletely closed sodium channels. Because this allows the neurons to fire more easily, or even spontaneously, the muscles innervated by such neurons may go into spasms (low-calcium tetany). High calcium ion concentrations lead to the opposite condition, neurons becoming less excitable and more difficult to fire. This occurs with certain anesthetics, cocaine use, and poisoning by certain pesticides (e.g., chlorinated hydrocarbons).

SYNAPSES

Although neurons may come very close to one another, they do not touch and remain separated by a tiny space or gap. The same is true of neurons and the cells of such effectors as muscles, the **neuromuscular junction.** The neuron-neuron and neuron-effector junctions are called **synapses** (Figure 19.10). Neurons that end at a specific synapse are called **presynaptic,** whereas those that begin there are called **postsynaptic.** Some neurons are presynaptic for one synapse and postsynaptic for another.

At most neuron-neuron synapses, nerve impulses cross **chemical synapses** by means of specific **neurotransmitters** that are released by the presynaptic neurons into the synaptic cleft. These chemical messengers (Chapter 18) diffuse across the synaptic cleft and bind to specific receptor proteins on the membrane of the postsynaptic cell. The presynaptic terminals are generally knoblike, and neurotransmitters are synthesized there continuously and stored in the tiny **synaptic vesicles.** The ATP needed for the synthesis is provided by the mitochondria in the presynaptic knobs.

Neuromuscular junctions are special kinds of chemical synapses. The axons of motor neurons branch, and each forms a neuromuscular junction with a muscle cell or fiber (Chapter 21). Each motor neuron and the set of muscle fibers (cells) with which it forms junctions is called a

FIGURE 19.9

Very rapid conduction occurs in large myelinated axons, even though myelin insulates the axon where it is present. Such conduction also requires less energy than the continuous conduction along unmyelinated neurons. Depolarization skips along the myelinated axons from one node of Ranvier to the next because only at those sites is the membrane in direct contact with the interstitial fluid.

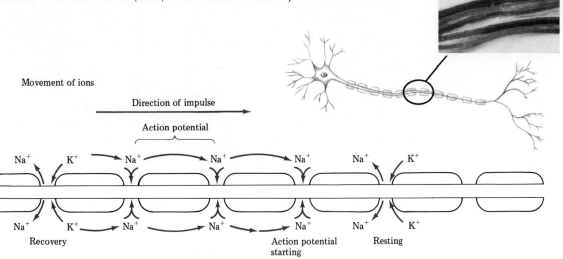

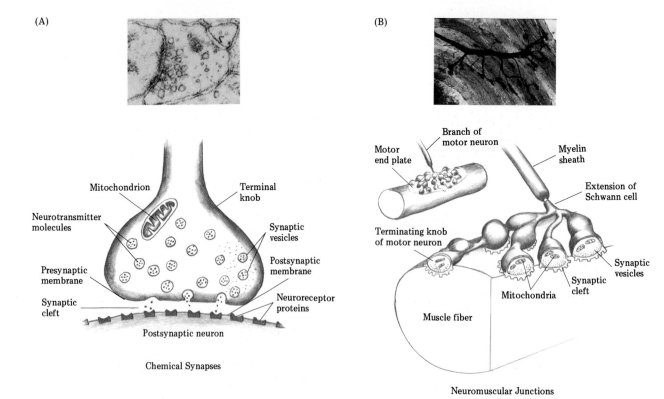

(A)

Mitochondrion

Neurotransmitter molecules

Presynaptic membrane

Synaptic cleft

Terminal knob

Synaptic vesicles

Postsynaptic membrane

Neuroreceptor proteins

Postsynaptic neuron

Chemical Synapses

(B)

Motor end plate

Branch of motor neuron

Myelin sheath

Extension of Schwann cell

Terminating knob of motor neuron

Synaptic vesicles

Synaptic cleft

Mitochondria

Muscle fiber

Neuromuscular Junctions

FIGURE 19.10

Types of synapses. (A) The arrival of the action potential at the terminal knob of the axon allows the entry of calcium ions. These cause some synaptic vesicles to fuse to the plasma membrane and release their specific neurotransmitters into the synaptic cleft.

After diffusing quickly across the cleft, a neurotransmitter binds to specific receptor proteins, and either allows an impulse to be initiated in the postsynaptic cell or inhibits it. Therefore, because transmission in chemical synapses takes place by means of neurotransmitters, it is unidirectional, from a presynaptic neuron to a postsynaptic cell.

(B) The branching ends of a motor axon each form a neuromuscular junction with a different muscle fiber. Acetylcholine is the neurotransmitter involved in these special forms of chemical synapses. All of the muscle fibers in the motor unit contract simultaneously when a motor neuron fires.

motor unit. Impulses in a motor neuron cause the simultaneous contraction of all muscle fibers in that motor unit. Other types of neuron-effector junctions are described later in this chapter.

In chemical synapses, only presynaptic neurons are able to produce and release neurotransmitters, and only postsynaptic cells are able to receive them. Therefore, transmission is unidirectional in chemical synapses, with the synapse acting as a gate. For example, when you step on a sharp object, you become aware of the pain after the stimulus passes through a number of neurons. However, the message to lift your foot off the sharp object must travel over a different set of neurons.

Although many neurons secrete more than one kind of neurotransmitter, "wiring" may be very specific. A typical neuron has a variety of receptors in its plasma membrane, and its ability to respond to a particular neurotransmitter depends upon the exact nature of its receptors. Among the numerous neurotransmitters are **acetylcholine, norepinephrine, serotonin,** and several amino acids. Acetylcholine is the neurotransmitter for motor neurons that arise in the spinal cord, and it is involved with vertebrate neuromuscular synapses. Norepinephrine is released by many neurons in both the CNS and the PNS. Serotonin and other amine neurotransmitters are especially important in the brain.

Neurotransmitters seldom remain long in the synaptic cleft. If they did, then the postsynaptic cell might be constantly active because its receptors would usually be bound to the transmitters. Although removal of the neurotransmitters is essential for normal neural functioning, it only rarely happens because the neurotransmitters simply diffuse away (Figure 19.11). Rather, in the case of acetylcholine, for example, a specific enzyme (acetylcholinesterase) breaks down the chemical messenger and clears the cleft. Amine neurotransmitters are often taken back into the axon knob by active transport, and glial cells remove some amino acid transmitters.

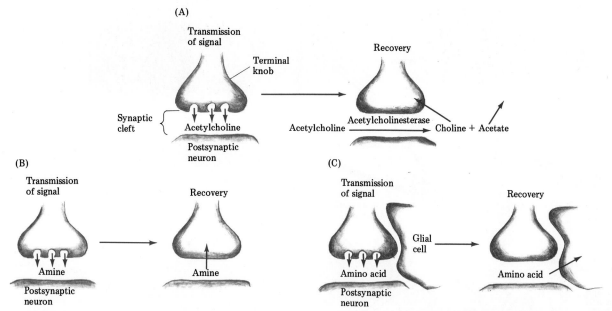

FIGURE 19.11
Clearing the synapse quickly of neurotransmitters is important for the proper functioning of neurons. Although neurotransmitters occasionally just diffuse away, one of the three processes shown is far more common: (A) An enzyme in the synaptic cleft may break down the neurotransmitter (e.g., acetycholine); (B) reuptake by the axon knob may occur by active transport (e.g., amines); or (C) glial cells may take up the neurotransmitter (e.g., amino acids).

Various substances interfere with the process of clearing the synapse of neurotransmitters. For example, certain nerve gases and pesticides (organophosphates) inactivate acetylcholinesterase. Because acetylcholine increases with each successive action potential, spasms may result from the excessive stimulation of muscle cells. Asphyxiation may occur if this happens in the larynx or diaphragm. Some drugs that affect mood do so at least partially by their effects on the synaptic levels of amine-type neurotransmitters.

Postsynaptic neurons may have receptors for more than one kind of neurotransmitter. In so-called excitatory synapses, the ability of a postsynaptic membrane to generate an action potential is increased when a neurotransmitter binds to its receptor proteins. The opposite happens in inhibitory synapses, where the generation of an action potential is made less likely. It is not the neurotransmitter that determines whether a synapse is excitatory or inhibitory, but rather the postsynaptic membrane. A particular neurotransmitter may be excitatory at some synapses and inhibitory at others.

Postsynaptic membranes may be in contact with a few to tens of thousands of presynaptic membranes (Figure 19.12). **Summation** is necessary to make sense out of the numerous excitatory and inhibitory messages. For example, action potentials may be generated if excitatory events exceed inhibitory ones. However, no impulse is triggered until the threshold is reached. One way this happens is for several excitatory signals to reach the postsynaptic cell simultaneously (spatial summation). Another possibility is for rapid and repeated stimulation through a single excitatory synapse (temporal summation).

Many pathogens, toxic substances, and drugs affect the synthesis, storage, and activity of neurotransmitters. For example, the toxin of the bacterium that causes **tetanus** (*Clostridium tetani*) blocks inhibitory impulses that would normally reach spinal motor neurons. Because those neurons then receive only a flow of excitatory impulses, uncontrolled muscle spasms and seizures result. (Popularly, tetanus has been called *lockjaw* because the jaw muscles are often affected.) A related bacterium (*C. botulinum*) releases a powerful toxin (only 60 billionths of a gram kills a human) that produces **botulism** by blocking the release of acetylcholine at the synapse.

(A)

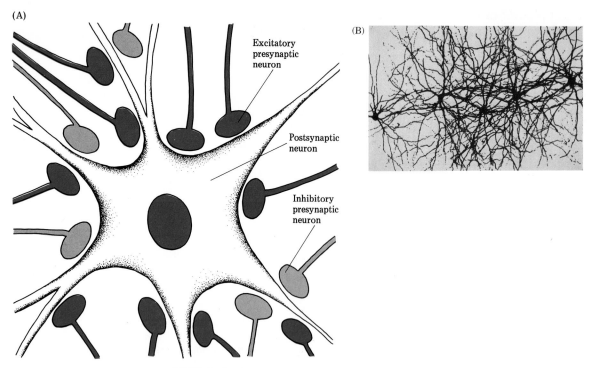

Excitatory
presynaptic
neuron

Postsynaptic
neuron

Inhibitory
presynaptic
neuron

(B)

FIGURE 19.12

(A) Scanning electron micrograph of a cell
body and numerous axon knobs converging
on it. **(B)** Simplified network of extensions
from just five spinal motor neurons. Some
messages are stimulatory and some are
inhibitory, necessitating integration, and
often summation.

REFLEXES AND BASIC CIRCUITS

Mammalian nervous systems are composed of billions of neurons, each
of which typically synapses with many other cells. Obviously, proper
homeostatic responses to changing stimuli require neural organization.
This section briefly explains some simple neural circuits or pathways;
descriptions of entire nervous systems occur later.

Reflexes

A **reflex** is an automatic, stereotyped, predetermined response to a
given stimulus, such as the familiar pupillary reflex or knee jerk reflex
(Figure 19.13). Reflexes are functional units of the nervous system that
depend only upon the anatomic relationships of the neurons involved.
Typically, they involve only part of the body rather than the body as a
whole. Much of the behavior of some simple animals is controlled by
reflexes. However, reflexes are also important in the vertebrates, and
many physiological mechanisms involve them. For example, a change in
body temperature causes the body's thermostat in the hypothalamus to
mobilize the necessary homeostatic responses to return the body temperature to normal (Chapter 2). Other reflexes are involved in the maintenance of breathing and heart rates, regulation of blood pressure, and
salivation.

The simple knee jerk reflex requires only two neurons and one synapse. The stretching of a muscle fiber when the tendon of the knee is
tapped sends an impulse to the spinal cord over a sensory neuron. After
crossing a single synapse there, the impulse travels back to the same leg
muscle over a motor neuron. The familiar kick occurs when the muscle
contracts.

Most reflexes are more complicated, such as the withdrawal reflexes
when we touch something very hot or sharp. In these cases sensory
neurons synapse with one or more **association neurons** in the spinal cord
before a message travels to an effector over a motor neuron. Most

neurons in the central nervous system are association neurons and they are very important in the integration of responses. For example, the association neuron that received the impulse from the sensory neuron and transmitted it to the motor neuron may also send a message to the brain via other association neurons in the spinal cord. The brain may then decide on other actions, such as yelling in pain, but the reflex itself occurred without involvement of the brain. However, some reflex actions can be modified consciously. Urination in babies is a reflex response that occurs when the bladder stretches to a certain point and a sphincter muscle relaxes and allows urine to flow through the urethra (Chapter 17). We gradually learn to either urinate before the bladder fills or, within limits, to inhibit the reflex when it occurs at an inconvenient time or place.

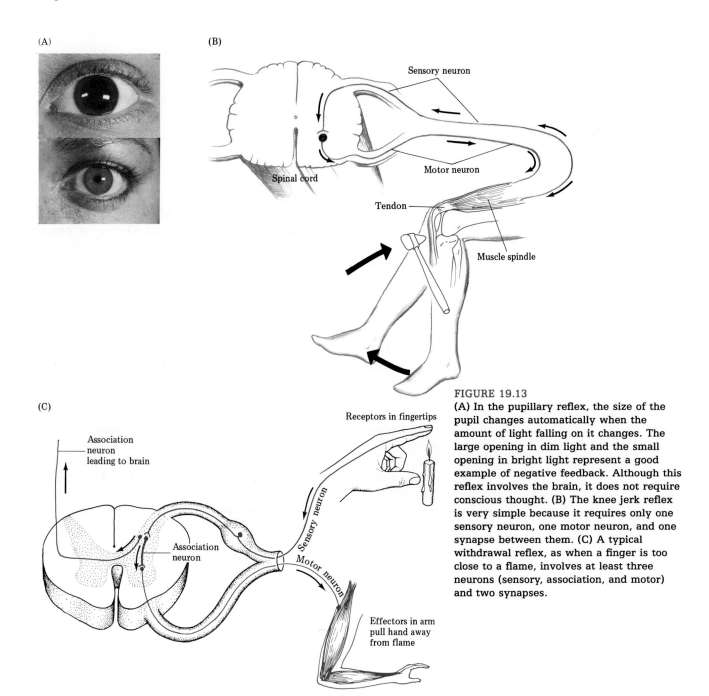

FIGURE 19.13

(A) In the pupillary reflex, the size of the pupil changes automatically when the amount of light falling on it changes. The large opening in dim light and the small opening in bright light represent a good example of negative feedback. Although this reflex involves the brain, it does not require conscious thought. **(B)** The knee jerk reflex is very simple because it requires only one sensory neuron, one motor neuron, and one synapse between them. **(C)** A typical withdrawal reflex, as when a finger is too close to a flame, involves at least three neurons (sensory, association, and motor) and two synapses.

FIGURE 19.14
(A) Convergent circuits allow one neuron to receive impulses from more than one other neuron. (B) In divergent circuits, one presynaptic neuron synapses with more than one postsynaptic cell. The complex combinations that often occur may facilitate summation. (C) Reverberating circuits are important in processes as diverse as breathing and thinking. Axon collaterals assist in generating repeated impulses, either in a neuron itself or along a particular pathway.

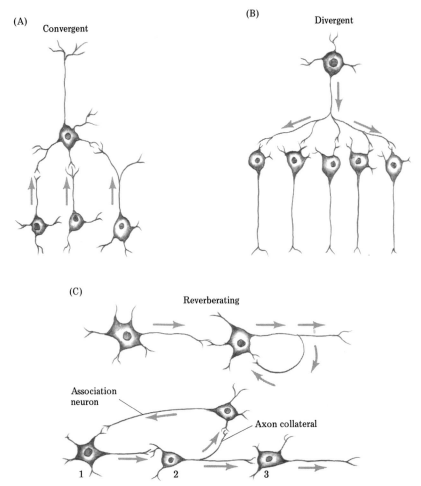

(A) Convergent

(B) Divergent

(C) Reverberating

Association neuron

Axon collateral

1   2   3

Simple Circuits

In **convergent circuits,** a postsynaptic neuron is controlled by impulses from two or more presynaptic neurons (Figure 19.14). For example, an association neuron in the spinal cord may receive impulses from other CNS association neurons as well as from a sensory neuron entering the cord. The various incoming impulses must be integrated before the association neuron fires and stimulates a motor neuron. In **divergent circuits,** a single presynaptic neuron communicates with more than one postsynaptic neuron, and sometimes with tens of thousands of such cells. This is often important in summation that was described earlier. The **reverberating circuits** that involve axon collaterals allow a neuron to stimulate itself or to trigger repeated new impulses through a pathway. All of these basic circuits, as well as more complicated combinations of them, operate in controlling our many physiological functions and behaviors.

THE VERTEBRATE NERVOUS SYSTEM

Cephalization reaches its peak in the vertebrates, which have large, complex brains and numerous sensory structures concentrated in, but not limited to, the head. In this section you will first explore the involuntary ANS portion of the PNS that controls many aspects of the internal environment and then examine the general traits of vertebrate brains.

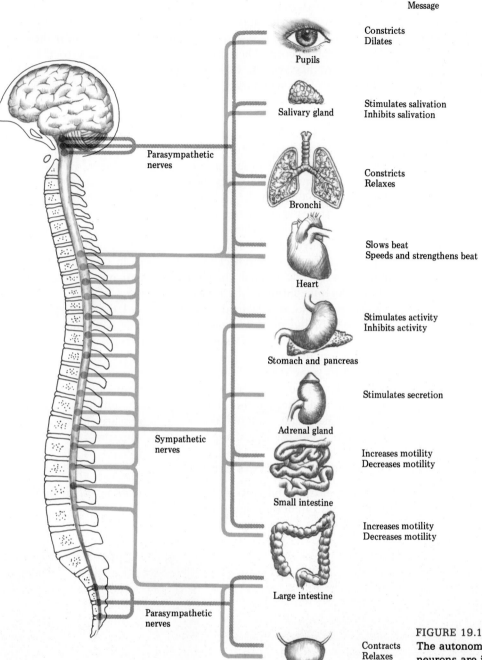

Message

Constricts
Dilates
Pupils

Stimulates salivation
Inhibits salivation
Salivary gland

Parasympathetic
nerves

Constricts
Relaxes
Bronchi

Slows beat
Speeds and strengthens beat
Heart

Stimulates activity
Inhibits activity
Stomach and pancreas

Stimulates secretion
Adrenal gland

Sympathetic
nerves

Increases motility
Decreases motility
Small intestine

Increases motility
Decreases motility
Large intestine

Parasympathetic
nerves

Contracts
Relaxes
Bladder

FIGURE 19.15

The autonomic nervous system. Pairs of neurons are involved in the control of involuntary functions by both subdivisions. The presynaptic neurons of the sympathetic system synapse with the postsynaptic neurons in various ganglia. Presynaptic neurons of the parasympathetic system extend from the lower brain and lower spinal cord to the target organ, where they synapse with short postsynaptic neurons. In general, the neurons from the two subdivisions have opposite effects on the same target organ. The sympathetic system generally mobilizes energy and aids the body in responding to stress; the parasympathetic system assists in restoring the body to resting states after the stressful situation.

Of course, both the PNS and the CNS work together toward the end of maintaining homeostasis.

The Autonomic Nervous System (ANS)

Both subdivisions of the ANS operate through pairs of neurons, whose effects on the same target organ are often, but not always, antagonistic (Figure 19.15). In the **sympathetic nervous system** subdivision, the cell bodies of the presynaptic neurons lie in the middle region of the spinal cord. These neurons synapse with the postsynaptic neurons in various ganglia. The postsynaptic neurons then extend to such target organs as the involuntary muscles and glands. In contrast, the cell bodies of the **parasympathetic nervous system** lie in the lower brain or near the lower end of the spinal cord. Their axons extend all the way to the target organs, where they synapse with short postsynaptic neurons.

The effects of the sympathetic neurons contribute to the fight-or-flight response (Chapter 18), such as an accelerated heartbeat and more forceful contraction, increased blood flow to skeletal and cardiac muscles, increased blood glucose, increased blood pressure, and dilated pupils of the eyes. The parasympathetic system helps to restore body functions to normal after a stressful situation, and it is dominant in ordinary, quiet activities. The initial emergency response is triggered largely by the nervous system, after which it is augmented by the secretion of epinephrine and norepinephrine by the adrenal medulla.

The neurotransmitter released by the presynaptic axons of both subdivisions is acetylcholine. However, the neurotransmitters secreted by the postsynaptic neurons reflect the differences in function of the two subdivisions, acetylcholine in the parasympathetic portion and epinephrine and norepinephrine in the sympathetic system.

When the ANS was named, it was believed that the system was autonomous or independent from the CNS. It is now known that the hypothalamus and other CNS structures help to regulate the ANS. Also, although it usually functions automatically, the ANS can be consciously controlled to some degree through techniques such as **biofeedback.** In biofeedback, various auditory, visual, or other stimuli aid a person in monitoring and altering the status of certain autonomic functions. Trained humans have altered their blood pressure, heart rates and rhythm, blood glucose levels, and brain wave patterns.

Overview of Vertebrate Brains

The three major subdivisions of the vertebrate brain are the **hindbrain, midbrain,** and **forebrain** (Figure 19.16).

FIGURE 19.16

The three major regions of the vertebrate brain, using a cross-sectional view of the human brain as an example. The hindbrain is shown in a darker tint, the midbrain in a lighter tint, the diencephalon portion of the forebrain in light gray, and the telencephalon in dark gray.

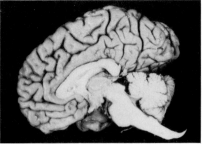

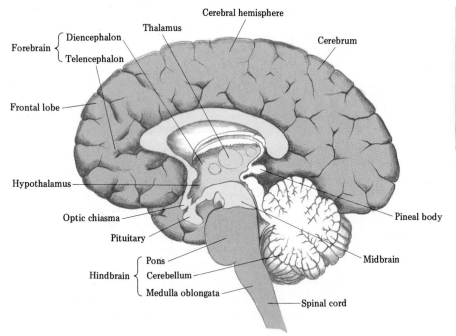

The portion of the hindbrain closest to the spinal cord is the **medulla oblongata.** Its masses of gray matter are important in the control of circulation, breathing, swallowing, vomiting, sneezing, and coughing. Many important tracts of white matter also pass through the medulla. Sensory information from the right side of the body goes to the left cerebral hemisphere and vice versa. The nearby **cerebellum,** the second-largest part of the human brain, is involved in the control of body movements and equilibrium. Its size and complexity is correlated roughly with the complexity of muscular activity in an animal group. It receives impulses from the muscles, tendons, joints, and inner ear, as well as from the cerebral cortex, regarding the appropriate type of movement. The **pons** contains areas of gray matter and tracts of white matter that link the medulla to the cerebellum and cerebrum.

The midbrain structures play a variety of roles in vertebrates. Many complex behaviors of fishes and amphibians are controlled by this region, which is the most prominent part of their brains. The midbrain is also important in aspects of vision in all vertebrates, although many visual functions are performed by the cerebrum in the higher vertebrates.

The forebrain contains two subdivisions, the **diencephalon** and **telencephalon.** The diencephalon contains the **thalamus,** the **hypothalamus** and the associated posterior pituitary (Chapter 18), and the pineal gland. The general functions of the diencephalon structures are to control arousal and various emotional responses and to integrate and relay information from the sensory structures. The size of the cerebrum portion of the telencephalon varies greatly in different groups of vertebrates (Figure 19.17). The cerebrum, which is usually divided into two hemispheres, is quite small in fish, somewhat larger in amphibians, and still larger in reptiles. It is the major area in birds, although its surface is relatively smooth. The mammalian cerebrum is very large and its thin outer layer is highly convoluted, which greatly increases its surface area and adds to the potential for complex behavior. Many functions controlled by other parts of the brain in other vertebrates are performed by the cerebrum in mammals.

THE HUMAN CENTRAL NERVOUS SYSTEM

The soft and fragile brain and spinal cord are protected both by the bones of the skull and vertebrae and by the three layers of connective

FIGURE 19.17
Comparison of the brains of various vertebrates. Olfaction, or the sense of smell, is the dominant sense in most lower vertebrates. A major trend in vertebrate evolution is the increase in the size of the cerebrum and in the surface area of the cerebral cortex.

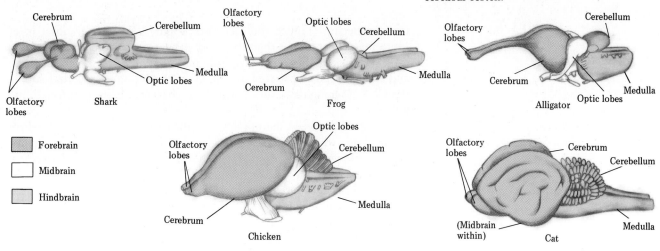

(A)

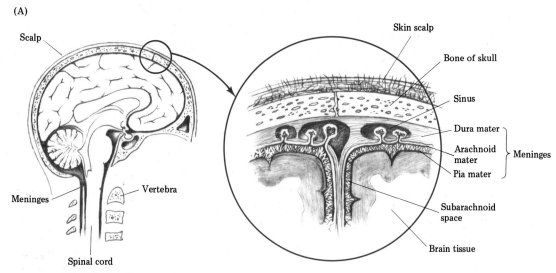

FIGURE 19.18
The CNS is well protected by bone, the meninges, cerebrospinal fluid, and body tissues in general. (A) Detailed view showing the skull and meninges. (B) Cerebrospinal fluid is produced continuously, circulates around the CNS, and is reabsorbed into the blood.

(B)

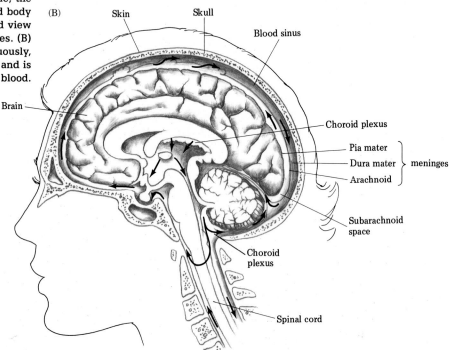

tissue collectively called the **meninges** (Figure 19.18). The outer layer (dura mater) is tougher than the more delicate cobweblike middle layer (arachnoid). The thin inner layer (pia mater) adheres closely to the brain and spinal cord, and it contains many of the blood vessels that supply the CNS. **Meningitis** occurs when these coverings become infected and inflamed.

The space between the middle and inner meninges contains **cerebrospinal fluid,** which further protects the CNS by cushioning it against physical blows. After forming continuously in special capillary networks that extend from an area of the inner meninges (choroid plexuses), the cerebrospinal fluid circulates through the CNS and is reabsorbed into the blood. A blockage of the flow of cerebrospinal fluid, such as by a tumor, may allow it to accumulate, press against, and damage neural tissue (e.g., hydrocephalus).

Spinal Cord

Far less than a walking stick in length (or meter stick, for scientists), the oval **spinal cord** controls many reflex actions and transmits impulses to and from the brain (Figure 19.19). The inner part of the cord contains a butterfly-shaped region of gray matter made up of nerve cell bodies, dendrites, unmyelinated axons, glial cells, and blood vessels. The surrounding white matter consists of myelinated axons arranged in tracts that transmit impulses up and down the cord.

(A)

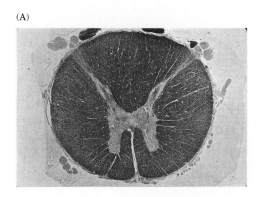

(B)

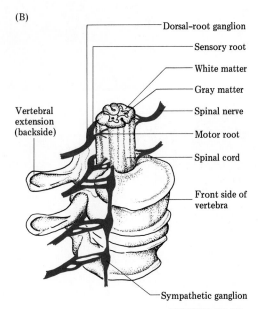

Dorsal-root ganglion

Sensory root

White matter

Gray matter

Spinal nerve

Motor root

Spinal cord

Front side of vertebra

Vertebral extension (backside)

Sympathetic ganglion

FIGURE 19.19
(A) The spinal cord is about the diameter of a finger and is well protected by the meninges and vertebrae. Two grooves or fissures partially divide it into halves. White matter surrounds the inner gray matter, the opposite arrangement as in the cerebrum. (B) A pair of spinal nerves (part of the somatic system of the PNS) emerges from the space between adjacent vertebrae. Spinal nerves are connected to the cord by two so-called roots, and the ganglion or swelling on the dorsal root contains the cell bodies of the sensory neurons.

The Brain

The person whose brain is shown in Figure 19.20 once loved, hated, was afraid, was fearless, cried, and was happy. How can the jellylike mass that weighs a mere 1.4 kg (3 lb) cause those and so many more emotions while it also receives, integrates, and directs the innumerable impulses that control our physiological functions? We are still finding out, using far lesser computers in the process.

The **brain** is a demanding structure, a mere 2 percent of the body weight that requires about 20 percent of the total oxygen used by the body. The brain weight of children is relatively greater than it is in adults, and up to 50 percent of the child's oxygen may be used to support the brain's activities. Stopping the blood flow to the brain for as little as 10 seconds may cause loss of consciousness. Irreversible damage to many brain neurons may occur from only 3 to 5 minutes of blood flow deprivation. A **stroke** is any serious disruption of the brain's circulation, such as by hemmorhage or a blood clot. The severity of damage depends upon the extent and location of the area affected.

Although **phrenology** is as phony as it is old, the notion that function is localized in the brain has been confirmed (Figure 19.21 and Table 19.1). There is no evidence, however, that surface bumps and depressions on the skull represent unusual development or relative lack of brain function in the underlying **cortex,** or outer part of the cerebrum. Experimental neurologists, superficially resembling acupuncturists inserting needles, probe the brain with tiny electrodes while attempting to learn about localization of function. Because the brain itself has no pain

FIGURE 19.20
External view of the top of the human brain, showing the deep cleft that separates the two cerebral hemispheres. The extensive convolutions of the cerebral cortex greatly increase its surface area. The overlying meninges and the extensive network of blood vessels that provide the abundant oxygen and nutrients, and carry away metabolic wastes, were removed to provide a clear view of the cerebrum.

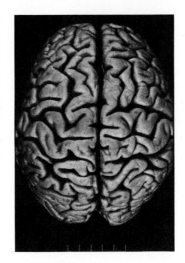

TABLE 19.1
Structures and Functions of the Human Brain

Structure	Functions
Hindbrain	
Medulla	Vital centers within its reticular formation regulate breathing, heartbeat, and blood pressure; reflex centers control coughing, sneezing, swallowing, and vomiting; relays messages to and from the brain
Pons	Helps regulate respiration; links various parts of the brain
Cerebellum	Helps maintain coordinated movement, posture, muscle tone, and equilibrium
Midbrain	
Midbrain	Mediates visual and auditory reflexes; integrates information on posture and muscle tone
Forebrain	
Diencephalon	
Thalamus	Main relay center between brain and spinal cord; helps sort and integrate incoming messages before relaying them to cerebral centers
Hypothalamus	Control center for body temperature, hunger, fluid balance; involved in some emotional and sexual responses; helps control ANS; secretes releasing hormones that regulate pituitary gland
Telencephalon	
Cerebrum	Center of intellect, memory, language, judgment, personality, and consciousness; receives and interprets sensory information; controls motor functions; links various parts of brain

(A)

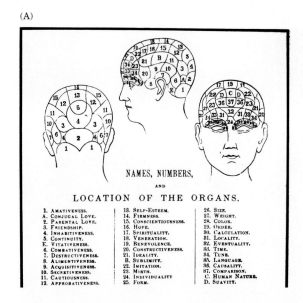

NAMES, NUMBERS,
AND
LOCATION OF THE ORGANS.

1. AMATIVENESS.	13. SELF-ESTEEM.	26. SIZE.
A. CONJUGAL LOVE.	14. FIRMNESS.	27. WEIGHT.
2. PARENTAL LOVE.	15. CONSCIENTIOUSNESS.	28. COLOR.
3. FRIENDSHIP.	16. HOPE.	29. ORDER.
4. INHABITIVENESS.	17. SPIRITUALITY.	30. CALCULATION.
5. CONTINUITY.	18. VENERATION.	31. LOCALITY.
E. VITATIVENESS.	19. BENEVOLENCE.	32. EVENTUALITY.
6. COMBATIVENESS.	20. CONSTRUCTIVENESS.	33. TIME.
7. DESTRUCTIVENESS.	21. IDEALITY.	34. TUNE.
8. ALIMENTIVENESS.	B. SUBLIMITY.	35. LANGUAGE.
9. ACQUISITIVENESS.	22. IMITATION.	36. CAUSALITY.
10. SECRETIVENESS.	23. MIRTH.	37. COMPARISON.
11. CAUTIOUSNESS.	24. INDIVIDUALITY.	C. HUMAN NATURE.
12. APPROBATIVENESS.	25. FORM.	D. SUAVITY.

(B)

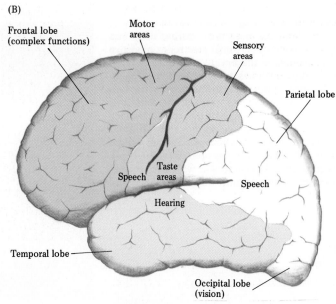

FIGURE 19.21

(A) A phrenologist's model of the skull. Phrenologists believe erroneously that surface bumps and depressions reflect brain development beneath, and therefore that one's capabilities and potentials can be "read" by examining the head's surface. (B) Localization of function in the brain is evident, however, as shown in this current "map" of some of the main sensory and motor regions of the cerebral cortex.

receptors, subjects can be conscious and can report their sensations as cells are stimulated. Not only are major events recalled in presumably complete and accurate detail, but so are seemingly irrelevant experiences (e.g., the afternoon weather conditions on this date 12 years ago). Interestingly, recall by electrode stimulation seems to occur in real time (i.e., it is not speeded up or condensed). This may present a practical problem in using electrodes in your brain on the next exam!

As for every body system, much knowledge also comes from observing patients with brain injury or disease, and then studying the brain carefully after death. Although aspects of the study of "brain waves" are debated, the use of **electroencephalograms** (EEGs) is almost routine (Figure 19.22). EEGs are measurements of the electrical activity of the brain, and they show clearly that the brain is active continuously, even during sleep.

A blow to the back of the head often causes the sensation of light, and serious injury or removal of that region causes blindness. From those and other observations, it is clear that that region (posterior occipital lobe) contains the visual centers. Similarly, a blow to the head just above the ear causes a sensation of sound, and serious injury or removal of such areas (lateral temporal lobes) on both sides causes deafness. These two examples demonstrate that what we sense does not depend entirely upon the nature of the stimulus but, rather, upon the integrator that receives the sensory input. The pressure of a blow to the head is sensed as light if the messages go to one area and as sound if they terminate in another region.

The large fissure (central sulcus) that crosses the top of each hemisphere partially separates some motor areas (in the frontal lobe) from some sensory areas (in the parietal lobe). Remember how peculiar and distorted you looked when you stood in front of a trick mirror in an amusement park? Figure 19.23 shows how we might look if our body features were formed in proportion to the size of the brain areas that control them. It is clear, for example, that the size of the motor area in the cortex is proportional to the extent and complexity of the muscular action involved rather than to the quantity of muscle tissue itself.

FIGURE 19.22
(A) A person being "wired" for the study of brain wave patterns during sleep and wakefulness. Small electrodes measure changes in the brain's electrical activity. Only local anesthesia is needed for most brain surgery because the brain senses no pain in itself. As a result, the conscious patient can verbalize the sensations experienced. (B) Some normal and abnormal EEG patterns. In an epileptic seizure (about 1 in 200 Americans has epilepsy), one brain section stimulates another, which stimulates another, and so on until the neurons fatigue.

(A)

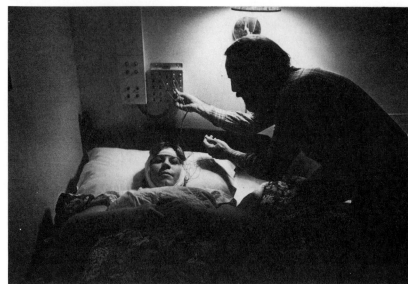

(B)

Amplifiers

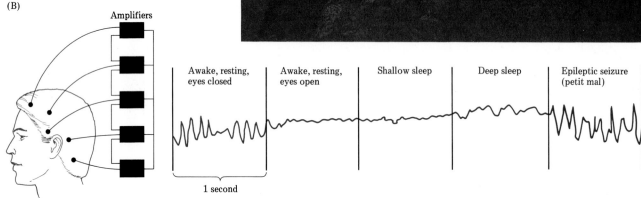

Awake, resting, eyes closed

Awake, resting, eyes open

Shallow sleep

Deep sleep

Epileptic seizure (petit mal)

1 second

Note that large sections of the human cortex in Figure 19.21B are unlabeled. Maps like these almost cover a rat's cortex, cover a large part of a dog's cortex, and a moderate proportion of a monkey's cortex. The extensive regions of the human cortex not involved in the primary sensory or motor functions are hardly empty, however. These are the **association areas** responsible for the "higher" faculties of learning, memory, reasoning, imagination, and personality. Remember all the questions you had about memory before you forgot them?

The fact that the brain is not bilaterally symmetrical in all of its functions comes from the study of people with so-called **split brains.** In such people the right and left hemispheres are disconnected from one another. Other evidence comes from victims of damage to one or the other hemisphere.

Language is usually a function of the left hemisphere only. For example, suppose that a familiar object such as a key is placed in the *right* hand of a blindfolded, split-brain person. As usual, nerve impulses about the presence of the object travel to the primary sensory areas of the *left* hemisphere. The association areas surrounding the sensory cortex process the information, and the person "knows" what the object is and can name it. However, if the object had been placed in the person's *left* hand, naming the object would not be possible. Even so, the person knows what the object is, can pick out the key by touch from a group of dissimilar objects, and can demonstrate how keys work. The person can say "key" when the blindfold is removed because visual information, unlike that from the hands, is received about equally by both hemispheres. Because vision is their prime way of recognizing objects, people

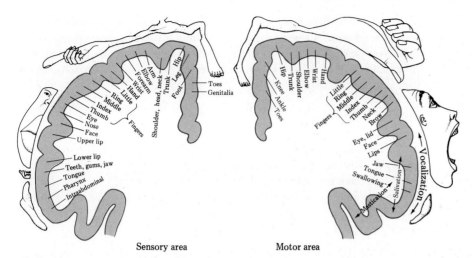

Sensory area   Motor area

FIGURE 19.23
We would appear quite different if the parts of our bodies shown were constructed in proportion to the areas of the cerebral cortex devoted to them. Also note that the motor "person" would not look exactly like the sensory "person." Much cortical tissue is devoted to structures capable of complex movements, even though they are relatively small, such as the tongue and fingers.

with split brains usually conduct everyday activities without difficulty. Unlike adults, young children who sustain damage to the speech area of the left hemisphere can often regain language abilities.

Similar experiments show that many mental activities are localized in one hemisphere or another. The right hemisphere is associated with the appreciation of music (nonsymbolic sounds) and of objects in space (as opposed to time). On the other hand, the left hemisphere is involved mainly with mental activities that can be expressed in words, are logical, and are well ordered in time.

SUMMARY

1. **Stimuli** are detectable changes within organisms and in their environments. Even simple homeostatic responses usually involve the sensing of a stimulus by a **receptor** and the transmission of the information to an **integrator**. A response requires further transmission to an **effector**.

2. In vertebrates, the **brain** and **spinal cord** constitute the **central nervous system** (CNS). The **peripheral nervous system** (PNS) includes the sensory receptors, the **nerves** that link them with the CNS, and the nerves that link the CNS with effectors. The two divisions of the PNS are the **somatic system** (voluntary motor system) and the **autonomic nervous system** (ANS) that regulates the internal environment. The ANS is subdivided further into the **sympathetic system** and **parasympathetic system**.

3. The **neuron,** the anatomical and functional unit of the nervous system, typically consists of a cell body with several short **dendrites** on one side and one longer **axon** on the other. Axons may branch to form **collaterals.**

4. Nerves are bundles of parallel axons in the PNS that are held together by connective tissue; **tracts** are similar bundles within the CNS. The cell bodies of neurons are clustered in **ganglia** in the PNS and in **nuclei** in the CNS. Some nerves are primarily sensory, others are mainly motor, and still others are mixed.

5. **Glial cells** support and nourish neurons and help to maintain a proper ionic environment for them. They assist neurons in the transmis-

sion process although they do not generate or conduct **impulses** themselves. **Schwann cells** produce myelin, which insulates some axons. Nerve impulse velocity is faster in large **myelinated axons** than in small, unmyelinated fibers. **Multiple sclerosis** occurs when there is patchy, irregular destruction of myelin along CNS neurons.

6. In the CNS, the cell bodies of neurons constitute the **gray matter;** masses of myelinated axons make up the **white matter.**

7. The plasma membrane of resting neurons separates the negatively charged interiors from the positively charged exteriors. Sodium and potassium ions are actively pumped out of and into neurons, respectively. The nerve impulse or **action potential** is a traveling wave of depolarization along the membrane. Following a very short **refractory period,** a neuron is ready to fire again.

8. A neuron will fire only if its **threshold** is reached, and it does so in an **all-or-none** manner. An above-threshold stimulus elicits the same response as a threshold-level stimulus.

9. Neuron-neuron and neuron-effector junctions are called **synapses.** In **chemical synapses, neurotransmitters** are necessary for transmission across the synaptic cleft from the presynaptic to the postsynaptic cell, and transmission is thus unidirectional. **Neuromuscular junctions** are special types of chemical synapses.

10. Clearing the chemical synapse quickly of neurotransmitters is important for normal neural functioning. Neurotransmitters are generally broken down by enzymes, reabsorbed by axons, or removed by glial cells.

11. Because **postsynaptic neurons** may have receptors for more than one kind of neurotransmitter, they may be excited or inhibited. The postsynaptic membrane, not the neurotransmitter, determines whether a synapse is excitatory or inhibitory. A particular neurotransmitter may be excitatory at some synapses and inhibitory at others. **Summation** is necessary to integrate the numerous excitatory and inhibitory impulses at synapses.

12. **Reflexes** (e.g., knee jerk and pupillary reflexes) are automatic, predetermined, and stereotyped responses to a given stimulus. Although only one sensory and one motor neuron are necessary in very simple reflexes, most reflexes also involve one or more association neurons.

13. Simple circuits include **convergent circuits** (postsynaptic neuron controlled by impulses from two or more **presynaptic neurons**), **divergent circuits** (single presynaptic neuron communicates with more than one postsynaptic neuron), and **reverberating circuits** (axon collaterals restimulating a neuron or triggering repeated impulses through a pathway).

14. Sponges have no nervous systems and respond to stimuli only at the cellular level. **Nerve nets,** like those in *Hydra,* constitute the simplest nervous systems. In addition to their nerve nets, starfish also have nerve rings with radial branches. Flatworms have a ladderlike network of nerves, with ganglia in the head. Earthworms and arthropods have nerve cords, as well as numerous small body ganglia and larger ganglia in their heads. **Cephalization** is the trend toward the increased concentration of neural tissue in the head, and is especially evident in **bilaterally symmetrical animals.**

15. The autonomic nervous system of vertebrates operates through pairs of neurons whose effects on the same target organ are often antagonistic. The sympathetic nerves emerge from the middle of the spinal cord, and their general effects are to help mobilize the body for reaction to stress. The parasympathetic nerves that emerge from the lower brain and the lower spinal cord generally dominate during ordinary times. Through **biofeedback**, some control of autonomic functions is possible.

16. The vertebrate brain is divided into three regions—the **hindbrain, midbrain,** and **forebrain.** The structures and functions of these regions are summarized concisely in Table 19.1.

17. CNS structures are protected by bones (skull and vertebrae), the **meninges, cerebrospinal fluid,** and other body structures. The meninges contain three layers: tough outer layer, delicate middle layer, and thin and vascular inner layer; **meningitis** occurs when these become infected and inflamed. Cerebrospinal fluid originates in special capillaries, flows through the CNS, and is reabsorbed into the blood; blockage of its flow may damage the CNS.

18. The spinal cord controls many reflex actions and transmits impulses to and from the brain. White matter surrounds the central gray matter.

19. For its size, the brain uses a disproportionately large percentage of the body's oxygen and blood glucose. A **stroke** is any serious disruption of the brain's circulation.

20. Although localization of function is well established, there is no evidence for **phrenology.** Methods of studying the brain include the use of tiny electrodes, the observation of the effects of brain damage and of the damaged brain after death, and the analysis of **electroencephalograms.**

21. The sizes of the sensory and motor areas of the **cerebral cortex** are in proportion to the complexity of the associated body areas rather than with the actual size of the body structures. The large **association areas** are responsible for faculties such as learning, memory, and judgment.

22. The two cerebral hemispheres are not bilaterally symmetrical in function. For example, language and logical mental activities are involved primarily with the left hemisphere. Because of this, people with **split brains** may experience difficulties in performing certain mental functions.

REVIEW QUESTIONS

1. Describe the general flow of information through the nervous system from an external or internal stimulus to an animal's response.

2. Distinguish between the structures and functions of the central and peripheral nervous systems, including the divisions and subdivisions of the PNS.

3. Describe the structures of a typical neuron and explain the functions of each.

4. Describe a nerve and distinguish it from a neuron.

5. Distinguish between tracts and nerves and between ganglia and nuclei.

6. Where are glial cells found and what are their functions?

7. Describe the structure of a Schwann cell, including the node of Ranvier, and explain the source and function of myelin.

8. Compare the function and location of white and gray matter in various parts of the CNS.

9. Explain the nerve impulse by comparing the polarization of the membranes of resting neurons first with ones in which action potentials are occurring, and then with ones during refractory periods.

10. State and explain the all-or-none law in terms of a neuron's threshold level.

11. Describe chemical synapses, including neuromuscular junctions.

12. Explain the origin and function of common vertebrate neurotransmitters.

13. Explain the importance of the rapid clearance of chemical synapses of neurotransmitters, and describe the common methods by which this happens.

14. Describe some examples of transmission problems at the synapse caused by pathogens, drugs, and/or toxic substances.

15. Explain what determines whether a synapse is excitatory or inhibitory.

16. Describe and explain a reflex that involves one sensory, at least one association, and one motor neuron.

17. Compare and contrast convergent, divergent, and reverberating neural circuits.

18. Compare the nervous systems of typical radially symmetrical and bilaterally symmetrical animals.

19. Describe the trends, including cephalization, in the structures of the nervous systems of simple to more complex invertebrates.

20. Compare and contrast the structures, general functions, and neurotransmitters in the sympathetic and parasympathetic subdivisions of the ANS.

21. What is biofeedback?

22. Describe the locations and major functions of the following brain structures: medulla oblongata, pons, cerebellum, midbrain, thalamus, hypothalamus, and cerebrum.

23. Describe how CNS structures are protected.

24. Describe the meninges and explain their functions.

25. Where does cerebrospinal fluid form, what does it do, where does it go, and what may happen if its flow is blocked?

26. What are the major functions of the spinal cord?

27. What are strokes, what are some of their causes, and why are they so potentially serious?

28. Describe some common methods used to study human brain function.

29. Why are there differences in "real" human bodies and those constructed theoretically based upon the sizes of the sensory or motor areas of the cerebral cortex that are associated with them?

30. Draw a general map of the major sensory and motor regions of the cerebral cortex.

31. Where are association areas in the brain, and what are their functions?

32. Explain the behavior of people with split brains.

CHAPTER 20

RECEPTORS: SENSORY SYSTEMS

::

The senses link us with the external environment and with our inner conditions. Because of this, the senses play an extremely useful role in homeostasis, providing the initial information upon which responses are determined. Sensory information is especially useful in the integrative functions of the endocrine and nervous systems (Chapters 18 and 19), and in our ability to respond to changing conditions in appropriate ways (Chapter 21). While few animals surpass humans in the variety of well-developed senses available to them, many animals exceed us in the development of particular senses. Thus, we are unaware of some hues of flowers seen by certain insects, and of the ultrasonic echoes used by bats in navigating at night. Also, we are not fully aware of the extremely rich variety of environmental smells that are so important in the lives of many animals. In this chapter you will explore how we and other animals sense our environments and process the information received.

Looking, listening, and smelling, the prairie dog is vigilant at the opening of its burrow. Badgers could rush across the bare zone at any time, or a large snake could slowly crawl near, blending almost invisibly into the background. Circling overhead, a hawk could swoop down with fatal quickness. This chapter is about the receptors or sensory structures used by humans and other animals to survive and maintain homeostasis.

SENSORY INFORMATION

It is very evident that multiple sensory inputs are important to the survival of the prairie dog shown in the chapter opener. Obviously, any suspicious movements, subtle odors, and unusual sounds that might indicate the nearness of a predator are important. However, as with humans, survival also depends upon the array of unconscious information about the heart rate and blood pressure, the concentrations of oxygen, carbon dioxide, and glucose in the blood, and muscle tone. The simplest way to classify **receptors** is to call them **exteroceptors** if they provide information about the external world, and **interoceptors** if they sense the internal environment.

In simple animals, the cell membrane is the receptor, and much responsiveness depends upon the inherent irritability of the cytoplasm. Specialized receptors are common in the complex invertebrates and vertebrates. Sometimes various accessory cells work with the receptors. Thus, the light-sensitive rods and cones in the retina of our eyes are the receptors, and the cells of the cornea, iris, lens, and other structures support the functions of the sensory cells.

Although receptors were once classified simply as exteroceptors or as interoceptors, a somewhat more precise classification names receptors by the type of energy or condition to which they respond. Thus, **photoreceptors** respond to light energy, **phonoreceptors** to sound, **chemoreceptors** to certain chemicals, and **thermoreceptors** to temperatures. **Mechanoreceptors** respond to touch, pressure, movement, and gravity. **Proprioceptors** (from the Latin for "self") are the specific mechanoreceptors found within muscles, tendons, and joints that provide animals with a sense of the position and condition of their body parts. If you close your eyes and raise an arm toward your face, your proprioceptors will probably allow you to touch your nose very precisely with the tip of a finger. In addition to the above senses, many animals also sense pain and some, such as certain fish, detect changes in the electrical energy of their environment.

Many receptors are found in epithelial tissue, no surprise when we recall that such tissue is commonly in contact with both the external and internal environments. Some sensory structures serve several purposes, and apparently a different type of receptor does not exist for each known sense.

In this chapter, the discussions about invertebrate and vertebrate receptors occur together in the sections on each sense.

HOW RECEPTORS WORK

Technically, exteroceptors do not provide animals with information about the external world *per se* but only about changes in the receptor. Even then, the receptor itself does not determine completely what animals sense, as you can experience by trying the activity in Figure 20.1.

Undoubtedly you have at some point received a blow to the head and noticed that you "saw stars." (Closing your eyes and tapping the lid of one eye may give you somewhat the same sensation less painfully.) Although the immediate stimulation of a blow to the head or eye is pressure, an impulse reaching the vision area of the brain is interpreted as light, whatever the nature of the stimulus. Similarly, an ear infection may stimulate the auditory nerve and you may "hear" the condition as a ringing sound. The various inputs to the brain from most interoceptors never stir our consciousness. Thus, receptors (e.g., in the hypothalamus

Part 1

Part 2

FIGURE 20.1
In vertebrates at least, a change in a stimulus is often more important than a set level of a stimulus. For example, set up the containers of water as shown. Leave your hands in the warm and cold water for at least five minutes before you place both hands into the room-temperature water. Do your two hands provide you with the same information about the room-temperature water?

or carotid artery) sense changes in the blood but we experience no such sensations consciously.

Receptor cells are either neuron endings or specialized cells in close contact with sensory neurons. Receptors are specialized to respond to one or more particular forms of energy, which they convert into electrical energy. This produces a state of depolarization that spreads down the dendrite until it reaches an area of the cell body near the axon. If the threshold is reached, an action potential is generated and the information ultimately reaches the central nervous system (CNS). Receptors and associated cells generally indicate the intensity of stimuli by changing the rate at which they produce action potentials, firing more rapidly as the intensities of stimuli increase.

Remember how the initial discomfort from a small stone in your shoe stopped after a time? The diminished response to a constant stimulus is called **adaptation.** This commonly occurs when the receptor produces a smaller action potential and/or becomes less responsive to stimulation. Many receptors are sensitive to changes in stimuli rather than to set levels of stimuli. For example, although you may smell a particular odor when you enter a room (e.g., fish or onions cooking in a kitchen), it often "disappears" after a few minutes. Similarly, you feel a hat, clothing, or glasses when you first put them on, but you soon stop noticing the pressure. However, such receptors as those for pain or cold adapt very slowly, and may continue to produce action potentials as long as the stimulus persists.

Some stimuli affect more than one kind of receptor, as when a flame stimulates both heat and pain receptors. In fact, many intense stimuli such as loud sounds, bright lights, and heavy pressure also result in the sensation of pain. Also, perhaps you have noticed goose bumps appear briefly on your skin when you enter a hot shower or bath (Figure 20.2). If so, the hot water was intense enough to stimulate the cold receptors first, which are generally closer to the skin surface than the heat receptors.

Sensory overload would occur quickly if we were conscious of every sight, scent, touch, or sound that affected our receptors. The sensory structures work closely with the CNS, and parts of the latter "shelter" our consciousness from the enormous quantity and variety of information about our environments.

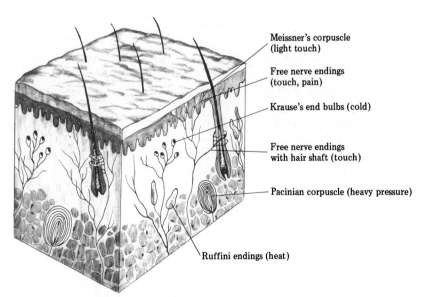

Meissner's corpuscle (light touch)

Free nerve endings (touch, pain)

Krause's end bulbs (cold)

Free nerve endings with hair shaft (touch)

Pacinian corpuscle (heavy pressure)

Ruffini endings (heat)

FIGURE 20.2
Many receptors are found in epithelial tissue, as this cross-sectional view of the human skin shows. Not all receptors in the same category, such as thermoreceptors, respond in the same way. For example, bathwater that feels warm to your hand may feel hot when you sit in it.

FIGURE 20.3
Increasingly, public facilities are equipped for those with sensory handicaps. For example, the park visitor in (A) is able to read trail signs in Braille. In other cases, the senses of animals, such as those of the guide dog in (B), supplement and complement those of humans.

(A)

(B)

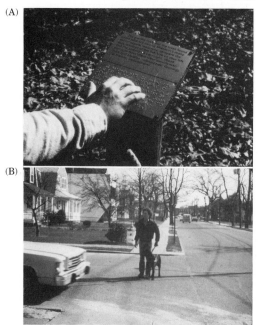

MECHANORECEPTORS

Mechanoreceptors respond to forces such as pressure, stretching, and twisting. Some mechanoreceptors in contact with the external environment are shown in Figure 20.2. Others are located internally, such as the stretch receptors in the alveoli and diaphragm that are important in breathing (Chapter 14), and those in the bladder that are involved in urination (Chapter 17). As the following subsections illustrate, many mechanoreceptors are **hair cells.** These sensory cells often indicate directionality by causing an impulse in the associated neuron when deflected one way but inhibiting the neuron when a reverse deflection occurs.

Tactile Receptors

Our touch-sensitive wrapper, the skin, contains about 500,000 touch and pressure receptors. Other tactile receptors lie at the bases of hairs and are stimulated indirectly when the bristles or hairs are moved. Hairs on the face and around the genitalia are the most highly innervated.

Human tactile receptors are not evenly distributed, being especially concentrated on the tongue, lips, and fingertips. The latter are important in manual dexterity and, for the blind, in using the Braille alphabet (Figure 20.3). Touch is important in many forms of behavior, such as the erogenous zones in sexual activities. Research has also shown the immense psychological importance of physical contact during the infancy of many mammals. A pat on the back, a slap on the face, and the nature of a handshake tell us much about another's feelings as do many other aspects of "body language."

The sensation of deep pressure involves nerve endings surrounded by layers of connective tissue interspersed with fluid, **Pacinian corpuscles.** These onionlike receptors are found in the skin and in the lining of the vertebrate gut. Adaptation of these receptors occurs quickly, although the initial deformation of the layers, or a change in level of deformation, leads to action potentials.

Pain is an important and very complicated protective sense caused by cellular damage, potential damage, or overstimulation of various receptors. We have millions of pain receptors, mainly free nerve endings, that are involved in "simple" pain. So-called **referred pain** appears to come from an area other than where it originates. For example, pain in the liver may be felt in the shoulders, and certain heart problems (e.g., angina pectoris) may lead to pain in the left arm or shoulder as well as in the chest. Referred pain occurs because pain fibers from different parts of the body lead to the same general area of the CNS. Another puzzling form of pain is the "phantom limb" phenomenon that occurs when an amputated limb may seem to itch or hurt for some time even though it is missing.

Headaches are one of the most common forms of pain (Table 20.1). About 90 percent of adult Americans experience headaches, and perhaps 20 percent have headaches severe enough to disrupt their lives. Because headaches obviously occur in someone's head, and because humans differ so physically and psychologically, it is difficult to develop an exact classification of headaches.

Studies of pain often use **placebos,** or substances that do not contain an active ingredient for the process being investigated (e.g., sugar pills or starch tablets). Placebos may be injected or taken orally as liquids or tablets. Many details about how placebos work are unknown, although their effects may be very powerful. For example, one study found that placebos relieved serious postoperative pain and that of chronic cancer

TABLE 20.1
Common Forms of Headaches

Type	Description
Migraine	Often attacks one side of the head, or starts there. From Greek roots, meaning "half head." Arteries within the head dilate, pulling on surrounding nerves.
Cluster	Sharp pains (often in the temple area or behind the eye), followed by pain-free or nearly pain-free period, followed by more pain, and so on. May result from excess histamine released when a person is emotionally upset. The histamine may cause dilation of blood vessels, congestion in the nose and sinuses, and watery eyes.
Tension	May occur from long periods of muscle contraction brought on by work or emotional tension.
Chemical	May result from allergies, alcohol, or various other substances.
Sinus	Excessive pressure on surrounding tissues by clogged sinuses may cause severe headaches.
Organic	Headaches are symptoms of some serious diseases or body damage (e.g., eye disorders).

in about 40 percent of patients, whereas "real" morphine-based medications relieved pain in about 60 percent of patients.

Chemicals control pain in several ways, such as by depressing conduction of nerve impulses or by raising the pain threshold. The cerebral cortex helps us localize painful stimuli, especially if touch receptors are stimulated at the same time. However, the thalamus is apparently the part of the brain that interprets most pain. Pain may also be relieved for some by the methods described below, although the detailed mechanisms are still unclear.

Hypnotism was apparently discovered in the 1780s, but was not named until the 1840s. Although the word *hypnosis* came from the Greek word for sleep, and despite the fact that some hypnotized people seem to be in a trance, hypnosis is not true sleep. (More information about hypnosis can be found in psychology texts.) A suggestion during hypnosis may cause the incoming pain sensations to be "turned off," and for some people hypnosis can relieve pain from surgery or dental work. High motivation to ignore pain, for example on the part of athletes, may do the same. Pain can also be relieved by being distracted, as in an emergency or even by a good book, movie, or television show.

The **spinal gate theory** appears to explain many aspects of pain. Nerve impulses travel many times faster in myelinated or fast fibers than in nonmyelinated or slow fibers (Chapter 19). Stimulation of some fast fibers of the spinal cord by an electrode leads to pressure or temperature sensations, but more intense stimulation of the slow fibers may cause pain. One "spinal gate" appears to be in the upper spinal cord or lower brain, and Figure 20.4 shows how it may work. Note that the same impulses travel over the nonmyelinated fibers in each case, and whether the impulses are painful or not depends upon how active the myelinated fibers are at the time. If the fast fibers are cut or damaged, almost any stimulation of the slow fibers causes pain. If the fast fibers are firing rapidly, no pain is felt even if the slow fibers are also firing rapidly. Activity in the myelinated fibers inhibits or "closes the gate" on painful stimulations.

Acupuncture has been used in China for about 5000 years. Hundreds of specific points, when pierced by sterilized needles or bamboo slivers, appear to help relieve pain in patients with a wide variety of problems.

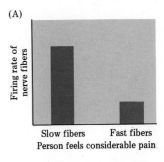

(A)

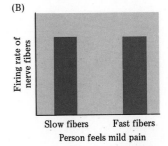

(B)

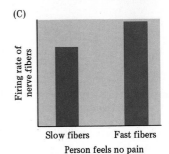

(C)

FIGURE 20.4
A "spinal gate" theory to explain pain.

FIGURE 20.5
Some acupressure points for sinus
headaches.

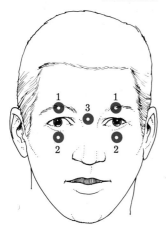

FIGURE 20.5
Some acupressure points for sinus
headaches.

Like hypnosis, acupuncture may work partially through the placebo effect. That is, if one believes strongly enough that acupuncture works, and that a particular practitioner is skillful and is sincerely trying to help, then the cerebral cortex may adjust the spinal gate to suppress slow-fiber firing. If at the same time an inserted object stimulates the firing of myelinated fibers, then the relative amount of nonmyelinated firing is reduced further.

Although references to **acupressure** can be found in American medical journals as far back as the 1830s, the practice started to become widely known and used only in the 1970s. Acupressure depends upon force applied by the fingertips to strategic spots and is preferred over acupuncture by many Western physicians (Figure 20.5).

Occasionally, otherwise normal individuals feel no pain. In these rare cases, a brain center may have turned off the spinal gate permanently. While this situation may sound ideal, the protective value of pain is lost.

Gravity Receptors and Equilibrium

Sensing gravity is important to animals, especially if they move quickly. **Statocysts** help many invertebrates reassume their normal positions if they are displaced (Figure 20.6). Statocysts are usually epidermal infoldings that are lined with receptor cells with hairs. They also contain one or more sand grains or calcium carbonate particles. Gravity pulls on the particles, and that stimulates the hair cells. Movement bends the hair cells or causes the particles to stimulate other hair cells. An experiment on crustaceans dramatically demonstrated the function of statocysts. When iron filings were substituted for the sand grains in statocysts, the animals swam upside down when a strong magnet was held above them.

"The better to hear with" may come to mind when we think about the ear (Figure 20.7). However, as important as hearing is for some vertebrates, the basic function of the vertebrate ear is to maintain **equilibrium,** our sense of balance or the ability to maintain an orientation between opposing forces. Many ears also contain gravity receptors. Fishes do not use their ears for hearing at all, and many vertebrates have no outer or middle ears. However, all vertebrates possess complex inner ears of interconnected canals and sacs. In the higher vertebrates, the main sacs are the **saccule** and **utricle,** and the canals are called the **semicircular canals.** The nearby cochlea that is involved in hearing is discussed in the next section. Equilibrium in many vertebrates also depends upon visual and proprioceptive information, and pressure-sensitive cells on the soles of the feet are also important in mammals.

The sensory cells in the saccule and utricle are hair cells tipped by

(A)

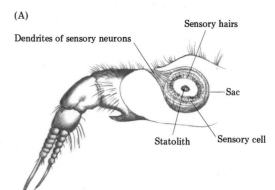

Dendrites of sensory neurons

Sensory hairs

Sac

Statolith

Sensory cell

(B)

FIGURE 20.6
Some statocysts from crustaceans (A). In flies (B), the hind wings or halteres are modified into paired organs of equilibrium. Numerous mechanoreceptors at the base of each haltere detect changes in the fly's cuticle during flight. The impulses sent to the CNS help the fly make the necessary corrections and movements.

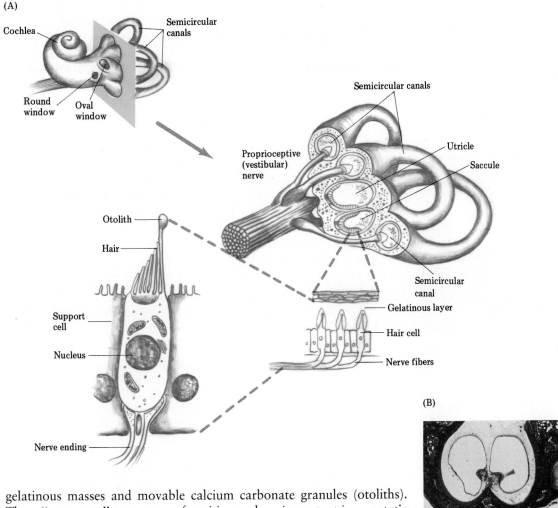

(A)

Cochlea

Semicircular canals

Round window

Oval window

Semicircular canals

Proprioceptive (vestibular) nerve

Utricle

Saccule

Semicircular canal

Gelatinous layer

Otolith

Hair

Support cell

Nucleus

Hair cell

Nerve fibers

Nerve ending

(B)

FIGURE 20.7

The structures involved in hearing and in equilibrium are closely associated in the vertebrate ear, as shown in this human example. (A) The saccules, utricles, and semicircular canals are the organs of equilibrium (vestibular apparatus). The sensory hair cells in the sacs and ampullae of the canals are similar in that distortion may lead to a nerve impulse. However, the ear stones that occur in the utricles and saccules are absent in the ampullae (B).

gelatinous masses and movable calcium carbonate granules (otoliths). These "ear stones" are organs of position and are important in our **static sense of balance.** Because the receptor cells lie in different planes in the two sacs, a wide variety of sensory impulses to the brain is possible as the head tilts or body movements accelerate or decelerate. The sliding of the ear stones and the deformations of the gelatinous layers cause the hair cells to move, and these movements initiate nerve impulses. Movements bend some hair cells more than others, and each type of body motion leads to a distinct pattern of nerve impulses.

Each fluid-filled (endolymph) semicircular canal originates in the utricle and lies at right angles to the other two, like the surfaces making the corners of a box. A group of hair cells similar to those in the sacs but lacking ear stones are found in the swelling at the base of each canal. As the body turns and the liquid in the semicircular canals moves, the sensory hair cells are stimulated. Not only do we sense rotation but the stimulation also often causes reflex movements of the eyes and head in a direction opposite to the original rotation. This sense is often called our **dynamic sense of balance,** as opposed to our static sense of balance, discussed below.

Proprioception

The semicircular canals and the cerebellum work together to control movement and balance, our **proprioceptive sense.** Most adults think of movement, such as that of a leg, as a simple event, forgetting how long it took us to learn this task (Figure 20.8). However, over 50 muscles are

FIGURE 20.8
The joy of taking the first step is evident in this young child. When locomotion is possible, we are able to expand our sensory horizons considerably.

involved in the forward movement of a leg, and managing all these muscles requires a complex control system, including "knowing" at all times where body parts are and what they are doing.

Humans lead mainly a two-dimensional life (forward and backward, left and right) and are relatively unaccustomed to vertical movements. **Motion sickness** may result from excessive rotary movement and rapid vertical movements. Visual and psychological factors may compound the problem, and nausea, dizziness or **vertigo,** headache, cold sweat, and vomiting are common symptoms. Impulses from the semicircular canals excite the area in the medulla that controls nausea and vomiting. **Ménière's disease** occurs when too much endolymph is produced in the cochlea and the equilibrium organs. Tinnitis, vomiting, vertigo, and hearing loss are common symptoms. Severe vertigo is disabling and may last for several hours or even days. Infections, allergies, arteriosclerosis, or poisons may lead to overproduction of endolymph.

Thus, although we mainly think of muscles, tendons, and joints as effectors (Chapter 21), parts of them also provide us with important sensory information (Figure 20.9). In fact, our proprioceptors are among our most continuously active receptors. We even feel the lack of proprioception, as when an arm or leg "goes to sleep." Complicated acts are impossible without the information from the nerve endings that detect muscle movements, the stretching of tendons (which attach muscle to bone), and the movements of ligaments in joints.

PHONORECEPTION

Although phonoreception is a form of mechanoreception, it is treated separately because of its conscious importance to the higher vertebrates. Among arthropods, phonoreceptors are found in cicadas, crickets, some moths, and spiders. Although sound is relatively unimportant in the lives of fishes and reptiles, it plays a major role in the reproductive behaviors of some amphibians, and the springtime chorus from frogs, toads, and tree frogs can be almost deafening.

The mechanoreceptor cells in the **cochlea** of the inner ear are especially important to birds and mammals (Figure 20.10). The human coiled cochlea, looking somewhat like a snail's shell, consists of three canals separated from one another by thin membranes. Two of the fluid-filled canals (vestibular and tympanic ducts) are connected with one another at the apex of the cochlea. The phonoreceptor, the **organ of Corti,** is found in the middle fluid-filled (cochlear) canal. Each organ of Corti consists of over 20,000 hair cells arranged in rows along the length of the cochlea. The **basilar membrane** upon which the hair cells rest separates the middle canal from the neighboring outer canal. Another membrane (tectorial membrane) overhangs the cilia of the hair cells.

While some humans can wiggle their ears slightly, many mammals can move their ears extensively. The function of the outer and middle ear structures in terrestrial vertebrates is to collect and alter sound waves (mechanical energy) in air to pressure waves in the cochlear fluid. For example, the sound waves that are funnelled by our cartilaginous ear flaps into the ear canal set the **eardrum** or tympanic membrane vibrating. These vibrations are transmitted to and amplified by the three smallest bones of the body, named the *hammer* (malleus), *anvil* (incus), and *stirrup* (stapes) because of their shapes. The vibrations of these levers, in turn, are transmitted to a membrane, the **oval window,** that sets the fluid in one of the cochlea canals in motion. Because fluids are almost incompressible, each inward movement of the oval window leads to an

(A)

FIGURE 20.9
Proprioception. (A) The beauty and skill of
the gymnast's movements are possible
because proprioceptors are sending to the
brain a constant stream of information about
the relative position and tension of the
muscles, tendons, and joints. (B) The major
vertebrate proprioceptors are found in
muscles (muscle spindles) and in tendons
(Golgi end organs). Joint receptors also
detect movement in ligaments.

(B)

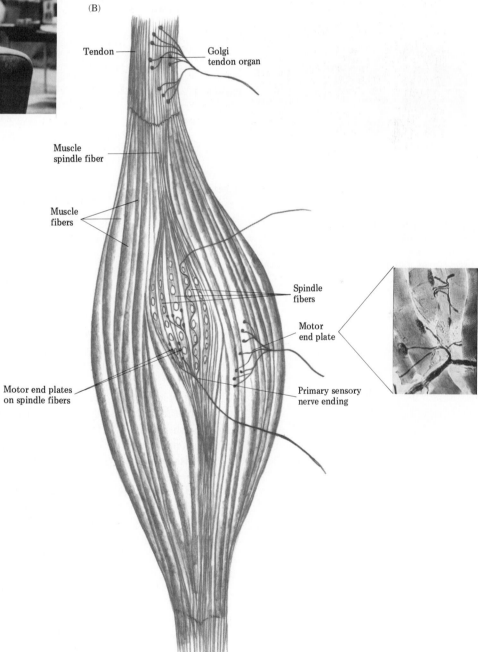

Tendon

Golgi
tendon organ

Muscle
spindle fiber

Muscle
fibers

Spindle
fibers

Motor
end plate

Motor end plates
on spindle fibers

Primary sensory
nerve ending

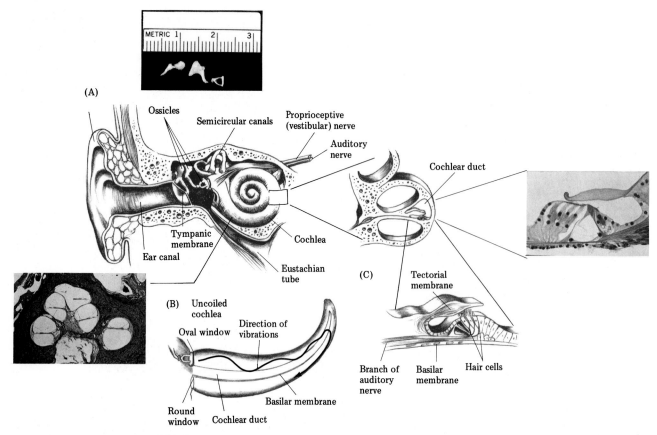

FIGURE 20.10

The structures of the mammalian ear, using the human example. (A) Cross-section view through the outer, middle, and inner ear structures. (B) The structure of the cochlea uncoiled. (C) Details of the organ of Corti.

outward movement of another membrane, the **round window** at the other end of the duct. These pulsations move the basilar membrane and cause the sensory hair cells of the organ of Corti to rub against the overhanging membrane. The distortion of the hair cells initiates nerve impulses in the dendrites of the nerve (cochlear) lying at the base of the hair cells.

The fibers of the basilar membrane are of different lengths, thus resembling the strings of a piano or harp. Sounds of different pitches or frequencies of sound waves set up fluid waves that stimulate particular groups of hair cells. Based upon where the auditory impulses come from, the brain infers the pitch of the sound. The patterns to the stimulation of hair cells lead to the determination of a sound's quality (e.g., overtones and harmonics). Thus, the brain can distinguish between sounds of the same pitch made by different musical instruments. Loud sounds lead to more intense stimulation of the hair cells, and that increases the number of impulses per second in the auditory nerve.

Human ears can detect sounds from about 20 to 20,000 cycles per second (or hertz, Hz), a range of over 10 octaves. Although human voices range from about 300 to 3000 cycles per second, our greatest acuity seems to occur between about 1000 and 2000 Hz. A high piano note is about 4000 Hz. Various vertebrates can hear frequencies well above the human limits, as when the sound of a dog whistle falls silent upon human ears.

Hearing is perhaps 10 times more sensitive than vision when the energy of audible sound waves is compared with the energy of visible light waves. In fact, if hearing were more efficient, it might result in our experiencing a constant buzzing or hissing sensation. In general, there is little fatigue in the hearing neurons, including those in the integrative brain centers. However, see the essay "Noise Pollution" for information about a growing problem.

Noise Pollution

The word *noise* is derived from the Latin term for nausea. Noise pollution is a growing and serious problem in modern society and excessive noise may cause progressive hearing loss and deafness. I SAY, NOISE POLLUTION MAY CAUSE . . . . Over 20 million Americans have hearing problems, and the number is increasing rapidly. Many of these problems were caused by environmental noise, and over half of the United States population is now affected adversely by noise. Table B20.1 gives the sound levels of common events in decibels. Because the decibel scale is logarithmic, information on the relative sound intensity is given.

The most serious problems occur in people whose work exposes them frequently to sounds above 80 decibels, such as many military and industrial personnel. Some noisy industries have solved the problem of high worker turnover by hiring mainly deaf people. Long-term exposure to excessively amplified music is another problem, and sounds above about 180 decibels can even kill. One technical problem in developing noise protection devices is how to protect people from loud sounds while also allowing them to hear normal sounds.

Noise is stressful, and hearing losses are not its only effect. Another common effect is to constrict blood vessels, which increases blood pressure and contributes to heart disease. Other effects include dilation of the pupils, increased heartbeat, tensed muscles, and gastrointestinal problems. If noise affects normal sleep regularly, one's general health is worsened. Excessive noise may also affect the fetus during pregnancy.

FIGURE B20.1

TABLE B20.1
Common Noise Levels

Examples	Decibels	Relative Sound Intensity	Description and Effects
Jet takeoff (close range)	150	1,000,000,000,000,000	Eardrum ruptures
Deck of aircraft carrier	140	100,000,000,000,000	
Military personnel carrier	130	10,000,000,000,000	
Thunderclap, jet takeoff (65 m)	120	1,000,000,000,000	Pain threshold
Auto horn (1 m), riveting, live rock band, steel mill	110	100,000,000,000	
Outboard engines, power mowers, farm tractor, jackhammer, blender, motorcycle (8 m), printing plant, subway, jet (300 m)	100	10,000,000,000	Serious hearing damage after long-term exposure
Diesel truck, busy city street	90	1,000,000,000	
Garbage disposal, average factory, freight train (14 m), noisy office, dishwasher	80	100,000,000	Hearing damage after long-term exposure
Freeway traffic (15 m), vacuum cleaner	70	10,000,000	
Typical suburb, average restaurant	60	1,000,000	
Living-room conversation, quiet suburb (daytime)	50	100,000	Quiet
Library	40	10,000	
Quiet rural area (nighttime)	30	1,000	
Whisper, rustling leaves	20	100	Very quiet
Breathing	10	10	
	0–1	1	Threshold of hearing

Because numerous structures are involved in hearing, damage to or malformations in any of several parts may result in deafness or difficulty in hearing. Thus, the ear canal may fill with wax, the eardrum may tear or become inflamed, the middle ear bones may fuse, or there may be damage to the hair cells, auditory nerve, or sound-perceiving brain centers.

The **eustachian tube** connects the air-filled, middle-ear cavity with the pharynx, allowing the air pressure within the middle ear to equal the outside air pressure. Sudden changes in outside air pressure may temporarily hinder hearing by making the eardrum bulge and tauten. Similar problems and an uncomfortable feeling also occur when the nasal swelling that results from a cold keeps the slitlike opening of the soft-walled eustachian tube in the pharynx closed. Yawning or swallowing often help "pop" your ears.

The middle ear may become infected, usually by pathogens arriving through the eustachian tube. For example, vigorous nose blowing or holding one's nose when sneezing may force pathogens into the eustachian tube. Pus may fill the middle-ear cavity and scar tissue may form, both of which interfere with hearing. Pain, pressure, tinnitus, and difficulty in hearing are common symptoms, and drainage is one remedy.

CHEMORECEPTORS: TASTE AND SMELL

Chemical messengers are important in the control of many internal processes (e.g., hormones and neurotransmitters), as they are in many forms of behavior. This section treats the chemical senses of **taste** (gustation) and **smell** (olfaction).

In many aquatic animals the chemical sense is generalized over the body surface. Because it typically involves simple nerve endings of certain bipolar cells, it is often also nonspecific. Frogs, for example, will react in much the same way when any number of different chemical compounds touch their skins. In terrestrial vertebrates, in contrast, both chemical senses are restricted to small areas of moist epithelial tissue.

One general problem in studying the chemical senses is our poor vocabulary of taste and smell words. (Try to think of some besides "yucky!") Therefore, we often resort to saying that something smells or tastes like something else.

Taste

The taste hairs of insects are well known, and while some of these sensitive chemoreceptors are in the mouthparts, they are also on the terminal segments of the legs in flies. For example, houseflies can distinguish between water, sugar, and salts. A thirsty fly extends its retractable proboscis and drinks when one sensitive hair touches water. It feeds when another hair touches something sweet, but it moves on if another hair touches a salty object. As in many other receptors, the sensory neurons are bipolar cells. A number of other arthropods (e.g., crabs) also taste by sticking their feet into substances.

The sense of taste in mammals involves many thousands of **taste buds** that are found mostly on small elevations or papillae, on the tongue and soft palate (Figure 20.11). Don't worry too much if you "burn your tongue" on hot food because the taste buds are replaced every few days. Each oval taste bud contains several taste receptors—epithelial cells with microvilli on their free surfaces. A hairlike projection extends through a pore to the external surface of the taste bud. Each taste receptor is

(A)

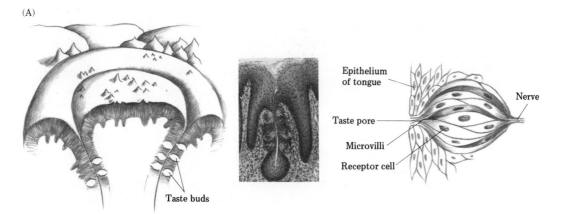

Epithelium of tongue

Nerve

Taste pore

Microvilli

Receptor cell

Taste buds

(B)

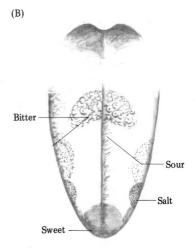

Bitter

Sour

Salt

Sweet

FIGURE 20.11
(A) Cross-section of a typical human taste bud. (B) The general distribution of sensitivity to the traditional four basic tastes.

innervated by more than one neuron, and some neurons connect more than one taste cell.

That substances must be in solution to be tasted can be demonstrated by a simple experiment. With your eyes closed, note how long it takes you to identify a crystal of sugar or salt, without your knowing which, that a friend drops onto the tip of your tongue after you have dried it with a towel. Some taste cells can detect certain chemicals in concentrations of only 1 part in 2 or 3 million parts of water.

Despite the large number of distinct tastes, it is commonly thought that there are only four basic tastes: sweet, salty, sour, and bitter. Although the greatest sensitivity to each primary taste is typically in the areas shown in Figure 20.11B, many papillae and individual taste buds may respond to more than one primary taste. Also, different sugars (e.g., glucose, lactose, and sucrose) produce distinct tastes. Some substances elicit one taste sensation in one location but another taste upon contact with other taste cells. Further, the same substance may produce different tastes in different people.

The sense of taste is also complicated by the fact that flavor depends upon our sense of smell, as everyone who has had clogged nasal passages can verify. The taste buds are unaffected by a cold or the flu, but the vapors from food may not reach the olfactory receptors. The texture and temperature of food also commonly affect how it tastes. For example, hot food generally releases more odorous vapors than cold food and it also affects the thermoreceptors on the tongue.

You may also wish to try a simple experiment to see whether different foods can be identified as easily by taste alone as when one can also smell them. First cut very small pieces of onion and apple and then mash them until their texture is similar. Place these on the tongue of a friend whose eyes are closed and whose nose is held tightly shut to see whether the foods can be distinquished.

Smell

It is relatively easy to explore the taste sensitivity of the tongue. A blindfolded subject needs only to report on each sensation as various substances are placed precisely on different regions of the tongue. Learning about smell is far more difficult, because the **olfactory epithelium** is a small, relatively inaccessible area in the roof of the nasal cavity (Figure 20.12). Also, olfaction involves sensing chemicals in air, and gases are far more difficult to manipulate precisely than liquids are. Finally, olfactory adaptation occurs very quickly, with receptors losing about half of their sensitivity within a second or so after stimulation. This is nice if a

(A)

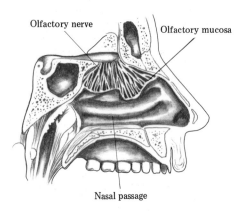

Olfactory nerve
Olfactory mucosa
Nasal passage

(B)

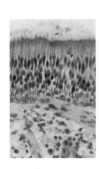
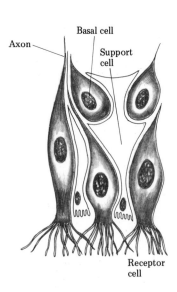

Axon
Basal cell
Support cell
Receptor cell

FIGURE 20.12
(A) General location of the olfactory epithelium and (B) detailed views. As with the sense of taste, substances must be in solution to be sensed.

substance has an offensive odor but it makes comparing perfumes difficult.

Densely packed olfactory neurons, with hairlike extensions on their dendrites, are interspersed among nonsensory epithelial cells. The axons extend upward through tiny pores in the skull to form the olfactory nerve. When an appropriate chemical makes contact with a specialized macromolecule in the plasma membrane of the receptor cell, an impulse is generated. This reaction, which is somewhat like that between enzyme and substrate, probably occurs because of an effect upon the permeability of the chemoreceptor membrane to sodium.

The sense of smell is generally far more sensitive than the sense of taste. Some substances can be detected in concentrations of only 1 part per tens of billions of parts of air. As with taste, substances must be in solution to be smelled, and the olfactory epithelium is mucus covered. Basic odors are also more numerous than basic tastes, although there is no complete agreement on their exact number. As with tastes, mixtures of the primary sensations produce the large number of different odors that we perceive. Thus, chemoreception is somewhat analogous to our ability to perceive an enormous number of hues made from just three primary colors.

THERMORECEPTORS

Many invertebrates are sensitive to changes in environmental temperatures, although relatively little is known about their thermoreceptors. Probably free nerve endings in the body covering are involved, and perhaps the CNS itself. The antennae of some insects contain thermoreceptors, which may be sensitive to changes of less than 0.5°C. This probably helps mosquitoes, other blood-sucking insects, and ticks find you and other hosts.

Pits in the heads of snakes such as rattlesnakes contain thermoreceptors useful in detecting endothermic prey (Figure 20.13). These sensitive cells can detect the presence of a mouse, for example, as far as about 50 cm (20 in.) away.

In endotherms, changes in environmental temperatures are sensed by both free nerve endings and specialized thermoreceptors. Human skin

FIGURE 20.13
In the pit located between the eye and the nostril of this rattlesnake are several thousand sensitive thermoreceptors useful in detecting endothermic prey. In the dark, pit viper snakes typically orient themselves so that the pits on both sides of their heads detect the same amount of heat.

contains about 150,000 cold receptors and about 16,000 heat receptors. Sensory information about temperature is received and integrated by the hypothalamus. Chapter 2 describes in detail how both ectotherms and endotherms maintain their body temperatures homeostatically.

How temperature changes cause the thermoreceptors to send impulses is not entirely clear, but probably the permeability of the plasma membrane is changed. When your skin touches large cold or warm objects, heat conduction is important and, barefooted, you have probably noticed how concrete or tile floors feel colder than rugs. Although the floor and the rug are at the same temperature, the floor feels colder because we sense the higher rate of heat loss to it.

Points or electrodes with known temperatures can be used to map the cold and heat receptors in the skin. One can map the pressure and pain receptors of some animals, but only humans can report temperature sensations. (Imagine trying this with Rover: "Bark once if it feels cold and twice if it feels warm.") However, mapping the thermoreceptors is only the first step, as it is with other skin receptors. Next, one must remove pieces of the mapped skin, slice them thinly, and look for the receptors where a subject reports feeling cold, pressure, or pain. Any volunteers?

PHOTORECEPTORS

Photoreception occurs when light energy causes light-sensitive cells to initiate a nerve impulse. **Vision** is only one form of photoreception and usually requires an eye with a **lens** to focus light onto photoreceptors. Vision involves image formation, and that requires eyes more complex than those shown in Figure 20.14. As in all photoreceptors, the simple "eyes" or **ocelli** of jellyfish and planarian flatworms contain pigments that absorb light energy.

Several more complex eyes are shown in Figure 20.15. While the compound eyes of arthropods such as insects and crustaceans are useful in detecting movement, most do not form images even though they have lenses. The eyes of shellfish such as scallops and abalones form simple images, somewhat as pinhole cameras do, although they have no lens. The eye of the octopus, however, has a lens and forms an image. Although cephalopod and vertebrate eyes function similarly, having so-called camera-type eyes, they are analogous and not homologous. That is, they evolved independently and developed from different structures.

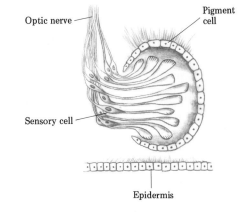

(A) Simple invertebrate eye
(Ocellus of jellyfish)

(B) Ocellus of planarian worm

FIGURE 20.14
(A) Like all ''eyes,'' the ocelli of jellyfish contain special pigments. (B) Although planarians have light-sensitive cells scattered over their surfaces, the ocelli are in the head. The bowl-shaped pigment cells allow light to enter mainly from above and the front, and thus assist the flatworm in detecting the direction of the source of light.

The Human Eye

Possibly no animal sees color better than humans, although many insects may detect motion better. Some insects also see ultraviolet light that we cannot see, so flowers may look different to these animals. Most humans consider vision to be their most important sense, and you need only close your eyes and try to perform routine tasks to appreciate the value of sight. The major features of the human eye are shown in Figure 20.16.

Comparisons are often made between vertebrate eyes and cameras, and there are some similarities (Figure 20.17). For example, both have a lens that can be focused for different distances, a mechanism for controlling the amount of entering light (the eye's **iris**), a light-sensitive material (the eye's **retina** and the camera's film), and a light-tight container. Both also form upside-down images and both require care and occasional repair. However, there are also important differences, one of the most prominent of which is the focusing mechanism. We focus by changing the shape of our lens, not the distance between the lens and the object being viewed as in most modern cameras.

FIGURE 20.15
Some complex eyes of invertebrates. (A) The surface of a large compound eye of an insect. The eyes of most arthropods do not form precise images but they are useful in detecting light intensity and movement. (B) The scallop eye forms a fuzzy image even though it has no lens to focus light. (C) The octopus eye resembles the vertebrate eye functionally but it evolved independently.

(A)　　　　　　　　(B)

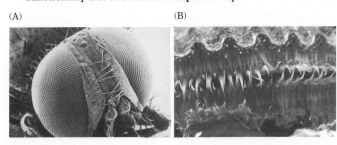

(C)

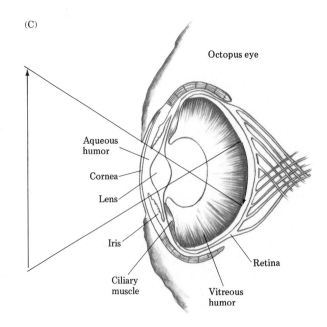

(A)

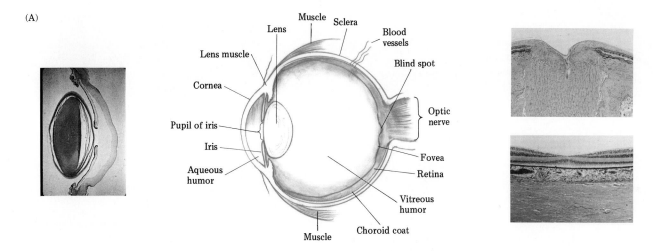

(B)

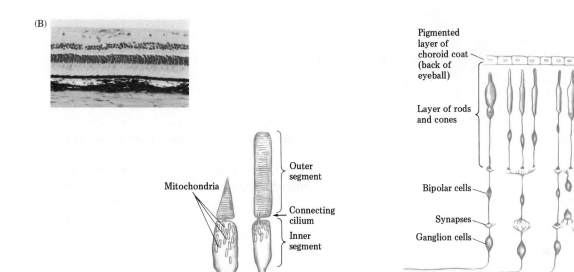

FIGURE 20.16

(A) Cross-section of the human eye. (B) Detailed view of the retina. Light must pass through the associated ganglion and bipolar cells before it reaches the visual receptors. Of the latter, there are about 20 times as many rods as cones. The cones are most concentrated around the fovea, and the rods around the periphery of the retina.

How is it that the upside-down image on our retina is "seen" as right side up? Learning seems to play a major role, as shown by experiments in which humans wear special lenses that make the retinal image right side up. At first disoriented, the subjects adjust in a few days and the world again looks right side up. However, the world appears upside down again for a time when the subjects remove the special lenses.

The eyeball is padded by fat deposits and is largely surrounded by bone. Also, the thin, transparent **conjunctiva** layer of cells protects the front surface of the eye and is continuous with the eyelids. The eye is kept moist by tears from the **lacrimal glands.** Tears are spread over the surface when we blink, which we do mostly unconsciously every few seconds. The rate of blinking increases in dusty environments, when the eyes are tired, and during stress. Tears contain antibacterial substances that protect the eyes against infections. Excess fluid drains into the nasal sinuses through a duct on the inner corner of the lower eyelid (Figure 20.18).

(A)

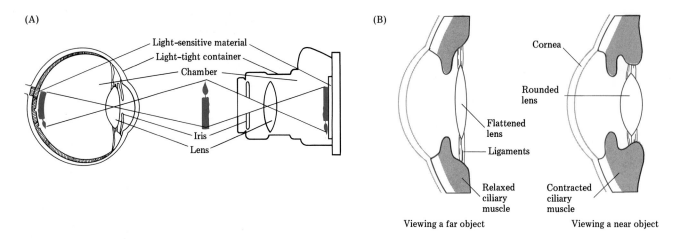

Light-sensitive material
Light-tight container
Chamber
Iris
Lens

(B)

Cornea
Rounded lens
Flattened lens
Ligaments
Relaxed ciliary muscle
Contracted ciliary muscle

Viewing a far object   Viewing a near object

(C)

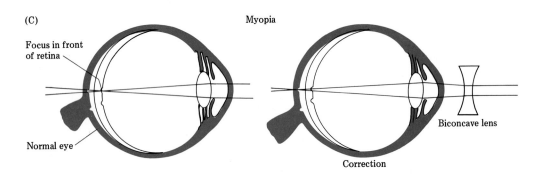

Myopia

Focus in front of retina

Normal eye

Biconcave lens

Correction

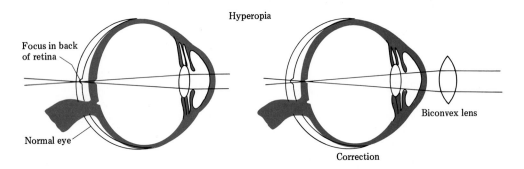

Hyperopia

Focus in back of retina

Normal eye

Biconvex lens

Normal eye

Correction

FIGURE 20.17

(A) Some eye-camera comparisons. (B) Accommodation is accomplished by the lens, although the cornea and the humors are also important in the refraction of light. (C) About 50 percent of Americans wear eyeglasses, usually for nearsightedness (caused by an eyeball that is too long) or for farsightedness (caused by an eyeball that is too short). Corrections for the uneven curvature of the cornea and/or the lens (astigmatism) may also be necessary. To a person with uncorrected astigmatism, some of the lines on charts like the one in (D) may appear darker than others.

(D)

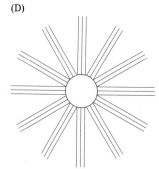

The outer of the eye's three layers, the **sclera,** is the tough, opaque, and fibrous "white" of the eye that helps to protect and maintain the shape of the eye. The muscles that move the eye in its socket are also attached to the sclera, and although each muscle can move alone, the movement of all muscles must be coordinated to aim the eyes properly. **Strabismus** is a general term for the eyes either aiming inward (cross-

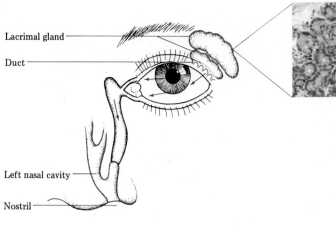

Lacrimal gland

Duct

Left nasal cavity

Nostril

FIGURE 20.18
The lacrimal or tear glands. Also, glands at the inner edge of the eyelids secrete an oily fluid that keeps the eyelids from sticking together. Styes result from inflammation of these glands caused by infection. Conjunctivitis causes the eyes to appear red and bloodshot, giving the problem its common name of "pink eye."

eyed) or outward (wall-eyed). If strabismus is not corrected, the brain may ignore one of the two images it receives from the eyes and the person becomes functionally blind in one eye. This "lazy eye" condition is called **amblyopia.** The thinner and transparent front surface of the sclera is the **cornea,** through which light enters.

The middle **choroid** layer contains nonreflecting dark pigment that acts as a light shield and prevents light from scattering. It also contains the eye's major blood circulation. The visible part of this layer is the iris, whose color depends upon the nature and amount of pigment present. Tiny sets of antagonistic muscles in the iris allow the diameter of the opening, or **pupil,** to change sixfold. This means that about 40 times as much light enters the eye when the pupil is large in dim light than when it is small in bright light.

If there is much pigment, or **melanin,** in the iris, the eye color is brown. Blue-eyed people have little pigment, and albinos lack pigment altogether. Their eyes appear pink because the blood shows through the choroid's vessels. It takes time for melanin to be produced postnatally, and all but albino infants are generally born with blue eyes.

The transparent, elastic lens is located behind the iris. The lens and the cornea are responsible for focusing or refracting light onto the retina, the innermost layer. The cornea accounts for about 70 percent of the refraction, and the lens for the remainder. **Humors** help give the eyeball its shape and supply nutrients for the lens and cornea. The watery aqueous humor fills the space between the cornea and the lens, and the more viscous vitreous humor fills the larger chamber between the lens and the retina. Particles are often in the humors, but although annoying, these "floaters" seldom disrupt vision. Far more serious, however, is excessive pressure of the humors which causes **glaucoma,** a major cause of blindness in the world. The high pressure injures blood vessels and causes retina and neuron damage. Blurred vision, poor side vision ("tunnel vision"), and seeing halos or rainbows around or near objects are common symptoms. Glaucoma tests are easy and are part of most eye examinations. Excess humor fluid may be withdrawn in serious cases, and special eye drops may help minor cases.

By changing the curvature of the lens, the eye can **accommodate** or focus for near or far vision. The relaxation or contraction of small muscles (ciliary) causes the lens to be flattened or rounded for distance or close vision, respectively. Accommodation is often more difficult with age because the lens becomes less elastic and somewhat larger (presbyopia). Bifocal or trifocal eyeglasses usually correct such problems, just as lenses do for **nearsightedness** (**myopia**) or for **farsightedness** (**hyperopia**). The uneven curvature of the cornea and/or lens, called **astigmatism,** may also lead to blurred vision, eye fatigue, and chronic headaches if uncorrected, because rays of light come to a focus at different points. Naturally, difficulties or blindness occur if the cornea or lens becomes cloudy, a condition known as **cataracts.** However, corneal transplants are usually quite successful because tissue rejection is uncommon (Chapter 16).

Both **ophthalmologists** and **optometrists** can examine eyes for visual problems and prescribe corrective lenses. However, because only ophthalmologists are M.D.s, only they can diagnose and treat eye diseases. Corrective lenses are ground by **opticians,** who may also assist in selecting and fitting frames.

The retina is the light-sensitive inner layer, with its numerous **rods** (over 100 million) and **cones** (about 6 million). Note in Figure 20.16 that incoming light must pass through several layers of connecting neurons (ganglion and bipolar cells) before it reaches the two types of receptor cells, which were named for their shapes. The axons of the associated neurons unite at a point in the back of the eye to form the **optic nerve,** which connects the eyeball with the brain. Because light striking this region forms no images, it is called the "blind spot" (Figure 20.19).

The cones, which are responsible for perceiving color and fine detail, are highly concentrated in the **fovea.** When bright light strikes this small depressed area, which it does when we look directly at an object, we have our keenest vision. Rods have a much lower threshold to light than cones and are involved in noncolor and motion vision. Most receptors around the periphery of the retina are rods. Because of this, we often see better in dim light by not looking directly at an object but, rather, slightly to one side of it. In this way, the light strikes the rods with their low light threshold rather than the region of concentrated cones with their high light threshold.

A simple experiment will demonstrate these facts about the photoreceptors. While you stare *straight ahead,* have a friend move a colored object with writing on it slowly into view from the side. Note that you can detect the object's presence before you can identify its color. The object must be almost directly in front of you before you can also read the writing on it.

Acuity is measured by various charts that a person usually views from a distance of 20 ft (6.1 m). The ability to read certain letters with "normal" eyes is called *20/20 vision.* Someone who can read letters at only 20 ft that a normal person can read at 100 ft (30.5 m) is said to have 20/100 vision.

In some animals, the layer of pigmented epithelium that lies directly behind the rods and cones moves. For example, this layer may be close to the rods by day, reducing their sensitivity somewhat, and withdrawn at night, increasing their sensitivity. In some mammals, such as cats, a **tapetum** or reflecting layer occurs immediately behind the retina. This layer, which causes the eyes of some animals to "glow" in the dark, increases night vision. Light that passes the rods initially without being absorbed is reflected back past the rods, thus providing another opportunity for it to be absorbed. Humans do not possess a tapetum.

FIGURE 20.19

The blind spot can be detected and its shape mapped roughly by use of these marks. Hold this page at arm's length directly in front of you, close your left eye, and stare *only* at the circle. Slowly bring the book toward your face. When the plus mark disappears the light is falling on the blind spot, where the optic nerve joins the eyeball and where rods and cones are absent. Continue moving the book closer until the spot reappears. Reverse the procedure (make sure the circle is on your left) to find the blind spot in your left eye.

The position of the eyes is important in depth perception (Figure 20.20). Because their eyes are relatively close together, with a large overlap in their visual fields, humans and some other vertebrates have excellent **binocular vision** that allows them to accurately judge distance and depth. Predators and fast, tree-dwelling mammals, such as monkeys, tend to have good three-dimensional (stereoscopic) vision. Animals that are hunted, such as many herbivores, often have **monocular vision,** with little or no overlap between the two visual fields. By having their eyes on the sides of their heads, however, such animals have a wide total visual field.

The fact that most people are right- or left-handed is well known. Most people are also right- or left-footed, which can be determined by observing which foot you use automatically when you run toward and kick a ball. Similarly, one eye is usually dominant over the other. To discover your eyedness, tear a small circle from a folded piece of paper. With both eyes open, look through the hole at a distant object. Close your left eye without moving the paper; if you still see the same view, you are right-eyed.

The Chemistry of Vision

The primary visual pigment **rhodopsin,** or "visual purple," is found in the rods of land vertebrates, marine fishes, and some invertebrates. A slightly different pigment is found in freshwater fishes. Frog larvae also contain this related pigment, although adults possess rhodopsin. Rhodopsin consists of a protein (opsin) and a vitamin A derivative (retinal). A single quantum of light can be absorbed by one molecule of rhodopsin, and this may lead to an impulse from a rod cell. The ability to see in dim light depends upon the relative rates of synthesis and breakdown of rhodopsin. The synthesis process is relatively slow, and the rhodopsin concentration is low when the eye is exposed to bright light. This is why it is so difficult to see when you first enter a darkened room from bright sunlight. Lengthy exposure to very bright light may cause temporary blindness, for example, the snow blindness familiar to Eskimos and skiers. After an hour or so of dark adaptation, when rhodopsin is synthesized, the sensitivity of the eye to light increases about 1 million-fold. The resynthesis of rhodopsin requires vitamin A, and a person whose diet lacks enough vitamin A may suffer from **night blindness,** the relative inability to see in dim light or adjust to dim light after exposure to bright light, such as an oncoming headlight.

Many questions about the chemistry of color vision remain. Humans, other primates, and cats are exceptions among mammals, most of which are at least partially color-blind. However, birds, lizards, frogs, and many fishes have color vision. In animals with color vision there are three types of cones and three different pigments (red, green, and blue), each of which is sensitive to a particular range of wavelengths. The brain interprets color based upon the rates at which various cone types respond. For example, we sense yellow when red-sensitive and green-sensitive cones are stimulated. Color intensity depends on the number of nerve impulses produced when light hits the cones. Color blindness (often an inherited sex-linked trait; see Chapter 9) occurs when one or more of the cone types is absent.

When you see a movie, you are really seeing many still pictures each second. The ability to fuse separate images into a continuous sequence depends upon "visual persistence" in the retina. Because the breakdown products of rhodopsin are not converted by an enzyme immediately, an image persists, fuses with the next, and so on.

FIGURE 20.20
Depth perception depends upon several factors, one of which is the degree to which the two visual fields overlap. The monocular vision that is common in such prey animals as herbivores gives them a wide field of vision for detecting predators. Predatory animals usually have good binocular vision. To get a rough feel for what monocular vision is like, when each eye sees a different view, try common activities while holding a short tube (e.g., from a role of paper towels or toilet paper) pointed out from each eye.

M: monocular
B: binocular

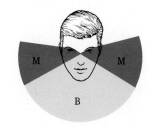

Man

Horse

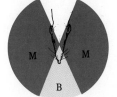

Rat

Processing Visual Information

The way the rods synapse with bipolar cells is important, partly because bipolar cells can sum the inputs from several rods. It is also important in our sensing of motion, because slight differences in the arrival at a bipolar cell of impulses from different rods indicate that we are viewing a moving object or are moving past an object. The arrangement by which many cones synapse individually with bipolar cells helps to give us visual acuity, or the ability to distinguish close points.

Ganglion cells do not simply receive impulses from the bipolar cells and transmit them to the optic nerve, but they also analyze and alter the impulses first. The optic nerves in mammals feed mainly to areas on the underside of the brain (lateral geniculate bodies), where additional "filtering" of information occurs (Figure 20.21). From there, fibers go to the visual cortex at the back of the cerebrum for final and more complex processing. Note that the optic nerves do not connect the eye and the brain in a simple manner. Instead, fibers from the left side of each retina enter the left side of the brain, and fibers from the right side of each eye go to the right side of the brain.

Optic chiasma viewed from below

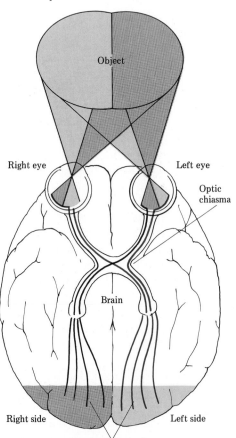

FIGURE 20.21
The human visual system. Note that the optic nerves that leave the point of crossing (optic chiasm) carry "outside" information from one retina and "inside" messages from the other.

SUMMARY

1. **Receptors** are sensory structures useful to animals in maintaining homeostasis. **Exteroceptors** link animals with their external environment, and **interoceptors** inform them of their internal conditions. Exteroceptors are commonly named after the form of energy or condition to which they respond, such as **photo-, phono-, chemo-, thermo-, mechano-,** and **proprioceptors.** Some stimuli affect more than one kind of receptor.

2. Receptors are either neuron endings or specialized cells in close contact with sensory neurons. Receptors respond to an appropriate form of energy by converting it into a nerve impulse. The intensity of a stimulus is usually indicated by the rate at which impulses are generated. The intense stimulation of many receptors results in pain.

3. **Adaptation** is the diminished response to a constant stimulus. Many receptors are sensitive to changes in stimuli rather than to set levels of stimuli. Adaptation to odor is especially fast, whereas it occurs very slowly in response to pain and to cold.

4. Mechanoreceptors respond to forces such as pressure, stretching, and twisting. Many mechanoreceptors are **hair cells** or are neurons located at the bases of hairs. Other mechanorecptors respond to being deformed (e.g., tactile and pressure receptors).

5. **Statocysts** are hair cells important to many animals in sensing gravity and maintaining **equilibrium.** Equilibrium is the sense of balance or the ability to maintain an orientation between opposing forces. The basic function of the vertebrate **ear** is to maintain equilibrium. The receptors are hair cells found in the **saccule, utricle,** and **semicircular canals.**

6. Proprioceptors involve mechanoreceptors found in muscles, tendons, and joints. These sensory cells provide animals with a sense of the position and condition of their body parts.

7. Phonoreception is a form of mechanoreception that is especially important in birds and mammals. The receptor cells are hair cells located along the **basilar membrane** within the **cochlea.** Air waves set the **eardrum** vibrating, and these vibrations are transmitted to and amplified by three tiny bones. One bone transmits the vibrations to the incompressible fluid in the outer cochlear canals. Those pulsations move particular regions of the basilar membrane and cause deformations in certain hair cells. The impulses in the associated nerve then travel to the brain.

8. The chemical sense in many aquatic animals is generalized over the body surface and is relatively nonspecific. In terrestrial vertebrates, the chemoreceptors are concentrated in small areas of moist epithelial tissue.

9. The **taste** receptors of several arthropods (e.g., flies and crabs) are found on their feet. In mammals, taste involves receptors in **taste buds** found mostly on the tongue. Four basic tastes are often recognized (sweet, salty, sour, and bitter), and there is some differential sensitivity to these tastes on different parts of the tongue.

10. The **olfactory epithelium** is a small region deep in the nasal cavity. As with taste, substances must be in solution to be smelled. The sense

of smell is generally far more sensitive than the sense of taste, and the number of basic odors is probably more numerous.

11. Thermoreceptors inform animals about both the external and internal temperature environments. Both free nerve endings and specialized receptors are involved in sensing temperature.

12. Photoreception occurs when light energy causes appropriate receptors to initiate nerve impulses. **Vision,** the formation of images of the external world, generally requires a complex eye with a **lens** to focus light onto photoreceptors. Simple eyes are called **ocelli.** Specialized pigments (e.g., the **rhodopsins** of cepahalopods and vertebrates) are important in photoreception.

13. Although cephalopod and vertebrate eyes function similarly, they developed independently in evolution. Both are camera-type eyes. The human eye is well protected by bone, fat deposits, and a tough outer layer. Light enters through the transparent **cornea.** Small muscles in the **iris** control the amount of light that passes through the **pupil.** The lens and cornea refract light onto the **retina. Humors** help maintain the shape of the eyeball but may cause **glaucoma** if their pressure is excessive.

14. The light-sensitive cells in the retina include the **cones,** important in color vision and acuity, and the far more numerous **rods,** important in vision in dim light and in detecting motion. The highest concentration of cones is in the **fovea.** Some mammals that are primarily nocturnal have a **tapetum** immediately behind the retina.

15. Depth perception is facilitated by **binocular vision.** Predators and arboreal mammals tend to have binocular vision, whereas many prey species possess the wide fields of vision afforded them by **monocular vision.**

16. The relative rates of synthesis and breakdown of rhodopsin, the primary visual pigment, determine the ability to see in dim light. Light breaks down rhodopsin, and its synthesis process is relatively slow. Although many vertebrates have color vision, most mammals are at least partially color-blind. There are three types of cones and three different pigments, each sensitive to a particular range of wavelengths.

17. Light entering the eye passes through ganglion and bipolar cells before it reaches the retina. Bipolar cells can sum the inputs of several rods, but one-to-one synapses between cones and bipolar cells are common. Although the visual cortex interprets light, some filtering of visual information occurs in ganglion cells and in other brain cells before it reaches the cerebrum.

REVIEW QUESTIONS

1. Distinguish between interoceptors and exteroceptors, and state at least two examples of each.

2. Why is it advantageous for many receptors to be found in epithelial tissue?

3. Explain why people "see stars" when an eye is hit.

4. Describe sensory adaptation and explain how it occurs.

5. Describe the variety of hair cells involved in mechanoreception, and explain how they generate impulses.

6. Describe at least three ways by which animals sense touch and pressure.

7. Explain pain and describe some of its causes.

8. Distinguish between the static and the dynamic senses of balance, and describe the structures and processes involved in each.

9. What are proprioceptors and of what value to animals are they?

10. Describe the structure of the organ of Corti and explain how it and associated structures work in phonoreception.

11. Compare the relative sensitivities of the senses of vision and hearing, and of the senses of taste and smell.

12. What are the structural and functional features shared by the senses of taste and smell?

13. Distinguish among the structures and functions of ocelli, compound eyes, and camera-type eyes.

14. In what ways are cephalopod and vertebrate eyes similar to and different from cameras?

15. Describe the typical path of light through the human eye.

16. Name and explain at least four common abnormalities of human vision.

17. Compare the functions of the rods with those of the cones.

18. Contrast binocular and monocular vision, and the animals that typically possess each way of seeing.

19. Explain how rhodopsin works.

CHAPTER 21

EFFECTORS AND SUPPORT: ACTION SYSTEMS

Sensing the external and internal environments and integrating those messages are only two steps in the maintenance of homeostasis. It is also important for animals to be able to act upon the information—for example, to walk, jump, run, or talk. In this chapter you will explore the workings of a variety of effectors, especially muscles. Glands, the body's other major category of effectors, are discussed in Chapter 18. By weight, muscles are the principal tissues in vertebrates. Muscles not only help us move but also assist in such tasks as pumping the heart and moving food through the body. You will also study the associated skeletal systems that assist effectors in performing their tasks.

In addition to being involved in locomotion and the operation of internal body parts, muscles are also important in verbal communication and the nonverbal communication commonly called body language. Only by using our muscles can we express our thoughts and feelings.

AN OVERVIEW OF EFFECTORS

From the simple cilia in many invertebrates to the complex muscle-and-bone structures of vertebrates, the array of animal **effectors** is diverse. Effectors are those structures that move motile animals through their environment, move the environment past sessile animals, or move or alter animal structures. Several such actions may occur simultaneously in animals. Thus, while effectors such as muscles assist mammals in walking or running, internal cilia are moving mucus and particles and phagocytic cells are engulfing pathogens (Figure 21.1). Most effectors are controlled by neurons and/or by various chemical messengers.

The structures of **cilia** are described in Chapter 3, and their functions in various animals and body systems are reviewed below. Cilia provide the means of locomotion for the larval stages of many invertebrates, which would otherwise be almost totally at the mercy of water currents. Many sessile animals use cilia to propel the environment past them, thus bringing food particles into contact with the mucus that entraps them. Sweeping cilia also propel nutrients to locations where they are ingested. Within mammals, beating cilia in the respiratory passages help to keep out particulate matter. Dust, pathogens, and other particles are continually trapped in mucus, and cilia move both to the mouth for elimination. Ciliated cells in the nasal passages do much the same. Internal cilia are also important in processes as diverse as excretion in flatworms (e.g., flame cells; Chapter 17) and propelling eggs through the mammalian oviduct (Chapter 6). Cilia on the external surfaces of flatworms and some mollusks assist the rippling movements of muscles in locomotion. However, multicellular animals that rely even partially on beating cilia for locomotion lead generally sluggish lives.

The hairlike structures on cells that are usually somewhat longer and less numerous than cilia are called **flagella.** Beating flagella on specialized cells maintain a flow of water through the bodies of sponges that brings in nutrients and oxygen and helps remove wastes. Flagella also propel the sperm of many animals.

Named after the familiar motion of amoebas, **amoeboid movement** is important to many animals, including humans. In sponges, amoeboid cells perform many functions, such as aiding in digestion, carrying food from one area to another, secreting protective structures, and forming reproductive cells. The important process of cells moving from one area to another during embryonic development is accomplished by means of amoeboid movement. Phagocytes also depend upon amoeboid movement to locate and ingest pathogens, other foreign material, and cellular debris (Chapter 16). Two of the same proteins that are involved in muscle action (actin and myosin) are involved in amoeboid movement.

MUSCLE TYPES

Most animals possess **muscle cells,** contractile units specialized to move other structures. About half of our body weight is made up of our approximately 600 different muscles. Muscles are classified into three groups based upon differences in the structures of their cells (Figure 21.2).

The simplest muscle type, **smooth muscle,** is found throughout the animal kingdom, and it was the first to evolve. In vertebrates, smooth muscles are part of many internal organs that are not under voluntary control, such as in the walls of blood vessels and in the organs of the digestive tract. The long, spindle-shaped cells are uninucleate, and the

FIGURE 21.1
Strong leg muscles are obviously necessary for this huge bull bison to plod through the deep snow as it tries to eek out an existence during the harsh winter. Strong neck muscles allow it to move its head from side to side to push the snow from the meager vegetation needed to fuel its "inner fires," so necessary in thermoregulation. Internally, effectors are at work constricting and relaxing blood vessels, causing breathing movements and the heartbeat, and innumerable other actions. Although the animal may not survive the winter, as it has so many previously, its survival indicates that homeostasis is currently being maintained. If the individual-preserving processes are successful, then the animal's effectors may be useful in such species-preserving activities as mating.

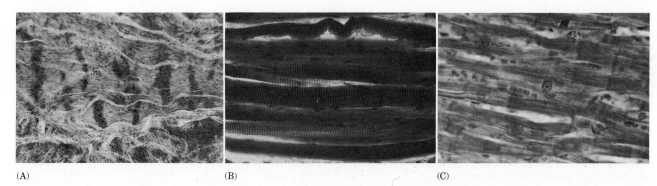

(A)   (B)   (C)

FIGURE 21.2
Three types of vertebrate muscle cells. (A) These smooth muscle cells are found in the wall of the urinary bladder. They perform their functions relatively slowly, do not tire easily, and are not under conscious control. (B) These striated muscle cells from the human arm act faster and more forcefully than smooth muscles, and they are under voluntary control. (C) Cardiac muscle, found only in vertebrate hearts, has some traits of both smooth and striated muscles.

contractile proteins in the cytoplasm are not easily seen under the light microscope. Smooth muscles act relatively slowly and do not tire quickly.

The most common muscles in vertebrates are the **striated muscles,** or voluntary muscles. These are also commonly called *skeletal muscles* because they move structures such as bones and thus provide the basis for locomotion. The cells of striated muscles are fused into large units called **muscle fibers.** Muscle fibers are bound by delicate connective tissue and entire muscles are surrounded by tough sheaths of connective tissue. Muscle fibers contain many nuclei and their contractions are both more powerful and faster than those of the same mass of smooth muscle cells. The striated muscles were so named because their contractile proteins are often arranged regularly in sequences of bands or striations.

Cardiac muscle occurs only in vertebrate hearts. The fibers possess striations but they branch profusely and are mixed with noncontractile cells. Cardiac muscle contracts with an inherent rhythm, even in the absence of nervous stimulation (Chapter 15). Like striated muscle (of which it could be considered a variety), cardiac muscle acts quickly and forcefully but, like smooth muscle, it is not under voluntary control and does not tire quickly. Cardiac muscle needs abundant oxygen and stops functioning if deprived of oxygen for as little as 30 seconds.

MUSCLE TRAITS

Muscles either move skeletal parts to which they are attached or change the size and/or shape of a soft tissue (e.g., when the heart beats). An important and basic trait of muscle cells and fibers is that they contract actively and relax passively. That is, they contract when they are stimulated and relax when unstimulated. Muscles cannot be stimulated to relax and, thus, they exert their forces in only one direction. Because of this, complex acts and movements often involve two sets of muscles that are called **antagonists** because the muscles have opposite effects, one relaxing while the other contracts (Figure 21.3). However, even "relaxed" muscle is generally contracted somewhat, so-called **muscle tone.** Muscles under slight tension respond more rapidly and contract more forcefully than those that are relaxed completely.

Many examples of antagonistic muscles are found in the joints of arthropods and vertebrates. Each joint is bent by one or more **flexors** and is straigthened by **extensors.** For example, the two ends of vertebrate striated muscles are typically attached to two different bones. Some striated muscles, however, connect one part of the skin with another, such as the muscles of facial expression, or connect a bone with the skin. Other terms for and examples of common antagonists are shown in Table 21.1 and Figure 21.4.

FIGURE 21.3
The appendages of arthropods (A) and vertebrates (B) are moved by antagonistic sets of muscles. When flexors contract while extensors relax, the limb bends; the opposite actions lead to straightening. Note that the antagonists of arthropods are attached to their external skeleton, whereas those of vertebrates are connected to the bones of their internal skeleton. (C) Both biceps and triceps of the human arm have their origins on the shoulder, which remains relatively fixed during arm movements. The insertions of these antagonistic muscles are on the bones of the forearm, which move as the arm flexes and extends. Other sets of antagonistic muscles move body parts toward or away from the center of the body, raise or lower body parts, decrease or increase the size of openings, or rotate body parts one way or another. (D) The tendons in the human hand attach the finger bones to muscles in the forearm. We would not have the same manual dexterity if the muscles were attached directly to the finger bones, as on the right.

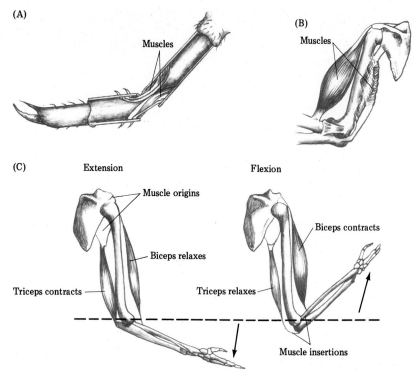

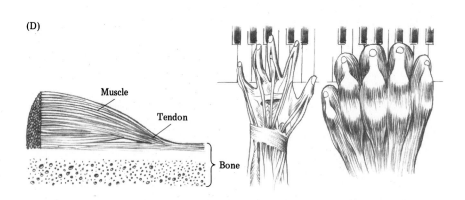

TABLE 21.1
Antagonistic Muscles

Names	Actions	Examples
Adductors	Move parts of the body toward the center line of the body	Move the arms toward and away from the midline of the body
Abductors	Move parts of the body away from the center line of the body	
Levators	Raise body parts	Move the lower jaw
Depressors	Lower body parts	
Sphincters	Decrease the size of openings	Control the size of anal opening
Dilators	Increase the size of openings	
Pronators	Rotate downward-backward	Turn the palm down and up
Supinator	Rotate upward-forward	

In addition to antagonists, there are muscle **synergists,** muscles that contract at the same time as the muscle or prime mover that exerts the major pulling force. The function of synergists is to aid the prime mover either by exerting more pull or by stabilizing the body part that is being moved.

Many of the muscles that act upon bones are attached to them indirectly by means of inelastic **tendons.** Tendons allow muscles to act at some distance from the bones they move. When you clench your fist, you can see the tendons of some of the arm muscles that move the fingers tense in your wrist. When a muscle contracts, one end, the **origin,** remains relatively fixed and the other end, the **insertion,** moves.

Because muscles are grouped into bundles covered by a connective-tissue sheath, individual bundles can act separately. Adjacent bundles can also slide easily over one another as muscles contract. Bundles of muscle fibers often extend the length of striated muscles. Near the ends of muscles, the sheaths taper and gather to form the tendon. As we age there is an increased amount of connective tissue and decreased muscle strength.

Each fiber contains filaments called **myofibrils** (Figure 21.5). The ATP energy source for contraction of the myofibrils is provided by the numerous mitochondria in the cytoplasm (called *sarcoplasm* in muscles) within which the myofibrils occur. Myofibrils, in turn, consist of long rows of individual functional units called **sarcomeres.** The numerous sarcomere units give skeletal muscle its striated appearance. Finally, each sarcomere consists of two thin actin filaments for every thicker and longer myosin filament. Sarcomeres are the contractile units of striated muscles.

Sliding Filaments

Muscle contraction is caused by the sliding movements of **actin** and **myosin** filaments within the sarcomere (Figure 21.6). About 90 percent of muscle protein consists of actin and myosin, although other proteins (tropomyosin and troponin) are also important in muscle actions. The globular protruding "heads" of myosin molecules serve as hooks that attach to actin molecules. The "bridges" between the two filaments form, break, and reform during the sliding action that reduces the combined lengths of the filaments without shortening the filaments themselves. Relaxation involves the detachment of the bridges between actin and myosin, something that requires ATP. In rigor mortis, the stiffening of muscles in dead animals, the bridges bind actin and myosin because, obviously, no ATP is being produced to cause their separation. The changes in the striation patterns can be observed as contraction and relaxation occur.

This contractile mechanism is initiated when a nerve impulse reaches a neuromuscular junction (Chapter 19). The depolarization of the plasma membrane of the muscle fiber (sarcolemma) at the synapse is followed by a wave of depolarization over the surface of the muscle fiber. The nerve fiber and the muscle fibers that its branches supply constitute a **motor unit.** Fine, precise control of the eyeball muscles is achieved with little power because the motor units contain just a few muscle fibers. In contrast, the motor units of large muscles, such as the arm's biceps, may contain thousands of fibers, resulting in powerful contractions when many of these motor units contract at once. In response to a nerve impulse, all components of a motor unit contract simultaneously. Relaxation involves a reversal of the events of contrac-

FIGURE 21.4
Examples of antagonistic muscle actions. Note that striated muscles are not usually located within the parts they move. For example, the biceps is found in the upper arm but it acts to move the lower arm.

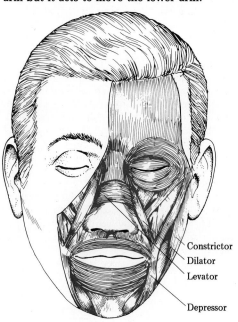

Constrictor
Dilator
Levator

Depressor

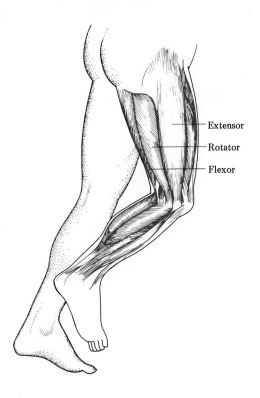

Extensor
Rotator
Flexor

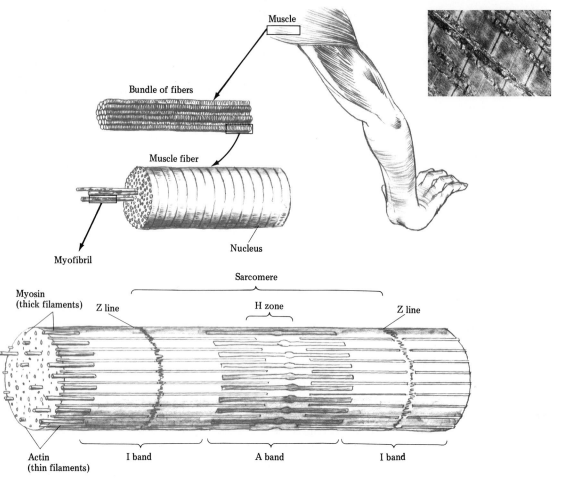

FIGURE 21.5

The structure of vertebrate striated muscle. Muscles consist of bundles of fibers that, in turn, contain myofibrils, with their functional units called sarcomeres. Actin and myosin are the two types of protein filaments.

tion. Movements of calcium ions are important in both contraction and relaxation.

Various poisons, nerve gases, and pesticides affect the neurotransmitters, some preventing their release. This is true of the **botulism** toxin, one of the most powerful natural poisons known. This substance, produced by the bacterium *Clostridium botulinum,* prevents muscle stimulation and causes paralytic food poisoning (Chapter 12). The **tetanus** toxin, produced by a related bacteria (*C. tetani*), acts in a different way. Like the poison strychnine, it suppresses inhibition of muscle antagonists. As a result, antagonists contract at the same time and limbs are locked in a state of spastic paralysis. Some other chemicals (e.g., curare) prevent the neurotransmitter from reaching the muscle fiber, causing paralysis and sometimes death if breathing or other vital muscular movements are inhibited.

Muscular dystrophy and **myasthenia gravis** are diseases in which the neuromuscular junction does not function properly. In muscular dystrophy, something seems to inhibit the action of a neurotransmitter on the muscle fiber, whereas myasthenia gravis appears to involve abnormal neuron endings.

Gradations and Speeds of Response

From threading a needle to driving a car, our striated muscles are capable of a wide range of graded responses. However, a given muscle fiber operates in an all-or-none fashion, and is either contracted or relaxed. (See Chapter 19 for a discussion of a similar situation in the

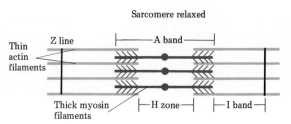

Sarcomere relaxed

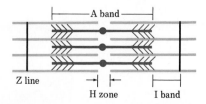

Sarcomere contracted

FIGURE 21.6
The sliding filament model of vertebrate striated muscle contraction. The thick myosin filaments are unanchored in the center of the sarcomere. As contraction occurs, the actin filaments (which are attached to the so-called Z line) slide past the myosin filaments.

action of neurons.) One reason for graded responses is that different numbers of muscle fibers may be involved. The strong, rapid action of throwing a heavy rock involves many more motor units than turning a page in a book. The frequency with which motor units are stimulated is also important.

In vertebrates, striated muscles are often divided into **red muscle** and **white muscle,** depending upon the relative presence of the red protein pigment myoglobin (which is similar to the hemoglobin in erythrocytes). Red muscle fibers are specialized for moderate, relatively slow activity, whereas white fibers are specialized for intense, rapid muscle activity. For example, long-distance runners have a higher proportion of red fibers than weight lifters, who in turn have a higher percentage of white fibers in their muscles. Like hemoglobin, myoglobin can bind reversibly with oxygen, and oxidative phosphorylation of fatty acids typically provides most of the energy for red muscle (Chapter 4). These striated muscles are designed for relatively continuous use that requires endurance. In contrast, the typical fuel for white muscles is glucose or glycogen, which they break down by anaerobic glycolysis. This results in the accumulation of lactic acid and a so-called oxygen debt. The heavy breathing after exercising is one way mammals "repay" the oxygen debt and use oxygen to remove the accumulated lactic acid.

While it is true of many mechanical devices that the more they are used, the faster they wear out, this is not the case with muscles. Well-used muscles **hypertrophy,** or become larger and stronger. Neither the number of muscle cells nor the number of bundles increases, but the number of myofibrils per cell may. Muscles that are not used regularly **atrophy,** or decrease in size and strength, partly because noncontractile connective tissue replaces the unused muscle cells.

As with neurons, muscle fibers have a very short refractory period when they cannot respond to another stimulus. If stimuli are applied too fast, a muscle may be unable to relax between stimuli and may remain contracted (tetanus). If this occurs for long, the muscle will fatigue and its strength of contraction will diminish even when it receives constant stimulation. Muscle fatigue from heavy use results from the depletion of fuels and the accumulation of lactic acid. Usually, however, the neuromuscular junction "tires" (produces insufficient neurotransmitters) before the muscle itself does.

For information on effectors in other animals, see the essay "Comparative Effectors."

Comparative Effectors

While most animals possess muscles and many also have exocrine and endocrine glands (Chapter 18), some effectors are found in only certain groups of animals. For example, cnidarians such as jellyfish and hydras have large numbers of **stinging cells** (nematocysts) on their arms (Figure B21.1). These are threadlike structures with basal spines that are coiled tightly within a capsule. When potential prey touch the trigger, the stinging cell fires by turning inside out, and it penetrates or entangles the animal's body. Poison may also help to subdue the prey, which is then swallowed.

Blending into the background is obviously useful in avoiding detection (Figure B21.2). In squid, several kinds of fish, chameleons, and some other animals, pigment-bearing **chromatophores** help animals change their color and/or pattern, sometimes within seconds. Most commonly, cellular pigment granules either become concentrated (paling the animal) or dispersed (darkening it), or some combination thereof (producing

a blotchy pattern). Under nervous and/or hormonal control, microfilaments move the granules. In other cases, muscular contraction or relaxation cause changes in the shapes of chromatophores. In still other instances, amoeboid chromatophores move closer to or further away from the animal's surface.

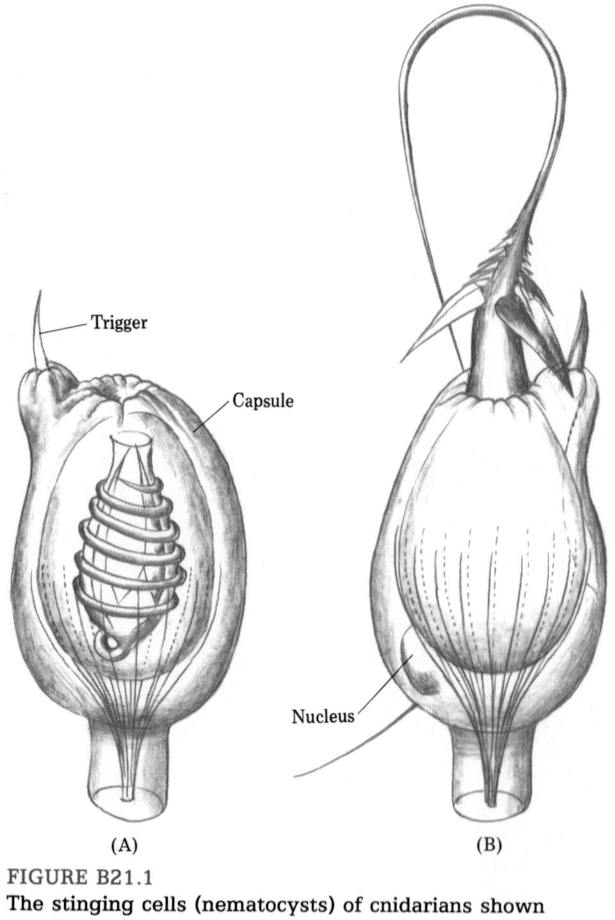

(A)

(B1)

(B2)

(B3)

FIGURE B21.2
(A) This fish matches the color and pattern of its background very precisely. **(B)** In the most common type of chromatophore, the concentration of pigment in the center of the cell causes a paling of the color, whereas the dispersal of the pigment leads to a darker color. The light micrographs of melanophores on a fish scale show the pigment dispersed (1), during aggregation (2), and completely aggregated (3). The time between (1) and (3) is 20 sec.

Trigger

Capsule

Nucleus

(A)   (B)

FIGURE B21.1
The stinging cells (nematocysts) of cnidarians shown "loaded" (A) and triggered (B).

SKELETAL SYSTEMS

For locomotion, the evolution of muscles provided a significant advance over the use of cilia. Another major evolutionary step was the development of a **skeleton.** Most animals require a skeleton against which muscles can pull, as well as to support them against the pull of gravity. See the essay "Comparative Skeletal Systems" for information about the supportive systems of invertebrates.

Vertebrates have a supportive and protective **endoskeleton,** or internal scaffolding, to which muscles attach (Figure 21.7). Such skeletons also maintain the soft tissues against the pull of gravity. The adult human skeleton contains about 206 bones, most of which are found in the appendicular skeleton. Our entire skeleton makes up almost 20 percent of our body weight.

Two kinds of connective tissue make up the endoskeleton, **cartilage** and **bone.** Cartilage or "gristle" consists of widely spaced cells within a

(Continues on p. 465.)

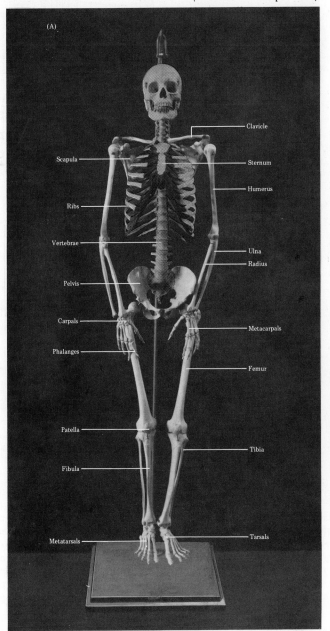

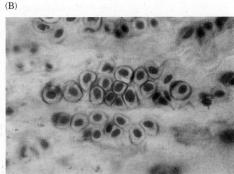

FIGURE 21.7

(A) Bone, as in this human skeleton, is a major part of the vertebrate endoskeleton, to which the striated or skeletal muscles are attached. (B) Cartilage, the other important component of the endoskeleton, consists of cells scattered in a rubbery matrix. (C) Collagen is the principal protein in cartilage, and some is also found in bone.

Comparative Skeletal Systems

The simplest type of skeleton is a **hydrostatic skeleton** that consists of incompressible body fluids (Figure B21.3). Hydrostatic skeletons are common in various soft-bodied invertebrates, such as the largely sessile *Hydra*. As when a water-filled balloon is squeezed, contraction of circular muscle cells lengthens the body. The antagonistic contraction of longitudinal muscle cells shortens the body.

Although only relatively crude mass movements are possible with a hydrostatic skeleton, an earthworm uses one to move quite effectively, for example, to extend from its burrow on damp evenings to reach the bits of decaying vegetation upon which it feeds. Alternatively, it may extend its posterior end in order to defecate its "castings." The wall of each earthworm segment consists of both circular and longitudinal muscles. Con-

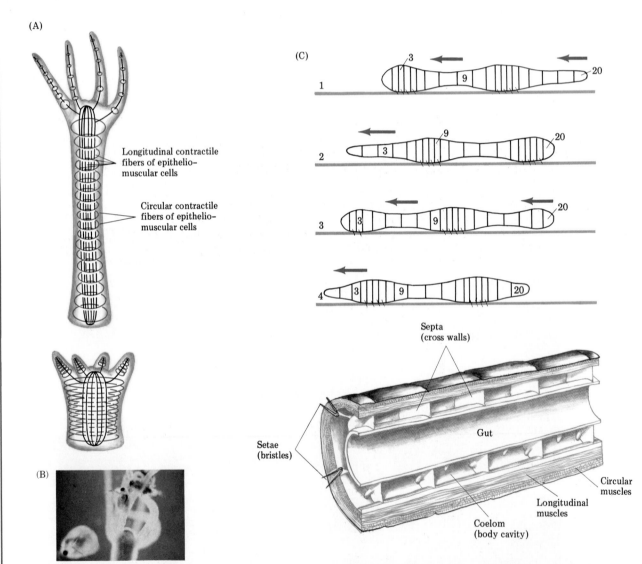

FIGURE B21.3

(A) The basic movements of *Hydra* involve circular and longitudinal muscle cells that act antagonistically against its hydrostatic skeleton. Its movements are relatively crude, although it does move by somersaulting on occasion and it can capture food and carry the food to its mouth with its flexible arms (B). The earthworm (C) moves by alternately contracting and relaxing muscles in the walls of

its body segments, which push against its hydrostatic skeleton. Bristles anchor the bulging segments (due to contraction of the longitudinal muscles), while the contraction of circular muscles allows other segments to elongate. The sectional view of the earthworm body shows the circular and longitudinal muscles, the bristles, and the liquid-filled cavities of the several segments that are separated from one another by partitions (septa).

traction of circular muscles elongates the segments, whereas contraction of longitudinal muscles shortens them and causes their wall to bulge outward. By narrowing some segments, earthworms extend themselves and then anchor those segments in soil with bristles. The contraction of other segments moves them toward the first, which can then be extended again.

The tube feet of starfish and some of their relatives also operate on hydrostatic principles. So do some body parts of vertebrates, such as the penis and clitoris when they become erect due to increases in hydrostatic pressure. However, for most animals who do more than drag themselves along, other skeletal systems are necessary.

The nonliving external covering or cuticle of arthropods such as insects, crustaceans, and spiders is hardened into an **exoskeleton.** The thick inner portion of the tough but pliable cuticle consists of protein and chitin (a polysaccharide), and is found only in arthropods. Muscles are attached to the inner surface of the exoskeleton, and the thinner and more flexible exoskeleton in joints facilitates movements. Although exoskeletons provide support for muscles as well as protection for the soft inner structures, their rigidity limits growth and they must be shed and replaced periodically.

The mechanism of insect flight differs from that of vertebrates such as bats and birds—another case of

(A)

FIGURE B21.4

(A) Most insects, like these dragonflies, have two pairs of wings. Like many insect wings, those of dragonflies are thin and membranous. However, some insect wings are parchmentlike (e.g., front wings of grasshoppers) or thick and horny (e.g., front wings of beetles); those of butterflies and moths are covered with fine scales. (B) The mechanisms of insect flight. Dragonflies reach speeds of about 40 kilometers per hour. Although their flight is slower (about 10 km/hr), migrating monarch butterflies are capable of long continuous flights.

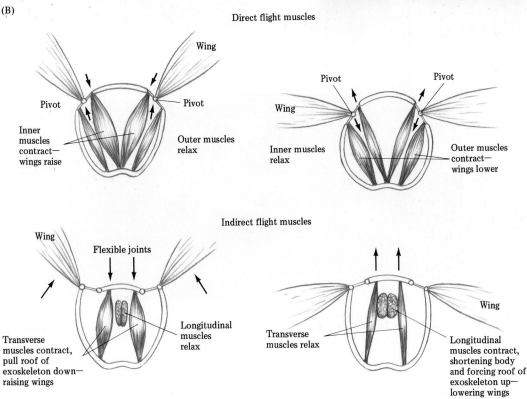

analogy but not homology. Insect wings are composed of cuticle and are outgrowths of the body wall. Although some insects are flightless, and true flies have but a single pair of wings (the halteres replacing the hindwings), most insects have two pairs of wings (Figure B21.4). A complex set of **flight muscles** in the insect's midbody (thorax) controls wing movements. Many familiar insects such as dragonflies, butterflies, and grasshoppers have so-called direct flight muscles. These insert directly on the base of the wings and raise, lower, or tilt the wings directly. Other so-called indirect flight muscles attach to the thorax walls rather than directly to the wings. These more powerful muscles are typical of flies and midges. The longitudinal and dorsoventral (top-to-bottom) sets of muscles exert a lever action on the wing by changing the shape of the body wall.

The exoskeletons of mollusks usually consist of one or a few shells (Figure B21.5). The shell serves to protect the animal and is often a point of attachment for muscles. Both muscular and hydrostatic mechanisms are important in the movements of these animals. For example, clams and snails extend the foot hydraulically, by engorgement with fluid. A clam's foot extended into sand or mud can be enlarged and used as an anchor. Longitudinal muscles contract to shorten the foot and to pull the body toward it. The foot of free-swimming forms may be modified into mobile structures for swimming. Jet propulsion is used by several mollusks, as when scallops clap their shells together quickly to expel water. The expulsion of water through a funnel allows squid and their relatives to use jet propulsion far more precisely.

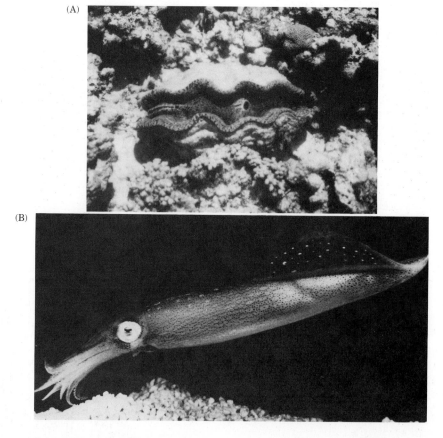

(A)

(B)

FIGURE B21.5

(A) The massive shell of this giant clam obviously protects the soft body parts inside. The several sets of muscles that allow the clam to open and shut its two hinged shells are attached to the inner surfaces of the shells. The movements of the foot depend upon both muscular and hydrostatic actions. Scallops can move fairly quickly but jerkily by alternately opening their shells and snapping them shut to expel water quickly. You are consuming the muscles when you eat scallops. (B) Jet propulsion is far more precise in squid, which are able to turn the funnel in various directions and to control their speed by the force with which water is expelled.

rubbery matrix of proteins (mainly collagen) and polysaccharides. Collagen fibers occur in all directions in cartilage and contribute to its strength and resiliency. The texture of cartilage varies from the hard discs between the vertebrae to the more elastic cartilage of the outer ear. Hyaline cartilage, the most common type, is found in the trachea, the ends of the long bones, and other sites. The skeletons of some fishes, such as sharks and rays, are composed entirely of cartilage, as are the embryonic skeletons of vertebrates.

Most people think of bone as rather passive, probably because the bones that we see are usually dead and dry. However, the rapid growth of a child and the mending of broken bones shows the dynamic nature of this connective tissue. Have you ever wondered how the age, sex, race, and medical history of a deceased person can be determined from the person's skeleton? Forensic or "legal" physicians and coroners often make such judgments, as do archaeologists and physical anthropologists. Bone is so durable and its growth patterns so well known that it has been extremely important in unraveling human prehistory and vertebrate evolution (Chapter 24).

The living cells that secrete the materials making up bone are **osteocytes** (Figure 21.8). Fine cytoplasmic threads pass through tiny tubes to connect the osteocytes that are scattered in circular layers throughout the hard material that they secrete. Collagen is the major protein in bone, as it is in cartilage. The calcium salts in bone help to maintain the blood calcium level homeostatically (Chapter 18). Because blood calcium levels are maintained at the expense of bone, any condition that lowers blood calcium will affect bone structure. For example, pregnant and lactating females provide much calcium to the fetus and infant, respectively, and such women may develop weak bones if their diet is calcium poor. People with rickets have weak, poorly calcified bones caused by a lack of vitamin D, a substance needed for the proper absorption of calcium from the digestive tract (Chapter 12).

Because nutrients and wastes do not diffuse readily through the matrix of bones, many interconnecting canals house blood vessels and neurons. Neurons in the fibrous connective-tissue sheath (periosteum) that covers bone make it very sensitive, as is evident when you bang your shins.

Types of Bone

Bone is organized somewhat as steel-reinforced concrete is. The tough and fibrous collagen is analogous to the flexible steel rods, and the minerals are like the cement. The detailed structure of bone varies with its function and position (Figure 21.9). To the unaided eye, bone is classified as either "compact" or "spongy," although both types are merely arrangements of the same elements and are similar chemically. For example, shafts of long bones consist of hard compact cylinders that surround cavities filled with **bone marrow.** This hollow-tube architecture provides a strong but lightweight structure important for efficient locomotion. The cavities in the long bones lessen the weight of the skeleton with only a slight reduction in strength.

Bones are commonly grouped on the basis of their shape and structure into four classes (Figure 21.10). **Long bones** are composed externally of a shaft and knoblike ends and internally by two types of bone marrow. **Yellow marrow** is fat-storing tissue, and **red marrow** is the hematopoietic tissue that produces many blood cells (Chapter 15). Most of the bones of children are filled with red marrow, but as we grow much of the red marrow changes to yellow marrow. In adults, red marrow is found only

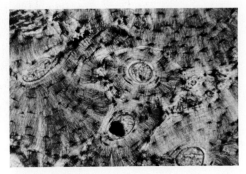

FIGURE 21.8
The major features of bone. Osteocytes secrete bone, which in humans is about two-thirds inorganic salts (including about 98 percent of the body's calcium) and one-third collagen, the material that yields gelatin when bones are boiled.

FIGURE 21.9
Compact and spongy bone. Compact bone makes up most of the shaft of this leg bone, whereas spongy (cancellous) bone is prominent in the head portion. Spongy bone, with numerous internal struts and ties, is especially important to birds. Such bone in wings is not only strong but light and it also provides space for the important air sacs (Chapter 14). The skeleton of the frigate bird, which has a 2-meter wingspread, weighs a mere 115 grams.

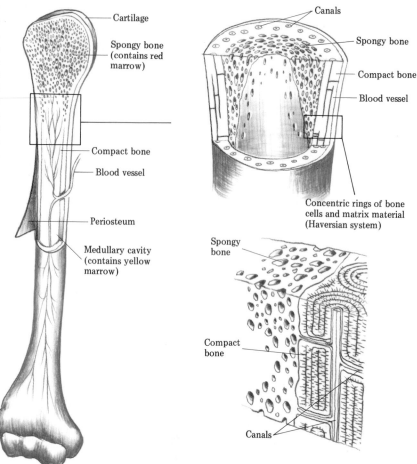

in some skull bones, the vertebrae, sternum, the ribs, and in the upper arm and upper leg bones. **Short bones,** such as those in the wrists and ankles, are usually cube shaped and contain a thin layer of compact bone over an inner mass of spongy bone. Some skull bones, the ribs, and the shoulder blade are **flat bones** that contain a central layer of spongy bone sandwiched between outer layers of compact bone. **Irregular bones** have complex shapes but, like short and flat bones, have the core of spongy bone surrounded by a thin layer of compact bone.

Bone Growth

In humans, for example, the embryonic skeleton consists first of collagen fibers (Figure 21.11). Later, cartilage is deposited between the fibers, and **ossification** or bone growth usually begins by the eighth week. The change of cartilage to bone involves the enlargement and death of the cartilage cells and the growth of blood vessels and connective tissue into the spaces left behind. Some of these cells become osteocytes and others become hematopoietic tissue. Early growth is especially obvious in the central region of the long bones, with ossification of the tips usually occurring after birth. The growth zone is below the tip and above the shaft and, in essence, the tip of the bone grows farther and farther away from the middle of the shaft. If the growth zone were elsewhere, it could interfere with the articulation of the bone to the bone next to it. As bones increase in thickness and diameter their marrow cavities enlarge and specialized cells (osteoclasts) remove material from the central part of the shaft. These "bone eater" cells are large, multinucleate cells that resorb bone by secreting digestive enzymes.

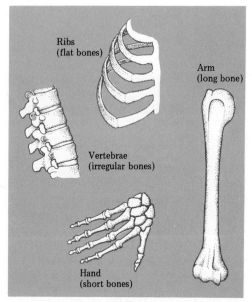

FIGURE 21.10
Classification of bone according to shape and structure. What additional examples of each bone type can you name?

(A)

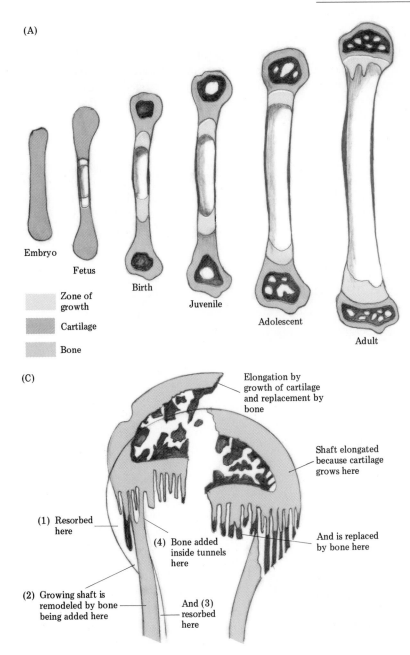

Embryo

Fetus

Birth

Juvenile

Adolescent

Adult

Zone of growth

Cartilage

Bone

(C)

Elongation by growth of cartilage and replacement by bone

Shaft elongated because cartilage grows here

(1) Resorbed here

(4) Bone added inside tunnels here

And is replaced by bone here

(2) Growing shaft is remodeled by bone being added here

And (3) resorbed here

FIGURE 21.11

(A) The growth of a long bone in humans from embryo to adult. These bones ossify first near the center of the shaft and later in the tips, as is clear in the view of the developing bone in a fetal finger, shown in (B). (C) Details of the process by which such bones grow. Because the growth zone is below the tip, the articulation with a neighboring bone is not disrupted. (D) The cartilaginous surface of this ball joint is clear during its formation.

(B)

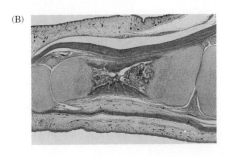

(D)

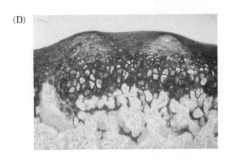

The final growth of most bones occurs between about the ages of 13 and 25 years. Although no further increase in length normally occurs thereafter, new bone may appear in broken or fractured bones (Figure 21.12). The scar tissue (mainly connective tissue) that forms first at the site of the injury changes to cartilage, and specialized cells from the adjacent periosteum gradually form bone that replaces the cartilage. The ability of bones to heal is usually excellent in children, but declines in middle and old age. Healing occurs best if the bone's ends are immobilized, properly aligned (using splints, casts, or bone pins), and brought into close contact.

Many factors affect the size and shape of bones. Heredity is a major factor in early development, but mechanical stress, muscle tension, hormone levels, and diet (especially calcium and vitamin D) are also important. As with muscles, we "use 'em or lose 'em." That is, bones subjected to heavy use usually grow thick and strong, whereas bones not used for long periods atrophy, such as those in a cast or in people with polio.

Simple

Displaced

Compound

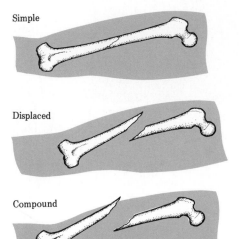

FIGURE 21.12

Three common types of bone fracture.

Bone Movement

If everything but bones were to disappear, most of the typical vertebrate skeleton would collapse because most bones are separate units held together by nonbony materials. Bones fit together to form **joints,** which are often held together by tough fibrous **ligaments** (Figure 21.13).

The ends of two movable bones do not contact each other directly but are covered with a smooth, slippery cap of cartilage. The joint surfaces are encased in capsules with slippery linings (bursae) made of ligaments and filled with **synovial fluid** that is secreted continuously by the membrane lining the cavity. The secretion of synovial fluid typically decreases as animals age, and the cartilaginous surfaces of the bones tend to become ossified. As a result, the joints become stiffer and movement is more difficult. Some common types of chronic joint problems are outlined in Table 21.2.

FIGURE 21.13

(A) Ligaments hold bones together, as in this example of the human hand. (B) Some of the major types of movable joints in the mammal skeleton. (C) The sutures of the skull also hold bones together but they are largely immovable. (D) Details of a synovial joint. Although too little synovial fluid causes problems, as in older mammals, excessive fluid also causes joints to become swollen and painful.

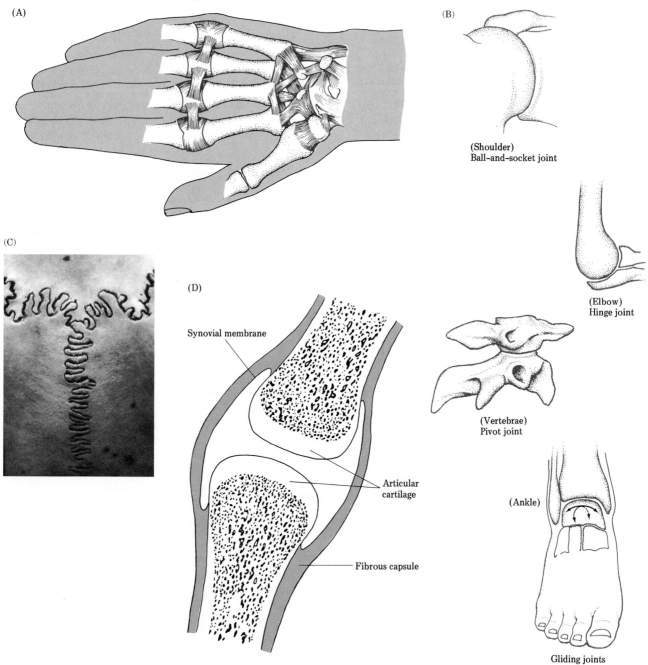

(A)

(B)

(Shoulder)
Ball-and-socket joint

(Elbow)
Hinge joint

(C)

(D)

Synovial membrane

Articular cartilage

Fibrous capsule

(Vertebrae)
Pivot joint

(Ankle)

Gliding joints

TABLE 21.2
Joint Diseases

Disease	Characteristics	Treatment
Arthritis	General term for inflammation of the joints. Fatigue, fever, morning stiffness, swollen and sometimes deformed joints, especially in the hands, knees, and feet. Rheumatoid arthritis affects about 1 of 50 Americans. May be caused by a chronic infection and/or an autoimmune response. Osteoarthritis results from the degeneration of the cartilage that covers the bones at their articulating surfaces.	Rest, heat, aspirin, antiinflammatory drugs during severe attacks; surgically replacing damaged joints with plastic ones
Bursitis	Infection or irritation of a bursa (e.g., tennis elbow)	Antiinflammatory drugs
Gout	Inflammation of the joints due to deposition of uric acid in the cartilage of joints.	Dietary control; various drugs

SUMMARY

1. **Effectors** are structures such as **muscles** or **cilia** that move organisms or their parts. Cilia and **flagella** are important in the locomotion of small, simple animals; the movement of sperm; or the movement of mucus and particles within organisms. **Amoeboid** cells are involved with various internal processes, such as phagocytosis.

2. Most animals possess contractile units called muscles. The three types of muscles differ in the structure of their cells. Muscles either change the shape and/or size of soft tissues or move skeletal parts to which they are attached. For larger animals, muscles are far more versatile for locomotion than cilia are.

3. **Smooth muscle** cells were the earliest to evolve, and they are found in all animals. Smooth muscles contract slowly, do not tire readily, and are not under conscious control.

4. **Striated muscles** are the most common muscles in vertebrates. They contract rapidly and forcefully, tire relatively easily, and are under voluntary control. In addition to other functions, striated muscles move skeletal units and provide the basis for intricate animal movements.

5. **Cardiac muscle** is found only in vertebrate hearts. It has an inherent rhythm of contraction and shares some traits with both smooth and striated muscles.

6. Many important movements of skeletal muscles involve **antagonists,** such as **flexors** and **extensors.** The **origin** of such muscles is often to a relatively fixed bone, whereas the **insertion** is usually to a movable bone. Skeletal muscles are often connected to bones by means of **tendons,** which allow muscles to act at some distance from the bones they move.

7. Striated muscles contain bundles of **muscle fibers,** each of which contains **myofibrils.** Myofibrils consist of many individual **sarcomeres,** which give the muscle its striated appearance. Sarcomeres are the contractile units of such muscles.

8. With the assistance of other proteins, the sliding movements of **actin** and **myosin** filaments within sarcomeres cause the contractions of striated muscles.

9. **Motor units** consist of nerve fibers and the muscle fibers that they innervate. All components of a motor unit contract simultaneously when stimulated. Precise control occurs when an axon connects only a few muscle fibers. More powerful contractions occur when nerve fibers connect with many muscle fibers.

10. Although individual muscle fibers operate in an all-or-none manner, gradations of response may occur when different numbers of fibers are involved and as the frequency of stimulation changes.

11. Not all striated muscles contract at the same rate. For example, vertebrate **red muscle** is specialized for relatively slow sustained action, and **white muscle** for rapid, intense contraction.

12. Some effectors are found in only certain animals, such as the **stinging cells** of cnidarians. **Chromatophores** assist animals in several groups to change colors and/or patterns to match their backgrounds.

13. Most animals require a **skeleton** against which muscles can pull. The simplest skeletons are **hydrostatic,** consisting of incompressible body fluids. Antagonistic circular and longitudinal muscles are common in such animals as cnidarians and earthworms. Hydrostatic mechanisms sometimes affect only certain body parts, often working with muscles.

14. **Exoskeletons** consist of nonliving material secreted by animals, such as the cuticles of arthropods and the shells of mollusks. Muscles are attached to the inner surfaces of the exoskeleton.

15. Insect flight depends either upon **flight muscles** connected directly to the wings or upon muscles that move the wings indirectly by changing the shape of the body wall. Flight in birds involves antagonistic muscles, where the origin is on the keel and the insertion is on the wing bones.

16. The **endoskeletons** of vertebrates consist of **cartilage** and/or **bone.** Some fishes have a skeleton entirely of cartilage, but most vertebrates possess both types of connective tissue. Cartilage consists of cells in a rubbery matrix made up mainly of collagen. In bone, **osteocytes** are scattered in a matrix composed largely of calcium salts.

17. Some bone is **compact** and other bone is **spongy,** although the basic composition is similar. The hollow areas in many bones contain **bone marrow** where many blood cells form.

18. In humans, **ossification** begins at about the eighth week of embryonic development. The growth zone of the long bones is below the tip and above the shaft. Thus, bones grow away from the middle of the shaft but not at the tips where growth could interfere with articulation.

19. Bones fit together to form **joints,** often being held together by **ligaments. Synovial fluids** lubricate some joints between movable bones. Arthritis, bursitis, and gout are common joint problems.

REVIEW QUESTIONS

1. Define *effector* and name at least six examples.

2. Distinguish between the structures and functions of cilia and flagella.

3. Describe the importance of amoeboid movements to animals.

4. Distinguish among the locations, structures, and functions of smooth, striated, and cardiac muscles.

5. Distinguish between the origin and insertion of skeletal muscles.

6. What are antagonists and how do they function?

7. How do the functions and locations of ligaments differ from those of tendons?

8. Describe the following structures in striated muscles and explain their functions: myofibril, fiber, and sarcomere.

9. Explain the sliding filament model of striated muscle contraction.

10. What is a motor unit and what is the significance of its size?

11. In light of the all-or-none manner in which muscle fibers contract, explain the graded responses of muscles.

12. Distinguish between red and white striated muscle.

13. What is a hydrostatic skeleton and how does it function?

14. Distinguish between an exoskeleton and an endoskeleton.

15. Compare the muscular actions involved in the flight of insects and of birds.

16. Contrast the structures and functions of cartilage and bone.

17. Distinguish between compact and spongy bones.

18. Where is bone marrow found and what is its function?

19. Describe the process of ossification during human embryonic development.

20. How and where do the long bones grow?

21. Describe at least three types of joints.

22. What is synovial fluid?

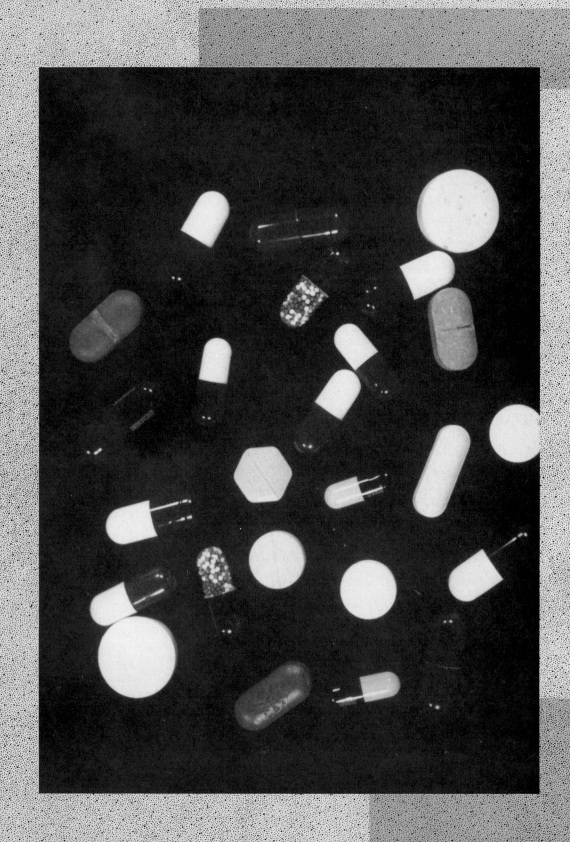

CHAPTER 22

SUBSTANCE ABUSE: A CHALLENGE TO HOMEOSTASIS

Substance abuse is a common and major threat to homeostasis. Humans have probably used mind-affecting substances since long before written history. There is an old Persian story about three men who had reached a walled town at night to find the gates locked. One man was an alcoholic, another was addicted to opium, and the third used hashish. The alcoholic wanted to break down the gate, the opium smoker thought that they should lie down and sleep until morning, but the hashish addict suggested that they pass through the keyhole!

The growing of drug plants is big business, as is their processing, transporting, and marketing. Thousands of species of plants contain psychoactive compounds, as well as numerous other substances useful in medicine and industry. The misuse of many of these chemicals is of major concern because society as a whole, not merely the drug abuser, is affected in many ways.

SUBSTANCE ABUSE

This chapter examines some of the diverse effects of **stimulants, depressants,** and **hallucinogens** (Figure 22.1). As their names imply, stimulants ("uppers") generally speed body processes and depressants ("downers") slow them. Hallucinogens affect how the body uses sensory information, although some also have stimulant and depressant effects.

Interest in and knowledge about brain chemistry has increased in recent years as we have sought to better understand human thought and behavior. Such studies were also stimulated by the rise of the drug culture and by the increasingly diverse range of psychoactive substances being produced, including the many so-called designer drugs. One major com-

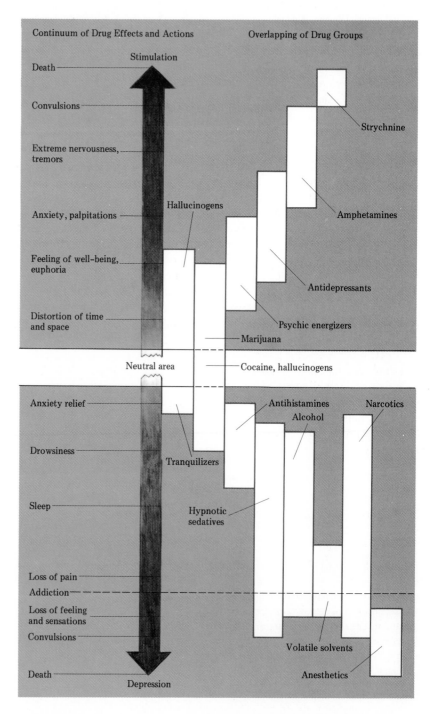

FIGURE 22.1
Some major categories of abused substances
and their general effects.

plication is that some compounds cause unusual behavior in one person but alleviate an abnormal condition in another.

Many modern societies are highly oriented toward chemicals. Think of the advertising you hear and see for chemical substances. "Trouble sleeping? Use Closelid before going to bed." "When you have a super-headache, take two Cranium Painiums—they're far stronger and faster than Head-Deads." "Too much partying? Chew two Excess-Aids and you'll feel better in minutes."

A change in threshold, or **tolerance,** often develops with repeated substance use, meaning that a given quantity of a compound has less and less effect. Larger amounts of the substance are needed to achieve the same effect initially caused by a small amount, and the increased substance intake may damage many organs. However, the problems of **withdrawal illness** ("cold turkey") may also occur if heavy substance use stops suddenly. Common symptoms of withdrawal include muscular pain, cramps, and twitching; aching bones and joints; nausea, vomiting, perspiration, and diarrhea; restlessness and trembling; gooseflesh; dilated pupils; watery eyes; increased blood pressure; loss of appetite; and sometimes delusions, hallucinations, delirium, and/or convulsions.

Substance use may also lead to physical and/or psychological **dependence,** when the users may crave and their cells may require frequent substance inputs. At the extreme, someone's personality, motivation, values, and interests may change and revolve around substance usage. The terms **addicted** and "hooked" describe this condition, which may also involve tolerance (Figure 22.2).

Many substances show **synergistic effects** ("1 + 1 = 5") when they are used together. For example, alcohol is one of the most abused substances, and its use with other compounds is especially common. Unpredictable and sometimes tragic results occur from such mixing, and a common but major problem results when alcohol is mixed with barbiturates, such as those found in some sleeping pills. The uncertain effects of "street drugs" occur partly because they are sometimes mixed with other ingredients. A user accustomed to adulterated drugs may experience an overdose when "lucky" enough to secure a relatively pure product.

Some substances are difficult to classify absolutely as stimulants, depressants, or hallucinogens because their effects vary from person to person and depend upon the mood and condition of the user. For example, marijuana is defined legally as a narcotic but is often grouped with the hallucinogens. Therefore, thus far in this discussion, the general term *substance* has been used.

STIMULANTS

Stimulants include a wide range of substances that cause both metabolic stimulation and mood elevation. Some stimulants are synthetic, and others are derived from certain plants, as is summarized in Table 22.1.

Psychic energizers, amphetamines, and antidepressants may combat depression and increase the rate of many metabolic processes. Some of the actions of these substances on the central nervous system (CNS) are to change the levels of synaptic neurotransmitters and various chemicals in the brain. For example, the release of some stimulatory neurotransmitters may be increased, or the release of some inhibitory neurotransmitters may be reduced. Either the neurotransmitter itself, enzymes important to it, presynaptic processes, or postsynaptic events may be affected.

FIGURE 22.2
(A) Addiction. Have you experienced tolerances for certain substances? Dependence? What nonchemical side effects may disturb the well-being of an addict? (B) Drug sale viewed with infrared night-vision device.

(A)

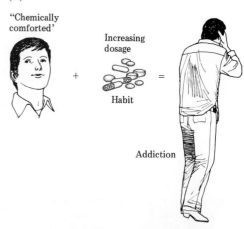

(B)

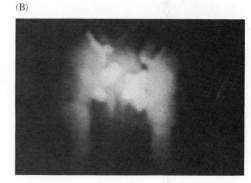

TABLE 22.1
Common Stimulants

Substance	Origin and Characteristics	Usual Method of Use	Potential for Dependence		Overall Potential for Abuse
			Physical	Psychological	
Amphetamines (e.g., Benzedrine, Dexedrine, Methedrine)	Manufactured	Swallowing or injection	Low	High	High
Cocaine	Leaves of coca tree (not cacao); white, fluffy, odorless powder	Sniffing; injection; chewing leaves	Moderate	High	High
Caffeine (e.g., coffee, tea, many soft drinks, stay-awake pills, many pain relievers)	Coffee berries, tea leaves, maté leaves, kola nuts	Swallowing liquid or pill	Low	Moderate	Low
Nicotine (e.g., cigarettes, cigars, chewing tobacco)	Tobacco leaves	Smoking or chewing	Low	High	Moderate

DEPRESSANTS

Depressants include a wide variety of substances that depress the action of the CNS (Table 22.2).

The term *narcotic* is confusing because it is used to describe several things: (1) a CNS depressant, (2) an addicting substance, and (3) a substance controlled by Federal narcotic laws. The action of depressants on the CNS may take several forms. Analgesics relieve pain, sedatives relieve anxiety and relax muscles, hypnotics increase drowsiness and sleep, and others produce euphoria or an exaggerated sense of well-being and contentment. Many actions of depressants involve the cerebral cortex, thalamus, hypothalamus, and brainstem (Chapter 19). The levels of various brain chemicals (e.g., serotonin) and neurotransmitters may also be altered, and depressants often constrict the pupil of the eye and depress the cough center. The body temperature is commonly reduced, partly due to evaporative cooling from increased perspiration, and some depressants stimulate the centers in the medulla that control nausea and vomiting. Overdoses commonly affect the breathing center in the medulla and lead to respiratory failure.

Abuse of depressants often comes from the euphoric state that may result. Euphoria may include feelings of peace and tranquillity, elevation of mood, and relief from fear and apprehension. After the euphoria passes, the user may become apathetic and drift into a state of sleep.

Potential of Tolerance	Withdrawal Symptoms	Possible Complications		Death by Overdose
		Physical	Psychological	
Yes	Depression, apathy, fatigue	Loss of appetite, needle-caused infections, blood vessel diseases, shaking, blackouts	Intoxication, mental illness, antisocial behavior, irritability and restlessness, hallucinations, talkativeness, aggressiveness, insomnia, delusions	Convulsions, coma, brain hemorrhage
Low	Usually minor	Loss of appetite, nasal membrane destruction from sniffing, loss of coordination, convulsions, brain damage	Intoxication, mental illness, excited state followed by depression, hallucinations, insomnia	Convulsions, respiratory failure
Yes	Headache, intestinal upset	Increased urine and salivary flow	Sometimes restlessness, insomnia	Convulsions, respiratory failure
Yes	Nervousness	Lung (and other) cancer, cardiovascular problems, cough, bronchitis, emphysema, deadened taste buds, decreased skin temperature, hypertension	Reduced alertness (from CO poisoning)	Tremors, convulsions, respiratory failure

Persons addicted to depressants frequently are pale, suffer from malnutrition and constipation, and often have a low or absent sex drive. Because of haste or a general disregard for themselves, addicts are often careless about sterilizing their equipment, and bacterial and viral infections (e.g., hepatitis and AIDS) often result. Withdrawal symptoms may be severe, and some addicts continue substance usage more to avoid withdrawal illness than to achieve a euphoric state. Society is also affected, because addicts must often resort to crime in order to support their habits.

Several depressants listed in Table 22.2 are **opiates.** Opium is the air-dried juice of the Oriental opium poppy obtained by cutting the unripe flower pods (Figure 22.3). The juice hardens into dark, sticky masses that taste bitter and smell sweet. Morphine, heroin, codeine, and many other substances are derived from opium, and as opium is refined the strength of its derivatives is often increased. For example, morphine is about ten times stronger than opium, and heroin is two to four times stronger than morphine. Because of their potency, morphine and heroin are usually diluted or "cut" before use. Many other depressants are synthetic (e.g., methadone and Demerol).

Why do opiates such as heroin and morphine combine with receptors on brain cells? Apparently, the receptors are where some naturally occurring substances found in the brain and pituitary gland combine. These morphinelike **endorphins** ("endogenous morphines") are neurotransmit-

FIGURE 22.3
Oriental opium poppy (*Papaver somniferum*) flowers and pods. Slashes on the pods are made for the extraction of raw opium.

ters that inhibit the production of nerve impulses associated with the perception of pain. Purified human endorphin is similar in potency to morphine. Perhaps the effectiveness of acupuncture is due in part to the release of these natural painkillers (Chapter 20). However, endorphins have also been found in parts of the brain involved with emotions rather than with pain. Although much about endorphins is unclear, they are known to induce a condition similar to schizophrenia when injected into rats. Elevated levels of endorphins have been found in some human schizophrenics, perhaps indicating one biochemical basis for this widespread mental illness.

As with other abusable substances, the effects of any depressant depend upon the individual, the setting, the dosage, and the method of

TABLE 22.2
Common Depressants

Substance	Origin and Characteristics	Usual Method of Use	Potential for Dependence		Potential for Tolerance
			Physical	Psychological	
Opium	Oriental opium poppy	Smoking	High	High	High
Heroin	Opium derivative; whitish to brownish powder	Injection, sometimes smoked with tobacco	High	High	High
Morphine	Opium derivative; white powder or tablets	Injection	High	High	High
Codeine	Opium derivative	Swallowing	High	High	High
Alcohol	Fermenting grapes, other fruits and grains	Swallowing	High	High	High
Barbiturates (e.g., phenobarbital, Nembutal, Seconal, Tuinal, Amytal)	Manufactured	Swallowing pills or capsules	High	High	High
Tranquilizers (e.g., reserpine, phenothiasines, Valium, Librium)	Manufactured	Swallowing pills or capsules	High	Moderate	Low
Volatile solvents (e.g., glue, gasoline, paint thinner, lighter fluid, carbon tetrachloride, benzine, toluene, cleaning fluid)	Manufactured	Inhaling	Low	Minimal to moderate	High

usage. Depressants may be swallowed (e.g., alcohol and codeine), inhaled (e.g., "snorting" heroin or sniffing volatile solvents), smoked (e.g., mixing heroin with tobacco), or injected into a muscle ("skin popping" or "muscling") or into a vein ("mainlining").

The major difference between the action of hypnotics and sedatives is the degree of CNS depression. Hypnotics generally have a greater depressant effect than sedatives and often lead to sleep after administration. Alcohol is probably the oldest hypnotic sedative used by humans and is by far the most abused depressant today. Other hypnotic-sedative substances include alcohol derivatives (e.g., chloral hydrate and paraldehyde), bromides, barbiturates, tranquilizers, and anesthetics. Synergistic effects are common, especially with alcohol, and because the lethal

Withdrawal Symptoms	Possible Complications		Death by Overdose	Overall Potential for Abuse
	Physical	Psychological		
Vomiting, diarrhea, aches, shaking, perspiration	Constipation, loss of appetite and weight, temporary impotence or sterility, loss of coordination	Intoxication, antisocial behavior, drowsiness, dulled senses	Coma, respiratory failure	High
As above	As above; infections from contaminated needles	As above	As above	High
As above	As above	As above	As above	High
As above but less severe	Constipation, loss of appetite	Drowsiness, dulled senses	Possible but unlikely	Low
As with opiates but with hallucinations, delusions, and illusions	Impaired reaction time and coordinations, sometimes obesity, brain and liver damage, stomach irritation, malnutrition	Drowsiness, impaired judgment and emotional control, often aggressive behavior, intoxication, mental illness, delirium tremors	Coma, respiratory failure	High
Convulsions, delirium tremors	Impaired reaction time and coordination, weight loss and reduced appetite, slurred speech	Drowsiness, impaired judgment and emotional control, irritability	Coma, respiratory failure, shock, low blood pressure	High
	Sometimes dryness of mouth, blurred vision, skin rash, tremor, occasional jaundice	Sometimes drowsiness	Possible but uncommon	Low
	Impaired coordination; often serious liver, heart, and kidney damage; irritated mucous membranes; anemia; bone marow depression; slowed speech	Similar to barbiturates and narcotics, disorientation, intoxication, drowsiness, insomnia	Coma; liver, kidney, or respiratory failure	Moderate

dose of many depressants varies from person to person, or from time to time in the same individual, synergistic interactions are potentially very dangerous.

Tolerance to barbiturates by the CNS develops quickly, and the chronic user may increase the dose to a lethal level without realizing the danger. Barbiturates are often used in suicides, although many deaths attributed to suicide are probably accidental overdoses. Alcohol intensifies the effects of barbiturates, and its use often accompanies accidental overdoses.

Histamines, natural chemicals found in most cells, produce marked and often harmful, effects in very small concentrations (Chapter 16). By dilating small blood vessels they may cause a rise in skin temperature, a drop in blood pressure, and headaches. Antihistamines block the actions of histamines but also lead to drowsiness or sleep at high dosage levels. Therefore, antihistamines are often considered with other depressants.

Because of its common use and abuse, **alcohol** is a depressant in a class by itself. About 70 percent of adult Americans consume alcoholic beverages to some extent, and alcoholism affects many millions of people directly and indirectly. Alcohol is the product of the fermentation of sugar by yeast, and the alcohol in beverages is mainly ethanol or ethyl alcohol. The methyl alcohol (methanol or wood alcohol) used in many products such as fuels and antifreezes is a deadly poison that can cause blindness or death in even small amounts. Isopropyl or rubbing alcohol is also too poisonous to be consumed.

In addition to alcohol and water, many alcoholic beverages contain flavoring and coloring agents. Although alcoholic beverages have little nutritive value, the caloric content is relatively high (Table 22.3). However, because these calories are "empty" or without other important nutrients, chronic alcoholics often suffer from malnutrition, and many of the physical ailments of alcoholics are complicated by this condition.

All alcoholic beverages have basically the same effects on the body, but various beverages differ in alcoholic content. Beers and ales average about 4–8 percent alcohol by volume, and the comparable figure for wines is about 12–21 percent. The volume of distilled beverages such as whiskey, brandy, rum, vodka, and gin average 40–50 percent alcohol. Figure 22.4 shows the equivalent alcohol content of common beverages. Each of these drinks produces an average blood alcohol content of 0.03 percent, and the average person oxidizes one of these drinks in about one hour.

Although most alcohol is absorbed in the duodenum, some is absorbed in the stomach. The presence of food in the stomach slows absorption. Carbon dioxide is one of many factors that speeds the passage of alcohol into the small intestine. Thus, an alcoholic beverage

TABLE 22.3
Caloric and Nutritional Content of Alcoholic Beverages

Calories (kcal) and Nutrients	Type and Quantity of Beverage		
	Beer (340 g or 12 oz)	Whiskey (57 g or 2 oz)	Wine (227 g or 8 oz)
Calories, total	171.0	140.0	275.0
Calories, alcohol	114.0	140.0	240.0
Protein (g)	2.0	0.0	0.0
Fat (g)	0.0	0.0	0.0
Carbohydrates (g)	12.0	0.0	8.5

4-oz (113 g) glass
of table wine
(12 percent alcohol)

12-oz (340 g) glass
of light beer
(4 percent alcohol)

4-oz (113 g) glass
of champagne
(24 proof, or
12 percent alcohol)

1-oz (28 g) of vodka,
taken with ice
(100 proof, or
50 percent alcohol)

1.25-oz (35 g)
"shot" of whiskey
(80 proof, or
40 percent alcohol)

1.5-oz (43 g) glass
of aperitif liquor
(25 percent alcohol)

FIGURE 22.4
Equivalent drinks in terms of alcohol content. An average person can oxidize one of these drinks in a one-hour period.

with a carbonated mixer is more potent than the beverage mixed with water, and the extra "kick" of champagne is due to dissolved CO_2. Table 22.4 shows the relationship among body size, number of drinks, and the resulting blood alcohol concentration.

Intoxication occurs because the brain centers depressed first are those affecting control and inhibition. Sensory perception is also affected, in part because the flow of neurotransmitters across synapses is affected, and sight is often one of the first senses affected by alcohol. The stimulant effect of alcohol is illusory and occurs because the restraining factors of one's personality are depressed. The use and storage of information is greatly diminished by alcohol, but small amounts may aid digestion by reducing nervous tension (which also often reduces the quantity of food consumed), and by stimulating the release of stomach acid. However, large amounts of alcohol, along with the increased amount of hydro-

TABLE 22.4
Blood-Alcohol Levels

Body Weight (lb)	Number of Drinks[a]											
	1	2	3	4	5	6	7	8	9	10	11	12
100	0.038	0.075	0.113	0.150	0.188	0.225	0.263	0.300	0.338	0.375	0.413	0.450
120	0.031	0.063	0.094	0.125	0.156	0.188	0.219	0.250	0.281	0.313	0.344	0.375
140	0.027	0.054	0.080	0.107	0.134	0.161	0.188	0.214	0.241	0.268	0.295	0.321
160	0.023	0.047	0.070	0.094	0.117	0.141	0.164	0.188	0.211	0.234	0.258	0.281
180	0.021	0.042	0.063	0.083	0.104	0.125	0.146	0.167	0.188	0.208	0.229	0.250
200	0.019	0.038	0.056	0.075	0.094	0.113	0.131	0.150	0.169	0.188	0.206	0.225
220	0.017	0.034	0.051	0.068	0.085	0.102	0.119	0.136	0.153	0.170	0.188	0.205
240	0.016	0.031	0.047	0.063	0.078	0.094	0.109	0.125	0.141	0.156	0.172	0.188

Under 0.05	0.05 to 0.10	0.10 to 0.15	Over 0.15
	Driving becomes increasingly dangerous Legally drunk in some states	Driving is dangerous Legally drunk in many states	Driving is very dangerous Legally drunk in all states
Driving may not be seriously impaired			

[a] One drink equals 1 oz (28 g) of 100-proof liquor or 12 oz (340 g) of beer.

chloric acid, irritate the stomach lining. Blood vessels in the brain dilate, leading in part to the headache associated with a hangover. The lack of restful sleep affects many physiological and psychological traits, causing, for example, fatigue and grouchiness.

Liver ailments affect many alcoholics, because about 90 percent of the alcohol taken into the body is metabolized by the liver. The use of all nutrients is affected by a faulty liver, contributing further to the malnutrition problems of alcoholics (Chapter 13). **Cirrhosis,** the increase in connective tissue in the liver, is about six times more common among alcoholics than among the general population. Alcohol also affects the kidney and the body's water balance. In time there is a shift of intracellular water and associated ions such as potassium into the extracellular fluids, causing both the insatiable thirst and the bloated, "beer belly" look common to heavy drinkers and alcoholics. Ironically, the greatest damage from alcohol probably occurs in those best at "holding their liquor." That is, most organs do not develop tolerance as rapidly as the brain does.

Alcohol plays a major role in homicides and automobile accidents. Alcohol often makes drivers *think* that their driving ability is improved, but blood alcohol levels as low as 0.03 percent may alter reaction time and impair judgment. At 40 miles per hour (64 km/hr), for example, an automobile driver who has been drinking moderately requires about 36 ft (11 m) more distance to stop than a nondrinking driver does. The distances increase to about 60 ft (18 m) and 90 ft (27 m) at 50 mph (80 km/hr) and 60 mph (96 km/hr), respectively. In addition to impaired coordination and slowed reaction time, increased aggressiveness is a common effect of alcohol and many other depressants.

HALLUCINOGENS

I noted with dismay that my environment was undergoing progressive change. Everything seemed strange and I had the greatest difficulty in expressing myself. My visual fields wavered and everything appeared deformed as in a faulty mirror. I was overcome by a feeling that I was going crazy, the worst part being that I was clearly aware of my condition.

So wrote Albert Hofmann in 1943 after he accidentally ingested **LSD** that he had synthesized earlier. LSD is derived from the ergot fungus of rye, a black substance that grows on the grain. Pure LSD is colorless, tasteless, and odorless but is so potent that 1 g (1/28 oz) makes about 10,000 doses. For this obvious reason, LSD is usually mixed with other substances by users.

Many studies of substance effects use placebos in order to separate real and imagined effects. **Double-blind experiments** are useful for the

FIGURE 22.5

Some results of studies of LSD taken by volunteers. What are some of the main effects of LSD? How long did the effects last? Were they dose-related?

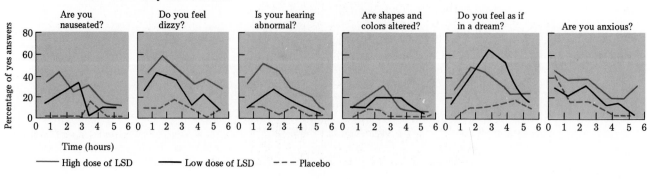

same reason and also often involve placebos. In these experiments neither the subject nor the person administering the substance (such as a nurse or technician) knows what the substance is. Figure 22.5 shows some typical results of double-blind experiments of LSD taken by volunteers.

Hallucinogens may distort sensory information without greatly disturbing the user's consciousness, and the terms **hallucination, illusion,** and **delusion** are often confused. Although everyone probably sees illusions, such as those in Figure 22.6, hallucinations are not usually experienced by most people. With illusions, objects are often perceived in variable ways or as different from the way they really are. In hallucinations, there is no object to be perceived incorrectly or differently, although objects may trigger them. Whereas most people see illusions similarly, hallucinations may be unique to the individual. Delusions usually involve people's feelings more than their senses, and the "persecution complex," when people feel incorrectly that others are trying to injure them, is a well-known example. People deluded by the false belief that nothing could harm them have jumped from high windows, shot themselves, or walked in front of moving vehicles.

Some hallucinogens seem to produce psychoticlike behaviors, and *psychedelic* (mind-expanding or mind-realizing) is another term used for some hallucinogens. Certain hallucinogens produce **synesthesia,** or mixing of the senses. One sensory input may be translated into another, such as when one "sees" high C or "tastes" purple. There is some evidence of chromosome damage from the use of certain hallucinogens, and some may also have teratogenic effects (Table 22.5). However, whether the chromosome damage found among LSD users at higher rates than among nonusers is due directly to LSD or to an impurity used to cut the drug is yet unclear.

Two natural sources of hallucinogens are shown in Figure 22.7. **Mescaline** comes from the small peyote cactus (*Lophophora williamsii*) that grows in northeastern Mexico and the southwestern United States. The small part that extends above the ground is cut off and is eaten fresh, chewed after it has dried ("mescal" button), or boiled, and the broth is drunk. The only legal use for mescaline in the United States is by members of the Native American Church for religious purposes. **Psilocybin** comes from a mushroom (*Psilocybe mexicana*) that grows in central Mexico. Hundreds of other hallucinogenic compounds occur in plants and in some animals, and many are produced synthetically.

Marijuana, the most commonly used hallucinogen, is derived from the leaves and flowering tops of female hemp or *Cannabis sativa* plants (Figure 22.8). The potency of cannabis depends upon the plant, the parts used, and the way it is prepared and stored. Hashish is a more concentrated preparation than marijuana. Although often grouped with hallucinogens, cannabis has a wide range of effects, some resembling those

FIGURE 22.6
A variety of visual illusions. (A) Does the hat look taller or wider? (B) How many different ways can you see the object? (C) Which person looks taller? What illusions involving other senses have you experienced?

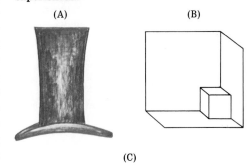

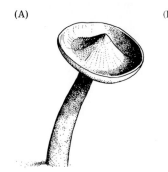 

FIGURE 22.7
(A) The mushroom from which psilocybin is secured and (B) the peyote cactus, which yields mescaline.

TABLE 22.5
Common Hallucinogens

Substance	Origin and Characteristics	Usual Method of Use	Potential for Dependence	
			Physical	Psychological
LSD	Semisynthetic derivative of ergot fungus	Swallowing	Low	Moderate
Mescaline	Peyote cactus	Chewed or swallowed; sometimes injected	Low	Moderate
Psilocybin	Mushroom	Swallowing	Low	Moderate
Marijuana	Hemp; synthetics	Smoking, chewing, swallowing, or sniffing	Low	Moderate to high

FIGURE 22.8

Cannabis sativa, from which marijuana and hashish are prepared. When the supply of hemp was cut off from the Philippines in World War II, *Cannabis* was cultivated widely in the United States to produce fibers for rope.

of stimulants and depressants. The effects of even large dosages of cannabis are milder and more easily controlled than those of most hallucinogens. Wakefulness is common during intoxication with most hallucinogens, but cannabis intoxication often ends in sedation. At low dosages, marijuana produces some of the same effects as alcohol, such as early excitement and a later sedated phase, reduced short-term memory, and slowed information processing. Alcohol used with marijuana produces synergistic effects, and as with many substances, the effects of cannabis depend greatly upon the user's state, environment, and expectations.

Hallucinogens and some other substances may lead to a **flashback,** the experiencing of some of the substance's effects at a later time. If the most recent or other important experience with the substance ("trip") was bad (a "bummer"), the flashback may be terrifying. Obviously, when a substance is used, some effects are expected, but part of the problem of flashbacks is that they occur at unpredictable times, days to months later.

SUMMARY

1. Three of the major classes of abused substances or psychoactive drugs are the **stimulants, depressants,** and **hallucinogens.** Stimulants generally accelerate many body functions, and depressants usually depress them. While some hallucinogens may have some stimulant or depressant effects, they primarily affect mental processes.

Potential for Tolerance	Withdrawal Symptoms	Possible Complications		Overall Potential for Abuse
		Physical	Psychological	
Moderate	Flashbacks common	Nausea, impaired coordination, aches, pains, dilated pupils, increased blood pressure and strength of reflexes	Hallucinations, delusions, anxiety, precipitation or intensification of an existing psychosis, flashbacks, alteration of sensory input	Moderate
Moderate	Flashbacks possible	Nausea and vomiting, cramps, sweating, increased blood pressure, muscle twitching, high pulse rate	Hallucinations	Moderate
Moderate	Flashbacks possible	As above	Hallucinations	Moderate
Low	Minimal	Increased appetite, sometimes impaired coordination, dilated pupils, reduced body temperature, sometimes nausea, bronchitis, speech difficulties	Hallucinations, altered sensory inputs, sometimes impaired judgment, reduced inhibitions, sometimes a sense of security even in the face of real danger	Moderate

2. **Tolerance** is an increase in threshold that results from repeated substance use. **Withdrawal illness** occurs if heavy and chronic substance use stops suddenly. **Dependence** occurs when a person craves or needs a substance. Both tolerance and dependence may be physical and/or psychological. **Addiction** usually involves both tolerance and dependence.

3. Many substances interact **synergistically,** often in unpredictable ways, and this sometimes results in death or serious bodily damage.

4. Some stimulants are derived from plants and others are synthetic. Medically, stimulants are commonly prescribed to increase some metabolic functions or to combat depression. The effects of stimulants on the CNS largely involve neurotransmitters and the levels of various chemicals in the brain.

5. Depressants include narcotics, analgesics, sedatives, and hypnotics. Through their actions on several regions of the brain, various depressants produce euphoria. Overdoses may be fatal if the rates of functions such as breathing are reduced excessively.

6. **Endorphins** are natural, morphinelike neurotransmitters produced in the brain. Apparently the brain receptors involved in the actions of **opiates** are the endorphin-producing sites. Endorphins are involved in functions such as pain perception and in various emotions.

7. **Alcohol** (ethyl) is a very widely abused depressant, and it often also acts synergistically with other abused substances. Alcohol contains calories but few other useful nutrients, and many problems of alcoholics

are complicated by malnutrition. Most alcohol is absorbed from the duodenum, although some enters the body quickly from the stomach. Food slows and CO_2 speeds the absorption of alcohol.

8. Hallucinogens include a wide range of natural (e.g., **mescaline, psilocybin,** and **marijuana**) and synthetic (e.g., **LSD**) compounds. Some hallucinogens produce **synesthesia,** or a mixing of sensory inputs (e.g., hearing tastes).

9. **Double-blind experiments** are particularly important in the study of real vs. imagined effects of substances. In these experiments, neither the subject nor the person administering the compound knows what the substance is (e.g., whether it is a placebo).

10. **Illusions,** or misperceptions of objects, are experienced by most people in a similar manner. However, **hallucinations,** which require no object, may be unique to each individual. Substance abuse may also affect **delusions,** or false beliefs or emotions.

11. When substances are used, the user expects some results. **Flashbacks** sometimes occur at later times unassociated with substance use. If the initial experience was unpleasant, the unexpected flashback may be serious.

12. Humans differ greatly, as does one individual from time to time. This variation greatly complicates making generalizations about the effects of substances. Effects are often not only dose related but also involve the user's past experience, expectations, and immediate environment.

REVIEW QUESTIONS

1. How do you differentiate among the terms *medicine, drug, narcotic, chemical,* and *substance*?

2. List all nonfood chemicals you use and try to discover the nature of their ingredients. Do you consider yourself to be a heavy chemical user? Would you like to stop using any chemical listed? Are there unlisted chemicals that you would like to use?

3. What advertisements for chemicals do you see? Do these advertisements show or imply that controlled experiments were done on the product?

4. Summarize the difference between dependence and tolerance. Can tolerance be psychological as well as physical? Can tolerance occur without dependence? Dependence without tolerance?

5. Do you think physical or psychological dependence is usually more difficult to treat? Physical or psychological tolerance?

6. How can chemical use by one individual affect relatives and friends? Society in general?

7. Why are double-blind experiments helpful in determining the true effect of a chemical?

8. Many experiments on the effects of substances on humans use volunteers. Do you think that experimental results from volunteers are representative of others? Do you think that experiments should be done on nonvolunteers (i.e., on people unaware that they are involved in an experiment)?

9. Describe at least three examples of synergistic interactions among substances.

10. Compare and contrast the major effects of stimulants and depressants.

11. How do the effects of hallucinogens differ from those of stimulants and depressants?

12. Distinguish among illusions, delusions, and hallucinations, using specific examples.

CHAPTER 23

FITNESS

Many people define good health as the absence of disease or ailments, although overlooked in this simple definition is the idea that health and fitness include the abilities to resist disease, to function sometimes at above-normal levels without harm, and to repair injuries quickly and completely. Although adequate exercise is as important in fitness as proper eating and sleeping, many people are "out of shape," partially because technology has reduced the need for labor and exercise. Obviously, the optimum level of fitness varies considerably from person to person, and the athlete, farmer, student, child, and construction worker each require a different level of fitness. In this chapter, you will examine some basic facts about fitness.

Sales of jogging shoes and exercise equipment are booming, as are sales of books and home videos of one fitness program or another. The operation of fitness centers, modern "sweat shops," is also big business, and programs are tailored to all age groups.

BASIC CONCEPTS

We need energy to support our basal metabolic rate (BMR) and our activities (Chapter 12). The contraction of a muscle produces a force, and most energy needs above the BMR are for moving large muscles. **Work** occurs when a force is applied over a distance. Note that time is not a part of the formula Work = Force × Distance. Therefore, both experienced ditch diggers and average office workers digging large holes perform about the same work and use about the same amount of energy. However, ditch diggers probably perform and recover faster than office workers.

Three important aspects of fitness are **strength, endurance,** and **flexibility,** as well as such specific motor skills as speed, reaction time, agility, balance, power, and coordination. Strength, the ability to mobilize power for a particular task, is often measured in two ways (Figure 23.1). **Isotonic** or movement strength is measured by the greatest weight that can be moved and **isometric** or holding strength is measured by the greatest force that can be exerted against a fixed resistance. Endurance, the length of time that muscular power can be used, is often expressed as the ability to continue muscular action when resistance is submaximal.

Different mechanisms account for increases in muscular strength and muscular endurance. Because muscular strength is proportional to the cross-sectional area of a muscle, it increases as a muscle hypertrophies. The strength of associated tendons and ligaments also increases. Further, the number of used muscle fibers increases as muscles are used. Improved muscular endurance depends heavily upon increases in blood supply and mitochondria.

Flexibility is the ability of a joint or series of joints to move within a normal range (Figure 23.2). Good flexibility not only allows you to move more easily but also decreases your chance of injuries.

Isotonic strength

Isometric strength

FIGURE 23.1
Isotonic and isometric examples of muscular strength.

Don't bounce! Stretch slowly and hold for 20-30 seconds.

Shoulder stretch

A. Extend your arms over your head, hands together, palms forward.

B. Force your arms backward until you feel a slight pain.

C. Hold this position for 20-30 seconds, then relax.

D. Repeat five times.

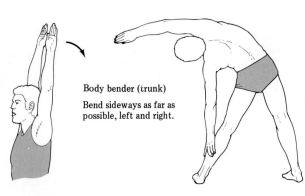

Body bender (trunk)

Bend sideways as far as possible, left and right.

Side leg raise (muscles on inside of leg)

A. Lie on your side, arms extended overhead. Rest your head on your lower arm. Extend your legs fully, one on top of the other.

B. Raise the top leg vertically, and hold it 20-30 seconds.

C. Return to starting position. Repeat five times. Then repeat on other side.

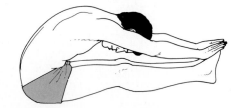

Leg stretcher

A. Sit with legs stretched out in front of you.

B. Bend at waist and reach for your toes. Keep your knees straight.

C. Stretch as far as you can and hold that position for 20-30 seconds.

D. Repeat five times.

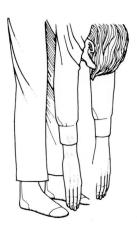

FIGURE 23.2
Some simple exercises to promote flexibility. If done regularly and at a sufficiently high level, these exercises, like most, also improve strength and endurance. In toe-touching activities, it is important not to "bounce," as this may damage some muscle fibers.

CARDIORESPIRATORY FITNESS

Although we often think of stress in negative terms, "good stress" in terms of fitness means a useful challenge to one's physiological capacity. Exercise programs are designed to provide the stress needed to increase one's physical performance. Because active tissues need oxygen and nutrients and produce metabolic wastes, the cardiovascular (Chapter 15) and respiratory systems (Chapter 14), hereafter abbreviated CRS for **cardiorespiratory system,** are especially important in fitness. See the essay "The Step Test" for a simple way to measure some aspects of your CRS fitness.

CRS fitness is the ability to deliver the oxygen needed by working tissues, especially the body's large muscles. Heavy exercise may result in a 20-fold increase in ventilation or oxygen consumption and a 7-fold increase in cardiac output (Figure 23.3). Obviously, anything that improves the ability of the heart to pump more blood contributes to CRS fitness. Because the action of muscles, including the heart, is to contract, a force is needed to return them to the relaxed state (Chapter 21). For example, skeletal muscles are arranged in antagonistic pairs, such as in the arm, where contraction of the triceps returns the flexed biceps to its noncontracted state.

The heart is not part of an antagonistic-pair system and this fact is a key to CRS fitness. Thus, the way to stretch heart muscles to their noncontracted state following diastole is to fill the chambers with blood. This requires increasing the return of venous blood to the heart and slowly to increase the force and volume of each contraction. However, all CRS conditioning programs eventually reach a plateau at a particular

FIGURE 23.3
Some typical effects of heavy exercise on various body functions of a well-conditioned person.

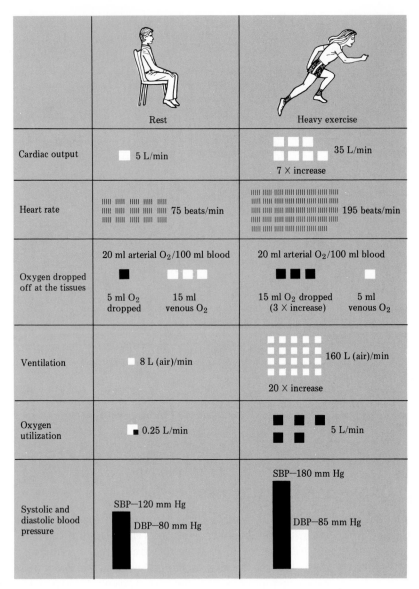

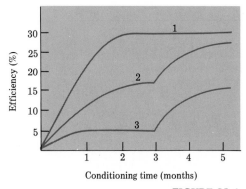

FIGURE 23.4

Typical results of various exercise programs for humans in low to medium initial levels of fitness: (1) very high activity for 5 months, (2) high activity for 3 months followed by very high activity for 2 months, (3) moderate activity for 3 months followed by high activity for 2 months.

stage (Figure 23.4). Figure 23.5 shows that the blood flow to many organs changes during exercise, and that such changes are usually dependent upon the intensity and pattern of exercising.

Many CRS fitness programs involve high-intensity exercise (so-called "no pain, no gain" regimes) such as cross-country skiing, running, or rowing. In such programs, the heart rate during exercising is usually quite high (e.g., about 150 beats per minute), and even conditioned individuals should be "puffing." After age 40 years, one might begin such a program with an exercising heart rate at a somewhat lower level (e.g., perhaps 110–120 beats per minute). Proponents of these programs suggest that if the heart rate is not sufficiently high, then additional stress is needed for CRS fitness to improve. However, exercise resulting in a heart rate that is too high (e.g., 175 beats per minute) may tire individuals too quickly to do much good. When an individual is well conditioned, the heart rate is often about 10 beats per minute less than it was before conditioning (Figure 23.6). Many believe that, in general, the slower the heart rate, the longer the heart will probably last, and that a well-conditioned heart will usually respond to unusual stress better than a nonconditioned one.

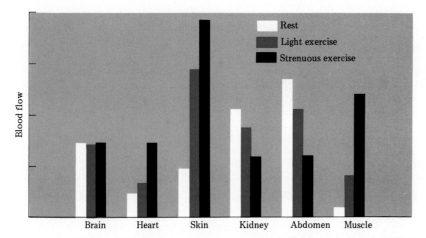

FIGURE 23.5
Effects of exercise on blood flow to various parts of the body. To what organs does blood flow increase? Decrease? Remain constant? How do you explain the blood flow to the various organs?

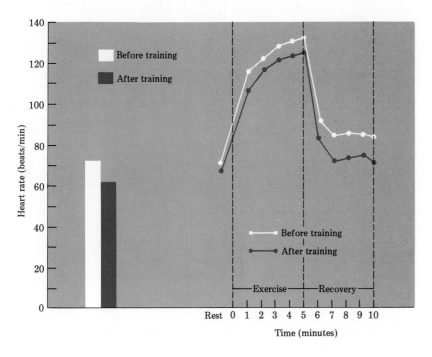

FIGURE 23.6
Effect of exercise on heart rates. Note that after conditioning, recovery time is faster and the heart rate is lower than before training.

Three factors that an exercise program should consider are frequency, intensity, and duration. Many exercise physiologists believe that 3 or 4 periods of high-intensity **interval exercise** a week are needed to cause a major change in CRS fitness. This schedule of alternating exercise and rest often produces rapid progress. Proponents suggest that exercise periods should last between 15 and 90 minutes, depending upon one's general health, physical condition, and age.

WARMING UP AND GETTING WARM

Warming up before strenuous exercise helps improve performance and prevent injuries (Figure 23.7). Nerve impulses travel faster and muscles stretch more easily when you warm up. However, the best warm-up time needed to increase ability to an optimum level depends upon one's physical condition and the activity.

Working muscles produce heat that causes the body temperature to rise (Figure 23.7B). Because excess heat is lost to the air (Chapter 2),

(A)

1. Side leg–raise, 10 times

2. Jogging in place, 1 minute

3. Trunk rotation, 20 times

4. Half knee bend, 10 times

5. Arm circles, 15 each way

6. Jumping jacks, 20 times

(B)

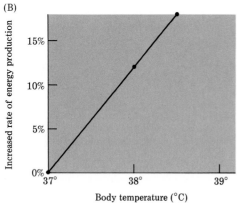

FIGURE 23.7

Warming up. (A) Some familiar warm-up exercises. (B) Typical increases in energy production and body temperature during warm-up exercising. (C) Common percentages of increase in running speed as a result of warming-up time. For those in poor shape just beginning an exercise program, 15 minutes of warm-up activities may seem difficult enough.

(C)

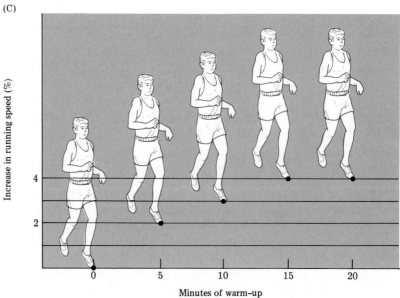

The Step Test

CAUTION: If you are ill, are recovering from an illness, or are otherwise disabled, do not attempt this or other activities suggested in this chapter.

This exercise has four positions and is repeated for four minutes (Figure B23.1). One cycle should be completed in two seconds. A partner should hold the chair or bench steady, give you signals, and be timekeeper. Conduct the activity according to the outline below. Use the pulse count in step 8 to find your approximate "recovery index": Poor—164 or higher, fair—143–163, good—126–142, excellent—125 or below. (Note: These values are common for young adults but may not be valid for others.)

Outline of the Step Test:
(1) exercise for four minutes,
(2) rest for one minute,
(3) count your pulse for 30 seconds,
(4) rest for 30 seconds,
(5) count your pulse for 30 seconds,
(6) rest for 30 seconds,
(7) count your pulse for 30 seconds, and
(8) add the three 30-second pulse readings.

Repeating this activity at intervals during an exercise program will provide one index of the level of improvement in your CRS fitness.

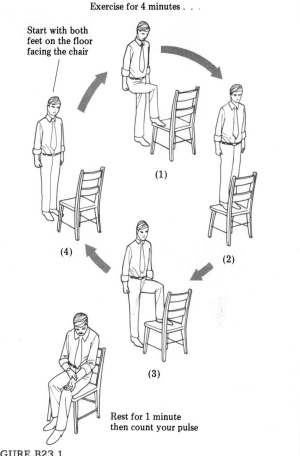

Exercise for 4 minutes . . .

Start with both feet on the floor facing the chair

(1)

(2)

(3)

(4)

Rest for 1 minute then count your pulse

FIGURE B23.1
The basic step-test procedure.

many people believe that it is always wise to expose as much skin to the air as possible. This is not always wise because when it is very hot and humid your body may gain heat from the environment. Light-colored clothing reflects some sunlight and aids evaporative cooling by absorbing perspiration rather than allowing large sweat droplets to fall from the body. Lightweight, loose-fitting clothing aids cooling by allowing more air to circulate than heavy, tight-fitting clothing.

According to some specialists, medium-intensity activities such as swimming, jogging, or tennis may not always provide enough stress to achieve full CRS fitness, and, they suggest, normal walking, golf, gardening, and other low-intensity activities contribute only very slowly to CRS fitness (Table 23.1). However, walking or similar relatively low-intensity activities for 30–60 minutes a day are obviously superior to "vigorous" TV watching or other sedentary activities. Any exercise uses calories, and weight maintenance is important for good health.

After about age 30, conditioning is more difficult than earlier but also usually more important. Hypertension often occurs with aging, but adequate exercise may minimize or even reverse this trend. Stroke is a leading cause of death of Americans, and high blood pressure is often associated with strokes.

TABLE 23.1

Comparison of Typical Effects of Various Activities on CRS Fitness[a]

	Endurance	Strength	Flexibility
Archery		•	
Badminton	•		•
Basketball	•		•
Bicycle riding	•	•	
Calisthenics		•	•
Canoeing and rowing	•	•	
Gymnastics		•	•
Handball	•		•
Hiking	•	•	
Jogging	•		
Long, brisk walking	•		
Mountain climbing	•	•	
Skiing	•	•	
Swimming	•		
Tennis	•		•
Wrestling	•	•	•
Weight lifting		•	

[a] • = very helpful.

Before beginning a conditioning program, everyone should obtain a physical examination, and some physicians believe that an exercise stress test is important (Figure 23.8). However, other physicians worry about the risk of some such tests triggering a heart attack, and there is also disagreement about how close to one's maximum heart rate a person should perform. It is easy to overexert in any conditioning program, especially in the beginning, and the essay "Exercise Aches and Pains" discusses some common problems and treatments.

OXYGEN DEBT

Any exercise causes an increased demand for oxygen by cells, and both the breathing rate and the tidal volume increase. If oxygen needs exceed available oxygen in the tissues, muscles function anaerobically (Chapters 4 and 21). This may continue until people reach their maximum **oxygen debt tolerance** (ODT). The higher one's ODT, the greater one's work capacity. Activities such as sprints are usually anaerobic, whereas dis-

FIGURE 23.8

An exercise stress test often involves peddling a stationary bicycle against resistance or walking on a treadmill. While exercising, one's blood pressure is measured, and a continuous electrocardiogram may be monitored for cardiac irregularities.

tance races tend to be aerobic or steady-state events. People in the latter events often experience a "second wind" when their cardiorespiratory systems can keep up with the oxygen requirements, and the heart rate and oxygen uptake level off. Aerobic work generally occurs only in submaximal tasks, although highly conditioned people such as marathon runners may maintain a steady state up to about 85 percent of their maximum oxygen uptake (Figure 23.9).

Lactic acid is one metabolic waste that leads to muscle fatigue. When exercise stops, oxygen uptake (heavy breathing) continues for several minutes, lactic acid is reconverted to glycogen, and the oxygen debt is "paid." Blood glucose levels during exercise are maintained largely by the liver (Chapter 13), and glycogen is converted to glucose, as are some lipids and amino acids. After exercise, the level of glycogen is gradually replenished (Figure 23.10).

DIET, EXERCISE, AND HEALTH

Exercise contributes to weight maintenance or reduction in several ways. A kilogram (2.2 lb) of body weight is equivalent to about 7700 kcal, and to lose 1 kg of body weight, your food intake must be 7700 kcal less than the energy used, or you must use 7700 kcal more than your food intake (Table 23.2). One of the best exercises is to push yourself away from the table at mealtime.

Regular exercise often also tends to lower the "appestat" in the hypothalamus, and the appestat of sedentary people is frequently too

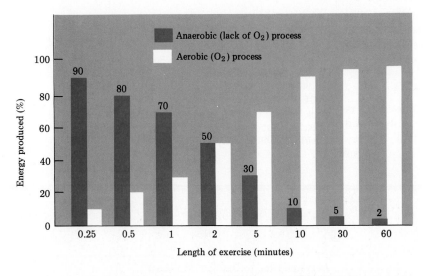

FIGURE 23.9
Aerobic and anaerobic processes related to length of exercise.

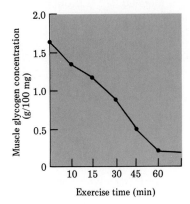
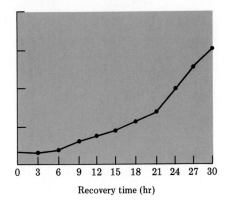

FIGURE 23.10
Effect of exercise and recovery times on muscle glycogen.

Exercise Aches and Pains

At one time or another most people suffer from one or more of the problems described below. Many of these problems are reduced by proper warm-up activities, because cold and tight muscles and joints are more likely to be injured than those that are warmed up and stretched.

Sore and Stiff Muscles

A new activity often causes muscles to stretch more than they used to, and the microscopic breaks in muscle fibers may cause pain. Usually there is no swelling, and no blood vessels are broken. Soreness can often be relieved by massaging or applying heat to the sore muscles and/or by doing light exercise to increase blood flow to the affected area.

Pulled Muscles

Because pulled muscles involve larger muscle fibers and broken blood vessels, they are more serious than simple soreness or stiffness. Common treatment for this problem includes the use of ice packs to slow the movement of blood to the area of broken blood vessels, pressure wraps to help close the broken vessels, elevating the injury to help drain fluids, and rest. After several days, heat may be applied to increase circulation and speed healing, which may take several weeks to complete.

Charley Horses and Bruises

A sharp blow to a muscle bruises some muscle fibers and breaks some blood vessels. Charley horses and bruises are treated like pulled muscles.

Muscle Cramps and Spasms

Physiologists do not agree about why muscles, especially leg muscles, suddenly and painfully contract. However, the buildup of waste materials and/or the loss of certain substances through heavy sweating seem to be involved in some cases. Slowly stretching a cramped muscle until the pain stops often helps. If pain returns, the muscle can be released slowly and then stretched again. When cramping has stopped, heat helps keep the muscle relaxed. Cramping may happen while muscles are being exercised but more often occurs after exercising.

(A)   (B)

Tests the strength of the upper back muscles

Tests the strength of the lower back

FIGURE B23.2
Back muscle tests for the upper (A) and lower (B) back muscles. Have someone hold down your body at the positions indicated by the arrows.

TABLE 23.2
Using Calories

Food	Kilocalories	Average Time to Oxidize (Minutes)[a]				
		Reclining	Walking	Bicycling	Running	Swimming
Apple, 1 large	100	66	28	16	5	8
Bacon, 2 strips	96	64	27	16	5	8
Beer, 1 glass	115	76	32	19	6	9
Bread and butter	78	60	15	10	4	6
Cookie, 1 plain	15	12	3	2	1	1
Donut, 1 medium	125	96	24	15	4	8
Egg, fried	110	73	31	18	6	9
Hamburger, medium, with bun	350	233	100	58	19	30

[a] Average values for a person weighing 70 kg (154 lb).

Back Disorders

Low-back pain is a problem for about half of all Americans. Although some problems are due to spinal disorders, about 80 percent are muscular problems. Figure B23.2 shows how to test the strength of some of your back muscles.

Many back problems result from weak abdominal muscles and/or inflexible hamstring muscles and tendons (Figure B23.3). Either or both conditions may cause the pelvis to tilt incorrectly and the lower back to curve improperly. Hamstring flexibility is maintained or increased by doing "toe touches" regularly, and bent-knee situps improve the strength and endurance of the abdominal muscles.

Joint Injuries

There are two main types of joint injuries: **sprains** and **dislocations**. Sprains most often affect ankle, knee, and thumb joints. The twisting of a joint may stretch or tear ligaments and tendons, and break blood vessels, causing internal bleeding and swelling. A bad sprain should be X-rayed to be sure no bones are broken. Treatment is often similar to that for pulled muscles.

Dislocations, which occur when the end of a bone is moved out of place, are most common in shoulder, finger, and thumb joints. A physician should reposition a dislocated bone and prescibe other treatment.

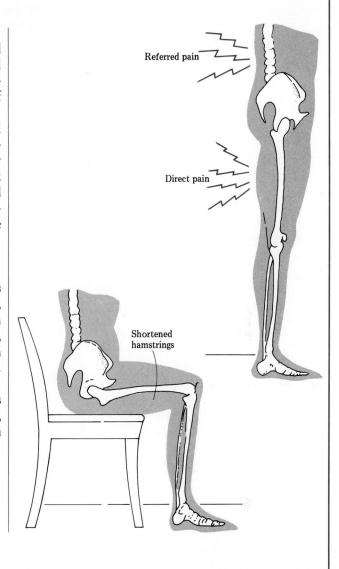

FIGURE B23.3
Shortening or contracture of the hamstring muscles often occurs as a result of prolonged sitting and a sedentary lifestyle.

Food	Kilocalories	Average Time to Oxidize (Minutes)				
		Reclining	Walking	Bicycling	Running	Swimming
Malted milk shake	502	386	97	61	26	41
Orange juice, 1 glass	120	92	23	15	6	9
Pecan pie, 1 piece	700	466	200	117	39	58
Potato chips, 10 medium	115	76	32	19	6	9
TV dinner, chicken	542	417	104	66	28	44

high, leading to weight gain. Interestingly, the appetite of some people on exercise programs has increased, with no weight loss occurring, and sometimes even weight gain. However, regular exercise reduces the amount of body fat, even if the body weight stays the same or increases slightly, and most people who exercise look and feel better in general.

We in the United States have one of the highest rates of heart disease in the world, and sedentary people generally have more heart problems than those who are active. Among those who exercise regularly, cardiovascular diseases tend to occur later in life, if at all, and to be less severe than in sedentary individuals. All the ways that exercise helps prevent or delay heart disease are not known, but one major factor is that exercise helps lower blood cholesterol levels. General CRS fitness and moderate exercise before and after meals speed the removal of plasma lipids after eating (Figure 23.11).

FIGURE 23.11
Common events following a high-fat meal among different individuals.

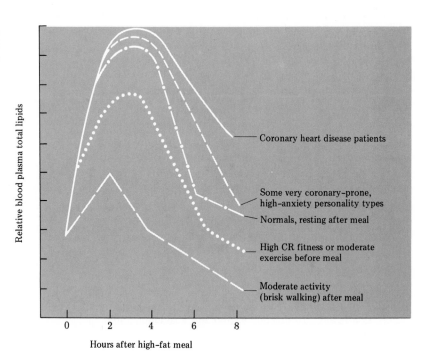

Relative blood plasma total lipids

Coronary heart disease patients

Some very coronary-prone, high-anxiety personality types

Normals, resting after meal

High CR fitness or moderate exercise before meal

Moderate activity (brisk walking) after meal

Hours after high-fat meal

SUMMARY

1. **Work** is a force applied over a distance, and most bodily work above the basal metabolic rate involves contractions of large muscles.

2. Three major aspects of **cardiorespiratory fitness** are **strength** (the ability to mobilize power for a particular task), **endurance** (the length of time that muscular power can be used), and **flexibility** (the ability of a joint or series of joints to move within a normal range).

3. Two kinds of muscular strength are commonly recognized: **isotonic** (movement strength) and **isometric** (holding strength). Strength is proportional to the size of muscles and to the number of muscle fibers used. The number of mitochondria in cells and the vascular supply to cells affect endurance.

4. Like all muscles, cardiac muscle hypertrophies with use and atrophies with disuse. Exercise not only increases the heart rate and cardiac

output but also increases the ventilation rate, the oxygen dropped off at and used by tissues, and systolic and diastolic blood pressures.

5. Exercise programs vary greatly in type and intensity but high-intensity and **interval exercise** regimes generally lead to the fastest improvement in fitness. **Warming-up** activities increase the rate of energy production, body temperature, and performance ability.

6. Fitness programs should be started only after an appropriate physical examination, which may include an exercise stress test.

7. Muscles function anaerobically when oxygen needs exceed the oxygen available in the tissues. In general, the higher one's **oxygen debt tolerance,** the greater one's work capacity. High-endurance activities such as long-distance running may approach an aerobic or steady-state condition. During exercise, the liver converts glycogen and some other stored nutrients to blood glucose. When exercise stops, the lactic acid that contributed to muscle fatigue is reconverted to glycogen.

8. A kilogram of body weight is roughly equal to 7700 kilocalories. Weight loss occurs if energy utilization exceeds energy intake. Regular exercise generally lowers the appestat, reduces the amount of body fat, and makes people feel better. Weight loss usually also results, and exercise helps prevent or delay cardiovascular problems.

REVIEW QUESTIONS

1. Define *work* scientifically and state at least two examples.

2. Distinguish between isotonic and isometric muscular strength.

3. Describe at least two reasons that muscles get stronger as they are used.

4. Define *cardiorespiratory fitness.*

5. Describe some typical changes in body functions resulting from heavy exercise.

6. How does exercising the large muscles help move venous blood and lymphatic fluid?

7. Define *interval exercise* and describe an example.

8. What is an exercise stress test?

9. Does longer-term exercising involve primarily aerobic or anaerobic processes? Why?

10. How are exercise, diet, and health interrelated?

PART FIVE

LIVING IN THE ENVIRONMENT

Both the past and the future are unclear in many respects, and the present isn't always easy to understand either. However, in this part you will examine our evolutionary origins, the current status of life, and some possible scenarios for the future.

Weatherwise, it's partly sunny in the region of Africa where humans probably evolved. In some other ways, however, it's a dismal day for many humans and other organisms there and elsewhere on Spaceship Earth because of overpopulation, hunger and starvation, habitat destruction, and environmental contamination.

CHAPTER 24

THE HUMAN PAST

About 200 million years ago the first small mammals coexisted with the ancestral dinosaurs. For millions of years mammals were dominated by the abundant, diverse, well-adapted, and sometimes quite large dinosaurs. When the dinosaurs became extinct, the mammals diversified and the apes and humans evolved from a common ancestor. You will explore the evolutionary development of Homo sapiens in this chapter; Chapter 1 deals with some general principles of evolution. In part because of our intense interest in ourselves, we probably know as much about our own evolution as about that of any species, although many unanswered questions remain.

Searching for one's roots interests many people, and the search for human origins in general attracts scientists and nonscientists alike. While many fossil human teeth, bones, and artifacts have been discovered, and the search for more is very active, the story is still incomplete and probably always will be.

TABLE 24.1

Traits Shared by Humans and Anthropoid Apes

1. Large, complex brains
2. Chest (pectoral) mammary glands
3. Hands have four fingers and an opposable thumb
4. Flattened face replaces muzzle
5. Single uterus
6. Usually one or two young born at a time
7. Prolonged and elaborate parental care of young
8. Menstrual cycles
9. Reduced sense of smell; prime emphasis upon vision
10. Tactile hairs on nose are lost
11. Collar bone (clavicle) braces the shoulder against the rib cage, giving the arms strength for brachiation (swinging)
12. Tendency to live in groups in which complex social interactions occur

MAJOR RELATIONSHIPS

Humans have been called "naked apes" in part because of the many traits we share with the higher modern **primates** (Figure 24.1 and Table 24.1). **Anthropoid** (from the Greek for "human") primates include the monkeys, apes, and humans; **hominids** (from the Latin for "man") include humans and their immediate ancestors.

These similarities have long suggested to biologists that humans are primates and that primates descended from a common ancestor. However, there are numerous other forms of evidence, such as the blood chemistry of humans and other anthropoids being similar, including common blood types such as ABO and Rh. Rh, in fact, was named after the rhesus monkey, in whose body the antigen was first found. The chromosomes of humans and apes, especially the chimpanzee, are similar. Humans and some other primates share many diseases, including syphilis. The sequences of amino acids in the proteins of humans and the anthropoid apes are generally quite close, and the differences are used as a "molecular clock" to estimate the time of divergence of various groups from one another. Actually, there are fewer differences between humans and the gorilla and chimpanzee than there are between the apes and the lower primates (Figure 24.2 and Table 24.2).

Walking humans are bipedal, whereas other primates are largely quadrupedal (four-legged) most of the time. Free from locomotor needs, humans arms and hands may be used for tool making and fine manipulative skills. Some other animals use tools but none so expertly and extensively as humans. The basketlike human pelvis provides support for many internal organs and assists us in our upright locomotion by providing the attachment point for various leg muscles. The anthropoid brain is only about one-third the size of the large human brain, which is housed within a dome-shaped cranium. Our ability for thought and language far surpasses that of other primates, due in part to the very extensive gray matter on the surface of the cerebral hemispheres (Chapter 19).

THE HUMAN FAMILY TREE

Humans did not descend *from* the apes, as some once contended erroneously, but rather *with* the apes (Figure 24.3). Modern apes, like humans, are recent products of evolution, and all these forms descended from a common, extinct ancestral mammal. See the essay "Mammalian Evolution" for more information.

Unfortunately, the fossil record of humans and other primates is incomplete and, of course, behavior, speech, and social organization are not fossilized directly but must be inferred from fossil remains and

FIGURE 24.1

The primates are one of over a dozen major groups (orders) of mammals. Of particular interest to humans are the anthropoid primates, a group that includes us and the chimpanzee (A), gorilla (B), orangutan (C), and gibbon (D).

(A)   (B)   (C)   (D)

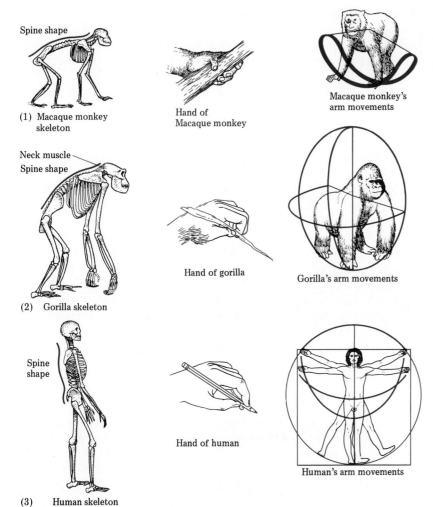

(1) Macaque monkey skeleton

Hand of Macaque monkey

Macaque monkey's arm movements

(2) Gorilla skeleton

Hand of gorilla

Gorilla's arm movements

(3) Human skeleton

Hand of human

Human's arm movements

FIGURE 24.2
Some comparative traits of three modern anthropoids. Although some other primates can walk bipedally, their gait is rather ungainly. Also, although the higher primates have opposable thumbs, the musculature differs. To appreciate the difference, try threading a needle or buttoning clothes with fingers flexed like those of the ape.

TABLE 24.2
Differences Between Humans and Anthropoid Apes

Trait	Anthropoid Apes	Humans
Thumb	Relatively small	Larger, used more
Growth rate	Slow	Slower
Sexual maturity	About 8 years	About 13 years
Infant dependency	About 2 years	About 6–8 years
Female receptivity to sexual intercourse	At ovulation	Continuous
Long bones	Legs shorter than arms	Legs longer than arms
Feet	Relatively short ankle and long toes; generally archless	Relatively long ankle and short toes; arched
Trunk	Long compared with leg length	Short compared with leg length
Vertebral column	Straight or curved backward uniformly	Alternating forward and backward curves; S-shaped
Skull-vertebral column joint	At back of skull	Almost in center of base of skull
Canine teeth	Large	No larger than premolars
Jaws	Long, large	Short
Face	Long, protruding in front of brain	Short, flatter, under brain

FIGURE 24.3
"RETURN OF THE FOREFATHER. MR. G-G-G-O-O-O-Rilla" is the caption of this cartoon published in the late 1800s during the height of the controversy over Darwinian theory. Darwin correctly associated humans with primates although many at the time (including Alfred Wallace, who codiscovered the concept of natural selection) believed that humans did not evolve as other organisms did.

FIGURE 24.4
Modern prosimians include the tree shrew (A), lemur (B), and loris (C). Note that some have large eyes that are close together on their flattened faces, giving them good binocular vision.

(A)

(B)

(C)

artifacts. Therefore, our reconstruction of human evolutionary history is hypothetical and, like all areas of science, is open to change based upon new findings, methods, and/or interpretations. Understandably, various experts viewing the same fossils devise somewhat different theories.

Before looking at the fossil record, it is important to note the recency of humans and to recognize that we are newcomers on Earth's block (Table 24.3). Many dinosaur species became extinct several million years after the early primates appeared. Although science fiction novels and movies sometimes indicate otherwise, humans and dinosaurs never coexisted.

The small, extinct insectivores from which primates evolved probably had generalized traits somewhat like modern tree shrews (Figure 24.4). Tree shrews, lemurs, lorises, and tarsiers are called **prosimians** ("premonkeys") and these primates are found today only in Madagascar and tropical Africa and Asia. They were abundant worldwide 55 million years ago and lived in North America until about 25 million years ago. Although their brains are small, prosimians have grasping hands and opposable thumbs, and some have nails rather than claws.

Exactly how and when the anthropoids diverged from prosimians is not certain. However, *Dryopithecus* that lived about 20 million years ago and the more recent *Ramapithecus* are two ape ancestors (Figure 24.5).

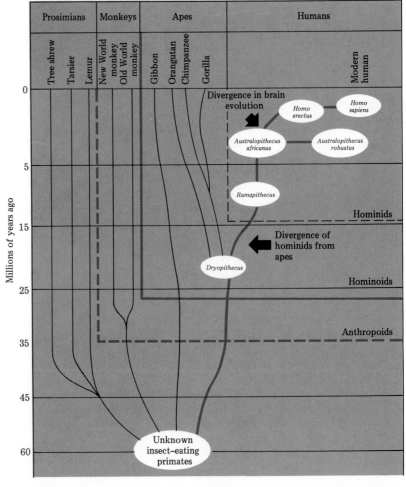

(A)

FIGURE 24.5
(A) Chart of the possible ancestry of the primates. (B) Comparison of the teeth and skulls of some extinct and modern anthropoids. (C) Reconstruction of some stages of human evolution.

The several species of *Australopithecus* were slightly over a meter (about 4 ft) tall, weighed up to about 20 kg (44 lb), and had gorilla-sized brains (350–450 cm³). The fact that the hole through the skull within which the spinal cord passes to connect to the brain faced almost downward is one indication that *Australopithecus* stood and moved erect. Their rounded jaws and their teeth suggest an omnivorous diet. Also, like the pelvis of modern humans, that of *Australopithecus* was flat and broad, with a rear-projecting flange. The latter anchors the large buttock muscles that balance the trunk on the erect lower limbs. The internal organs of quadrupeds are suspended from the back, whereas they are cradled by the pelvis in *Australopithecus* and *Homo*.

Why walk erect, and of what benefit are reduced canines? There are no certain answers but it is inferred by many that prehumans sometimes left arboreal life in the forest to forage in savannas, or areas of scattered

TABLE 24.3
Earth's Time Scale

Origin of	Millions of Years Ago (approximate)
Earth	4500
Life	3200
Vertebrates	420
Mammals	210
Primates	65
Anthropoids	36
Hominids	6
Australopithecus	5.5
Genus *Homo*	2
Homo sapiens	0.5

(B)

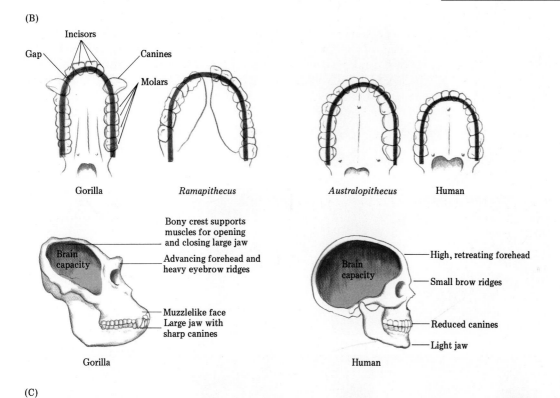

(C)

Dryopithecus   *Ramapithecus*   *Australopithecus*   *Homo erectus*   Early *Homo sapiens*   Neanderthal human   Cro-Magnon human   Modern human

Mammalian Evolution

Mammals evolved from reptilian ancestors even before birds did. The shrew-sized ancestral mammals that coexisted with the dinosaurs during the Mesozoic era were apparently insectivorous and nocturnal. With the demise of most of the dinosaurs, the mammals underwent a rapid adaptive radiation and diversified to occupy most habitats. The Mesozoic climate was generally warm and consistent, but Earth's climate became cooler and more variable toward the end of that era. Endothermy and the ability to utilize efficiently the numerous new forms of flowering plants that were developing probably contributed to the ultimate success of the mammals.

Modern mammals fall into three groups (Figure B24.1). The **monotremes** are egg-laying forms, such as the platypus and echidna. **Marsupial** mammals, such as kangaroos and koalas, possess pouches within which the development of the young, which are born in an extremely early stage, is completed. All other mammals are called **placental** because they complete their embryonic development within a uterus, connected to the mother by the placenta (Chapter 6).

(A)

(B1)   (B2)

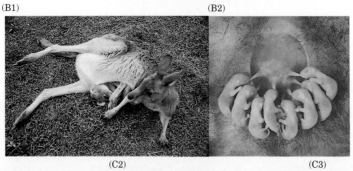

(C1)   (C2)   (C3)

FIGURE B24.1

Modern mammals. (A) The monotremes, such as this echidna, are the only mammals to lay eggs. Although, as with other mammals, the monotremes possess milk-producing glands, nipples are absent, and the young suck milk secreted onto the fur. (B) (1) This ''joey'' of the red kangaroo is almost too large to spend much more time in its mother's pouch. (2) Marsupials, like these opossum young, are insect-sized at birth, and they use their forelimbs to crawl slowly into their mother's pouch. There, they fix on a teat and complete their development while nursing. (C) Placental mammals are by far the most diverse modern mammals, occupying land habitats, like this elk (1); aerial habitats, like this bat (2); and aquatic habitats, like this dolphin (3).

trees, shrubs, and grasses (Chapter 25). An erect posture would have facilitated finding food and sighting enemies, as well as freeing the hands for defense if an animal were caught in the open. The large eyes and excellent binocular vision undoubtedly also contributed to survival, and the relatively unspecialized teeth probably allowed these primates to eat a variety of foods. These adaptations coupled with the small number of offspring probably allowed parents to devote much time to caring for and teaching their young.

Some biologists suggest that an upright posture developed before tool making but others believe that tool-making preceded bipedal posture. Whichever the case, tool users work best walking upright. Although culture itself cannot be fossilized, tools are a common and useful trace of human activity. Tools first appeared about 2 million years ago and during the course of human evolution they became more and more refined (Figure 24.6). When tools eventually made possible the hunting of big game, humans with the best learning and manipulative abilities were better adapted. Tools could also be used to crush food before it was eaten. Game hunting made possible some division of labor, social organization, and better communication. Some have suggested that if aggressive behavior enhanced survival, then it may have become part of the human makeup.

Because tools were found associated with the fossils of some very early members of the genus *Homo*, one species was named *Homo habilis* (from the Latin for "able man"). *H. habilis* fossils have been found widely in Africa, and the species lived for about half a million years (Figure 24.7). The species was a little larger than *Australopithecus* and, like its apparent ancestor, it still regularly climbed trees.

About 1.7 million years ago, another species of human, *Homo erectus* ("upright man"), also appeared in Africa. About 1 million years ago, *H. erectus* emigrated into Europe and Asia, and this species persisted until about 500,000 years ago. Almost as tall as modern humans, *H. erectus* had a brain size (700–1300 cm^3) about twice that of *Australopithecus* and *H. habilis*. The skull was larger and heavier than that of modern humans, and *H. erectus* possessed massive teeth and prominent brow ridges. *H. erectus* was apparently a good hunter, ate mostly meat, and built campsites. Although *H. erectus* "domesticated" fire, it may have

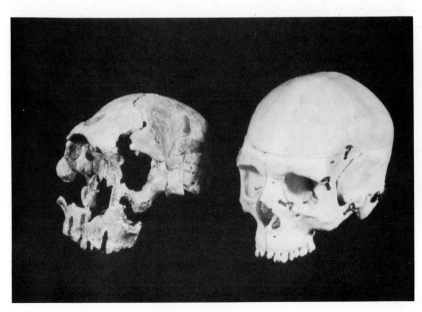

FIGURE 24.6

The evolution of tool use. What trends occur in the tools? How do you think the tools were made?

Age (years)

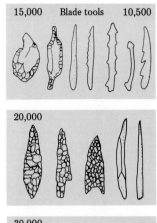

15,000   Blade tools   10,500

20,000

30,000

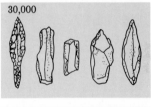

40,000   Flake tools

} *Homo sapiens*

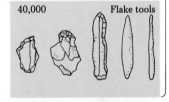

45,000   Flake tools

75,000–150,000   Core tools

} *Homo sapiens*
  Homo erectus

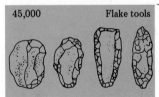

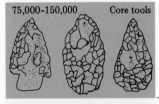

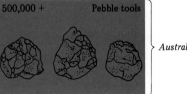

500,000 +   Pebble tools

} *Australopithecus*

FIGURE 24.7

Comparison of the skulls of *Homo habilis*, the handyman, and *Homo sapiens*, the wise man.

been a "fire gatherer" rather than a fire maker. Fire permitted use of a wider variety of food and probably allowed larger populations to grow. Caves were occupied, but it is not known whether other permanent homes were built or whether clothing was worn. At least a rudimentary form of language was probable. The species lived longer in Asia than in Africa, and it includes the famous "Java man" and "Peking man" that lived about 250,000 years ago.

At some time within about the last 500,000 years, *Homo sapiens* ("wise man") originated in Africa or Eurasia. Today, no other animal species is as widely distributed as humans, with the possible exception of body lice and similar organisms that "tag along" with humans wherever they go.

One extinct early race, the **Neanderthal** humans, lived in Europe, Asia, and parts of Africa from about 70,000 to 32,000 years ago. The Neanderthals were the classic cavepeople; they were short, with heavy brows, retreating foreheads, large jaws with heavy teeth, stocky bodies, a lumbering gait, and stooped posture. According to some, the latter trait may have been caused by severe rickets due to low sunlight during the Ice Ages. The Neanderthals wore clothing, were good hunters, and made diverse tools. They also practiced crude surgery, kept records, created some art, and conducted formal burials.

The Neanderthals were replaced relatively quickly by another group of *H. sapiens,* the **Cro-Magnon** humans, who left behind their artistic legacy deep inside caves (Figure 24.8). They were apparently the first to make tools of bone, antler, and ivory.

The major changes in hominids during the past few hundred-thousand years were in the brain and skull traits. Mainly, the brain enlarged to about 300 cm^3 larger than in *H. erectus,* to an average of about 1400–1450 cm^3. Of course, brain size is only a crude measure of an organ whose complexity should be studied microscopically and biochemically. Whales and elephants have larger brains than humans, although not in proportion to their total body sizes.

Although many discovery sites for early *H. sapiens* are in Europe, it is unlikely that humans originated in a small geographical area (Figure 24.9). Early humans were probably distributed widely in breeding communities in Europe, Asia, Africa, and India, and the separation of modern humans into geographical races probably occurred quite early. Also, various Asian populations moved to Pacific islands and to Australia. At least 13,000 years ago, when the ice from the last glacial period was retreating, some Asian groups also moved across the Bering Strait to reach the Americas.

HUMAN RACES

Modern humans belong to one species, and can and do interbreed, therefore sharing a common gene pool. However, humans live in various populations separated from others by certain physical and/or cultural barriers. Over time, with more breeding occurring within rather than between populations, some groups have come to differ physically from others. Human races apparently originated in basically the same way as did the races of other species.

When human populations were very small and the barriers between them more important, **genetic drift** could have occurred by chance alone. An example of how this occurs is provided by the Dunkers, a devout Protestant group who immigrated to Pennsylvania from Germany in the early 1700s. Marriage outside the church is not allowed. One American

FIGURE 24.8
Paintings by Cro-Magnons are found in caves, usually deep inside, at numerous European locations. The paintings were made over a period of about 20,000 years, and some are only 8,000–10,000 years old. Many depict the large game animals that were familiar to our immediate human ancestors.

FIGURE 24.9
Distribution sites for modern humans and various ancestors.

Dryopithecus and *Pliopithecus*
Apes
8–18 million years ago

Australopithecus
Human-apes
2–5 million years ago

Homo erectus
Early human
400,000–700,000 years ago

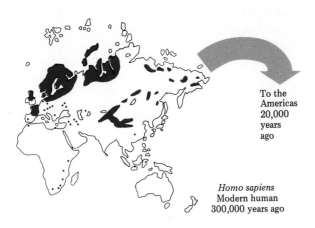

To the
Americas
20,000
years
ago

Homo sapiens
Modern human
300,000 years ago

(A)

(B)

(C)

Dunker community consists of about 300 people, and its size has been nearly the same for several generations. This makes it an ideal population to look for the effects of genetic drift. Table 24.4 compares the frequency of various genes for blood groups among Americans in general, the Dunker community, and their homeland population.

Genetic drift affects the frequencies of apparently nonadaptive traits when the number of human parents in any one generation is a few hundred or less. Before the development of agriculture about 11,000 years ago, most human populations were probably within this size range. Prehistoric hunters and gatherers probably lived in small groups that ranged in size from a few families to tribes of a few hundred people.

Today, most human populations are large and there are far fewer major geographical and cultural barriers between them. In large populations when individuals mate at random, the frequencies of genes not influenced obviously by natural selection tend to remain constant from one generation to another. Chance is the main factor that determines the frequencies of such genes. In large populations, chance changes in gene frequencies tend to balance one another.

Two types of human variations are *individual* and *geographic*. The individual variations among members of a population (polymorphism, or intrapopulation diversity) may be continuous (e.g, weight) or discontinuous (e.g., either-or traits such as blood types). Although each population shows some polymorphism, the polymorphisms within the populations of Sri Lanka, Ecuador, Yugoslavia, Canada, and Morocco all differ, or show geographic variation. Typically, the closer populations are geographically, the more similar are their polymorphisms. Species that show interpopulation diversity are called **polytypic,** and the subgroups within a polytypic species that differ in the incidence of certain genes are called *races* or *subspecies*. The number of races within a species is often open to debate, and the human species is no exception. Some biologists recognize three races, whereas others describe 30 or more (Figure 24.10).

In the past, human races were described mainly on the basis of physique, skin color, hair texture and color, and other external characteristics. Now, internal traits such as blood types and the chemical nature of proteins are also used (Figure 24.11). However, at the biochemical level the differences among human races are usually less pronounced than the external traits. Whether external or internal traits are considered, the intrapopulation variation generally exceeds the average interpopulation differences.

Two major factors involved in race formation are isolation and selection. In humans, cultural isolation may be as significant as geographical isolation. Isolated populations may be subject to different selective pressures and in time may come to differ from other populations. Because "pure" human races do not exist, races are usually defined statistically, and no person is "typical" of his or her race in all traits. Yet,

TABLE 24.4
Some Comparative Frequencies of Genes for Blood Types

Population	Blood Type Alleles (%)		
	OO	AA	BB
Dunkers	60	38	2
West Germans	64	29	7
Americans, generally	70	26	4

(A)

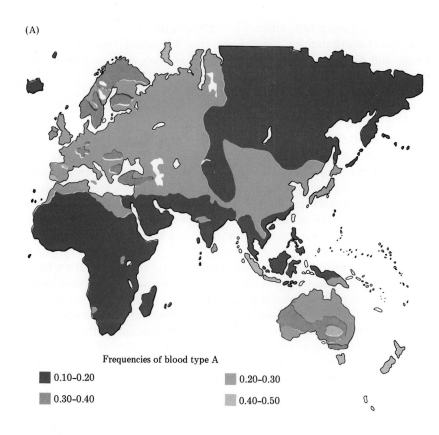

Frequencies of blood type A

■ 0.10–0.20   ▨ 0.20–0.30

▦ 0.30–0.40   ▨ 0.40–0.50

FIGURE 24.11
Internal traits used in classifying human groups include blood types (A) and blood proteins (B). The distribution of blood type A is estimated to be that existing before the year 1500. The blood protein varieties shown in (B) are of albumin.

(B)

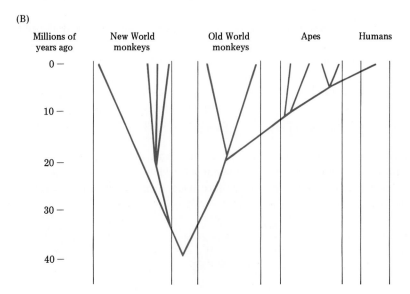

racism often uses the erroneous view of a "type specimen" for each race. Obvious to many, but not all, is the fact that human races are not equivalent to language, cultural, or religious groups or to social status groups.

Today, geographic variation in humans is largely irrelevant in terms of natural selection. With the aid of technological and medical advances, most humans may now adapt to their environment by behavior rather than by physical strength, for example. The biological basis of behavior is in the brain and is little affected by natural selection on most superficial external traits. Relatively few humans now live under "natural" conditions, and members of any race are about as likely to reproduce in one environment as in another.

SKIN COLOR

Variations in different populations of the same species result from the interaction of the local environment and the gene pool of the population. The environment determines which genes among the range in the population's DNA pool will be expressed and to what degree. One trait often used in designating human races that is affected by the environment in a significant way is skin color. Many dozens of human skin colors have been described, ranging from almost white to jet black (Figure 24.12).

In recent centuries, exploration, exploitation, and colonization have "blurred" many human traits through racial mixing. Prior to that, most darkly pigmented people lived around the equator, and lightly pigmented people lived elsewhere (Figure 24.13). The degree of pigmentation regulates the amount of calciferol (vitamin D) synthesized by the action of ultraviolet (UV) light. It has been suggested that skin color is related mainly to the amount of light in the environment. By screening out many UV rays, dark skin pigments prevent the excessive synthesis of calciferol. Light-skinned individuals would be favored in temperate areas where there is less intense light for the synthesis of normal amounts of calciferol. The only northerly humans with dark skin are the Eskimos. However, because native Eskimos ate fish livers (once their only dietary source rich in vitamin D), they apparently got adequate calciferol. Although this theory is attractive in many ways, further work is needed to confirm it.

Because *H. sapiens* evolved first in the tropics, it is likely that early humans were darkly pigmented. Natural selection may have favored dark skin not only in terms of calciferol synthesis but also because a dark color may have provided better camouflage against human and nonhuman enemies.

RECENT AND FUTURE HUMAN EVOLUTION

Humans have changed very little physically during the past 40,000 years. Facial features have become somewhat more delicate, and the size of the molars has decreased. However, there have been no major changes in

FIGURE 24.12
It is difficult for a modern drugstore to stock flesh-colored Band-Aids for all its customers.

Lightest
Medium–light
Medium
Medium–dark
Darkest

FIGURE 24.13
Worldwide distribution of native skin pigmentation. What are the main functions of calciferol, and what are symptoms of deficiencies and excesses? Where on Earth would you expect the highest and the lowest levels of ultraviolet light? What calciferol-related problems might a light-skinned person have in the tropics? A dark-skinned person in temperate regions?

cranial size or skull contours. The most important changes have been behavioral rather than physical. See the essay "Of Humans and Boats."

One of the most important behavioral changes occurred about 11,000 years ago, when agriculture developed independently in the Near East, Asia, and Latin America (Figure 24.14). As the knowledge spread outward, more and more populations gave up or modified their hunting-and-gathering life-styles. Another major technological breakthrough was the development of metallurgy about 6000–7000 years ago.

Will new species of *Homo* evolve in the future? Although we cannot know for certain, many biologists think not, for two main reasons. First, because humans already live in most habitats on Earth, the development of new traits or adaptive radiation following further geographic expansion is unlikely. Second, little geographic isolation and few reproductive barriers exist between populations today.

In the past, mutation, genetic recombination, reproductive isolation, and natural selection were important in human speciation. Although each factor still operates, we also have the anatomical, physiological, medical, and genetic knowledge to chart much of our evolutionary course. For example, many people with certain genetic syndromes once did not survive to produce offspring. Will the fact that some such individuals now reproduce and perpetuate their deleterious genes threaten human existence? It appears that with proper medical care, there is little such threat to our future (Chapter 10).

One important trend in human evolution has been an increase in brain size. Will human brains continue to enlarge? Again, probably not, but if they do, it could complicate the birth process if this enlargement happened prenatally unless there were also concurrent evolutionary changes in the size of the female pelvis and/or a shorter pregnancy. Also, it is not certain that a larger brain would be of benefit, since we apparently do not use all of the gray matter we already have.

Some have suggested that manual dexterity would be increased by an extra thumb, and peripheral vision improved by protruding eyes. In these cases our expanding technology seems to make such evolutionary "advances" unnecessary, and it is very uncertain that we could successfully apply animal husbandry principles or artificial selection to ourselves

Of Humans and Boats

Legends and science fiction writers have produced many humanlike creatures, sometimes smaller or larger than the "real" thing—elves, titans, leprechauns, nymphs, and fairies. Centaurs, harpies, mermaids, minotaurs, satyrs, sphinxes, and tritons are only partly human.

In today's world of space exploration, we often hear speculation concerning intelligent life elsewhere in the universe. Would such life be like us (except perhaps for such details as more pointed ears, and bushy eyebrows that point up, as depicted in some TV programs and movies) or unimaginably exotic? If it exists, the important thing for such life, as on Earth, is to survive and reproduce.

Some who suggest that intelligent life elsewhere would be very similar to humans point to the analogy of boat design. They argue that bowl-shaped or saucer-shaped boat designs would soon be abandoned for most purposes, as would square or star-shaped boats. The conclusion is that "regular shaped" boats would result wherever they were invented not because of coincidence but because they work. Do you agree with this position? What forms of intelligent life do you think might exist elsewhere in space?

FIGURE 24.14
Sheep and goats have long been domesticated in the Near East and southwestern Asia. Domestic animals and the rising human population made possible by agriculture, other technological breakthroughs, and better medicine and health had, and continue to have, a devastating effect on Earth's habitats and on its other inhabitants.

anyway. However, the idea that humans be given genes to produce a multicompartmental stomach (a cow has four) may have more merit because we might then digest the cellulose that makes up so much of plant material, and thereby perhaps eliminate most human hunger. Technology has led to increased food production, but the population is growing almost too rapidly for that to make much difference on a per capita basis. Immense numbers of humans are chronically hungry, millions starve annually, and we are in the midst of the greatest famine in human history in terms of the numbers of people affected.

Although natural selection still operates on humans, most populations are not challenged physically today as they were in the past by such agents as disease and temperature extremes. As the cartoon character Pogo said, "We have met the enemy and he is us." Perhaps our evolutionary future depends most upon our ability to manage Spaceship Earth and its inhabitants (Chapter 25).

SUMMARY

1. Humans belong to the mammalian group called **primates,** specifically **anthropoid** primates, which also includes monkeys and apes. Humans and their immediate ancestors are called **hominids.** Humans share many external and internal traits with the other anthropoids but we also differ from them in several important respects.

2. Other living primates include the **prosimians,** such as lemurs and lorises. The ancestor to primates may have had generalized traits similar to those of the modern tree shrews.

3. Early mammals and birds evolved from reptilian ancestors while the dinosaurs reigned during the Mesozoic era. Upon the extinction of the dinosaurs, mammals became the dominant animals, and they now occupy most terrestrial, aquatic, and aerial habitats. Humans and dinosaurs did not coexist.

4. Both humans and modern apes evolved from a common, extinct, insectivorous mammalian ancestor. The fossil record of humans has been studied extensively, but it is still incomplete and many of our notions about our evolutionary history are only inferences.

5. Some of the earliest hominids were species of *Australopithecus,* which apparently spent part of their lives walking upright in a savanna environment. Members of the genus *Homo* appeared about 2 million years ago in Africa. *H. habilis* used tools and *H. erectus,* which was still larger, perfected tools further and also domesticated fire.

6. *Homo sapiens* originated about 500,000 years ago. The stocky **Neanderthals,** an early race, became extinct about 32,000 years ago and were replaced by our immediate ancestors, the **Cro-Magnons.**

7. Throughout human evolution there has been a general increase in brain size. Among other effects, this has resulted in humans being born at a very immature stage. The long period of dependency, however, provides the time during which culture is transmitted.

8. Humans spread to all habitats on Earth and became very **polytypic.** When populations were small and when geographical and/or cultural barriers to interbreeding were of major importance, **genetic drift** may have accounted for some variation.

9. Skin color is one of many external traits used to designate human **races.** Some authorities recognize three human races (e.g., Mongoloid, Negroid, and Caucasoid) while others suggest a far larger number. One theory postulates that skin color is an adaptation to calciferol production in various levels of ultraviolet light. Increasingly, internal as well as external traits are used to designate human groups.

REVIEW QUESTIONS

1. Distinguish among primates, prosimians, anthropoids, and hominids.

2. Summarize the major ways that humans resemble and differ from anthropoid apes.

3. Distinguish among the traits of egg-laying, marsupial, and placental mammals.

4. What was the approximate time span between the occurrence of *Australopithecus* and the origin of the genus *Homo*? Between *Australopithecus* and *H. sapiens* specifically?

5. What major trends have occurred in humans during their evolution?

6. Describe how you would infer culture from the typically small fragments of human bones, teeth, and artifacts.

7. What traits in a fossil would have to be present for you to call it a human?

8. Describe some examples of cultural and geographical factors that separated humans in the past and, to a lesser degree, continue to do so.

9. Describe how the calciferol-ultraviolet light theory is used to explain variations in human skin color.

10. How many human races do you recognize, and what traits are most important in your classification system?

11. In what ways do you think humans will change in the future?

CHAPTER 25

HUMAN ECOLOGY: LIVING IN THE ENVIRONMENT

The biosphere is that portion of Earth's land, water, and gaseous envelope where life occurs. It is a thin and fragile layer, seldom more than 16 km (10 mi) thick, and continued life therein depends upon our understanding of ecology, the interrelationships between organisms and their living and nonliving environments. This chapter examines the major principles of ecology, the habitats occupied and used by humans, and the role of humans within the biosphere.

Knowing where we are going is as important as knowing where we came from. We can only investigate and speculate about our past, but theoretically, our future is under our control. However, to plan for continued survival on Earth, we must have a firm understanding of life on Earth and, especially, of the effects of our actions on it and on ourselves.

FIGURE 25.1

Levels of organization of life. Which levels are of particular interest to an anatomist and physiologist? To an ecologist? What analogies do you see between cells and organisms, between tissues and populations, and between organs and communities?

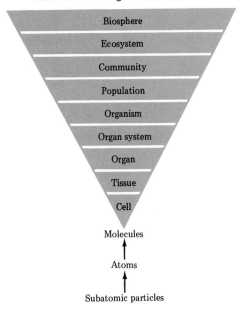

Biosphere

Ecosystem

Community

Population

Organism

Organ system

Organ

Tissue

Cell

Molecules

↑

Atoms

↑

Subatomic particles

ECOSYSTEMS

Life can be examined at many levels, and most chapters in this text deal with the "lower" levels, such as cells and organs (Figure 25.1). Most ecologists are also interested in cells, tissues, and organs but their particular concern is with the "higher" levels of organization. A **population** is a group of interacting organisms of the same species, and a **community** is an interacting group of different kinds of organisms. The functional unit in ecology is the **ecosystem** because it includes all of the interactions of communities with both their living (biotic) and their physical (abiotic) environments. Ecosystems may be large and natural, such as deserts, forests, grasslands, and lakes, or small and artificial, such as an aquarium or a potted houseplant.

The sun is the source of energy for nearly all ecosystems (Chapter 1). Through the process of photosynthesis, autotrophs such as green plants not only produce their own food but also release oxygen (Figure 25.2). Thus, heterotrophs such as animals and decomposers such as most bacteria and fungi depend upon autotrophs for both nutrients and oxygen. Because of this important role in ecosystems, autotrophs have traditionally been called **producers,** although they might better be called *converters.* **Consumers** include the animals that feed upon plants, the **herbivores;** those that eat flesh, the **carnivores;** and the **omnivores,** which eat both producers and consumers.

ENERGY FLOW

Implied in the description above is the concept of **food chains** (Figure 25.3). However, most consumers eat more than one kind of prey, and in turn are eaten by more than one kind of predator. Therefore, ecosystems contain many **food webs,** by means of which the lives of many kinds of organisms are intertwined (Figure 25.4).

Not all the energy "produced" by a green plant in its lifetime is passed along to the consumers that eat it. Obviously, much energy is used by the plant in its growth, development, and metabolism, and some is lost to the environment as dead leaves, roots, stems, or flowers. Similarly, not all of the energy taken in by a herbivore during its lifetime is available to the carnivore that consumes it. Ecologists often use the **ten percent principle** to generalize about this energy flow. In other words, only about 10 percent of the energy incorporated into an organism at one feeding or **trophic level** is passed along to the organism that consumes it (Figure 25.5). This explains why so little energy is available beyond the fourth or fifth "link" in a typical food chain.

Americans and many others in developed nations eat "high" on food chains, a luxury most humans on Earth cannot afford. Often, the animals that eventually will be eaten by humans eat more grain than ultimately would feed humans directly. Also, most of those in developing nations

FIGURE 25.2

(*above*) Some basic processes in an ecosystem. Animals utilize the oxygen produced by green plants during the process of photosynthesis. In turn, autotrophs utilize the CO_2 released by heterotrophs. The energy at one feeding level may be used by organisms at higher trophic levels.

FIGURE 25.3

(*right*) A food chain. Name examples of herbivores, carnivores, and third-order consumers. What organisms are "consumers" of all of the above? Why are the links in a food chain usually limited to five or less?

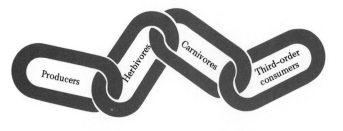

Producers Herbivores Carnivores Third-order consumers

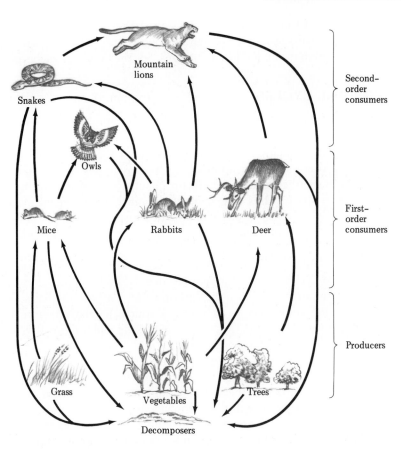

Second-order consumers

First-order consumers

Producers

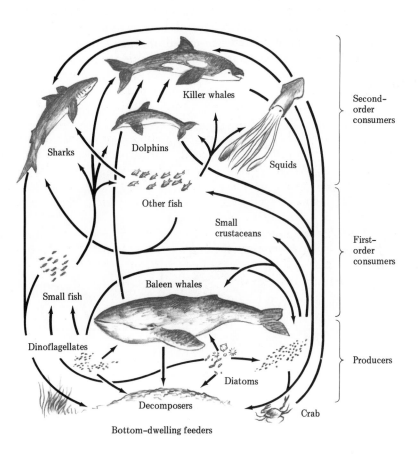

Second-order consumers

First-order consumers

Producers

FIGURE 25.4
A terrestrial and an aquatic food web.

FIGURE 25.5
This energy pyramid in an aquatic food chain illustrates the ten percent principle.

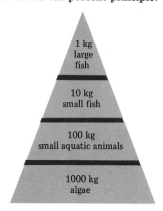

1 kg
large
fish

10 kg
small fish

100 kg
small aquatic animals

1000 kg
algae

can ill afford to maintain many pets. American dogs and cats, which, like humans, also usually feed high on food chains, may well eat better than many humans in developing nations. See the essay "Algaeburgers?"

MINERAL CYCLES

Energy flows but does not cycle through ecosystems. That is, without a constant input of energy from the sun, the power of life, most ecosystems would collapse quickly. It is a different story for the material substances that life processes depend upon. Earth has essentially a finite quantity of minerals, neither gaining significant quantities from nor losing large amounts to space. Thus, the minerals in most living organisms were once in other organisms, and when the organisms die, the minerals will return to the environment or become incorporated into still other organisms.

Although minerals cycle incessantly between the living and nonliving worlds, the cycle of each has unique details. Figure 25.6 shows one extremely important pattern, the **carbon cycle.** Carbon, the basis for all organic compounds, has as its ultimate source a relatively rare compound, carbon dioxide (CO_2), which makes up only about 0.03–0.04 percent of the atmosphere.

Many mineral cycles occur quite rapidly, such as some in lawns and grasslands. As the vegetation dies back each year, decomposers act quickly to break down the material and make constituent minerals available for further growth the following year. In other cases, minerals may remain unused by organisms for long periods. For example, most **fossil fuels** are the remains of organisms that died about 300 million years ago. Until the petroleum, coal, or natural gas is mined and used, the carbon it contains is largely unavailable for use by other organisms. Additionally, huge quantities of minerals lie in marine sediments for very long periods, recycling only when mined or when vulcanism or crustal movements expose them to weathering processes.

FIGURE 25.6
The carbon cycle simplified. How do your activities affect this mineral cycle?

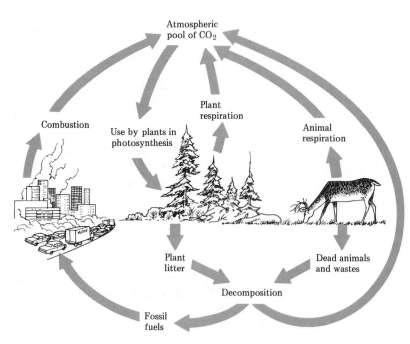

Atmospheric
pool of CO_2

Combustion

Use by plants in
photosynthesis

Plant
respiration

Animal
respiration

Plant
litter

Dead animals
and wastes

Decomposition

Fossil
fuels

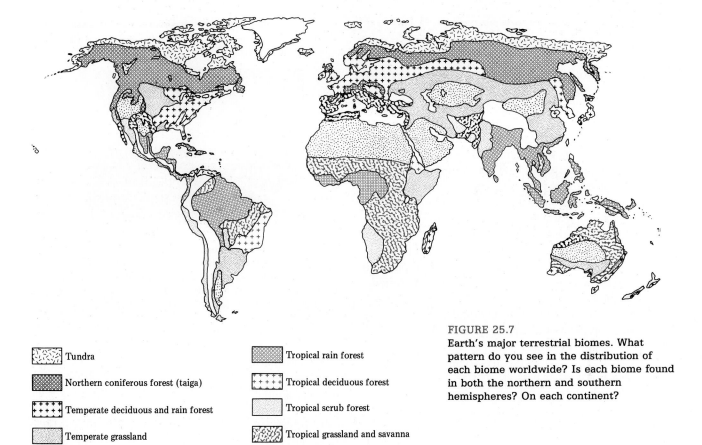

Tundra

Northern coniferous forest (taiga)

Temperate deciduous and rain forest

Temperate grassland

Chaparral

Desert

Tropical rain forest

Tropical deciduous forest

Tropical scrub forest

Tropical grassland and savanna

Mountains (complex zonation)

FIGURE 25.7

Earth's major terrestrial biomes. What pattern do you see in the distribution of each biome worldwide? Is each biome found in both the northern and southern hemispheres? On each continent?

SOME MAJOR TERRESTRIAL ECOSYSTEMS

The major large terrestrial ecosystems or **biomes** are shown in Figure 25.7. However, to paraphrase a familiar saying, no ecosystem is an island unto itself. A forest may contain meadows, lakes, bogs, and rivers, each of which may be viewed as an ecosystem. Also, each ecosystem is partially dependent upon what is happening in other ecosystems. Because humans live in, travel to, and use materials from all biomes, it is important to understand biomes in order to act responsibly in today's interdependent world.

Notice in Figure 25.7 that many biomes are named after the makeup of the major plant community. This is done not only because plants are so relatively easy to observe but also because to a high degree, plants determine the nature of the animal life in each region.

Somes biomes are far larger than others, although size is not always an index of importance to humans. For example, **chaparral** is not found extensively in the United States but it is a dominant biome in Southern California, and about 1 of every 11 Americans lives in or near chaparral (Figure 25.8). Many millions more live in or near this biome around the Mediterranean and in the Middle East. Misunderstandings about this biome contribute to millions of dollars worth of damage annually due to fires, floods, and mud slides.

FIGURE 25.8

Many homes in Southern California are in or close to chaparral. Periodic fires in this biome cause much property damage, as do floods and mud slides that result when rains fall onto the bare slopes. The chaparral biome requires fire, although fire conflicts with most human interests in this ecosystem.

Algaeburgers?

Like it or not, many humans may be forced to eat lower on food chains as the number of humans increases and the amount of good agricultural land declines. Already, many people are eating more plant material than they suspect. Soybeans and other high-protein plant products are often used as meat extenders. However, soybean plants are not completely utilized. Like most crops, many parts of soybean plants have no known use, even though some parts can be fed to animals or plowed back into the soil to increase its organic content. But small algae are relatively easy to grow by **hydroponics,** the culture of crops in water without soil (Figure B25.1). Many algae even grow well in polluted water and could be used, for example, to feed astronauts on long space trips.

Although most familiar agricultural plants produce only one crop per year, algae grow quickly and can be harvested regularly. Wheat yields of 2.5–6 metric tons/hectare/year (1–3 tons/acre) are common, and sugar beets yield up to 30 MT/ha/yr (12 t/a). By comparison, algae grown hydroponically have yielded 75 MT/ha/yr (30 t/a).

Protein is essential for human health (Chapter 12). Although the protein content of plant crops is usually

FIGURE B25.1

These traditional crops are being grown hydroponically—that is, their roots are immersed in water with optimal levels of minerals. Temperature and light are also controlled carefully in such experimental facilities. How economical hydroponics will be on a large scale remains to be seen.

lower than that from animals, growing plant protein is far more efficient than growing animal protein. For example, annual yields of soybean protein average 2.5 MT/ha/yr (1 t/a) compared with less than 0.2 MT/ha/yr (100 lb/a) from farm animals. Algae yield up to 30 MT/ha/yr (12 t/a). Is it becoming unethical to eat steak?

Arid Biomes

The driest **deserts** receive less than 25 cm (10 in.) of precipitation annually, often in a few sudden downpours (Figure 25.9). Although some high-altitude deserts are quite cold in winter, most deserts are warm to hot during the day, with large temperature decreases at night.

Desert plants often have either shallow roots to tap sparse rainfall before it evaporates, or extensive, deep roots; some have both root types. Water loss is reduced by the plants' having small, light-colored leaves and/or spines to inhibit animals from eating them. Many plants complete their life cycles very quickly during the occasional, short, favorable seasons, and their seeds or spores lie dormant for years until proper conditions return.

Desert animals also face many temperature and moisture challenges to homeostasis. Like many plants, desert animals are often light colored and either leave during the harshest season or become nocturnal or dormant. Some desert mammals are excellent examples of Bergman's and Allen's principles (Chapter 2).

The generally regular winter rains allow chaparral plants to grow somewhat larger and closer together than desert plants. However, many adaptations of the chaparral shrubs, small trees, and annual plants are similar to those of desert plants.

Some arid cities in the Sunbelt have undergone very rapid population increases in recent decades. This has been true for all of Southern California and for cities such as Las Vegas, Nevada; Phoenix and Tucson, Arizona; Albuquerque and Las Cruces, New Mexico; and Dallas, El

FIGURE 25.9
Typical desert scene in the southwestern United States. Ironically, many features of deserts are formed by water erosion even though total precipitation is very sparse. Note the wide spacing of the plants and their generally small, light-colored leaves.

Paso, Fort Worth, and San Antonio, Texas. Local water supplies are often inadequate, and the water tables and underground aquifers are often reduced, sometimes requiring massive and expensive water importation programs. Lawns are outlawed in some areas, and people are urged to use native vegetation for landscaping. See the essay "Pave It, Plant It, or Preserve It?"

Grasslands

Because they are Earth's "breadbaskets," **grasslands** are extremely important to humans (Figure 25.10). Many major crops are grasses, such as wheat, oats, corn, and rice. Due to human use of these areas for growing crops and raising animals, only remnants of native grasslands survive and most herds of large grazing herbivores are gone or reduced.

In North America, the grasslands may be divided roughly into the **plains** and the **prairies.** Grasses in the western plains are generally shorter and more widely spaced than those in the eastern prairies, where richer soil and more precipitation occurs. Of course, many other plants besides grasses grow in these regions. Areas where widely spaced trees occur with grasses are called **savannas,** and it was in such ecosystems that humans apparently originated (Chapter 24). These areas, especially in Africa, once supported very large populations of animals. Drought, habitat destruction, and poaching threaten the few areas that still house some of Earth's most spectacular assemblages of wildlife (Figure 25.11).

Forests

Much of the eastern United States, Western Europe, and eastern Asia is covered by **temperate deciduous forests** (Figure 25.12). Oaks, maples, poplars, hickories, and beeches are familiar examples of trees that shed their leaves in the autumn. Some of the areas where they grow experience harsh winters but warm, humid summers. The 76–152+ cm (30–50 in.) of precipitation generally is spread throughout the year, often with snow in the winter. Because trees lead to significant stratification of light, temperature, and humidity, many microclimates are available to forest organisms, and species diversity is far higher than in arid lands.

Many major cities exist in this forest biome. Humans also use these areas for both summer and winter recreation, and make extensive use of forest products.

The **northern coniferous forest** or **taiga** occurs in a broad band across the northern parts of North America, Europe, and Asia. Both precipitation and average annual temperatures are lower than in the deciduous forests. Some areas are dominated by only one or two evergreen species, such as fir or spruce. Human use of this biome is often limited to mining and timber operations.

FIGURE 25.10
Few relatively untouched grasslands such as this remain in the United States. Most grasslands have long been converted to agriculture.

FIGURE 25.11
On the savanna at the base of Mt. Kilimanjaro in East Africa, one can still see diverse species of large animals. Only through human commitment and diligence will such herds withstand the onslaughts of poaching, habitat destruction, and even tourism.

FIGURE 25.12
Forests: (A) Temperate deciduous forest, (B) taiga, and (C) tropical rainforest.

(A)

(B)

(C)

Pave It, Plant It, or Preserve It?

Think of the enormous amounts of land paved for streets, highways, and parking lots and occupied by homes, stores, schools, and businesses. Sometimes the land involved was "wasteland," if there is waste in nature, that had to be filled or drained prior to construction. However, in many cases the easiest land to build on is high-quality agricultural land because it is already cleared and relatively flat. Thousands of hectares of such land are lost daily in the United States alone. Many farmers want to keep their land but sometimes lose it to a highway construction project, or because of high taxes or attractive offers by developers. American farmers make up less than 0.1 percent of Earth's population, yet their crops feed about one-fourth of the world's people. How should we preserve prime agricultural land and keep a sufficient amount of land undeveloped for use as parks, greenbelts, or natural areas?

In many societies throughout history, women, children, aliens, the elderly and infirm, and various minorities had few if any legal rights. Although such is still the case in many cultures, today many of these groups and their supporters have won basic rights. It seems a natural and noble thing to do if all people are truly to become equal. However, what are we doing about birds and trees and lakes and oceans, which cannot speak for themselves about the "insults" of pollution or habitat destruction? Do they have legal rights or is this only a human privilege?

In his book *Should Trees Have Standing?* (William Kaufman), Professor Christopher D. Stone argues that we should give legal rights to organisms and environments, as we do to corporations. Humans might sue for damages or seek an injunction against actions detrimental to the environment or organisms. Does an area of rock polished by a glacier have a right not to be destroyed by highway construction, and does fragile tundra vegetation have the right not to be destroyed by human hikers or oil pipelines? What about the right of a redwood tree to die naturally rather than at the hands of a chainsaw? Do waters, air, and land have legal bases to remain free from toxic wastes?

What is your opinion on Stone's thesis? If you agree with it, join an environmental organization and act on your beliefs.

FIGURE 25.13
Slash-and-burn practices have long been used in the tropics by small, scattered populations of humans who have moved on to other areas after a year or two. The generally poor soil rarely supports extensive long-term agriculture, and the second-growth jungle that grows on abandoned lands is often quite different from the original forest.

Tropical rainforests occupy enormous areas but are often sparsely settled by humans. The diversity of life is high and astounding, in part due to the abundant precipitation, high temperatures, and stratification. Human use of rainforests is increasing rapidly, sometimes for timber trees and mineral resources but often simply for the land itself. Small groups of humans have long used a slash-and-burn technique of agriculture and moved from one area to another every few years (Figure 25.13). Now, however, because of the high rate of population growth in many tropical countries, vast areas of rainforest are being cleared. When not done properly, this deforestation results in major and perhaps irreversible changes in soil and nutrient cycles that make such areas unfit not only for long-term agriculture but also for the regrowth of the original type of forest itself. Technically, a **jungle** is a second-growth rainforest. Some scientists are very concerned about deforestation changing Earth's weather patterns and the oxygen and CO_2 content of the atmosphere.

Tundra

This biome, common in the far north and in many high-altitude locations, is characterized by long, cold winters and a cool, short growing season (Figure 25.14). Because the ground remains frozen much of the year in the Arctic (permafrost), the area is very wet during the summer even though precipitation is relatively low. Although hardy, the lichens, mosses, annuals, and small shrubs of the tundra are also fragile.

Disturbances are very slow to heal, much as in arid lands, a fact humans would do well to remember as they use tundra regions increasingly for oil, other minerals, or recreation.

AQUATIC ECOSYSTEMS

Being air breathers, humans obviously do not live in water, but we all depend upon water in numerous ways. Although we hear about droughts in one area and floods elsewhere, Earth's water supply is essentially finite. While we may not run out of water, we often use it for too many purposes in particular locations. Most of Earth's water is salty, and one of our largest problems is keeping the far smaller quantities of fresh water unpolluted for use and reuse.

As in terrestrial ecosystems, sunlight is a **lake's** source of energy. In the upper (euphotic) zone, photosynthesis occurs and the oxygen released by the process becomes dissolved in the water. At a certain depth (compensation point), the amount of oxygen used by cellular respiration of all organisms present equals the amount of oxygen produced by autotrophs. Below that depth most organisms are heterotrophic, depending upon the "rain" of organic material falling from above.

Temperature has a profound effect upon both organisms and their environments (Chapter 2). Being most dense at about 4°C (30°F), water at that temperature sinks beneath water that is either colder or warmer (Figure 25.15). As a result, many lakes undergo regular seasonal changes, including the spring and fall **overturns** that mix oxygen and nutrients.

FIGURE 25.14
Tundra is found in both high altitude and high latitude locations. Tundra plants are small, species diversity is relatively low, and many organisms must either complete their life cycles quickly, leave during the winter, or become dormant for long periods.

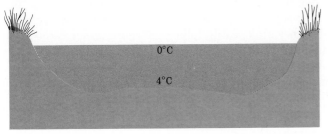

Winter

FIGURE 25.15
Many temperate lakes undergo two overturns annually. What effects might overturns have upon lake ecology?

Spring and fall overturns

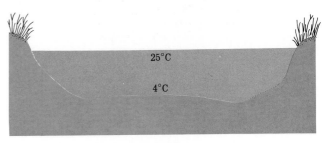

Summer

Just as forests may be called young or mature, lakes also age (Figure 25.16). When young, many lakes are clear, but because there are few nutrients these lakes do not support much life. As lakes age, the quantity and diversity of life generally increase as sediments and organic matter accumulate. Eventually, lakes may be too nutrient rich and oxygen poor and, at that stage, are called **eutrophic.** Although it is a natural process, eutrophication is being hastened by such human activities as allowing sewage, detergents, and fertilizers to enter lakes. In the water, these substances accelerate the growth of certain algae and waterweeds. These growths decrease light penetration and dissolved oxygen, add to the undecomposed sediment when they die, interfere with recreational use of the lake, and often produce unpleasant odors and tastes. Perhaps more water pollution problems are caused by cultural or human-induced eutrophication than by the addition of toxic pollutants.

POLLUTANTS IN ECOSYSTEMS

Most people today have a generalized awareness of the threat of poisonous substances such as many industrial wastes, pesticides, heavy metals such as mercury, oil, and other chemicals entering lakes, rivers, the ocean, or underground water (aquifers). Although some of these chemicals are toxic enough to kill many organisms on contact, others become involved in **biological magnification.** Whereas energy decreases with each transfer in a food chain, the concentration of various chemicals may increase (Figure 25.17). Most organisms cannot excrete pollutants as quickly as they breathe or ingest them. For example, birds at high trophic levels have suffered reproductive failure or have laid eggs with shells so thin that they broke easily, killing the young. Some carnivorous mammals accumulate some toxins in their fat deposits and die slowly as they use these lipid stores during hibernation. Because humans are rather high on many food chains, there is a good chance that many of our wastes will "return to haunt us" physiologically.

NICHES AND HABITATS

Each species occupies a different **niche** in its **habitat.** An organism's niche may be thought of as its "profession," and habitat as its "address," or a description of where it lives. In this sense, niche is the role the organism plays in its ecosystem, such as what it eats and what eats it. In human terms we might think of the hypothetical Smith family of 229

FIGURE 25.16

Eutrophication. (A) A eutrophic lake and (B) some stages in the process of eutrophication. How does cultural eutrophication differ from natural eutrophication?

(A)

(B)

(1)   (2)

(3)   (4)

Main Street. Although all the Smiths share one habitat (address), the parents occupy different niches (interactions with their environment) from each other and from the various niches of their children.

It is generally predicted that if two related species were to occupy the same niche, then they might compete for many of the same resources. Because the two species would not be equally adapted, one would eventually replace the other or their niche requirements would change to allow for coexistence. For example, the organisms might come to occupy slightly different habitats, eat differently, or be active at different times.

We often think of interactions between organisms as contestlike, as when we discuss "competition" for survival between a coyote and its main prey, rabbits. Certainly a "race" of sorts for survival may occur but not in the same terms as in human athletic competitions. However, it follows from the concept of niches that the greatest competition occurs between closely related organisms. Thus, coyotes compete mainly with coyotes and rabbits mainly with other rabbits. Similarly, the greatest threat to human existence is human behavior.

BUILT ENVIRONMENTS

Most people live in built environments most of the time. Perhaps 80 percent of humans live in the **urban ecosystem** on less than 2 percent of Earth's land. Maybe that is why it is so nice to get away from it all when we can. However, many modern campgrounds and resorts are about as crowded, noisy, and polluted as the environments left behind. In addition, the demand on many wilderness areas is so great that special permits to enter are required in order to maintain the area as wilderness.

Urban environments have been developing for at least 11,000 years and, like humans themselves, know few climatic or geographic boundaries. As populations and cities grew, so did urban sprawl, producing metropolises. Today, some metropolitan areas are growing together to form **megalopolises.** Several megalopolises in the United States have been given such tentative names as *Chipitts* (Chicago-to-Pittsburgh) and *Bowash* (Boston-to-Washington).

If cities are ecosystems, they are most unusual in that they have few producers and decomposers relative to the large number of consumers. Most materials needed by populations such as humans, starlings, rats, cockroaches, and pigeons must be imported. For example, enormous quantities of water, food, and fossil fuels must enter cities daily. This is in addition to immense quantities of other materials needed to construct and operate homes, offices, schools, businesses, factories, and transportation systems. In using these materials, cities produce huge amounts of trash, industrial waste, air pollutants, toxic materials, and sewage. According to some estimates, the pets alone in a typical city of 1 million humans produce over 500 metric tons of feces and 3 million liters of urine daily! Obviously, all of this strains other ecosystems quite distant from the city.

Although most materials are recycled in nature, recycling is the exception in cities, and this makes urban areas quite unstable. Most wastes—newspapers, bags, boxes, bottles, cans, aluminum foil, cellophane, tires, plastic, and many "disposable" objects—are exported, further straining other ecosystems (Figure 25.18).

While some people are managing and attempting to improve old cities, others are planning or building new ones, and Figure 25.19 shows two of many ideas for the design of new communities. See the essay "Asphalt Jungles and Favorite Places."

FIGURE 25.17
Biological magnification illustrated by the concentration of pesticide (dots) in a food chain. The quantity (biomass) of organisms at each trophic level is indicated by the size of the rectangles.

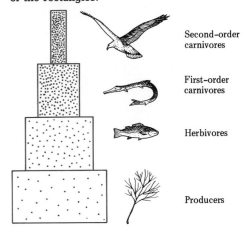

Second-order carnivores

First-order carnivores

Herbivores

Producers

FIGURE 25.18
Many people are "down in the dumps" about the enormous problem of dealing with solid waste. Where do the things you use daily come from? (Looking at the labels on products may help you answer this question partially.) How do you affect ecosystems elsewhere in the country by using these products? Elsewhere in the world? What happens to water after you have used it? To your fecal, urinary, and other wastes? What is the environmental consequence of your use of energy?

Those in developed nations have an impact on Earth 30–50 times that of individuals in developing nations. Does your very existence threaten the survival of some humans elsewhere? How could you reduce your personal impact on your immediate environment? On other ecosystems?

FIGURE 25.19
Two plans for new communities. What are the advantages and disadvantages of each? What other plans can you suggest?

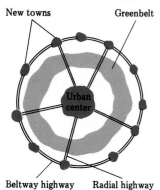

Ring-of-cities plan

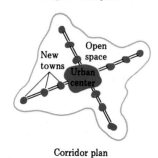

Corridor plan

TYPES OF ORGANISM-TO-ORGANISM INTERACTIONS

As shown in Table 25.1, predators and parasites harm their prey and hosts, respectively. Other types of interactions may benefit one species but not harm another species, or both interacting species may benefit.

Predation

Many people have mixed feelings while watching a house cat catch a bird or while seeing a film of a cheetah bring down a "poor little antelope." Others shrug and accept the fact that "that's nature" and that all animals eat something and are, in turn, consumed by something. While some side with the prey ("underdogism"?), others feel that many **predators** are overly persecuted. Many species of large wild cats and dogs are, in fact, in danger of extinction. Actually, we humans are Earth's major predators.

By eating mainly slow, young, old, injured, and deformed prey, predators actually help keep the prey population healthy. That is, healthy, prime prey individuals generally escape predation, whereas those animals least likely to provide "good" genes to the population often succumb to predators. Similarly, sickly or abnormal predators are less likely to survive and reproduce than healthy and normal animals.

Parasitism

Ticks and fleas (ectoparasites) and tapeworms and liver flukes (endoparasites) are not the most appealing of animals to most people. However, **parasitism** is a successful life-style and there are hundreds of thousands of species of parasites. Some parasites kill their hosts, especially if the host is weakened by disease or malnutrition. However, because the death of the host may also mean death for the parasite, there is a selective advantage for parasites who do not "overdo a good thing." To use an economic analogy, parasites live on interest and predators on capital.

Parasites often have complex life cycles that involve more than one host and one environment (Figure 25.20). Combating some such parasites may involve reducing the **vector** or carrier of one stage of the parasite. Thus, killing appropriate snails, the vectors of flukes, may reduce the parasite problem.

Social parasites benefit from other organisms behaviorally rather than by means of physical attachment. The cowbird, for example, builds no nest of its own but lays its eggs in the nests of other species. Eagles sometimes cause ospreys to drop their catch, which they swoop down to capture in their talons. Perhaps this is not regal behavior for our national symbol, but human social parasites do much worse.

TABLE 25.1
Interspecific Interactions

Type of Interaction	Effect on Population Growth and Survival When Two Populations A and B Are Interacting[a]	
	A	B
Predation	+	−
Parasitism	+	−
Commensalism	+	0
Mutualism	+	+

[a] + = positive effect, − = negative effect, 0 = no major effect.

Asphalt Jungles and Favorite Places

Many people, if not most, pay little real attention to their environment, a problem that you can solve partially by trying some of the activities suggested in this essay. For example, walk along city streets for several blocks and observe the built environment carefully. What functions were served by the structures you saw? Was there visual and other homogeneity along the streets? Was the route pleasant to walk along? How do new and old constructions differ? What open space was available and how was it used? What structures were designed for vehicles rather than pedestrians? What would make the street scenes more orderly, memorable, functional, or aesthetically pleasing? How would traveling the same route by bicycle, car, or bus affect your perceptions? What assumptions might you make about the personalities of the people who built, occupy, or are changing the environment you observed?

What attracts people to cities? What are some advantages and disadvantages of living there? What attracts people to suburban or rural areas, and what are the advantages and disadvantages of living in those areas?

Describe your favorite place and explain why you selected it when you share your views with others. Include the shapes, colors, textures, temperatures, odors, and other traits. What environments make you feel relaxed, excited, or depressed, and why? Are your feelings about your environments the same now as they were when you were 5 years old? 15 years old? Will they be the same when you are 65 or 85 years old? What determines whether we feel free, secure, and "at home" vs. feeling constrained, insecure, and formal? Why, as children, did we enjoy clubhouses, tree houses, "caves," "forts," and tents? What is your "personal space" at home, at school, and at work?

Commensalism

You are involved in a **commensalistic** association with a tree when you sit in its shade on a hot day, or even when you stand back and marvel at its beauty. That is, you benefit but with no harm or help received by the tree. Similarly, the tree may simply provide support for a climbing vine or a nesting place for a bird.

Mutualism

As the name implies, both interacting species benefit in **mutualistic** associations. Lichens that are common on rocks and tree trunks are

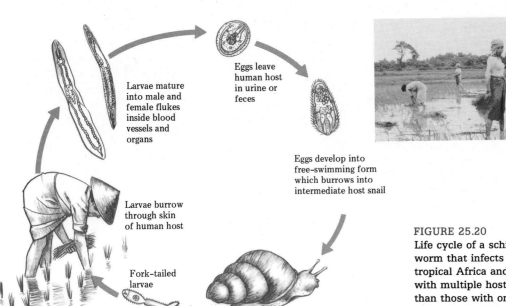

Larvae mature into male and female flukes inside blood vessels and organs

Eggs leave human host in urine or feces

Eggs develop into free-swimming form which burrows into intermediate host snail

Larvae burrow through skin of human host

Fork-tailed larvae

Intermediate host snail

FIGURE 25.20

Life cycle of a schistosome fluke, a parasitic worm that infects millions of humans in tropical Africa and Asia. Why are parasites with multiple hosts more difficult to control than those with one host? Why are endoparasites often more difficult to control than ectoparasites?

FIGURE 25.21
(A) The algae in a lichen are producers whereas the fungal members are consumers. **(B)** This electron micrograph clearly shows the two members of this common mutualistic association.

(A)

(B)

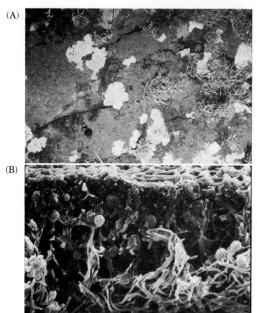

composed of algae and fungi (Figure 25.21). The autotrophic algae produce food not only for themselves but also for the fungi. Because the fungi are typically more resistant than algae to drying and to extreme temperatures and other harsh features of the environment, they thus protect the more delicate algae, allowing both to "live happily ever after."

The associations between many flowers and their various animal pollinators (e.g., bees, hummingbirds, and bats) are also mutualistic. In some cases the relationships are essential or obligate for the two species, but in other instances the organisms can live independently, although sometimes not as well.

POPULLUTION

The term **popullution** was coined to describe the relationship between human population and environmental pollution. In addition to "soiling our nest," the growing human population has resulted in over half of humans being malnourished (Chapter 12). The following common suggestions for combating the hunger problem are not without their own problems, which you should try to identify as each possibility is described. Slowing population growth or even reducing Earth's population is, of course, an obvious solution.

Increasing Farmlands

Only about one-fifth of Earth's surface is covered by land, and about 90 percent of that is unsuitable for traditional agriculture. Converting more land to agricultural use is possible through practices such as deforestation and irrigation, even though this would lead to habitat changes and many biological effects; see the essay "Endangered Species and Habitats."

In the past, getting more food from water has meant exploitation, to the point of making many species of fish and marine mammals nearly extinct. Now, there is growing interest in the scientific "farming of the sea," or **aquaculture.** Already, this is quite successful for certain algae, fish, shellfish, and crustaceans.

Green Revolution

The 1970 Nobel Peace Prize was given to Dr. Norman Borlaug, an American, for his pioneering work on "miracle seeds" and the "**Green Revolution.**" These new crops produce more and are often resistant to temperature extremes, drought or flooding, and to many pests. Some new strains grow fast enough to produce two crops per year. However, many of the new crops require irrigation and fertilization in order to reach their full potential, and many developing nations cannot afford either technology. Both also have many widespread environmental effects.

Changing Living Habits

Everyone probably enjoys green lawns, except when they have to be mowed, but few think of the enormous quantities of fertilizer and pesticides used on them and in parks, on golf courses, and along highways. Could these substances be used instead to raise more crops? Perhaps many of us do live by more than bread alone, but is it ethical to use rich land to grow flowers or golf greens in light of the millions of starving humans?

Endangered Species and Habitats

Don't feel sorry about the chartreuse-winged, yellow-tufted marsh warbler, because it does not exist. But suppose you learned that it once had and that human encroachments upon its habitat had made it extinct? Would you feel sad, even though you had never seen one? Perhaps most humans who contribute to save-the-whale funds have never seen a majestic live whale in nature, yet they would feel truly sorry if those great marine mammals ceased sharing Earth with us. Would you?

A tiny fish called the snail darter almost stopped completion of a large and expensive American dam because the species was on the endangered list. Although the "dam" humans won, the fish may not have lost, depending upon the long-term success of transplantation and breeding programs. Some take the position, "Who cares anyway, whether a few small fish live, without sport or recreational value? And who cares whether Cherokee Indian ancestral lands and grave sites are covered by water, or whether a stretch of wild river is lost, or whether some fertile farmland is under water? Just think of the electricity generated, the floods prevented, and the recreational opportunities provided."

What is a species, a wild river, or historical site worth to you? What do you understand a cost-benefit analysis to mean in biological and environmental terms? Would the extinction of an endangered species *really* matter to you?

Mechanization could also increase crop yields in many areas, but machines require fuel, and world supplies are dwindling (Figure 25.22). Would less fuel use by developed nations result in that fuel becoming available and affordable for agriculture in developing nations?

Some experts estimate that over one-quarter of food in many developed nations is wasted. How much food do you waste? Would that food have been available for those in need had it not been wasted?

Many people like pets, and Americans own many millions of cats, dogs, horses, gerbils, and other species. These pets eat large amounts of food, some of which might have been used by humans, or fed to agricultural animals, or the resources used in its processing applied to human food production. Notice how much space in a supermarket is devoted to animal feed and to products for caring for pets.

There have been about 300 wars during the last 500 years, killing about 35 million humans. However, at today's rate of population increase, that number would be replaced in less than six months. About 600,000 Americans have died in wars, a number replaced in three days of human reproduction today. Many wonder when and at what level this growth rate will stop. Will famine, overcrowding, pollution, nuclear war, or human choice cause it to level off?

FIGURE 25.22

Humanpower is common in agriculture in developing nations (A), whereas agriculture in developed nations is very mechanized (B).

(A)

(B)

SUMMARY

1. Although all biologists are interested in organisms, the levels of biological organization of especial interest to ecologists are **populations** (intraspecific interactions), **communities** (interspecific interactions), and **ecosystems** (interactions of communities and their environments). The thin biosphere includes all of Earth's ecosystems.

2. The sun provides the energy for most biological processes. Autotrophs utilize sunlight in the process of photosynthesis, and various **consumers** such as **herbivores, carnivores,** and **omnivores** utilize **producers** for energy and materials. All organisms and their wastes are ultimately degraded by **decomposers.**

3. Many organisms within particular ecosystems are interconnected by **food chains** and **food webs**. Because only about 10 percent of the energy available at one feeding level is incorporated into the next **trophic level**, most food chains contain few links.

4. Materials cycle through ecosystems, but energy does not. Each **mineral cycle** has unique traits, although many are interrelated. Photosynthesis and cellular respiration involve the very important **carbon** and **oxygen cycles**, among others.

5. **Biomes** are large ecosystems such as deserts, grasslands, forests, rivers, and oceans. Terrestrial biomes are commonly named after the dominant plants.

6. Some common traits of **desert** and **chaparral** plants include small, light-colored leaves, spines, and shallow and/or deep roots. Annual plants often complete their life cycles very quickly.

7. **Grasslands** are Earth's major agricultural biome. Drier grasslands in the United States are called **plains;** moister regions with taller, more closely spaced plants are known as **prairies. Savannas** are regions of grasslands with scattered trees.

8. Forest biomes exist in many forms. **Temperate deciduous forests** occur generally in warmer and wetter regions than the **taiga,** and they have a higher species diversity. **Tropical rainforests** are being destroyed quickly, although vast areas remain.

9. Most **lakes** have an upper lighted zone, wherein oxygen is produced by autotrophs, and a deeper dark zone dominated by heterotrophs. Annual changes in temperature are important in the typical spring and fall **overturns.**

10. **Biological magnification** occurs when a substance becomes more concentrated with each transfer in a food chain. The phenomenon is of particular significance to organisms such as humans that are high on food chains.

11. The role that an organism plays in its **habitat** is called its **niche.** The greater the similarity between the niches of two species, the greater the competition.

12. **Urban ecosystems** are unusual in that they are composed mainly of consumers, with few producers. Other ecosystems produce materials for import to cities, and they are the recipients of many of the city's wastes and pollutants. New cities may or may not be the solution.

13. Interspecific interactions are symbolized $+,-$ in **predation** and **parasitism;** $+,0$ in **commensalism;** and $+,+$ in **mutualism.** Predator-prey and parasite-host populations generally keep one another healthy in the sense that only the best-adapted individuals usually survive. The life cycles of many parasites involve one or more **vectors** and often more than one habitat. **Social parasites** affect hosts behaviorally rather than physically. Mutualistic interactions generally involve organisms with very different needs, such as autotrophs and heterotrophs.

14. The relationship between population and pollution has been termed **popullution.** Some commonly suggested solutions include increasing farmlands through deforestation and irrigation, utilizing aquaculture and Green Revolution crops, and changing living habits.

REVIEW QUESTIONS

1. Distinguish between producers and consumers. Why would the term *converter* be better technically than the commonly used term *producer*?

2. How do you depend upon producers for energy, shelter, building materials, medications, and other purposes?

3. State examples of occasions when you are a second-, third-, fourth-, or fifth-order consumer.

4. Describe and explain the 10 percent principle.

5. Compare and contrast the movements of energy and materials through ecosystems.

6. Draw a simplified oxygen cycle. How is the oxygen cycle related to the carbon cycle?

7. Why are most terrestrial biomes named after the makeup of the dominant plants?

8. What is chaparral and where is it found?

9. Describe some typical adaptive traits of desert plants and animals.

10. Distinguish among plains, prairies, and savannas.

11. Compare and explain the differences in species diversity in the various types of forests.

12. How do you think deforestation of the tropical rainforests might alter conditions on Earth?

13. Explain the spring and fall overturns in lakes, and describe their ecological significance.

14. In what ways do you contribute to eutrophication?

15. State an example of biological magnification and explain the phenomenon.

16. Use particular examples to distinguish between niches and habitats.

17. Distinguish between a metropolis and a megalopolis. What additional megalopolises are likely in the near future?

18. How do urban ecosystems differ from such natural biomes as grasslands and forests?

19. Describe some examples of human social parasitism.

20. Describe a situation when a commensalistic interaction might become a case of parasitism or of mutualism.

21. Many apparent solutions for human hunger and pollution problems create other concerns. What are some potential drawbacks to increasing farmlands through deforestation and irrigation, aquaculture, and increased utilization of Green Revolution crops?

22. Would it affect you if agricultural chemicals were not used on lawns and parks? If florist shops were illegal? If only native plants could be grown?

23. How do you view the prospect of most populations quickly changing their basic living habits in major ways (e.g., accepting algae-burgers)?

COMMON BIOLOGICAL PREFIXES AND SUFFIXES:
Finding Your Way Through the Scientific Word Jungle

Prefix or Suffix	Meaning	Examples
a-	not, without, no	atypical
ab-	from, away	abnormal, abductor
-ad-	to, toward	addiction, adduct
aden-	gland	adenoids
albi-	white	albino, albicans
-alg	pain	analgesic, neuralgia
amyl-	starch	amylase
an-	not, without	anesthesia
ana-	up, upon	anabolism
anti-	against, as opposed to	antitoxin, antibody
arth-	joint	arthritis
-ase	designates an enzyme	amylase, lipase
auto-	self	autonomic, autoimmune
bi-	two, twice	biceps, bifocal
-bio-	life	biology, antibiotic
calc-	stone	calcification
-calor-	heat	calorimeter, kilocalorie
-card-	heart	cardiac, electrocardiogram
cata-, kata-	down	catabolism
-ceph-	head	cephalic, encephalitis
cerebro-	brain	cerebrospinal
chrom-	color	chromosome
-cidal	killing	bactericidal, spermicide
circum-	around	circumcision, circular
-cis	cut	excise, incision
c-, com-, con-	with, together	coagulate, compress, consult
contra-	against	contraception
corpus-	body	corpuscle, corpus luteum
cort-	shell	cortex
cut-	skin	cutaneous, cuticle
cyst-	hollow, bladder	blastocyst, cystitis
-cyte, cyto-	cell	cytoplasm, lymphocyte
-derm-	skin	hypodermic, dermatology
di-	two, twice	diploid, digest

Prefix or Suffix	Meaning	Examples
dia-	through, between	diabetes, diaphragm
dors-	back	dorsal
dys-	bad, difficult	dysmenorrhea, dysentery
-ectomy	cut out, remove	appendectomy, tonsillectomy
em-, en-, endo-	in, into	embolism, endotherm
-emia	blood	anemia, leukemia
end-	within	endometrium, endotherm
entero-	intestine	gastroenteritis
epi-	on, above, upon	epidermis, epiglottis
erythro-	red	erythrocyte
eu-	good	eutrophic, eugenics
extra-	outside of, beyond	extrasensory, extraterrestrial
-fer-	to carry, transport	afferent, efferent
-fract-	break	fracture, refraction
gastr-	stomach	gastric, gastrointestinal
-gen-	to produce, begin; origin	glycogen, genetics, psychogenic
glyc-	sugar, sweet	glycogen, glycolysis
-gnosis	knowledge	diagnosis
-gram	record	electrocardiogram, encephalogram
-graph	write	cardiograph
gyne-	woman	gynecology
hemi-	half	hemisphere
hemo-	blood	hemorrhage, hemoglobin
hetero-	different, other	heterozygous, heterotroph, heterosexual
homo-	alike, same	homozygous, homotherm
hydro-	water	hydrophobic, hydrocephalus
hyper-	over, more than	hypersecretion, hypertension, hypertrophy
hypo-	under, less than	hypothyroidism, hypothalamus
inter-	between, together	intercostal, intercourse
intra-	within	intravenous, intraspecific
ir-	not	irregular
-itis	inflammation	appendicitis, arthritis
kin-	to move	kinetic
-lac-	milk	lactation, prolactin
leuk-	white	leukocyte, leukemia
lip-	fat	lipid, lipase
-logy	science, knowledge	physiology, pathology
lumb-	loin	lumbar, lumbago
lymph-, lympho-	lymph	lymphatic, lymphocyte
-lysin, -lysis, -lytic	dissolve, destroy	hemolysis
macro-	large	macroscopic, macrophage
mal-	bad	malignant, malnutrition
-meg-, mega-	large, great	acromegaly
mens-	month	menopause, menstruation
-meter, -metry	measure	millimeter, optometry
micro-	small	microorganism, microscope
mono-	one	monocyte, monosaccharide
myo-	muscle	myosin, myoglobin
neo-	new	neonate, neoplasm
nephr-	kidneys	nephron, nephritis
neur-	nerve	neurology, neuritis
-oid	like, similar	ameboid, lymphoid
-ole	small	arteriole, bronchiole
-oma	swelling, tumor	carcinoma, sarcoma

Prefix or Suffix	Meaning	Examples
op-, -oph	eye	optic, ophthalmologist
-opia	sight	hyperopia, myopia
-oscopy	look inside	laparoscopy
-osis	condition, process	phagocytosis, pinocytosis
os-, oste-, osteo-	bone	osteocyte, osteology
ot-	ear	otosclerosis
-otomy	cut into	hysterotomy
ov-	egg	oviduct, ovum
-par-	birth, bear	postpartum
para-	near by, beside	paramedic, parathyroid
path-	disease; sick, suffering	pathogenic, pathology
-pend-	hang down	appendicitis, pendulous
per-	through	perception, perforate
peri-	around, near	pericardium, periosteum
phago-	to eat	phagocyte, phagocytosis
-phil-	loving	basophil, hydrophilic
phleb-	vein	phlebitis
-plasm-	substance	cytoplasm, plasmolysis
pneumo-	air, lung	pneumonia
-pod-	foot	pseudopod, podocyte
poly-	many	polypeptide, polysaccharide
post-	after, behind	posterior, postpartum
pre-	in front of, before	precursor, prenatal, precancerous
proprio-	one's own	proprioceptor
proto-	first	protoplasm
pseudo-	false	pseudopod, pseudoscience
-psycho-	mind	psychosomatic, psychosis, somatopsychological
pulmo-	lung	pulmonary
-renal	kidney	adrenal
-rrhea	flow	diarrhea
salp-	tube	salpingectomy, salpingitis
sarco-	flesh, muscle	sarcoplasm
-sclera-	hard	arteriosclerosis, sclera
-soma-	body	psychosomatic, somatotypology
-some	body	chromosome
-therm-	heat	ecotherm, thermometer
-thrombo-	clot, coagulation	thrombophlebitis, thrombosis
-tome, -tomy	to cut	hysterectomy
-tox-	poison	antitoxin, toxemia
trans-	across, by way of	transfusion, transport, transurethral
tri-	three	tricuspid, trimester
-trophic	feeding	autotrophic
-tropic	attracted to	thyrotropic
-ule	small	venule
-ur	urine	uric acid, urology
vas-	duct, vessel	vascular, vasectomy
vaso-	blood vessel	vasodilation
vit-	life	vital, vitamin

MEASUREMENTS AND GRAPHS

MEASUREMENT

Of the two major systems of measurement used worldwide, most countries use the metric system. Only the United States and a few other nations use systems based on the English system. Most of the latter countries are in the process of converting to the metric system, which is used by scientists in all countries. This resource will help you convert from metric to U.S. customary units and vice versa.

Weight Measures

1 g = 0.035 oz   1 oz = 28.35 g
1 kg = 2.2 lb, 35.27 oz   1 lb = 0.45 kg, 453.6 g
1 metric ton (MT) = 1.1 ton   1 ton = 0.91 MT, 907 kg

Linear Measures

1 cm = 0.39 in.   1 in. = 2.54 cm, 25.4 mm
1 m = 3.28 ft, 39.37 in.   1 ft = 0.3 m, 30.48 cm
1 m = 3.2 ft   1 yd = 0.91 m, 91.4 cm
1 km = 0.62 mi, 3273 ft   1 mi = 1.6 km, 1609 m

Fluid Measures

1 ml (cm^3) = 0.03 oz   1 oz = 3 ml (cm^3), 0.3 L
1 L = 2.1 pt   1 pt = 0.47 L
1 L = 1.06 qt   1 qt = 0.95 L
1 L = 0.27 gal   1 gal = 3.79 L

Temperature Measures

To convert temperature scales (see Figure R2.1):

$$\text{Fahrenheit to Celsius } °C = \frac{5}{9}(°F - 32)$$

$$\text{Celsius to Fahrenheit } °F = \frac{9}{5}(°C) + 32$$

GRAPHS

Most of the graphs in this text look like those in Figure R2.2. The "low ends" of the measurement scales are in the lower left corner of the graph.

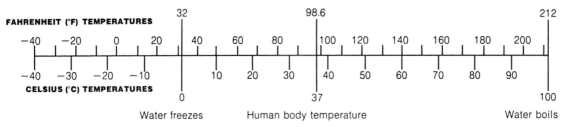

FIGURE R2.1

Comparison of Fahrenheit and Celsius temperature scales.

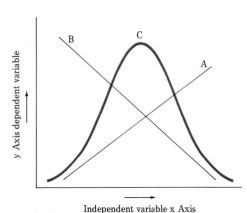

FIGURE R2.2

Common types of graphs. Line A shows a direct relationship (increases in both variables), and line B shows an inverse relationship (decrease in one variable as the other variable increases). Line C is a normal distribution.

The values increase as you go to the right along the horizontal axis (x axis, or abscissa) and as you go up along the vertical axis (y axis, or ordinate). The units along the x axis reflect values of the independent variable, which is usually the one under the control of the experimenter or observer, and those along the y axis reflect the dependent variable. The latter are so named because they vary with ("depend upon") changes in the independent variable. Thus, for example, plant growth (plotted on the y axis) often increases up to a point as the temperature, light, water, or "richness" of the soil increases (plotted on the x axis).

Line A in Figure R2.2 represents a direct relationship, when one variable increases as the other also increases. For example, all other things being equal, the more you eat (independent variable), the more you weigh (dependent variable). Line B shows an inverse relationship. Thus, again assuming that all other things are equal (such as your diet and health), the more you exercise (independent variable), the less you weigh (dependent variable). Finally, Line C is the familiar normal distribution, or bell-shaped curve. For example, we find such a curved line when the y axis represents the numbers of people and the x axis represents height (or weight). Most people are near the middle, but some are very short or very tall (or are lighter or heavier) than the average. Curved lines like this also commonly depict homeostatic adaptations graphically. That is, the peak of the curve is often the optimum or the average response to an environmental variable, and the other portions of the line represent the range of tolerance. Thus, one of your digestive enzymes functions best at a particular temperature and less well at lower or higher temperatures.

SELF-EVALUATION QUESTIONS

Use the objective questions presented for each chapter as a quick check on your understanding of the major terms and concepts. Try the questions honestly before looking at the answers at the end of this resource!

While multiple choice questions are familiar to most, the comparison items may be new. There are many dichotomies and other comparisons to be made in biology, and these questions show two columns of short statements or phrases to be compared. Basically, you are asked to judge which item implies the greater (e.g., larger, faster, or older), for example. In some cases the two items are the same, or essentially so, and these instances call for "c" answers. There is not much to read but often much to think about, and one value of the comparison items is that they allow you to evaluate your understanding of *relationships,* not nitty-gritty *details.* That is, you do not have to know the details about rates, dates, sizes, amounts, and so forth, but rather the relationships between the two phrases. Some students view these statements in a true-false manner, in that one phrase is often true when the other is false.

SELF-EVALUATION QUESTIONS

Chapter 1—Biology: The Study of Life

Multiple Choice. *Choose the one best answer for each question.*

1. The time of Darwin's work was ___. (a) 1775–1825 (b) 1825–1875 (c) 1875–1925 (d) 1925–1975
2. The person who devised the basis for our current system of scientific names was ___. (a) Jean Baptiste Lamarck (b) René Descartes (c) Charles Darwin (d) Carolus Linnaeus
3. The use-and-disuse theory of evolution was developed by ___. (a) Adam Sedgewick (b) Jean Baptiste Lamarck (c) Georges Buffon (d) Charles Darwin
4. ___ developed the theory of uniformitarianism. (a) Charles Darwin (b) Charles Lyell (c) Thomas Malthus (d) creationists
5. Darwin was quite surprised when he learned that ___ had developed a theory of natural selection independently. (a) Charles Lyell (b) Thomas Malthus (c) Gregor Mendel (d) Alfred Wallace
6. Which is not a typical example of a decomposer? (a) virus (b) mold (c) bacterium (d) mushroom

7. A bacteriophage is a type of ____. (a) simple plant (b) prokaryote (c) virus (d) bacteria
8. A disease-causing agent is called a(n) ____. (a) prokaryote (b) antibody (c) phage (d) pathogen

Comparison Items. *Study each pair of statements and use the following key: (a) the statement on the left suggests the greater quantity, (b) the statement on the right suggests the greater quantity, or (c) the two statements are essentially the same.*

9. Changing nature of organisms according to creationism views | Changing nature of organisms according to evolutionary views
10. Arithmetic progression represented by 6, 8, 10, etc. | Geometric progression represented by 6, 8, 10, etc.
11. Resemblance of human population growth to arithmetic progression | Resemblance of human population growth to geometric progression
12. Importance of natural selection in the development of the finches on the Galápagos Islands | Importance of artificial selection in the development of the finches of the Galápagos Islands
13. Control of the dependent variable by an experimenter | Control of the independent variable by an experimenter
14. Ability of heterotrophs to synthesize their own food | Ability of autotrophs to synthesize their own food
15. Typical amount of meat in the diet of an herbivorous mammal | Typical amount of meat in the diet of an omnivorous mammal
16. Frequency of internal membranes in eukaryotic cells | Frequency of internal membranes in prokaryotic cells

Chapter 2—Homeostasis and Adaptation: An Introduction

Multiple Choice. *Choose the one best answer for each question.*

1. Processes that keep body functions within their range of tolerance are called ____ feedback. (a) negative (b) positive (c) negative and positive (d) runaway
2. The human thermostat is found in the ____. (a) heart (b) blood (c) skin (d) brain
3. In countercurrent circulation, ____. (a) venous blood is cooled (b) arterial blood is cooled (c) venous and arterial blood are cooled (d) venous and arterial blood are warmed
4. Negative feedback systems may be symbolized by ____. (a) $-,-$ (b) $-,+$ (c) $+,+$

Comparison Items. *Study each pair of statements and use the following key: (a) the statement on the left suggests the greater quantity, (b) the statement on the right suggests the greater quantity, or (c) the two statements are essentially the same.*

5. Corrective feedback within the range of tolerance | Corrective feedback outside the range of tolerance

6. Vasodilation in the skin when you are hot / Vasodilation in the skin when you are cold
7. Rate at which air removes body heat / Rate at which water removes body heat
8. Constancy of body temperature of endotherms / Constancy of body temperature of ectotherms
9. Amount of oxygen cold water can dissolve / Amount of oxygen warm water can dissolve
10. Frequency of blood returning to the heart in veins / Frequency of blood returning to the heart in arteries
11. Average body size of mammal species X in a cold region / Average body size of mammal species X in a warm region
12. Typical ear size of mammal species X in a cold region / Typical ear size of mammal species X in a warm region
13. Metabolic rate when exposed to optimum temperatures / Metabolic rate when exposed to above- and below-optimum temperatures

Chapter 3—Structural Organization for Homeostasis

Multiple Choice. *Choose the one best answer for each question.*

1. What is the correct sequence of body organization, from smallest to largest unit? (a) cell-system-organ-tissue (b) cell-organ-tissue-system (c) cell-tissue-system-organ (d) cell-tissue-organ-system
2. The fundamental unit that possesses the properties of life is the ____. (a) gene (b) tissue (c) cell (d) organism

Questions 3–6 refer to the typical cell shown.
(a) 1 (b) 2 (c) 3 (d) 4

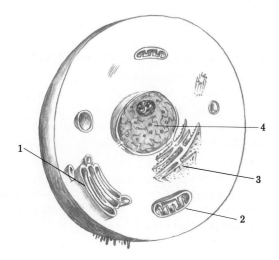

3. Help make proteins
4. Contains the genes
5. Powerhouse of the cell
6. Packaging organelle

7. The study of cells is called ____. (a) microscopy (b) pathology (c) histology (d) cytology

8. The structure of a cell is most closely related to its ___.
   (a) function (b) size (c) age (d) number of nucleoli

9. Endocrine glands secrete ___. (a) into ducts (b) into body fluids
   (c) into the nucleus (d) through the skin

10. Diffusion occurs only ___. (a) across a selectively permeable
    membrane (b) in the presence of concentration differences
    (c) in true solutions (d) in living cells

11. The U-shaped tube contains solutions of salt water separated by a
    membrane impermeable to salt but permeable to water. In which
    side of the tube will the water level rise? (a) A (b) B (c) both sides
    (d) neither side

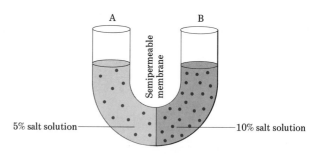

12. Osmosis ___. (a) is not important in mature cells (b) is a type of
    diffusion (c) occurs when molecules move in the opposite direc-
    tion from those moving by diffusion (d) occurs only in human
    cells

13. Chemicals ___ enter or leave cells by diffusion. (a) never
    (b) sometimes (c) always

14. During interphase of cell division, chromosomes are ___. (a) not
    visible (b) not present (c) present only (d) countable only

15. Through meiosis, ___. (a) the haploid chromosome number oc-
    curs in the offspring (b) gametes receive only one-half of each
    chromosome (c) the chromosome numbers vary among the mem-
    bers of a population (d) gametes are produced in a way that re-
    duces the chromosome number by one-half

16. Identical sets of genetic material normally occur in ___ somatic
    cells. (a) no two (b) closely similar (c) all

17. The heart is an example of ___. (a) a tissue (b) a highly special-
    ized cell (c) an organ system (d) an organ

18. Which of the following is not an example of connective tissue?
    (a) cartilage (b) bone (c) muscle (d) blood

Comparison Items. *Study each pair of statements and use the following
key: (a) the statement on the left suggests the greater quantity, (b) the
statement on the right suggests the greater quantity, or (c) the two
statements are essentially the same.*

19. Details of cell structures visi-   Details of cell structures visible
    ble through light microscopes   through electron microscopes

20. Ability to consciously control   Ability to consciously control
    smooth muscles   striated muscles

21. Speed of activity of most   Speed of activity of most skeletal
    smooth muscles   muscles

22. Frequency of tendons attach-   Frequency of ligaments attaching
    ing muscles to bones   muscles to bones

23. Frequency of cell divisions in mature epithelial tissue

Frequency of cell divisions in mature nervous tissue

24. Typical quantity of stored wastes in Golgi bodies

Typical quantity of stored wastes in ribosomes

25. Usual number of cilia per ciliated cell

Usual number of flagella per flagellated cell

26. Usual length of cilia

Usual length of flagella

27. Frequency of plastids in plant cells

Frequency of plastids in animal cells

28. Number of chromosomes in diploid cells

Number of chromosomes in haploid cells

29. Cellular work performed in moving materials by diffusion

Cellular work performed in moving materials by active transport

30. Particles taken into a cell by phagocytosis

Fluids taken into a cell by phagocytosis

31. Increase in surface area as an object grows

Increase in volume as an object grows

Chapter 4—Chemical Organization for Homeostasis

Multiple Choice. *Choose the one best answer for each question.*

1. Which of the following substances is inorganic? (a) protein (b) water (c) sugar (d) oil
2. Which of the following is not a protein? (a) glycogen (b) hemoglobin (c) enzyme (d) insulin
3. Which of the following elements is more common in Earth's crust than in humans? (a) oxygen (b) hydrogen (c) nitrogen (d) iron
4. Which of the following is not a carbohydrate? (a) steroid (b) cellulose (c) glycogen (d) disaccharide
5. Which of the following is the main cellular fuel? (a) lactose (b) galactose (c) glucose (d) fructose
6. Cells are mainly a combination of water and ____. (a) minerals (b) proteins (c) carbohydrates (d) lipids
7. The largest part of most diets is made up of ____. (a) nucleic acids (b) proteins (c) carbohydrates (d) lipids
8. The protein "alphabet" consists of ____ amino acids. (a) 18 (b) 20 (c) 46 (d) untold thousands of
9. Chemically, enzymes are ____. (a) nucleic acids (b) proteins (c) carbohydrates (d) lipids
10. Enzymes involved in chemical reactions ____. (a) are used up during reactions (b) decompose during reactions (c) react more readily as reactions progress (d) are not used up during reactions
11. Enzymes control the ____ of chemical reactions. (a) speed (b) direction (c) nature of the end products (d) temperature
12. The second law of thermodynamics states that in an energy transformation, the amount of usable energy ____. (a) increases (b) remains the same (c) decreases (d) disappears
13. Isotopes differ from other forms of an element in their number of ____. (a) electrons (b) protons (c) neutrons (d) orbitals

Comparison Items. *Study each pair of statements and use the following key: (a) the statement on the left suggests the greater quantity, (b) the statement on the right suggests the greater quantity, or (c) the two statements are essentially the same.*

14. Frequency of carbon in organic compounds / Frequency of carbon in inorganic compounds
15. Quantity of organic matter in organisms / Quantity of inorganic matter in organisms
16. Usual size and complexity of organic compounds / Usual size and complexity of inorganic compounds
17. Amount of water added during hydrolysis reactions / Amount of water added during condensation reactions
18. Frequency of sugar names ending in *-ase* / Frequency of enzyme names ending in *-ase*
19. Size of most carbohydrate molecules / Size of most protein molecules
20. Number of different kinds of carbohydrates / Number of different kinds of proteins
21. Frequency of electrons being gained through oxidation / Frequency of electrons being gained through reduction
22. Numerical index of an acid substance / Numerical index of a basic substance
23. Strength of ionic bonds / Strength of covalent bonds
24. Frequency of photosynthetic autotrophs / Frequency of photosynthetic heterotrophs

Chapter 5—The Reproductive System

Multiple Choice. *Choose the one best answer for each question.*

1. The testes normally descend into the scrotum at about ____. (a) six months into pregnancy (b) the time of birth (c) the age of 3 months (d) the time of puberty
2. Embryologically, the male penis and the female ____ develop from similar tissue. (a) hymen (b) vagina (c) clitoris (d) ovary
3. Testosterone is to male as ____ is to female. (a) progesterone (b) LH (c) estrogen (d) FSH
4. The pituitary hormone that stimulates the Leydig cells is to male as ____ is to female. (a) progesterone (b) LH (c) estrogen (d) FSH
5. Gamete-producing organs are called ____. (a) gonads (b) zygotes (c) hernias (d) glans
6. The movement of eggs from the ovaries to the uterus proceeds through the ____. (a) cervix (b) vagina (c) placenta (d) fallopian tubes
7. The corpus luteum is formed from the action of ____. (a) FSH (b) LH (c) progesterone (d) estrogen
8. Ovulation generally occurs ____. (a) during menstruation (b) midway through the menstrual cycle (c) immediately after menstruation (d) just prior to ovulation
9. In females, FSH causes ____. (a) development of follicles (b) production of the corpus luteum (c) disintegration of the ovum (d) menstruation
10. After ovulation, the ruptured follicle ____. (a) disappears after all its cells disintegrate (b) passes as waste material through the oviduct (c) mends itself and begins the maturation of another egg (d) differentiates into a temporary endocrine gland
11. In young females, cervix is to uterus as ____ is to vagina. (a) immature ovary (b) fallopian tube (c) hymen (d) a labial lip
12. Menopause usually occurs between the ages of ____. (a) 35 and 40 (b) 40 and 45 (c) 45 and 50 (d) 50 and 55

Comparison Items. *Study each pair of statements and use the following key: (a) the statement on the left suggests the greater quantity, (b) the statement on the right suggests the greater quantity, or (c) the two statements are essentially the same.*

13. Size of sperm	Size of eggs
14. Temperature of epididymis	Temperature of prostate
15. Amount of testosterone made in seminiferous tubules	Amount of testosterone made in interstitial cells
16. Length of urethra in males	Length of urethra in females
17. Acid nature of semen	Basic nature of semen
18. Typical number of years males produce sperm	Typical number of years females produce eggs
19. Effect of progesterone on thickening the endometrium	Effect of estrogen on thickening the endometrium
20. Number of viable sperm in sterile males	Number of viable sperm in impotent males
21. Frequency of sperm in the urethra	Frequency of sperm in a ureter
22. Secretion of progesterone by the pituitary	Secretion of progesterone by the corpus luteum
23. Stimulatory effect of high levels of progesterone upon the secretion of LH	Inhibitory effect of high levels of progesterone upon the secretion of LH
24. Similarity of offspring and parents in asexual reproduction	Similarity of offspring and parents in sexual reproduction
25. Muscular tissue in the myometrium	Muscular tissue in the endometrium
26. Frequency of polar bodies in spermatogenesis	Frequency of polar bodies in oogenesis

Chapter 6—Pregnancy and Birth

Multiple Choice. *Choose the one best answer for each question.*

1. Fertilization usually occurs ____ of the fallopian tubes. (a) in the upper third (b) in the middle third (c) in the lower third (d) outside

2. The average length of pregnancy is ____ weeks. (a) 34–36 (b) 36–38 (c) 38–40 (d) 40–42

3. Living low-birth-weight infants usually weigh ____ kilograms. (a) about 0.25 (b) less than 2.5 (c) 2.5–3.0 (d) over 3.0

4. Monozygotic twins are ____ of the same sex. (a) always (b) sometimes (c) usually (d) never

5. The term *imprinting* refers to ____. (a) the effect of repeated learning (b) the importance of an initial major stimulus (c) a way to measure fetal brain activity (d) sequential patterning of learned behavior

6. Agents that cause birth defects are called ____. (a) pathogens (b) mutagens (c) teratogens (d) abortogens

7. Developmental changes in the relative lengths of the head and body represent an example of the ____ developmental direction. (a) cephalodistal (b) cephalocaudal (c) proximodistal (d) proximocaudal

8. When a child is born, its blood ____. (a) flows for the first time (b) reverses its flow through the heart (c) ceases to pass from one atrium to the other (d) carries wastes for the first time

9. Lactotropic hormone stimulates the ____. (a) corpus luteum to secrete (b) mammary glands to form milk (c) endometrium to thicken (d) secondary sex traits to form

10. Heart is to mesoderm as lungs are to ____. (a) endoderm (b) ectoderm (c) pachyderm (d) mesoderm

11. Which maternal disease listed is most likely to adversely affect the developing human? (a) tetanus (b) smallpox (c) rubella (d) influenza

12. Following fertilization, implantation usually occurs ____. (a) almost immediately (b) within 12–24 hours (c) within 1–3 days (d) after 3 days

13. The chorionic gonadotropic hormone is secreted by the ____. (a) placenta (b) ovary (c) uterus (d) pituitary

Comparison Items. *Study each pair of statements and use the following key: (a) the statement on the left suggests the greater quantity, (b) the statement on the right suggests the greater quantity, or (c) the two statements are essentially the same.*

14. Mitotic cell divisions in the early embryo	Meiotic cell divisions in the early embryo
15. Typical size of the zygote	Typical size of the morula
16. Harmful effects of most agents early in pregnancy	Harmful effects of most agents late in pregnancy
17. Number of years since human sperm were observed first	Number of years since human eggs were observed first
18. Intrauterine fertilizations	Extrauterine fertilizations
19. Rate of development of the extremities of the fetus	Rate of development of the midline structures of the fetus
20. Probable concentration of a body contaminant in maternal blood	Probable concentration of a body contaminant in fetal blood
21. Frequency of fraternal multiple births	Frequency of identical multiple births
22. Frequency of multiple births in Mongoloids	Frequency of multiple births in Negroids
23. Amount of FSH in lactating females	Amount of FSH in nonlactating females
24. Frequency of monozygotic twins in younger mothers	Frequency of monozygotic twins in older mothers
25. Quantity of hormones released by the corpus luteum early in pregnancy	Quantity of hormones released by the corpus luteum late in pregnancy
26. Average birth weight of infants born to smokers	Average birth weight of infants born to nonsmokers
27. Proportion of the body length occupied by the head in the early fetus	Proportion of the body length occupied by the head in the late fetus
28. Relative development of altricial young	Relative development of precocial young
29. Frequency of acrosomes in sperm	Frequency of acrosomes in eggs

| 30. | Frequency of intrauterine implantations in ectopic pregnancies | Frequency of extrauterine implantations in ectopic pregnancies |

Chapter 7—Contraception

Multiple Choice. *Choose the one best answer for each question.*

1. The operation to sterilize males is called ___. (a) hysterotomy (b) spermectomy (c) vasectomy (d) gametectomy
2. The rhythm method of birth control is ___. (a) based on the assumption that ovulation can be predicted (b) considered to be the most reliable of all birth control methods (c) a type of dancing prescribed for pregnant women (d) especially useful near the beginning and end of a woman's reproductive life
3. The present world population is about ___. (a) 250 million (b) 500 million (c) 5 billion (d) 8.2 billion
4. Which of the following contraceptive methods has the lowest reliability? (a) spermicides (b) postcoital douche (c) condom (d) diaphragm

Comparison Items. *Study each pair of statements and use the following key: (a) the statement on the left suggests the greater quantity, (b) the statement on the right suggests the greater quantity, or (c) the two statements are essentially the same.*

5.	FSH content of most birth control pills	Estrogen content of most birth control pills
6.	Blood testosterone levels in vasectomized males	Blood testosterone levels in castrated males
7.	Frequency of salpingectomy for strictly contraceptive purposes	Frequency of hysterectomy for strictly contraceptive purposes
8.	Average body temperature before ovulation	Average body temperature after ovulation
9.	Typical failure rate of the rhythm method	Typical failure rate of oral contraceptives
10.	Chance that an embolus is located where it formed	Chance that a thrombus is located where it formed
11.	Frequency of early abortions using dilatation and evacuation	Frequency of early abortions using vacuum aspiration
12.	Percentage of Earth's population composed of Americans	Percentage of Earth's nonrenewable resources used by Americans

Chapter 8—Sexually Transmitted Diseases

Multiple Choice. *Choose the one best answer for each question.*

1. Gonorrheal infections in females generally begin in the ___. (a) vagina (b) cervix (c) uterus (d) oviducts
2. Which form of STD listed is caused by a virus? (a) monilial vaginitis (b) genital herpes (c) gonorrhea (d) cystitis

3. Cystitis is an inflammation of the ___. (a) vagina (b) urethra (c) anus (d) bladder
4. *Trichomonas* is a ___. (a) bacterium (b) virus (c) protozoan (d) fungus
5. Septicemia occurs when ___. (a) a person has both syphilis and gonorrhea at the same time (b) a person is reinfected with an STD pathogen (c) bacteria enter the bloodstream (d) someone is allergic to STD treatment

Comparison Items. *Study each pair of statements and use the following key: (a) the statement on the left suggests the greater quantity, (b) the statement on the right suggests the greater quantity, or (c) the two statements are essentially the same.*

6. Frequency of syphilis infections / Frequency of gonorrhea infections
7. Physiological seriousness of most cases of syphilis / Physiological seriousness of most cases of gonorrhea
8. Ease of detecting most cases of STD in males / Ease of detecting most cases of STD in females
9. Time between infection and the appearance of symptoms in most cases of syphilis / Time between infection and the appearance of symptoms in most cases of gonorrhea
10. Use of eyedrops in neonates to prevent damage from syphilis / Use of eyedrops in neonates to prevent damage from gonorrhea
11. Frequency of damage by syphilis outside of the urinary-genital system in adults / Frequency of damage by gonorrhea outside of the urinary-genital system in adults
12. Frequency of chancres in cases of syphilis / Frequency of chancres in cases of gonorrhea
13. Number of common STDs caused by species of bacteria / Number of common STDs caused by types of viruses
14. Use of antibiotics in treating bacterial infections / Use of antibiotics in treating viral infections
15. STDs spread by contacts between people / STDs spread by contacts with non-human objects
16. Chance of Pill users getting gonorrhea / Chance of non-Pill users getting gonorrhea
17. Frequency of gonorrhea infections in the United States / Frequency of genital herpes infections in the United States

Chapter 9—Basic Human Genetics

Multiple Choice. *Choose the one best answer for each question.*

1. Down syndrome is correlated with ___. (a) an extra chromosome (b) an XXX condition (c) a poor uterine environment (d) a defective gene
2. Hermaphrodite means that ___. (a) organisms can only reproduce asexually (b) both gonad types are found in a single individual (c) specialized reproductive organs are lacking (d) sexual reproduction is not possible
3. Normal females have ___ Barr bodies. (a) 0 (b) 1 (c) 2 (d) 3
4. Humans with Turner syndrome are symbolized ___. (a) XXX (b) XXY (c) XO (d) XYY

5. Consanguineous matings tend to produce ____. (a) very strong hybrid offspring (b) heterozygous offspring (c) homozygous offspring (d) matched pairs of chromosomes

6. Which statement concerning the offspring of the first generation is most likely if *Aa* mates with *aa*? (a) All the offspring will exhibit the dominant phenotype. (b) All the offspring will exhibit the recessive phenotype. (c) The dominant phenotype will show up in 75 percent of the offspring. (d) The recessive phenotype will appear in half of the offspring.

7. If an individual's genotype were *AaBbCc* and independent assortment and segregation of chromosomes were operating through meiosis, then ____ types of gametes could be formed. (a) 3 (b) 6 (c) 8 (d) 9

8. Mendel conducted his research on peas in what time period? (a) 1750–1800 (b) 1800–1850 (c) 1850–1900 (d) 1900–1950

9. Which of the following represents alleles? (a) *AA* (b) *AB* (c) *Cd* (d) *ef*

10. If a woman who is heterozygous for color blindness and a man with normal vision have children, ____ percent of their sons would be expected to be color-blind. (a) 0 (b) 25 (c) 50 (d) 75

11. Which listed trait is typically least dependent upon a person's heredity? (a) eye color (b) weight (c) skin pigmentation (d) hair color and texture

12. Alleles that are dominant in one sex but recessive in another are ____. (a) sex linked (b) sex influenced (c) sex limited (d) unknown in humans

13. Symbols used in discussing intermediate inheritance are ____. (a) AB^c (b) *AB* (c) *Aa* (d) *aa*

14. One or more genes canceling out the effects of other genes is known as ____. (a) codominance (b) nondisjunction (c) intermediate inheritance (d) epistasis

15. The individual with which listed genotype will exhibit the phenotype of a homozygous recessive condition? (a) *ss* (b) *Ss* (c) *sS* (d) *SS*

Comparison Items. *Study each pair of statements and use the following key: (a) the statement on the left suggests the greater quantity, (b) the statement on the right suggests the greater quantity, or (c) the two statements are essentially the same.*

16. Frequency of Barr bodies in male cells | Frequency of Barr bodies in female cells

17. Size of X chromosome | Size of Y chromosome

18. Frequency of Klinefelter syndrome in females | Frequency of Klinefelter syndrome in males

19. Frequency of XO adults | Frequency of YO adults

20. Number of pairs of autosomes | Number of pairs of sex chromosomes

21. Chance that a hemophiliac inherited the causal gene from his mother | Chance that a hemophiliac inherited the causal gene from his father

22. Frequency of color-blind males | Frequency of color-blind females

23. Proportion of genes operating at a particular time | Proportion of genes not operating at a particular time

24. Number of X-linked traits

    Number of Y-linked traits

25. Chance of finding a sex chromosome in a gamete

    Chance of finding a sex chromosome in a liver cell

26. Chance of discontinuous variation in polygenic inheritance

    Chance of continuous variation in polygenic inheritance

27. Seriousness of most cases of sickle-cell anemia

    Seriousness of most cases of sickle-cell trait

28. Probable number of males conceived

    Probable number of females conceived

29. Average IQ of children with phenylketonuria

    Average IQ of children with hemophilia

30. *In vitro* fertilization during artificial insemination

    In vitro fertilization during artificial fertilization

31. Use of amniocentesis to predetermine a child's sex

    Use of amniocentesis to discover a child's sex prenatally

Chapter 10—Hereditary Mechanisms

Multiple Choice. *Choose the one best answer for each question.*

1. Transformation occurs when ___. (a) one form of RNA becomes another form (b) external genetic information is assimilated by an organism (c) DNA strands duplicate themselves (d) genetic information is used to make proteins

2. The monomers of DNA are ___. (a) amino acids (b) base pairs (c) sugar-phosphate groups (d) nucleotides

3. If ACG represents one DNA codon, then ___ represents its complement. (a) ACG (b) GCA (c) TGC (d) CGT

4. Which of the following is the most accurate description of gene action, one gene-one ___? (a) polypeptide (b) amino acid (c) protein (d) enzyme

5. There are ___ amino acids. (a) 3 (b) 4 (c) 20 (d) an infinite variety of

6. Codons contain ___ bases. (a) 2 (b) 3 (c) hundreds to thousands (d) an untold number of

7. Transcription involves making ___ from a ___ template. (a) one DNA strand; DNA (b) two DNA strands; DNA double helix (c) mRNA; DNA (d) tRNA; mRNA

8. The form of RNA that picks up appropriate amino acids for use in synthesizing proteins is ___ RNA. (a) transcribed (b) transfer (c) messenger (d) ribosomal

9. The organelle most directly involved with protein synthesis is the ___. (a) ribosome (b) Golgi body (c) rough endoplasmic reticulum (d) chromosome

10. A human-machine combination is known as a ___. (a) bionic being (b) clone (c) cyborg (d) mutagen

Comparison Items. *Study each pair of statements and use the following key: (a) the statement on the left suggests the greater quantity, (b) the statement on the right suggests the greater quantity, or (c) the two statements are essentially the same.*

11. Frequency of circular chromosomes in eukaryotes

    Frequency of circular chromosomes in prokaryotes

12. Ability of a virus to metabolize independently | Ability of a bacterium to metabolize independently
13. Sugar-phosphate in the DNA "ropes" | Sugar-phosphate in the DNA "rungs"
14. Amount of DNA in the nucleoplasm | Amount of DNA in the cytoplasm
15. Number of strands of DNA | Number of strands of RNA
16. Typical beneficial effects of mutations | Typical detrimental effects of mutations
17. Chance of mutations in somatic cell lines affecting future generations | Chance of mutations in germ cell lines affecting future generations
18. Probable role of regulatory proteins in stimulating transcription | Probable role of regulatory proteins in inhibiting transcription
19. Frequency of clones of cells produced asexually | Frequency of clones of cells produced sexually
20. Frequency of artificial insemination *in vivo* | Frequency of artificial insemination *in vitro*
21. Amount of genetic manipulation in eugenics | Amount of genetic manipulation in euthenics

Chapter 11—Growth and Development

Multiple Choice. *Choose the one best answer for each question.*

1. Slight differences in identical twins supports the hypothesis that ____. (a) genetic traits are influenced by many genes (b) single genes may produce multiple effects (c) the environment affects the expression of genetic traits (d) they developed from separate fertilized eggs

2. People who are either very thin, very fat, very short, or very bright are ____. (a) pathological (b) unnatural (c) immature (d) deviants from the norm

3. The growth spurt during adolescence is ____. (a) greater and earlier in females than in males (b) greater in females but earlier in males (c) smaller and later in females (d) smaller but earlier in females

4. According to Sheldon, one is least likely to find a(n) ____ at a social event. (a) extrovert (b) ectomorph (c) mesomorph (d) endomorph

5. The part of the body that enlarges least from birth to maturity is (are) the ____. (a) appendages (b) head (c) muscles (d) abdomen

6. An excellent control of maturational variables in studies of the environment's effect on children's physical and psychological growth and development is the use of ____. (a) matched groups of children (b) introspection (self-analysis) techniques (c) both direct observation and questionnaires (d) the co-twin control method

7. The typical growth curve is shaped somewhat like the letter ____. (a) J (b) W (c) S (d) M

8. The average age of menarche in the United States is at ____ years. (a) 14–15 (b) 13–14 (c) 12–13 (d) 11–12

9. Cryonics is the study of ____. (a) bumps and depressions on the skull (b) fingerprint patterns (c) substances at low temperatures (d) the relationship between physical and psychological traits

10. Gerontology is the study of ____. (a) senescence and aging (b) the relationship between fingerprint patterns and physical disorders (c) patterns of concordance (d) somatotypes

Comparison Items. *Study each pair of statements and use the following key: (a) the statement on the left suggests the greater quantity, (b) the statement on the right suggests the greater quantity, or (c) the two statements are essentially the same.*

11. Average life expectancy of females	Average life expectancy of males
12. Usual effect of the environment on a person's height	Usual effect of the environment on a person's weight
13. Typical range of variation in the traits of children	Typical range of variation in the traits of adults
14. Average age of puberty in females	Average age of puberty in males
15. Average age of menarche in 1900	Average age of menarche today
16. Frequency of administering treatment in euthanasia	Frequency of withholding treatment in euthanasia
17. Chance that blindness will cause somatopsychological problems	Chance that blindness will cause psychosomatic problems
18. Chance that an ectomorph is tall and thin	Chance that an ectomorph is short and fat
19. Degree of relationship when the correlation between two factors is 0.2	Degree of relationship when the correlation between two factors is −0.8
20. Typical similarity in height of monozygotic twins reared apart	Typical similarity in height of dizygotic twins reared together (same sex)
21. Length of time an acute condition lasts	Length of time a chronic condition lasts

Chapter 12—Nutrition: Fueling the Body

Multiple Choice. *Choose the one best answer for each question.*

1. Proteins are built from ____. (a) vitamins (b) monosaccharides (c) amino acids (d) enzymes
2. Per weight, which of the following foods contains the most energy? (a) lipids (b) starches (c) sugars (d) proteins
3. Vitamins are ____. (a) inorganic compounds obtained almost entirely from food (b) enzymes produced to aid digestion (c) mainly amino acids obtained from plants (d) various organic compounds
4. The vitamin that is made when ultraviolet light strikes the skin is ____. (a) A (b) B_1 (c) C (d) D
5. Synergism refers to ____. (a) the effect of one agent minus the effect of another (b) the product of the effects of two agents (c) one agent intensifying the effect of another (d) the effect of one agent being canceled by the effect of another
6. Most cases of food poisoning result from ____. (a) food additives (b) chemical contaminants in food (c) toxins produced by microorganisms in food (d) hypervitaminosis

Comparison Items. *Study each pair of statements and use the following key: (a) the statement on the left suggests the greater quantity, (b) the statement on the right suggests the greater quantity, or (c) the two statements are essentially the same.*

7.	Average male BMR	Average female BMR
8.	Caloric needs per kilogram of body weight in children	Caloric needs per kilogram of body weight in adults
9.	Calories needed per hour for running	Calories needed per hour for climbing stairs
10.	Energy content of a gcal	Energy content of a kcal
11.	Quantity of calcium needed by the body	Quantity of iodine needed by the body
12.	Quantity of vitamins needed by the body	Quantity of minerals needed by the body
13.	Water-soluble nature of vitamin C	Fat-soluble nature of vitamin C
14.	Frequency of night blindness cause by a vitamin A deficiency	Frequency of night blindness caused by a vitamin B deficiency
15.	Chance of rickets caused by a deficiency of vitamin B	Chance of rickets caused by a deficiency of vitamin D
16.	Cycling of minerals needed by organisms	Cycling of energy needed by organisms
17.	Average number of calories needed by a sedentary person to support the basal metabolic rate	Average number of calories needed by a sedentary person to support activities
18.	Increase in the surface area of a cell as it grows	Increase in the volume of a cell as it grows
19.	Typical amount of water consumed directly daily	Typical amount of water consumed in food daily

Chapter 13—Nutrition: The Digestive Process

Multiple Choice. *Choose the one best answer for each question.*

1. The epiglottis ____. (a) prevents air from entering the esophagus (b) prevents nutrients from entering the trachea (c) is the sound-producing organ (d) producs important digestive enzymes
2. The products of chemical digestion are absorbed largely through the mucosa of the ____. (a) rectum (b) colon (c) duodenum (d) stomach
3. Before blood carrying the products of chemical digestion returns to the heart, it first passes through capillary networks in the ____. (a) gallbladder (b) pancreas (c) hypothalamus (d) liver
4. The lack of bile usually interferes most with the absorption of ____. (a) starches (b) lipids (c) sugars (d) proteins
5. Which of the following is usually most important in the sensation of hunger? (a) the amount of food in the stomach (b) a hormone from the liver (c) the level of amino acids in the blood (d) the blood glucose level
6. An important end product of carbohydrate digestion is (are) ____. (a) glucose (b) glycogen (c) glycerol (d) amino acids

7. Small fat molecules are absorbed into ＿＿. (a) capillaries in the stomach wall (b) the gallbladder (c) lymph in villi (d) bacteria in the colon

8. Glycogen reserves can be converted to glucose by the ＿＿. (a) liver (b) pancreas (c) hypothalamus (d) intestinal bacteria

9. An enzyme in saliva begins to digest ＿＿. (a) fats (b) proteins (c) oils (d) carbohydrates

10. Most water absorption occurs in the ＿＿. (a) mouth (b) stomach (c) small intestine (d) large intestine

11. The appendix opens into the ＿＿. (a) stomach (b) duodenum (c) colon (d) rectum

12. Bile is made by the ＿＿. (a) liver (b) gallbladder (c) pancreas (d) duodenum

13. A tooth's major blood vessels and nerve endings are found in the ＿＿. (a) enamel (b) pulp cavity (c) periodontal membrane (d) dentin

14. A zymogen is ＿＿. (a) a substance that speeds chemical reactions without being used up in the process (b) the most potent form of digestive enzyme released by the duodenum (c) an inactive form of a digestive enzyme (d) a major form of malnutrition caused by a caloric deficiency

Comparison Items. *Study each pair of statements and use the following key: (a) the statement on the left suggests the greater quantity, (b) the statement on the right suggests the greater quantity, or (c) the two statements are essentially the same.*

15. Acidity of intestinal juices	Alkalinity of intestinal juices
16. Importance of carbohydrate deficiency in causing kwashiorkor	Importance of protein deficiency in causing kwashiorkor
17. Absorption of alcohol from the stomach	Absorption of alcohol elsewhere in the gastrointestinal tract
18. Typical seriousness of constipation	Typical seriousness of diarrhea
19. Extracellular actions of human digestive enzymes	Intracellular actions of human digestive enzymes
20. Usual loss of teeth to dental caries	Usual loss of teeth to periodontal diseases
21. Number of deciduous teeth	Number of permanent teeth
22. Importance of the stomach in digestion	Importance of the stomach in food storage
23. Length of the small intestine	Length of the large intestine

Chapter 14—Respiration: Gas Exchange

Multiple Choice. *Choose the one best answer for each question.*

1. The digestive and breathing passages cross in the ＿＿. (a) esophagus (b) larynx (c) pharynx (d) trachea

2. Diffusion occurs only ＿＿. (a) across a selectively permeable membrane (b) when there are concentration differences (c) in solutions (d) across wet respiratory membranes

For questions 3–5, the approximate content of the air for each listed gas is ___ percent. (a) <1 (b) 20 (c) 60 (d) 80

3. oxygen
4. nitrogen
5. carbon dioxide

6. Puncture of the thoracic wall but not of the lung itself will cause ___. (a) collapse of the lung (b) inflation of the lung (c) no important interference with breathing (d) a decreased breathing rate
7. Exhaled air contains about ___ percent oxygen. (a) 0 (b) 6 (c) 11 (d) 16
8. The function of the eustachian tube is to ___. (a) control the size of the alveoli (b) determine whether air enters the trachea or the esophagus (c) equalize air pressure on the two sides of the eardrum (d) control the position of the diaphragm
9. The respiratory center is in the ___. (a) hypothalamus (b) medulla oblongata (c) lung (d) wall of the diaphragm and pleural cavity
10. Tracheae are most common in ___. (a) tadpoles (b) mammals (c) birds (d) insects

Comparison Items. *Study each pair of statements and use the following key: (a) the statement on the left suggests the greater quantity, (b) the statement on the right suggests the greater quantity, or (c) the two statements are essentially the same.*

11. Average breathing rate of neonates	Average breathing rate of adults
12. Surface area-to-volume ratio of a small animal	Surface area-to-volume ratio of a large animal
13. Length of time most people can go without air	Length of time most people can go without food
14. Weight of air consumed in an average day	Weight of water and food consumed in an average day
15. Rigidity of the trachea	Rigidity of the esophagus
16. Diameter of the bronchioles	Diameter of the bronchi
17. Healthiness of breathing through the nose	Healthiness of breathing through the mouth
18. Average temperature of environmental air	Average body temperature
19. Importance of the diaphragm in breathing	Importance of other muscles in breathing
20. Percentage of nitrogen in inhaled air	Percentage of nitrogen in exhaled air
21. Relative importance of high blood CO_2 in stimulating breathing	Relative importance of low blood oxygen in stimulating breathing
22. Inspiratory reserve volume	Expiratory reserve volume
23. Chance that the diaphragm is domed during inhalation	Chance that the diaphragm is domed during exhalation
24. Importance of oxygen in causing the bends	Importance of nitrogen in causing the bends
25. Affinity of oxygen to hemoglobin	Affinity of carbon monoxide to hemoglobin

26. Sound produced when air enters the trachea | Sound produced when air exits the trachea
27. Average vital capacity of females | Average vital capacity of males
28. Movement of two substances in the same direction in countercurrent circulation | Movement of two substances in different directions in countercurrent circulation
29. Efficiency of respiration in mammals | Efficiency of respiration in birds
30. pH number of acidic rain | pH number of basic rain

Chapter 15—Circulation: Internal Transport

Multiple Choice. *Choose the one best answer for each question.*

1. The part of the circulatory system in most direct contact with each cell is (are) the ____. (a) lymph (b) plasma (c) hematocrit (d) blood cells
2. CO is harmful because it ____. (a) saturates the plasma (b) forms a stable compound with hemoglobin (c) prevents the passage of erythrocytes through blood vessels (d) reduces the number of nitrogen bubbles in the blood
3. Hemoglobin is involved mainly with ____. (a) blood clotting (b) hormone formation (c) transport of nutrients (d) oxygen transport
4. Plasma makes up about ____ percent of the blood of humans living near sea level. (a) 35 (b) 45 (c) 55 (d) 65
5. Iron is essential because it is ____. (a) a major clotting agent (b) in the nucleus of erythrocytes (c) part of hemoglobin (d) useful in combating infections
6. A function of the spleen is to ____. (a) be stimulated by anger (b) cause reverse peristalsis in lymph vessels (c) secrete a blood-stimulating hormone (d) control blood volume
7. Leukocytes kill bacteria by ____. (a) outbreeding them (b) poisoning them with metabolic wastes (c) ingesting them (d) depriving them of oxygen
8. Hematocrit means the proportion of ____ in the blood. (a) formed elements (b) serum (c) plasma (d) proteins
9. Average adults contain about ____ liters of blood. (a) 0.5–1.0 (b) 3.0–4.0 (c) 5.0–6.0 (d) 7.0–8.0
10. Common heavy metals that enter the body tend to become concentrated in the ____. (a) spleen (b) thymus (c) bone marrow (d) lymph nodes
11. Arteries always carry ____. (a) oxygenated blood (b) deoxygenated blood (c) blood to the heart (d) blood from the heart
12. Heart murmur is due to a ____. (a) nonfunctional atrium (b) leaky valve (c) coronary thrombosis (d) small aorta
13. The ventricles provide the main force ____. (a) that propels blood through the body (b) that closes the valves of the aorta (c) needed to return venous blood (d) that moves blood out of the atria
14. In ____ systems, the blood that leaves capillaries flows into veins that terminate in other capillaries before it returns to the heart. (a) systemic (b) pulmonary (c) portal (d) cardiac
15. During diastole, ____ are relaxed. (a) only the ventricles (b) only the atria (c) both the ventricles and atria (d) neither the ventricles nor the atria

16. The longest part of the cardiac cycle is ___. (a) diastole (b) ventricular systole (c) atrial systole (d) the cardiac output
17. The pacemaker is located in the ___. (a) right atrium (b) left atrium (c) right ventricle (d) left ventricle
18. Impulses from the pacemaker cause the ___. (a) ventricles and contract (b) atria to contract (c) both the ventricles and atria to contract (d) heart to relax
19. The AV node is found in the ___. (a) aorta (b) atria (c) ventricles (d) wall between the atria and ventricles

Comparison Items. *Study each pair of statements and use the following key: (a) the statement on the left suggests the greater quantity, (b) the statement on the right suggests the greater quantity, or (c) the two statements are essentially the same.*

20. Ability of erythrocytes to move independently	Ability of leukocytes to move independently
21. Presence of nuclei in mature erythrocytes	Presence of nuclei in mature leukocytes
22. Frequency of erythrocytes per unit of blood	Frequency of leukocytes per unit of blood
23. Frequency of leukocytes per unit of blood	Frequency of thrombocytes per unit of blood
24. Importance of leukocytes in blood clotting	Importance of thrombocytes in blood clotting
25. Size of average leukocytes	Size of average thrombocytes
26. Number of erythrocytes in humans living at low altitudes	Number of erythrocytes in humans living at high altitudes
27. Volume of blood in humans living at low altitudes	Volume of blood in humans living at high altitudes
28. Amount of water in blood plasma	Amount of other chemicals in blood plasma
29. Average hematocrit in anemic humans	Average hematocrit in nonanemic humans
30. Quantity of oxygen carried by the blood plasma	Quantity of CO_2 carried by the blood plasma
31. Chance that a thrombus is found where it formed	Chance that an embolus is found where it formed
32. Speed of lymph flow at rest	Speed of blood flow at rest
33. Speed of lymph flow at rest	Speed of lymph flow while walking
34. Affinity of fetal hemoglobin for oxygen	Affinity of adult hemoglobin for oxygen
35. Percentage of lymphocytes	Percentage of neutrophils
36. Average number of leukocytes in victims of leukemia	Average number of leukocytes in nonvictims of leukemia
37. Blood pressure in arteries	Blood pressure in veins
38. Thickness of the right ventricle wall	Thickness of the left ventricle wall
39. Thickness of the atria walls	Thickness of the ventricle walls
40. Diastolic blood pressure	Systolic blood pressure
41. Frequency of valves in arteries	Frequency of valves in veins
42. Neonate's heartbeat rate	Adult's heartbeat rate
43. Similarity of arteries and lymphatic vessels	Similarity of veins and lymphatic vessels

Chapter 16—Disease, Cancer, and Defense

Multiple Choice. *Choose the one best answer for each question.*

1. Disease-causing organisms are called ___. (a) teratogens (b) mutagens (c) pathogens (d) carcinogens
2. Rh problems may develop when an ___ fetus is being carried by an ___ woman. (a) Rh^+; Rh^+ (b) Rh^+; Rh^- (c) Rh^-; Rh^+ (d) Rh^-; Rh^-
3. Normally, the ___ are the most common form of leukocytes. (a) neutrophils (b) monocytes (c) eosinophils (d) basophils
4. Macrophages develop from the ___. (a) neutrophils (b) monocytes (c) eosinophils (d) basophils
5. Interferons ___. (a) are the main ingredients in common vaccines (b) are produced by the body in response to virus infections (c) are obtained from pathogenic bacteria (d) decrease the body's ability to resist a second virus infection
6. So-called universal donors have blood type ___. (a) A (b) B (c) AB (d) O
7. No plasma antibodies are found in those with blood type ___. (a) A (b) B (c) AB (d) O
8. Metastasis is said to occur when ___. (a) malignant tumors become benign (b) cancer cells are totally destroyed (c) cancer cells enter the bloodstream (d) tumors enlarge

Comparison Items. *Study each pair of statements and use the following key: (a) the statement on the left suggests the greater quantity, (b) the statement on the right suggests the greater quantity, or (c) the two statements are essentially the same.*

9. Production of antibodies by B cells | Production of antibodies by T cells
10. Poisonous nature of toxoids | Poisonous nature of toxins
11. Cellular immunity involved with immediate-type hypersensitivity | Humoral immunity involved with immediate-type hypersensitivity
12. Allergic reactions due to histamines | Allergic reactions due to antihistamines
13. Number of antibodies formed upon exposure to antigens | Number of antigens formed upon exposure to antibodies
14. Frequency of Rh^+ Americans | Frequency of Rh^- Americans
15. Typical damage done to the firstborn child by an untreated Rh incompatibility | Typical damage done to subsequent children by an untreated Rh incompatibility
16. An antibody as an example of a nonspecific defense mechanism | An antibody as an example of a specific defense mechanism
17. Increase in porosity of capillaries due to histamine | Decrease in porosity of capillaries due to histamine
18. Stimulatory effect of fever to most pathogens | Inhibitory effect of fever to most pathogens
19. Y-shaped nature of most antigens | Y-shaped nature of most antibodies
20. Formation of granulocytes in red bone marrow | Formation of agranulocytes in red bone marrow

21. Frequency of lymphocytes in blood	Frequency of lymphocytes in lymph
22. Involvement of B cells in phagocytosis	Involvement of T cells in phagocytosis
23. Formation of plasma cells from B cells	Formation of plasma cells from T cells
24. Importance of antibodies in humoral defense mechanisms	Importance of antibodies in cellular defense mechanisms
25. Release of lymphokines from B cells	Release of lymphokines from T cells
26. Typical duration of natural immunity	Typical duration of passive immunity
27. Typical concern about benign tumors	Typical concern about malignant tumors
28. Percentage of Americans who died of cancer in 1900	Percentage of Americans who die from cancer today
29. B-cell deficiency in those with AIDS	T-cell deficiency in those with AIDS

Chapter 17—Excretion and Osmoregulation

Multiple Choice. *Choose the one best answer for each question.*

1. Which of the following substances is (are) normally resorbed completely by the kidney tubules? (a) CO_2 (b) minerals (c) nitrogen compounds (d) glucose
2. The average adult is about ____ water by weight. (a) one-fourth (b) one-third (c) half (d) two-thirds
3. Which of the following is not a function of the kidney? (a) controlling the chemical composition of the feces (b) regulating blood volume (c) regulating mineral concentrations (d) helping maintain the body's pH balance
4. All but about ____ percent of the water that passes through the nephrons is resorbed. (a) 0.03 (b) 1 (c) 20 (d) 99
5. The thirst center is located in the ____. (a) hypothalamus (b) kidneys (c) mouth (d) pituitary
6. Antidiuretic hormone is released by the ____. (a) hypothalamus (b) kidneys (c) cerebrum (d) pituitary
7. Urination generally occurs when the bladder holds about ____ ml of urine. (a) 50–300 (b) 300–800 (c) 800–1200 (d) 1200–1700
8. The main form of nitrogenous waste produced by humans is ____. (a) nitrogen oxides (b) ammonia (c) urea (d) uric acid
9. Our external environment is typically ____ relative to our tissue fluids. (a) hypotonic (b) hypertonic (c) isotonic (d) ginandtonic

Comparison Items. *Study each pair of statements and use the following key: (a) the statement on the left suggests the greater quantity, (b) the statement on the right suggests the greater quantity, or (c) the two statements are essentially the same.*

10. Average daily water loss from the lungs	Average daily water loss from perspiration
11. Intracellular body water	Extracellular body water
12. Normal number of ureters	Normal number of urethras

13. Variety of fluids carried in the female urethra — Variety of fluids carried in the male urethra
14. Stimulatory effect of diuretics upon urine production — Inhibitory effect of diuretics upon urine production
15. Length of time most humans can go without water — Length of time most humans can go without food
16. Amount of medulla tissue on the kidney's outer edge — Amount of cortex tissue on the kidney's outer edge
17. Permeability of the descending tubule to water — Permeability of the ascending tubule to water
18. Loss of Na^+ from the descending tubule — Loss of Na^+ from the ascending tubule
19. Normal quantity of Na^+ in blood plasma — Normal quantity of Na^+ in urine
20. Effect of drinking large amounts of water on stimulating release of ADH — Effect of dehydration on stimulating release of ADH
21. Quantity of urine produced by those with diabetes mellitus — Quantity of urine produced by those with diabetes insipidus
22. Average length of the male urethra — Average length of the female urethra
23. Toxicity of ammonia — Toxicity of uric acid
24. Range of tolerance for a euryhaline animal — Range of tolerance for a stenohaline animal
25. Frequency of Malpighian tubules in flatworms — Frequency of Malpighian tubules in insects

Chapter 18—Regulation: The Endocrine System

Multiple Choice. *Choose the one best answer for each question.*

1. Hormones are organic compounds that are _____. (a) required in the diet (b) used to speed chemical reactions (c) inherited from one's parents (d) coordinators of cellular actions
2. Endocrine glands secrete into _____. (a) ducts (b) the blood (c) the brain (d) the hypothalamus
3. The major gland involved with the body's immune response is the _____. (a) pituitary (b) thymus (c) parathyroid (d) hypothalamus
4. Insufficient dietary iodine causes a(n) _____. (a) enlarged thyroid (a) small thyroid (c) acromegaly (d) giant
5. An important effect of thyroxin is to _____. (a) decrease the metabolic rate of the liver (b) prevent bulging eyes (c) prevent hypothyroidism by maintaining a proper level of blood iodine (d) maintain the metabolic rate of most cells
6. Because Mary lacks endocrine cells in her pancreas, we would expect to find _____. (a) low blood glucose (b) normal blood glucose but high urine glucose (c) high blood glucose and high urine glucose (d) high blood glucose but low urine glucose
7. Mary's condition probably requires _____. (a) insulin to promote glucose movement into cells (b) insulin to slow the movement of glucose into cells (c) tablets of gastric juice to regulate carbohydrate digestion (d) glycogen to increase the blood glucose level

For questions 8–12, choose from the following:
(a) hypothalamus (b) pineal (c) parathyroid (d) pituitary

8. Controls calcium metabolism
9. Produces growth hormone
10. Contains the body's thermostat
11. Produces hormones that are stored in the posterior pituitary
12. Affects the onset of puberty

13. Secretions from the _____ are most important in the body's fight-or-flight response to a stressful situation. (a) pituitary (b) adrenal (c) thyroid (d) hypothalamus
14. A secretion released by one organism that influences the behavior of another organism is called a(n) _____. (a) melatonin (b) endorphin (c) pheromone (d) prostaglandin

Comparison Items. *Study each pair of statements and use the following key: (a) the statement on the left suggests the greater quantity, (b) the statement on the right suggests the greater quantity, or (c) the two statements are essentially the same.*

15. Frequency of acromegaly developing before puberty | Frequency of acromegaly developing after puberty
16. Necessity of adrenal cortex for life | Necessity of adrenal medulla for life
17. Chance of cretinism developing from a hypothyroid condition | Chance of cretinism developing from a hyperthyroid condition
18. Number of tropic hormones released by the pituitary | Number of tropic hormones released by the hypothalamus
19. Necessity of the thyroid gland for life | Necessity of the parathyroid glands for life
20. Effect of oxytocin upon milk production | Effect of oxytocin upon milk secretion
21. Effect of ADH in causing diabetes insipidus | Effect of ADH in causing diabetes mellitus
22. Chance that hormones function as first messengers | Chance that hormones function as second messengers
23. Sterility from hyposecretion of follicle-stimulating hormone | Sterility from hypersecretion of follicle-stimulating hormone
24. Frequency of diabetes mellitus in males | Frequency of diabetes mellitus in females

Chapter 19—Regulation: Neurons and the Nervous System

Multiple Choice. *Choose the one best answer for each question.*

1. Which of the following is the correct order in a normal reflex?
   (a) receptor, effector, sensory neuron, integrator, motor neuron
   (b) receptor, integrator, sensory neuron, motor neuron, effector
   (c) effector, sensory neuron, motor neuron, integrator, receptor
   (d) receptor, sensory neuron, integrator, motor neuron, effector
2. The cerebellum _____. (a) releases hormones (b) contains the pons (c) controls coordination (d) is filled with cerebrospinal fluid
3. The basic arrangement of the autonomic nervous system is that _____. (a) the parasympathetic stimulates and the sympathetic division inhibits (b) the sympathetic stimulates and the parasympathetic division inhibits (c) on any one organ the two divisions

have opposite effects (d) on any one organ the two divisions have additive effects

4. The brain center that controls many basic physiological processes is the ___. (a) hypothalamus (b) cerebellum (c) cerebrum (d) midbrain

Questions 5–8 refer to the diagram, in which the letters show parts of the nervous system that are blocked.
KEY: (a) block is at A (b) block is at B (c) block is at C (d) no block

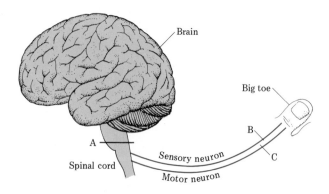

5. The person can move the toes but cannot feel the movement.
6. The person can feel the toe being stimulated but cannot move it.
7. When the skin of the toe is stimulated the toe moves and the person knows that it is moving.
8. When the skin of the toe is stimulated the toe moves but the person does not know that it is moving.

9. The meninges surround and protect ___. (a) the brain only (b) the spinal cord only (c) both the brain and the spinal cord (d) neither the brain nor the spinal cord

Comparison Items. *Study each pair of statements and use the following key: (a) the statement on the left suggests the greater quantity, (b) the statement on the right suggests the greater quantity, or (c) the two statements are essentially the same.*

10. Frequency of chemical control in organisms | Frequency of nervous control in organisms
11. Speed of nerve impulses in myelinated fibers | Speed of nerve impulses in non-myelinated fibers
12. Frequency of neurotransmitters passing from axons to dendrites | Frequency of neurotransmitters passing from dendrites to axons
13. Number of dendrites in "typical" neurons | Number of axons in "typical" neurons
14. Amount of white matter composed of neuron cell bodies | Amount of white matter composed of neuron fibers
15. Response of a neuron to a threshold stimulus | Response of a neuron to an above-threshold stimulus
16. Number of sensory neurons in the nervous system | Number of association neurons in the nervous system
17. Control of breathing by the medulla oblongata | Control of breathing by the cerebellum

18. Number of spinal nerves   Number of cranial nerves
19. Gray matter located in the inner spinal cord   White matter located in the inner spinal cord
20. Sympathetic nerves originating from the upper and lower portions of the spinal cord   Parasympathetic nerves originating from the upper and lower portions of the spinal cord
21. Speed of electricity   Speed of the nerve impulse
22. Relative activity of most radially symmetrical animals   Relative activity of most bilaterally symmetrical animals

Chapter 20—Regulation: Sensory Systems

Multiple Choice. *Choose the one best answer for each question.*

1. Proprioceptors are most concerned with ____. (a) balance (b) touch (c) pain (d) temperature
2. The back of the tongue is sensitive mainly to ____. (a) sweet (b) bitter (c) salt (d) sour
3. Which receptor listed has the simplest structure? (a) pain (b) pressure (c) temperature (d) chemical
4. Headaches that attack one side of the head or begin there are called ____. (a) tension (b) chemical (c) cluster (d) migraine
5. The amount of light entering the eye is controlled by the ____. (a) lens (b) iris (c) cornea (d) retina
6. If the light-sensitive cells in the retina are induced to send an impulse because of increased pressure on the eyeball, the resulting sensation will be of ____. (a) pressure (b) pain (c) light (d) vertigo
7. Your blind spot is where ____. (a) the optic nerve leaves the retina (b) there are only cones on the retina (c) only rods occur on the retina (d) the lens has become cloudy
8. Astigmatism occurs when ____. (a) there are no rods (b) the cones are absent (c) people wear inverted glasses (d) the cornea or lens is not curved evenly
9. The eustachian tube helps ____. (a) both eyes focus properly (b) your sense of balance (c) keep the pressure even on both sides of the eardrum (d) lower the pitch of very high sounds
10. The front of the eye is protected by the ____. (a) melanin (b) conjunctiva (c) choroid coat (d) optic nerve
11. Difficulty in accommodating is called ____. (a) presbyopia (b) hyperopia (c) myopia (d) amblyopia
12. A sound of 20 decibels has a sound intensity ____ times that of a threshold sound. (a) $\frac{1}{20}$ (b) 20 (c) 100 (d) 1000
13. Balance and hearing are centered, respectively, in the ____. (a) cochlea and semicircular canals (b) semicircular canals and cochlea (c) cochlea and middle-ear bones (d) middle-ear bones and semicircular canals
14. A placebo is the ____. (a) same as hypnosis (b) use of a substance inactive in the process being studied (c) insertion of sterilized needles into the body (d) application of pressure to certain sites to relieve pain

Comparison Items. *Study each pair of statements and use the following key: (a) the statement on the left suggests the greater quantity, (b) the statement on the right suggests the greater quantity, or (c) the two statements are essentially the same.*

15. Speed of adaptation to odors — Speed of adaptation to pain

16. Number of cold receptors in the skin — Number of heat receptors in the skin

17. Number of pressure receptors in the skin — Number of pain receptors in the skin

18. Depth of cold receptors in the skin — Depth of heat receptors in the skin

19. Referred pain felt in an injured organ — Referred pain felt away from an injured organ

20. Importance of an absolute fiber firing rate in the sensation of pain — Importance of a relative fiber firing rate in the sensation of pain

21. Sensitivity of the tip of the tongue to sour tastes — Sensitivity of the tip of the tongue to salty tastes

22. Sensitivity of our sense of smell — Sensitivity of our sense of taste

23. Typical effect of taste on our sense of smell — Typical effect of smell on our sense of taste

24. Amount of refraction by the cornea — Amount of refraction by the lens

25. Frequency of a flat, stretched lens when viewing distant objects — Frequency of a thickened lens when viewing distant objects

26. Frequency of an abnormally long eyeball in hyperopia — Frequency of an abnormally short eyeball in hyperopia

27. Medical training of an ophthalmologist — Medical training of an optometrist

28. Number of rods in the retina — Number of cones in the retina

29. Light threshold for rods — Light threshold for cones

30. Density of rods around the periphery of the retina — Density of cones around the periphery of the retina

31. Involvement of rods in color vision — Involvement of cones in color vision

32. Size of pupil in bright light — Size of pupil in dim light

33. Right-side-up images on your retina — Upside down images on your retina

34. Frequency of excessive intraocular pressure causing glaucoma — Frequency of excessive intraocular pressure causing cataracts

35. Importance of "ear stones" to our dynamic sense of balance — Importance of "ear stones" to our static sense of balance

36. Binocular vision when eyes are close together on an animal's face — Binocular vision when eyes are widely separated on the sides of an animal's head

37. Importance of vitamin A in preventing night blindness — Importance of vitamin C in preventing night blindness

Chapter 21—Effectors and Support: Action Systems

Multiple Choice. *Choose the one best answer for each question.*

1. If one muscle relaxes when a second muscle contracts, the two muscles are called ___. (a) abductors (b) depressors (c) sphincters (d) antagonists

2. When a muscle fiber is stimulated, it ____. (a) contracts in proportion to the strength of the stimulus (b) contracts completely or not at all (c) relaxes (d) contracts if it is smooth but relaxes if it is cardiac muscle

For questions 3–6, choose from the following:
(a) abductor (b) levator (c) sphincter (d) supinator

3. Raises the lower jaw
4. Rotates the palm upward and outward
5. Closes the opening between the stomach and the small intestine
6. Moves the legs outward sideways

7. The main organic component of muscle is ____. (a) lipid (b) carbohydrate (c) protein (d) water
8. The main organic component of bone is ____. (a) collagen (b) calcium (c) glycogen (d) myoglobin
9. Which statement is true? (a) Levels of calcium in the blood and in the bones are unrelated. (b) The calcium levels in bone and blood are identical. (c) Levels of bone calcium are maintained at the expense of blood calcium. (d) Levels of blood calcium are maintained at the expense of bone calcium.

Comparison Items. *Study each pair of statements and use the following key: (a) the statement on the left suggests the greater quantity, (b) the statement on the right suggests the greater quantity, or (c) the two statements are essentially the same.*

10. Striations in cardiac muscle — Striations in smooth muscle
11. Contraction rate of most smooth muscles — Contraction rate of most striated muscles
12. Rate at which most smooth muscles tire — Rate at which most skeletal muscles tire
13. Specialization of red skeletal muscle fibers for intense activity — Specialization of white skeletal muscle fibers for intense activity
14. Voluntary control over striated muscles — Voluntary control over nonstriated muscles
15. Ability of muscles to push — Ability of muscles to pull
16. Frequency of bones attached to bones by tendons — Frequency of bones attached to bones by ligaments
17. Assistance given to antagonists by synergists — Assistance given to synergists by antagonists
18. Movement involving a muscle's origin — Movement involving a muscle's insertion
19. Number of finger-moving muscles found in the finger — Number of finger-moving muscles found in the forearm
20. Size of hypertrophied muscles — Size of atrophied muscles
21. Number of bones in the axial skeleton — Number of bones in the appendicular skeleton
22. Number of bones in the body — Number of muscles in the body
23. Amount of inorganic material in bone — Amount of organic material in bone
24. Amount of yellow marrow in children — Amount of yellow marrow in adults

25. Blood vessels in bone | Blood vessels in cartilage
26. Hematopoietic regions in red marrow | Hematopoietic regions in yellow marrow
27. Growth at the tips of bones | Growth within the middle regions of bones
28. Quantity of synovial fluid secreted by younger adults | Quantity of synovial fluid secretion by older adults
29. Number of muscle fibers in the motor units of leg muscles | Number of muscle fibers in the motor units of eyeball muscles
30. Exoskeletons in arthropods | Endoskeletons in arthropods

Chapter 22—Substance Abuse: A Challenge to Homeostasis

Multiple Choice. *Choose the one best answer for each question.*

1. Because of synergistic effects, the net effect of several agents occurring together is ___. (a) greater than the combination of the independent effects of each agent (b) less than the combination of the independent effects of each agent (c) the sum of the effects of each agent acting independently (d) zero or very low because the agents usually cancel each other's effect
2. Which of the following is not a stimulant? (a) cocaine (b) amphetamine (c) nicotine (d) marijuana
3. Which substance is not derived from opium? (a) morphine (b) cocaine (c) codeine (d) heroin
4. Although alcohol has many effects, it is usually classified as a ___. (a) hallucinogen (b) stimulant (c) depressant (d) barbiturate
5. Mixing of the senses is known as ___. (a) a hallucination (b) an illusion (c) a delusion (d) synesthesia
6. Which of the following is not a hallucinogen? (a) histamine (b) psilocybin (c) mescaline (d) marijuana
7. Endorphins are ___. (a) produced in the brain (b) the most active ingredients in opiates (c) the main class of natural hallucinogens (d) designer stimulants

Comparison Items. *Study each pair of statements and use the following key: (a) the statement on the left suggests the greater quantity, (b) the statement on the right suggests the greater quantity, or (c) the two statements are essentially the same.*

8. Chance for physical dependence from use of common stimulants | Chance for psychological dependence from use of common stimulants
9. Potency of opium | Potency of heroin
10. Alcohol content (by volume) of most wines | Alcohol content (by volume) of most beers and ales
11. Absorption rate of alcohol from an empty stomach | Absorption rate of alcohol from a stomach that also contains food
12. Chance that flashbacks occur from heavy use of stimulants | Chance that flashbacks occur from heavy use of hallucinogens
13. Chance that delusions involve one's feelings | Chance that delusions involve one's senses
14. Chance for physical dependence from use of common hallucinogens | Chance for psychological dependence from use of common hallucinogens

15. Involvement of a change in threshold in tolerance	Involvement of a change in threshold in dependence
16. Methanol content of alcoholic beverages	Ethanol content of alcoholic beverages
17. Amount of alcohol absorbed by the stomach	Amount of alcohol absorbed by the duodenum

Chapter 23—Fitness

Multiple Choice. *Choose the one best answer for each question.*

1. Work equals ____. (a) force plus distance (b) force divided by distance (c) force times distance (d) force plus distance times time
2. The length of time that power is used is ____. (a) strength (b) endurance (c) work (d) physiological reserve
3. During heavy exercise, blood flow to the ____ decreases. (a) skin (b) brain (c) muscles (d) kidneys

Comparison Items. *Study each pair of statements and use the following key: (a) the statement on the left suggests the greater quantity, (b) the statement on the right suggests the greater quantity, or (c) the two statements are essentially the same.*

4. Muscle movement involved in isometric exercise	Muscle movement involved in isotonic exercise
5. Number of antagonistic muscles in the legs	Number of antagonistic muscles in the heart
6. Increase in blood flow to the brain during heavy exercise	Increase in blood flow to the heart during heavy exercise
7. Speed of aerobic metabolism	Speed of anaerobic metabolism
8. Size of a hypertrophied muscle	Size of an atrophied muscle
9. Increase in ventilation (above normal rate) during heavy exercise	Increase in cardiac output (above normal rate) during heavy exercise
10. Improvement in endurance from jogging	Improvement in endurance from weight lifting
11. Value of early application of heat to a pulled muscle	Value of early application of heat to a sore or stiff muscle
12. Relative increase in diastolic pressure during heavy exercise	Relative increase in systolic pressure during heavy exercise
13. Presence of a "hamstring" muscle in the back	Presence of a "hamstring" muscle in the leg

Chapter 24—The Human Past

Multiple Choice. *Choose the one best answer for each question.*

1. Which statement is important to the theory of natural selection? (a) The traits an organism acquires during its lifetime are passed to its offspring. (b) Organisms with certain traits are better able to survive and reproduce. (c) Organisms change their traits in order to survive better. (d) Only complex organisms survive when their environments change.

2. Part of Darwin's theory is that ____. (a) humans developed from modern apes and monkeys (b) acquired traits of parents are always inherited by their offspring (c) organisms with unfavorable variations have no right to survive (d) organisms with favorable traits are most likely to survive

3. Which of the following is not a fossil? (a) extinct clam (b) petrified wood (c) hardened imprint of a raindrop (d) dinosaur footprint

4. Primates appear to have developed from small, extinct ____. (a) carnivores (b) herbivores (c) insectivores (d) omnivores

5. The scientific name of modern humans is ____. (a) *Homo Sapiens* (b) *homo sapiens* (c) *Homo sapiens* (d) *homo Sapiens*

6. Which of the following is the oldest genus? (a) *Australopithecus* (b) *Dryopithecus* (c) *Homo* (d) *Ramapithecus*

7. The only vitamin that humans can synthesize in their bodies is vitamin ____. (a) A (b) B (c) C (d) D

8. Which of the following is the most primitive group of primates? (a) anthropoids (b) humans (c) prosimians (d) hominids

9. Egg-laying mammals are called ____. (a) placental (b) monotremes (c) marsupials (d) primates

Comparison Items. *Study each pair of statements and use the following key: (a) the statement on the left suggests the greater quantity, (b) the statement on the right suggests the greater quantity, or (c) the two statements are essentially the same.*

10. Types of geological processes today | Assumed types of geological processes in the past

11. Amount of melanin visible in skin exposed to UV light | Amount of melanin visible in skin not exposed to UV light

12. Differences between humans and the gorillas and chimpanzees | Differences between gorillas and the chimpanzees and lower apes

13. Chance that organisms adapt immediately to environmental changes in order to survive | Chance that organisms survive environmental changes because they are already adapted

14. Number of years since Darwin wrote *On the Origin of Species* | Number of years since Darwin wrote *The Descent of Man*

15. Length of time organisms lived on Earth | Length of time no organisms lived on Earth

16. Typical intrapopulation breeding | Typical interpopulation breeding

17. Chance of genetic drift in small populations | Chance of genetic drift in large populations

18. Differences between human races in biochemical traits | Differences between human races in external traits

Chapter 25—Human Ecology: Living in the Environment

Multiple Choice. *Choose the one best answer for each question.*

1. Cultural eutrophication means the ____. (a) slowing of lake aging (b) speeding of lake aging (c) addition of large amounts of toxic substances to lakes (d) filling of lakes to produce more industrial land

2. Aquaculture is the ____. (a) water portion of the biosphere (b) "farming" of water (c) water content of organisms (d) loss of water by agricultural plants

3. In which of the following organisms that belong to a food chain would you expect to find the highest level of an environmental contaminant? (a) snake (b) petunia caterpillar (c) petunia (d) toad

4. Which graph illustrates the human population growth? (a) A (b) B (c) C (d) D

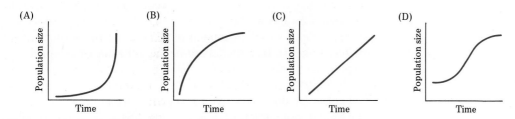

5. Water is most dense at ____°C. (a) 0 (b) 2 (c) 4 (d) 6

6. Organisms are composed of matter and use energy for their activities. Matter and energy ____. (a) both circulate and move in cycles from nonliving to living, back to nonliving, and so on (b) both pass from nonliving to living things but are then lost and never used again (c) differ in that matter depends upon CO_2 and energy upon O_2 (d) differ in that matter moves in cycles but energy does not

For questions 7–9, refer to the figure illustrating a food web.

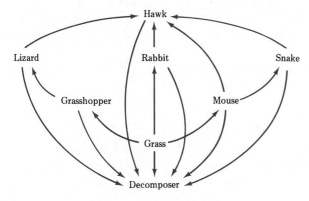

7. The hawk is a third-order consumer if its eats the ____. (a) grass (b) mouse (c) rabbit (d) snake

8. The producer is the ____. (a) grass (b) hawk (c) grasshopper (d) lizard

9. The decomposers function to ____. (a) return energy to the air for use by producers (b) release minerals from dead organisms for use by other organisms (c) increase the heat in the ecosystem (d) decrease the mineral supply

10. One winter, seven sailors were shipwrecked on a barren Arctic island that had neither soil nor vegetation. A crate of corn flakes and one containing seven hens was also cast ashore with them. In order to survive as long as possible, the sailors should ____. (a) feed the corn flakes to the hens as long as they last and then kill and eat the hens (b) kill and eat the hens and then eat the corn flakes (c) feed the corn flakes to the hens and then eat the eggs that the hens produce (d) eat the corn flakes, giving none to the hens, and then eat the hens when they die of starvation

11. Suppose bear eat salmon, salmon eat crustaceans, and crustaceans eat algae. Then 1000 kg of algae would support about how many kilograms of bear? (a) 1000 (b) 100 (c) 10 (d) 1

12. The *basic functional unit* in ecology is the ____. (a) organism (b) biosphere (c) ecosystem (d) population

13. Which of the following is not a fossil fuel? (a) coal (b) peat (c) uranium (d) petroleum

14. The best example of an ecological community is ____. (a) a meadow (b) one oak tree (c) all the people in Los Angeles (d) all the whales on Earth

15. An example of social parasitism is ____. (a) mistletoe on trees (b) liver flukes in humans (c) egg-laying behavior of cowbirds (d) fleas on a dog

16. An example of commensalism is ____. (a) leaves and caterpillars (b) lichens (c) vines on a tree (d) lice on a bird

17. The human population on Earth is now about ____. (a) 50 million (b) 500 million (c) 5 billion (d) 50 billion

Comparison Items. *Study each pair of statements and use the following key: (a) the statement on the left suggests the greater quantity, (b) the statement on the right suggests the greater quantity, or (c) the two statements are essentially the same.*

18. Number of people in a mega-lopolis | Number of people in a metropolis

19. Typical interspecific competition | Typical intraspecific competition

20. Precipitation in prairies | Precipitation in plains

21. Latitude of major coniferous forests | Latitude of major deciduous forests

22. Number of species in a population | Number of species in a community

23. Amount of water on Earth a century ago | Amount of water on Earth today

24. Typical number of species in a food web | Typical number of species in a food chain

25. Usual quantity of precipitation in grasslands | Usual quantity of precipitation in savannas

ANSWERS TO SELF-EVALUATION QUESTIONS

Chapter 1—Biology: The Study of Life

1. b 2. d 3. b 4. b 5. d 6. a 7. c 8. d 9. b 10. a 11. b 12. a 13. b 14. b 15. b 16. a

Chapter 2—Homeostasis and Adaptation: An Introduction

1. a 2. d 3. b 4. b 5. a 6. a 7. b 8. a 9. a 10. a 11. a 12. b 13. b

Chapter 3—Structural Organization for Homeostasis

1. d 2. c 3. c 4. d 5. b 6. a 7. d 8. a 9. b 10. b 11. b 12. b 13. b 14. a 15. d 16. c 17. d 18. c 19. b 20. b 21. b 22. a 23. a 24. a 25. a 26. b 27. a 28. a 29. b 30. a 31. b

Chapter 4—Chemical Organization for Homeostasis

1. b 2. a 3. d 4. a 5. c 6. b 7. c 8. b 9. b 10. d 11. a 12. c 13. c 14. a 15. b 16. a 17. a 18. b 19. b 20. b 21. b 22. b 23. b 24. a

Chapter 5—The Reproductive System

1. b 2. c 3. c 4. d 5. a 6. d 7. b 8. b 9. a 10. d 11. c 12. c 13. b 14. b 15. b 16. a 17. b 18. a 19. a 20. b 21. a 22. b 23. b 24. a 25. a 26. b

Chapter 6—Pregnancy and Birth

1. a 2. c 3. b 4. a 5. b 6. c 7. b 8. c 9. b 10. a 11. c 12. d 13. a 14. a 15. c 16. a 17. a 18. b 19. b 20. b 21. a 22. b 23. b 24. c 25. a 26. b 27. a 28. b 29. a 30. b

Chapter 7—Contraception

1. c 2. a 3. c 4. b 5. b 6. a 7. a 8. b 9. a 10. b 11. b 12. b

Chapter 8—Sexually Transmitted Diseases

1. b 2. b 3. d 4. c 5. c 6. b 7. a 8. a 9. a 10. b 11. a 12. a 13. a 14. a 15. a 16. a 17. b

Chapter 9—Basic Human Genetics

1. a 2. b 3. b 4. c 5. c 6. d 7. c 8. c 9. a 10. c 11. b 12. b 13. a 14. d 15. a 16. b 17. a 18. b 19. a 20. a 21. a 22. a 23. b 24. a 25. c 26. b 27. a 28. a 29. b 30. b 31. b

Chapter 10—Hereditary Mechanisms

1. b 2. d 3. c 4. a 5. c 6. b 7. c 8. b 9. a 10. c 11. b 12. b 13. a 14. a 15. a 16. b 17. b 18. a 19. a 20. a 21. a

Chapter 11—Growth and Development

1. c 2. d 3. d 4. b 5. b 6. d 7. c 8. c 9. c 10. a 11. a 12. b 13. b 14. b 15. a 16. b 17. a 18. a 19. b 20. a 21. b

Chapter 12—Nutrition: Fueling the Body

1. c 2. a 3. d 4. d 5. c 6. c 7. a 8. a 9. b 10. b 11. a 12. b 13. a 14. a 15. b 16. a 17. a 18. b 19. a

Chapter 13—Nutrition: The Digestive Process

1. b 2. c 3. d 4. b 5. d 6. a 7. c 8. a 9. d 10. d 11. c 12. a 13. b 14. c 15. b 16. b 17. b 18. b 19. a 20. b 21. b 22. b 23. a

Chapter 14—Respiration: Gas Exchange

1. c 2. b 3. b 4. d 5. a 6. a 7. d 8. c 9. b 10. d 11. a 12. a 13. b 14. a 15. a 16. b 17. a 18. b 19. a 20. c 21. a 22. a 23. b 24. b 25. b 26. b 27. b 28. b 29. b 30. b

Chapter 15—Circulation: Internal Transport

1. a 2. b 3. d 4. c 5. c 6. d 7. c 8. a 9. c 10. c 11. d 12. b 13. a 14. c 15. c 16. a 17. a 18. b 19. d 20. b 21. b 22. a 23. b 24. b 25. a 26. b 27. b 28. a 29. b 30. b 31. a 32. b 33. b 34. a 35. b 36. a 37. a 38. b 39. b 40. b 41. b 42. a 43. b

Chapter 16—Disease, Cancer, and Defense

1. c 2. b 3. a 4. b 5. b 6. d 7. c 8. c 9. a 10. b 11. b 12. a 13. a 14. a 15. b 16. b 17. a 18. b 19. b 20. a 21. a 22. b 23. a 24. a 25. b 26. a 27. b 28. b 29. b

Chapter 17—Excretion and Osmoregulation

1. d 2. d 3. a 4. b 5. a 6. d 7. b 8. c 9. a 10. b 11. a 12. a 13. b 14. a 15. b 16. b 17. a 18. b 19. c 20. b 21. b 22. a 23. a 24. a 25. b

Chapter 18—Regulation: The Endocrine System

1. d 2. b 3. b 4. a 5. d 6. c 7. a 8. c 9. d 10. a 11. a 12. b 13. b 14. c 15. b 16. a 17. a 18. a 19. b 20. b 21. a 22. a 23. a 24. b

Chapter 19—Regulation: Neurons and the Nervous System

1. d 2. c 3. c 4. a 5. b 6. c 7. d 8. a 9. c 10. a 11. a 12. a 13. a 14. b 15. c 16. b 17. a 18. a 19. a 20. b 21. a 22. b

Chapter 20—Regulation: Sensory Systems

1. a 2. b 3. a 4. d 5. b 6. c 7. a 8. d 9. c 10. b 11. a 12. c 13. b 14. b 15. a 16. a 17. b 18. b 19. b 20. b 21. b 22. a 23. b 24. a 25. a 26. b 27. a 28. a 29. b 30. a 31. b 32. b 33. b 34. a 35. b 36. a 37. a

Chapter 21—Effectors and Support: Action Systems

1. d 2. b 3. b 4. d 5. c 6. a 7. c 8. a 9. d 10. a 11. b 12. b 13. a 14. a 15. b 16. b 17. a 18. b 19. b 20. a 21. b 22. b 23. a 24. b 25. a 26. a 27. a 28. a 29. a 30. a

Chapter 22—Substance Abuse: A Challenge to Homeostasis

1. a 2. d 3. b 4. c 5. d 6. a 7. a 8. b 9. b 10. a 11. a 12. b 13. a 14. b 15. a 16. b 17. b

Chapter 23—Fitness

1. c 2. b 3. d 4. b 5. a 6. b 7. a 8. a 9. a 10. a 11. b 12. b 13. b

Chapter 24—The Human Past

1. b 2. d 3. c 4. c 5. c 6. b 7. d 8. c 9. b 10. c 11. a 12. b 13. b 14. a 15. a 16. a 17. a 18. b

Chapter 25—Human Ecology: Living in the Environment

1. b 2. b 3. a 4. a 5. c 6. d 7. d 8. a 9. b 10. b 11. d 12. c 13. c 14. a 15. c 16. c 17. c 18. a 19. b 20. a 21. a 22. b 23. c 24. a 25. b

GLOSSARY

ABO blood group A group of blood types based upon the presence or absence of the A and B antigens.

Abortion *(a-BOR-shun)* The premature loss or removal of the embyro or fetus; may be spontaneous or induced.

Absorption *(ab-SORP-shun)* The taking up of liquids by solids or of gases by solids or liquids.

Abstinence Not engaging in an act, such as coitus.

Accommodation *(a-kom-o-DA-shun)* A change in the curvature of the eye lens for focusing on objects at various distances.

Acetylcholine *(ass-see-til-KO-leen)* A neurotransmitter liberated at synapses in the central nervous system; stimulates striated muscle contraction.

Acid *(AS-id)* A proton donor; excess hydrogen ions produce a pH less than 7; acids have a sour taste and unite with bases to form salts.

Acid rain Precipitation with a pH below about 5; may damage vegetation, aquatic life, and soil organisms.

Acidosis A serious condition in which there are reduced amounts of the normal alkaline substances in blood.

Acquired immune deficiency syndrome An infectious, perhaps always fatal viral disease that attacks T cells and thus reduces the body's ability to defend itself against pathogens; abbreviated AIDS.

Acromegaly *(ak-ro-MEG-uh-lee)* Thickened bone and enlargement of cartilaginous structures (e.g., nose, jaws); caused by hypersecretion of growth hormone in an adult.

Acrosome *(AK-ruh-some)* A caplike organelle at the tip of a sperm that helps the sperm penetrate an egg.

ACTH Abbreviation for adrenocorticotropic hormone; see *Addison's disease*.

Actin *(AK-tin)* The contractile protein that makes up thin myofilaments in muscle fibers, see *myosin*.

Action potential An impulse of negativity or rapid change in the membrane potential along an excitable cell such as a neuron.

Activation energy The energy required for a molecule to undergo a specific chemical reaction.

Active immunity Immunity resulting from antibodies formed by a person in response to an antigen; see *passive immunity*.

Active transport The movement of a substance across the cell membrane against a concentration gradient by a process that requires energy, which is typically supplied by the expenditure of ATP.

Acuity *(a-KYOO-i-tee)* Clearness or sharpness, usually of vision.

Acupressure Application of pressure to select points for the relief of pain.

Acupuncture *(AK-u-punk-chur)* Insertion of sterilized needles to select points for the relief of pain.

Acute *(a-KYOOT)* Having rapid onset, serious symptoms, and a short course; not chronic.

Adaptation (1) Changes in the eye's pupil in response to light variations. (2) The ability of an organism to adjust to an environmental change by means of its anatomy, physiology, and/or behavior.

Addiction Mental or physical condition requiring continuing use of a drug.

Addison's disease A deficiency in the secretion of adrenocortical hormones; results in weakness, loss of appetite, low blood sugar, and characteristic skin pigmentation.

Adenoid The pharyngeal tonsils.

ADH Abbreviation for <u>a</u>nti<u>d</u>iuretic <u>h</u>ormone.

Adipose Referring to fatty tissue or to the fat itself.

Adrenal *(a-DREE-nal)* A gland on top of the kidney; contains an outer cortex and an inner medulla.

Adrenalin A secretion of the adrenal medulla; also called *epinephrine;* responsible for the physiological changes associated with the alarm response.

Adrenocorticotropic hormone Hormone, produced by the anterior pituitary, which affects the adrenal cortex; abbreviated ACTH.

Aerobic A biological process that requires free oxygen.

Afterbirth The placenta and associated structures expelled shortly after the birth process.

Agglutination *(a-gloot-i-NAY-shun)* Clumping of blood cells; an antigen-antibody reaction; an immunity response.

Aging Progressive decline in the body's homeostatic adaptive abilities.

Agranulocytes Leukocytes that lack cytoplasmic granules and usually have spherical nuclei (e.g., lymphocytes and monocytes); develop from lymphoid tissue.

AIDS Abbreviation for <u>a</u>cquired <u>i</u>mmune <u>d</u>eficiency <u>s</u>yndrome.

Air pollution Impairment of air quality due to the presence of harmful unnatural substances or forms of energy or the adverse increase or reduction in a natural component.

Air sac Common name for an alveolus.

Albinism Nonpathological absence (partial or total) of pigment in skin, hair, and eyes.

Alcohol Colorless, volatile, flammable liquid; C_2H_6O; beverage alcohol is ethanol (ethyl alcohol) that is produced through fermentation of various materials.

Aldosterone *(al-do-STER-on)* The main mineralocorticoid hormone produced by the adrenal cortex; regulates blood sodium and potassium.

Alkalosis A condition due to an excess of alkalies or withdrawal of acid or chlorides from the blood.

Allele *(uh-LEEL)* One of two or more alternative forms of a gene.

Allen's principle Principle that the size of the extremities of endotherms living in warm areas tends to be larger than those of relatives living in cold regions; see *Bergman's principle.*

Allergen *(AL-er-jen)* A substance that can induce an allergy; usually antigenic.

Allergy A hypersensitivity to an environmental substance, manifested as hay fever, skin rash, or asthma.

All-or-none-principle   The idea that muscle fibers and neurons respond either completely or not at all.

Altricial *(al-TRISH-ul)*   Condition of young being produced in a very immature and helpless condition, requiring much parental care; contrast with *precocial*.

Alveolus *(al-VEE-oh-lus)*   A small cavity; the thin-walled air sacs within the lungs where gas exchange occurs with capillaries; a milk-secreting sac in a mammary gland.

Amblyopia   Dimness of vision or partial loss of sight; usually due to the brain rejecting images from the affected eye; "lazy eye" condition.

Amenorrhea   Absence or suppression of menstruation.

Amino acid *(uh-MEE-noh)*   One of 20 nitrogen-containing compounds that may link to form peptide chains of protein molecules.

Ammonia   The simplest nitrogen compound that is soluble in water; NH_3.

Amniocentesis *(AM-nee-oh-sen-TEE-sis)*   Removal of amniotic fluid and cells by inserting a needle through the abdomen into the amniotic cavity; a technique used to detect genetic or metabolic abnormalities in a fetus.

Amnion *(AM-nee-on)*   The thin, innermost extraembryonic membrane that encloses a fluid-filled sac around the embryo and fetus.

Amoeboid motion *(uh-MEE-boyd)*   The movement of a cell or its plasma membrane by means of the slow flow of cellular contents.

Anaerobic   Life processes not requiring free oxygen.

Anaphase   The mitotic stage in which the chromatids separate and move to opposite poles.

Anaphylactic shock   A violent allergic reaction; may result in death through respiratory and cardiac failure; an immediate-type hypersensitivity.

Anatomy   The internal structures of the body, or their study, and the relationship of the parts to one another.

Androgen *(ANN-droh-jen)*   Any male sex hormone (e.g., testosterone); stimulates and maintains the male reproductive system and secondary sex traits.

Anemia *(uh-NEE-mee-uh)*   A decrease in the number of functional red cells or their hemoglobin content.

Aneurysm *(AN-yoo-rizm)*   A saclike enlargement of a blood vessel caused by a weakening of the wall.

Angina pectoris *(an-JI-na PEK-to-ris)*   A strong chest pain related to reduced coronary circulation that may or may not involve heart or artery disease.

Animalia   The kingdom that includes animals.

Anorexia nervosa *(an-o-REK-see-uh ner-VO-sa)*   Loss of appetite and sometimes unusual patterns of eating.

Anoxia *(an-OX-see-uh)*   Deficiency of oxygen.

ANS   Abbreviation for <u>a</u>utonomic <u>n</u>ervous <u>s</u>ystem.

Antagonist   A muscle or element of the nervous system that acts in opposition to another.

Anthropoid   A higher primate; monkey, ape, or human.

Antibiotic   A substance that inhibits or kills microorganisms.

Antibody *(AN-tee-bod-ee)*   A protein produced by lymphocytes in response to a foreign macromolecule or antigen and released into the bloodstream.

Anticoagulant *(an-ti-co-AG-yoo-lant)*   A substance that is able to reduce, delay, or prevent blood clotting.

Antidiuretic *(an-ti-di-yoo-RET-ik)*   Substance that inhibits urine formation.

Antidiuretic hormone (ADH)   A peptide hormone synthesized in the hypothalamus, stored in the posterior pituitary, and released when the blood becomes too concentrated; promotes reabsorption of water in the kidneys.

Antigen *(ANN-ti-jen)*   A substance (usually a protein or polysaccharide) that stimulates lymphocytes to proliferate and secrete specific antibodies that bind with it.

Antigenic determinant   Localized regions on the surface of an antigen that antibodies chemically recognize.

Antihistamine   A compound that counteracts the effects of histamine.

Antioxidant   A substance that inhibits oxidation.

Antitoxin   An antibody produced in response to the presence of a toxin released by a microorganism.

Anus   The distal end and outlet of the rectum.

Aorta   The main systemic artery; emerges from the left ventricle to carry oxygenated blood to all body regions except the lungs.

Aphasia   Defect or loss of ability to speak or write.

Apocrine glands   Exocrine glands that open into the hair follicles rather than onto the skin surface.

Appendicitis   Inflammation of the appendix.

Appendicular skeleton   The bones of the limbs.

Appendix   A wormlike extension of the large intestine.

Appestat   The hunger center in the hypothalamus.

Aquaculture   The growing and harvesting of such crops as algae, fish, and shellfish in bodies of water.

Areola   The pigmented ring around the nipple of the breast.

Arithmetic progression   Population growth characterized by the addition of a constant number of individuals during a unit interval of time (e.g., 3, 5, 7, 9, etc.); contrast with *geometric progression*.

Arteriole *(ar-TEER-ee-ol)*   A small artery, especially one next to a capillary.

Arteriosclerosis   A degeneration and hardening of artery walls.

Artery   A vessel that carries blood from the heart to body tissues.

Arthritis   A chronic, disabling inflammation of the joints.

Artificial fertilization   *In vitro* fertilization.

Artificial insemination   Introduction of semen into the female reproductive tract by artificial means for *in vivo* fertilization.

Artificial resuscitation   Any means of restoring breathing to a nonbreathing person (e.g., mouth-to-mouth methods).

Artificial selection   The selective breeding of domesticated organisms to produce descendants with desired traits; individuals without the desired traits are usually not allowed to reproduce.

Asbestos   A mineral that separates into long, thin, flexible fibers; often retained in the lungs where it may produce serious damage.

Ascorbic acid   Vitamin C.

Asexual reproduction   Reproduction involving one parent and no union of gametes; produces genetically identical offspring.

Association area   An area of the cerebral cortex concerned with particular sensory, emotional, motor, and intellectual processes.

Association neuron   A neuron in the central nervous system that links a sensory neuron with a motor neuron or synapses with many other neurons in the CNS.

Asthma   Difficulty in breathing due to spasms of bronchiolar musscles.

Astigmatism An irregularity of the lens or cornea of the eye, causing the image to be out of focus.

Asymptomatic A disease or condition without obvious external symptoms.

Atherosclerosis *(ATH-ur-oh-sklur-OH-sis)* A cardiovascular disease in which plaques develop on the inner walls of arteries, decreasing their interior size.

Atmosphere (1) The pressure of air at sea level. (2) Air.

Atom The smallest quantity of an element that can retain the chemical properties of the element; composed of a central nucleus containing protons and neutrons, and electrons that circle the nucleus in specific orbitals.

Atomic number The number of protons in the nucleus of an atom; unique for each element.

ATP Abbreviation for adenosine triphosphate *(uh-DEN-oh-sin try-FOS-fate);* a compound in which energy for metabolism is stored in high-energy bonds.

Atrioventricular node The compact mass of conducting cells located in the right atrial wall of the heart.

Atrioventricular valve A valve located between an atrium and a ventricle of the heart.

Atrium *(AY-tree-um)* A cavity or sinus; the upper heart chambers that receive blood returning to the heart.

Atrophy A decrease in size of a part due to nonuse or abnormal nutrition; contrast with *hypertrophy.*

Australopithecus A genus of early humans or hominids that gave rise to the genus *Homo*; lived from about 5 million to 1.3 million years ago.

Autoimmunity Condition due to failure of the immune system to distinguish between the body's own proteins and foreign proteins; a delayed-type hypersensitivity to one's own tissues.

Autonomic Self-controlling or independent.

Autonomic nervous system The part of the peripheral nervous sytem that controls the involuntary body functions (e.g., heart, glands, smooth muscles); subdivided into the sympathetic and parasympathetic divisions, whose effects oppose one another; abbreviated ANS.

Autopsy The examination of the body after death.

Autosome *(AW-tuh-some)* A chromosome other than a sex chromosome.

Autotroph A self-nourishing organism, such as a green plant; an organism able to synthesize organic molecules from simple inorganic materials; contrasts with *heterotroph.*

Axial skeleton The bones of the cranium and vertebral column.

Axon *(AKS-on)* A typically long process that conducts nerve impulses away from the cell body toward target cells.

B cell A lymphocyte that develops in the bone marrow and later secretes a specific antibody upon exposure to a suitable antigen; involved in humoral immunity.

Bacillus A rod-shaped bacterium.

Bacteria Various members of the kingdom Monera; includes many species of economic, medical, and ecological importance; commonly categorized by shape, bacilli (rods), cocci (dots), and spirilla (spirals).

Bacteriophage *(bak-TEER-ee-oh-faj)* A virus that infects a bacterium; phage.

Barr body An inactivated X chromosome; the dense body lies along the inside of the nuclear envelope in the cells of females.

Bartholin's gland A small mucous gland located on the side of the vaginal opening.

Basal ganglia Large groups of neuron bodies deep in the cerebral hemispheres.

Basal metabolic rate The minimum rate of metabolism needed to maintain the body's vital processes when no food is being digested and when the body is at rest; abbreviated BMR.

Base A proton acceptor; excess OH ions producing a pH greater than 7, alkaline; opposite of *acid*.

Basilar membrane The inner membrane in the cochlea of the inner ear upon which the organ of Corti is found.

Basophil *(BAY-so-fil)* A white blood cell with a pale nucleus and large, dense granules.

Bends Gas bubbles in the blood due to a sudden reduction in environmental pressure.

Benign Not malignant.

Bergman's principle Principle that the body size of endotherms living in cold areas tend to be larger than relatives living in warmer areas; see *Allen's principle*.

Beriberi Malnutrition due to a deficiency of thiamin in the diet.

Bilateral symmetry A body form with a central longitudinal plane dividing the body into two equal but opposite halves; characteristic of motile and higher animals.

Bile A solution of organic salts secreted by the liver and stored temporarily in the gallbladder; emulsifies fats in the small intestine.

Bilirubin A bile pigment produced from the hemoglobin of red blood cells.

Binocular vision Vision through both eyes; the overlapping fields of vision usually result in good depth perception.

Bioconcentration Biological magnification.

Biofeedback A process in which people learn to receive some sensory information regarding visceral functions and to control those functions consciously to some extent.

Biological magnification The buildup in concentration of retained substances in successively higher feeding levels in a food chain or web.

Biome *(BY-ome)* A large terrestrial ecosystem characterized by distinctive organisms and climatic conditions (e.g., desert, tundra, forest, or grassland).

Bionics The science of designing replacement body parts.

Biopsy Removal of material from the living body for examination.

Biosphere *(BY-oh-sfeer)* The areas on and near Earth's surface (air, land, and water) where organisms live.

Biotechnology The utilization of recombinant DNA to develop new products or organisms with altered traits; includes genetic engineering.

Biotic potential The maximum rate of population increase possible under ideal conditions, if all females bred as often as possible, and all individuals survived past their reproductive age.

Bipolar cells Retinal neurons that relay the input via the optic nerves.

Black lung disease A group of lung diseases caused by excessive dust (especially from coal) or other particles that accumulate in the alveoli and disrupt normal breathing; some particles are carcinogenic; debilitating and often fatal.

Blastocyst   An early stage in embryonic development; consists of a hollow ball of cells with an inner cell mass at one end that gives rise to the embryo.

Blastula   *(BLAS-tyoo-la)*   A hollow ball of cells produced by cleavage of a zygote.

Blind spot   The area in the retina lacking light receptor cells; the location of the optic nerve.

Blood   The fluid that flows through the heart, arteries, capillaries, and veins; consists of a liquid or plasma portion and a portion of formed elements or blood cells.

Blood-brain barrier   A specialized capillary network that restricts the passage of most substances from blood into the cerebrospinal fluid and brain, thus helping to maintain the constancy of the brain's environment.

Blue baby   A neonate whose foramen ovale has not closed properly.

BMR   The abbreviation for basal metabolic rate.

Bolus   *(BO-lus)*   A soft, round mass of food that passes through the digestive tract.

Bone   A type of connective tissue in which the cells are surrounded by a mineral and organic matrix; often contains a central marrow region; commonly categorized by structure (e.g., compact or spongy) or by shape (e.g., long or flat).

Bone marrow   Soft, spongy material in the cavities of bone; see *red* and *yellow marrow.*

Botulism   An often fatal type of food poisoning caused by the bacterium, *Clostridium botulinum,* usually in improperly canned food, especially meat.

Bowman's capsule   A double-walled, cup-shaped sac of cells that surrounds the glomerulus at the end of each nephron in the kidney.

Brain   A mass of nerve tissue usually located in the head in higher animals but sometimes found elsewhere in lower forms.

Brainstem   The brain exclusive of the cerebrum and cerebellum; includes the medulla, pons, midbrain, thalamus, and hypothalamus.

Breathing center   A region of the medulla oblongata and the pons.

Bronchitis   Inflammation of the bronchi; characterized by a persistent cough.

Bronchus   One of a pair of large respiratory tubes that branch from the lower end of the trachea into each lung.

Bronchiole   A smaller division of a bronchus; culminates in the alveoli.

Budding   An asexual means of reproduction in which outgrowths from the parent form, and either pinch off to live independently or remain attached to form colonies of genetically similar organisms.

Buffer   A substance that minimizes changes in hydrogen-ion concentration or pH when bases or acids are added to a solution.

Bypass operation   Procedure by which blocked arteries are removed from the heart and replaced by healthy vessels, often veins from the person's own body to reduce rejection.

Caesarean section   Incision through the abdominal and uterine walls for delivery of a fetus; done when birth through natural passages is impossible or dangerous.

Calciferol   A form of vitamin D.

Calcitonin   *(kal-see-TO-nin)*   One of the major hormones from the thyroid gland; inhibits the release of calcium from and accelerates calcium absorption by bone.

Calorie   A unit of heat. The small or gram calorie (gcal) is the amount of heat required to increase the temperature of 1 gram of water 1°C. The large or kilocalorie (kcal) is the amount of heat needed to raise 1 kilogram of water 1°C.

Calorimeter A device used to measure the caloric content of nutrients.

Cancer A tumor or cell mass resulting from uncontrolled cell division; often tends to move, invade other tissues, and stimulate new growths.

Cannabis The hemp plant from which marijuana is derived.

Capillary *(KAP-ill-air-ee)* A microscopic blood vessel that connects an artery and a vein; consists of a single layer of endothelial cells, through which exchanges occur between blood and interstitial fluid.

Carbohydrate An organic compound containing carbon, hydrogen, and oxygen in the ration of 1:2:1 (e.g., sugar, glycogen, cellulose, and starch).

Carcinogen *(kar-SIN-oh-jen)* Any agent capable of causing cancer.

Carcinoma A cancer found in epithelial tissue.

Cardiac cycle The pumping cycle of the heart, consisting of an alternation of contraction (systole) and relaxation (diastole).

Cardiac muscle The specialized striated and involuntary muscle of the heart.

Cardiac output The quantity of blood ejected each minute by each ventricle.

Caries Decay and death of a tooth or bone associated with inflammation and the formation of abscesses in the periosteum and surrounding tissues.

Carnivore An animal-eating organism.

Carrier An individual who possesses a particular gene, pathogen, or parasite.

Cartilage *(KAR-til-ij)* Connective tissue consisting of fibrous material and cells embedded in a matrix of firm, gelatinlike substance; usually with minimal nerve or blood supply.

Castration The removal of the gonads.

Catalyst A substance that accelerates the rate of a chemical reaction without becoming part of the end product or being used up in the process; enzymes are biological catalysts.

Cataract Loss of transparency of the crystalline lens of the eye and/or its capsule.

Caudal Pertaining to any taillike structure or to the tail region.

Cell The basic structural and functional unit of organisms; contains various organelles in its cytoplasm and nucleoplasm.

Cell cycle The sequence of events of growth and division in the life of a cell.

Cell membrane The selectively permeable outermost membrane of a cell that regulates exchanges between the cell and its environment; also called *plasma membrane*.

Cell theory The principle that cells are the fundamental units of organization in organisms and that all cells today arise from preexisting cells.

Cellular immunity Immunity provided by sensitized T cells.

Cellular respiration The process by which energy is released from organic compounds within cells; oxygen is consumed as a reactant with an organic fuel.

Cementum The bony material covering the root of a tooth.

Central nervous system The brain and spinal cord; abbreviated CNS.

Centrifuge A device used to separate heavier from lighter substances (e.g., blood cells from plasma).

Centriole *(SEN-tree-ol)* A small cytoplasmic organelle lying outside the nuclear membrane; divides and organizes the spindle fibers during cell division.

Centromere *(SEN-troh-mere)* The position where spindle fibers connect sister chromatids during cell division.

Cephalic Pertaining to the head.

Cephalization The evolutionary trend through which neural organization and specialization has become localized in the head end of animals.

Cephalocaudal The developmental direction that proceeds from the head to the lower extremities; compare with *proximodistal.*

Cerebellum The part of the brain that integrates information about body position and motion, maintains equilibrium, and coordinates muscular activities.

Cerebrospinal fluid The fluid filling the ventricles of the brain, central canal of the spinal cord, and a space of the meninges.

Cerebrum The forebrain that occupies the upper cranium, consisting of two hemispheres united by the corpus callosum. It processes and coordinates sensory inputs and motor responses; the primary association center of the brain.

Cervical cap A small contraceptive device placed immediately over the cervix.

Cervicitis Inflammation of the cervix.

Cervix The lower part of the uterus.

Chancre A firm, open, painless ulcer at the site of syphilis infection; usually heals without a scar.

Chancroid A bacteria-caused sexually transmitted disease.

Chaparral *(SHAP-uh-RAL)* Evergreen, shrubby vegetation typical of Mediterranean-type climates of long, hot, dry summers and mild, rainy winters.

Chemical bond The attractive force between the atoms in a molecule.

Chemical reaction The making or breaking of chemical bonds.

Chemoreceptor A sensory cell or organ that responds to specific chemical stimuli by initiating a nerve impulse; includes taste and smell receptors.

Chemotherapy Treatment of a disease or condition (especially cancer) with appropriate drugs.

Chlorophyll *(KLOR-oh-fill)* Various steroid green pigments found in chloroplasts; the receptors of light energy in photosynthesis.

Chloroplast A plastid in an autotroph cell containing chlorophyll, and the location where photosynthesis occurs.

Cholesterol *(kol-ESS-teh-rol)* A steroid found in many tissues.

Choline One of the B vitamins; involved in carbohydrate and fat metabolism.

Chorion *(KOR-ee-on)* The outermost embryonic and fetal membrane; contributes to the structure of the placenta.

Chorionic gonadotropin A hormone secreted by the placenta beginning about 14 days after fertilization; it maintains the corpus luteum and stimulates the placenta to secrete estrogen and progesterone.

Choroid The middle vascular coat of the eyeball.

Chromatid *(KROH-muh-tid)* One of two identical halves or daughter strands of a duplicated chromosome that are joined by a single centromere.

Chromatin *(KROH-muh-tin)* The stainable complex of dispersed genetic material observed between periods of cell division; consist of the DNA and proteins of which chromosomes are composed.

Chromatography   A method of separating chemical substances by their differential movement.

Chromatophore   A pigment-containing cell; the distribution of the pigment is important in adaptive color changes in various animals.

Chromosome   *(KROH-muh-som)*   A structure in a cell that contains the genes; located within the nucleus of eukaryotes but circular and free in prokaryotes.

Chyme   The semifluid mixture of partly digested food and digestive secretions found in the stomach and small intestine.

Cilium   *(SILL-ee-um)*   One of many short hairlike cytoplasmic projection on the free surface of some cells; may move cells or substances past cells.

Circumcision   Removal of the foreskin from the glans penis.

Cirrhosis   A liver disease that results in scar tissue and disruption of normal liver functions.

Cleavage   The succession of rapid cell divisions without growth that change a zygote into a multicellular, hollow blastula.

Climacteric   A syndrome of traits associated with the "change of life" (e.g., menopausal years) in females and males.

Clitoris   *(KLI-to-ris)*   A small concentration of erogenous tissue that is part of the female genitalia.

Clone   A population of cells derived by common descent from a single ancestral cell; also used for an organism produced asexually.

Closed transport system   A circulatory system in which blood is enclosed within vessels throughout the body; contrast with *open transport system.*

CNS   Abbreviation for central nervous system.

Coccus   A spherical bacterium.

Cochlea   *(KOH-klee-uh)*   A spirally coiled tubular cavity of the inner ear; contains the organ of Corti.

Codominance   A situation in which the effects of both alleles at a particular locus are apparent in the phenotype of a heterozygote.

Codon   A series of three nucleotides in DNA or in RNA specifying a particular amino acid in protein synthesis or a termination signal; the basic unit of the genetic code.

Coenzyme   A nonprotein organic molecule that plays an accessory role in enzymatic reactions; includes many vitamins.

Cofactor   A nonprotein molecule or ion that is necessary for the proper functioning of an enzyme.

Coitus   Introduction of sperm into the female body to facilitate fertilization; sexual intercourse or copulation.

Coitus interruptus   Withdrawal of the penis from the vagina prior to ejaculation during sexual intercourse.

Collagen   A tough, fibrous protein found in bones, tendons, and other connective tissues.

Collateral   A branch of a nerve fiber or blood vessel.

Collecting duct   The duct that collects fluids from the distal convoluted tubules of nephrons and discharges them into the renal pelvis.

Colon   The major portion of the large intestine.

Color blindness   The inability to distinguish colors (e.g., shades of reds and greens); lack or abnormality in one or more photopigments of the cones.

Colostrum   *(ko-LOS-trum)*   The first milklike material secreted at the end of pregnancy.

Coma   A deep state of unconsciousness from which one cannot be aroused easily.

Commensalism *(kuh-MEN-sul-iz-um)* The interspecific interaction in which one species benefits from an association with another species that is unaffected by the association.

Community A group of interacting species in a common environment.

Compact bone Bone tissue with a dense matrix.

Competition The intraspecific or interspecific interaction between organisms attempting to use the same limited resources.

Complete digestive system A tubular digestive tract with both an oral and an anal opening; contrast with *incomplete digestive system.*

Compound A molecule composed of two or more kinds of atoms in definite ratios held together by chemical bonds; degradable by ordinary chemical processes into its components.

Conception The fertilization of an egg.

Concordance The degree of similarity between twins and siblings in specific traits.

Condensation reaction A chemical reaction in which two molecules join to form a larger molecule, splitting off a molecule of water in the process; the reactions in which monomers link to form polymers; also called *dehydration synthesis.*

Condom A sheath for the penis worn during coitus to prevent conception or infection.

Conduction The transfer of heat through a medium; see *convection* and *radiation.*

Cone A photoreceptor neuron in the retina concerned with the perception of color and the discrimination of detail.

Congenital Existing at birth; usually refers to a physical or mental abnormality.

Coniferous forest A forest composed primarily of cone-bearing trees, usually with evergreen leaves (e.g., pine, fir, spruce); see *taiga.*

Conjunctiva The thin membrane that lines the eyelids and the exposed surface of the eyeball.

Connective tissue Supporting tissue formed of fibrous and intercellular substances with relatively few cells; includes blood, cartilage, bone, and adipose cells.

Consanguineous mating Mating between close relatives.

Constipation Excessive dehydration of the feces in the large intestine, resulting in difficult defecation.

Consumer A heterotroph incapable of synthesizing its own food; secures its nutrients from living or freshly killed organisms; includes herbivores, carnivores, and omnivores.

Contact dermatitis A delayed-type hypersensitivity reaction caused by contact between various chemicals and the skin.

Contagious Communicable; able to be transmitted directly from one person to another.

Continuous variation Variation in traits whose measurements form a spectrum that is continuous from one extreme to another; often exhibits a normal distribution; typical of traits affected by many genes.

Contraception Means of birth control.

Control A standard of comparison in a scientific experiment; a replicate of the experiment in which a factor being studied is omitted; see *variable.*

Convection The transfer of heat by the movement of a fluid or gas; see *conduction* and *radiation.*

Convergent circuit The arrangement in which the synaptic knobs of several presynaptic neurons terminate on one postsynaptic neuron.

Convulsion A violent, uncontrolled muscle spasm, or a series of them; sometimes repeated rapidly and/or accompanied by unconsciousness.

Cornea The transparent front of the sclera and the eye's principal focusing structure.

Corona radiata A layer of follicular cells that surrounds the egg at ovulation.

Corpus albicans The ovarian scar tissue that forms after the corpus luteum degenerates.

Corpus callosum A mass of transverse fibers connecting the cerebral hemispheres.

Corpus luteum *(KOR-pus LOO-tee-um)* A yellow glandular mass in the ovary formed by cells of an ovarian follicle that has matured and discharged its ovum; secretes progesterone.

Corpuscle A cell-like structure without a nucleus, such as an erythrocyte; a small mass or body.

Corrective feedback Negative feedback; the major mechanism for homeostatic control.

Cortex The outer layer of a structure (e.g., cerebral, adrenal, and kidney cortexes); see *medulla*.

Co-twin method The study of the traits of identical twins during their lifetimes.

Cough A reflex triggered by irritation of the throat or trachea; the controlling brain center is in the medulla oblongata.

Countercurrent exchange The movement of one substance in the opposite direction to that of another, maximizing exchange rates; also applies to energy exchanges, such as heat.

Covalent bond *(koh-VAL-ent)* A strong chemical bond in which two atoms share one or more electrons.

Cowper's gland A small gland that joins the urethra near the point where the urethra enters the penis; produces a mucous secretion that forms part of semen.

Creationism The belief that species were created in essentially their present form; contrast with *evolution*.

Cretin *(KREE-tin)* Victim of severe congenital thyroid deficiency, causing physical and mental retardation.

Cro-magnon An essentially modern form of tool-using *Homo sapiens*; creators of well-known cave paintings.

Crop An enlarged temporary storage organ in the digestive tract of some animals (e.g., birds and earthworms).

Crossing-over The process during meiosis in which homologous chromosomes exchange chromatid segments; contributes to genetic recombination.

Cross-linkage A temporary or permanent linkage between molecules.

Cryonics The study of materials under very low temperature conditions.

Cryptorchidism Failure of the testes to descend into the scrotum normally.

Cunnilingus Oral stimulation of a female's genitals.

Curettage Scraping the interior of an organ (usually the uterus) with a spoonlike instrument (curette).

Cyanobacteria Members of the kingdom Monera; various photosynthetic prokaryotes; formerly called *blue-green algae*.

Cyanocobalamin Vitamin B_{12}; essential for the manufacture of red blood cells.

Cyborg A human-machine combination.

Cystic fibrosis An inherited childhood disease.

Cystitis Inflammation of the urinary bladder.

Cytokinesis The division of a cell into two daughter cells after nuclear division.

Cytology The study of cell structure.

Cytoplasm *(SIGH-toh-plaz-um)* The living material within a cell, excluding the nucleus.

Cytoskeleton Complex internal network of cells consisting of protein microfilaments and microtubules; helps maintain cell shape, anchors the organelles, and is involved with cell motility.

Daughter cell Either of the two cells created when one cell divides.

Dead air A volume of air in the breathing system that is not available for gas exchange and that is not moved by normal breathing; see *residual air.*

Deamination Removal of an amino group ($-NH_2$) from an organic compound (e.g., an amino acid).

Decibel A unit for measuring the loudness of sounds; the logarithmic decibel scale measures sound intensities relative to the intensity of the faintest audible sound.

Deciduous That which eventually falls out, such as our early teeth; not permanent. Also applied to the shedding of leaves at a certain season.

Decomposer An organism that breaks down nonliving organic matter into smaller molecules; includes bacteria and fungi.

Decomposition reaction A chemical reaction in which larger molecules are broken down into smaller, simpler molecules; contrast with *synthesis reaction.*

Defecation The discharge of feces from the rectum.

Deficiency disease A disorder caused by lack of an essential nutrient.

Delayed-type hypersensitivity Cellular hypersensitivity reactions; allergic responses due to sensitized T cells.

Delusion A false belief that is difficult to correct by reason.

Dendrite *(DEN-dryt)* A process extending from the cell body of a neuron, typically numerous and branched, that conducts a nerve impulse toward the cell body.

Denature To alter a protein, as by heat, so as to destroy its properties.

Dentin The osseous tissues of a tooth, enclosing the pulp cavity.

Dependence Change in function, resulting from prolonged use of a chemical substance, characterized by a necessity for the substance for the maintenance of function.

Dependent variable The observed or measured results of an experiment whose changes depend upon the changes in the independent variable being controlled by the experimenter.

Depressant General term for chemical substances whose effects include a slowing of many metabolic functions; contrast with *stimulant.*

Dermatitis A general term for inflammation of the skin.

Dermatoglyphics The medical study of fingerprint, palm, and footprint patterns as an aid to diagnosing a variety of ailments.

Desert A relatively barren biome where precipitation is irregular and generally less than about 25 cm per year.

Detritus Dead organic matter or organic wastes.

Development The appearance of new structures and/or functions; closely associated with *growth.*

Developmental directions The tendency for the head and the central part of the body to develop faster than the lower and distal part of the body, respectively (*cephalocaudal* and *proximodistal*).

Diabetes insipidus *(die-uh-BEE-tez in-SIP-i-dus)* A disease caused by a lack of antidiuretic hormone from the pituitary; characterized by chills, extreme thirst, and elimination of very large amounts of urine.

Diabetes mellitus   The inability to utilize carbohydrates and fats properly due to insufficient insulin.

Dialysis   The process of separating smaller particules from larger particles by the difference in their rates of diffusion.

Diaphragm   (1) The dome-shaped partition between the thoracic and abdominal cavities. (2) A membrane used to cover the cervix to prevent conception.

Diarrhea   A disruption of the normal dehydrating mechanism of the large intestine that results in frequent, watery feces.

Diastole   *(Die-ASS-tuh-lee)*   The stage of the cardiac muscle in which the heart muscle is relaxed, allowing the chambers to fill with blood.

Diencephalon   The part of the brain consisting primarily of the thalamus and hypothalamus.

Diffusion   *(deh-FU-shun)*   The net movement of materials from their higher concentration to areas of their lower concentration. The process results from the random movement of individual molecules, and it tends to distribute molecules uniformly.

Digestion   The breakdown of the large organic molecules in food to smaller molecules that can be absorbed into cells and used as the raw materials for synthetic purposes or degraded to yield energy.

Dihybrid cross   A cross in which the parents differ in two traits being investigated; extends to a mating of their offspring to produce an F_2 generation.

Dilatation and evacuation   Dilatation of the cervix and evacuation of the uterine contents; abbreviated D and E.

Dilate (dilatate)   To expand or swell.

Diploid   *(DIP-loid)*   Containing two sets of chromosomes (2_n); typical of somatic cells; in contrast with *haploid*.

Disaccharide   *(DY-SAK-uh-ryd)*   A double sugar composed of two chemically bonded monosaccharide or simple sugar molecules.

Discontinuous variation   Variation in traits that fall into two or more nonoverlapping classes; characteristic of either-or traits.

Dislocation   Out of place; especially a bone out of its socket.

Distal   Farthest away from the center, the median line, or the trunk; opposite of *proximal*.

Diuretic   Any agent that increases urine secretion.

Divergent circuit   The arrangement in which the synaptic knobs of one presynaptic neuron terminate on several postsynapic neurons; see *convergent circuit*.

Dizygotic twins   Twins that develop from two different zygotes; fraternal.

DNA   Abbreviation for deoxyribonucleic acid *(DEE-ox-ee-rye-boh-noo-KLAY-ik)*. The genetic material present in chromosomes; the double-stranded, helical molecule contains genetic information coded in specific sequences of its two complementary nucleotides wound in a double helix.

Dominant allele   The allele that is fully expressed in the phenotype.

Dorsal   Pertaining to the back, or upper surface; opposite of *ventral*.

Dosage compensation   The genetic mechanism that compensates for genes that, as a result of their location on the X chromosome, exist in two doses in females and one dose in males.

Double-blind experiment   A research technique in which neither the subject nor the experimenter knows which subject has been exposed to the variable being studied.

Double helix The DNA molecule, which consists of two strands wound helically or spirally around each other.

Douche A stream of material directed against a part or into a cavity.

Down syndrome Mental and physical retardation associated with trisomy of chromosome pair 21; once called *mongolism*.

Drumstick A nuclear extension of some female neutrophils.

Duodenum The upper portion of the small intestine; the principle site of food digestion.

Dynamic equilibrium The maintenance of body position; especially the head, in response to sudden movements such as acceleration, deceleration, and rotation; contrast with *static equilibrium*.

Dysentery Disruption of colon function caused by pathogens, parasites, or chemicals; characterized by diarrhea.

Dysmenorrhea Painful or difficult menstruation.

Eardrum The thin membrane (tympanic) that separates the external from the middle ear; receives vibrations and transmits them to the ear bones or ossicles.

Eccrine Exocrine glands that open onto the skin surface.

Ecology *(ee-KOL-uh-jee)* The study of the interactions between organisms and other organisms and with their physical environments.

Ecosystem *(EE-koh-sis-tum)* The interactions of the organisms within a community with each other and with their physical environments, such as in a pond or forest.

Ectoderm *(EK-toh-durm)* The outermost of the three embryonic germ layers; gives rise to the outer epithelium (skin, hair, and nails) and the nervous system.

Ectomorph A body build characterized by linearity, fragility, large surface area, and thin muscles and subcutaneous tissues.

Ectopic Out of the normal location (e.g., attachment of an embryo to a structure other than the endometrium).

Ectotherm *(EK-toh-therm)* An animal that derives its body temperature primarily from its environment; contrast with *endotherm*.

Edema Excessive fluid in tissues.

Effector A structure that performs an activity after neural or hormonal stimulation; usually a muscle or a gland.

Egestion The excretion from the body of material that is indigestible.

Ejaculation Release of seminal fluid from the male urethra.

Electrocardiogram A recording of the electric current produced by the contraction of cardiac muscle; abbreviated ECG.

Electroencephalogram A recording of the electric currents developed in the brain; abbreviated EEG.

Electrolyte A substance that dissolves into ions in an aqueous solution.

Electron The small, nearly weightless subatomic particle that bears a negative electric charge equal to the positive charge of the proton.

Electron microscope A microscope using beams of electrons and magnetic focusing to allow small specimens to be viewed.

Electrophoresis The movement of charged particles suspended in a liquid on various media under the influence of an applied electric field.

Element One of the hundred or so types of matter, natural or synthetic, composed of atoms all of which have the same number of protons in the nucleus and the same number of orbiting electrons; substances that cannot be broken down to any other substance by ordinary chemical means.

Elephantiasis Enlargement of certain body parts due to edema that results from chronic infections of filarial worms.

Embolism   Obstruction or closure of a vessel by a transported blood clot, a mass of pathogens, or other foreign material.

Embryo   The early developmental stages after conception; in humans, between the second and eighth weeks, inclusive.

Emergency response   The sum of the body's physiological reactions to stress; often includes short-term positive feedback (e.g., increase in blood pressure and blood sugar above normal); the general adaptation syndrome.

Emphysema   An abnormal swelling or inflation of air passages and alveoli.

Emulsify   To break down large fat globules to smaller, uniformly distributed particles; the increased surface area that results facilitates their digestion.

Enamel   The white, compact, and very hard substance that covers and protects the dentin of the crown of the tooth.

Encephalitis   Inflammation of the brain.

Endocrine   *(EN-doh-krin)*   Secreting internally; applied to ductless glands that secrete hormones into extracellular space, from which they diffuse into blood or lymph; contrasts with *exocrine*.

Endocrinology   The study of the structure and function of the endocrine glands.

Endocytosis   *(EN-do-sigh-TOE-sis)*   The process by which a cell engulfs particles and solutes; the plasma membrane surrounds the substance and pinches off to form an intracellular vesicle; includes phagocytosis and pinocytosis; contrast with *exocytosis*.

Endoderm   *(EN-doh-durm)*   The innermost of the embryonic germ layers; gives rise to epithelium that lines the digestive tract and its outgrowths, most of the respiratory tract, the urinary bladder, and some endocrine glands.

Endometrium   *(EN-doh-MEE-tree-um)*   The glandular, inner uterine lining shed during menstruation; contrast with *myometrium*.

Endomorph   A body build characterized by a relatively large accumulation of fat and digestive viscera.

Endoplasmic reticulum   *(EN-doh-plaz-mik reh-TIK-yoo-lum)*   An extensive system of membranes in the cytoplasm of the cell; called "rough" if it contains many ribosomes and "smooth" otherwise.

Endorphin   *(en-DOR-fin)*   A neuropeptide hormone produced in the brain and anterior pituitary that inhibits pain and performs various other functions.

Endoskeleton   An internal supportive system, usually of cartilage and bone; common in vertebrates and echinoderms; contrast with *exoskeleton*.

Endothelium   *(EN-doh-THEEL-ee-um)*   Flat cells that line blood and lymphatic vessels, the heart, and some body cavities.

Endotherm   *(EN-doh-thurm)*   An animal that derives its body temperature primarily from its metabolism; mainly a bird or mammal; contrast with *ectotherm*.

Endurance   The length of time that muscular power is used.

Energy   The capacity to do work by moving matter against an opposing force; see *kinetic* and *potential energy*.

Entropy   A quantitative measure of disorder or randomness in a system; the amount of energy in a closed system that is not available for doing work.

Enuresis   Partial or complete involuntary urination after about age 3 years; bed wetting.

Enzyme   *(EN-zyme)*   A biological catalyst; a protein that facilitates chemical reactions without being used up or becoming part of the end product.

Eosinophil White blood cells produced in the bone marrow; nonphagocytic cells that make up about 2 percent of leukocytes.

Epidemiology The study of the occurrence and distribution of disease in human populations.

Epidermis *(EP-eh-DER-mis)* The outermost layer of body cells; of ectodermal origin.

Epididymis A complexly coiled sperm storage vessel next to the testis.

Epiglottis The lidlike structure that covers the opening to the larynx.

Epilepsy A disturbance of the brain's electrical activity that results in impairment or loss of consciousness and abnormal movements and sensations.

Epinephrine A secretion of the adrenal medulla; adrenalin.

Episode An occurrence of serious air pollution.

Epistasis The process by which one gene alters the expression of another gene that is inherited independently at another locus.

Epithelium *(ep-eh-THEEL-ee-um)* The tissue that covers an exposed surface or lines a tube or cavity.

Equilibrium The maintenance of a state of balance between two opposing forces; see *dynamic* and *static equilibrium.*

Erogenous zone A body region that when properly stimulated produces sexual excitement.

Erythroblasts The cells from which red blood cells develop.

Erythrocyte Red blood cell.

Escherichia coli A common bacterium in the digestive tract; used widely in genetic engineering.

Esophagus The muscular tube leading from the pharynx to the stomach.

Essential amino acid An amino acid required for metabolism that cannot by synthesized by the body in adequate amounts or at a necessary rate; must be present in the diet.

Estivation A state of dormancy caused by dryness and heat; contrast with *hibernation.*

Estrogen *(ES-troh-jen)* One of a group of steroid female sex hormones produced mainly in the ovary; they maintain the female reproductive system and develop female secondary sex traits.

Eugenics Improving the human condition by the scientific, humane, and moral use of genetic knowledge.

Eukaryote Organisms whose cells possess a membrane-bounded nucleus and membrane-bounded organelles; includes protists, fungi, plants, and animals; contrast with *prokaryote.*

Eunuch A male suffering from a lack or deficiency of androgens; traits may include fat deposition characteristic of the female, a femalelike voice, and absence of facial hair.

Euphoria A feeling of well-being, not always well founded.

Eury- tolerance A wide range of tolerance to a condition (e.g., temperature, light, or salinity); contrast with *steno- tolerance.*

Eustachian tube A tube from the pharnyx admitting air into the middle ear and thus equalizing pressure on either side of the tympanic membrane.

Euthanasia Allowing a person to die without the use of extraordinary life-support.

Euthenics Improving the human condition by altering or improving environmental conditions (e.g., better nutrition, education, and public health).

Eutrophication   The process whereby a body of water becomes enriched with nutrients, increases in productivity, and accumulates organic matter; called *cultural eutrophication* when the nutrients result from human activity.

Evaporation   The process by which a liquid changes to a gas; the loss of highly kinetic molecules from wet surfaces to air cools animals or their parts.

Evolution   A process of competition, mutation, hybridization, and selection that results in the continuous genetic adaptation to environmental conditions; descent with modification.

Excess deaths   The number of deaths above that expected for a time period; often applied to the effects of environmental contaminants.

Excitement stage   The stage in the sexual response cycle characterized by swelling in erectile tissues, secretion of various moistening and lubricating fluids, increased myotonia, and changes in the position and state of various organs following appropriate sensory stimulation.

Excrete   To eliminate waste products from the body.

Exercise stress test   An examination to determine one's fitness to engage in a particular exercise program without physiological damage.

Exocrine   *(EX-oh-krin)*   Denotes a gland whose secretion is released through a duct (e.g., sweat and digestive glands); contrasts with *endocrine.*

Exocytosis   The expelling of cellular particles and solutes following fusion of vesicles with the plasma membrane; contrast with *endocytosis.*

Exoskeleton   An external supportive system, as in shelled mollusks and in arthropods; contrast with *endoskeleton.*

Experiment   A test of a hypothesis; generally implies the use of controls.

Expiratory reserve   The volume of air one can forcibly exhale beyond the tidal exhalation.

Expressivity   The degree of phenotype manifestation of a given gene.

Extensor   A muscle that straightens a limb or part of a limb; contrast with *flexor.*

Exteroceptor   A sense organ excited by stimuli from outside the body.

Facilitated diffusion   Carrier-assisted diffusion through specific channels in a cell membrane; contrast with *active transport.*

Facilitation   The promotion or hastening of a natural process, particularly in the transmission of nerve impulses.

Fallopian tube   Either of two slender oviducts that carry the ova from the ovary to the uterus.

Farsightedness   A condition in which close vision is difficult due to abnormal shortening of the eyeball; hyperopia.

Fast fibers   Myelinated (insulated) nerve fibers.

Fat   A lipid molecule composed of glycerol and fatty acids; called *oils* in the liquid state.

Fatty acids   Various chains of carbon and hydrogen atoms insoluble in water; may be unattached or bound to glycerol.

Feces   Material (stool) discharged from the bowel, made up of food residue, bacteria, and secretions.

Feedback system   System in which the increase of a product leads to a decrease in its rate of production, or a deficiency of a product leads to an increase in its rate of production; a major mechanism for homeostatic control.

Fellatio   Oral stimulation of the penis.

Fertilization   The fusion of haploid gametes to form a diploid zygote.

Fetal hemoglobin   Hemoglobin found in the fetus that has a higher affinity for oxygen than adult hemoglobin.

Fetus   The developing human *in utero* from about eight weeks to birth.

Fibrillation   Irregular twitching movements of individual or small groups of muscle fibers, preventing effective action by an organ or a muscle.

Fibrin   *(FY-brin)*   The filamentlike coagulated protein forming the substance of a blood clot.

Field of vision   The area that can be seen by an eye without moving it.

Filter feeder   A heterotroph, generally aquatic, that sifts food particles from its environment.

Flagellum   *(fluh-JEL-um)*   A long, hairlike, motile organelle protruding from the surface of a cell.

Flame cell   A cup-shaped cell with a tuft of cilia in its depression that regulates the content of extracellular fluid and participates in the excretion of wastes and in osmoregulation.

Flashback   Experiencing at a later, and usually unexpected time, an earlier event (e.g., the responses to a psychoactive agent).

Flexibility   The ability of a joint or series of joints to move within a certain range.

Flexor   A muscle that bends a limb or part of a limb; contrast with *extensor.*

Fluoridation   The addition of fluorine compounds to drinking water to reduce the incidence of tooth decay.

Fluorosis   Excessive ingestion of fluorine, often resulting in mottled tooth enamel.

Folic acid   A B vitamin, having a role in red blood cell production.

Follicle   *(FOL-eh-kul)*   A small secretory sac or cavity; a spherical chamber in the ovary that contains an oocyte and secretes estrogens.

Follicle-stimulating hormone   A hormone released by the anterior pituitary that stimulates the growth of ovarian follicles; abbreviated FSH.

Food additive   A substance intentionally added to food to preserve, enrich, or color the food or enhance its appeal; contrast with *food contaminant.*

Food chain   A portion of a food web; usually a sequence from producers to consumers to decomposers.

Food contaminant   A foreign substance in food not added intentionally (e.g., pesticide residue or pathogens); contrast with *food additive.*

Food poisoning   The development of an adverse reaction to pathogens or their products in food (e.g., botulism).

Food web   The total interlocking set of food chains in a community.

Foramen ovale   The oval window between the right and left atria present in the fetus, by means of which blood entering the right atrium may enter the aorta without passing through the lung.

Forebrain   The anterior division of the vertebrate brain; includes the cerebrum, thalamus, and hypothalamus.

Forensic   Pertaining to legal proceedings.

Foreplay   The sensory activities primarily during the excitement stage of the sexual response cycle.

Foreskin   The prepuce, especially of the penis and clitoris.

Formed elements   The cells and platelets of the blood.

Fossil   Any evidence of past life (e.g., bone, cast of a shell, or track).

Fossil fuel   The altered remains of once-living organisms from a previous geological period that are burned to release energy (e.g., coal, crude oil, and natural gas).

Fovea A small depression in the center of the retina that contains mostly cones and provides for keenest vision.

Fraternal twins Twins that develop from separately fertilized eggs; dizygotic; contrast with *identical twins.*

Free radicals A highly reactive atomic or molecular unit with an unpaired electron.

Frontal lobotomy Surgical isolation of the prefrontal areas of the cerebrum.

FSH Abbreviation for follicle-stimulating hormone.

Fungi A kingdom of relatively simple heterotrophic organisms, which include yeast, molds, rusts, and mushrooms; a number are pathogenic.

G6PD enzyme An enzyme involved in carbohydrate metabolism; people lacking it may develop a form of anemia (hemolytic).

Gallbladder The bile-storing sac located between the lobes of the liver.

Gallstone A crystal of bile salts that forms in the gallbladder or bile duct and disrupts liver function.

Gamete *(GAM-eet)* A sperm or egg; a haploid reproductive cell.

Gamma globulin A plasma protein carrying most of the blood-borne antibodies.

Ganglion A mass of neuron cell bodies found outside the central nervous system.

Gangrene Death of a mass of cells.

Gastric juice The secretions of stomach glands; contains water, mucus, enzymes, hydrochloric acid, and other substances; strongly acidic.

Gastrointestinal tract The stomach, intestines, and associated organs involved in digestion; abbreviated G.I. tract.

gcal Abbreviation for gram calorie; see *calorie.*

Gene The basic unit of heredity; a sequence of DNA nucleotides on a chromosome that encodes a protein.

General adaptation syndrome The emergency response.

Genetic code The arrangement by which the 64 possible codons each specify an amino acid or act as a termination signal in protein synthesis.

Genetic drift The chance fluctuation of allele frequencies over time within a population, especially a small, isolated population.

Genetic mosaic An individual composed of cells of different genotypes.

Genetics The study of heredity.

Genitalia The external and internal reproductive organs.

Genome The total genetic makeup of an organism or of a species.

Geometric progression The growth of a population by a constant proportion during each unit of time (e.g., 3, 6, 12, 24, etc.); contrast with *arithmetic progression.*

Genotype *(JEE-noh-type)* The genetic combination that controls a trait or set of traits; see *phenotype.*

Germ cell A gamete or a cell that gives rise directly to a gamete; see *somatic.*

Germ layer One of the three basic embryonic tissue layers (ectoderm, endoderm, and mesoderm); each gives rise to certain tissues or structures as the organism develops.

Gerontology The study of old age.

Giantism Abnormal development of the body or its parts.

Gill A respiratory organ of aquatic animals; usually a filamentous, vascular process with a large surface area.

Gingivitis Inflammation of the gums.

Gizzard A structure used by various animals (e.g., earthworms and birds) to grind food, sometimes with the aid of ingested stones or sand.

Glans The cap of a structure (e.g., glans penis, glans clitoris).

Glaucoma An eye disorder caused by increased pressure due to an excess of fluid within the eye.

Glial cell *(GLEE-ul)* A nonconducting supportive, protective, and insulative cell of the nervous system.

Glomerular filtrate The filtrate or filtered blood that enters the proximal convoluted tubule of the kidney after passing through Bowman's capsule.

Glomerulus *(glum-AR-yoo-lus)* A cluster of capillaries enclosed by Bowman's capsule in the kidney; serves as the site of filtration.

Glottis The slitlike opening into the larynx between the vocal cords.

Glucagon *(GLOO-ka-gon)* A pancreatic hormone that stimulates the breakdown of glycogen to glucose in the liver, thereby increasing the blood glucose level.

Glucocorticoid One of a group of adrenal cortical hormones that protect against stress and affect protein and carbohydrate metabolism.

Glucose A simple sugar or monosaccharide; the main blood sugar.

Glycerol An alcohol-like substance that combines with fatty acids to form a fat or oil.

Glycogen *(GLY-koh-jen)* A complex polysaccharide that serves as a food reserve; can be broken down readily into glucose; the animal equivalent of starch; found mainly in the liver and muscles.

Goiter An enlargement of the thyroid gland.

Golgi body *(GOAL-jee)* An organelle that functions as a collecting and packaging center for cellular products manufactured for export; consists of flat sacs, tubules, and vesicles.

Gonad The male (testis) or female (ovary) gamete-producing organ.

Gonadotropin *(go-NAD-oh-TROH-pin)* Any hormone that affects the gonads, especially FSH and LH.

Gonococcus The coccus bacterium (diplococcus) that causes gonorrhea *(Neisseria gonorrhoeae)*.

Gonorrhea A bacteria-caused sexually transmitted disease.

Gout A condition characterized by excessive uric acid in the blood; uric acid crystals may form deposits in joints and in the kidneys.

Graafian follicle The last stage in follicle development in the ovary; contains the maturing oocyte.

Graded exercise Exercise requiring increasing levels of intensity.

Granulocyte A leukocyte that has granules in the cytoplasm and lobed nuclei; develops from red bone marrow (e.g., neutrophils, eosinophils, and basophils).

Granuloma inguinale A bacteria-caused sexually transmitted disease.

Grassland A biome characterized by grasses and other plants but generally treeless; usually having higher precipitation than deserts but lower than forests; see *plain, prairie,* and *savanna.*

Gray matter Parts of the central nervous system composed mainly of neuron cell bodies.

Green gland The excretory organ of certain crustaceans.

Green Revolution Describes the increased yields from scientifically bred or selected crops (usually grains), generally in association with large amounts of fertilizer, water, and pesticide usage.

Growth The process in which cells become increasingly specialized and enlarged in size and number; see *development.*

Growth curve The curve resulting when growth is plotted against time on a graph; often S-shaped for many traits.

Growth hormone An anterior pituitary hormone that aids in growth and development by stimulating protein synthesis and affecting other metabolic processes.

Habitat The place where an organism, population, or community lives.

Hair cell A type of mechanoreceptor found in various parts of different animals (e.g., vertebrate inner ear and arthropod statocyst), which detects some type of motion.

Hallucination *(ha-loo-si-NAY-shun)* A false perception having no basis in reality and not accounted for by an external stimulus.

Hallucinogen An agent capable of producing hallucinations.

Haploid *(HAP-loid)* Having a single set of chromosomes (1*n*), as normally present in mature gametes; in contrast to *diploid*.

Heart murmur Sounds produced by blood flowing through faulty valves between the atria and the ventricles.

Heimlich maneuver A first-aid procedure for choking that employs a quick, upward thrust against the diaphragm to force air out of the lungs with sufficient force to eject any lodged material.

Hematocrit *(hee-MAT-oh-krit)* The volume of blood cells (especially red blood cells) packed by centrifugation in a blood sample.

Hematopoietic tissue *(hem-a-to-poy-EH-tik)* Tissues that produce red blood cells.

Hemizygous A gene present in a single dose.

Hemoglobin *(HEE-moh-gloh-bin)* The red, iron-containing protein pigment of red blood cells that transports oxygen.

Hemolytic anemia A decrease in hemoglobin due to a destruction of red blood cells.

Hemorrhage Bleeding, especially when it is profuse.

Hemorrhoid *(HEM-oh-royd)* A dilated or varicose blood vessel (usually a vein) in the anal region; piles.

Hemophilia A group of hereditary diseases characterized by blood that does not clot properly; "bleeder's" disease.

Hepatic Pertaining to the liver.

Hepatitis Acute (serum) or chronic (infectious) viral inflammation of the liver.

Herbivore A plant-eating organism.

Heredity The transmission of inherited traits from parent to offspring.

Heritability A measure of the degree to which a phenotype is genetically influenced.

Hermaphrodite *(her-MAF-roh-dite)* An organism that possesses both male and female sex organs; may or may not be self-fertilizing.

Hernia The protrusion or projection of an organ or part of an organ through the wall cavity that normally contains it.

Herpes genitalis A virus-caused sexually transmitted disease.

Herpes simplex The virus that causes oral infections such as cold sores and fever blisters.

Heterotroph *(HET-ur-oh-troph)* An organism that cannot make its own food from inorganic materials and therefore must live on other organisms, wastes, or decaying matter; contrasts with *autotroph*.

Heterozygous *(HET-ur-oh-ZY-gus)* Possessing two different alleles for a given trait at the corresponding loci of homologous chromosomes; opposite of *homozygous*.

Hibernation The passage of an unfavorable period (usually winter conditions) in a state of dormancy by certain animals; usually refers to endotherms; contrast with *estivation*.

Hindbrain   The lower or posterior portion of the brain just above the spinal cord; includes the cerebellum, pons, and medulla oblongata.

Histamine *(HISS-tuh-meen)*   A substance produced by injured and mast cells; causes contraction of smooth muscle, increases fluid loss from blood vessels, and is largely responsible for inflammatory responses in allergies.

Histology   The microscopic study of tissues.

Homeostasis *(HOME-ee-oh-STAY-sis)*   The tendency to maintain a stable internal environment despite changes in the external environment; typically involves feedback mechanisms.

Hominid   A primate group composed of modern humans and related extinct forms.

Homo   The genus of humans. *H. habilis* ("able man") lived earlier than *H. erectus* ("upright man"); modern humans are *H. sapiens* ("wise man").

Homologous chromosome *(home-OL-uh-gus)*   One chromosome of a pair in diploid cells that carry equivalent genes.

Homozygous *(HOME-oh-ZY-gus)*   Possessing an identical pair of alleles for a given trait at the corresponding loci of homologous chromosomes; opposite of *heterozygous*.

Homunculus   A miniature human imagined by early biologists to be present in a gamete.

Hormone *(HOR-mone)*   A substance produced in cells in one part of the body that is transported by the bloodstream to other parts of the body, where it regulates and coordinates the activities of these parts; a chemical messenger.

Host   An organism that harbors or sustains a pathogen, parasite, or other symbiotic organism.

Humor   Any fluid or semifluid substance in the body.

Humoral immunity *(HYOO-mur-al)*   Immunity residing in the blood and provided by antibodies.

Hunger center   A region of the hypothalamus; the appestat.

Huntington disease   Progressive mental deterioration due to a particular autosomal dominant gene.

Hyaline membrane disease   A condition characterized by difficult breathing in neonates, especially low-birth-weight infants.

Hyaluronidase   An enzyme present in the acrosome of the sperm that functions to digest the follicle cells around the ovum to permit fertilization.

Hybrid   The result of the mating of genetically dissimilar individuals or species.

Hydrocarbon *(HY-droh-kar-bon)*   An organic compound consisting only of carbon and hydrogen atoms.

Hydrocephalus   Accumulation of fluid in the head due to blocked circulation of cerebrospinal fluid.

Hydrogen bond   A weak bond between two molecules that is formed when a hydrogen atom is shared between two atoms.

Hydrolysis *(hy-DROL-eh-sis)*   The splitting of a molecule into parts by the addition of H^+ and OH^- ions derived from water; opposite of *condensation reaction* or *dehydration synthesis;* an essential process of digestion.

Hydrophilic *(HY-droh-FIL-ik)*   A trait of charged molecules in which they interact readily with water molecules; "water loving;" contrast with *hydrophobic*.

Hydrophobic   Refers to a molecule that resists wetting but dissolves readily in organic solvents; not containing polar groups; "water fearing;" contrast with *hydrophilic.*

Hydroponics   The growth of plants in a water and mineral solution.

Hydrostatic skeleton   Support provided by fluids under pressure.

Hymen   A fold of mucous membrane that commonly partially covers the entrance to the vagina for a time in the life of a female.

Hyperopia   Farsightedness.

Hypersenstivity   Abnormally increased sensitivity; ability to react to the presence of allergens in amounts innocuous to normal individuals.

Hypertension   Abnormally high blood pressure.

Hypertonic   Refers to a solution that contains a higher concentration of solute particles; having an osmotic pressure greater than that of a solution with which it is compared; contrast with *hypotonic* and *isotonic.*

Hypertrophy   Excessive growth due to increase in cell size; contrast with *atrophy.*

Hyperventilation   Rapid and deep breathing.

Hypervitaminosis   A condition due to an excess of one or more vitamins.

Hypnosis   A socially induced, unusual state of consciousness in which a person is very susceptible to suggestion.

Hypochondriac   A person who imagines an affliction when none is present.

Hypothalamus   A region on the floor of the forebrain containing various centers of the autonomic nervous system that control many neural and endocrine functions (e.g., thirst, hunger, temperature).

Hypothermia   Abnormally low body temperature.

Hypothesis   A statement that provides a testable explanation for a particular phenomenon; ranges from a tentative explanation to one that is generally accepted as highly probable in light of established facts; see *theory.*

Hypotonic   Refers to a solution that contains a lower concentration of solute particles; having an osmotic pressure less than that of a solution with which it is compared; contrast with *hypertonic* and *isotonic.*

Hysterectomy   The surgical removal of the uterus.

Hysterotomy   Incision of the uterus.

Identical twins   Twins arising from the same zygote; monozygotic; contrast with *fraternal twins.*

Illusion   A distorted or incorrect perception of a physically present object.

Immediate-type hypersensitivity   A fast-acting allergic reaction involving humoral antibodies.

Immune system   The widely dispersed tissues that respond to antigens.

Immunity   The state of being resistant to injury by foreign proteins, poisons, parasites, and pathogens due to the presence of antibodies.

Implantation   The attachment of the developing embryo (blastocyst) to the uterine endometrium.

Impotence   Inability to copulate (e.g., inability to maintain an erection); weakness.

Imprinting   A form of rapid learning by which a young animal forms a strong social attachment to an object within a short time after birth or hatching.

Incest   Sexual intercourse between closely related persons.

Inclusion   Any of a number of nonliving structures occurring within the cytoplasm.

Incomplete digestive system   A digestive sac with only a single opening for both the ingestion of nutrients and the egestion of wastes.

Incomplete dominance   The condition in heterozygotes in which the phenotype is intermediate between the two homozygotes; also called *intermediate inheritance.*

Incontinence *(in-KON-ti-nens)*   Inability to control the release of urine or feces due to loss of sphincter control.

Incubation period   The interval between exposure to a pathogen and the development of a disease.

Independent assortment   The principle that segregation of alternative alleles at one locus during gametogenesis is independent of the segregation of alleles at other loci; Mendel's second law; only true for genes located on different chromosomes or far apart on one chromosome.

Independent variable   The condition under the control of the experimenter; contrast with *dependent variable.*

Inert   Having no action.

Infertility   Inability to conceive or to cause conception.

Inflammation   The reactions of tissues to injury, foreign substances, or pathogens; often involves pain, increased temperature, redness, and the accumulation of white blood cells, especially phagocytic macrophages.

Ingestion *(in-JEST-shun)*   The intake of food and fluids.

Inguinal hernia   The protrusion of a part of the viscera into the inguinal canal; a common form of hernia.

Inner ear   The area of the ear that contains several receptors responsible for hearing and the sense of equilibrium (e.g., cochlea and semicircular canals).

Inorganic   Chemicals in which carbon is not the principal element.

Insensible perspiration   Perspiration that occurs at even low temperatures, of which we are unaware.

Insertion   The place of attachment of a muscle to the bone that it moves; contrast with *origin.*

Inspiratory reserve   The volume of air one can forcibly inhale beyond the tidal inhalation.

Insulin   A peptide hormone produced by the islet cells of the pancreas that is essential for the proper metabolism of glucose; promotes glycogen formation and lowers the blood glucose level.

Integrator   A general term used for a neural structure (e.g., brain, spinal cord, ganglion) that receives an input and "decides" what to do with information, which may include ignoring it or sending a message to association or motor neurons.

Interferon   A cellular protein produced by virus-infected cells that assists other cells in resisting viruses by inhibiting the replication of viruses.

Intermediate inheritance   Incomplete dominance.

Interoceptor   A receptor sensitive to stimuli arising inside the body.

Interphase   The period between two mitotic or meiotic cell divisions in which a cell grows and its DNA replicates.

Interstitial *(IN-tur-STISH-ul)*   Refers to certain cells located between other cells (e.g., those between the seminiferous tubules that produce testosterone) or to fluids that fill spaces between cells.

Interval exercise   Alternating exercise and rest.

Intestinal juice   A complex mixture of digestive enzymes produced by the mucosa of the small intestine plus the secretions that enter from the pancreas and bile duct.

Intrauterine device A semipermanent contraceptive device consisting of plastic or metal of variable shape inserted into the uterus; abbreviated IUD.

In utero Within the uterus.

Invertebrate An animal without a backbone.

In vitro In glass, as in a test tube.

In vivo In the living body.

Ion *(EYE-on)* An atom or molecule bearing an electric charge, either positive (cation) or negative (anion), due to an unequal number of electrons and protons.

Ionic bond Electrostatic bond; chemical bond formed as a result of the mutual attraction of ions of opposite charge.

Iris The colored extension of the ciliary body in the eye, with the pupil at its center.

Islets of Langerhans Scattered cells in the pancreas that secrete the hormones insulin and glucagon.

Isometric Contraction of a muscle in which shortening or lengthening is prevented.

Isotonic (1) Refers to two solution that have equal concentrations of solute particles; contrast with *hypotonic* and *hypertonic*. (2) A muscle contraction in which fibers shorten but the tension remains the same.

Isotope *(EYE-so-tope)* An alternative form of a chemical element that differs from other forms of the same element in the number of neutrons in the nucleus. All isotopes behave the same chemically; some are unstable and emit radiation.

IUD Abbreviation for intrauterine device.

Jaundice A condition characterized by yellowness of the skin; whites of eyes, mucuous membranes, and body fluids; caused by the deposition of bile pigments following liver disease, excessive destruction of erythrocytes, or obstruction of the bile duct.

Joint The place of junction of two or more bones; often movable.

Jungle A second-growth tropical forest.

Karyotype *(KAR-ee-oh-type)* The physical chromosome traits of an individual.

kcal Abbreviation for kilocalorie; see *calorie*.

Keratin A fibrous, insoluble protein found in hair, nails, and other tough epidermal tissues.

Kidney The principle excretory and osmoregulatory organ(s) of an animal's body.

Kidney stone A mineral stone that may accumulate in and sometimes block the central cavity of the kidney.

Kinetic energy Energy of motion; contrast with *potential energy*.

Kingdom One of the primary divisions of organisms (Monera, Protista, Fungi, Plantae, and Animalia).

Klinefelter syndrome An abnormal condition in which the constitution of the sex chromosome is XXY; the testes of the phenotypic males are poorly developed and sterility is common.

Kwashiorkor A childhood disease resulting from a deficiency of dietary protein.

Labium A lip or liplike structure (e.g., labium majorum and labium minorum).

Labor The process of birth consisting of increasing contractions of the uterine muscle.

Lacrimal *(LAK-ri-mal)* Pertaining to tears.

Lactation The secretion and ejection of milk by the mammary glands.

Lacteal *(lak-TEEL)* A small lymph vessel within the core of an intestinal villus that takes up digested fat molecules.

Lactic acid A metabolic waste that causes muscle fatigue.

Lactotropic hormone A hormone secreted by the anterior pituitary that stimulates milk secretion.

Lanugo Fine hair found on the fetus.

Laparotomy An operation in which an incision is made through the abdominal wall and into the abdominal cavity.

Large intestine The portion of the gastrointestinal tract extending from the end of the small intestine to the anus; includes the cecum, colon, rectum, and anal canal.

Larynx The cartilaginous structure at the entrance of the trachea that functions in sound production; the voice box.

Laryngitis Inflammation of the mucous membrane lining the larynx.

Laws of thermodynamics See *thermodynamics*.

LBW Abbreviation for low-birth-weight infants.

Lens A transparent part of the eye that focuses rays of light onto the retina.

Letdown reflex The release of milk stored in the breast as a result of suckling.

Lethal gene A gene whose phenotypic effect results in the death of the bearer.

Leukemia A malignant disease of the blood-forming organs characterized by a rapid and abnormal increase in the number of leukocytes plus many immature cells in the blood.

Leukocyte A white blood cell.

Leukocytosis An increase in the number of leukocytes, characteristic of many infections and other disorders.

Leukopenia A deficiency of leukocytes.

Leydig cell An interstitial cell in the testis that secretes testosterone.

Libido *(li-BEE-do)* The sex drive.

Ligament *(LIG-uh-ment)* The fibrous connective tissue that joins bones together at a joint.

Light microscope A microscope using beams of light to allow specimens to be viewed; contrast with *electron microscope*.

LH Abbreviation for luteinizing hormone.

Lipid A category of water-insoluble, nonpolar organic molecules, wholly or mainly formed of carbon, hydrogen, and oxygen; includes fats, oils, waxes, phospholipids, and steroids.

Liver The large, glandular, vascular organ that functions in nutrient storage, the biochemical alteration of nutrient molecules, the production of bile and various blood proteins, and detoxification.

Locus The location of a gene or transcription unit on a chromosome.

Loop of Henle The hairpin loop of the nephron of the kidney between the proximal and distal convoluted tubules; significant in osmoregulation.

Low-birth-weight infants Infants weighing less than 2.5 kg or 5.5 lb at birth; formerly termed *premature*.

Lung Any of a variety of animal structures that function in the exchange of gases between the atmosphere and the bloodstream.

Luteinizing hormone A pituitary gonadotropic hormone required for the completion of gametogenesis in both sexes and for the formation and support of the corpus luteum during the ovarian cycle; abbreviated LH.

Lymph The colorless fluid derived from blood by filtration through capillary walls in the tissues; contains white blood cells, some of which enter from the tissue fluid and others of which are made in the lymph nodes.

Lymph node A mass of spongy lymphatic tissue in which lymphocytes are generated, antibodies are produced, the spread of pathogens is impeded or stopped, and debris and dead cells are degraded.

Lymphocyte A type of white blood cell formed in lymph nodes; responsible for the immune response; exists as B or T cells.

Lymphogranuloma venereum A rickettsia-caused sexually transmitted disease.

Lymphokine A substance produced by an activated T cell.

Lymphotoxin A lymphokine that directly harms or destroys a pathogen.

Lyon hypothesis The theory that dosage compensation is due to inactivation of one of the X chromosomes in all somatic cells of a female.

Lysosome *(LYE-so-som)* An organelle containing digestive enzymes released when the structure ruptures; important in recycling cellular debris.

Macromolecule Any large biological polymer (e.g., carbohydrate, protein, lipid, nucleic acid).

Macronutrient A nutrient required in relatively large quantities; contrast with *micronutrient.*

Macrophage A large phagocytic cell found in the circulating blood as a monocyte and in various areas of the body as a "wandering" cell.

Malignant Diseases that worsen and cause death, especially in the invasion and spreading of cancer.

Malpighian tubule A hollow, tubular, blind structure that empties into an insect gut and functions in excretion of nitrogenous wastes and osmoregulation.

Mammary gland A milk-producing gland; a breast.

Marasmus Extreme malnutrition with marked weight loss.

Marfan syndrome A genetic disorder characterized by tall, thin stature and excessively long fingers and toes.

Marijuana A psychoactive derivative of the hemp plant *(Cannabis sativa).*

Marrow The soft tissue in the cavity of bones; important as a site of blood cell formation; see *red* and *yellow marrow.*

Marsupial *(mar-SOOP-ee-ul)* A group of mammals whose young typically complete their embryonic development inside a maternal pouch (e.g., kangaroo, koala, opossum); contrast with *monotreme* and *placental mammals.*

Mass peristalsis Strong peristaltic contractions that move large amounts of material long distances.

Mastoiditis Inflammation of the mastoid bone cells.

Masturbation Obtaining a sexual climax by self-stimulation and manipulation.

Matrix The intercellular liquid, jellylike, or solid substance of a tissue; frequently contains an interlacing network of microscopic fibers.

Matter Anything that has mass and occupies space.

Mechanoreceptor A receptor that detects mechanical deformation of itself or of an adjacent cell; includes touch, pressure, vibration (hearing, equilibrium), and proprioception.

Meditation Methods of achieving higher consciousness and/or alteration of physiological response (e.g., yoga and transcendental meditation).

Medulla The inner part of a structure (e.g., adrenal or kidney); contrast with *cortex.*

Medulla oblongata The most posterior part of the brain lying next to the spinal cord; the hindbrain.

Megalopolis The region resulting from the physical merger of two or more large metropolitan areas (metropolises).

Meiosis *(my-OH-sis)* The two successive nuclear divisions in which a diploid cell (2*n*) forms four haploid daughter cells (1*n*); reduction division; compare with *mitosis*.

Melanin A dark pigment found in some cells, such as those in the skin.

Melatonin A pineal gland hormone that functions in sexual development.

Menarche The onset of menstruation; the first period.

Ménière's disease A disease caused by upset in sodium metabolism; characterized by intense dizziness and vertigo.

Meninges The tough, protective, connective tissues that enclose and protect the brain and spinal cord.

Meningitis Inflammation of the meninges resulting from a head injury or as a complication of various infectious diseases.

Menopause The cessation of menstrual cycles at the end of a female's reproductive lifetime, and the associated physical, physiological, and behavior changes.

Menstrual·cycle *(MEN-stroo-ul)* A cyclic process whose function is to prepare the endometrium for the possible implantation of a young embryo.

Menstrual extraction Use of suction to quickly remove the loosened endometrium that would normally be expelled over a period of days during menstruation.

Menstruation The cyclic sloughing of the blood-enriched endometrium that normally recurs at about 28-day intervals in the absence of pregnancy during the reproductive lifetime of the female.

Mescaline An alkaloid psychoactive derivative of the peyote cactus.

Mesoderm *(MEZ-oh-durm)* The middle of the three primary embryonic germ layers, lying between the ectoderm and the endoderm; gives rise to muscle, bone, and other connective tissue, the circulatory system, the peritoneum, and most reproductive and excretory structures.

Mesomorph A body type characterized by large muscles and relatively great strength.

Metabolism *(meh-TAB-ol-liz-um)* The sum of the physical and chemical processes within a living cell or organism.

Metaphase The stage of cell division characterized by the chromosomes coming to lie in the spindle's equatorial plane.

Metastasis *(meh-TAS-teh-sis)* The transfer of disease from one organ or body part to another that is not connected to it; refers especially to the spread of cancer cells.

Microclimate The physical conditions immediately surrounding an organism, not those measured in a standard weather station.

Microfilament A fine protein thread composed of actin and myosin; its contractile ability is important in muscle contraction, cell motility, and changes in cell shape.

Micronutrient A nutrient required in relatively small quantities (e.g., vitamins and some minerals); contrast with *macronutrient*.

Microtubule A long, hollow protein cylinder; important in moving the chromosomes during cell division, providing internal structure to cilia and flagella, and in maintaining cell shape.

Microvillus A tiny fingerlike extension of the plasma membrane of certain cells (e.g., epithelium); greatly increases a cell's surface area.

Midbrain The middle portion of the brain concerned with visual and auditory reflexes.

Middle ear The air-filled chamber in a cranial bone containing the ear ossicles that conduct vibrations from the tympanic membrane to the oval window.

Midwife A paramedical person specializing in the delivery of babies.

Mineral An inorganic chemical compound.

Mineralocorticoid A steroid hormone of the adrenal cortex that regulates sodium and potassium levels in the blood.

Miscarriage A spontaneous expulsion of an embryo or fetus from the uterus before full term.

Mitochondrion A spherical or elongated organelle that contains the electron transmitter system and certain other enzymes, produces most ATP, and is the site of cellular respiration.

Mitosis *(my-TOE-sis)* A form of nuclear division in somatic cells in which the duplicated chromosomes separate to form two genetically identical daughter nuclei; usually accompanied by cytokinesis; compare with *meiosis.*

Molecule The smallest unit of an element or compound having the composition and properties of the substance; two or more atoms held together by chemical bonds.

Monera A kingdom of prokaryotes such as bacteria and cyanobacteria.

Mongolism An older term for Down syndrome, or trisomy-21.

Monilial vaginitis A fungus-caused type of vaginitis.

Monoclonal antibody A specific antibody produced by laboratory-cloned B cells hybridized with cancerous cells.

Monocular vision Vision through only one eye; see *binocular vision.*

Monocyte A type of leukocyte; a large, highly mobile cell (macrophage) whose main function is phagocytosis.

Monomer A subunit capable of linking with other like molecules to form a polymer.

Mononucleosis An infectious disease characterized by weakness, sore throat, swollen glands, and abnormal numbers of blood cells.

Monosaccharide A sugar not composed of smaller sugar subunits (e.g., glucose, fructose).

Monotreme An egg-laying mammal (e.g., echidna, platypus); contrast with *marsupial* and *placental mammals.*

Monozygotic twins Identical twins; result from a single zygote.

Mons veneris The pad of coarse skin and fatty tissue overlying the pubic bone in a female; typically covered by a triangular mass of short, curly hair after puberty.

Morula The solid ball of cells that results from the early cleavage divisions of the zygote.

Motile Capable of movement from place to place; contrast with *sessile.*

Motor neuron A neuron that innervates an effector (e.g., muscle, gland), and impulses from which stimulate the effector; contrast with *sensory neuron.*

Motor unit All the skeletal muscle fibers stimulated by a single motor neuron.

Mucosa The mucous membrane lining passages and cavities communicating with the outside of the body.

Mulatto A person with one white and one black parent.

Multiple allele A set of alleles that contains more than two contrasting members of a given locus.

Multiple sclerosis A chronic disease caused by degeneration of myelin sheaths of the central nervous system.

Muscle An organ specialized to produce voluntary or involuntary movements of the body or its parts; composed of muscle fibers.

Muscle fiber An individual cell of a muscle; typically long and multinucleated; its numerous myofibrils are capable of contraction when stimulated.

Muscle insertion The attachment of a muscle to a bone that moves; contrast with *muscle origin*.

Muscle origin The fixed skeletal attachment of a muscle; contrast with *muscle insertion*.

Muscle tissue Tissue specialized to produce motion; important traits include excitability, extensibility, contractility, and elasticity; see *cardiac, smooth,* and *striated muscle*.

Muscle tone A partial, sustained contraction of skeletal muscle in response to activation of stretch receptors.

Muscular dystrophy A hereditary disease characterized by chronic weakening, atrophy, or paralysis of a muscle, muscle group, or entire limb.

Mutagen Any agent capable of inducing a mutation or change in DNA; includes various pathogens, chemicals, and forms of radiation.

Mutation *(myoo-TAY-shun)* A permanent, heritable change in a cell's DNA (e.g., alteration in nucleotide sequence, change of gene position, insertion of foreign sequences, gene duplication or loss); contributes to genetic diversity.

Mutualism An interspecific interaction that benefits both species.

Myasthenia gravis A chronic disease that primarily affects the voluntary muscles of the face, mouth, and eyes.

Myelin *(MY-eh-lin)* The fatty, insulative sheath around certain axons in the central nervous system and in some peripheral nerves; composed of the membranes of Schwann cells.

Myocardial infarction A heart attack; death of a portion of the heart muscle caused by blockage of a coronary artery.

Myocardium Heart muscle.

Myofibril A contractile microfilament within muscle, composed of actin and myosin.

Myoglobin A red, iron-containing, oxygen-storing protein pigment in muscle tissue.

Myometrium *(my-oh-MEE-tree-um)* The smooth muscle coat of the uterus; contrast with *endometrium*.

Myopia A visual defect whereby objects can be seen distinctly only when very close to the eyes; nearsightedness.

Myosin *(MY-oh-sin)* A contractile protein involved in cell structure and movement; especially important in muscle fibers; see *actin*.

Myotonia *(my-oh-TOE-nee-uh)* Increased muscular irritability and tendency to contract.

Myxedema A condition that results from a deficiency of thryoxin in adults; characterized by a low metabolic rate.

Narcotic Originally, addictive, pain-relieving, and sleep-inducing substances, such as opiates and various synthetic drugs; now often defined loosely as various psychoactive agents or in legal terms.

Natural active immunity Immunity acquired through the production of antibodies against a pathogen or other antigen that the body encounters.

Natural selection The process whereby organisms with various heritable traits experience differential survival and reproduction, resulting in the perpetuation of those traits and organisms best adapted to a particular environment; the basis for evolutionary change; contrast with *artificial selection.*

Neanderthal An extinct race or subspecies of humans *(Homo sapiens neanderthalensis).*

Nearsightedness Myopia; contrast with *farsightedness.*

Negative feedback A homeostatic control mechanism in which the output is used to constantly modify the input to keep both factors within certain limits; contrast with *positive feedback.*

Neonate A newborn infant.

Neoplasm A new and abnormal formation of tissue, as a cancer or tumor.

Nephridial organ An excretory organ of various invertebrates; typically paired organs consisting of a ciliated funnel draining wastes from the coelom through a twisted duct to the exterior.

Nephritis Kidney inflammation.

Nephron The microscopic anatomical and functional unit of the kidney; involved in filtration and selective reabsorption of blood components; consists of a Bowman's capsule, an enclosed glomerulus, and a long tubule.

Nerve A bundle of axons and associated cells held together by connective tissue and located outside of the central nervous system; contrast with *tract.*

Nerve impulse A wave of transitory membrane depolarization or excitement transmitted through a neuron, across a synapse, or along a nerve pathway; an action potential.

Nerve net The netlike system of neurons in some invertebrates dispersed through epithelium; permits only diffuse responses to stimuli.

Nervous tissue A tissue that transmits nerve impulses that function to coordinate homeostasis.

Neuromuscular junction The place where neurons and muscle fibers unite.

Neuron *(NYOOR-on)* A nerve cell; the structural unit of the nervous system; includes a cell body, dendrite(s), and axon.

Neurosecretory cell Any cell of the nervous system that produces a hormone; applied especially to the hypothalamus.

Neurotransmitter A short-lived, hormonelike chemical released by a terminus of an axon into a synaptic cleft, where it stimulates a dendrite during the transmission of a nerve impulse from one neuron to another (e.g., acetylcholine, norepinephrine).

Neutron An electrically uncharged particle in the atomic nucleus of most elements.

Neutrophil A phagocytic white blood cell formed in the bone marrow.

Niacin A B vitamin; used to treat pellagra.

Niche *(nich)* The functional role of an organism or a species in an ecosystem.

Night blindness The impaired ability of the eyes to adapt to vision in the dark; caused by a vitamin A deficiency.

Nitrogen narcosis The bends.

Node of Ranvier *(ron-VEE-a)* A constriction in a myelin sheath at the location of a gap between successive Schwann cells.

Nondisjunction The failure of homologous chromosomes to separate at metaphase, resulting in one daughter cell receiving both and the other daughter cell neither of the chromosomes involved.

Nonself Foreign cells or substances that enter the body or altered body cells that are not recognized as self (e.g., cancer cells).

Nonspecific defense system Defenses against pathogens that do not involve antibodies; includes the skin, mucous membranes and their secretions, and normal flora, phagocytes, interferon, and inflammatory responses.

Nonspecific urethritis Inflammation of the urethra that cannot be attributed to a specific cause.

Norepinephrine A hormone produced by the adrenal medulla.

Northern coniferous forest The biome dominated by conifers (e.g., spruce, fir) found at relatively high latitudes and altitudes; taiga.

Nuclear envelope The double membrane surrounding the eukaryote nucleus, the outermost of which is continuous with the endoplasmic reticulum.

Nucleic acid A nucleotide polymer that forms the basis of heredity and protein synthesis (e.g., DNA, RNA).

Nucleolus A spherical organelle within the cell nucleus that is the site of rRNA synthesis.

Nucleoplasm The structures and protoplasm found in the cell's nucleus; see *cytoplasm.*

Nucleotide The repeating subunit of DNA or RNA that consists of a nitrogenous base and a phosphate group linked to a 5-carbon sugar molecule.

Nucleus (1) The membrane-enclosed organelle in the cell that contains the chromosomal material. (2) In atoms, the positively charged core consisting typically of neurons and protons. (3) A cluster of neuron bodies in the central nervous system.

Nutrient Any substance required for maintenance and growth.

Obesity The condition of being excessively fat, or roughly over 20 percent above average.

Obstetrician A physician who specializes in delivering babies and diagnosing and treating the diseases and disorders of women near the time of childbirth.

Ocellus A small simple eye or eyespot; one element in the compound eye of an arthropod.

Olfaction The sense of smell; the process of smelling.

Omnivore *(OM-neh-vor)* An organism that can eat both plant and animal material.

Oncogenic Capable of producing a cancer, neoplasm, or tumor.

Oocyte A cell that gives rise to an ovum by meiosis.

Oogenesis *(OH-uh-JEN-eh-sis)* The process of forming mature eggs in the ovary.

Oogonium The primordial cell from which the egg arises; grows to become a primary oocyte.

Open transport system A circulatory system in which blood vessels open into intercellular spaces, allowing blood to bathe organs directly; contrast with *closed transport system.*

Ophthalmologist An M.D. who specializes in examining eyes, prescribing corrective lenses and/or exercises, and diagnosing and treating eye diseases and disorders.

Opiate A drug derived from opium (e.g., heroin, morphine).

Optic chiasma The X-shaped structure on the underside of the brain; formed by the optic nerves from the eyes and the optic tracts to the brain, where nerve fibers from both eyes are distributed to both sides of the brain.

Optic disc The region where the optic nerve and the main blood vessels enter and leave the eye; the blind spot.

Optic nerve   The main nerve to the eye, arising in the brain.

Optician   A technician who grinds lenses according to a prescription from an ophthalmologist or optometrist.

Optometrist   A person trained to examine eyes for visual problems (e.g., myopia, hyperopia, astigmatism) and to prescribe corrective lenses and/or exercises.

Orbital   A shell, level, or volume of space surrounding the atomic nucleus in which an electron will be found most of the time.

Organ   A body structure composed of several different tissues grouped together in a structural and functional unit.

Organ of Corti   The sensory organ within the cochlear duct that functions to detect sound.

Organ system   A collection of organs working together to serve the same function.

Organelle   *(OR-gan-EL)*   A small specialized part of a cell (e.g., mitochondrion, ribosome, vesicle).

Organic   A chemical compound that contains carbon.

Orgasm   A state of highly emotional excitement that occurs at the climax of sexual intercourse or masturbation.

Origin   The more fixed point of attachment of a muscle to the skeletal structure; contrast with *insertion*.

Orthodontia   The use of braces to bring crooked teeth back into line.

Osmoconformer   An organism, especially aquatic, that does not regulate the osmotic potential of its body tissues, but allows it to change with that of its environment; contrast with *osmoregulator*.

Osmoregulation   The homeostatic regulation of body fluid volume and composition.

Osmoregulator   An organism, especially aquatic, that regulates the osmotic potential of its body fluids homeostatically; contrast with *osmoconformer*.

Osmosis   *(oz-MOH-sis)*   The diffusion of water molecules across a selectively permeable membrane.

Osmotic potential   The tendency for water to move across a selectively permeable membrane from one solution to another due to differences in the solute concentrations.

Osmotic pressure   The total concentration of solutes in a solution.

Ossicle   Any small bone (e.g., the three tiny ear bones).

Ossification   Formation of bone; especially the replacement of cartilage by bone.

Osteocyte   A mature bone cell that has ceased its bone-building activity.

Osteoporosis   A disease characterized by a widening of the normal channels within bone, resulting in increased porosity and softness.

Outer ear   The ear flap and ear canal.

Oval window   The membrane separating the middle ear from the inner ear to which vibrations are imparted by the ossicles.

Ovary   The female gonad; produces eggs and the hormones estrogen and progesterone.

Overturn   Vertical mixing of a body of water brought about by seasonal changes in temperature.

Oviduct   The tube through which an egg passes from the ovary to the uterus; a fallopian tube.

Ovotestis   An abnormal, sterile, gonadlike structure with some traits of both a testis and an ovary.

Ovulation   The discharge of a mature ovum from a follicle of the ovary.

Ovum   The female gamete; an egg cell.

Oxidation   The combining of oxygen and food in the tissues; the loss of an electron by an atom or molecule; occurs simultaneously with reduction (gain of an electron).

Oxidative water   Water released from food by chemical digestion.

Oxygen debt   The amount of oxygen required to oxidize the excess lactic acid accumulated in muscles during heavy exercise.

Oxygen dissociation curve   A line on a graph that relates the percent saturation of hemoglobin with oxygen and the gaseous pressure of oxygen.

Oxytocin   A hormone produced in the hypothalamus and secreted by the pituitary; regulates uterine contractions and acts on the breasts to stimulate milk release.

Ozone O_3; filters ultraviolet radiation at high altitudes but often harmful to organisms and materials near Earth's surface.

Pacemaker   The sinoatrial node that initiates the heartbeat and regulates the rate of contraction of the heart.

Pacinian corpuscle   Small oval pressure receptors in the skin, tendons, and internal organs.

Palate   The roof of the mouth, having both hard (bony) and soft sections.

Pancreas   A large exocrine (digestive enzymes) and endocrine gland (insulin and glucagon); located near the duodenum.

Pap smear   A cell sample taken from the cervix used to determine the presence or absence of uterine cancer.

Parasite   *(PAR-uh-site)*   An organism that lives in (endo-) or on (ecto-) another organism, at whose expense it gains nourishment and often protection.

Parasympathetic   A subdivision of the autonomic nervous system; fibers originate in the brain and the pelvic region of the spinal cord and innervate primarily the internal organs; operates antagonistically with the *sympathetic system.*

Parathormone   The hormone of the parathyroid glands; decreases blood phosphate and increases blood calcium levels.

Parathyroid   A pea-sized gland found within the substance of the thyroid gland; mainly regulates the metabolism of calcium and phosphorus.

Parkinson's disease   A chronic progressive paralysis involving rhythmic tremor, masklike facial appearance, and rigidity and slowness of muscle action.

Parthenogenesis   Asexual reproduction in which an egg develops without fertilization.

Particulate   A solid air pollutant (e.g., dust, smoke, ash, pollen).

Passive immunity   Temporary protection against an infectious disease by the injection of antibodies taken from an animal or another human who was exposed to the disease; see *active immunity.*

Passive transport   Movement of fluids, solutes, or other substances, especially across a membrane, without the expenditure of energy (e.g., diffusion, osmosis); contrast with *active transport.*

Pathogen   A disease-causing organism or virus.

Pathology   The study of diseases and the changes in structure and function produced by pathogens.

Pedigree   The ancestral history of an individual.

Pellagra   Deficiency of niacin; symptoms are diarrhea, dermatitis, and mental changes.

Pelvic inflammatory disease   One of a group of diseases that affect the uterus, oviducts, or ovaries; caused by a variety of pathogens; abbreviated PID.

Penis The male copulatory organ; delivers sperm into the female reproductive tract during coitus.

Peptide A chain of two or more amino acids linked by peptide bonds.

Peptide bond A type of covalent chemical bond that links amino acids together in peptides, polypeptides, and proteins; forms by dehydration synthesis.

Perineum The area between the anus and the external genitalia.

Periodontal membrane The membrane surrounding the root of the tooth and binding it to the adjacent bone.

Peripheral nervous system All of the neurons and nerve fibers outside of the central nervous system; includes the cranial and spinal nerves, the autonomic nervous system, and sensory and motor neurons.

Peristalsis *(PER-is-TAL-sis)* Successive rhythmic waves of involuntary muscular contraction and relaxation in the walls of hollow, tubular organs (e.g., ureter, oviduct, parts of the digestive tract); serves to move the tube's contents forward.

Peritoneum The membrane that lines the body cavity and forms the external covering of the organs within it.

Peritonitis Inflammation of the peritoneum.

Pernicious anemia A decreased red blood cell count due to vitamin B_{12} deficiency.

pH The symbol used in expressing hydrogen ion concentration; a number from 0 to 14 on the acid (below 7) and base (above 7) scale.

Phagocytosis *(FAY-go-sigh-TOE-sis)* The engulfing of microorganisms, other cells, and foreign particles by a cell such as a white blood cell.

Pharynx *(FAH-rinks)* The muscular tube in the throat that connects the mouth cavity and the esophagus.

Phenocopy The simulation by an individual of traits characteristic of another genotype resulting from physical or chemical influences in the environment.

Phenotype *(FEE-nuh-type)* The realized expression of the genetic makeup of an organism; see *genotype*.

Phenylketonuria A genetic disease of children involving serious brain damage and mental retardation; characterized by high blood levels of the amino acid phenylalanine; abbreviated PKU.

Pheromone A substance released by an exocrine gland of one organism that influences the physiology or behavior of another organism of the same species.

Phlebitis Inflammation of a vein.

Phocomelia A birth deformity of the arms or legs.

Phonoreceptor A receptor of sound energy.

Phospholipid A phosphorylated lipid; a major component of cell membranes; consist of a polar, hydrophilic head and a nonpolar, hydrophobic tail.

Photochemical smog Complex air pollutants resulting from the interaction of sunlight with such contaminants as hydrocarbons and nitrogen oxides.

Photoreceptor A receptor of light energy.

Photosynthesis The synthesis of organic compounds from carbon dioxide and water, using the energy of light captured by chlorophyll.

Phrenology An unsupported theory that one can learn about an individual's personality and future by noting bumps and depressions on the skull.

Physiological reserve The ability to repair injuries quickly and completely, to function for a time at above-normal levels without permanent harm, and to resist diseases.

Physiology The study of body functions.

PID Abbreviation for pelvic inflammatory disease.

"Pill" An oral contraceptive.

Pineal A small endocrine gland in the brain; secretes melatonin.

Pinocytosis *(PEE-no-sigh-TOE-sis)* The engulfing of liquids and their dissolved solutes by cells.

Pitch The quality of sound dependent upon the frequency of the vibrations.

Pituitary An endocrine gland with numerous secretions located on the underside of the brain; controlled through releasing hormones by the hypothalamus; consists of an anterior and a posterior lobe in humans.

PKU Abbreviation for phenylketonuria.

Placebo A substances having no active ingredient in the process being studied; administered as a control in testing the effect of another substance.

Placenta *(pluh-SEN-tuh)* A structure formed from the union of the uterine mucosa with the extraembryonic membranes of the embryo, by means of which the embryo and fetus receive nutrients and oxygen and eliminate wastes.

Placental mammal Mammals that form placentas; contrast with *marsupials* and *monotremes.*

Plain A relatively dry grassland, characterized by small, widely spaced grasses; see *prairie.*

Plantae The eukaryote kingdom that includes plants.

Plaque (1) A deposit of material (polysaccharide, bacteria, mucus, detritus) on the surface of a tooth. (2) A deposit of lipid, fibrous material, and often calcium salts, on the inner wall of a blood vessel.

Plasma *(PLAZ-muh)* The fluid part of the blood within which the cells are suspended; contains mostly water but also dissolved nutrients and wastes, hormones, and various proteins.

Plasma cell An antibody-producing cell resulting from the multiplication and differentiation of a B lymphocyte that has interacted with an antigen.

Plasma membrane The outermost selectively permeable membrane through which all nutrients entering the cell and all waste products or secretions leaving it pass; cell membrane.

Plasmid A small fragment of extrachromosomal DNA that replicates independently of the main chromosome.

Plastid One of various membrane-bounded organelles that is the site of photosynthesis (e.g., chloroplast), pigmentation, or starch storage.

Plateau stage The stage of the sexual response cycle between the excitement and orgasm stage; characterized by swelling of the vagina, labial lips, breast, penis, and testes.

Platelet A small blood cell involved in blood clotting.

Pleura *(PLOOR-a)* The tough, connective tissue membrane that covers the lungs and lines the walls of the thorax and diaphragm.

Pleurisy Inflammation of the pleura.

Plexus A network of nerves, veins, or lymphatic vessels.

Polar A charged portion of a molecule capable of forming hydrogen bonds with water.

Polar body   A small nonfunctioning cell that forms during the meiotic divisions of oogenesis; consists mainly of a nucleus with little cytoplasm.

Polygenic inheritance   The condition when two or more pairs of genes affect the same trait.

Polymer   *(POL-eh-mur)*   A large molecule formed by the linkage of many simpler, identical subunits (e.g., starch is a polymer of glucose).

Polypeptide   A polymer consisting of many linked amino acids; not as complex as a protein.

Polysaccharide   A sugar polymer; a carbohydrate composed of many linked monosaccharide subunits (e.g., starch, cellulose, glycogen).

Pons   A broad mass of nerve fibers on the ventral surface of the hindbrain anterior to the medulla.

Population   A group of interacting organisms of the same species.

Popullution   Refers to the relationships between the size and density of the human population, resource use, and environmental pollution and damage.

Portal system   A group of veins that drain one region and lead to a capillary bed in another organ rather than directly to the heart.

Positive feedback   Processes in which the output constantly enhances the input; runaway feedback; contrast with *negative feedback*.

Potential energy   Energy stored by matter as a result of its location or spatial arrangement; contrast with *kinetic energy*.

Power   Work performed per unit of time.

Prairie   A relatively moist grassland, characterized by tall, closely spaced grasses; see *plain*.

Precapillary sphincter   Rings of smooth muscles in arterioles, capable of regulating blood flow into capillaries.

Precocial   Refers to a relatively high degree of development and of the capability for independent activity at the time of hatching or birth; contrast with *altricial*.

Precursor   A chemical substance or object from which another substance or object forms.

Predation   The interspecific interaction in which an organism of one species (the predator) captures and consumes an organism of another species (the prey).

Preformed water   Free water in food released by digestion, especially physical digestion.

Prematurity   Refers to low-birth-weight infants (less than 2.5 kg or 5.5 lb).

Prepuce   Foreskin; the loose tissue covering the glans of the penis or clitoris.

Presbyopia   Loss of elasticity of the eye lens and resulting difficulty in focusing.

Primary sex traits   The characteristics of the gonads.

Primate   A group of mammals that includes prosimians and anthropoids (e.g., tarsiers, monkeys, humans).

Proctitis   Inflammation of the rectum.

Producer   An organism (autotroph) capable of synthesizing its own food from inorganic substances.

Progesterone   *(pro-JES-ter-one)*   The steroid hormone secreted by the corpus luteum of the ovary and by the placenta; helps to regulate menstrual cycles and to maintain the uterus during pregnancy.

Prokaryote   Organisms in the kingdom Monera (e.g., bacteria and cyanobacteria); no nucleus or membrane-bound organelles are present and the single chromosome is circular; contrast with *eukaryote.*

Prolactin   A hormone of the anterior pituitary that stimulates milk secretion and helps maintain the corpus luteum.

Proliferation   Rapid, repeated reproduction of new parts, especially cells.

Prophase   An early stage in mitosis characterized by nuclear disorganization, the formation of a microtubule spindle, the shortening and thickening of the chromosomes, and their movement toward the equator of the spindle.

Prophylaxis   The prevention of disease.

Proprioceptor   A sensory receptor located in the muscles, tendons, and joints; senses the body's position and movements.

Prosimian   Nonanthropoid primates (e.g., tarsier, lemur, loris, tree shrew).

Prostaglandin   *(PROS-tuh-GLAN-din)*   One of a group of compounds (modified fatty acids) secreted by all or most tissues, with many hormonelike effects.

Prostate   A large accessory sex gland in males that surrounds the upper urethra and produces part of the seminal fluid.

Protein   A large molecule composed of chains of amino acids joined by peptide bonds; one of the principal types of compounds present in all cells; some form structures and many are enzymes.

Protista   The eukaryote kingdom that includes organisms such as protozoans and algae; includes mostly unicellular forms or multicellular lines derived from them.

Proton   An elementary particle present in the nuclei of all atoms that has a positive electrical charge and a mass similar to that of a neutron; a hydrogen ion.

Protozoa   Single-celled animallike organisms; members of the kingdom Protista.

Proximodistal   Refers to the developmental direction in which the structures along the midline of the body develop prior to those in the extremities; compare with *cephalocaudal.*

Psilocybin   A hallucinogen found naturally in a particular mushroom.

Psychoactive   Pertains to substances that primarily alter one's emotional state, such as hallucinogens.

Psychogenic   Referring to an emotional or psychological origin.

Psychosomatic   Pertaining to a mental state affecting a physical condition.

Puberty   *(PYOO-ber-tee)*   The period of life when the gonads become functional and when the secondary sex traits develop.

Pubic lice   Small insects (often erroneously called "crabs") that cause pubic irritation.

Pulp cavity   The central cavity of the tooth containing the nerves and blood vessels.

Pulse   The rhythmic expansion and recoil of the arteries caused by blood flow from the left ventricle; pulse rate corresponds to the heart rate.

Punnett square   A cross-multiplication square used to calculate the types and expected frequencies of offspring from a mating.

Pupil   The opening of the center of the iris that admits light into the eye.

Pure-breeding   Individuals homozygous for all genes being considered.

Pus   A product of inflammation that consists of tissue debris, dead or living pathogens, and dead phagocytes in a fluid excretion.

Pyridoxine   Vitamin B$_6$.

Rabies   A viral disease transmitted by the bite of an infected animal.

Race   A variant form of a species characterized by a set of distinguishing traits.

Radial symmetry   The regular arrangement of parts around a central axis such that any plane passing through the central axis divides the organism into halves that are essentially mirror images; contrast with *bilateral symmetry.*

Radiation   (1) The transfer of heat as waves; contrast with *conduction* and *convection.* (2) Energetic radiation that produces ions, some of which may harm tissues or be mutagenic; includes X-rays, gamma rays, and particles from radioisotopes.

Radioactivity   The release of radiant energy as a result of decay in the nucleus of unstable atoms.

Radioisotope   An unstable isotope of an element that decays spontaneously, emitting radiation.

Radiotherapy   The treatment of diseases by use of X-rays, radioisotopes, or other forms of radiation.

Range of tolerance   The limits within which a cell or organism can survive.

RBC   Abbreviation for r̲ed b̲lood c̲ell.

RDA   Abbreviation for r̲ecommended d̲aily a̲llowance.

Reabsorb   To take back a substance that was released previously.

Receptor   A nerve ending specialized to receive a stimulus.

Recessive allele   An allele whose phenotypic effect is masked in a heterozygote by a dominant allele; alleles that produce their effect when homozygous.

Recombinant DNA   A new cell or individual is produced by combining fragments of DNA from two species; a key part of genetic engineering technology.

Recombination   The formation of new gene combinations, such as through crossing-over, the reassortment of chromosomes during meiosis and fertilization, mutation, or by the transfer of genes into cells.

Recommended daily allowance   The estimate of minimal or optimal daily intake levels for various nutrients; abbreviated RDA.

Rectum   The terminal portion of the large intestine used for the temporary storage of feces.

Red blood cell   Erythrocyte; abbreviated RBC; function is to transport gases, especially oxygen.

Red marrow   Regions within certain bones (e.g., ribs, sternum, vertebrae) where red blood cells are produced; contrast with *yellow marrow.*

Red muscle   Muscle fibers specialized for relatively slow or moderate activity; contrast with *white muscle.*

Reduction   The addition of electrons or hydrogen atoms to a substance; contrast with *oxidation.*

Referred pain   Pain that is felt somewhere apart from its origin.

Reflex   An inborn, automatic, involuntary response to an appropriate stimulus; a functional unit of the nervous system that consists of a sensory neuron, usually one or more association neurons, and a motor neuron.

Refractory period   *(ree-FRAK-tor-ee)*   The brief time when a neuron or muscle fiber cannot respond to a second stimulus.

Regeneration   The replacement of parts of an organism that were lost due to an injury.

Releasing hormone   A peptide hormone produced by the hypothalamus that stimulates or inhibits secretion of specific hormones by the anterior pituitary.

Replication   *(rep-lih-KAY-shun)*   The synthesis of DNA in which the double helix opens, the two strands separate, and each is used as a template for producing a new complementary strand.

Residual air   Air that stays in the lungs after the maximum possible exhalation; see *dead air.*

Resolution stage   The stage of the sexual response cycle following orgasm, when body conditions return to normal.

Respiration   The process by which cells use oxygen, produce CO_2, and conserve the energy of nutrients in biologically useful form (e.g., ATP); the act or function of breathing.

Respiratory center   A region in the medulla oblongata that regulates the rate and depth of breathing.

Respiratory pigment   A pigment molecule in blood that assists in carrying gases (e.g., hemoglobin).

Resting membrane potential   The normal difference in charge or electric potential that exists across a neuron membrane at rest.

Restriction enzyme   A bacterial enzyme that cleaves a DNA molecule at a specific base sequence; used widely in recombinant DNA technology.

Retina   The innermost of the three layers of the eyeball; contains the light-sensitive rods and cones and several layers of other neurons.

Retinopathy   Disease of the retina.

Reverberating circuit   An arrangement of closed chains of neurons that, when stimulated by a single stimulus, continue to generate impulses from collaterals of their cells.

Rh factor   An inherited antigen found on the surface of the red blood cells of certain humans (Rh^+) but absent in others (Rh^-).

Rhodopsin   Visual purple; a pigment found in rod cells that decomposes when light strikes it and creates an impulse.

Rhythm method   A contraceptive method relying on abstinence from sexual intercourse for an appropriate period of time before and after predicted ovulation in the female.

Riboflavin   Vitamin B_2.

Ribonucleic acid   A class of nucleic acids (mRNA, tRNA, rRNA) that takes part in cytoplasmic protein synthesis.

Ribosome   A small organelle found either free in the cytoplasm or attached to the outer surface of the endoplasmic reticulum; carries out protein synthesis.

Rickets   *(RIK-ets)*   A condition of malnutrition in children that is characterized by soft, deformed bones; caused by disruption of calcium metabolism due to inadequate vitamin D.

Rickettsia   An intracellular parasite.

RNA   Abbreviation for <u>r</u>ibo<u>n</u>ucleic <u>a</u>cid.

Rod   A slender, sensory cell in the retina involved in black-and-white vision; sensitive to dimmer light than a cone.

Root canal   Extension of the pulp cavity into the one or more projections of the tooth in the bony socket.

Roughage   The undigestible part of foods, usually plant fibers, that helps stimulate digestive secretions and movement by distension of the intestine.

Round window   The membranous structure of the inner ear at the opposite end of the cochlea from the oval window.

Rubella   A highly contagious viral infection most common in humans 4–19 years old; formerly called *German measles*.

Ruminant   A hooved grazing mammal that digests cellulose in a four-part stomach with the assistance of microorganisms.

Saccharide   A carbohydrate containing one or more simple sugar units; a compound consisting of carbon, hydrogen, and oxygen in which the ratio of hydrogen to oxygen is 2 : 1, or $(CH_2O)_n$.

Sacculus   One of two organs of static equilibrium in the inner ear; see *utriculus*.

Saline   Salt (especially sodium chloride) solution; a solution of salts isotonic with body fluids.

Saliva   The alkaline secretion of the salivary glands; contains water, enzymes, minerals, and other substances.

Salpingectomy   The cutting or tying of the oviducts, especially for contraceptive purposes.

Salpingitis   Inflammation of the oviducts.

Salt gland   A specialized structure in various vertebrates for concentrating and excreting salts ingested with food and fluids.

Sarcoma   A tumor of connective tissue, often highly malignant.

Sarcomere   The contractile unit of striated muscle; consists of repeating bands of the proteins actin and myosin.

Satiety center   *(sa-TYE-e-tee)*   A region of the hypothalamus that inhibits the desire for food.

Saturated fat   A fat containing fatty acids with no double-bonded carbon atoms; usually solid at room temperature; contrast with *unsaturated fat*.

Savanna   *(suh-VAN-uh)*   A tropical or subtropical grassland with scattered tree or shrubs.

Scanning electron microscope   An electron microscope used to study the surfaces of objects, abbreviated SEM; contrast with *transmission electron microscope*.

Schwann cell   A cell that forms myelin and an insulating sheath around a nerve fiber.

Scientific method   Any of several relatively formal methods of research used by scientists to investigate phenomena; typically involves identifying a problem, gathering relevant data, and formulating and testing hypotheses, often through the use of controlled experimentation.

Sclera   The fibrous, outermost protective covering of the eyeball; whitish except for the clear region (cornea) in the front of the eye.

Sclerosis   A hardening and loss of elasticity of the tissues.

Scrotum   An external pouch of loose skin containing the testes.

Scurvy   A deficiency disease caused by lack of vitamin C.

Sebaceous gland   An oil-producing gland.

Second messenger   A chemical signal that relays a hormonal message from the cell's surface to its interior.

Secondary sex trait   An external trait that differs between the sexes (e.g., mammary glands, body hair, voice, fat deposition); not directly involved in reproduction.

Secretion   A substance produced and released by a cell, especially a useful product as opposed to waste material.

Segmentation   (1) Movements of the smooth muscles in the walls of the digestive tract that serve to churn the contents. (2) The serial repetition of similar body parts.

Segregation   The separation of homologous chromosomes or of alternative alleles into different gametes during meiosis; Mendel's first law; contrast with *independent assortment*.

Selectively permeable membrane A membrane that permits passage of water and some solutes but excludes other solutes.

Self Cells and substances that the body recognizes as its own; nonforeign body materials to which antibodies are not produced; contrast with *nonself*.

SEM Abbreviation for scanning electron microscope.

Semen The sperm-bearing secretion discharged from the penis during male orgasm; the alkaline fluid also contains fructose that nourishes the sperm.

Semicircular canal An inner-ear structure (three-part chamber) concerned with dynamic equilibrium; contrast with *utriculus* and *sacculus*.

Seminal vesicle One of a pair of pouches that lies above the prostate gland; produces part of seminal fluid or stores sperm.

Seminiferous tubule *(SEM-in-IF-er-us)* A thin, tubular, coiled structure in which the spermatozoa are produced; constitutes most of the testis.

Senescence *(se-NES-ens)* The process of growing old; the period of old age; changes, especially deterioration, associated with aging.

Sensory nerve A nerve that conducts sensory information from a receptor to the central nervous system; contrast with *motor nerve*.

Septicemia The serious condition caused by bacteria living in the blood stream.

Serendipity The discovery of useful or valuable information not sought for.

Serotonin A vasoconstricting hormone released by thrombocytes in damaged tissues, thus aiding in the control of bleeding.

Serum The liquid that separates from blood after coagulation; blood plasma without cells and the protein fibrinogen.

Serum sickness Illness following the injection of a serum into someone who is allergic to an ingredient in the serum.

Sessile *(SES-il)* Animals not free to move in their environment; attached; contrast with *motile*.

Set point The optimum for a particular function at a particular time.

Sex chromosome The X or the Y chromosome; see *autosome*.

Sex hormone A steroid hormone from a gonad that has a variety of reproductive and related functions (e.g., estrogen and testosterone).

Sex-influenced trait A trait in which dominance of an allele depends on the sex of the bearer; a trait more common in one sex than another.

Sex-limited trait A trait expressed in only one of the sexes.

Sex-linked trait A trait that is determined by a gene located on a sex chromosome, usually the X chromosome.

Sex preselection The use of a method (e.g., timing of coitus) or a technology (e.g., centrifuging semen) to increase the likelihood of an X- or a Y-carrying sperm fertilizing an egg and producing a child of a sex desired by the parents.

Sex reversal Any procedure to alter some of the primary and secondary sex traits of an individual through surgery and hormone treatment.

Sexual intercourse Coitus or copulation.

Sexual reproduction Reproduction that involves meiosis and fertilization; contrast with *asexual reproduction*.

Sexual response cycle The sequence of anatomic and physiological changes that occur during sexual behavior; includes the excitement, plateau, orgasm, and resolution stages.

Sexually transmitted disease Any disease spread by sexual contact (e.g., syphilis, gonorrhea, AIDS, herpes genitalis); abbreviated STD; known formerly as *venereal disease,* VD.

Shock   A state resulting from circulatory collapse (weak heart action and low blood pressure).

Sickle-cell anemia   A serious form of anemia seen in individuals homozygous for an autosomal, codominant gene; red blood cells containing abnormal hemoglobin assume a sickle shape.

Sickle-cell trait   A generally mild anemia seen in individuals heterozygous for the sickling gene.

Simian line   The large crease(s) across the upper palm.

Singleton   A single child born in a particular pregnancy.

Sinoatrial node   A small mass of conducting tissue in the right atrium that initiates cardiac cycles; the pacemaker.

Sinus   An air-filled, mucous membrane-lined cavity in the skull; any cavity in bone or tissues.

Skeletal muscle   Striated or voluntary muscle; contrast with *smooth* and *cardiac muscle.*

Slow fiber   An uninsulated or nonmyelinated axon.

Small intestine   The region of the intestine with the smaller diameter in which the major digestion and absorption of food occurs; includes the duodenum, jejunum, and ileum.

Smegma   Glandular secretions and cellular debris found under the labia minora around the glans clitoris and under the male prepuce.

Smooth muscle   The involuntary nonstriated muscle tissue in the walls of visceral organs and blood vessels; contrast with *striated* and *cardiac muscles.*

Sneeze   A reflex triggered by the irritation of the nasal passages; controlled by a brain center in the medulla oblongata.

Social parasite   An organism that benefits from another organism behaviorally rather than by means of a physical attachment; see *parasite.*

Solute   *(SOL-yoot)*   A dissolved substance.

Solvent   A substance, usually liquid, in which other substances (solutes) can be dissolved.

Somatic   Pertaining to the body exclusive of the gametes or the cells that give rise to gametes; see *germ cell.*

Somatic nervous system   The voluntary nervous system; the neurons of the peripheral nervous system that control skeletal muscle; contrast with *autonomic nervous system.*

Somatopsychological   Pertains to body traits affecting psychological traits.

Somatotypology   The study of the relationship between physical and psychological traits.

Specific defense system   The immune response whereby the body produces specific antibodies in response to each antigen it encounters; contrast with *nonspecific defense system.*

Sperm   The mature male gamete; spermatozoa.

Spermatocyte   A diploid cell formed from spermatogonia that becomes modified into mature sperm.

Spermatogenesis   The process by which spermatogonia develop into sperm.

Spermatogonium   An unspecialized diploid cell on the wall of a testis that undergoes meiotic division to become a spermatocyte and later a sperm.

Spermicide   Any agent that kills sperm.

Sphincter   A circular muscle capable of closing a tubular opening by constriction.

Sphygmomanometer An instrument for measuring arterial blood pressure.

Spinal cord The mass of nerve tissue located in the vertebral canal; the portion of the central nervous system that extends downward from the brain stem and from which the 31 pairs of spinal nerves originate.

Spinal gate theory The theory that the relative firing rate of fast (myelinated) and slow (uninsulated) fibers explains many aspects of pain sensation.

Spindle fiber One of a group of microtubules that together make up the spindle that is involved with cell division.

Spiracle One of several external openings of an insect tracheal respiratory system.

Spirillum A coiled or spiral-shaped bacterium.

Spirochete A spiral-shaped bacterium; often used to refer to the syphilis pathogen.

Spleen A vascular organ in which lymphocytes are produced and red blood cells are destroyed; assists in maintaining blood volume.

Split brain Lack of connections between the cerebral hemispheres.

Spongy bone Bone tissue containing spaces among the irregular network of matrix.

Spontaneous abortion A miscarriage.

Sprain A tear, rupture, or marked stretching of a muscle, ligament, or joint with dislocation.

Starch A polysaccharide (e.g., glycogen).

Static equilibrium The sense of the head's or body's position in space when it is not in motion; contrast with *dynamic equilibrium*.

Statocyst A sense organ of equilibrium, orientation, and movement found in many invertebrates; consists of a fluid-filled chamber with a small, dense object that stimulates cilia when it moves.

STD Abbreviation for <u>s</u>exually <u>t</u>ransmitted <u>d</u>isease.

Steno- tolerance A narrow range of tolerance; contrast with *eury-tolerance*.

Sterility Inability to produce proper numbers of viable gametes or to conceive or produce offspring.

Sterilization Rendering an individual incapable of reproduction (e.g, castration, vasectomy, salpingectomy, hysterectomy).

Sternum The breastbone.

Steroid One of a group of lipids; includes cholesterol, sex hormones, and hormones from the adrenal cortex.

Stillbirth Birth of a dead fetus.

Stimulant A substance that increases physiological activity, produces a state of alertness, elevates mood, diminishes appetite, and enhances physical performance.

Stimulus An agent or change that causes a reaction in an organism or in any of its parts.

Stomach The muscular sac of the gastrointestinal tract between the esophagus and the duodenum; produces acidic gastric juice that sterilizes ingested material and initiates some digestion; primarily a storage organ, in which limited absorption occurs.

Stomate *(STOH-mayt)* An opening in a leaf surface (between two guard cells), through which gases are exchanged with the environment and transpiration occurs.

Strabismus A condition in which the eyes do not both focus at the same point; crossed eyes.

Strength The ability to mobilize power for a particular task.

Streptococcus A spherical-shaped bacteria that occurs in chains.

Stress External or internal stimuli that initiate negative feedback or homeostatic responses, or the body condition that results from responding to such stimuli.

Stretch receptor A sensory receptor stimulated by a stretch of the containing tissue.

Striated muscle A skeletal or voluntary muscle; sometimes used to include cardiac muscle; named from the striped appearance, which reflects the arrangement of its contractile units; contrast with *smooth muscle*.

Stroke The total set of symptoms resulting from a cerebral vascular disorder; apoplexy.

Stroke volume The volume of blood that leaves each ventricle per heartbeat.

Stye Inflammation of one or more of the oil glands of the eyelids.

Substrate (1) A substance upon which an enzyme works. (2) The surface upon which an organism lives.

Summation The increasing state of excitation leading to a nerve impulse or a muscle contraction in response to rapidly repeated subthreshold stimulation (temporal) or to the simultaneous arrival of impulses at numerous synapses (spatial).

Surface area-to-volume relationship As an object grows, its volume increases at a faster rate than does its surface area.

Surfactant A chemical that reduces the surface tension of water.

Suture A type of joint, especially in the skull, where bone surfaces are closely united and relatively immovable.

Symbiosis Refers to a variety of interactions between two species that live in close contact (e.g., mutualism, parasitism, commensalism).

Sympathetic A subdivision of the autonomic nervous system that prepares the body to respond to emergency situations; typically operates antagonistically with the parasympathetic subdivision.

Synapse *(SIN-aps)* The point of near contact between the tip of an axon and the surface of another excitable cell. The gap is bridged by neurotransmitter molecules, which may result in excitatory or inhibitory responses.

Synaptic vesicle A membrane-bound sac at the ends of axons that contain neurotransmitters that are released to cross a synapse.

Syndrome A group of symptoms characteristic of a particular disease or abnormality.

Synergism A cooperative or multiplicative action of two or more agents upon the same function.

Synergist A muscle that assists the action of a prime mover.

Synesthesia The condition in which a stimulus of one sense is perceived as another type of sensation.

Synovial fluid The viscous lubricating fluid normally present in small amounts in a joint.

Synthesis reaction The formation of a more complex molecule from simpler ones.

Syphilis A sexually transmitted disease caused by a bacterium *(Treponema pallidum)*.

Systole Heart muscle contraction, especially that of the ventricles; contrast with *diastole*.

Taiga *(TYE-guh)* The coniferous or boreal forest, usually at relatively high latitudes or altitudes; characterized by long, harsh winters and short, cool summers.

Tapetum The layer behind the retina that reflects light back through the retina, thus aiding vision in various vertebrates.

Target cell A cell acted upon by a specific hormone.

Taste bud A chemoreceptor on the tongue, palate, or pharynx able to detect one or more of the four basic tastes.

T cell A lymphocyte that has matured in the thymus during late fetal life and infancy; responsible for cellular immunity.

Telencephalon The front or most anterior portion of the brain; includes the cerebrum and associated structures.

Telophase The last stage of nuclear division, during which the chromosomes become reorganized into two nuclei.

TEM Abbreviation for transmission electron microscope.

Temperate deciduous forest A midlatitude forest biome dominated by broad-leaved deciduous species (e.g., oak, maple, birch).

Tendon A tough, fibrous connective tissue that attaches muscle to bone.

10 percent rule The generalization that about 10 percent of the chemical energy available at one feeding level is transferred in usable form to the next feeding level in a food chain.

Teratogen An agent that results in an abnormal embryo.

Testicle An organ consisting of the testis surrounded by several layers of connective tissue.

Testis The male gonad; produces sperm and sex hormones.

Testosterone The major androgen or steroid sex hormone in a male.

Tetanus *(TET-un-us)* A bacteria-caused disease *(Clostridium tetani)* characterized by painful, sustained muscular contractions, including those of the jaw (lockjaw).

Tetany A nervous condition characterized by intermittent tonic muscular contractions of the extremities.

Thalamus The part of the forebrain just posterior to the cerebrum; acts as a relay center to the cerebrum for most impulses from the sense organs, the cerebellum, and other parts of the nervous system.

Theory A well-tested hypothesis that fits existing data, explains how processes or events are thought to occur, and is a basis for predicting future events or discoveries.

Thermal pollution The adverse effects of excess heat in the environment (e.g., lowering the quantity of dissolved oxygen water can hold and speeding metabolic rates of aquatic organisms).

Thermodynamics The study of transformations of energy. The *first law* of thermodynamics states that energy cannot be created nor destroyed; the total energy of the universe remains constant. The *second law* of thermodynamics states that with each transformation of energy, some energy becomes unusable; the entropy, or degree of disorder, tends to increase.

Thermoreceptor A sensory structure that responds to environmental temperature changes.

Thiamin Vitamin B_1.

Thirst center The hypothalamic center that controls fluid intake into the body.

Thorax The chest region; contains the heart and lungs.

Threshold The lowest level of stimulation that can evoke a response in a particular muscle, muscle fiber, nerve, or neuron.

Thrombocyte A blood platelet, which functions in blood clotting.

Thrombus A blood clot obstructing a vessel; remains where it formed; see *embolism.*

Thymus An endocrine gland in the lower neck or upper chest; functions in producing T cells important in cellular immunity.

Thyroid A bilobed endocrine gland straddling the ventral side of the trachea; secretes thyroxin.

Thyroid-stimulating hormone   A peptide hormone of the anterior pituitary that stimulates growth and secretion by the thyroid gland; abbreviated TSH.

Thyrotropic hormone   A peptide hormone from the hypothalamus that stimulates the anterior pituitary to release thyroid-stimulating hormone.

Thyroxin   The main thyroid hormone, whose function is to elevate the basal metabolic rate and influence growth.

Tidal air   The air volume normally inhaled and exhaled while at rest.

Tinnitus   Ringing in the ears.

Tissue   A group of similar cells that perform a common function or group of functions.

Tolerance   The condition in which a larger dose of a chemical substance must be used to produce effects equivalent to those achieved initially with a smaller dose.

Tonsil   An aggregation of lymphoid tissue near the mouth or pharynx.

Torpor   A state of decreased metabolism or activity, often in response to low or high environmental temperatures.

Toxemia   Coma and convulsions that may develop during or immediately after pregnancy related to the presence of protein in the urine and to hypertension.

Toxin   A poisonous substance derived from an organism.

Toxoid   A toxin treated to decrease its toxicity but that is still able to cause antibody production.

Trace element   Required elements normally found in minute traces in tissues (e.g., fluoride, copper, and manganese).

Tracer   A radioactive particle or group of particles that can be detected as it travels through the body.

Trachea   *(TRAY-kee-uh)*   The air tube from the pharynx to the bronchi; the windpipe; supported by cartilaginous rings.

Tracheae   The tubules that make up the breathing system of insects.

Tract   A collection of axons traveling through the white matter of the brain and spinal cord that has specific anatomical locations; compare with *nerve*.

Transcription   The transfer of information from a DNA molecule into an RNA molecule.

Transformation   The transfer of naked DNA from one organism to another; the incorporation of external genetic material into a cell.

Transfusion reaction   An adverse reaction caused by giving incompatible blood.

Translation   The transfer of information from an RNA molecule into a polypeptide; the assembly of a protein on ribosomes, with mRNA directing the order of amino acids.

Transmission electron microscope   An electron microscope used to examine internal details of objects; abbreviated TEM; contrast with *scanning electron microscope*.

Transplant   The placement of a tissue or organ from one part of the body into another, or from one individual to another.

Trichomonas vaginalis   A protozoan pathogen that causes a common form of vaginitis.

Trimester   A three-month period of pregnancy.

Triradius   An area where three ridges come together on fingerprints, palmprints, or toeprints.

Trisomy   An individual having one extra chromosome of a set.

Trophic   Refers to feeding or nutrition, such as the level in a food web; see *producer, consumer,* and *decomposer*.

Trophoblast The thin-walled side of a blastocyst that gives rise to the placenta and the membranes surrounding the embryo.

Tropic hormone A hormone, especially one from the anterior pituitary or hypothalamus, that influences a different endocrine gland.

Tropical rainforest A complex and large forest biome located mainly in equatorial regions; characterized by warm temperatures, high precipitation, and little climatic variation.

Tubal ligation A method of female sterilization in which each oviduct is cut and the free ends tied.

Tumor A swelling or enlargement in otherwise normal tissue; usually caused by uncontrolled growth of a transformed cell.

Tundra The cold, treeless biome at high latitudes or altitudes.

Turbinate A bony protuberance in the nasal passageways covered with a highly vascularized mucous membrane that may swell and cause nasal congestion.

Turner syndrome Individuals who are phenotypic females (usually XO) but with rudimentary sexual organs and immature secondary sex traits.

Type A behavior Exhibited by people who are achievement-oriented, energetic, tense, perfectionistic, and impatient and who find it hard to relax.

Type B behavior Exhibited by people who have personality traits generally opposite those of Type A individuals.

Ulcer *(UL-ser)* An open lesion of the skin or mucous membranes, with loss of substance and necrosis of the tissue.

Umbilical cord *(um-BIL-i-kal)* The long cord containing blood vessels that links the fetus to the placenta and transports nutrients and wastes.

Unsaturated fat A fat that contains one or more double covalent bonds between the carbon atoms of its fatty acids; usually liquid at room temperature; contrast with *saturated fat.*

Urea A soluble nitrogenous waste excreted by most adult amphibians and mammals; contrast with *ammonia* and *uric acid.*

Ureter *(YOO-re-ter)* A long thin tube leading from the kidney to the urinary bladder.

Urethra *(yoo-REE-thra)* Duct from the urinary bladder to the exterior; also carries semen in males.

Uric acid A relatively insoluble nitrogenous waste; excreted by land snails, insects, birds, and some reptiles; crystals sometimes form in joints and in the kidneys, causing gout; contrast with *ammonia* and *urea.*

Uniformitarianism The view that geological processes now modifying Earth's crust have acted in essentially the same way throughout geologic time, although possibly at different rates.

Urinalysis Examination of the urine.

Urinary bladder A hollow, muscular organ situated in the pelvic cavity; receives urine from the ureters and releases it into the urethra.

Urine The liquid waste filtered from the blood by the kidney and stored in the bladder prior to elimination.

Uterus *(YOO-te-rus)* The chamber in which the embryo develops and receives nourishment prior to birth; the womb; the site of changes during the menstrual cycle.

Utriculus Inner ear receptor for static body balance.

UV Abbreviation for ultraviolet light.

Uvula A fleshy appendage that projects from the soft palate.

Vaccine A preparation of pathogens or substances derived from them administered to a person in order to establish resistance to an infection.

Vacuum aspiration   Use of suction to remove a structure or substance.

Vagina *(va-JYE-na)*   The female organ that receives sperm from the male penis during intercourse; the birth canal and the channel for the menstrual flow.

Vaginitis   An infection or inflammation of the vagina.

Variable   In experiments, a condition being manipulated or measured (e.g., temperature, light, humidity) and a condition that varies from one organism to another (e.g., growth rate, size, color); see *dependent* and *independent variables.*

Varicose vein   A distended, knotted vein.

Vas deferens   The duct carrying sperm from a testis to the urethra.

Vasectomy   Cutting and ligation or removal of a portion of the vas deferens for contraceptive purposes.

Vasoconstriction   The narrowing of blood vessels, especially small arteries; contrast with *vasodilation.*

Vasodilation   Widening (relaxation) of blood vessels, especially small arteries; contrast with *vasoconstriction.*

Vasomotion   The general term for the constriction or dilation of the blood vessels.

VD   Abbreviation for venereal disease.

Vector   (1) An organism or agent that carries and transmits pathogens or parasites from one host to another. (2) A virus or plasmid DNA into which a gene is integrated and transferred into a cell.

Vegetarian   One who prefers not to eat meat or also, sometimes, animal products (e.g., milk, eggs, cheese).

Vein   A blood vessel carrying blood from tissues to the heart.

Vena cava   One of the two major veins that returns blood to the heart from the body.

Venereal disease   One of a group of diseases caused primarily by sexual activities; abbreviated VD. Now more commonly called *sexually transmitted disease* (STD).

Venereal warts   Virus-caused sexually transmitted infections.

Ventricle *(VEN-treh-kul)*   One of two lower muscular chambers of the heart; receives blood from an atrium and pumps it to a lung or to the body tissues.

Venule *(VEN-ule)*   A small blood vessel that collects blood from a capillary and delivers it to a vein.

Vernix   Greasy glandular secretion that covers the fetus.

Vertebrate   A backboned animal.

Vertigo   Dizziness, especially the feeling that one's surroundings are moving.

Vesicle   Small intracellular, membrane-bound sac in which various substances are stored and transported.

Villus *(VIL-us)*   A short fingerlike projection from certain cell membrane surfaces, such as those that line the duodenum.

Virulence   The ability to produce disease.

Virus   A group of pathogens visible only with the electron microscope; incapable of reproduction or growth apart from the living cells they infect.

Vital capacity   The total volume of air displaced when one breathes in as deeply as possible and then breathes out as completely as possible; sum of inspiratory reserve, tidal air, and expiratory reserve.

Vitamin   An organic molecule necessary in trace amounts; secured from food or synthesized by intestinal bacteria.

Vitamin A A fat-soluble vitamin necessary for proper vision and the growth of bones, teeth, and epithelium.

Vitamin B complex Several water-soluble vitamins that are converted to coenzymes needed for proper energy metabolism.

Vitamin B$_1$ Thiamin; a deficiency is beriberi.

Vitamin B$_2$ Riboflavin.

Vitamin B$_6$ Pyridoxine; essential for the utilization of several amino acids.

Vitamin B$_{12}$ Cyanocobalamin; a vitamin needed for proper formation of red blood cells.

Vitamin C Ascorbic acid; essential for the formation of connective tissue.

Vitamin D A fat-soluble vitamin involved in calcium and phosphorus metabolism.

Vitamin E An essential vitamin whose exact function in humans is unclear.

Vitamin K A fat-soluble vitamin needed for proper blood clotting.

Vocal cords Cords of muscle and connective tissue stretching across the larynx; vibrate when air passes over them and are capable of producing sound.

Vulva Region of the female's external genitalia.

WBC Abbreviation for white blood cell.

White blood cell Leukocyte; abbreviated WBC; involved in protection and phagocytosis.

White matter A region in the central nervous system consisting mainly of nerve fibers.

White muscle Muscle fibers specialized for rapid, intense activity; contrast with *red muscle*.

Wind chill The fact that moving air at one temperature cools the body as much as still air at another, cooler temperature.

Windpipe The trachea.

Withdrawal illness The symptoms that occur when humans discontinue use of a chemical substance to which they have become addicted.

Work The application of a force over a distance.

X chromosome One of the sex chromosomes; associated with sex determination; normal human females have two Xs and normal human males have one X and one Y.

X-linked inheritance Transmission of genes found on the X chromosome.

XYY syndrome A genetic abnormality of the sex chromosomes in phenotypic males.

Y chromosome One of the sex chromosomes; all males have at least one Y chromosome.

Yellow marrow Fat storage areas in bone marrow; contrast with *red marrow*.

Y-linked inheritance Transmission of genes found on the Y chromosome.

Zona pellucida A jellylike layer that surrounds an ovum.

Zygote The fertilized egg; the diploid product of the union of haploid gametes in conception.

Zymogen An inactive form of an enzyme.

PHOTO CREDITS

Part One Opener: Black man, Albert J. Copley/Visuals Unlimited; Indian woman, CHIMA/Visuals Unlimited; Egyptian man, Nada Pecnik/Visuals Unlimited; Chinese girls, Michael G. Gabridge/Visuals Unlimited; Amoebas, John D. Cunningham/Visuals Unlimited

Chapter 1 Opener: John D. Cunningham/Visuals Unlimited

Chapter 1: 1.1 John D. Cunningham/Visuals Unlimited; 1.2A Historic VU/Visuals Unlimited; 1.3A John D. Cunningham/Visuals Unlimited; 1.3B C.P. Hickman/Visuals Unlimited; 1.3C E.F. Anderson/Visuals Unlimited; 1.5 William Ormerod, Jr./Visuals Unlimited; 1.7 John D. Cunningham/Visuals Unlimited; 1.8 Len Rue, Jr./Visuals Unlimited; 1.9A Leonard R. Rue III/Visuals Unlimited; 1.9B Leonard R. Rue III/Visuals Unlimited; 1.10 C.P. Hickman/Visuals Unlimited; 1.12A K.G. Murti/Visuals Unlimited; 1.12B David M. Phillips/Visuals Unlimited; 1.12C John D. Cunningham/Visuals Unlimited; 1.14A(1) David M. Phillips/Visuals Unlimited; 1.14A(2) Sherman Thomson/Visuals Unlimited; 1.14B(1) David M. Phillips/Visuals Unlimited; 1.14B(2) Stan Elems/Visuals Unlimited; 1.14C(1) Harold E. Sweetman/Visuals Unlimited; 1.14C(2) David M. Phillips/Visuals Unlimited; 1.14D(1) John D. Cunningham/Visuals Unlimited; 1.14D(2) David Newman/Visuals Unlimited; B1.1 Christine L. Case/Visuals Unlimited; B1.2 Granger Collection; B1.3B K.G. Murti/Visuals Unlimited; B1.4A K.G. Murti/Visuals Unlimited; B1.4B K.G. Murti/Visuals Unlimited; B1.4C K.G. Murti/Visuals Unlimited; B1.4D K.G. Murti/Visuals Unlimited

Part Two Opener: Brazilian Indians, G. Prance/Visuals Unlimited; Eskimos, Nada Pecnik/Visuals Unlimited; African girls, Forest W. Buchanan/Visuals Unlimited

Chapter 2 Opener: Bruce Berg/Visuals Unlimited

Chapter 2: 2.1A G. Prance/Visuals Unlimited; 2.1B Tom W. French/Visuals Unlimited; 2.1C Albert J. Copley/Visuals Unlimited; 2.7 John D. Cunningham/Visuals Unlimited; 2.9A John D. Cunningham/Visuals Unlimited; 2.9B John D. Cunningham/Visuals Unlimited; 2.9C John D. Cunningham/Visuals Unlimited; 2.12A John D. Cunningham/Visuals Unlimited; 2.12B Leonard L. Rue III/Visuals Unlimited; 2.12C Leonard L. Rue III/Visuals Unlimited; 2.15A Leonard L. Rue III/Visuals Unlimited; 2.15B John D. Cunningham/Visuals Unlimited; 2.16A Leonard L. Rue III/Visuals Unlimited; 2.16B Tom J. Ulrich/Visuals Unlimited; 2.21A John D. Cunningham/Visuals Unlimited; 2.21B Leonard L. Rue III/Visuals Unlimited; 2.23B Leonard L. Rue III/Visuals Unlimited; 2.23C Leonard L. Rue III/Visuals Unlimited; 2.24(1) Dennis Paulson/Visuals Unlimited; 2.24(2) Bayard H. Brattstrom/Visuals Unlimited; B2.1A John D.

Cunningham/Visuals Unlimited; B2.1B Leonard L. Rue III/Visuals Unlimited; B2.1C Leonard L. Rue III/Visuals Unlimited

Chapter 3 Opener: All by David M. Phillips/Visuals Unlimited

Chapter 3: 3.1 Granger Collection; 3.2A John D. Cunningham/Visuals Unlimited; 3.2B John D. Cunningham/Visuals Unlimited; 3.2C David M. Phillips/Visuals Unlimited; 3.2D David M. Phillips/Visuals Unlimited; 3.3(1) K.G. Murti/Visuals Unlimited; 3.3(2) David M. Phillips/Visuals Unlimited; 3.3(3) David M. Phillips/Visuals Unlimited; 3.3(4) David M. Phillips/Visuals Unlimited; 3.3(5) K.G. Murti/Visuals Unlimited; 3.3(6) K.G. Murti/Visuals Unlimited; 3.3(7) K.G. Murti/Visuals Unlimited; 3.3(8) David M. Phillips/Visuals Unlimited; 3.3(9) David M. Phillips/Visuals Unlimited; 3.3(10) David M. Phillips/Visuals Unlimited; 3.3(11) Manfred Schliwa/Visuals Unlimited; 3.3(12) David M. Phillips/Visuals Unlimited; 3.3(13) K.G. Murti/Visuals Unlimited; 3.3(14) David M. Phillips/Visuals Unlimited; 3.3(15) David M. Phillips/Visuals Unlimited; 3.4B K.G. Murti/Visuals Unlimited; 3.5 Martha Powell/Visuals Unlimited; 3.6 D. H. Oscar/Visuals Unlimited; 3.9 David M. Phillips/Visuals Unlimited; 3.10B David M. Phillips/Visuals Unlimited; 3.17A Kwang Jeon/Visuals Unlimited; 3.18B John D. Cunningham/Visuals Unlimited; 3.18B(1) John D. Cunningham/Visuals Unlimited; 3.18B(2) John D. Cunningham/Visuals Unlimited; 3.18B(3) John D. Cunningham/Visuals Unlimited; 3.19 David M. Phillips/Visuals Unlimited; 3.23A Stan Elems/Visuals Unlimited; 3.23B Fred Hossler/Visuals Unlimited; 3.23C Fred Hossler/Visuals Unlimited; 3.24(1) Fred Hossler/Visuals Unlimited; 3.24(2) John D. Cunningham/Visuals Unlimited; 3.24(3) John D. Cunningham/Visuals Unlimited; 3.24(4) John D. Cunningham; 3.24(5) David M. Phillips/Visuals Unlimited; 3.25A John D. Cunningham/Visuals Unlimited; 3.25B John D. Cunningham/Visuals Unlimited; 3.26A C. P. Hickman/Visuals Unlimited; 3.26B John D. Cunningham/Visuals Unlimited; 3.26C John D. Cunningham/Visuals Unlimited; 3.27 David M. Phillips/Visuals Unlimited; 3.28 John D. Cunningham/Visuals Unlimited

Chapter 4 Opener: George Musil/Visuals Unlimited

Chapter 4: 4.4B Harold E. Sweetman/Visuals Unlimited; 4.9A Science VU-NASA/Visuals Unlimited; 4.9B William A. Banaszewski/Visuals Unlimited; 4.10B Peggy Starborn/Visuals Unlimited

Part Three Opener: David M. Phillips/Visuals Unlimited

Chapter 5 Opener: All by David M. Phillips/Visuals Unlimited

Chapter 15 Opener: John D. Cunningham/Visuals Unlimited

Chapter 15: 15.2A David M. Phillips/Visuals Unlimited; 15.2C David M. Phillips/Visuals Unlimited; 15.3A David M. Phillips/Visuals Unlimited; 15.3B David M. Phillips/Visuals Unlimited; 15.4A David M. Phillips/Visuals Unlimited; 15.7(2) David M. Phillips/Visuals Unlimited; 15.7(3) David M. Phillips/Visuals Unlimited; 15.8C John D. Cunningham/Visuals Unlimited; 15.11C John D. Cunningham/Visuals Unlimited; 15.17 Science VU-Cardiac Control Systems/Visuals Unlimited; 15.18(2) Science VU-EPA/Visuals Unlimited; B15.1 John D. Cunningham/Visuals Unlimited

Chapter 16 Opener: John D. Cunningham/Visuals Unlimited

Chapter 16: 16.1A John D. Cunningham/Visuals Unlimited; 16.1B George Musil/Visuals Unlimited; 16.1C Science VU-NIH/Visuals Unlimited; 16.1D Science VU-NIH/Visuals Unlimited; 16.1E David Newman/Visuals Unlimited; 16.2 Fred Hossler/Visuals Unlimited; 16.4A(1) John D. Cunningham/Visuals Unlimited; 16.4A(2) John D. Cunningham/Visuals Unlimited; 16.4A(3) John D. Cunningham/Visuals Unlimited; 16.4A(4) John D. Cunningham/Visuals Unlimited; 16.5A(5) John D. Cunningham/Visuals Unlimited; 16.4A(6) W.J. Johnson/Visuals Unlimited; 16.4A(7) David M. Phillips/Visuals Unlimited; 16.4B David M. Phillips/Visuals Unlimited; 16.4C David M. Phillips/Visuals Unlimited; 16.5(1) John D. Cunningham/Visuals Unlimited; 16.5(2) David M. Phillips/Visuals Unlimited; 16.9 E.F. Anderson/Visuals Unlimited; 16.10B John D. Cunningham/Visuals Unlimited; 16.11 Science VU-UUMC/Visuals Unlimited; 16.12 Science VU-NIH/Visuals Unlimited; 16.13 Science VU-CDC/Visuals Unlimited; 16.14(1) David M. Phillips/Visuals Unlimited; 16.14(2) John D. Cunningham/Visuals Unlimited

Chapter 17 Opener: John D. Cunningham/Visuals Unlimited

Chapter 17: 17.1A John D. Cunningham/Visuals Unlimited; 17.1B John D. Cunningham/Visuals Unlimited; 17.1C Len Rue, Jr./Visuals Unlimited; 17.2A Stan Elems/Visuals Unlimited; 17.2B John D. Cunningham/Visuals Unlimited; 17.5B(2) John D. Cunningham/Visuals Unlimited; 17.6B John D. Cunningham/Visuals Unlimited; 17.6C John D. Cunningham/Visuals Unlimited; 17.7 David M. Phillips/Visuals Unlimited; 17.10B Science VU-NIH/Visuals Unlimited; B17.2B Albert J. Copley/Visuals Unlimited; B17.4 Patrice Ceisel/Visuals Unlimited

Chapter 18 Opener: David M. Phillips/Visuals Unlimited

Chapter 18: 18.6(1) John D. Cunningham/Visuals Unlimited; 18.6(2) John D. Cunningham/Visuals Unlim-

ited; 18.9A John D. Cunningham/Visuals Unlimited; 18.10A Fred Hossler/Visuals Unlimited; 18.11(1) John D. Cunningham/Visuals Unlimited; 18.11(2) Fred Hossler/Visuals Unlimited; 18.11(3) John D. Cunningham/Visuals Unlimited; 18.12A(1) John D. Cunningham/Visuals Unlimited; 18.13B(1) John D. Cunningham/Visuals Unlimited; B18.1(2) John D. Cunningham/Visuals Unlimited; B18.2 Mansell Collection

Chapter 19 Opener: George Musil/Visuals Unlimited

Chapter 19: 19.4A John D. Cunningham/Visuals Unlimited; 19.5A John D. Cunningham/Visuals Unlimited; 19.6A Cedric S. Raine/Visuals Unlimited; 19.9 John D. Cunningham/Visuals Unlimited; 19.10A David Fuller/Visuals Unlimited; 19.10B John D. Cunningham/Visuals Unlimited; 19.12B Science VU-NIH/Visuals Unlimited; 19.13A(1) Frank T. Awbrey/Visuals Unlimited; 19.13A(2) Frank T. Awbrey/Visuals Unlimited; 19.16 Fred Hossler/Visuals Unlimited; 19.19A John D. Cunningham/Visuals Unlimited; 19.21 George Musil/Visuals Unlimited; 19.21A John D. Cunningham/Visuals Unlimited; 19.22A Stuart Bratesman/Visuals Unlimited

Chapter 20 Opener: Leonard Lee Rue III/Visuals Unlimited

Chapter 20: 20.3A William Palmer/Visuals Unlimited; 20.3B John D. Cunningham/Visuals Unlimited; 20.6B John D. Cunningham/Visuals Unlimited; 20.7B John D. Cunningham/Visuals Unlimited; 20.8 John D. Cunningham/Visuals Unlimited; 20.9A H. Oscar/Visuals Unlimited; 20.9B John D. Cunningham/Visuals Unlimited; 20.10A Stan Elems/Visuals Unlimited; 20.10B John D. Cunningham/Visuals Unlimited; 20.10C John D. Cunningham/Visuals Unlimited; 20.11A John D. Cunningham/Visuals Unlimited; 20.12B John D. Cunningham/Visuals Unlimited; 20.13 John D. Cunningham/Visuals Unlimited; 20.15A S. Meola/Visuals Unlimited; 20.15B John D. Cunningham/Visuals Unlimited; 20.16A(1) John D. Cunningham/Visuals Unlimited; 20.16A(2) John D. Cunningham/Visuals Unlimited; 20.16A(3) John D. Cunningham/Visuals Unlimited; 20.16B John D. Cunningham/Visuals Unlimited; 20.18 John D. Cunningham/Visuals Unlimited; B20.1 John D. Cunningham/Visuals Unlimited

Chapter 21 Opener: Bruce Berg/Visuals Unlimited

Chapter 21: 21.1 William J. Weber/Visuals Unlimited; 21.2A John D. Cunningham/Visuals Unlimited; 21.2B John D. Cunningham/Visuals Unlimited; 21.2C John D. Cunningham/Visuals Unlimited; 21.5 D.W. Fawcett/Visuals Unlimited; 21.7B John D. Cunningham/Visuals Unlimited; 21.7C David M. Phillips/Visuals Unlimited; 21.8 John D. Cunningham/Visuals Unlimited; 21.11B John D. Cunningham/Visuals Unlimited; 21.11D David M. Phillips/Visuals Unlimited; 21.13C Michael Gabridge/Visuals Unlimited; B21.2A John D. Cunningham/Visuals Unlimited; B21.2B(1) Manfred Schliwa/Visuals Unlimited; B21.2B(2) Manfred Schliwa/Visuals Unlimited; B21.2B(3) Manfred Schliwa/Visuals Unlimited; B21.3B Tom E. Adams/Visuals Unlimited; B21.4A John D. Cunningham/Visuals Unlimited; B21.5A Eliot C. Williams/Visuals Unlimited; B21.5B H. Oscar/Visuals Unlimited

Chapter 22 Opener: John D. Cunningham/Visuals Unlimited

Chapter 22: 22.2B Science VU-NASA/Visuals Unlimited; 22.3 E.F. Anderson/Visuals Unlimited; 22.7B E.F. Anderson/Visuals Unlimited; 22.8 William J. Weber/Visuals Unlimited

Chapter 23 Opener: William Ormerod, Jr./Visuals Unlimited

Chapter 23: 23.1(1) Albert J. Copley/Visuals Unlimited; 23.1(2) Hank Andrews/Visuals Unlimited; 23.7A(1) John D. Cunningham/Visuals Unlimited; 23.8 Science VU-EPA/Visuals Unlimited

Part Five Opener: Science VU-NASA/Visuals Unlimited

Chapter 24 Opener: Nada Pecnik/Visuals Unlimited

Chapter 24: 24.1A John D. Cunningham/Visuals Unlimited; 24.1B John D. Cunningham/Visuals Unlimited; 24.1C John D. Cunningham/Visuals Unlimited; 24.1D John D. Cunningham/Visuals Unlimited; 24.3 The Granger Collection; 24.4A Milton H. Tierney, Jr./Visuals Unlimited; 24.4B John D. Cunningham/Visuals Unlimited; 24.4C John D. Cunningham/Visuals Unlimited; 24.7 Frank T. Awbrey/Visuals Unlimited; 24.8 John D. Cunningham/Visuals Unlimited; 24.10A Russell L. Ciochon/Visuals Unlimited; 24.10B John D. Cunningham/Visuals Unlimited; 24.10C John D. Cunningham/Visuals Unlimited; 24.14 Rudolf Arndt/Visuals Unlimited; B24.1A John D. Cunningham/Visuals Unlimited; B24.1B(1) Milton H. Tierney, Jr./Visuals Unlimited; B24.1B(2) John D. Cunningham/Visuals Unlimited; B24.1C(1) John D. Cunningham/Visuals Unlimited; B24.1C(2) John D. Cunningham/Visuals Unlimited; B24.1C(3) John D. Cunningham/Visuals Unlimited

Chapter 25 Opener: Science VU-WHO/Visuals Unlimited

Chapter 25: 25.8 John D. Cunningham/Visuals Unlimited; 25.9 John D. Cunningham/Visuals Unlimited; 25.10 Richard Thom/Visuals Unlimited; 25.11 Pat Armstrong/Visuals Unlimited; 25.12A John D. Cunningham/Visuals Unlimited; 25.12B David L. Pearson/Visuals Unlimited; 25.12C John D. Cunningham/Visuals Unlimited; 25.13 Ghillean Prance/Visuals Unlimited; 25.14 John D. Cunningham/Visuals Unlimited; 25.16 John D. Cunningham/Visuals Unlimited; 25.18 John D. Cunningham/Visuals Unlimited; 25.20 Virginia Crowl/Visuals Unlimited; 25.21A John D. Cunningham/Visuals Unlimited; 25.21B Vernon Ahmadjian/Visuals Unlimited; 25.22A Jon Turk/Visuals Unlimited; 25.22B John D. Cunningham/Visuals Unlimited; B25.1 David Newman/Visuals Unlimited

INDEX

Page numbers in italic type indicate references to illustrations. Also see the Glossary.